D0722028

Handbook of Plastics Technologies

The Complete Guide to Properties and Performance

Charles A. Harper   Editor-in-Chief

Technology Seminars, Inc., Lutherville, Maryland

McGRAW-HILL

New York   Chicago   San Francisco   Lisbon   London   Madrid
Mexico City   Milan   New Delhi   San Juan   Seoul
Singapore   Sydney   Toronto

The McGraw-Hill Companies

Cataloging-in-Publication Data is on file with the Library of Congress.

Copyright © 2006 by The McGraw-Hill Companies, Inc. All rights reserved. Printed in the United States of America. Except as permitted under the United States Copyright Act of 1976, no part of this publication may be reproduced or distributed in any form or by any means, or stored in a data base or retrieval system, without the prior written permission of the publisher.

1 2 3 4 5 6 7 8 9 0 DOC/DOC 0 1 0 9 8 7 6

ISBN 0-07-146068-3

The sponsoring editor for this book was Kenneth P. McCombs and the production supervisor was Pamela A. Pelton. It was set in Times roman by J. K. Eckert & Company, Inc. The art director for the cover was Handel Low.

Printed and bound by RR Donnelley.

McGraw-Hill books are available at special quantity discounts to use as premiums and sales promotions, or for use in corporate training programs. For more information, please write to the Director of Special Sales, McGraw-Hill Professional, Two Penn Plaza, New York, NY 10121-2298. Or contact your local bookstore.

CONTENTS

Chapter 6 Nanomanufacturing with Polymers
Daniel Schmidt, Joey Mead, Carol Barry
Julie Chen

Chapter 7 Plastics Joining
Edward M. Petrie

Chapter 8 Plastics Recycling and Biodegradable Plastics
Susan E. Selke

Chapter 9 Plastics and Elastomers: Automotive Applications
K. Sehanobish and Tom Traugott

CONTRIBUTORS

Anne-Marie Baker   *University of Massachusetts, Lowell, MA* (CHAPS. 1, 2)

Carol M. F. Barry   *University of Massachusetts, Lowell, MA* (CHAP. 1, 6)

Julie Chen   *University of Massachusetts, Lowell, MA* (CHAP. 6)

Aubert Y. Coran   *Consultant, Longboat Key, FL* (CHAP. 4)

Rudolph D. Deanin   *University of Massachusetts, Lowell, MA* (CHAPS. 3, 5)

Joey L. Mead   *University of Massachusetts, Lowell, MA* (CHAPS. 1, 2, 6)

Edward M. Petrie   *Consultant, Cary, NC* (CHAP. 7)

Daniel Schmidt   *University of Massachusetts, Lowell, MA* (CHAP. 6)

Kalyan K. Sehanobish   *Dow Chemical Company, Auburn Hills, MI* (CHAP. 9)

Susan E. Selke   *Michigan State University, East Lansing, MI* (CHAP. 8)

Tom Traugott   *Dow Chemical Company, Auburn Hills, MI* (CHAP. 9)

PREFACE

It is a pleasure to present to my reading audience this new book in my series, entitled *Handbook of Plastics Technologies*. This new book, an extension of my Materials Science and Engineering Series, is a thorough, comprehensive, and completely up-to-date treatment of the ever more important and critical field of plastics. Prepared by a leading team of professional experts in this field, it will be a useful and practical addition to the bookshelves of both my established and new readers. Like the other books in my series, this *Handbook of Plastics Technologies* provides the broad array of practical information, data, and guidelines necessary to easily understand and use plastics to best advantage in all types of products and applications. This book will be useful to the wide spectrum of readers ranging from product designers to researchers to plastics application and marketing people.

Like the other books in my series, this book has been organized for easy reader convenience. First, a chapter is presented covering the fundamental and introductory aspects of plastics—a basic understanding of plastics types, categories, and forms. Next, a set of chapters is offered on the individual plastics categories, namely, thermoplastics, thermosets, and elastomers. Following that is a chapter covering plastics additives—the myriad ways in which plastics can be modified for specific end-use or application requirements. Then, a chapter is included on nanotechnology, the increasingly vital and exciting area in modern plastics technology. Finally, a set of chapters is included to cover three other areas of great importance in the successful application of plastics, namely the myriad techniques for joining of plastic parts, the increasingly critical area of plastics recycling by conventional means and by use of biodegradable plastics, and lastly, the successful use of plastics in the large and very critical area of automotive applications.

Needless to say, a book of this caliber could only have been achieved with a group of outstanding contributing authors, such as I have been extremely fortunate to have in preparing this book. I would like to take this opportunity to express my thanks to them. They have not only been great from the technology viewpoint, but they have been great people with whom to work. My thanks to all of them and the organizations that they represent. And my special thanks to Dr. Anne-Marie Baker for her work with the University of Massachusetts-Lowell team.

Charles A. Harper
Technology Seminars, Inc.
Lutherville, Maryland

CHAPTER 1
INTRODUCTION TO POLYMERS AND PLASTICS

Carol M. F. Barry, Anne-Marie Baker, Joey L. Mead
University of Massachusetts
Lowell, Massachusetts

1.1 INTRODUCTION

Plastics are an important part of everyday life; products made from plastics range from sophisticated products, such as prosthetic hip and knee joints, to disposable food utensils. One of the reasons for the great popularity of plastics in a wide variety of industrial applications is the tremendous range of properties exhibited by plastics and their ease of processing. Plastic properties can be tailored to meet specific needs by varying the atomic composition of the repeat structure, by varying molecular weight and molecular weight distribution. The flexibility can also be varied through the presence of side chain branching, via the lengths and polarities of the side chains. The degree of crystallinity can be controlled through the amount of orientation imparted to the plastic during processing, through copolymerization, blending with other plastics, and through the incorporation of an enormous range of additives (fillers, fibers, plasticizers, stabilizers). Given all of the avenues available for tailoring any given polymer, it is not surprising that the variety of choices available to us today exist.

Polymeric materials have been used since early times even though their exact nature was unknown. In the 1400s, Christopher Columbus found natives of Haiti playing with balls made from material obtained from a tree. This was natural rubber, which became an important product after Charles Goodyear discovered that the addition of sulfur dramatically improved the properties; however, the use of polymeric materials was still limited to natural-based materials. The first true synthetic polymers were prepared in the early 1900s using phenol and formaldehyde to form resins—Baekeland's Bakelite. Even with the development of synthetic polymers, scientists were still unaware of the true nature of the materials they had prepared. For many years, scientists believed they were colloids—a substance that is an aggregate of molecules. It was not until the 1920s that Herman Staudinger showed that polymers were giant molecules or macromolecules. In 1928, Carothers developed linear polyesters and then polyamides, now known as nylon. In the 1950s, Ziegler and Natta's work on anionic coordination catalysts led to the development of polypropylene, high-density, linear polyethylene, and other stereospecific polymers. More recent developments include Metallocene catalysts for preparation of stereospecific polymers and the use of polymers in nanotechnology applications.

77777

Materials are often classified as either metals, ceramics, or polymers. Polymers differ from the other materials in a variety of ways but generally exhibit lower densities, thermal conductivities, and moduli. Table 1.1 compares the properties of polymers to some representative ceramic and metallic materials. The lower densities of polymeric materials offer an advantage in applications where lighter weight is desired. The use of additives allows the compounder to develop a host of materials for specific application. For example, the addition of conducting fillers generates materials from insulating to conducting. As a result, polymers may find application in EMI shielding and antistatic protection.

Polymeric materials are used in a vast array of products. In the automotive area, they are used for interior parts and in under-the-hood applications. Packaging applications are a large area for thermoplastics, from carbonated beverage bottles to plastic wrap. Application requirements vary widely but, luckily, plastic materials can be synthesized to meet these varied service conditions. It remains the job of the part designer to select from the array of thermoplastic materials available to meet the required demands.

1.2 POLYMER STRUCTURE AND SYNTHESIS

A polymer is prepared by stringing together a low molecular weight species (monomer; e.g., ethylene) into an extremely long chain (polymer; in the case of ethylene, the polymer is polyethylene) much as one would string together a series of bead to make a necklace (see Fig. 1.1). The chemical characteristics of the starting low molecular weight species will determine the properties of the final polymer. When two low different molecular

TABLE 1.1   Properties of Selected Materials[48]

Material	Specific gravity	Thermal conductivity, Joule cm/(°C cm^2 s)	Electrical resistivity, $\mu\Omega$ cm	Modulus, MPa
Aluminum	2.7	2.2	2.9	70,000
Brass	8.5	1.2	6.2	110,000
Copper	8.9	4.0	1.7	110,000
Steel (1040)	7.85	0.48	17.1	205,000
Al$_2$O$_3$	3.8	0.29	>10^{14}	350,000
Concrete	2.4	0.01	–	14,000
Borosilicate glass	2.4	0.01	>10^{17}	70,000
MgO	3.6	–	10^5 (2000° F)	205,000
Polyethylene (H.D.)	0.96	0.0052	10^{14}–10^{18}	350–1,250
Polystyrene	1.05	0.0008	10^{18}	2,800
Polymethyl methacrylate	1.2	0.002	10^{16}	3,500
Nylon	1.15	0.0025	10^{14}	2,800

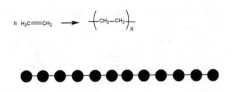

FIGURE 1.1 Polymerization.

weight species are polymerized, the resulting polymer is termed a *copolymer*—for example, ethylene vinylacetate. This is depicted in Fig. 1.2. Plastics can also be classified as either *thermoplastics* or *thermosets*. A thermoplastic material is a high molecular weight polymer that is not crosslinked. It can exist in either a linear or branched structure. Upon heating, thermoplastics soften and melt, allowing them to be shaped using plastics processing equipment. A thermoset has all of the chains tied together with covalent bonds in a three-dimensional network (crosslinked). Thermoset materials will not flow once crosslinked, but a thermoplastic material can be reprocessed simply by heating it to the appropriate temperature. The different types of structures are shown in Fig. 1.3. The properties of different polymers can vary widely; for example, the modulus can vary from 1 MN/m^2 to 50 GN/m^2. For a given polymer, it is also possible to vary the properties simply by varying the microstructure of the material.

FIGURE 1.2 Copolymer structure.

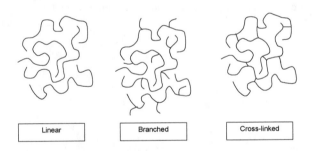

FIGURE 1.3 Linear, branched, and cross-linked polymer structures.

There are two primary polymerization approaches: *step-reaction polymerization* and *chain-reaction polymerization.*[1] In step-reaction (also referred to as *condensation polymerization*), reaction occurs between two polyfunctional monomers, often liberating a small molecule such as water. As the reaction proceeds higher molecular weight species are produced as longer and longer groups react together. For example, two monomers can react to form a dimer then react with another monomer to form a trimer. The reaction can be described as n-mer + m-mer → (n + m)mer, where n and m refer to the number of monomer units for each reactant. Molecular weight of the polymer builds up gradually with time, and high conversions are usually required to produce high molecular weight polymers. Polymers synthesized by this method typically have atoms other than carbon in the backbone. Examples include polyesters and polyamides.

Chain-reaction polymerizations (also referred to as *addition polymerizations*) require an initiator for polymerization to occur. Initiation can occur by a free radical, an anionic, or a cationic species. These initiators open the double bond of a vinyl monomer, and the reaction proceeds as shown above in Fig. 1.1. Chain-reaction polymers typically contain only carbon in their backbone and include such polymers as polystyrene and polyvinyl chloride.

Unlike low molecular weight species, polymeric materials do not possess one unique molecular weight but rather a distribution of weights as depicted in Fig. 1.4. Molecular weights for polymers are usually described by two different average molecular weights, the number average molecular weight, $\overline{M_n}$, and the weight average molecular weight, $\overline{M_w}$. These averages are calculated using the equations below:

$$\overline{M_n} = \sum_{i=1}^{\infty} \frac{n_i M_i}{n_i} \tag{1.1}$$

$$\overline{M_w} = \sum_{i=1}^{\infty} \frac{n_i M_i^2}{n_i M_i} \tag{1.2}$$

where n_i is the number of moles of species i, and M_i is the molecular weight of species i. The processing and properties of polymeric materials are dependent on the molecular weights of the polymer as well as the molecular weight distribution. The molecular weight of a polymer can be determined by a number of techniques including light scattering, solution viscosity, osmotic pressure, and gel permeation chromatography.

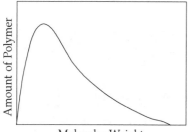

FIGURE 1.4   Molecular weight distribution.

1.3 SOLID PROPERTIES OF POLYMERS

1.3.1   Glass Transition Temperature (T_g)

Polymers come in many forms, including plastics, rubber, and fibers. Plastics are stiffer than rubber yet have reduced low-temperature properties. Generally, a plastic differs from a rubbery material due to the location of its glass transition temperature (T_g). A plastic has a T_g above room temperature, while a rubber has a T_g below room temperature. T_g is most clearly defined by evaluating the classic relationship of elastic modulus to temperature for polymers as presented in Fig. 1.5.

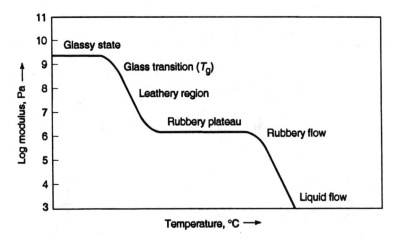

FIGURE 1.5   Relationship between elastic modulus and temperature.

At low temperatures, the material can best be described as a glassy solid. It has a high modulus, and behavior in this state is characterized ideally as a purely elastic solid. In this temperature regime, materials most closely obey Hooke's law:

$$\sigma = E\varepsilon \tag{1.3}$$

where σ is the stress being applied, and ε is the strain. Young's modulus, E, is the proportionality constant relating stress and strain.

In the leathery region, the modulus is reduced by up to three orders of magnitude from the glassy modulus for amorphous polymers. The temperature at which the polymer behavior changes from glassy to leathery is known as the glass transition temperature, T_g. The rubbery plateau has a relatively stable modulus until further temperature increases induce rubbery flow. Motion at this point does not involve entire molecules but, in this region, deformations begin to become nonrecoverable as permanent set takes place. As temperature is further increased, the onset of liquid flow eventually takes place. There is little elastic recovery in this region, and the flow involves entire molecules slipping past each other. This region models ideal viscous materials, which obey Newton's law:

$$\sigma = \eta\dot{\varepsilon} \tag{1.4}$$

In the case of a thermosetting material, the rubbery plateau is extended until degradation and no liquid flow will occur.

1.3.2   Crystallization and Melting Behavior (T_m)

In its solid form, a polymer can exhibit different morphologies, depending on the structure of the polymer chain as well as the processing conditions. The polymer may exist in a random unordered structure termed *amorphous*. An example of an amorphous polymer is polystyrene. If the structure of the polymer backbone is a regular, ordered structure, then the polymer can tightly pack into an ordered crystalline structure, although the material will generally be only semicrystalline. Examples are polyethylene and polypropylene. The exact makeup and architecture of the polymer backbone will determine whether the polymer is capable of crystallizing. This microstructure can be controlled by different synthetic methods. As mentioned above, the Ziegler-Natta catalysts are capable of controlling the microstructure to produce stereospecific polymers. The types of microstructure that can be obtained for a vinyl polymer are shown in Fig. 1.6. The isotactic and syndiotactic structures are capable of crystallizing because of their highly regular backbone, while the atactic form would produce an amorphous material. The amount of crystallinity actually present in the polymer depends on a number of factors, including the rate of cooling, crystallization kinetics, and the crystallization temperature. Thus, the extent of crystallization can vary greatly for a given polymer and can be controlled through processing conditions.

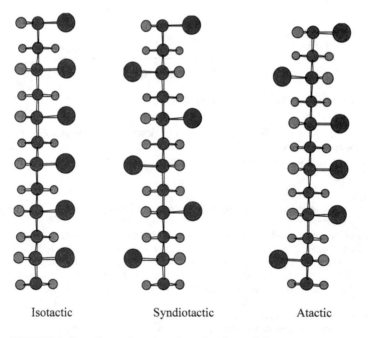

Isotactic   Syndiotactic   Atactic

FIGURE 1.6   Isotactic, syndiotactic, and atactic polymer chains.

1.4 MECHANICAL PROPERTIES

The mechanical behavior of polymers is dependent on many factors, including polymer type, molecular weight, and test procedure. Modulus values are obtained from a standard tensile test with a given rate of crosshead separation. In the linear region, the slope of a stress-strain curve will give the elastic or Young's modulus, E. Typical values for Young's modulus are given in Table 1.2. Polymeric material behavior may be affected by other fac-

TABLE 1.2 Comparative Properties of Thermoplastics [49,50]

Material	Heat deflection temperature @ 1.82 MPa (°C)	Tensile strength, MPa	Tensile modulus, GPa	Impact strength, J/m	Density, g/cm^3	Dielectric strength, MV/m	Dielectric constant @ 60 Hz
ABS	99	41	2.3	347	1.18	15.7	3.0
CA	68	37.6	1.26	210	1.30	16.7	5.5
CAB	69	34	.88	346	1.19	12.8	4.8
PTFE		17.1	.36	173	2.2	17.7	2.1
PCTFE		50.9	1.3	187	2.12	22.2	2.6
PVDF	90	49.2	2.5	202	1.77	10.2	10.0
PB	102	25.9	0.18	NB	0.91		2.25
LDPE	43	11.6	0.17	NB	0.92	18.9	2.3
HDPE	74	38.2		373	0.95	18.9	2.3
PMP		23.6	1.10	128	0.83	27.6	
PI		42.7	3.7	320	1.43	12.2	4.1
PP	102	35.8	1.6	43	0.90	25.6	2.2
PUR	68	59.4	1.24	346	1.18	18.1	6.5
PS	93	45.1	3.1	59	1.05	19.7	2.5
PVC—rigid	68	44.4	2.75	181	1.4	34.0	3.4
PVC—flexible		9.6		293	1.4	25.6	5.5

TABLE 1.2 Comparative Properties of Thermoplastics (*Continued*)[49,50]

Material	Heat deflection temperature @ 1.82 MPa (°C)	Tensile strength, MPa	Tensile modulus, GPa	Impact strength, J/m	Density, g/cm³	Dielectric strength, MV/m	Dielectric constant @ 60 Hz
POM	136	69	3.2	133	1.42	19.7	3.7
PMMA	92	72.4	3	21	1.19	19.7	3.7
Polyarylate	155	68	2.1	288	1.19	15.2	3.1
LCP	311	110	11	101	1.70	20.1	4.6
Nylon 6	65	81.4	2.76	59	1.13	16.5	3.8
Nylon 6,6	90	82.7	2.83	53	1.14	23.6	4.0
PBT	54	52	2.3	53	1.31	15.7	3.3
PC	129	69	2.3	694	1.20	15	3.2
PEEK	160	93.8	3.5	59	1.32		
PEI	210	105	3	53	1.27	28	3.2
PES	203	84.1	2.6	75	1.37	16.1	3.5
PET	224	159	8.96	101	1.56	21.3	3.6
PPO (modified)	100	54	2.5	267	1.09	15.7	3.9
PPS	260	138	11.7	69	1.67	17.7	3.1
PSU	174	73.8	2.5	64	1.24	16.7	3.5

tors such as test temperature and rates. This can be especially important to the designer when the product is used or tested at temperatures near the glass transition temperature, where dramatic changes in properties occur as depicted in Fig. 1.5. The time-dependent behavior of these materials is discussed below.

1.4.1 Viscoelasticity

Polymer properties exhibit time-dependent behavior, meaning that the measured properties are dependent on the test conditions and polymer type. Figure 1.7 shows a typical viscoelastic response of a polymer to changes in testing rate or temperature. Increases in testing rate or decreases in temperature cause the material to appear more rigid, while an increase in temperature or decrease in rate will cause the material to appear softer. This time-dependent behavior can also result in long-term effects such as stress-relaxation or creep.[2] These two time-dependent behaviors are shown in Fig. 1.8. Under a fixed displacement, the stress on the material will decrease over time, termed *stress relaxation*. This behavior can be modeled using a spring and dashpot in series as depicted in Fig. 1.9. The equation for the time dependent stress using this model is

$$\sigma(t) = \sigma_o e^{-t/\tau} \tag{1.5}$$

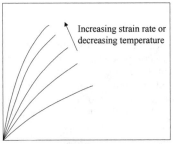

Increasing strain rate or decreasing temperature

Strain

FIGURE 1.7 Effect of strain rate or temperature on mechanical behavior.

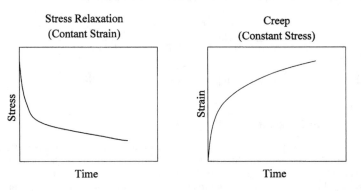

Stress Relaxation
(Contant Strain)

Creep
(Constant Stress)

Stress

Strain

Time

Time

FIGURE 1.8 Creep and stress relaxation behavior.

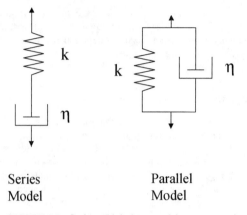

Series   Parallel
Model   Model

FIGURE 1.9 Spring and dashpot models.

where τ is the characteristic relaxation time (η/k). Under a fixed load, the specimen will continue to elongate with time, a phenomenon termed *creep*, which can be modeling using a spring and dashpot in parallel as seen in Fig. 1.9. This model predicts the time-dependent strain as

$$\varepsilon(t) = \varepsilon_o e^{-t/\tau} \tag{1.6}$$

For more accurate prediction of the time-dependent behavior, other models with more elements are often employed. In the design of polymeric products for long-term applications, the designer must consider the time-dependent behavior of the material.

If a series of stress relaxation curves is obtained at varying temperatures, it is found that these curves can be superimposed by horizontal shifts to produce a master curve.[3] This demonstrates an important feature in polymer behavior: the concept of time-temperature equivalence. In essence, a polymer at temperatures below room temperature will behave as if it were tested at a higher rate at room temperature. This principle can be applied to predict material behavior under testing rates or times that are not experimentally accessible through the use of shift factors (aT) and the equation below:

$$\ln a_T = \ln\left(\frac{t}{t_o}\right) = -\frac{17.44(T - T_g)}{51.6 + T - T_g} \tag{1.7}$$

where T_g is the glass transition temperature of the polymer.

1.4.2 Failure Behavior

The design of plastic parts requires the avoidance of failure without overdesign of the part, leading to increased part weight. The type of failure can depend on temperatures, rates, and materials. Some information on material strength can be obtained from simple tensile stress-strain behavior. Materials that fail at rather low elongations (1 percent strain or less) can be considered to have undergone brittle failure.[4] Polymers that produce this type of failure include general purpose polystyrene and acrylics. Failure typically starts at a defect where stresses are concentrated. Once a crack is formed, it will grow as a result of stress concentrations at the crack tip. Many amorphous polymers will also exhibit what are called *crazes*. Crazes appear to look like cracks, but they are load bearing, with fibrils of material bridging the two surfaces as shown in Fig. 1.10. Crazing is a form of yielding and, when present, can enhance the toughness of a material.

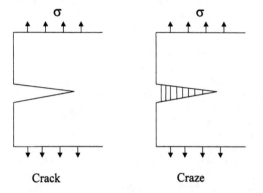

Crack   Craze

FIGURE 1.10   Cracks and crazes.

Ductile failure of polymers is exhibited by yielding of the polymer or slip of the molecular chains past one another. This is most often indicated by a maximum in the tensile stress-strain test or what is termed the *yield point*. Above this point, the material may exhibit lateral contraction upon further extension, termed *necking*.[5] Molecules in the necked region become oriented and result in increased local stiffness. Material in regions adjacent to the neck are thus preferentially deformed, and the neck region propagates. This process is known as *cold-drawing* (see Fig. 1.11). Cold drawing results in elongations of several hundred percent.

Under repeated cyclic loading, a material may fail at stresses well below the single-cycle failure stress found in a typical tensile test.[6] This process is called *fatigue* and is usually depicted by plotting the maximum stress versus the number of cycles to failure.

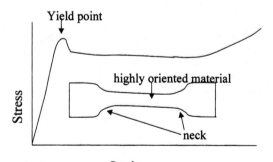

FIGURE 1.11   Ductile behavior.

Fatigue tests can be performed under a variety of loading conditions as specified by the service requirements. Thermal effects and the presence or absence of cracks are other variables to be considered when the fatigue life of a material is to be evaluated.

1.4.3 Effect of Fillers

The term *fillers* refers to solid additives, which are incorporated into the plastic matrix.[7] They are generally inorganic materials and can be classified according to their effect on the mechanical properties of the resulting mixture. Inert or extender fillers are added mainly to reduce the cost of the compound, whereas reinforcing fillers are added to improve certain mechanical properties such as modulus or tensile strength. Although termed inert, inert fillers can nonetheless affect other properties of the compound besides cost. In particular, they may increase the density of the compound, reduce the shrinkage, increase the hardness, and increase the heat deflection temperature. Reinforcing fillers typically will increase the tensile, compressive, and shear strengths, increase the heat deflection temperature, reduce shrinkage, increase the modulus, and improve the creep behavior. Reinforcing fillers improve the properties via several mechanisms. In some cases, a chemical bond is formed between the filler and the polymer; in other cases, the volume occupied by the filler affects the properties of the thermoplastic. As a result, the surface properties and interaction between the filler and the thermoplastic are of great importance. A number of filler properties govern their behavior, including the particle shape, the particle size and distribution of sizes, and the surface chemistry of the particle. In general, the smaller the particle, the greater the improvement in the mechanical property of interest (such as tensile strength).[8] Larger particles may give reduced properties compared to the pure thermoplastic. Particle shape can also influence the properties. For example, plate-like particles or fibrous particles may be oriented during processing, resulting in anisotropic properties. The surface chemistry of the particle is also important to promote interaction with the polymer and to allow for good interfacial adhesion. The polymer should wet the particle surface and have good interfacial bonding so as to obtain the best property enhancement.

Examples of inert or extender fillers include: china clay (kaolin), talc, and calcium carbonate. Calcium carbonate is an important filler, with a particle size of about 1 μm.[9] It is a natural product from sedimentary rocks and is separated into chalk, limestone, and marble. In some cases, the calcium carbonate may be treated to improve interaction with the thermoplastic. Glass spheres are also used as thermoplastic fillers. They may be either solid or hollow, depending on the particular application. Talc is a filler with a lamellar particle shape.[10] It is a natural, hydrated magnesium silicate with good slip properties. Kaolin and mica are also natural materials with lamellar structures. Other fillers include wollastonite, silica, barium sulfate, and metal powders. Carbon black is used as a filler primarily in the rubber industry, but it also finds application in thermoplastics for conductivity, for UV protection, and as a pigment. Fillers in fiber form are often used in thermoplastics. Types of fibers include cotton, wood flour, fiberglass, and carbon. Table 1.3 shows the fillers and their forms. An overview of some typical fillers and their effect on properties is shown in Table 1.4. Considerable research interest exists for the incorporation of nanoscale fillers into polymers. This aspect will be discussed in later chapters.

1.5 Rheological Properties

Viscosity is the resistance to flow. As shown in Table 1.5, polymer melts have viscosities of 100 to 1,000,000 Pa-s, whereas water has a viscosity of 0.001 Pa-s.[11] These high viscosities result from the long polymer chains and cause the polymer melt to exhibit laminar

TABLE 1.3 Forms of Various Fillers

Spherical	Lamellar	Fibrous
Sand/quartz powder	Mica	Glass fibers
Silica	Talc	Asbestos
Glass spheres	Graphite	Wollastonite
Calcium carbonate	Kaolin	Carbon fibers
Carbon black		Whiskers
Metallic Oxides		Cellulose
		Synthetic fibers

flow; that is, the melt moves in layers. Although, these melt layers may move at the same velocity, thereby producing plug flow, the melt layers typically flow at different the different velocities to provide shear. Changes in the cross-sectional area of the melt channel or drawing processes stretch or allow relaxation of the polymer chains, giving rise to elongation or extension.

The shear viscosity of polymer melts generally decreases with increasing shear rate. This pseudoplastic behavior contrasts with the shear-rate independent viscosity of fluids, such as water, solvents, and oligomers. The decrease in the viscosity of pseudoplastic fluids, however, does not occur immediately. At low shear rates, the polymer molecules flow as random coils, and the constant viscosity is called the *zero-shear rate viscosity* (η_o). With increasing shear rate, the polymer chains align in the direction of flow, and the viscosity decreases (Fig. 1.12). The shear rate corresponding to the onset of chain alignment or shear thinning increases with decreasing polymer molecular weight. When the viscosity decreases is proportional to the increase in shear rate, the viscosity can be modeled using:[12]

$$\eta = k\dot{\gamma}^{n-1} \tag{1.8}$$

where k is the consistency index and n is the power law index. The power law index is an indicator of a material's sensitivity to shear (rate), or the degree of non-Newtonian behavior. For Newtonian fluids $n = 1$, and for pseudoplastic fluids $n < 1$, with smaller values indicating greater shear sensitivity. Since shear rate varies considerably with the processing method (Table 1.6),[13] the degree of alignment, shear thinning, and material relaxation varies considerably with the process. Compression and rotational molding typically induce very little alignment of the polymer chains and thus produce low levels of orientation and retained stress. In contrast, the polymer chains are highly oriented during injection molded, and such parts exhibit high levels of residual stress.

As illustrated in Fig. 1.12, shear viscosity also decreases with temperature, since the polymer chains are more mobile. This temperature dependence of viscosity can be expressed using an Arrhenius equation:

$$\eta = A\exp\left(\frac{E_a}{RT}\right) \tag{1.9}$$

where A is a material constant, E_a is the activation energy (which varies with polymer and shear rate), R is a constant, and T is the absolute temperature. Since the activation energy depends on the difference between a polymer's processing and glass transition tempera-

TABLE 1.4 Effect of Filler Type on Properties[51]

	Glass fiber	Asbestos	Wollastonite	Carbon fiber	Whiskers	Synthetic fibers	Cellulose	Mica	Talc	Graphite	Sand/quartz powder	Silica	Kaolin	Glass spheres	Calcium carbonate	Metallic oxides	Carbon black
Tensile strength	++	+		+	+			+	O					+		+	+
Compressive strength	+								+		+			+	+		
Modulus of elasticity	++	++	++	++	+			++	+		+	+	+	+	+	+	+
Impact strength	+	−	−	−	−	++	+	+	−		−	−	−	−	+	−	+
Reduced thermal expansion	+	+		−	+			+	+		+	+	+			+	
Reduced shrinkage	+	+	+	+				+	+	+	+	+	+	+	+	+	+
Better thermal conductivity		+	+	+					+	+	+	+			+	+	+
Higher heat deflection temperature	++	+	+	++	+			+	+	+	+		+		+	+	
Electrical conductivity				+						+							+
Electrical resistance		+	+					++	+			+	++			+	
Thermal stability		+	+				+	+	+		+	+	+			+	+
Chemical resistance	+	+	+					+	O	+			+	+			
Better abrasion behavior				+				+	+	+			+				
Extrusion rate	+	+						+					+		+		
Machine abrasion	−	O			O	O	O		O	O	−			O	O		O
Price reduction	+	+	+				+	+	+	+	++	+	+	+	++	+	O

++ large influence, + influence, O no influence, − negative influence.

1.14

TABLE 1.5 Typical Viscosities

Material	Viscosity (Pa-s)
Air	10^{-5}
Water	10^{-3}
Polymer latexes	10^{-2}
Olive oil	10^{-1}
Glycerin	1
Polymer melts	$10^2 - 10^6$
Pitch	10^9
Plastics	10^{12}
Glass	10^{21}

TABLE 1.6 Typical Shear Rates for Selected Processes[52]

Process	Shear rate (s-1)
Compression molding	1–10
Calendering	10–100
Extrusion	100–1,000
Injection molding	1,000–10,000

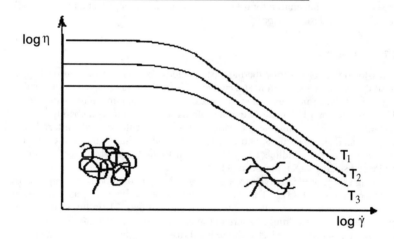

FIGURE 1.12 The effect of shear rate and temperature on viscosity, where $T_1 > T_2 > T_3$.

tures, materials such as polyethylene have activation energies less than 20 kJ/mol, whereas higher-temperature polymers, such as polycarbonate, exhibit activation energies that are greater than 50 kJ/mol. Pressure increases viscosity, but the effects are relatively insignificant when the processing pressures are less than 35 MPa (5,000 psi).[14] At higher pressures, the increase in viscosity is given by[15]

$$\eta = \eta_r \exp[\alpha_p (P - P_r)] \qquad\qquad (1.10)$$

where η_r is the viscosity at a reference P_r, and α_p is an empirical constant with values of 200 to 600 MPa^{-1}.

Shear viscosity increases with more rigid polymer structures, higher molecular weights, and additives such as fillers and fibers. Long chain branching and broader molecular weight distributions increase the shear sensitivity of viscosity. Blending two polymers can significantly alter polymer viscosity, but the effect depends on the two polymers. Additives such as lubricants typically decrease viscosity, whereas the effect of colorants and impact modifiers varies with type of additive.

In contrast, the effect of strain rate on extensional viscosity varies with the polymer structure. Branched polymers generally exhibit extensional thickening and a corresponding increase in viscosity. Linear polymers, such as LLDPE, undergo extensional thinning in which the viscosity decreases as the polymer sample necks. Generally, extensional viscosity is greater than shear viscosity and depends primarily on the molecular weight of the polymer

1.6 PROCESSING OF THERMOPLASTICS

Processing involves the conversion of the solid polymer into a desirable size and shape. There are a number of methods to shape the polymer, including injection molding, extrusion, thermoforming, blow molding, and rotational molding. The plastic material is heated to the appropriate temperature for it to flow, the material is shaped, and then it is cooled so as to preserve the desired shape.

1.6.1 Extrusion

In extrusion operations, a solid thermoplastic material is melted, forced through an orifice (die) of the desired cross section, and cooled. This method was adapted from metallurgists who use a similar form of extrusion to process molten aluminum and was first adapted in 1845 by Bewley and Brooman to extrude rubber around cable as a coating.[16] Extrusion processes are used to continuously produce film and sheet; shapes with uniform cross-sections, such as PVC pipe, tubes, and garden hose; profile with nonuniform cross-sections, such as PVC window moldings and gutters; synthetic fibers; polymer coatings for insulating wire and sealing paper, plastic, and metal packaging.

Although there are many types of extruders, the most common is the single-screw extruder (shown in Fig. 1.13).[17] This extruder consists of a screw in a metal cylinder or barrel. Electrical heater bands and fans that surround the barrel help bring the extruder to operating temperature during start-up and maintain barrel temperature during operation. One end of the screw is connected through a thrust bearing and gear box to a drive motor that rotates the screw in the barrel. The other end is free floating in the barrel. The barrel is connected to the feed throat, a separate "barrel section," with an opening called a feed port, and is connected to the feed hopper. A die adaptor is usually connected to the opposite end of the extruder. A breaker plate and a screen pack are sandwiched between the extruder and die adaptor. The breaker plate provides a seal between the extruder and die, converts the rotational motion of the melt (in the extruder) to linear motion (for the die), and supports the screen pack. The screen pack filters the melt, thereby prevent unmelted resin, degraded polymer, or other contaminants from producing defects in the extruded products and/or damaging the die.

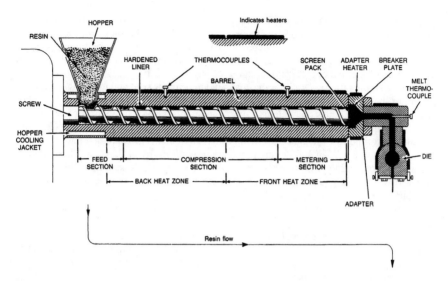

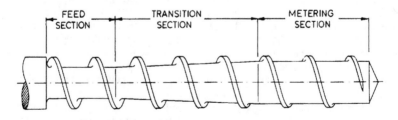

FIGURE 1.13   A single-screw extruder.

During extrusion, solid resin in the form of pellets or powder is fed from the hopper, through the feed port, and into the feed throat of the extruder. The solid resin falls onto the rotating screw and is packed into a solid bed in the first section of the screw (called the feed zone). The solid bed is melted as it travels through the middle section (transition zone) of the screw. The melt is mixed, and pressure is generated in the final section (metering zone) of the screw. Although the heater bands and cooling fans maintain the barrel at a set temperature profile, conduction from the barrel walls provides only 10 to 30 percent of the energy required to melt the resin. The remainder of the energy is generated from the frictional heat generated by the mechanical motion of the screw; this mechanism is called *viscous dissipation.*

Extruder screws are design to accommodate this pattern of packing, melting, and pressure generation. As illustrated in Fig. 1.14, the outside diameter of the screw, which is measured at the tops of the screw flights, remains constant.[18] The root diameter of the screw, however, changes. In the feed zone, the root diameter is small so that the large channel depth (i.e., distance between the outside and root diameters) can accommodate the packed solid resin particles. The root diameter of the transition or compression zone increases with the distance from the feed zone. This change in channel depth forces the solid

FIGURE 1.14   General-purpose extruder screw.[57]

into better contact with barrel wall, thereby promoting better melting. It also compresses the molten polymer in the screw channels. The root diameter becomes constant again in the metering zone, but the channel depth is very small. This geometry facilitates pressure generation and helps maintain the temperature of the polymer melt (i.e., polymers are poor conductors of thermal energy, and so thin melt layers have more uniform temperatures). The compression ratio (i.e., ratio of the channel depths in the feed and metering zones) and length of the transition zone significantly affect the melting in the single-screw extruders. Typically, extruder screws have length to diameter (L/D) ratios of about 30:1, with each zone requiring about one-third of the screw length. Barrier screws are used to improve melting performance while an assortment of mixing elements incorporated into the metering zone enhance mixing and the melt temperature uniformity of the melt. These include the addition of mixing pins on the barrel of the screw, ring barriers, and modified designs that involve very large screw diameters so as to force molten polymer through a small clearance between the mixing head and the inside of the barrel wall. Two stage-screws permit devolatilization of polymer melts, thereby eliminating entrapped moisture, air, and other volatiles from the melt.

Typical extruders have diameters of 25 to 150 mm, but this can vary from 20 to 600 mm (6 to 24 in). They typically operate at 1 to 2 rev/s (60 to 120 rpm) for large extruders and 1 to 5 rev/s (60 to 300 rpm) for small extruders.[19] Output varies as a function of processing parameters (particularly screw speed and pressure), the thermal and mechanical properties of the polymer, and the design and geometry of the screw. A 600-mm dia single-screw extruder is capable of delivering 29 metric tons of product an hour, whereas the smallest 20-mm dia single-screw extruders have a throughput capacity of 5 kg/h.[20] Operating pressures are typically 1 to 35 MPa (200 to 5000 psi).

Single-screw extruders account for 90 percent of all extruders, with the three types of twin-screw extruders making up the bulk of the remaining 10 percent. In nonintermeshing (tangential) extruders, the counter-rotating screws do not interlock with each other and convey the polymer using drag flow (i.e., like a single-screw extruder). These extruders permit tight control of heating and shear and so have been used for devolatilization, coagulation, reactive extrusion, and halogenation of polyolefins.[21] With intermeshing twin-screw extruders (Fig. 1.15), the flights of one screw fit into the channels of the other, and polymer is transferred from the channels of one screw to those of the other, thereby providing positive conveyance of the polymer and increased mixing. In counter-rotating, intermeshing twin-screw extruders, some material flows between the screws and the barrel wall, and the remainder is forced between the two screws. Polymer in co-rotating twin screws moves in a figure-eight pattern around the two screws, with little material flowing between the screws. The longer flow path produces longer extruder residence times than observed with counter-rotating, intermeshing twin-screw extruders but increases the degree of elongational flow and enhances mixing. Intermeshing twin-screw extruders are typically used in applications where mixing and compounding need to be accomplished, because the screws' elements can be rearranged (programmed) to suit a specific application. They are highly capable of dispersing small agglomerates such as carbon black and can be used, for example, to blend the components of duct tape adhesive as well as coat the finished adhesive onto the tape backing. Counter-rotating, intermeshing twin-screw extruders, which permit tight control of shear and residence time, are also employed for the extrusion of PVC pipes and profiles.[22] Although twin-screw extruders have relatively low pressure-generating capabilities, some materials can be compounded and formed directly if a gear pump is added to the end of the extruder.

Die designs depend on the product that will be formed. Typically, spiral flow and spider arm dies are used for blown film, tubing, and pipes. Crosshead dies are employed for tubing and wire coating. Wide dies with tee, coat hanger, and exponential are employed

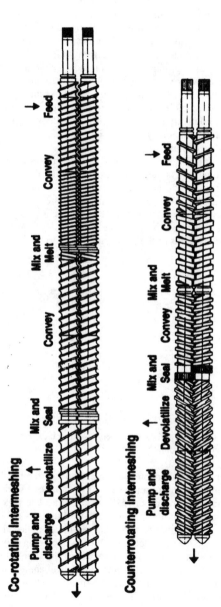

FIGURE 1.15 Intermeshing twin screws.[58]

for film, sheet, and extrusion coating. In die design, it is critical to avoid "dead spots" where the polymer melt can become stagnant and risk thermal degradation. It is also important that the polymer molecules be allowed to return to an equilibrium position to the greatest extent possible to minimize the orientation as a result of flow. Laminar flow is desired, and finite element analysis is used to design dies that enable laminar flow to the greatest extent. Multimanifold dies, such as plate dies, and feedblocks (along with film, sheet, and extrusion dies) combine melt streams from multiple single-screw extruders to produce co-extruded multilayer products. This common technique is used for producing multilayer packaging films, where each layer provides a particular feature. For example, garbage bags are often multilaminate constructions, as are packaging films where a PVDC layer may be incorporated for moisture or oxygen barrier properties, and HDPE may be used as a less-expensive, relatively strong, layer. EVA is a common "bonding layer" between different plastic layers. As many as eight or more extruders may be used to form highly specialized, multilayer films.

Common defects encountered with extrusion include effects associated with the viscoelastic nature of plastic melts. As the melt is extruded from the die for example, it may exhibit sharkskin melt fracture and extrudate (die) swell. Diagrams of these defects are shown in Fig. 1.16.[23] Sharkskin melt fracture occurs when the stresses being applied to the plastic melt exceed its tensile strength. Extrudate swell occurs due to the elastic component of the polymer melt's response to stress and is the result of the elastic rebound of the polymer as it leaves the constraints of the die channel prior to cooling.

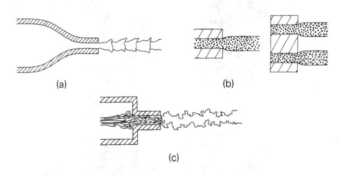

(a)   (b)

(c)

FIGURE 1.16 Common defects described from rod dies: (a) shark-skinning, (b) die swell, and (c) melt fracture.[23]

Pressure generated in the extruder forces the melt through the breaker plate, die adaptor, and die. The die forms the melt into the desired shape. Downstream equipment, such as a water bath cools the melt, and a puller draws the extrudate away from the die and through the water bath away from the die. Figure 1.17 illustrates the downstream equipment for tube extrusion. The annular tube exiting the die is pulled though a calibration unit, which maintains the outside diameter of the tube, while being cooled by a water bath. The puller stretches the molten tube, and a cutter slices the tube into preset lengths. In blown film extrusion (Fig. 1.18),[24] the melt forced though an annular die is expanded into a bubble using air blown through a hole in the die mandrel, stretched axially by take-up rolls, and cooled by forced convention. This biaxial orientation, thinning of the tube of film through the internal pressurization of the bubble, combined with the thinning of the film as it is stretched upwards, results in a strong, biaxially oriented film. Stretching continues until the freezing line is reached, at which point the film has cooled off to such an extent as to provide a high enough modulus to resist further deformation. Crystallization

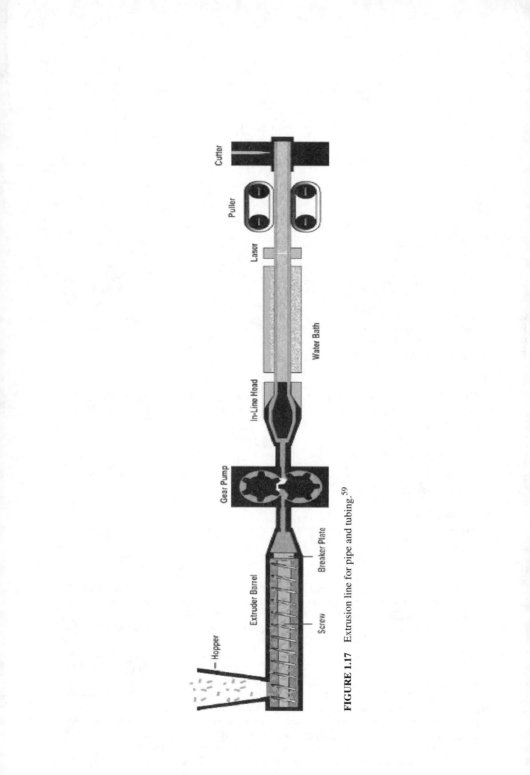

FIGURE 1.17 Extrusion line for pipe and tubing.[59]

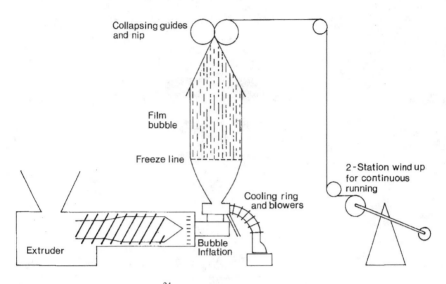

FIGURE 1.18 Blown film process.[24]

also enables the orientation to be maintained. A pair of collapsing rolls is used to flatten the bubble and allow the film to then be wound into a master roll for later converting processes such as slitting.

1.6.2 Injection Molding

Injection molding is a widely used process to produce parts with variable dimensions. An injection molding machine consists of the following four components:[25]

- Injection unit
- Control systems
- Drive system
- Clamping unit

The purpose of the injection unit is to heat and melt the polymer, inject the melt into the cavity, and apply pressure during the cooling phase. The most common type of injection molding machine is the reciprocating screw. In this type of machine, the screw rotates to plasticize the polymer, moving backward to deposit a volume of polymer melt ahead of the screw (shot). Once the correct shot size has been built, the screw then moves forward to inject the melt into the mold. Injection molding is a discontinuous process, and the clamping unit allows for the mold to open and close for part removal and to provide pressure as the cavity is filled. This is depicted in Fig. 1.19.

The purpose of an injection mold is to give the shape of the part (cavity), distribute the polymer melt to the cavities through a runner system, cool the part, and eject the part. During the injection molding cycle, the polymer flows from the nozzle on the injection unit through the sprue, then to the runners, which distribute the melt to each of the cavities. The entrance to the cavity is called the gate and is usually small so that the runner system can be easily removed from the part. A typical feed system for injection molding is shown

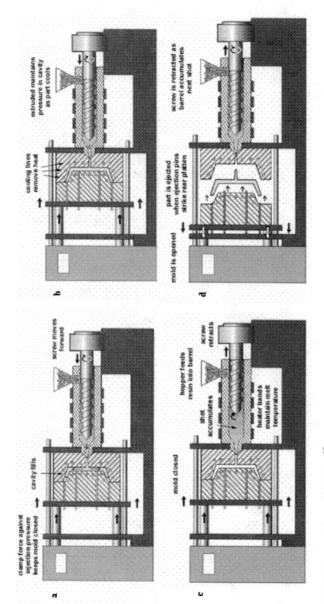

FIGURE 1.19 Injection molding.[60]

in Fig. 1.20. Figure 1.21 depicts a number of gate configurations. Molding conditions for a wide variety of thermoplastics are given in Table 1.7.

The molding process itself can have a large influence on the final properties of the part. The polymer chains undergo orientation in the flow direction during the mold filling phase of the injection cycle as shown in Fig. 1.22. The amount of orientation in the final part depends on how much orientation was induced during filling minus the amount that was been removed through molecular relaxation.[26] This is particularly true of the surface of the part, where hot material reaches the cool walls of the mold with rapid solidification coupled with the highest shear and induced orientation at the mold surface. Orientation can result in parts with anisotropic properties and should be accounted for during part design. Mechanical properties are thus typically higher in the direction of orientation.[27]

For semicrystalline polymers, the injection molding process parameters can have a large impact on the degree of crystallinity. As the cooling rate increases, the degree of crystallinity will decrease.[28] Cooling rate effects can cause a gradient of crystallinity across the thickness of a part where interior portions of the part may have higher crystallinity due to their slower cooling rates compared to the surface. The crystalline morphology will also be influenced by the cooling behavior. Slower cooling rates result in larger spherulites, while more rapid cooling rates result in a larger number of smaller spherulites.[29] A number of specialized injection molding processes also exist and are outlined below.

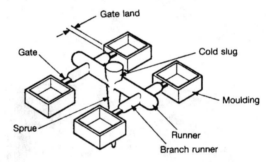

FIGURE 1.20   Injection molding feed system.[61]

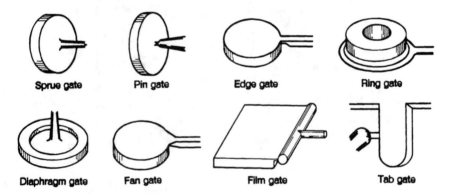

FIGURE 1.21   Injection molding gate types.[62]

TABLE 1.7 Injection Molding Guidelines for Unfilled Materials [53]

Material	Melt temp. range, °C	Mold temp. range, °C	Drying temp., °C	Drying times, hr
ASA	250–265	40–80	80–85	2–4
ABS	220–260	60–90	80–85	2–4
BDS	190–230	10–60	60	1
CA/CAB/CAP	160–230	40–80	55–85	3–4
FEP	300–380	200–240	150	2–4
HIPS/TPS	200–270	10–80	65–70	3–4
PA6	230–280	60–90	80–105	12–16
PA66	260–290	20–100	85–105	5–12
PA11/PA12	240–300	30–100	85	3–5
PBT	220–260	20–110	120–150	2.5–5.5
PC	280–320	80–120	120	2–4
PEBA	185–220	20–40	70–80	2–6
PEEL	195–255	10–70	90–120	10
PEEK	360–380	160–170	150	3
PE-HD	205–280	10–60	65	3
PE-LD	180–280	10–60	65	3
PE-LLD	160–280	10–60	65	3

TABLE 1.7 Injection Molding Guidelines for Unfilled Materials (*Continued*)[53]

Material	Melt temp. range, °C	Mold temp. range, °C	Drying temp., °C	Drying times, hr
PES	350–380	140–160	135–150	3–4
PET/PETP	265–295	120–140	135–165	2–4
PMMA	210–270	60–90	75	2–4
POM-H	190–215	40–120	110	2–3
POM-CO	175–220	40–120	110	2–3
PPO-M	260–300	60–110	100	2
PPS	300–360	135–160	150	3–6
PP	220–275	30–80	80	2–3
GPPS	200–250	10–80	70	2–3
PSU	350–380	100–150	135–150	3–4
PVDF	180–300	30–120	80	2–4
SAN	200–270	40–80	70–75	3–4
TPU/PUR	180–230	15–70	80	3
UPVC	185–205	30–60	65	2–3
PPVC	175–200	30–50	65	2
EVA	140–225	15–40	50–60	8

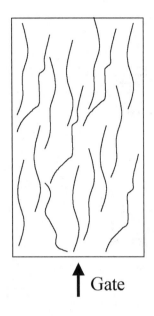

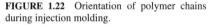

FIGURE 1.22 Orientation of polymer chains during injection molding.

Gate

1.6.2.1 Injection/Compression Molding. Injection/compression molding refers to the process wherein the cavity is not completely filled during injection of the resin.[25] In this process, the resin is injected while the mold is slightly open. The two halves of the mold then close, distributing the resin and filling the cavity. This process is useful for products that require high surface replication, such as compact discs or optical parts. Thin-walled parts can also be molded by this process, as the pressure losses are reduced, and there is less risk of premature resin solidification. Figure 1.23 illustrates this process.

1.6.2.2 Lost-Core Process. Products that are hollow or contain complex undercuts can be fabricated using the lost-core process as illustrated in Fig. 1.24. Core materials are typically low-melting alloys (around 150°C), that are removed by heating the part.[25] Before each molding cycle, a core is inserted into the mold, and the part is injection molded. The core is ejected with the part and then melted, resulting in the finished product. It is important that the core material melt at temperatures low enough that the plastic material is not affected by the heating cycle. Air manifolds for automotive and pump parts are often fabricated using this method.

1.6.2.3 Gas-Assisted Injection Molding. In gas-assisted injection molding the mold is partially filled with polymer, followed by a gas, which presses the polymer out to the surface of the mold, resulting in a hollow part. This process can be used for producing lighter-weight parts, often with reduced cycle times as a result of less material to cool. Thick-walled parts can be produced with fewer surface imperfections, such as sink marks, but equipment costs will be higher. Figure 1.25 shows the gas-assist injection molding method.

1.6.2.4 Coinjection Molding. Coinjection molding refers to a process whereby two materials are injected into the same cavity.[25] The first material is injected into the cavity and

step 1: injection

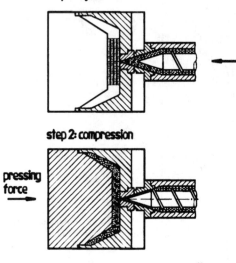

step 2: compression

pressing
force

FIGURE 1.23   Injection compression process.[63]

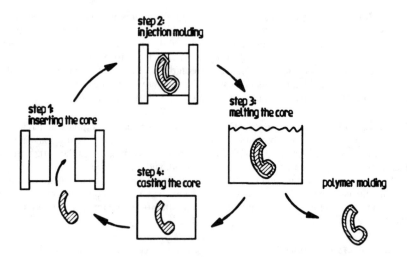

step 2:
injection molding

step 1:
inserting the core

step 3:
melting the core

step 4:
casting the core

polymer molding

FIGURE 1.24   Lost-core injection molding process.[64]

then followed by the second material as depicted in Fig. 1.26. In this process, the first material goes to the outside of the mold and forms the skin, and the second material forms the interior of the part. This is often referred to as *sandwich molding*. Materials may be injected either sequentially or simultaneously. Applications include the use of an expensive outer layer and a cheaper core material or with fiber reinforced materials, where a skin material is used for improved surface quality.

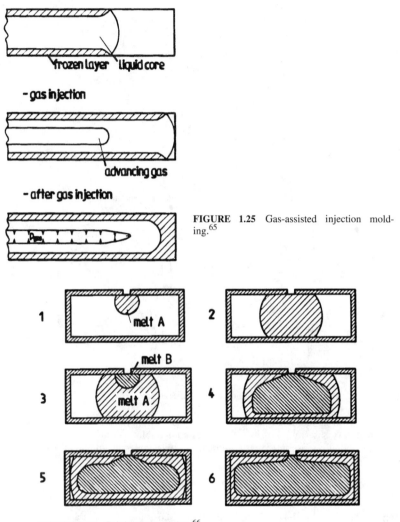

FIGURE 1.25 Gas-assisted injection molding.[65]

FIGURE 1.26 Coinjection molding.[66]

1.6.2.5 Two-Shot Injection Molding. Two-shot or overmolding refers to a process whereby either different colors or different materials are molded into one part. In this process, the first material or color is injected, then the mold is rotated, and the second shot is injected as depicted in Fig. 1.27. An alternative method is to use a retractable core.[67] In this case, the first material is injected, cooled to solidify, and then the core is retracted to allow injection of the second material as shown in Fig. 1.28. Bonding is accomplished through either strictly mechanical means or by adhesion between the two components through diffusion of the chains. This can result in parts with two materials combined with-

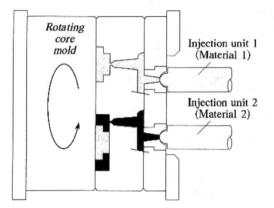

FIGURE 1.27 Two-shot molding with rotating mold.[67]

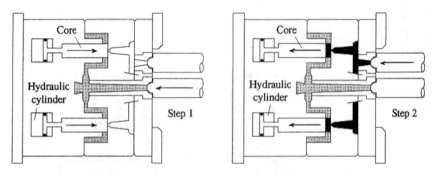

FIGURE 1.28 Two-shot injection molding with retractable core.[67]

out the need for an additional adhesive bonding step.[30] In the case where direct adhesion of the materials is desired, proper selection of compatible materials is required. Table 1.8 shows the bonding strength for a number of thermoplastic combinations for use in multicomponent injection molding.

1.6.3 Thermoforming

Thermoforming is the heating of a thermoplastic sheet until it is soft and stretchable and then forcing the hot sheet against the contours of a mold by mechanical (plug assist), vacuum, pressure, or a combination of all three. After cooling, the plastic sheet retains the mold's shape and detail.[31] Thermoforming is still a rapidly growing processing method because of the range of products that can be formed and the relatively low cost of required tooling and equipment.[32] Thermoformed products include dinnerware, cups, automotive parts, egg cartons, and blister packaging.[33]

There are a wide variety of processes for thermoforming. One-step processes include the following:[34]

- Drape forming
- Vacuum forming

TABLE 1.8 Bonding for Thermoplastic Combinations in Multicomponent Injection Molding [54]

Material	ABS	ASA	CA	EVA	PA 6	PA 6,6	PC	HDPE	LDPE	PMMA	POM	PP	PPO mod.	PS-GP	PS-HI	PBTP	TPU	PVC (soft)	SAN	TPR	PETP	PVAC	PSU	PC-PBTP	PC-ABS
ABS	+	+	+	+	+	+	+	−	−	+	−	−	−	N	N	+	+	+	+	N	+		+	+	+
ASA	+	+	+	+	+	+	+	−	−	+	−	−	−	N	−	+	+	+	+	N	+		+	+	+
CA	+	+	+	N				−	−		−	−	−	−	−	+	+	+	+	−					
EVA	+	+	N	+				+	+			+		+	+			−	+						
PA 6	+	+			+	+	+	N	N		−	N	−	−	−	+	+		+	−				+	+
PA 6,6	+	+			+	+	N	N	N		−	−	−	−	−	+	+		+	−	+			+	+
PC	+	+			+	N	+	−	−		−	−	−	−	−	+	+		+	−			+	+	+
HDPE	−	−	−	+	N	N	−	+	+	N	N	−	−	−	−	−	−	−	−	N	−		−	−	−
LDPE	−	−	−	+	N	N	−	+	+	N	N	+	−	N	−	−	−	−	−	N	−		−	−	−
PMMA	+	+						N	N	+	+	N	−	−	−	−		+	+		−			+	+
POM	−	−	−		−	−	−	N	N			−	−												
PP	−	−	−	+	N	−	−		+	N	−	+	−	−	−	−	−	−	−	+	−		−	−	−
PPO mod.	−	−	−		−	−	−	−	−	−	−	−	+	+	+	−	−	−	N	+	−		−	−	−

TABLE 1.8 Bonding for Thermoplastic Combinations in Multicomponent Injection Molding (*Continued*)[54]

Material	ABS	ASA	CA	EVA	PA 6	PA 6,6	PC	HDPE	LDPE	PMMA	POM	PP	PPO mod.	PS-GP	PS-HI	PBTP	TPU	PVC (soft)	SAN	TPR	PETP	PVAC	PSU	PC-PBTP	PC-ABS
PS-GP	N	N	−	+	−	−	−	−	N	−	−	−	+	+	+	−	−	−	−	−	−		−	−	−
PS-HI	N	−	−	+	−	−	−	−	−	−	−	−	+	+	+	−	−	−	−	N	−		−	−	−
PBTP	+	+	+		+	+	+	−	−	−	−	−	−	−	−	+	+	+	+	−	+		+	+	+
TPU	+	+	+		+	+	+	−	−			−	−	−	−	+	+	+	+	−	+			+	+
PVC (soft)	+	+	+	−				−	−	+		−	−	−	−	+	+	+	+					+	+
SAN	+	+	+	+	+	+	+	−	−	+		−	N	−	−	+		+	+					+	+
TPR	N	N	−		−	−	−	N	N			+	+	−	N					+	−		−	−	−
PETP	+	+					+	−	−	−	−		−	−	−	+				−			+	+	+
PVAC																						+			
PSU	+	+					+	−	−		−	−	−	−	−	+				−	+		+	+	+
PC-PBTP	+	+			+	+	+	−	−	+	−	−	−	−	−	+	+	+	+	−	+		+	+	+
PC-ABS	+	+			+	+	+	−	−	+	−	−	−	−	−	+	+	+	+	−	+		+	+	+

+ Good bonding; − poor bonding; N no bonding; blank, not evaluated.

- Pressure forming
- Free blowing
- Matched die molding

Drape forming, as shown in Fig. 1.29, involves either lowering the heated sheet onto a male mold or raising the mold into the sheet. Usually, either vacuum or pressure is used to force the sheet against the mold. In vacuum forming (Fig. 1.30), the sheet is clamped to the edges of a female mold, then vacuum is applied to force the sheet against the mold. Pressure forming is similar to vacuum forming except that air pressure is used to form the part (Fig. 1.31). In free blowing, the heated sheet is stretched by air pressure into shape, and the height of the bubble is controlled using air pressure. As the sheet expands outward, it cools into a free-form shape as shown in Fig. 1.32. This method was originally developed for aircraft gun enclosures. Matched die molding (Fig. 1.33) uses two mold halves to form the heated sheet. This method is often used to form relatively stiff sheets.

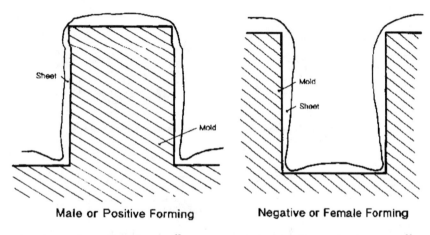

Male or Positive Forming

FIGURE 1.29 Drape-forming process.[68]

Negative or Female Forming

FIGURE 1.30 Vacuum-forming process.[66]

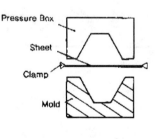

FIGURE 1.31 Pressure forming.[69]

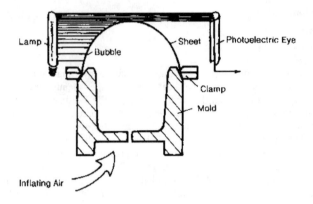

FIGURE 1.32 Free-blowing process.[69]

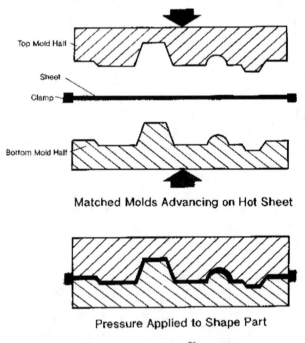

Matched Molds Advancing on Hot Sheet

Pressure Applied to Shape Part

FIGURE 1.33 Matched die thermoforming.[70]

Multistep forming is used in applications for thicker sheets or complex geometries with deep draw. In this type of thermoforming, the first step involves prestretching the sheet by techniques such as billowing or plug assist. After prestretching, the sheet is then pressed against the mold. Multistep forming includes the following:[35]

- Billow drape forming
- Billow vacuum forming
- Vacuum snap-back forming
- Plug assist vacuum forming
- Plug assist pressure forming
- Plug assist drape forming

Billow drape forming consists of a male mold pressed into a sheet prestretched by the billowing process (Fig. 1.34). A similar process is billow vacuum forming, wherein a female mold is used (Fig. 1.35). In vacuum snap-back forming, vacuum is used to prestretch the sheet, then a male mold is pressed into the sheet and, finally, pressure is used to force the sheet against the mold as seen in Fig. 1.36. In plug assist, a plug of material is used to prestretch the sheet. Either vacuum or pressure is then used to force the sheet against the walls of the mold as shown in Figs. 1.37 and 1.38. Plug assist drape forming is used to force a sheet into undercuts or corners (Fig. 1.39). The advantage of prestretching the sheet is more uniform wall thickness.

Materials suitable for thermoforming must be compliant enough to allow for forming against the mold, yet not produce excessive flow or sag while being heated.[36] Amorphous materials generally exhibit a wider process window than semicrystalline materials. Processing temperatures are typically 30 to 60°C above T_g for amorphous materials and usually just above T_m in the case of semicrystalline polymers.[37] Amorphous materials that are thermoformed include PS, ABS, PVC, PMMA, PETP, and PC. Semicrystalline materials

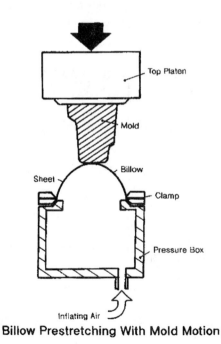

Billow Prestretching With Mold Motion

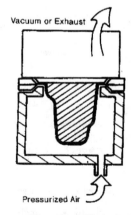

Vacuum/Pressure Forming

FIGURE 1.34 Billow drape forming.[71]

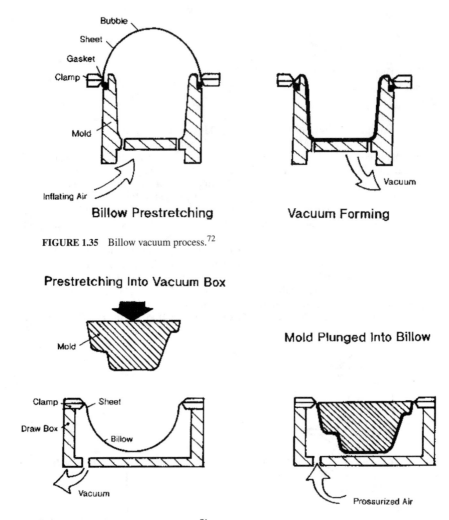

FIGURE 1.35   Billow vacuum process.[72]

FIGURE 1.36   Vacuum snap-back process.[71]

that can be successfully thermoformed include PE and nucleated PETP. Nylons typically do not have sufficient melt strength to be thermoformed. Table 1.9 shows processing temperatures for thermoforming a number of thermoplastics.

1.6.4   Blow Molding

Blow molding is a technique for forming nearly hollow articles and is very commonly practiced in the formation of PET soft-drink bottles. It is also used to make air ducts, surfboards, suitcase halves, and automobile gasoline tanks.[38] Blow molding involves taking a parison (a tubular profile) and expanding it against the walls of a mold by inserting pres-

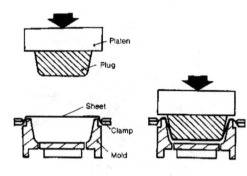

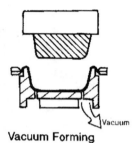

Plug Moving Into Hot Sheet   Plug Bottoming Out   Vacuum Forming

FIGURE 1.37   Plug assist vacuum forming.[73]

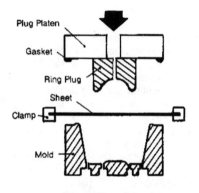

Ring Plug Moving
Into Hot Sheet

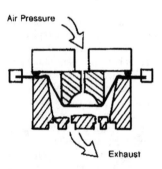

Pressure Balance
During Plugging

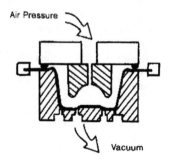

Vacuum/Pressure Forming

FIGURE 1.38   Plug assist pressure forming.[74]

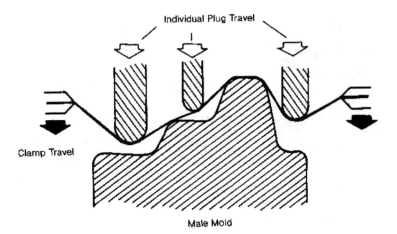

FIGURE 1.39 Plug assist drape forming.[74]

surized air into it. The mold is machined to have the negative contour of the final desired finished part. The mold, typically a mold split into two halves, then opens after the part has cooled to the extent that the dimensions are stable, and the bottle is ejected. Molds are commonly made out of aluminum, as molding pressures are relatively low, and aluminum has high thermal conductivity to promote rapid cooling of the part. The parison can either be made continuously with an extruder, or it can be injection molded; the method of parison production governs whether the process is called extrusion blow molding or injection blow molding. Figure 1.40 shows both the extrusion and injection blow molding processes.[39] Extrusion blow molding is often done with a rotary table so that the parison is extruded into a two-plate open mold, and the mold closes as the table rotates another mold under the extruder's die. The closing of the mold cuts off the parison and leaves the characteristic weld line on the bottom of many bottles as evidence of the pinch-off. Air is then blown into the parison to expand it to fit the mold configuration, and the part is then cooled and ejected before the position rotates back under the die to begin the process again. The blowing operation imparts radial and longitudinal orientation to the plastic melt, strengthening it through biaxial orientation. A container featuring this biaxial orientation is more optically clear, has increased mechanical properties, and reduced permeability, which is important in maintaining carbonation in soft drinks.

Injection blow molding has very similar treatment of the parison, but the parison itself is injection molded rather than extruded continuously. There is evidence of the gate on the bottom of the bottles rather than having a weld line where the parison was cut off. The parison can either be blown directly after molding while it is still hot, or it can be stored and reheated for the secondary blowing operation. An advantage of injection blow molding is that the parison can be molded to have finished threads. Cooling time is the largest part of this cycle and is the rate-limiting step. HDPE, LDPE, PP, PVC, and PET are commonly used in blow molding operations.

1.6.5 Rotational Molding

Rotational molding, also known as *rotomolding* or *centrifugal casting*, involves filling a mold cavity, generally with powder, and rotating the entire heated mold along two axes to

TABLE 1.9 Thermoforming Process Temperatures for Selected Materials[55]

Material	Mold and set temperature, °C	Lower processing limit, °C	Normal forming temperature, °C	Upper temperature limit, °C
ABS	85	127	149	182
Acetate	71	127	149	182
Acrylic	85	149	177	193
Butyrate	79	127	146	182
Polycarbonate	138	168	191	204
Polyester (PETG)	77	121	149	166
Polyethersulfone	204	274	316	371
Polyethersulfone-glass filled	210	279	343	382
HDPE	82	127	146	182
PP	88	129	154–166	166
PP-glass filled	91	129	204+	232
Polysulfone	163	190	246	302
Polystyrene	85	127	149	182
FEP	149	232	288	327
PVC -rigid	66	104	138–141	154

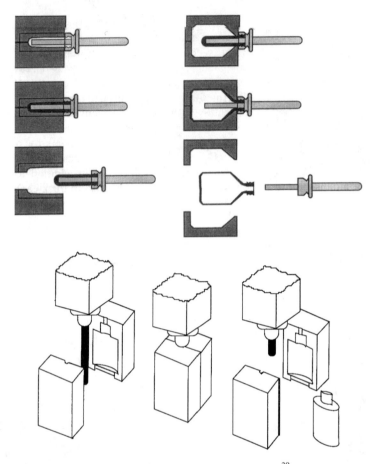

FIGURE 1.40 Extrusion and injection blow molding processes.[39]

uniformly distribute the plastic along the mold walls. This method is commonly used for making hollow parts, like blow molding, but is used either when the parts are very large (as in the case of kayaks, outdoor portable toilets, phone booths, and large chemical storage drums) or when the part requires very low residual stresses. Also, rotomolding is well suited, compared with blow molding, if the desired part design is complex or if it requires uniform wall thicknesses. Part walls produced by this method are very uniform as long as neither of the rotational axes corresponds to the centroid of the part design. The rotomolding operation imparts no shear stresses to the plastic, and the resultant molded article is therefore less prone to stress cracking, environmental attack, or premature failures along stress lines. Molded parts also are free of seams. Figure 1.41 shows a diagram of a typical rotational molding process.[40]

This is a relatively low-cost method, as molds are inexpensive and energy costs are low, thus making it suitable for short-run products. The drawback is that the heating and cooling times required are long, and therefore the cycle time is correspondingly long. High melt flow index PEs are often used in this process.

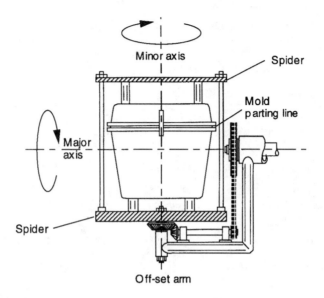

FIGURE 1.41 The rotational molding process.[40]

1.6.6 Foaming

The act of foaming a plastic material results in products with a wide range of densities. These materials are often termed *cellular plastics*. Cellular plastics can exist in two basic structures: closed-cell or open-cell. Closed-cell materials have individual voids or cells that are completely enclosed by plastics, and gas transport takes place by diffusion through the cell walls. In contrast, open-cell foams have cells that are interconnected, and fluids may pass easily between the cells. The two structures may exist together in a material so that it may be a combination of open and closed cells.

Blowing agents are used to produce foams, and they can be classified as either physical or chemical. Physical blowing agents include

- Incorporation of glass or resin beads (syntactic foams)
- Inclusion of an inert gas, such as nitrogen or carbon dioxide into the polymer at high pressure, which expands when the pressure is reduced
- Addition of low boiling liquids, which volatilize on heating, forming gas bubbles when pressure is released

Chemical blowing agents include

- Addition of compounds that decompose over a suitable temperature range with the evolution of gas
- Chemical reaction between components

The major types of chemical blowing agents include the azo compounds, hydrazine derivatives, semicarbazides, tetrazoles, and benzoxazines.[41] Table 1.10 shows some of the common blowing agents, their decomposition temperatures, and primary uses.

TABLE 1.10 Common Chemical Blowing Agents[56]

Blowing agent	Decomposition temp., °C	Gas yield, ml/g	Polymer applications
Azodicarbonamide	205–215	220	PVC, PE, PP, PS, ABS, PA
Modified azodicarbonamide	155–220	150-220	PVC, PE, PP, EVA, PS, ABS
4,4'-Oxybis(benzene-sulfohydrazide)	150–160	125	PE, PVC, EVA
Diphenylsulfone-3,3'-disulfohydrazide	155	110	PVC, PE, EVA
Trihydrazinotriazine	275	225	ABS, PE, PP, PA
p-Toluylenesulfonyl semicarbazide	228–235	140	ABS, PE, PP, PA, PS
5-Phenyltetrazole	240–250	190	ABS, PPE, PC, PA, PBT, LCP
Isatoic anhydride	210–225	115	PS, ABS, PA, PPE, PBT, PC

A wide range of thermoplastics can be converted into foams. Some of the most common materials include polyurethanes, polystyrene, and polyethylene. Polyurethanes are a popular and versatile material for the production of foams and may be foamed by either physical or chemical methods. In the physical reaction, an inert low-boiling chemical is added to the mixture, which volatilizes as a result of the heat produced from the exothermic chemical reaction to produce the polyurethane (reaction of isocyanate and diol). Chemical foaming can be done through the reaction of the isocyanate groups with water to produce carbamic acid, which decomposes to an amine and carbon dioxide gas.[42]

Rigid polyurethane foams can be formed by pour, spray, and froth.[43] Liquid polyurethane is poured into a cavity and allowed to expand in the pour process. In the spray method, heated two-component spray guns are used to apply the foam. This method is suitable for application in the field. The froth technique is similar to the pour technique except that the polyurethane is partially expanded before molding. A two-step expansion is used for this method using a low-boiling agent for preparation of the froth and a second higher-boiling agent for expansion once the mold is filled.

Polyurethane foams can also be produced by reaction injection molding or RIM.[44] This process combines low-molecular-weight isocyanate and polyol, which are accurately metered into the mixing chamber and then injected into the mold. The resulting structure consists of a solid skin and a foamed core.

Polystyrene foams are typically considered either as extruded or expanded bead.[45] Extruded polystyrene foam is produced by extrusion of polystyrene containing a blowing agent and allowing the material to expand into a closed cell foam. This product is used extensively as thermal insulation. Molded expanded polystyrene is produced by exposing polystyrene beads containing a blowing agent to heat.[46] If the shape is to be used as loosefill packaging, then no further processing steps are needed. If a part is to be made, the beads are then fused in a heated mold to shape the part. Bead polystyrene foam is used in thermal insulation applications, flotation devices, and insulated hot and cold drink cups.

Polyethylene foams are produced using chemical blowing agents and are typically closed-cell foams.[47] Cellular polyethylene offers advantages over solid polyethylene in terms of reduced weight and lower dielectric constant. As a result, these materials find application in electrical insulation markets. Polyethylene foams are also used in cushioning applications to protect products during shipping and handling.

1.7 REFERENCES

1. F. W. Billmeyer, *Textbook of Polymer Science,* 2nd ed., John Wiley & Sons, Inc., New York, 1971.
2. A.W. Birley, B. Haworth, and J. Batchelor, *Physics of Plastics,* Carl Hanser Verlag, Munich, 1992.
3. M.L. Williams, R.F. Landel, and J.D. Ferry, *J. Am. Chem. Soc.,* 77, 3701 (1955).
4. P.C. Powell, *Engineering with Polymers,* Chapman and Hall, London, 1983.
5. A.W. Birley, B. Haworth, and J. Batchelor, *Physics of Plastics,* Carl Hanser Verlag, Munich, 1992, pp. 283–284.
6. L.E. Nielsen and R.F. Landel, *Mechanical Properties of Polymers and Composites,* Marcel Dekker, New York, 1994, pp. 342–352.
7. A.W. Bosshard and H.P. Schlumpf, "Fillers and Reinforcements," in *Plastics Additives,* 2nd ed., R. Gachter and H. Muller, Eds., Hanser Publishers, New York, 1987, p. 397.
8. Brydson, J.A., *Plastics Materials,* 6th ed., Butterworth-Heinemann, Oxford, 1995, p. 122.
9. A.W. Bosshard and H.P. Schlumpf, "Fillers and Reinforcements," in *Plastics Additives,* 2nd ed. R. Gachter and H. Muller, Eds., Hanser Publishers, New York, 1987, p. 407.
10. A.W. Bosshard and H.P. Schlumpf, "Fillers and Reinforcements," in *Plastics Additives,* 2nd ed., R. Gachter and H. Muller, Eds., Hanser Publishers, New York, 1987, p. 420.
11. Sperling, L. H., *Introduction to Physical Polymer Science,* 2nd ed., John Wiley and Sons, New York (1992), 487.
12. W. Ostwald, *Kolloid Z.,* 36, 99 (1925).
13. Morton-Jones, D. H., *Polymer Processing,* Chapman Hall, New York (1989), 35.
14. Rauwendaal, C., Polymer Extrusion, 2nd ed., Hanser Publishers, New York (1990), 190.
15. Carreau, P. J., De Kee, D. C. R., and Chhabra, R. P., *Rheology of Polymeric Systems—Principles and Applications,* Hanser Publishers, New York (1997), 52.
16. Osswald, T.A., *Polymer Processing Fundamentals,* Hanser/Gardner Publications, New York, 1998, p. 67.
17. Brydson, J. A., *Plastics Materials,* 5th ed., London, England: Butterworths, 1989, p. 151.
18. C. Rauwendaal, *Polymer Extrusion,* 2nd ed., Hanser/Gardner Publications, Cincinnati, OH (1990), p. 24.
19. Osswald, T.A., *Polymer Processing Fundamentals,* Hanser/Gardner Publications, New York, 1998, p. 70.
20. *Encyclopedia of Polymer Science and Engineering,* 2nd ed., Vol. 6, Mark, Bilkales, Overberger, Menges, Kroschwitz, Eds., Wiley Interscience, 1986, p. 571.
21. White, J. L., "Simulation of Flow in Intermeshing Twin-Screw Extruders," in I. Manas-Zloczower and Z. Tadmor, *Mixing and Compounding of Polymers,* New York: Hanser Publishers, 1994, pp. 331–372.
22. Berins, M.L., *Plastics Engineering Handbook of the Society of the Plastics Industry,* 5th ed., Chapman and Hall, New York, 1991, p. 92.
23. Morton-Jones, D.H., *Polymer Processing,* Chapman and Hall, New York, 1989, pp. 107, 110, and 111.
24. Morton-Jones, D.H., *Polymer Processing,* Chapman and Hall, New York, 1989, p. 118.
25. G. Pötsch and W. Michaeli, *Injection Molding,* Hanser Publishers, Munich, Germany, 1995.

26. R.A. Malloy, *Plastic Part Design for Injection Molding,* Carl Hanser Verlag, Munich, 1994, p. 20.
27. G. Pötsch and W. Michaeli, *Injection Molding,* Hanser Publishers, Munich, Germany, 1995, p. 115.
28. G. Pötsch and W. Michaeli, *Injection Molding,* Hanser Publishers, Munich, Germany, 1995, p. 133.
29. G. Pötsch and W. Michaeli, *Injection Molding,* Hanser Publishers, Munich, Germany, 1995, p. 135.
30. R.A. Malloy, *Plastic Part Design for Injection Molding,* Carl Hanser Verlag, Munich, 1994, p. 120.
31. A.B. Strong, *Plastics: Materials and Processing,* Prentice-Hall, New Jersey, 1996.
32. J.L. Throne, *Technology of Thermoforming,* Carl Hanser Verlag, Munich, 1996.
33. M.L. Berins, *Plastics Engineering Handbook of the Society of the Plastics Industry,* 5th ed., Chapman and Hall, New York, 1991, p. 383.
34. J.L. Throne, *Technology of Thermoforming,* Carl Hanser Verlag, Munich, 1996, pp. 17–19.
35. J.L. Throne, *Technology of Thermoforming,* Carl Hanser Verlag, Munich, 1996, pp. 19–22.
36. A.W. Birley, B. Haworth, and J. Batchelor, *Physics of Plastics,* Carl Hanser Verlag, Munich, 1992, p. 229.
37. A.W. Birley, B. Haworth, and J. Batchelor, *Physics of Plastics,* Carl Hanser Verlag, Munich, 1992, p. 230.
38. Michaeli, W., *Plastics Processing, An Introduction,* Hanser/Gardner Publications, New York, 1992, p. 102.
39. Osswald, T.A., *Polymer Processing Fundamentals,* Hanser/Gardner Publications, New York, 1998, pp. 149 and 151.
40. Osswald, T.A., *Polymer Processing Fundamentals,* Hanser/Gardner Publications, New York, 1998, p. 176.
41. H. Hurnik, "Chemical Blowing Agents" in *Plastics Additives,* 4th ed., R. Gächter and H. Müller, Eds., Carl Hanser Verlag, Munich, 1993.
42. Berins, M.L., *Plastics Engineering Handbook of the Society of the Plastics Industry,* 5th ed., Chapman and Hall, New York, 1991, p. 553.
43. Berins, M.L., *Plastics Engineering Handbook of the Society of the Plastics Industry,* 5th ed., Chapman and Hall, New York, 1991, p. 555.
44. Berins, M.L., *Plastics Engineering Handbook of the Society of the Plastics Industry,* 5th ed., Chapman and Hall, New York, 1991, p. 559.
45. M.L. Berins, *Plastics Engineering Handbook of the Society of the Plastics Industry,* 5th ed., Chapman and Hall, New York, 1991, p. 593.
46. M.L. Berins, *Plastics Engineering Handbook of the Society of the Plastics Industry,* 5th ed., Chapman and Hall, New York, 1991, pp. 593-599.
47. M.L. Berins, *Plastics Engineering Handbook of the Society of the Plastics Industry,* 5th ed., Chapman and Hall, New York, 1991, pp. 600-605.
48. L.H. Van Vlack, *Elements of Materials Science and Engineering,* 3rd ed., Addison-Wesley, Reading, MA, 1975.
49. R.R. Maccani, "Characteristics Crucial to the Application of Engineering Plastics," in *Engineering Plastics,* Vol. 2, *Engineering Materials Handbook,* ASM International, Metals Park, OH, 1988, p. 69.
50. Berins, M.L., *Plastics Engineering Handbook of the Society of the Plastics Industry,* 5th ed., Chapman and Hall, New York, 1991, p. 48-49.
51. H.P. Schlumpf, "Fillers and Reinforcements", in *Plastics Additives,* 4th ed., R. Gächter and H. Müller, Eds., Carl Hanser Verlag, Munich, 1993.
52. Morton-Jones, D. H., *Polymer Processing,* Chapman Hall, New York (1989), 35.
53. T. Whelan and J. Goff, *The Dynisco Injection Molders Handbook,* 1st ed., Dynisco, ©T. Whelan and J. Goff, 1991.
54. http://www.mgstech.com/multishot_molding/materials/battenfeld_plastic_bonding_chart.gif.
55. Berins, M.L., *Plastics Engineering Handbook of the Society of the Plastics Industry,* 5th ed., Chapman and Hall, New York, 1991, p. 405.
56. H. Hurnik, "Chemical Blowing Agents," in *Plastics Additives,* 4th ed., R. Gächter and H. Müller, Eds., Carl Hanser Verlag, Munich, 1993.

57. C. Rauwendaal, *Polymer Extrusion,* 2nd ed., Hanser/Gardner Publications, Cincinnati, OH (1990), p. 24
58. *Twin Screw Report,* Somerville, NJ, American Leistritz Extruder Corp., (Nov., 1993).
59. http://www.ndhmedical.com/html/extrusion.htm, last accessed August 30, 2005.
60. G. Pötsch and W. Michaeli, *Injection Molding,* Hanser Publishers, Munich, Germany, 1995, p. 2.
61. N.G. McCrum, C.P. Buckley, and C.B. Bucknall, *Principles of Polymer Engineering,* 2nd ed., Oxford University Press, New York, 1997, p. 334.
62. N.G. McCrum, C.P. Buckley, and C.B. Bucknall, *Principles of Polymer Engineering,* 2nd ed., Oxford University Press, New York, 1997, p. 338.
63. G. Pötsch and W. Michaeli, *Injection Molding,* Hanser Publishers, Munich, Germany, 1995, p. 172.
64. G. Pötsch and W. Michaeli, *Injection Molding,* Hanser Publishers, Munich, Germany, 1995, p. 173.
65. G. Pötsch and W. Michaeli, *Injection Molding,* Hanser Publishers, Munich, Germany, 1995, p. 178.
66. G. Pötsch and W. Michaeli, *Injection Molding,* Hanser Publishers, Munich, Germany, 1995, p. 177.
67. R.A. Malloy, *Plastic Part Design for Injection Molding,* Carl Hanser Verlag, Munich, 1994, p. 396.
68. J.L. Throne, *Technology of Thermoforming,* Carl Hanser Verlag, Munich, 1996, p. 17.
69. J.L. Throne, *Technology of Thermoforming,* Carl Hanser Verlag, Munich, 1996, p. 18.
70. J.L. Throne, *Technology of Thermoforming,* Carl Hanser Verlag, Munich, 1996, p. 19.
71. J.L. Throne, *Technology of Thermoforming,* Carl Hanser Verlag, Munich, 1996, p. 20.
72. J.L. Throne, *Technology of Thermoforming,* Carl Hanser Verlag, Munich, 1996, p. 21.
73. J.L. Throne, *Technology of Thermoforming,* Carl Hanser Verlag, Munich, 1996, p. 22.
74. J.L. Throne, *Technology of Thermoforming,* Carl Hanser Verlag, Munich, 1996, p. 23.

CHAPTER 2
THERMOPLASTICS

Anne-Marie Baker, Joey L. Mead
University of Massachusetts
Lowell, Massachusetts

2.1 INTRODUCTION

Plastic materials encompass a broad range of materials. The effect of structure on the resulting properties was discussed more fully in Chap. 1. Here, we describe the details of the wide variety of plastic materials available for use. For a comprehensive listing of properties, the reader should refer to Chap. 1.

2.2 POLYMER CATEGORIES

2.2.1 Acetal (POM)

Acetal polymers are formed from the polymerization of formaldehyde. They are also given the name polyoxymethylenes (POMs). Polymers prepared from formaldehyde were studied by Staudinger in the 1920s, but thermally stable materials were not introduced until the 1950s, when DuPont developed Delrin.[1] Hompolymers are prepared from very pure formaldehyde by anionic polymerization as shown in Fig. 2.1. Amines and the soluble salts of alkali metals catalyze the reaction.[2] The polymer formed is insoluble and is removed as the reaction proceeds. Thermal degradation of the acetal resin occurs by unzipping with the release of formaldehyde. The thermal stability of the polymer is increased by esterification of the hydroxyl ends with acetic anhydride. An alternative method to improve the thermal stability is copolymerization with a second monomer, such as ethylene oxide. The copolymer is prepared by cationic methods[3] developed by Celanese and mar-

FIGURE 2.1 Polymerization of formaldehyde to polyoxymethylene.

keted under the trade name Celcon. Hostaform and Duracon are also copolymers. The presence of the second monomer reduces the tendency for the polymer to degrade by unzipping.[4]

There are four processes for the thermal degradation of acetal resins. The first is thermal or base-catalyzed depolymerization from the chain, resulting in the release of formaldehyde. End capping the polymer chain will reduce this tendency. The second is oxidative attack at random positions, again leading to depolymerization. The use of antioxidants will reduce this degradation mechanism. Copolymerization is also helpful. The third mechanism is cleavage of the acetal linkage by acids. It is therefore important not to process acetals in equipment used for PVC, unless it has been cleaned, due to the possible presence of traces of HCl. The fourth degradation mechanism is thermal depolymerization at temperatures above 270°C. It is important that processing temperatures remain below this temperature to avoid degradation of the polymer.[5]

Acetals are highly crystalline, typically 75 percent crystalline, with a melting point of 180°C.[6] Compared to polyethylene (PE), the chains pack closer together because of the shorter C-O bond. As a result, the polymer has a higher melting point. It is also harder than PE. The high degree of crystallinity imparts good solvent resistance to acetal polymers. The polymer is essentially linear with molecular weights (M_n) in the range of 20,000 to 110,000.[7]

Acetal resins are strong and stiff thermoplastics with good fatigue properties and dimensional stability. They also have a low coefficient of friction, and good heat resistance.[8] Acetal resins are considered similar to nylons but are better in fatigue, creep, stiffness, and water resistance.[9] Acetal resins do not, however, have the creep resistance of polycarbonate. As mentioned previously, acetal resins have excellent solvent resistance with no organic solvents found below 70°C; however, swelling may occur in some solvents. Acetal resins are susceptible to strong acids and alkalis as well as oxidizing agents. Although the C-O bond is polar, it is balanced and much less polar than the carbonyl group present in nylon. As a result, acetal resins have relatively low water absorption. The small amount of moisture absorbed may cause swelling and dimensional changes but will not degrade the polymer by hydrolysis.[10] The effects of moisture are considerable less dramatic than for nylon polymers. Ultraviolet light may cause degradation, which can be reduced by the addition of carbon black. The copolymers have generally similar properties, but the homopolymer may have slightly better mechanical properties, and higher melting point, but poorer thermal stability and poorer alkali resistance.[11] Along with both homopolymers and copolymers, there are also filled materials (glass, fluoropolymer, aramid fiber, and other fillers), toughened grades, and UV stabilized grades.[12] Blends of acetal with polyurethane elastomers show improved toughness and are available commercially.

Acetal resins are available for injection molding, blow molding, and extrusion. During processing, it is important to avoid overheating, or the production of formaldehyde may cause serious pressure buildup. The polymer should be purged from the machine before shutdown to avoid excessive heating during start-up.[13] Acetal resins should be stored in a dry place. The apparent viscosity of acetal resins is less dependent on shear stress and temperature than polyolefins, but the melt has low elasticity and melt strength. The low melt strength is a problem for blow molding applications. For blow molding applications, copolymers with branched structures are available. Crystallization occurs rapidly with post mold shrinkage complete within 48 hr of molding. Because of the rapid crystallization, it is difficult to obtain clear films.[14]

The market demand for acetal resins in the United States and Canada was 368 million lb in 1997.[15] Applications for acetal resins include gears, rollers, plumbing components, pump parts, fan blades, blow molded aerosol containers, and molded sprockets and chains. They are often used as direct replacements for metal. Most of the acetal resins are processed by injection molding, with the remainder used in extruded sheet and rod. Their low coefficient of friction make acetal resins good for bearings.[16]

2.2.2 Biodegradable Polymers

Disposal of solid waste is a challenging problem. The United States consumes over 53 billion lb of polymers a year for a variety of applications.[17] When the life cycle of these polymeric parts is completed, they may end up in a landfill. Plastics are often selected for applications based on their stability to degradation; however, this means degradation will be very slow, adding to the solid waste problem. Methods to reduce the amount of solid waste include either recycling or biodegradation.[18] Considerable work has been done to recycle plastics, both in the manufacturing and consumer area. Biodegradable materials offer another way to reduce the solid waste problem. Most waste is disposed of by burial in a landfill. Under these conditions, oxygen is depleted, and biodegradation must proceed without the presence of oxygen.[19] An alternative is aerobic composting. In selecting a polymer that will undergo biodegradation, it is important to ascertain the method of disposal. Will the polymer be degraded in the presence of oxygen and water, and what will be the pH level? Biodegradation can be separated into two types: chemical and microbial degradation. Chemical degradation includes degradation by oxidation, photodegradation, thermal degradation, and hydrolysis. Microbial degradation can include both fungi and bacteria. The susceptibility of a polymer to biodegradation depends on the structure of the backbone.[20] For example, polymers with hydrolyzable backbones can be attacked by acids or bases, breaking down the molecular weight. They are therefore more likely to be degraded. Polymers that fit into this category include most natural-based polymers, such as polysaccharides, and synthetic materials, such as polyurethanes, polyamides, polyesters, and polyethers. Polymers that contain only carbon groups in the backbone are more resistant to biodegradation.

Photodegradation can be accomplished by using polymers that are unstable to light sources or by the used of additives that undergo photodegration. Copolymers of divinyl ketone with styrene, ethylene, or polypropylene (Eco Atlantic) are examples of materials that are susceptible to photodegradation.[21] The addition of a UV absorbing material will also act to enhance photodegradation of a polymer. An example is the addition of iron dithiocarbamate.[22] The degradation must be controlled to ensure that the polymer does not degrade prematurely.

Many polymers described elsewhere in this book can be considered for biodegradable applications. Polyvinyl alcohol has been considered in applications requiring biodegradation because of its water solubility. However, the actual degradation of the polymer chain may be slow.[23] Polyvinyl alcohol is a semicrystalline polymer synthesized from polyvinyl acetate. The properties are governed by the molecular weight and by the amount of hydrolysis. Water soluble polyvinyl alcohol has a degree of hydrolysis near 88 percent. Water insoluble polymers are formed if the degree of hydrolysis is less than 85 percent.[24]

Cellulose-based polymers are some of the more widely available naturally based polymers. They can therefore be used in applications requiring biodegradation. For example, regenerated cellulose is used in packaging applications.[25] A biodegradable grade of cellulose acetate is available from Rhone-Poulenc (Bioceta and Biocellat), where an additive acts to enhance the biodegradation.[26] This material finds application in blister packaging, transparent window envelopes, and other packaging applications.

Starch-based products are also available for applications requiring biodegradability. The starch is often blended with polymers for better properties. For example, polyethylene films containing between 5 and 10 percent cornstarch have been used in biodegradable applications. Blends of starch with vinyl alcohol are produced by Fertec (Italy) and used in both film and solid product applications.[27] The content of starch in these blends can range up to 50 percent by weight, and the materials can be processed on conventional processing equipment. A product developed by Warner-Lambert call Novon is also a blend of polymer and starch, but the starch contents in Novon are higher than in the material by Fertec. In some cases, the content can be over 80 percent starch.[28]

Polylactides (PLAs) and copolymers are also of interest in biodegradable applications. This material is a thermoplastic polyester synthesized from ring opening of lactides. Lactides are cyclic diesters of lactic acid.[29] A similar material to polylactide is polyglycolide (PGA). PGA is also thermoplastic polyester but formed from glycolic acids. Both PLA and PGA are highly crystalline materials. These materials find application in surgical sutures and resorbable plates and screws for fractures, and new applications in food packaging are also being investigated.

Polycaprolactones are also considered in biodegradable applications such as films and slow-release matrices for pharmaceuticals and fertilizers.[30] Polycaprolactone is produced through ring opening polymerization of lactone rings with a typical molecular weight in the range of 15,000 to 40,000.[31] It is a linear, semicrystalline polymer with a melting point near 62°C and a glass transition temperature about –60°C.[32]

A more recent biodegradable polymer is polyhydroxybutyrate-valerate copolymer (PHBV). These copolymers differ from many of the typical plastic materials in that they are produced through biochemical means. It is produced commercially by ICI using the bacteria *Alcaligenes eutrophus*, which is fed a carbohydrate. The bacteria produce polyesters, which are harvested at the end of the process.[33] When the bacteria are fed glucose, the pure poly hydroxybutyrate polymer is formed, while a mixed feed of glucose and propionic acid will produce the copolymers.[34] Different grades are commercially available that vary in the amount of hydroxyvalerate units and the presence of plasticizers. The pure hydroxybutyrate polymer has a melting point between 173 and 180°C and a T_g near 5°C.[35] Copolymers with hydroxyvalerate have reduced melting points, greater flexibility, and impact strength, but lower modulus and tensile strength. The level of hydroxyvalerate is 5 to 12 percent. These copolymers are fully degradable in many microbial environments. Processing of PHBV copolymers requires careful control of the process temperatures. The material will degrade above 195°C, so processing temperatures should be kept below 180°C and the processing time kept to a minimum. It is more difficult to process unplasticized copolymers with lower hydroxyvalerate content because of the higher processing temperatures required. Applications for PHBV copolymers include shampoo bottles, cosmetic packaging, and as a laminating coating for paper products.[36]

Other biodegradable polymers include Konjac, a water-soluble natural polysaccharide produced by FMC; Chitin, another polysaccharide that is insoluble in water; and Chitosan, which is soluble in water.[37] Chitin is found in insects and in shellfish. Chitosan can be formed from chitin and is also found in fungal cell walls.[38] Chitin is used in many biomedical applications, including dialysis membranes, bacteriostatic agents, and wound dressings. Other applications include cosmetics, water treatment, adhesives, and fungicides.[39]

2.2.3 Cellulose

Cellulosic polymers are the most abundant organic polymers in the world, making up the principal polysaccharide in the walls of almost all of the cells of green plants and many fungi species.[40] Plants produce cellulose through photosynthesis. Pure cellulose decomposes before it melts and must be chemically modified to yield a thermoplastic. The chemical structure of cellulose is a heterochain linkage of different anhydroglucose units into high-molecular-weight polymer, regardless of plant source. The plant source however does affect molecular weight, molecular weight distribution, degrees of orientation, and morphological structure. Material described commonly as "cellulose" can actually contain hemicelluloses and lignin.[41] Wood is the largest source of cellulose, is processed as fibers to supply the paper industry, and is widely used in housing and industrial buildings. Cotton-derived cellulose is the largest source of textile and industrial fibers, with the combined result being that cellulose is the primary polymer serving the housing and clothing industries. Crystalline modifications result in celluloses of differing mechanical properties, and Table 2.1 compares the tensile strengths and ultimate elongations of some common celluloses.[42]

TABLE 2.1 Selected Mechanical Properties of Common Celluloses

Form	Tensile strength, MPa		Ultimate elongation, %	
	Dry	Wet	Dry	Wet
Ramie	900	1060	2.3	2.4
Cotton	200–800	200–800	12–16	6–13
Flax	824	863	1.8	2.2
Viscose rayon	200–400	100–200	8–26	13–43
Cellulose acetate	150–200	100–120	21–30	29–30

Cellulose, whose repeat structure features three hydroxyl groups, reacts with organic acids, anhydrides, and acid chlorides to form esters. Plastics from these cellulose esters are extruded into film and sheet and are injection molded to form a wide variety of parts. Cellulose esters can also be compression molded and cast from solution to form a coating. The three most industrially important cellulose ester plastics are cellulose acetate (CA), cellulose acetate butyrate (CAB), and cellulose acetate propionate (CAP), with structures as shown in Fig. 2.2.

These cellulose acetates are noted for their toughness, gloss, and transparency. CA is well suited for applications requiring hardness and stiffness, as long as the temperature and humidity conditions don't cause the CA to be too dimensionally unstable. CAB has the best environmental stress cracking resistance, low temperature impact strength, and dimensional stability. CAP has the highest tensile strength and hardness. A comparison of typical compositions and properties for a range of formulations is given in Table 2.2.[43] Properties can be tailored by formulating with different types and loadings of plasticizers.

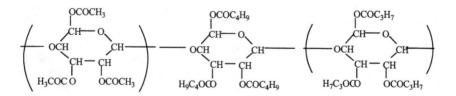

FIGURE 2.2 Structures of cellulose acetate, cellulose acetate butyrate, and cellulose acetate propionate.

TABLE 2.2   Selected Mechanical Properties of Cellulose Esters

	Cellulose acetate	Cellulose acetate butyrate	Cellulose acetate propionate
Composition, %			
Acetyl	38–40	13–15	1.5–3.5
Butyrl	–	36–38	–
Propionyl	–	–	43–47
Hydroxyl	3.5–4.5	1–2	2–3
Tensile strength at fracture, 23 °C, MPa	13.1–58.6	13.8–51.7	13.8–51.7
Ultimate elongation, %	6–50	38–74	35–60
Izod impact strength, J/m			
Notched, 23°C	6.6–132.7	9.9–149.3	13.3–182.5
Notched, –40°C	1.9–14.3	6.6–23.8	1.9–19.0
Rockwell hardness, R scale	39–120	29–117	20–120
Percent moisture absorption at 24 hr	2.0–6.5	1.0–4.0	1.0–3.0

Formulation of cellulose esters is required to reduce charring and thermal discoloration, and typically includes the addition of heat stabilizers, antioxidants, plasticizers, UV stabilizers, and coloring agents.[44] Cellulose molecules are rigid due to the strong intermolecular hydrogen bonding that occurs. Cellulose itself is insoluble and reaches its decomposition temperature prior to melting. The acetylation of the hydroxyl groups reduces intermolecular bonding and increases free volume, depending on the level and chemical nature of the alkylation.[45] CAs are thus soluble in specific solvents but still require plasticization for rheological properties appropriate to molding and extrusion processing conditions. Blends of ethylene vinyl acetate (EVA) copolymers and CAB are available. Cellulose acetates have also been graft-copolymerized with alkyl esters of acrylic and methacrylic acid and then blended with EVA to form a clear, readily processable thermoplastic.

CA is cast into sheet form for blister packaging, window envelopes, and file tab applications. CA is injection molded into tool handles, toothbrushes, ophthalmic frames, and appliance housings and is extruded into pens, pencils, knobs, packaging films, and industrial pressure-sensitive tapes. CAB is molded into steering wheels, tool handles, camera parts, safety goggles, and football noseguards. CAP is injection molded into steering wheels, telephones, appliance housings, flashlight cases, and screw and bolt anchors and is extruded into pens, pencils, toothbrushes, packaging film, and pipe.[46] Cellulose acetates are well suited for applications that require machining and then solvent vapor polishing, such as in the case of tool handles, where the consumer market values the clarity, toughness, and smooth finish. CA and CAP are likewise suitable for ophthalmic sheeting and injection molding applications that require many post-finishing steps.[47]

Cellulose acetates are also commercially important in the coatings arena. In this synthetic modification, cellulose is reacted with an alkyl halide, primarily methylchloride to yield methylcellulose or sodium chloroacetate to yield sodium cellulose methylcellulose (CMC). The structure of CMC is shown below in Fig. 2.3. CMC gums are water soluble and are used in food contact and packaging applications. Its outstanding film-forming properties are used in paper sizings and textiles, and its thickening properties are used in starch adhesive formulations, paper coatings, toothpaste, and shampoo. Other cellulose es-

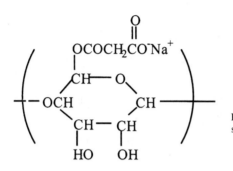

FIGURE 2.3 Sodium cellulose methylcellulose structure.

ters, including cellulosehydroxyethyl, hydroxypropylcellulose, and ethylcellulose, are used in film and coating applications, adhesives, and inks.

2.2.4 Fluoropolymers

Fluoropolymers are noted for their heat-resistance properties. This is due to the strength and stability of the carbon-fluorine bond.[48] The first patent was awarded in 1934 to IG Farben for a fluorine-containing polymer, polychlorotrifluoroethylene (PCTFE). This polymer had limited application, and fluoropolymers did not have wide application until the discovery of polytetrafluorethylene (PTFE) in 1938.[49] In addition to their high-temperature properties, fluoropolymers are known for their chemical resistance, very low coefficient of friction, and good dielectric properties. Their mechanical properties are not high unless reinforcing fillers, such as glass fibers, are added.[50] The compressive properties of fluoropolymers are generally superior to their tensile properties. In addition to their high temperature resistance, these materials have very good toughness and flexibility at low temperatures.[51]

A wide variety of fluoropolymers are available, including polytetrafluoroethylene (PTFE), polychlorotrifluoroethylene (PCTFE), fluorinated ethylene propylene (FEP), ethylene chlorotrifluoroethylene (ECTFE), ethylene tetrafluoroethylene (ETFE), polyvinylindene fluoride (PVDF), and polyvinyl fluoride (PVF).

2.2.4.1 Copolymers. Fluorinated ethylene propylene (FEP) is a copolymer of tetrafluoroethylene and hexafluoropropylene. It has properties similar to PTFE but with a melt viscosity suitable for molding with conventional thermoplastic processing techniques.[52] The improved processability is obtained by replacing one of the fluorine groups on PTFE with a trifluoromethyl group as shown in Fig. 2.4.[53]

FEP polymers were developed by DuPont, but other commercial sources are available, such as Neoflon (Daikin Kogyo) and Teflex (Niitechem, USSR).[54] FEP is a crystalline polymer with a melting point of 290°C, and it can be used for long periods at 200°C with good retention of properties.[55] FEP has good chemical resistance, a low dielectric con-

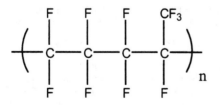

FIGURE 2.4 Structure of FEP.

stant, low friction properties, and low gas permeability. Its impact strength is better than PTFE, but the other mechanical properties are similar to PTFE.[56] FEP may be processed by injection, compression, or blow molding. FEP may be extruded into sheets, films, rods or other shapes. Typical processing temperatures for injection molding and extrusion are in the range of 300 to 380°C.[57] Extrusion should be done at low shear rates because of the polymer's high melt viscosity and melt fracture at low shear rates. Applications for FEP include chemical process pipe linings, wire and cable, and solar collector glazing.[58] A material similar to FEP, Hostaflon TFB (Hoechst), is a terpolymer of tetrafluoroethylene, hexafluoropropene, and vinylidene fluoride.

Ethylene chlorotrifluoroethylene (ECTFE) is an alternating copolymer of chlorotrifluoroethylene and ethylene. It has better wear properties than PTFE along with good flame resistance. Applications include wire and cable jackets, tank linings, chemical process valve and pump components, and corrosion-resistant coatings.[59]

Ethylene tetrafluoroethylene (ETFE) is a copolymer of ethylene and tetrafluoroethylene similar to ECTFE but with a higher use temperature. It does not have the flame resistance of ECTFE, however, and will decompose and melt when exposed to a flame.[60] The polymer has good abrasion resistance for a fluorine containing polymer, along with good impact strength. The polymer is used for wire and cable insulation, where its high temperature properties are important. ETFE finds application in electrical systems for computers, aircraft and heating systems.[61]

2.2.4.2 Polychlorotrifluoroethylene (PCTFE).   Polychlorotrifluoroethylene (PCTFE) is made by the polymerization of chlorotrifluoroethylene, which is prepared by the dechlorination of trichlorotrifluoroethane. The polymerization is initiated with redox initiators.[62] The replacement of one fluorine atom with a chlorine atom, as shown in Fig. 2.5, breaks up the symmetry of the PTFE molecule, resulting in a lower melting point and allowing PCTFE to be processed more easily than PTFE. The crystalline melting point of PCTFE at 218°C is lower than PTFE. Clear sheets of PCTFE with no crystallinity may also be prepared.

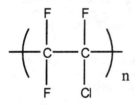

FIGURE 2.5   Structure of PCTFE.

PCTFE is resistant to temperatures up to 200°C and has excellent solvent resistance, with the exception of halogenated solvents or oxygen containing materials, which may swell the polymer.[63] The electrical properties of PCTFE are inferior to PTFE, but PCTFE is harder and has high tensile strength. The melt viscosity of PCTFE is low enough that it may be processing using most thermoplastic processing techniques.[64] Typical processing temperatures are in the range of 230 to 290°C.[65]

PCTFE is higher in cost than PTFE, somewhat limiting its use. Applications include gaskets, tubing, and wire and cable insulation. Very low vapor transmission films and sheets may also be prepared.[66]

2.2.4.3 Polytetrafluoroethylene (PTFE).   Polytetrafluoroethylene (PTFE) is polymerized from tetrafluoroethylene by free radical methods.[67] The reaction is shown below in Fig. 2.6. Commercially, there are two major processes for the polymerization of PTFE,

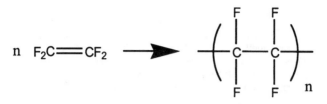

FIGURE 2.6 Preparation of PTFE.

one yielding a finer particle size dispersion polymer with lower molecular weight than the second method, which yields a "granular" polymer. The weight average molecular weights of commercial materials range from 400,000 to 9,000,000.[68] PTFE is a linear crystalline polymer with a melting point of 327°C.[69] Because of the larger fluorine atoms, PTFE takes up a twisted zigzag in the crystalline state, while polyethylene takes up the planar zigzag form.[70] There are several crystal forms for PTFE, with some of the transitions from one crystal form to another occurring near room temperature. As a result of these transitions, volume changes of about 1.3 percent may occur.

PTFE has excellent chemical resistance but may go into solution near its crystalline melting point. PTFE is resistant to most chemicals. Only alkali metals (molten) may attack the polymer.[71] The polymer does not absorb significant quantities of water, and it has low permeability to gases and moisture vapor.[72] PTFE is a tough polymer with good insulating properties. It is also known for its low coefficient of friction, with values in the range of 0.02 to 0.10.[73] PTFE, like other fluoropolymers, has excellent heat resistance and can withstand temperatures up to 260°C. Because of the high thermal stability, the mechanical and electrical properties of PTFE remain stable for long times at temperatures up to 250°C. However, PTFE can be degraded by high-energy radiation.

One disadvantage of PTFE is that it is extremely difficult to process by either molding or extrusion. PFTE is processed in powder form by either sintering or compression molding. It is also available as a dispersion for coating or impregnating porous materials.[74] PTFE has very high viscosity, prohibiting the use of many conventional processing techniques. For this reason, techniques developed for the processing of ceramics are often used. These techniques involve preforming the powder, followed by sintering above the melting point of the polymer. For granular polymers, the preforming is carried out with the powder compressed into a mold. Pressures should be controlled, as too low a pressure may cause voids, while too high a pressure may result in cleavage planes. After sintering, thick parts should be cooled in an oven at a controlled cooling rate, often under pressure. Thin parts may be cooled at room temperature. Simple shapes may be made by this technique, but more detailed parts should be machined.[75]

Extrusion methods may be used on the granular polymer at very low rates. In this case the polymer is fed into a sintering die that is heated. A typical sintering die has a length about 90 times the internal diameter. Dispersion polymers are more difficult to process by the techniques previously mentioned. The addition of a lubricant (15 to 25 percent) allows the manufacture of preforms by extrusion. The lubricant is then removed and the part sintered. Thick parts are not made by this process, because the lubricant must be removed. PTFE tapes are made by this process; however, the polymer is not sintered, and a nonvolatile oil is used.[76] Dispersions of PTFE are used to impregnate glass fabrics and to coat metal surfaces. Laminates of the impregnated glass cloth may be prepared by stacking the layers of fabric, followed by pressing at high temperatures.

Processing of PTFE requires adequate ventilation for the toxic gases that may be produced. In addition, PTFE should be processed under high cleanliness standards, because the presence of any organic matter during the sintering process will result in poor properties as a result of the thermal decomposition of the organic matter. This includes both poor visual qualities and poor electrical properties.[77] The final properties of PTFE are dependent on the processing methods and the type of polymer. Both particle size and molecular weight should be considered. The particle size will affect the amount of voids and processing ease, while crystallinity will be influenced by the molecular weight.

Additives for PTFE must be able to undergo the high processing temperatures required. This limits the range of additives available. Glass fiber is added to improve some mechanical properties. Graphite or molybdenum disulphide may be added to retain the low coefficient of friction while improving the dimensional stability. Only a few pigments are available that can withstand the processing conditions. These are mainly inorganic pigments such as iron oxides and cadmium compounds.[78]

Because of the excellent electrical properties, PTFE is used in a variety of electrical applications, such as wire and cable insulation and insulation for motors, capacitors, coils, and transformers. PTFE is also used for chemical equipment such as valve parts and gaskets. The low friction characteristics make PTFE suitable for use in bearings, mold release devices, and antistick cookware. Low-molecular-weight polymers may be used in aerosols for dry lubrication.[79]

2.2.4.4 Polyvinylindene Fluoride (PVDF). Polyvinylindene fluoride (PVDF) is crystalline with a melting point near 170°C.[80] The structure of PVDF is shown in Fig. 2.7. PVDF has good chemical and weather resistance, along with good resistance to distortion and creep at low and high temperatures. Although the chemical resistance is good, the polymer can be affected by very polar solvents, primary amines, and concentrated acids. PVDF has limited use as an insulator, because the dielectric properties are frequency dependent. The polymer is important because of its relatively low cost compared to other fluorinated polymers.[81] PVDF is unique in that the material has piezoelectric properties, meaning that it will generate electric current when compressed.[82] This unique feature has been utilized for the generation of ultrasonic waves.

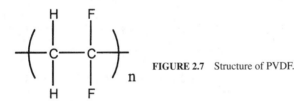

FIGURE 2.7   Structure of PVDF.

PVDF can be melt processed by most conventional processing techniques. The polymer has a wide range between the decomposition temperature and the melting point. Melt temperatures are usually 240 to 260°C.[83] Processing equipment should be extremely clean, as any contaminants may affect the thermal stability. As with other fluorinated polymers, the generation of HF is a concern. PVDF is used for applications in gaskets, coatings, wire and cable jackets, and chemical process piping and seals.[84]

2.2.4.5 Polyvinyl fluoride (PVF). Polyvinyl fluoride (PVF) is a crystalline polymer available in film form and used as a lamination on plywood and other panels.[85] The film is impermeable to many gases. PVF is structurally similar to polyvinyl chloride (PVC) except for the replacement of a chlorine atom with a fluorine atom. PVF exhibits low moisture absorption, good weatherability, and good thermal stability. Similar to PVC, PVF may give off hydrogen halides at elevated temperatures. However, PVF has a greater tendency to crystallize and better heat resistance than PVC.[86]

2.2.5 Nylons

Nylons were one of the early polymers developed by Carothers.[87] Today, nylons are an important thermoplastic, with consumption in the United States of about 1.2 billion lb in 1997.[88] Nylons, also known as polyamides, are synthesized by condensation polymerization methods, often an aliphatic diamine and a diacid. Nylon is a crystalline polymer with high modulus, strength, and impact properties, and low coefficient of friction and resistance to abrasion.[89] Although the materials possess a wide range of properties, they all contain the amide (-CONH-) linkage in their backbone. Their general structure is shown in Fig. 2.8.

FIGURE 2.8   General structure of nylons.

There are five main methods to polymerize nylon. They are

- Reaction of a diamine with a dicarboxylic acid
- Condensation of the appropriate amino acid
- Ring opening of a lactam
- Reaction of a diamine with a dicarboxylic acid
- Reaction of a diisocyanate with a dicarboxylic acid[90]

The type of nylon (nylon 6, nylon 10, etc.) is indicative of the number of carbon atoms in the repeat unit. Many different types of nylons can be prepared, depending on the starting monomers used. The type of nylon is determined by the number of carbon atoms in the monomers used in the polymerization. The number of carbon atoms between the amide linkages also controls the properties of the polymer. When only one monomer is used (lactam or amino acid), the nylon is identified with only one number (nylon 6, nylon 12).

When two monomers are used in the preparation, the nylon will be identified using two numbers (nylon 6,6, nylon 6,12).[91] This is shown in Fig. 2.9. The first number refers to the number of carbon atoms in the diamine used (a) and the second number refers to the number of carbon atoms in the diacid monomer (b + 2), due to the two carbons in the carbonyl group.[92]

The amide groups are polar groups and significantly affect the polymer properties. The presence of these groups allows for hydrogen bonding between chains, improving the interchain attraction. This gives nylon polymers good mechanical properties. The polar nature of nylons also improves the bondability of the materials, while the flexible aliphatic carbon groups give nylons low melt viscosity for easy processing.[93] This structure also gives polymers that are tough above their glass transition temperature.[94]

Nylons are relatively insensitive to nonpolar solvents; however, because of the presence of the polar groups, nylons can be affected by polar solvents, particularly water.[95] The presence of moisture must be considered in any nylon application. Moisture can cause

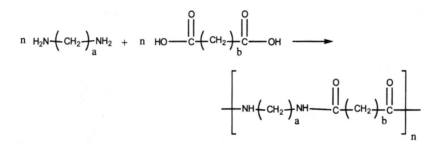

FIGURE 2.9   Synthesis of nylon.

changes in part dimensions and reduce the properties, particularly at elevated temperatures.[96] As a result, the material should be dried before any processing operations. In the absence of moisture, nylons are fairly good insulators but, as the level of moisture or the temperature increases, the nylons are less insulating.[97]

The strength and stiffness will be increased as the number of carbon atoms between amide linkages is decreased, because there are more polar groups per unit length along the polymer backbone.[98] The degree of moisture absorption is also strongly influenced by the number of polar groups along the backbone of the chain. Nylon grades with fewer carbon atoms between the amide linkages will absorb more moisture than grades with more carbon atoms between the amide linkages (nylon 6 will absorb more moisture than nylon 12). Furthermore, nylon types with an even number of carbon atoms between the amide groups have higher melting points than those with an odd number of carbon atoms. For example, the melting point of nylon 6,6 is greater than either nylon 5,6 or nylon 7,6.[99] Ring opened nylons behave similarly. This is due to the ability of the nylons with the even number of carbon atoms to pack better in the crystalline state.[100]

Nylon properties are affected by the amount of crystallinity. This can be controlled to a great extent in nylon polymers by the processing conditions. A slowly cooled part will have significantly greater crystallinity (50 to 60 percent) than a rapidly cooled, thin part (perhaps as low as 10 percent).[101] Not only can the degree of crystallinity be controlled, but also the size of the crystallites. In a slowly cooled material, the crystal size will be larger than for a rapidly cooled material. In injection molded parts where the surface is rapidly cooled, the crystal size may vary from the surface to internal sections.[102] Nucleating agents can be utilized to create smaller spherulites in some applications. This creates materials with higher tensile yield strength and hardness, but lower elongation and impact.[103] The degree of crystallinity will also affect the moisture absorption, with less crystalline polyamides being more prone to moisture pickup.[104]

The glass transition temperature of aliphatic polyamides is of secondary importance to the crystalline melting behavior. Dried polymers have T_g values near 50°C, while those with absorbed moisture may have T_gs in the range of 0°C.[105] The glass transition temperature can influence the crystallization behavior of nylons. For example, nylon 6,6 may be above its T_g at room temperature, causing crystallization at room temperature to occur slowly leading to post mold shrinkage. This is less significant for nylon 6.[106]

Nylons are processed by extrusion, injection molding, blow molding, and rotational molding, among other methods. Nylon has a very sharp melting point and low melt viscosity, which is advantageous in injection molding but causes difficulty in extrusion and blow molding. In extrusion applications a wide MWD is preferred, along with a reduced temperature at the exit to increase melt viscosity.[107]

When used in injection molding applications, nylons have a tendency to drool, due to their low melt viscosity. Special nozzles have been designed for use with nylons to reduce this problem.[108] Nylons show high mold shrinkage as a result of their crystallinity. Average values are about 0.018 cm/cm for nylon 6,6. Water absorption should also be considered for parts with tight dimensional tolerances. Water will act to plasticize the nylon, relieving some of the molding stresses and causing dimensional changes. In extrusion, a screw with a short compression zone is used, with cooling initiated as soon as the extrudate exits the die.[109]

A variety of commercial nylons are available, including nylon 6, nylon 11, nylon 12, nylon 6,6, nylon 6,10, and nylon 6,12. The most widely used nylons are nylon 6,6 and nylon 6.[110] Specialty grades with improved impact resistance, improved wear, or other properties are also available. Polyamides are used most often in the form of fibers, primarily nylon 6,6 and nylon 6, although engineering applications are also of importance.[111]

Nylon 6,6 is prepared from the polymerization of adipic acid and hexamethylenediamine. The need to control a 1:1 stoichiometric balance between the two monomers can be improved by the fact that adipic acid and hexamethylenediamine form a 1:1 salt that can be isolated. Nylon 6,6 is known for high strength, toughness, and abrasion resistance. It has a melting point of 265°C and can maintain properties up to 150°C.[112] Nylon 6,6 is used extensively in nylon fibers that are used in carpets, hose and belt reinforcements, and tire cord. Nylon 6,6 is used as an engineering resin in a variety of molding applications such as gears, bearings, rollers, and door latches because of its good abrasion resistance and self-lubricating tendencies.[113]

Nylon 6 is prepared from caprolactam. It has properties similar to those of nylon 6,6, but a lower melting point (255°C). One of the major applications is in tire cord. Nylon 6,10 has a melting point of 215°C and lower moisture absorption than nylon 6,6.[114] Nylon 11 and nylon 12 have lower moisture absorption and also lower melting points than nylon 6,6. Nylon 11 has found applications in packaging films. Nylon 4,6 has found applications in a variety of automotive applications due to its ability to withstand high mechanical and thermal stresses. It is used in gears, gearboxes, and clutch areas.[115] Other applications for nylons include brush bristles, fishing line, and packaging films.

Additives such as glass or carbon fibers can be incorporated to improve the strength and stiffness of the nylon. Mineral fillers are also used. A variety of stabilizers can be added to nylon to improve the heat and hydrolysis resistance. Light stabilizers are often added as well. Some common heat stabilizers include copper salts, phosphoric acid esters, and phenyl-β-naphthylamine. In bearing applications, self-lubricating grades are available that may incorporate graphite fillers. Although nylons are generally impact resistant, rubber is sometimes incorporated to improve the failure properties.[116] Nylon fibers do have a tendency to pick up a static charge, so antistatic agents are often added for carpeting and other applications.[117]

2.2.5.1 Aromatic Polyamides (Polyarylamides).

2.2.5.1 Aromatic Polyamides (Polyarylamides). A related polyamide is prepared when aromatic groups are present along the backbone. This imparts a great deal of stiffness to the polymer chain. One difficulty encountered in this class of materials is their tendency to decompose before melting.[118] However, certain aromatic polyamides have gained commercial importance. The aromatic polyamides can be classified into three groups: amorphous copolymers with a high T_g, crystalline polymers that can be used as a thermoplastic, and crystalline polymers used as fibers.

The copolymers are noncrystalline and clear. The rigid aromatic chain structure gives the materials a high T_g. One of the oldest types is poly(trimethylhexamethylene terephthalatamide) (Trogamid T®). This material has an irregular chain structure, restricting the material from crystallizing, but a T_g near 150°C.[119] Grilamid TR55® is another polyamide copolymer with a T_g about 160°C, and a lower water absorption and density than the Tro-gamid T.[120] The aromatic polyamides are tough materials and compete with polycarbon-ate, poly(methyl methacrylate) and polysulfone. These materials are used in applications requiring transparency. They have been used for solvent containers, flow meter parts, and clear housings for electrical equipment.[121] In the 1980s, a polyarylamide marketed as IXEF was introduced, and Solvay targets it for the injection molding market, particularly high-temperature automotive applications.[122]

An example of a crystallizable aromatic polyamide is poly-*m*-xylylene adipamide. It has a T_g near 85 to 100°C and a T_m of 235 to 240°C.[123] To obtain high heat deflection tem-perature, the filled grades are normally sold. Applications include gears, electrical plugs, and mowing machine components.[124]

Crystalline aromatic polyamides are also used in fiber applications. An example of this type of material is Kevlar®, a high-strength fiber used in bulletproof vests and composite structures. A similar material, which can be processed more easily, is Nomex®. It can be used to give flame retardance to cloth when used as a coating.[125]

2.2.6 Polyacrylonitrile

Polyacrylonitrile is prepared by the polymerization of acrylonitrile monomer using either free radical or anionic initiators. Bulk, emulsion, suspension, solution, or slurry methods may be used for the polymerization. The reaction is shown in Fig. 2.10.

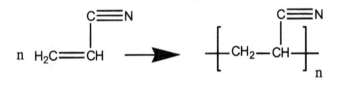

FIGURE 2.10   Preparation of polyacrylonitrile.

Polyacrylonitrile will decompose before reaching its melting point, making the materi-als difficult to form. The decomposition temperature is near 300°C.[126] Suitable solvents, such as dimethylformamide and tetramethylenesulphone, have been found for polyacry-lonitrile, allowing the polymer to be formed into fibers by dry and wet spinning tech-niques.[127]

Polyacrylonitrile is a polar material, giving the polymer good resistance to solvents, high rigidity, and low gas permeability.[128] Although the polymer degrades before melting, special techniques allowed a melting point of 317°C to be measured. The pure polymer is difficult to dissolve, but the copolymers can be dissolved in solvents such as methyl ethyl ketone, dioxane, acetone, dimethyl formamide, and tetrahydrofuran. Polyacrylonitrile ex-hibits exceptional barrier properties to oxygen and carbon dioxide.[129]

Copolymers of acrylonitrile with other monomers are widely used. Copolymers of vi-nylidene chloride and acrylonitrile find application in low-gas-permeability films. Styrene-acrylonitrile (SAN polymers) copolymers have also been used in packaging applications. Although the gas permeability of the copolymers is higher than for pure polyacrylonitrile, the acrylonitrile copolymers have lower gas permeability than many other packaging films. A number of acrylonitrile copolymers were developed for beverage containers, but the requirement for very low levels of residual acrylonitrile monomer in this application led to many products being removed from the market.[130] One copolymer currently available is Barex (BP Chemicals). The copolymer has better barrier properties than both polypropylene and polyethylene terephthalate.[131] Acrylonitrile is also used with butadiene and styrene to form ABS polymers. Unlike the homopolymer, copolymers can be processed by many methods, including extrusion, blow molding, and injection molding.[132]

Acrylonitrile is often copolymerized with other monomers to form fibers. Copolymerization with monomers such as vinyl acetate, vinyl pyrrolidone, and vinyl esters gives the fibers the ability to be dyed using normal textile dyes. The copolymer generally contains at least 85 percent acrylonitrile.[133] Acrylic fibers have good abrasion resistance, flex life, and toughness, and high strength. They have good resistance to stains and moisture. Modacrylic fibers contain between 35 and 85 percent acrylonitrile.[134]

Most of the acrylonitrile consumed goes into the production of fibers. Copolymers also consume large amounts of acrylonitrile. In addition to their use as fibers, polyacrylonitrile polymers can be used as precursors to carbon fibers.

2.2.7 Polyamide-imide (PAI)

Polyamide-imide (PAI) is a high-temperature amorphous thermoplastic that has been available since the 1970s under the trade name of Torlon.[135] PAI can be produced from the reaction of trimellitic trichloride with methylenedianiline as shown in Fig. 2.11.

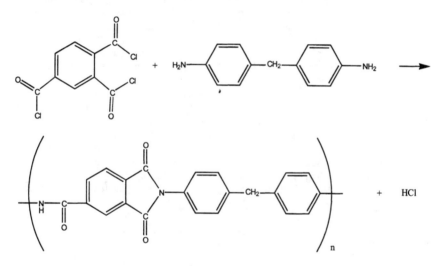

FIGURE 2.11 Preparation of polyamide-imide.

Polyamide-imides can be used from cryogenic temperatures to nearly 260°C. They have the temperature resistance of the polyimides but better mechanical properties, including good stiffness and creep resistance. PAI polymers are inherently flame retardant, with little smoke produced when they are burned. The polymer has good chemical resistance, but at high temperatures it can be affected by strong acids, bases, and steam.[136] PAI has a heat deflection temperature of 280°C, along with good wear and friction properties.[137] Polyamide-imides also have good radiation resistance and are more stable than standard nylons under different humidity conditions. The polymer has one of the highest glass transition temperatures, in the range of 270 to 285°C.[138]

Polyamide-imide can be processed by injection molding, but special screws are needed due to the reactivity of the polymer under molding conditions. Low-compression-ratio screws are recommended.[139] The parts should be annealed after molding at gradually increased temperatures.[140] For injection molding, the melt temperature should be near 355°C, with mold temperatures of 230°C. PAI can also be processed by compression molding or used in solution form. For compression molding, preheating at 280°C, followed by molding between 330 and 340°C with a pressure of 30 MPa, is generally used.[141]

Polyamide-imide polymers find application in hydraulic bushings and seals, mechanical parts for electronics, and engine components.[142] The polymer in solution has application as a laminating resin for spacecraft, a decorative finish for kitchen equipment, and as wire enamel.[143] Low coefficient of friction materials may be prepared by blending PAI with polytetrafluoroethylene and graphite.[144]

2.2.8 Polyarylate

Polyarylates are amorphous, aromatic polyesters. Polyarylates are polyesters prepared from dicarboxylic acids and bis-phenols.[145] Bis-phenol A is commonly used along with aromatic dicarboxylic acids, such as mixtures of isophthalic acid and terephthalic acid. The use of two different acids results in an amorphous polymer. However, the presence of the aromatic rings gives the polymer a high T_g and good temperature resistance. The temperature resistance of polyarylates lies between polysulfone and polycarbonate. The polymer is flame retardant and shows good toughness and UV resistance.[146] Polyarylates are transparent and have good electrical properties. The abrasion resistance of polyarylates is superior to that of polycarbonate. In addition, the polymers show very high recovery from deformation.

Polarylates are processed by most of the conventional methods. Injection molding should be performed with a melt temperature of 260 to 382°C, with mold temperatures of 65 to 150°C. Extrusion and blow molding grades are also available. Polyarylates can react with water at processing temperatures, and they should be dried prior to use.[147]

Polyarylates are used in automotive applications such as door handles, brackets, and headlamp and mirror housings. Polyarylates are also used in electrical applications for connectors and fuses. The polymer can be used in circuit board applications, because its high temperature resistance allows the part to survive exposure to the temperatures generated during soldering.[148] The excellent UV resistance of these polymers allows them to be used as a coating for other thermoplastics for improved UV resistance of the part. The good heat resistance of polyarylates allows them to be used in applications such as fire helmets and shields.[149]

2.2.9 Polybenzimidazole (PBI)

Polybenzimidazoles (PBIs) are high-temperature-resistant polymers. They are prepared from aromatic tetramines (for example, tetra amino-biphenol) and aromatic dicarboxylic acids (diphenylisophthalate).[150] The reactants are heated to form a soluble prepolymer that is converted to the insoluble polymer by heating at temperatures above 300°C.[151] The general structure of PBI is shown in Fig. 2.12.

The resulting polymer has high temperature stability, good chemical resistance, and nonflammability. The polymer releases very little toxic gas and does not melt when exposed to pyrolysis conditions. The polymer can be formed into fibers by dry-spinning processes. Polybenzimidazole is usually amorphous, with a T_g near 430°C.[152] Under certain

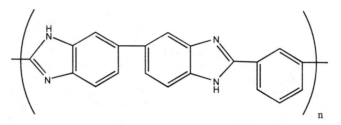

FIGURE 2.12   General structure of polybenzimidazoles.

conditions, crystallinity may be obtained. The lack of many single bonds and the high glass transition temperature give this polymer its superior high-temperature resistance. In addition to the high-temperature resistance, the polymer exhibits good low-temperature toughness. PBI polymers show good wear and frictional properties along with excellent compressive strength and high surface hardness.[153] The properties of PBI at elevated temperatures are among the highest of the thermoplastics. In hot, aqueous solutions, the polymer may absorb water, with a resulting loss in mechanical properties. Removal of moisture will restore the mechanical properties. The heat deflection temperature of PBI is higher than most thermoplastics, and this is coupled with a low coefficient of thermal expansion. PBI can withstand temperatures up to 760°C for short durations and exposure to 425°C for longer durations.

The polymer is not available as a resin and is generally not processed by conventional thermoplastic processing techniques, but rather by a high-temperature and high-pressure sintering process.[154] The polymer is available in fiber form, certain shaped forms, finished parts, and solutions for composite impregnation. PBI is often used in fiber form for a variety of applications such as protective clothing and aircraft furnishings.[155] Parts made from PBI are used as thermal insulators, electrical connectors, and seals.[156]

2.2.10   Polybutylene (PB)

Polybutylene polymers are prepared by the polymerization of 1-butene using Ziegler-Natta catalysts The molecular weights range from 770,000 to 3,000,000.[157] Copolymers with ethylene are often prepared as well. The chain structure is mainly isotactic and is shown in Fig. 2.13.[158]

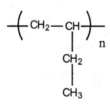

FIGURE 2.13   General structure for polybutylene.

The glass transition temperature for this polymer ranges from –17 to –25°C. Polybutylene resins are linear polymers exhibiting good resistance to creep at elevated temperatures and good resistance to environmental stress cracking.[159] They also show high impact strength, tear resistance, and puncture resistance. As with other polyolefins, polybutylene shows good resistance to chemicals, good moisture barrier properties, and good electrical insulation properties. Pipes prepared from polybutylene can be solvent welded, yet the polymer still exhibits good environmental stress cracking resistance.[160] The chemical resistance is quite good below 90°C but, at elevated temperatures, the polymer may dissolve in solvents such as toluene, decalin, chloroform, and strong oxidizing acids.[161]

Polybutylene is a crystalline polymer with three crystalline forms. The first crystalline form is obtained when the polymer is cooled from the melt. This first form is unstable and will change to a second crystalline form upon standing over a period of 3 to 10 days. The third crystalline form is obtained when polybutylene is crystallized from solution. The melting point and density of the first crystalline form are 124°C and 0.89 g/cm^3, respectively.[162] On transformation to the second crystalline form, the melting point increases to 135°C, and the density is increased to 0.95 g/cm^3. The transformation to the second crystalline form increases the polymer's hardness, stiffness, and yield strength.

Polybutylene can be processed on equipment similar to that used for low-density polyethylene. Polybutylene can be extruded and injection molded. Film samples can be blown or cast. The slow transformation from one crystalline form to another allows polybutylene to undergo post forming techniques, such as cold forming of molded parts or sheeting.[163] A range of 160 to 240°C is typically used to process polybutylene.[164] The die swell and shrinkage are generally greater for polybutylene than for polyethylene. Because of the crystalline transformation, initially molded samples should be handled with care.

An important application for polybutylene is plumbing pipe for both commercial and residential use. The excellent creep resistance of polybutylene allows for the manufacture of thinner wall pipes compared to pipes made from polyethylene or polypropylene. Polybutylene pipe can also be used for the transport of abrasive fluids. Other applications for polybutylene include hot melt adhesives and additives for other plastics. The addition of polybutylene improves the environmental stress cracking resistance of polyethylene and the impact and weld line strength of polypropylene.[165] Polybutylene is also used in packaging applications.[166]

2.2.11 Polycarbonate

Polycarbonate (PC) is often viewed as the quintessential engineering thermoplastic due to its combination of toughness, high strength, high heat-deflection temperatures, and transparency. The world wide growth rate, predicted in 1999 to be between eight and ten percent, is hampered only by the resin cost and is paced by applications where PC can replace ferrous or glass products. The polymer was discovered in 1898, and by 1958, both Bayer in Germany and General Electric in the United States had commenced production. Two current synthesis processes are commercialized, with the economically most successful one said to be the "interface" process, which involves the dissolution of bisphenol A in aqueous caustic soda and the introduction of phosgene in the presence of an inert solvent such as pyridine. The bisphenol A monomer is dissolved in the aqueous caustic soda then stirred with the solvent for phosgene. The water and solvent remain in separate phases. Upon phosgene introduction, the reaction occurs at the interface, with the ionic ends of the growing molecule being soluble in the catalytic caustic soda solution and the remainder of the molecule soluble in the organic solvent.[167] An alternative method involves transesterification of bisphenol A with diphenyl carbonate at elevated temperatures.[168] Both reactions are shown in Fig. 2.14. Molecular weights of between 30,000 and 50,000 g/mol can be obtained by the second route, while the phosgenation route results in higher-molecular-weight product.

The structure of PC, with its carbonate and bisphenolic structures, has many characteristics that promote its distinguished properties. The para-substitution on the phenyl rings results in a symmetry and lack of stereospecificity. The phenyl and methyl groups on the

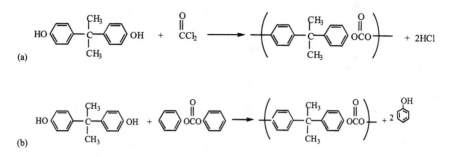

FIGURE 2.14 Synthesis routes for PC: (a) interface process and (b) transesterification reaction.

quartenary carbon promote a stiff structure. The ester-ether carbonate groups -OCOO- are polar, but their degree of intermolecular polar bond formation is minimized due to the steric hindrance posed by the benzene rings. The high level of aromaticity on the backbone, and the large size of the repeat structure, yield a molecule of very limited mobility. The ether linkage on the backbone permits some rotation and flexibility, producing high impact strength. Its amorphous nature with long, entangled chains contributes to the unusually high toughness. Upon crystallization, however, PC is brittle. PC is so reluctant to crystallize that films must be held at 180°C for several days to impart enough flexibility and thermal mobility required to conform to a structured three-dimensional crystalline lattice.[169] The rigidity of the molecule accounts for strong mechanical properties, elevated heat deflection temperatures, and high dimensional stability at elevated temperatures. The relatively high free volume results in a low-density polymer, with unfilled PC having a 1.22 g/cm^3 density. A disadvantage includes the need for drying and elevated-temperature processing. PC has limited chemical resistance to numerous aromatic solvents, including benzene, toluene, and xylene and has a weakness to notches. Selected mechanical and thermal properties are given in Table 2.3.[170]

TABLE 2.3 PC Thermal and Mechanical Properties

	Polycarbonate	30% glass-filled polycarbonate	Makroblend PR51, Bayer	Xenoy, CL101 GE
Heat deflection temperature, °C, method A	138	280	90	95
Heat deflection temperature, °C, method B	142	287	105	105
Ultimate tensile strength, N/mm^2	>65	70	56	>100
Ultimate elongation, %	110	3.5	120	>100
Tensile modulus, N/mm^2	2300	5500	2200	1900

Applications where PC is blended with ABS increase the heat distortion temperature of the ABS and improve the low-temperature impact strength of PC. The favorable ease of processing and improved economics make PC/ABS blends well suited for thin-walled electronic housing applications such as laptop computers. Blends with PBT are useful for improving the chemical resistance of PC to petroleum products and its low-temperature impact strength. PC alone is widely used as vacuum cleaner housings, household appliance housings, and power tools. These are arenas where PC's high impact strength, heat resistance, durability, and high-quality finish justify its expense. It is also used in safety helmets, riot shields, aircraft canopies, traffic light lens housings, and automotive battery cases. Design engineers take care not to design with tight radii where PC's tendency to stress crack could be a hindrance. PC cannot withstand constant exposure to hot water and can absorb 0.2 percent of its weight of water at 33°C and 65 percent relative humidity. This does not impair its mechanical properties but, at levels greater than 0.01 percent, processing results in streaks and blistering.

2.2.12 Polyester Thermoplastics

The broad class of organic chemicals called polyesters are characterized by the fact that they contain an ester linkage,

and may have either aliphatic or aromatic hydrocarbon units. As an introduction, Table 2.4 offers some selected thermal and mechanical properties as a means of comparing polybutylene terephthalate (PBT), polycyclohexylenedimethylene terephthalate (PCT), and poly(ethylene terephthalate) (PET).

2.2.12.1 Liquid Crystal Polymers (LCPs). Liquid crystal polyesters, known as liquid crystal polymers, are aromatic copolyesters. The presence of phenyl rings in the backbone of the polymer gives the chain rigidity, forming a rod-like chain structure. Generally, the phenyl rings are arranged in para linkages to give good rod-like structures.[171] This chain structure orients itself in an ordered fashion, both in the melt and in the solid state, as shown in Fig. 2.15. The materials are self-reinforcing with high mechanical properties, but

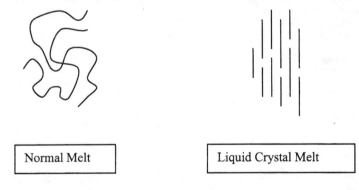

FIGURE 2.15 Melt configurations.

TABLE 2.4 Comparison of Thermal and Mechanical Properties of PBT, PCT, PCTA, PET, PEG, and PCTG.

	PBT unfilled	30% glass-filled PBT	30% glass-filled PCT	30% glass-filled PCTA	PET unfilled	30% glass-filled PET	PETG unfilled	PCTG unfilled
T_m, °C	220–267	220–267	–	285	212–265	245–265	–	–
Tensile modulus, MPa	1,930–3,000	8,960–10,000	–	–	2,760–4,140	8,960–9,930	–	–
Ultimate tensile strength, MPa	56–60	96–134	124–134	97	48–72	138–165	28	52
Ultimate elongation, %	50–300	2–4	1.9–2.3	3.1	30–300	2–7	110	330
Specific gravity	1.30–1.38	1.48–1.54	1.45	1.41	1.29–1.40	1.55–1.70	1.27	1.23
HDT, °C								
264 psi	50–85	196–225	260	221	21–65	210–227	64	65
66 psi	115–190	216–260	> 260	268	75	243–249	70	74

as a result of the oriented liquid crystal behavior, the properties will be anisotropic. The designer must be aware of this to properly design the part and gate the molds.[172] The phenyl ring also helps increase the heat distortion temperature.[173]

The basic building blocks for liquid crystal polyesters are *p*-hydroxybenzoic acid, terephthalic acid, and hydroquinone. Unfortunately, the use of these monomers alone gives materials that are difficult to process with very high melting points. The polymers often degraded before melting.[174] Various techniques have been developed to give materials with lower melting points and better processing behavior. Some methods include the incorporation of flexible units in the chain (copolymerizing with ethylene glycol), the addition of nonlinear rigid structures, and the addition of aromatic groups to the side of the chain.[175]

Liquid crystal polymers based on these techniques include Victrex (ICI), Vectra (Hoescht), and Xydar (Amoco). Xydar is based on terephthalic acid, *p*-hydroxybenzoic acid, and *p,p*′-dihydroxybiphenyl, while Vectra is based on *p*-hydroxybenzoic acid and hydroxynaphthoic acid.[176] These materials are known for their high temperature resistance, and particularly heat distortion temperature. The heat distortion temperature can vary from 170 to 350°C. They also have excellent mechanical properties, especially in the flow direction. For example, the tensile strength varies from 165 to 230 Mpa, the flexural strength varies from 169 to 256 Mpa, and the flexural modulus varies from 9 to 12.5 Gpa.[177] Filled materials exhibit even higher values. LCPs are also known for good solvent resistance and low water absorption compared to other heat-resistant polymers. They have good electrical insulation properties, low flammability with a limiting oxygen index in the range of 35 to 40, but a high specific gravity (about 1.40).[178] LCPs show little dimensional change when exposed to high temperatures and a low coefficient of thermal expansion.[179]

These materials can be high priced and often exhibit poor abrasion resistance due to the oriented nature of the polymer chains.[180] Surface fibrillation may occur quite easily.[181] The materials are processable on a variety of conventional equipment. Process temperatures are normally below 350°C, although some materials may need to be processed at higher temperatures. They generally have low melt viscosity as a result of their ordered melt and should be dried before use to avoid degradation.[182] LCPs can be injection molded on conventional equipment, and regrind may be used. Mold release is generally not required.[183] Part design for LCPs requires careful consideration of the anisotropic nature of the polymer. Weld lines can be very weak if the melt meets in a "butt" type of weld line. Other types of weld lines show better strength.[184]

Liquid crystal polymers are used in automotive, electrical, chemical processing, and household applications. One application is for oven and microwave cookware.[185] Because of their higher costs, the material is used only in applications where its superior performance justifies the additional expense.

2.2.12.2 Polybutylene Terephthalate (PBT). With the expiration of the original PET patents, manufacturers pursued the polymerization of other polyalkene terephthalates, particularly polybutylene terephthalate (PBT). The polymer is synthesized by reacting terephthalic acid with butane 1,4-diol to yield the structure shown in Fig. 2.16.

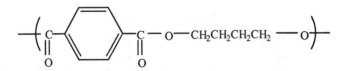

FIGURE 2.16 Repeat structure of PBT.

The only structural difference between PBT and PET is the substitution in PBT of four methylene repeat units rather than two present in PET. This feature imparts additional flexibility to the backbone and reduces the polarity of the molecule resulting in similar mechanical properties to PET (high strength, stiffness, and hardness). PBT growth is at least ten percent annually, in large part due to automotive exterior and under-the-hood applications such as electronic stability control and housings that are made out of a PBT/ASA (acrylonitrile/styrene/acrylic ester) blend. PBT/ASA blends are sold by BASF and GE Plastics Europe. Another development involving the use of PBT is coextrusion of PBT and a copolyester thermoplastic elastomer. This can then be blow molded into under-the-hood applications to minimize noise vibration. Highly filled PBTs are also making inroads into the kitchen and bathroom tile industries.[186] As with PET, PBT is also often glass fiber filled so as to increase its flexural modulus, creep resistance, and impact strength. PBT is suitable for applications requiring dimensional stability, particularly in water, and resistance to hydrocarbon oils without stress-cracking.[187] Hence, PBT is used in pump housings, distributors, impellers, bearing bushings, and gear wheels.

To improve PBT's poor notched impact strength, copolymerization with 5 percent ethylene and vinyl acetate onto the polyester backbone improves its toughness. PBT is also blended with PMMA, PET, PC, and polybutadiene to provide enhanced properties tailored to specific applications.

2.2.12.3 Polycyclohexylenedimethylene Terephthalate (PCT). Another polyalkylene terephthalate polyester of significant commercial importance is PCT—a condensation product of the reaction between dimethyl terephthalate and 1,4-cyclohexylene glycol as shown below in Fig. 2.17. This material is biaxially oriented into films and, while it is mechanically weaker than PET, it offers superior water resistance and weather resistance.[188] As seen in the introductory Table 2.4, PCT differentiates itself from PET and PBT with its high heat distortion temperature. As with PET and PBT, PCT has low moisture absorption, and its good chemical resistance to engine fluids and organic solvents lends it to under-the-hood applications such as alternator armatures and pressure sensors.[189]

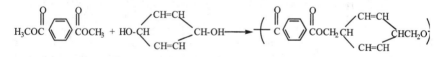

FIGURE 2.17 Synthesis route of PCT.

Copolymers of PCT include PCTA, an acid-modified polyester, and PCTG, a glycol-modified polyester. PCTA is used primarily for extruded film and sheet for packaging applications. PCTA has high clarity, tear strength, and chemical resistance, and when PCTA is filled, it is used for dual ovenable cookware. PCTG is primarily injection molded, and PCTG parts have notched Izod impact strengths similar to polycarbonate, against which it often competes. It also competes with ABS, another clear polymer. It finds use in medical and optical applications.[190]

2.2.12.4 Poly(ethylene Terephthalate) (PET). There are tremendous commercial applications for PET: as an injection-molding-grade material, for blow-molded bottles, and for oriented films. In 1998, the U.S. consumption of PET was 4,330 million lb, while domestic consumption of PBT was 346 million lb.[191] PET, also known as poly(oxyethylene oxyterephthaloyl), can be synthesized from dimethyl terephthalate and ethylene glycol by a two-step ester interchange process, as shown in Fig. 2.18.[192] The first stage involves a so-

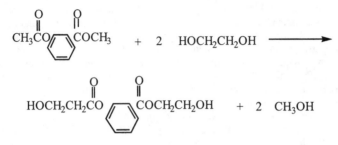

FIGURE 2.18   Direct esterification of a diacid (dimethyl terephthalate) with a diol(ethylene glycol) in the first stages of PET polymerization.

lution polymerization of one mole of dimethyl terephthalate with 2.1 to 2.2 moles of ethylene glycol.[193] The excess ethylene glycol increases the rate of formation of bis(2-hydroxyethyl) terephthalate. Small amounts of trimer, tetramer, and other oligomers are formed. A metal alkanoate, such as manganese acetate, is often added as a catalyst; this is later deactiviated by the addition of a phosphorous compound such as phosphoric acid. The antioxidant phosphate improves the thermal and color stability of the polymer during the higher-temperature second-stage process.[194] The first stage of the reaction is run at 150 to 200°C with continuous methanol distillation and removal.[195]

The second step of the polymerization, shown in Fig. 2.19, is a melt polymerization as the reaction temperature is raised to 260 to 290°C. This second stage is carried out under either partial vacuum (0.13 kPa)[196] to facilitate the removal of ethylene glycol or with an inert gas being forced through the reaction mixture. Antimony trioxide is often used as a polymerization catalyst for this stage.[197] It is critical that excess ethylene glycol be completely removed during this alcoholysis stage of the reaction so as to proceed to high-molecular-weight products; otherwise, equilibrium is established at an extent of reaction of less than 0.7. This second stage of the reaction proceeds until a number-average-molecular weight, M_n, of about 20,000 g/mol is obtained. The very high temperatures at the end of this reaction cause thermal decomposition of the end groups to yield acetaldehyde. Thermal ester scission also occurs, which competes with the polymer step-growth reactions. It is this competition, which limits the ultimate M_n, that can be achieved through this melt condensation reaction.[198] Weight-average molecular weights of oriented films are around 35,000 g/mol.

Other commercial manufacturing methods have evolved to a direct esterification of acid and glycol in place of the ester-exchange process. In direct esterification, terephthalic acid and ethylene glycol are reacted rather than esterifying terephthalic acid with methanol to produce the dimethyl terephthalate intermediate. The ester is easier to purify than the acid, which sublimes at 300°C and is insoluble. However, better catalysts and purer

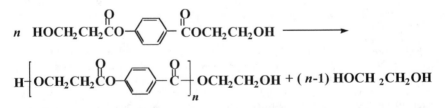

FIGURE 2.19   Polymerization of bis(2-hydroxyethyl) terephthalate to PET.

terephthalic acid offer the elimination of the intermediate use of methanol.[199] Generally, PET resins made by direct esterification of terephthalic acid contain more diethylene glycol, which is generated by an intermolecular ether-forming reaction between β-hydroxyethyl ester end groups. Oriented films produced from these resins have reduced mechanical strength and melting points as well as decreased thermo-oxidative resistance and poorer UV stability.[200]

The degree of crystallization and direction of the crystallite axis govern all of the resin's physical properties. The percentage of structure existing in crystalline domains is primarily determined through density measurements or by thermal means using a differential scanning calorimeter (DSC). The density of amorphous PET is 1.333 g/cm^3, while the density of a PET crystal is 1.455 g/cm^3.[201] Once the density is known, the fraction of crystalline material can be determined.

An alternate means of measuring crystallinity involves comparing the ratio of the heat of cold crystallization, ΔH_{cc}, of amorphous polymer to the heat of fusion, ΔH_f, of crystalline polymer. This ratio is 0.61 for an amorphous PET and a fully crystalline PET sample should yield a value close to zero.[202] After the sample with its initial morphology has been run once in the DSC, the heat of fusion determined in the next run can be considered as ΔH_{cc}. The lower the $\Delta H_{cc}/\Delta H_f$ ratio, the more crystalline the original sample was.

In the absence of nucleating agents and plasticizers, PET crystallizes slowly, which is a hindrance in injection molding applications, as either hot molds or costly extended cooling times are required. In the case of films, however, where crystallinity can be mechanically induced, PET resins combine rheological properties that lend themselves to melt extrusion with a well defined melting point, making them ideally suited for biaxially oriented film applications. The attachment of the ester linkage directly to the aromatic component of the backbone means that these linear, regular PET chains have enough flexibility to form stress-induced crystals and achieve enough molecular orientation to form strong, thermally stable films.[203]

Methods for producing oriented PET films have been well documented and will be only briefly discussed here. The process as described in the *Encyclopedia of Polymer Science and Engineering* usually involves a sequence of five steps which include[204]

- Melt extrusion and slot casting
- Quenching
- Drawing in the longitudinal machine direction (MD)
- Drawing in the transverse direction (TD)
- Annealing

Dried, highly viscous polymer melt is extruded through a slot die with an adjustable gap width onto a highly polished quenching drum. If very high output rates are required, a cascade system of extruders can be set up to first melt and homogenize the PET granules, then to use the next in-line extruder to meter the melt to the die. Molten resin is passed through filter packs with average pore sizes of 5 to 30 μm. Quenching to nearly 100 percent amorphous morphology is critical to avoid embrittlement; films that have been allowed to form spherulites are brittle and translucent, and are unable to be further processed.

The sheet is then heated to about 95°C (above the glass transition point of approximately 70°C), where thermal mobility allows the material to be stretched to three or four times its original dimension in the MD. This uniaxially oriented film has stress-induced crystals whose main axes are aligned in the machine direction. The benzene rings, however, are aligned parallel to the surface of the film in the <1,0,0> crystal plane. The film is then again heated, generally to above 100°C, and stretched to three to four times its initial dimension in the TD. This induces further crystallization, bringing the degree of crystallinity to 25 to 40 percent, and creates a film, which has isotropic tensile strength and elongation properties in the machine and transverse directions. The film at this point is thermally unstable above 100°C and must be annealed in the tenter frame to partially relieve the stress.

The annealing involves heating to 180 to 220°C for several seconds to allow amorphous chain relaxation, partial melting, recrystallization, and crystal growth to occur.[205] The resultant film is approximately 50 percent crystalline and possesses good mechanical strength, a smooth surface that readily accepts a wide variety of coatings, and good winding and handling characteristics. PET films are produced from 1.5 μm thick as capacitor films to 350 μm thick for use as electrical insulation in motors and generators.[206]

Due to the chemically inert nature of PET, films that are used in coatings applications are often treated with a variety of surface modifiers. Organic and inorganic fillers are often incorporated in relatively thick films to improve handling characteristics by roughening the surface slightly. For thin films, however, many applications require transparency, which would be marred by the incorporation of fillers. Therefore, an in-line coating step of either aqueous or solvent-based coatings is set up between the MD and TD drawing stations. The drawing of the film after the coating has been applied helps to achieve very thin coatings.

2.2.12.5 Polytrimethylene Terephthalate (PTT). PTT has been produced and marketed as three grades by Shell Chemicals under the trade name Corterra since the late 1990s, when Shell was able to develop a low-cost method of producing the starting raw material 1,3-propanediol. Corterra is used in the textile and carpet industries, which take advantage of its stain resistance, wearability as a result of high resilience and elastic recovery, color fastness, and soft hand.

2.2.12.6 Polyethylene Napthalate (PEN). Polyethylene napthalate (PEN) gained commercial importance in the late 1980s. Compared to PET, it has higher thermal resistance and tensile strength as well as better barrier properties and UV resistance. This is a result of the napthenic ring structures.[207] Both the T_g (124°C) and T_m (270 to 273°C) of PEN are higher compared to PET, while the crystallization rate of PEN is slower than PET. Currently, PEN is more expensive than PET, leading to the development of copolymers. Applications for PEN include fibers, films, and blow-molded products. Due to cost considerations, the markets for PEN blow molded products are generally in the medical arena.

2.2.13   Polyetherimide (PEI)

Polyetherimides (PEI) are a newer class of amorphous thermoplastics with high-temperature resistance, impact strength, creep resistance, and rigidity. They are transparent with an amber color.[208] The polymer is sold under the trade name of Ultem (General Electric) and has the structure shown in Fig. 2.20. It is prepared from the condensation polymerization of diamines and dianhydrides.[209]

The material can be melt processed because of the ether linkages present in the backbone of the polymer, but it still maintains properties similar to the polyimides.[210] The high-temperature resistance of the polymer allows it to compete with the polyketones, polysulfones, and poly(phenylene sulfides). The glass transition temperature of PEI is 215°C. The polymer has very high tensile strength, a UL temperature index of 170°C, flame resistance, and low smoke emission.[211] The polymer is resistant to alcohols, acids, and hydrocarbon solvents but will dissolve in partially halogenated solvents.[212] Both glass- and carbon-fiber-reinforced grades are available.[213]

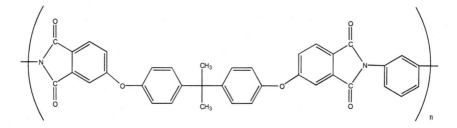

FIGURE 2.20   General structure of polyetherimide.

The polymer should be dried before processing, and typical melt temperatures are 340 to 425°C.[214] Polyetherimides can be processed by injection molding and extrusion. In addition, the high melt strength of the polymer allows it to be thermoformed and blow molded. Annealing of the parts is not required.

Polyetherimide is used in a variety of applications. Electrical applications include printed circuit substrates and burn-in sockets. In the automotive industry, PEI is used for under-the-hood temperature sensors and lamp sockets. PEI sheet has also been used to form an aircraft cargo vent.[215] The dimensional stability of this polymer allows its use for large flat parts such in hard disks for computers.

2.2.14   Polyethylene (PE)

Polyethylene (PE) is the highest-volume polymer in the world. Its high toughness, ductility, excellent chemical resistance, low water vapor permeability, and very low water absorption, combined the ease with which it can be processed, make PE of all different density grades an attractive choice for a variety of goods. PE is limited by its relatively low modulus, yield stress, and melting point. PE is used to make containers, bottles, film, and pipes, among other things. It is an incredibly versatile polymer with almost limitless variety due to copolymerization potential, a wide density range, a MW that ranges from very low (waxes have a MW of a few hundred) to very high (6×106), and the ability to vary MWD.

Its repeat structure is $(-CH_2CH_2-)_x$, which is written as polyethylene rather than polymethylene $(-CH_2)_x$, in deference to the various ethylene polymerization mechanisms. PE has a deceptive simplicity. PE homopolymers are made up exclusively of carbon and hydrogen atoms and, just as the properties of diamond and graphite (which are also materials made up entirely of carbon and hydrogen atoms) vary tremendously, different grades of PE have markedly different thermal and mechanical properties. While PE is generally a whitish, translucent polymer, it is available in grades of density that range from 0.91 to 0.97 g/cm^3. The density of a particular grade is governed by the morphology of the backbone: long, linear chains with very few side branches can assume a much more three-dimensionally compact, regular, crystalline structure. Commercially available grades are low-density PE (LDPE), linear low-density PE (LLDPE), high-density PE (HDPE), and ultra-high-molecular-weight PE (UHMWPE). Figure 2.21 demonstrates figurative differences in chain configuration that govern the degree of crystallinity, which, along with MW, determines final thermomechanical properties.

Four established production methods are (1) a gas phase method known as the Unipol process practiced by Union Carbide, (2) a solution method used by Dow and DuPont, (3) a slurry emulsion method practiced by Phillips, and (4) a high-pressure method.[216] Gener-

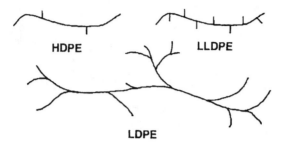

FIGURE 2.21   Chain configurations of polyethylene.

ally, yield strength and melt temperature increase with density, while elongation decreases with increased density.

2.2.14.1 Very-Low-Density Polyethylene (VLDPE).   This material was introduced in 1985 by Union Carbide, is very similar to LLDPE, and is principally used in film applications. VLDPE grades vary in density from 0.880 to 0.912 g/cm^3.[217] Its properties are marked by high elongation, good environmental stress cracking resistance, and excellent low-temperature properties, and it competes most frequently as an alternative to plasticized polyvinyl chloride (PVC) or ethylene-vinyl acetate (EVA). The inherent flexibility in the backbone of VLDPE circumvents plasticizer stability problems that can plague PVC, and it avoids odor and stability problems that are often associated with molding EVAs.[218]

2.2.14.2 Low-Density Polyethylene (LDPE).   LDPE combines high impact strength, toughness, and ductility to make it the material of choice for packaging films, which is one of its largest applications. Films range from shrink film, thin film for automatic packaging, heavy sacking, and multilayer films (both laminated and coextruded), where LDPE acts as a seal layer or a water vapor barrier.[219] It has found stiff competition from LLDPE in these film applications due to LLDPE's higher melt strength. LDPE is still very widely used, however, and is formed via free radical polymerization, with alkyl branch groups (given by the structure $-(CH_2)_x CH_3$) of two to eight carbon atom lengths. The most common branch length is four carbons long. High reaction pressures encourage crystalline regions. The reaction to form LDPE is shown in Fig. 2.22, where "n" approximately varies in commercial grades between 400 to 50,000.[220]

$$\text{n } CH_2 \!=\! CH_2 \xrightarrow[\substack{\text{small amounts of } O_2 \text{ or} \\ \text{organic peroxide present}}]{\substack{200°C \\ 20,000 \text{ - } 35,000 \text{ psi}}} -(CH_2CH_2)_n-$$

FIGURE 2.22   Polymerization of PE.

Medium-density PE is produced via the reaction above, carried out at lower polymerization temperatures.[221] The reduced temperatures are postulated to reduce the randomizing Brownian motion of the molecules, and this reduced thermal energy allows crystalline formation more readily at these lowered temperatures.

2.2.14.3 Linear Low-Density Polyethylene (LLDPE).   This product revolutionized the plastics industry with its enhanced tensile strength for the same density compared to

LDPE. Table 2.5 compares mechanical properties of LLDPE to LDPE. As is the case with LDPE, film accounts for approximately three-quarters of the consumption of LLDPE. As the name implies, it is a long linear chain without long side chains or branches. The short chains, which are present, disrupt the polymer chain uniformity enough to prevent crystalline formation and hence prevent the polymer from achieving high densities. Developments of the past decade have enabled production economies compared to LDPE due to lower polymerization pressures and temperatures. A typical LDPE process requires 35,000 psi, which is reduced to 300 psi in the case of LLDPE, and reaction temperatures as low as 100°C rather than 200 or 300°C are used. LLDPE is actually a copolymer containing side branches of 1-butene most commonly, with 1-hexene or 1-octene also present. Density ranges of 0.915 to 0.940 g/cm^3 are polymerized with Ziegler catalysts, which orient the polymer chain and govern the tacticity of the pendant side groups.[222]

TABLE 2.5 Comparison of Blown Film Properties of LLDPE and LDPE[*]

	LLDPE	LDPE
Density, g/cm^3	0.918	0.918
Melt index, g/10 min	2.0	2.0
Dart impact, g	110	110
Puncture energy, J/mm	60	25
Machine direction tensile strength, MPa	33	20
Cross direction tensile strength, MPa	25	18
Machine direction tensile elongation, %	690	300
Cross direction tensile elongation, %	740	500
Machine direction modulus, MPa	210	145
Cross direction modulus, MPa	250	175

[*]**Source:** *Encyclopedia of Polymer Science,* 2nd ed., vol. 6, Mark, Bikales, Overberger, Menges, and Kroschwitz, Eds., Wiley Interscience, 1986, p. 433.

2.2.14.4 High-Density Polyethylene (HDPE).

HDPE is one of the highest-volume commodity chemicals produced in the world. In 1998, the worldwide demand was 1.8×10^{10} kg.[223] The most common method of processing HDPE is blow molding, where resin is turned into bottles (especially for milk and juice), housewares, toys, pails, drums, and automotive gas tanks. It is also commonly injection molded into housewares, toys, food containers, garbage pails, milk crates, and cases. HDPE films are commonly found as bags in supermarkets, department stores, and as garbage bags.[224] Two commercial polymerization methods are most commonly practiced. One involves Phillips catalysts (chromium oxide), and the other involves Ziegler-Natta catalyst systems (supported heterogeneous catalysts such as titanium halides, titanium esters, and aluminum alkyls on a chemically inert support such as PE or PP). Molecular weight is governed primarily through temperature control, with elevated temperatures resulting in reduced molecular weights. The catalyst support and chemistry also play an important factor in controlling molecular weight and molecular weight distribution.

2.2.14.5 Ultra-High-Molecular-Weight Polyethylene (UHMWPE).

UHMWPE is identical to HDPE but, rather than having a MW of 50,000 g/mol, it typically has a MW of between 3×10^6 and 6×10^6. The high MW imparts outstanding abrasion resistance, high toughness (even at cryogenic temperatures), and excellent stress cracking resistance, but it does not generally allow the material to be processed conventionally. The polymer chains are so entangled, due to their considerable length, that the conventionally considered melt point doesn't exist practically, as it is too close to the degradation temperature—although an injection-molding grade is marketed by Hoechst. Hence, UHMWPE is often processed as a fine powder that can be ram extruded or compression molded. Its properties are taken advantage of in uses that include liners for chemical processing equipment, lubrication coatings in railcar applications to protect metal surfaces, recreational equipment such as ski bases, and medical devices.[225] A recent product has been developed by Allied Chemical that involves gel spinning UHMWPE into lightweight, very strong fibers that compete with Kevlar in applications for protective clothing.

2.2.15 Polyethylene Copolymers

Ethylene is copolymerized with many nonolefinic monomers, particularly acrylic acid variants and vinyl acetate, with EVA polymers being the most commercially significant. All of the copolymers discussed in this section necessarily involve disruption of the regular, crystallizable PE homopolymer and as such feature reduced yield stresses and moduli, with improved low-temperature flexibility.

2.2.15.1 Ethylene-Acrylic Acid (EAA) Copolymers. EAA copolymers, first identified in the 1950s, have enjoyed a renewed interest since 1974, when Dow introduced new grades characterized by outstanding adhesion to metallic and nonmetallic substrates.[226] The presence of the carboxyl and hydroxyl functionalities promotes hydrogen bonding, and these strong intermolecular interactions are taken advantage of to bond aluminum foil to polyethylene in multilayer extrusion-laminated toothpaste tubes and as tough coatings for aluminum foil pouches.

2.2.15.2 Ethylene-Ethyl Acrylate (EEA) Copolymers. EEA copolymers typically contain 15 to 30 percent by weight of ethyl acrylate (EA) and are flexible polymers of relatively high molecular weight suitable for extrusion, injection molding, and blow molding. Products made of EEA have high environmental stress cracking resistance, excellent resistance to flexural fatigue, and low-temperature properties down to as low as −65°C. Applications include molded rubber-like parts, flexible film for disposable gloves and hospital sheeting, extruded hoses, gaskets and bumpers.[227] Typical applications include polymer modifications where EEA is blended with olefin polymers (since it is compatible with VLDPE, LLDPE, LDPE, HDPE, and PP[228]) to yield a blend with a specific modulus, yet with the advantages inherent in EEA's polarity. The EA presence promotes toughness, flexibility, and greater adhesive properties. EEA blending can cost effectively improve the impact resistance of polyamides and polyesters.[229]

The similarity of ethyl acrylate monomer to vinyl acetate predicates that these copolymers have very similar properties, although EEA is considered to have higher abrasion and heat resistance, while EVA tends to be tougher and of greater clarity.[230] EEA copolymers are FDA approved up to 8 percent EA content in food contact applications.[231]

2.2.15.3 Ethylene-Methyl Acrylate (EMA) Copolymers. EMA copolymers are often blown into film with very rubbery mechanical properties and outstanding dart-drop impact strength. The latex-rubber-like properties of EMA film lend to its use in disposable gloves and medical devices without the associated hazards to people with allergies to latex rubber. Due to their adhesive properties, EMA copolymers, like their EAA and EEA counterparts, are used in extrusion coating, coextrusions, and laminating applications as heat-seal layers. EMA is one of the most thermally stable of this group, and as such it is commonly used to form heat and RF seals as well in multiextrusion tie-layer applications. This copolymer is also widely used as a blending compound with olefin homopolymers (VLDPE, LLDPE, LDPE, and PP) as well as with polyamides, polyesters, and polycarbonate to improve impact strength and toughness and to increase either heat seal response or to promote adhesion.[232] EMA is also used in soft blow-molded articles such as squeeze toys, tubing, disposable medical gloves, and foamed sheet. EMA copolymers and EEA copolymers containing up to 8 percent ethyl acrylate are approved by the FDA for food packaging.[233]

2.2.15.4 Ethylene-n-Butyl Acrylate (EBA) Copolymers. EBA copolymers are also widely blended with olefin homopolymers to improve impact strength, toughness, and heat sealability and to promote adhesion. The polymerization process and resultant repeat unit of EBA are shown in Fig. 2.23.

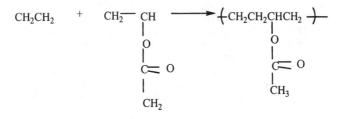

FIGURE 2.23 Polymerization and structure of EBA.

2.2.15.5 Ethylene-Vinyl Acetate (EVA) Copolymers. EVA copolymers are given by the structure shown in Fig. 2.24 and find commercial importance in the coating, laminating, and film industries. EVA copolymers typically contain between 10 and 15 mole percent vinyl acetate, which provides a bulky, polar pendant group to the ethylene and provides an opportunity to tailor the end properties by optimizing the vinyl acetate content. Very low vinyl-acetate content (approximately 3 mole percent) results in a copolymer that is essentially a modified low-density polyethylene,[234] with an even further reduced regular structure. The resultant copolymer is used as a film due to its flexibility and surface gloss. Vinyl acetate is a low-cost comonomer, which is nontoxic and allows for this copolymer to be

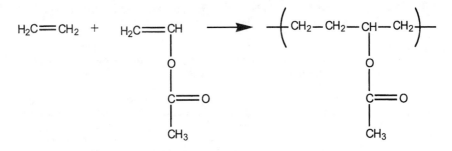

FIGURE 2.24 Polymerization of EVA.

used in many food packaging applications. These films are soft and tacky and therefore appropriate for cling-wrap applications (they are more thermally stable than the PVDC films often used as cling wrap) as well as interlayers in coextruded and laminated films.

EVA copolymers with approximately 11 mole percent vinyl acetate are widely used in the hot-melt coatings and adhesives arena, where the additional intermolecular bonding promoted by the polarity of the vinyl acetate ether and carbonyl linkages enhances melt strength while still enabling low melt-processing temperatures. At 15 mole percent vinyl acetate, a copolymer with very similar mechanical properties to plasticized PVC is formed. There are many advantages to an inherently flexible polymer for which there is no risk of plasticizer migration, and PVC-alternatives is the area of largest growth opportunity. These copolymers have higher moduli than standard elastomers and are preferable in that they are more easily processed without concern for the need to vulcanize.

2.2.15.6 Ethylene-Vinyl Alcohol (EVOH) Copolymers.   Poly(vinyl alcohol) is prepared through alcoholysis of poly(vinyl acetate). PVOH is an atactic polymer but, since the crystal lattice structure is not disrupted by hydroxyl groups, the presence of residual acetate groups greatly diminishes the crystal formation and the degree of hydrogen bonding. Polymers that are highly hydrolyzed (have low residual acetate content) have a high tendency to crystallize and for hydrogen bonding to occur. As the degree of hydrolysis increases, the molecules will very readily crystallize, and hydrogen bonds will keep them associated if they are not fully dispersed prior to dissolution. At degrees of hydrolysis above 98 percent, manufacturers recommend a minimum temperature of 96°C to ensure that the highest molecular weight components have enough thermal energy to go into solution. Polymers with low degrees of residual acetate have high humidity resistance.

2.2.15.7 Ethylene-Carbon Monoxide Copolymers (ECOs).   These polymers are random copolymers of ethylene and carbon monoxide, with properties similar to low-density polyethylene.[235] They are sold by Shell under the trade name Carilon. These polymers exhibit low water absorption and good barrier properties, but they are susceptible to UV degradation. They find application in packaging, fuel tanks, fuel lines, and in blends.

2.2.16   Modified Polyethylenes

The properties of PE can be tailored to meet the needs of a particular application by a variety of different methods. Chemical modification, copolymerization, and compounding can all dramatically alter specific properties. The homopolymer itself has a range of properties that depend on the molecular weight, the number and length of side branches, the degree of crystallinity, and the presence of additives such as fillers or reinforcing agents. Further modification is possible by chemical substitution of hydrogen atoms; this occurs preferentially at the tertiary carbons of a branching point and primarily involves chlorination, sulphonation, phosphorylination, and intermediate combinations.

2.2.16.1 Chlorinated Polyethylene (CPE).   The first patent on the chlorination of PE was awarded to ICI in 1938.[236] CPE is polymerized by substituting select hydrogen atoms on the backbone of either HDPE or LDPE with chlorine. Chlorination can occur in the gaseous phase, in solution, or as an emulsion. In the solution phase, chlorination is random, while the emulsion process can result in uneven chlorination due to the crystalline regions. The chlorination process generally occurs by a free-radical mechanism, shown in Fig. 2.25, where the chlorine free radical is catalyzed by ultraviolet light or initiators.

Interestingly, the properties of CPE can be adjusted to almost any intermediary position between PE and PVC by varying the properties of the parent PE and the degree and tacticity of chlorine substitution. Since the introduction of chlorine reduces the regularity

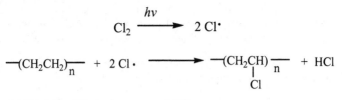

FIGURE 2.25 Chlorination process of CPE.

of the PE, crystallinity is disrupted and, at up to a 20 percent chlorine level, the modified material is rubbery (if the chlorine was randomly substituted). When the level of chlorine reaches 45 percent (approaching PVC), the material is stiff at room temperature. Typically, HDPE is chlorinated to a chlorine content of 23 to 48 percent.[237] Once the chlorine substitution reaches 50 percent, the polymer is identical to PVC, although the polymerization route differs. The largest use of CPE is as a blending agent with PVC to promote flexibility and thermal stability for increased ease of processing. Blending CPE with PVC essentially plasticizes the PVC without adding double-bond unsaturation prevalent with rubber-modified PVCs and results in a more UV-stable, weather-resistant polymer. While rigid PVC is too brittle to be machined, the addition of as little as three to six parts per hundred CPE in PVC allows extruded profiles such as sheets, films, and tubes to be sawed, bored and nailed.[238] Higher CPE content blends result in improved impact strength of PVC and are made into flexible films that don't have plasticizer migration problems. These films find applications in roofing, water and sewage-treatment pond covers, and sealing films in building construction.

CPE is used in highly filled applications, often using $CaCO_3$ as the filler, and finds use as a homopolymer in industrial sheeting, wire and cable insulations, and solution applications. When PE is reacted with chlorine in the presence of sulfur dioxide, a chlorosulfonyl substitution takes place, yielding an elastomer.

2.2.16.2 Chlorosulfonated Polyethylenes (CSPEs). Chlorosulfonation introduces the polar, cross-linkable SO_2 group onto the polymer chain, with the unavoidable introduction of chlorine atoms as well. The most common method involves exposing LDPE, which has been solubilized in a chlorinated hydrocarbon, to SO_2 and Cl in the presence of UV or high-energy radiation.[239] Both linear and branched PEs are used, and CSPEs contain 29 to 43 percent chlorine and 1 to 1.5 percent sulfur.[240] As in the case of CPEs, the introduction of Cl and SO_2 functionalities reduces the regularity of the PE structure, hence reducing the degree of crystallinity, and the resultant polymer is more elastomeric than the unmodified homopolymer. CSPE is manufactured by DuPont under the trade name Hypalon and is used in protective coating applications such as the lining for chemical processing equipment, as the liners and covers for waste-containment ponds, as cable jacketing and wire insulation, as spark plug boots, as power steering pressure hoses, and in the manufacture of elastomers.

2.2.16.3 Phosphorylated Polyethylenes. Phosphorylated PEs have higher ozone and heat resistance than ethylene propylene copolymers due to the fire retardant nature provided by phosphor.[241]

2.2.16.4 Ionomers. Acrylic acid can be copolymerized with polyethylene to form an ethylene acrylic acid copolymer (EAA) through addition or chain growth polymerization. It is structurally similar to ethylene vinyl acetate, but with acid groups off the backbone.

The concentration of acrylic acid groups is generally in the range of 3 to 20 percent.[242] The acid groups are then reacted with a metal containing base, such as sodium methoxide or magnesium acetate, to form the metal salt as depicted in Fig. 2.26.[243] The ionic groups can associate with each other, forming a cross-link between chains. The resulting materials are called ionomers in reference to the ionic bonds formed between chains. They were originally developed by DuPont under the trade name of Surlyn.

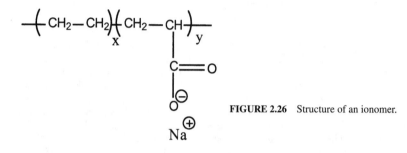

FIGURE 2.26   Structure of an ionomer.

The association of the ionic groups forms a thermally reversible crosslink that can be broken when exposed to heat and shear. This allows ionomers to be processed on conventional thermoplastic processing equipment while still maintaining some of the behavior of a thermoset at room temperature.[244] The association of ionic groups is generally believed to take two forms: multiplets and clusters.[245] Multiplets are considered to be a small number of ionic groups dispersed in the matrix, whereas clusters are phase-separated regions containing many ion pairs and also hydrocarbon backbone.

A wide range of properties can be obtained by varying the ethylene/methacrylic acid ratios, molecular weight, and the amount and type of metal cation used. Most commercial grades use either zinc or sodium for the cation. Materials using sodium as the cation generally have better optical properties and oil resistance, whereas those using zinc usually have better adhesive properties, lower water absorption, and better impact strength.[246]

The presence of the comonomer breaks up the crystallinity of the polyethylene so that ionomer films have lower crystallinity and better clarity compared to polyethylene.[247] Ionomers are known for their toughness and abrasion resistance, and the polar nature of the polymer improves both its bondability and paintability. Ionomers have good low-temperature flexibility and resistance to oils and organic solvents. Ionomers show a yield point with considerable cold drawing. In contrast to PE, the stress increases with strain during cold drawing, giving a very high energy to break.[248]

Ionomers can be processed by most conventional extrusion and molding techniques using conditions similar to other olefin polymers. For injection molding, the melt temperatures are in the range 210 to 260°C.[249] The melts are highly elastic due to the presence of the metal ions. Increasing temperatures rapidly decreases the melt viscosity, with the sodium and zinc based ionomers showing similar rheological behavior. Typical commercial ionomers have melt index values between 0.5 and 15.[250] Both unmodified and glass-filled grades are available.

Ionomers are used in applications such as golf ball covers and bowling pin coatings, where their good abrasion resistance is important.[251] The puncture resistance of films allows these materials to be widely used in packaging applications. One of the early applications was the packaging of fishhooks.[252] They are often used in composite products as an outer heat-seal layer. Their ability to bond to aluminum foil is also utilized in packaging applications.[253] Ionomers also find application in footwear for shoe heels.[254]

2.2.17 Polyimide (PI)

Thermoplastic polyimides are linear polymers noted for their high-temperature properties. Polyimides are prepared by condensation polymerization of pyromellitic anhydrides and primary diamines. A polyimide contains the structure -CO-NR-CO as a part of a ring structure along the backbone. The presence of ring structures along the backbone, as depicted in Fig. 2.27, gives the polymer good high-temperature properties.[255] Polyimides are used in high-performance applications as replacements for metal and glass. The use of aromatic diamines gives the polymer exceptional thermal stability. An example of this is the use of di-(4-amino-phenyl) ether, which is used in the manufacture of Kapton (Du Pont).

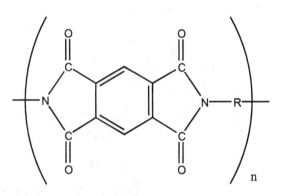

FIGURE 2.27 Structure of polyimide.

Although called thermoplastics, some polyimides must be processed in precursor form, because they will degrade before their softening point.[256] Fully imidized injection-molding grades are available, along with powder forms for compression molding and cold forming. However, injection molding of polyimides requires experience on the part of the molder.[257] Polyimides are also available as films and preformed stock shapes. The polymer may also be used as a soluble prepolymer, where heat and pressure are used to convert the polymer into the final, fully imidized form. Films can be formed by casting soluble polymers or precursors. It is generally difficult to form good films by melt extrusion. Laminates of polyimides can also be formed by impregnating fibers such as glass or graphite.

Polyimides have excellent physical properties and are used in applications where parts are exposed to harsh environments. They have outstanding high-temperature properties and their oxidative stability allows them to withstand continuous service in air at temperatures of 260°C.[258] Polyimides will burn, but they have self-extinguishing properties.[259] They are resistant to weak acids and organic solvents but are attacked by bases. The polymer also has good electrical properties and resistance to ionizing radiation.[260] A disadvantage of polyimides is their hydrolysis resistance. Exposure to water or steam above 100°C may cause parts to crack.[261]

The first application of polyimides was for wire enamel.[262] Applications for polyimides include bearings for appliances and aircraft, seals, and gaskets. Film versions are used in flexible wiring and electric motor insulation. Printed circuit boards are also fabricated with polyimides.[263]

2.2.18 Polyarylether Ketones

The family of aromatic polyether ketones includes structures that vary in the location and number of ketonic and ether linkages on their repeat unit and therefore include polyether ketone (PEK), polyether ether ketone (PEEK), polyether ether ketone ketone (PEEKK), as well as other combinations. Their structures are as shown in Fig. 2.28. All have very high thermal properties due to the aromaticity of their backbones and are readily processed via injection molding and extrusion, although their melt temperatures are very high—370°C for unfilled PEEK and 390°C for filled PEEK, and both unfilled and filled PEK. Mold temperatures as high as 165°C are also used.[264] Their toughness (surprisingly high for such high-heat-resistant materials), high dynamic cycles and fatigue resistance capabilities, low moisture absorption, and good hydrolytic stability lend these materials to applications such as parts found in nuclear plants, oil wells, high-pressure steam valves, chemical plants, and airplane and automobile engines.

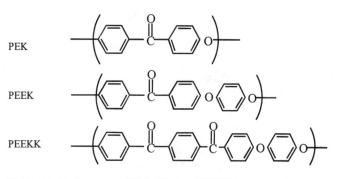

PEK

PEEK

PEEKK

FIGURE 2.28 Structures of PEK, PEEK, and PEEKK.

One of the two ether linkages in PEEK is not present in PEK, and the ensuing loss of some molecular flexibility results in PEK having an even higher T_m and heat distortion temperature than PEEK. A relatively higher ketonic concentration in the repeat unit results in high ultimate tensile properties as well. A comparison of different aromatic polyether ketones is given in Table 2.6.[265,266] As these properties are from different sources, strict comparison between the data is not advisable due to likely differing testing techniques.

Glass and carbon fiber reinforcements are the most important filler for all of the PEK family. While elastic extensibility is sacrificed, the additional heat resistance and moduli improvements allow glass- or carbon-fiber formulations entry into many applications.

PEK is polymerized either through self-condensation of structure (a) in Fig. 2.29, or via the reaction of intermediates (b) as shown below. Since these polymers can crystallize and tend therefore to precipitate from the reactant mixture, they must be reacted in high-boiling solvents close to the 320°C melt temperature.[267]

2.2.19 Poly(methylmethacrylate)

Poly(methyl methacrylate) is a transparent thermoplastic material of moderate mechanical strength and outstanding outdoor weather resistance. It is available as sheet, tubes, and rods, which can be machined, bonded, and formed into a variety of different parts. It is also available in bead form, which can be conventionally processed via extrusion or injection molding. The sheet form material is polymerized *in situ* by casting a monomer that

TABLE 2.6 Comparison of Selected PEK, PEEK, and PEEKK Properties

	PEK unfilled	30% glass-filled PEK	PEEK unfilled	30% glass-filled PEEK	PEEKK unfilled	30% glass-filled PEEKK
T_m, °C	323–381	329–381	334	334	365	–
Tensile modulus, MPa	3,585–4,000	9,722–12,090	–	8,620–11,030	4000	13,500
Ultimate elongation, %	50	2.2–3.4	30–150	2–3	–	–
Ultimate tensile strength, MPa	103	–	91	–	86	168
Specific gravity	1.3	1.47–1.53	1.30–1.32	1.49–1.54	1.3	1.55
Heat deflection temperature, °C, 264 psi	162–170	326–350	160	288–315	160	>320

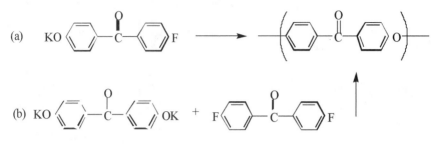

FIGURE 2.29   Routes for PEK synthesis.

has been partly prepolymerized by removing any inhibitor, heating, and adding an agent to initiate the free radical polymerization. This agent is typically a peroxide. This mixture of polymer and monomer is then poured into the sheet mold, and the plates are brought together and reinforced to prevent bowing to ensure that the final product will be of uniform thickness and flatness. This bulk polymerization process generates such high-molecular-weight material that the sheet or rod will decompose prior to melting. As such, this technique is not suitable for producing injection molding-grade resin, but it does aid in producing material that has a large rubbery plateau and has high enough elevated temperature strength to allow for bandsawing, drilling, and other common machinery practices as long as the localized heating doesn't reach the polymer's decomposition temperature.

Suspension polymerization provides a final polymer with low enough molecular weight to allow for typical melt processing. In this process, methyl methacrylate monomer is suspended in water, to which the peroxide is added along with emulsifying/suspension agents, protective colloids, lubricants, and chain transfer agents to aid in molecular weight control. The resultant bead can then be dried and is ready for injection molding, or it can be further compounded with any desired colorants, plasticizers, or rubber-modifier as required.[268] Number-average molecular weights from the suspension process are approximately 60,000 g/mol, while the bulk polymerization process can result in number average molecular weights of approximately 1 million g/mol.[269]

Typically, applications for PMMA optimize use of its clarity, with an up to 92 percent light transmission, depending upon the thickness of the sample. Again, because it has such strong weathering behavior, it is well suited for applications such as automobile rear-light housings, lenses, aircraft cockpits, helicopter canopies, dentures, steering wheel bosses, and windshields. Cast PMMA is used extensively as bathtub materials, in showers, and in whirlpools. [270]

Since the homopolymer is fairly brittle, PMMA can be toughened via copolymerization with another monomer (such as polybutadiene) or blended with an elastomer in the same way that high-impact polystyrene is, to enable better stress distribution via the elastomeric domain.

2.2.20   Polymethylpentene (PMP)

Polymethylpentene was introduced in the mid-1960s by ICI and is now marketed under the same trade name, TPX, by Mitsui Petrochemical Industries. The most significant commercial polymerization method involves the dimerization of propylene, as shown in Fig. 2.30.

As a polyolefin, this material offers chemical resistance to mineral acids, alkaline solutions, alcohols, and boiling water. It is not resistant to ketones or aromatic and chlorinated

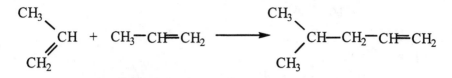

FIGURE 2.30   Polymerization route for polymethylpentene.

hydrocarbons. Like polyethylene and polypropylene, it is susceptible to environmental stress cracking[271] and requires formulation with antioxidants. Its use is primarily in injection molding and thermoforming applications, where the additional cost incurred compared to other polyolefins is justified by its high melt point (245°C), transparency, low density, and good dielectric properties. The high degree of transparency of polymethylpentene is attributed both to the similarities of the refractive indices of the amorphous and crystalline regions, as well as to the large coil size of the polymer due to the bulky branched four carbon side chain. The free-volume regions are large enough to allow light of visible-region wavelengths to pass unimpeded. This degree of free volume is also responsible for the 0.83 g/cm^3 low density. As typically cooled, the polymer achieves about 40 percent crystallinity, although with annealing can reach 65 percent crystallinity.[272] The structure of the polymer repeat unit is shown in Fig. 2.31.

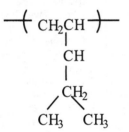

FIGURE 2.31   Repeat structure of polymethyl-pentene.

Voids are frequently formed at the crystalline/amorphous region interfaces during injection molding, rendering an often undesirable lack of transparency. To counter this, polymethylpentene is often copolymerized with hex-1-ene, oct-1-ene, dec-1-ene, and octadec-1-ene, which reduces the voids and concomitantly reduces the melting point and degree of crystallinity.[273] Typical products made from polymethylpentene include transparent pipes and other chemical plant applications, sterilizable medical equipment, light fittings, and transparent housings.

2.2.21   Polyphenylene Oxide

The term polyphenylene oxide (PPO) is a misnomer for a polymer that is more accurately named poly-(2,6-dimethyl-p-phenylene ether), and which in Europe is more commonly known as a polymer covered by the more generic term polyphenyleneether (PPE). This engineering polymer has high-temperature properties due to the large degree of aromaticity on the backbone, with dimethyl-substituted benzene rings joined by an ether linkage, as shown in Fig. 2.32.

The stiffness of this repeat unit results in a heat-resistant polymer with a T_g of 208°C and a T_m of 257°C. The fact that these two thermal transitions occur within such a short

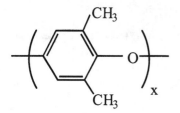

FIGURE 2.32   Repeat structure of PPO.

temperature span of each other means that PPO does not have time to crystallize while it cools before reaching a glassy state and as such is typically amorphous after process-ing.[274] Commercially available as PPO from General Electric, the polymer is sold in mo-lecular weight ranges of 25,000 to 60,000 g/mol.[275] Properties that distinguish PPO from other engineering polymers are its high degree of hydrolytic and dimensional stabilities, which enable it to be molded with precision, although high processing temperatures are re-quired. It finds application as television tuner strips, microwave insulation components, and transformer housings, which take advantage of its strong dielectric properties over wide temperature ranges. It is also used in applications that benefit from its hydrolytic sta-bility including pumps, water meters, sprinkler systems, and hot water tanks.[276] Its greater use is limited by the often-prohibitive cost, and General Electric responded by commer-cializing a PPO/PS blend marketed under the trade name Noryl. GE sells many grades of Noryl based on different blend ratios and specialty formulations. The styrenic nature of PPO leads one to surmise very close compatibility (similar solubility parameters) with PS, although strict thermodynamic compatibility is questioned due to the presence of two dis-tinct T_g peaks when measured by mechanical rather than calorimetric means.[277] The blends present the same high degree of dimensional stability, low water absorption, excel-lent resistance to hydrolysis, and good dielectric properties offered by PS, yet with the el-evated heat distort temperatures that result from PPO's contribution. These polymers are more cost competitive than PPO and are used in moldings for dishwashers, washing ma-chines, hair dryers, cameras, and instrument housings, and as television accessories.[278]

2.2.22   Polyphenylene Sulphide (PPS)

The structure of PPS, shown in Fig. 2.33, clearly indicates high temperature, high strength, and high chemical resistance due to the presence of the aromatic benzene ring on the backbone linked with the electronegative sulfur atom. In fact, the melt point of PPS is 288°C, and the tensile strength is 70 MPa at room temperature. The brittleness of PPS, due to the highly crystalline nature of the polymer, is often overcome by compounding with glass fiber reinforcements. Typical properties of PPS and a commercially available 40 percent glass-filled polymer blend are shown in Table 2.7.[279] The mechanical proper-ties of PPS are similar to other engineering thermoplastics such as polycarbonate and polysulphones except that, as mentioned, the PPS suffers from the brittleness arising from

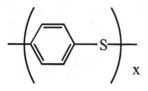

FIGURE 2.33   Repeat structure of polyphe-nylene sulphide.

TABLE 2.7   Selected Properties of PPS and GF PPS

Property, units	PPS	40% glass-filled PPS
T_g, °C	85	–
Heat distortion temperature, method A, °C	135	265
Tensile strength		
21 °C MPa	64–77	150
204 °C, MPa	33	33
Elongation at break, %	3	2
Flexural modulus, MPa	3,900	10,500
Limiting oxygen index, %	44	47

its crystallinity. But it does, however, offer improved resistance to environmental stress cracking.[280]

PPS is of most significant commercial interest as a thermoplastic, although it can be crosslinked into a thermoset system. Its strong inherent flame retardance puts this polymer in a fairly select class of polymers, including polyethersulphones, liquid crystal polyesters, polyketones, and polyetherimides.[281] As such, PPS finds application in electrical components, printed circuits, and contact and connector encapsulation. Other uses take advantage of the low mold shrinkage values and strong mechanical properties even at elevated temperatures. These include pump housings, impellers, bushings, and ball valves.[282]

2.2.23   Polyphthalamide (PPA)

Polyphthalamides were originally developed for use as fibers and later found application in other areas as high-temperature thermoplastics. They are semiaromatic polyamides based on the polymerization of terephthalic acid or isophthalic acid and an amine.[283] Both amorphous and crystalline grades are available. Solvay sells a semicrystalline grade polyphthalamide under the trade name Amodel®, available in both reinforced and nonreinforced grades, as a lower-cost, high-temperature plastic alternative to PPS and PEI. Amodel finds applications as automotive halogen lamp sockets and fog lamp assemblies, fuel system flanges, and fuel line connectors as well as vacuum cleaner impellers and lawn mower components. Polyphthalamides are polar materials with a melting point near 310°C and a glass transition temperature of 127°C.[284] The material has good strength and stiffness along with good chemical resistance. Polyphthalamides can be attacked by strong acids or oxidizing agents and are soluble in cresol and phenol.[285] Polyphthalamides are stronger, less moisture sensitive, and possess better thermal properties when compared to the aliphatic polyamides such as nylon 6,6. However, polyphthalamide is less ductile than nylon 6,6, although impact grades are available.[286] Polyphthalamides will absorb moisture, decreasing the glass transition temperature and causing dimensional changes. The material can be reinforced with glass and has extremely good high-temperature performance. Reinforced grades of polyphthalamides are able to withstand continuous use at 180°C.[287]

The crystalline grades are generally used in injection molding, while the amorphous grades are often used as barrier materials.[288] The recommended mold temperatures are 135 to 165°C, with recommended melt temperatures of 320 to 340°C.[289] The material should have a moisture content of 0.15 percent or less for processing.[290] Because mold temperature is important to surface finish, higher mold temperatures may be required for some applications.

Both crystalline and amorphous grades are available under the trade name Amodel (Amoco); amorphous grades are available under the names Zytel (Dupont) and Trogamid (Dynamit Nobel). Crystalline grades are available under the trade name Arlen (Mitsui).[291]

Polyphthalamides are used in automotive applications where their chemical resistance and temperature stability are important.[292] Examples include sensor housings, fuel line components, headlamp reflectors, electrical components, and structural components. Electrical components attached by infrared and vapor phase soldering are applications utilizing PPA's high-temperature stability. Switching devices, connectors, and motor brackets are often made from PPA. Mineral-filled grades are used in applications that require plating, such as decorative hardware and plumbing. Impact modified grades of unreinforced PPA are used in sporting goods, oil field parts, and military applications.

2.2.24 Polypropylene (PP)

Polypropylene is a versatile polymer used in applications from films to fibers, with a worldwide demand of over 21 million lb.[293] It is similar to polyethylene in structure except for the substitution of one hydrogen group with a methyl group on every other carbon. On the surface, this change would appear trivial, but this one replacement changes the symmetry of the polymer chain. This allows for the preparation of different stereoisomers, namely, syndiotactic, isotactic, and atactic chains. These configurations are shown in the introduction.

Polypropylene (PP) is synthesized by the polymerization of propylene, a monomer derived from petroleum products through the reaction shown in Fig. 2.34. It was not until Ziegler-Natta catalysts became available that polypropylene could be polymerized into a commercially viable product. These catalysts allowed the control of stereochemistry during polymerization to form polypropylene in the isotactic and syndiotactic forms, both capable of crystallizing into a more rigid, useful polymeric material.[294] The first commercial method for the production of polypropylene was a suspension process. Current methods of production include a gas phase process and a liquid slurry process.[295] New grades of polypropylene are now being polymerized using metallocene catalysts.[296] The range of molecular weights for PP is M_n = 38,000 to 60,000 and M_w = 220,000 to 700,000. The molecular weight distribution (M_n/M_w) can range from 2 to about 11.[297]

FIGURE 2.34 The reaction to prepare polypropylene.

Different behavior can be found for each of the three stereoisomers. Isotactic and syndiotactic polypropylene can pack into a regular crystalline array, giving a polymer with more rigidity. Both materials are crystalline. However, syndiotactic polypropylene has a lower T_m than the isotactic polymer.[298] The isotactic polymer is the most commercially used form, with a melting point of 165°C. Atactic polypropylene has a very small amount of crystallinity (5 to 10 percent), because its irregular structure prevents crystallization;

thus, it behaves as a soft flexible material.[299] It is used in applications such as sealing strips, paper laminating, and adhesives.

Unlike polyethylene, which crystallizes in the planar zigzag form, isotactic polypropylene crystallizes in a helical form because of the presence of the methyl groups on the chain.[300] Commercial polymers are about 90 to 95 percent isotactic. The amount of isotacticity present in the chain will influence the properties. As the amount of isotactic material (often quantified by an isotactic index) increases, the amount of crystallinity will also increase, resulting in increased modulus, softening point, and hardness.

Although in many respects polypropylene is similar to polyethylene, both being saturated hydrocarbon polymers, they differ in some significant properties. Isotactic polypropylene is harder and has a higher softening point than polyethylene, so it is used where higher stiffness materials are required. Polypropylene is less resistant to degradation, particularly high-temperature oxidation, than polyethylene, but it has better environmental stress cracking resistance.[301] The decreased degradation resistance of PP is due to the presence of a tertiary carbon in PP, allowing for easier hydrogen abstraction compared to PE.[302] As a result, antioxidants are added to polypropylene to improve the oxidation resistance. The degradation mechanisms of the two polymers are also different. PE cross-links on oxidation, while PP undergoes chain scission. This is also true of the polymers when exposed to high-energy radiation, a method commonly used to cross-link PE.

Polypropylene is one of the lightest plastics, with a density of 0.905.[303] The nonpolar nature of the polymer gives PP low water absorption. Polypropylene has good chemical resistance, but liquids such as chlorinated solvents, gasoline, and xylene can affect the material. Polypropylene has a low dielectric constant and is a good insulator. Difficulty in bonding to polypropylene can be overcome by the use of surface treatments to improve the adhesion characteristics.

With the exception of UHMWPE, polypropylene has a higher T_g and melting point than polyethylene. Service temperature is increased, but PP needs to be processed at higher temperatures. Because of the higher softening, PP can withstand boiling water and can be used in applications requiring steam sterilization.[304] Polypropylene is also more resistant to cracking in bending than PE and is preferred in applications that require tolerance to bending. This includes applications such as ropes, tapes, carpet fibers, and parts requiring a living hinge. Living hinges are integral parts of a molded piece that are thinner and allow for bending.[305] One weakness of polypropylene is its low-temperature brittleness behavior, with the polymer becoming brittle near 0°C.[306] This can be improved through copolymerization with other polymers such as ethylene.

Comparing the processing behavior of PP to PE, it is found that polypropylene is more non-Newtonian than PE and that the specific heat of PP is lower than polyethylene.[307] The melt viscosity of PE is less temperature sensitive than PP.[308] Mold shrinkage is generally less than for PE but is dependent on the actual processing conditions.

Unlike many other polymers, an increase in molecular weight of polypropylene does not always translate into improved properties. The melt viscosity and impact strength will increase with molecular weight but often with a decrease in hardness and softening point. A decrease in the ability of the polymer to crystallize as molecular weight increases is often offered as an explanation for this behavior.[309]

The molecular weight distribution (MWD) has important implications for processing. A PP grade with a broad MWD is more shear sensitive than a grade with a narrow MWD. Broad MWD materials will generally process better in injection molding applications. In contrast, a narrow MWD may be preferred for fiber formation.[310] Various grades of polypropylene are available tailored to particular application. These grades can be classified by flow rate, which depends on both average molecular weight and MWD. Lower-flow-rate materials are used in extrusion applications. In injection molding applications, low-flow-rate materials are used for thick parts, and high-flow-rate materials are used for thin-wall molding.

Polypropylene can be processed by methods similar to those used for PE. The melt temperatures are generally in the range of 210 to 250°C.[311] Heating times should be minimized to reduce the possibility of oxidation. Blow molding of PP requires the use of higher melt temperatures and shear, but these conditions tend to accelerate the degradation of PP. Because of this, blow molding of PP is more difficult than for PE. The screw metering zone should not be too shallow so as to avoid excessive shear. For a 60-mm screw, the flights depths are typically about 2.25 mm, and they are 3.0 mm for a 90-mm screw.[312]

In film applications, film clarity requires careful control of the crystallization process to ensure that small crystallites are formed. This is accomplished in blown film by extruding downwards into two converging boards. In the Shell TQ process, the boards are covered with a film of flowing, cooling water. Oriented films of PP are manufactured by passing the PP film into a heated area and stretching the film both transversely and longitudinally. To reduce shrinkage, the film may be annealed at 100°C while under tension.[313] Highly oriented films may show low transverse strength and a tendency to fibrillate. Other manufacturing methods for polypropylene include extruded sheet for thermoforming applications and extruded profiles.

If higher stiffness is required, short glass reinforcement can be added. The use of a coupling agent can dramatically improve the properties of glass filled PP.[314] Other fillers for polypropylene include calcium carbonate and talc, which can also improve the stiffness of PP.

Other additives such as pigments, antioxidants, and nucleating agents can be blended into polypropylene to give the desired properties. Carbon black is often added to polypropylene to impart UV resistance in outdoor applications. Antiblocking and slip agents may be added for film applications to decrease friction and prevent sticking. In packaging applications, antistatic agents may be incorporated.

The addition of rubber to polypropylene can lead to improvements in impact resistance. One of the most commonly added elastomers is ethylene-propylene rubber. The elastomer is blended with polypropylene, forming a separate elastomer phase. Rubber can be added in excess of 50 percent to give elastomeric compositions. Compounds with less than 50 percent added rubber are of considerable interest as modified thermoplastics. Impact grades of PP can be formed into films with good puncture resistance.

Copolymers of polypropylene with other monomers are also available, the most common monomer being ethylene. Copolymers usually contain between 1 and 7 weight percent of ethylene randomly placed in the polypropylene backbone. This disrupts the ability of the polymer chain to crystallize, giving more flexible products. This improves the impact resistance of the polymer, decreases the melting point, and increases flexibility. The degree of flexibility increases with ethylene content, eventually turning the polymer into an elastomer (ethylene propylene rubber). The copolymers also exhibit increased clarity and are used in blow molding, injection molding, and extrusion.

Polypropylene has many applications. Injection molding applications cover a broad range from automotive uses such as dome lights, kick panels, and car battery cases to luggage and washing machine parts. Filled PP can be used in automotive applications such as mounts and engine covers. Elastomer-modified PP is used in the automotive area for bumpers, fascia panels, and radiator grills. Ski boots are another application for these materials.[315] Structural foams, prepared with glass-filled PP, are used in the outer tank of washing machines. New grades of high-flow PPs are allowing manufacturers to mold high-performance housewares.[316] Polypropylene films are used in a variety of packaging applications. Both oriented and nonoriented films are used. Film tapes are used for carpet backing and sacks. Foamed sheet is used in a variety of applications including thermoformed packaging. Fibers are another important application for polypropylene, particularly in carpeting, because of its low cost and wear resistance. Fibers prepared from polypropylene are used in both woven and nonwoven fabrics.

2.2.25 Polyurethane (PUR)

Polyurethanes are very versatile polymers. They are used as flexible and rigid foams, elastomers, and coatings. Polyurethanes are available as both thermosets and thermoplastics. In addition, their hardnesses span the range from rigid material to elastomer. Thermoplastic polyurethanes will be the focus of this section. The term *polyurethane* is used to cover materials formed from the reaction of isocyanates and polyols.[317] The general reaction for a polyurethane produced through the reaction of a diisocyanate with a diol is shown in Fig. 2.35.

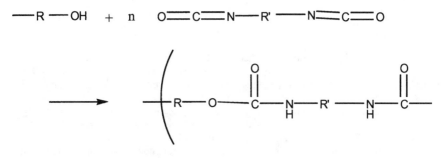

FIGURE 2.35   Polyurethane reaction.

$$[—A—B—A—B—A—B—]_n$$

FIGURE 2.36   Block structure of polyurethanes.

Polyurethanes are phase separated block copolymers as depicted in Fig. 2.36, where the A and B portions represent different polymer segments. One segment, called the hard segment, is rigid, while the other, the soft segment, is elastomeric. In polyurethanes, the soft segment is prepared from an elastomeric long-chain polyol, generally a polyester or polyether, but other rubbery polymers end-capped with a hydroxyl group could be used. The hard segment is composed of the diisocyanate and a short-chain diol called a *chain extender.* The hard segments have high interchain attraction due to hydrogen bonding between the urethane groups. In addition, they may be capable of crystallizing.[318] The soft elastomeric segments are held together by the hard phases, which are rigid at room temperature and act as physical cross-links. The hard segments hold the material together at room temperature but, at processing temperatures, the hard segments can flow and be processed.

The properties of polyurethanes can be varied by changing the type or amount of the three basic building blocks of the polyurethane: diisocyanate, short-chain diol, or long-chain diol. Given the same starting materials, the polymer can be varied simply by changing the ratio of the hard and soft segments. This allows the manufacturer a great deal of flexibility in compound development for specific applications. The materials are typically manufactured by reacting a linear polyol with an excess of diisocyanate. The polyol is end-capped with isocyanate groups. The end-capped polyol and free isocyanate are then reacted with a chain extender, usually a short--chain diol to form the polyurethane.[319]

There are a variety of starting materials available for use in the preparation of polyurethanes, some of which are listed below.

Diisocyanates

- 4,4′-diphenylmethane diisocyanate (MDI)
- Hexamethylene diisocyanate (HDI)
- Hydrogenated 4,4′-diphenylmethane diisocyanate (HMDI)

Chain Extenders

- 1,4 butanediol
- Ethylene glycol
- 1,6 hexanediol

Polyols

- Polyesters
- Polyethers

Polyurethanes are generally classified by the type of polyol used—for example, polyester polyurethane or polyether polyurethane. The type of polyol can affect certain properties. For example, polyether polyurethanes are more resistant to hydrolysis than polyester-based urethanes, while the polyester polyurethanes have better fuel and oil resistance.[320] Low-temperature flexibility can be controlled by proper selection of the long-chain polyol. Polyether polyurethanes generally have lower glass transition temperatures than polyester polyurethanes. The heat resistance of the polyurethane is governed by the hard segments. Polyurethanes are noted for their abrasion resistance, toughness, low-temperature impact strength, cut resistance, weather resistance, and fungus resistance.[321] Specialty polyurethanes include glass-reinforced products, fire-retardant grades, and UV-stabilized grades.

Polyurethanes find application in many areas. They can be used as impact modifiers for other plastics. Other applications include rollers or wheels, exterior body parts, drive belts, and hydraulic seals.[322] Polyurethanes can be used in film applications such as textile laminates for clothing and protective coatings for hospital beds. They are also used in tubing and hose in both unreinforced and reinforced forms because of their low-temperature properties and toughness. Their abrasion resistance allows them to be used in applications such as athletic shoe soles and ski boots. Polyurethanes are also used as coatings for wire and cable.[323]

Polyurethanes can be processed by a variety of methods, including extrusion, blow molding, and injection molding. They tend to pick up moisture and must be thoroughly dried prior to use. The processing conditions vary with the type of polyurethane; higher hardness grades usually require higher processing temperatures. Polyurethanes tend to exhibit shear sensitivity at lower melt temperatures. Post-mold heating in an oven, shortly after processing, can often improve the properties of the finished product. A cure cycle of 16 to 24 hr at 100°C is typical.[324]

2.2.26 Styrenics

The styrene family is well suited for applications where rigid, dimensionally stable molded parts are required. PS is a transparent, brittle, high-modulus material with a multitude of applications, primarily in packaging, disposable cups, and medical ware. When the mechanical properties of the PS homopolymer are modified to produce a tougher, more ductile blend, as in the case of rubber-modified high-impact grades of PS (HIPS), a far wider range of applications becomes available. HIPS is preferred for durable, molded items including radio, television, and stereo cabinets as well as compact disc jewel cases. Copolymerization is also used to produce engineering-grade plastics of higher performance as well as higher price, with acrylonitrile-butadiene-styrene (ABS) and styrene-acrylonitrile (SAN) plastics being of greatest industrial importance.

2.2.26.1 Acrylonitrile Butadiene Styrene (ABS) Terpolymer. As with any copolymers, there is tremendous flexibility in tailoring the properties of ABS by varying the ratios of the three monomers: acrylonitrile, butadiene, and styrene. The acrylonitrile component contributes heat resistance, strength, and chemical resistance. The elastomeric contribution of butadiene imparts higher impact strength, toughness, low-temperature property retention, and flexibility, while the styrene contributes rigidity, glossy finish, and ease of processability. As such, worldwide usage of ABS is surpassed only by that of the "big four" commodity thermoplastics (polyethylene, polypropylene, polystyrene, and polyvinyl chloride). Primary drawbacks to ABS include opacity, poor weather resistance, and poor flame resistance. Flame retardance can be improved by the addition of fire-retardant additives or by blending ABS with PVC, with some reduction in ease of processability.[325] As its use is widely prevalent as equipment housings (such as telephones, televisions, and computers), these disadvantages are tolerated. Figure 2.37 shows the repeat structure of ABS.

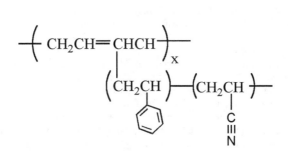

FIGURE 2.37 Repeat structure of ABS.

Most common methods of manufacturing ABS include graft polymerization of styrene and acrylonitrile onto a polybutadiene latex, blending with a styrene-acrylonitrile latex, and then coagulating and drying the resultant blend. Alternatively, the graft polymer of styrene, acrylonitrile, and polybutadiene can be manufactured separately from the styrene acrylonitrile latex and the two grafts blended and granulated after drying.[326]

Its ease of processing by a variety of common methods (including injection molding, extrusion, thermoforming, compression molding, and blow molding), combined with a good economic value for the mechanical properties achieved, results in widespread use of ABS. It is commonly found in under-the-hood automotive applications, refrigerator linings, radios, computer housings, telephones, business machine housings, and television housings.

2.2.26.2 Acrylonitrile-Chlorinated Polyethylene-Styrene (ACS) Terpolymer. While ABS itself can be readily tailored by modifying the ratios of the three monomers and by modifying the lengths of each grafted segment, several companies are pursuing the addition of a fourth monomer, such as alpha-methylstyrene for enhanced heat resistance and methylmethacrylate to produce a transparent ABS. One such modification involves using chlorinated polyethylene in place of the butadiene segments. This terpolymer, ACS, has very similar properties to the engineering terpolymer ABS, but the addition of chlorinated polyethylene imparts improved flame retardance, weatherability, and resistance to electrostatic deposition of dust without the addition of antistatic agents. The addition of the chlorinated olefin requires more care when injection molding to ensure that the chlorine does not dehydrohalogenate. Mold temperatures are recommended to be kept at between 190 and 210°C and not to exceed 220°C. As with other chlorinated polymers, such as polyvinyl chloride, residence times should be kept relatively short in the molding machine.[327]

Applications for ACS include housings and parts for office machines such as desktop calculators, copying machines, and electronic cash registers, as well as housings for television sets and video cassette recorders.[328]

2.2.26.3 Acrylic Styrene Acrylonitrile (ASA) Terpolymer. Like ACS, ASA is a specialty product with similar mechanical properties to ABS, but which offers improved outdoor weathering properties. This is due to the grafting of an acrylic ester elastomer onto the styrene-acrylonitrile backbone. Sunlight usually combines with atmospheric oxygen to result in embrittlement and yellowing of thermoplastics, and this process takes a much longer time in the case of ASA. Therefore, ASA finds applications in gutters, drain pipe fittings, signs, mail boxes, shutters, window trims, and outdoor furniture.[329]

2.2.26.4 General-Purpose Polystyrene (PS). PS is one of the four plastics whose combined usage accounts for 75 percent of the worldwide usage of plastics.[330] These four commodity thermoplastics are PE, PP, PVC, and PS. Although it can be polymerized via free-radical, anionic, cationic, and Ziegler mechanisms, commercially available PS is produced via free-radical addition polymerization. PS's popularity is due to its transparency, low density, relatively high modulus, excellent electrical properties, low cost, and ease of processing. The steric hindrance caused by the presence of the bulky benzene side groups results in brittle mechanical properties, with ultimate elongations only around 2 to 3 percent, depending on molecular weight and additive levels. Most commercially available PS grades are atatic, and in combination with the large benzene groups, result in an amorphous polymer. The amorphous morphology provides not only transparency but, in addition, the lack of crystalline regions means that there is no clearly defined temperature at which the plastic melts. PS is a glassy solid until its T_g of ~100°C is reached, whereupon further heating softens the plastic gradually from a glass to a liquid. Advantage is taken of this gradual transition by molders who can eject parts that have cooled to beneath the relatively high Vicat temperature. Also, the lack of a heat of crystallization means that high heating and cooling rates can be achieved, which reduces cycle time and also promotes an economical process. Lastly, upon cooling, PS does not crystallize the way PE and PP do. This gives PS low shrinkage values (0.004 to 0.005 mm/mm) and high dimensional stability during molding and forming operations.

Commercial PS is segmented into easy-flow, medium-flow, and high-heat-resistance grades. Comparison of these three grades is made in Table 2.8. The easy-flow grades are the lowest molecular weight, to which 3 to 4 percent mineral oil has been added. The mineral oil reduces melt viscosity, which is well suited for increased injection speeds while molding inexpensive thin-walled parts such as disposable dinnerware, toys, and packaging. The reduction in processing time comes at the cost of a reduced softening temperature and a more brittle polymer. The medium-flow grades are of slightly higher molecular weight and contain only 1 to 2 percent mineral oil. Applications include injection molded tumblers, medical ware, toys, injection-blow-molded bottles, and extruded food packaging. The high-heat-resistance plastics are of the highest molecular weight and have the

TABLE 2.8 Properties of Commercial Grades of General-Purpose PS[*]

Property	Easy-flow PS	Medium-flow PS	High-heat-resistance PS
M_w	218,000	225,000	300,000
M_n	74,000	92,000	130,000
Melt flow index, g/10 min	16	7.5	1.6
Vicat softening temperature, °C	88	102	108
Tensile modulus, MPa	3,100	2,450	3,340
Ultimate tensile strength, MPa	1.6	2.0	2.4

[*]*Source: Encyclopedia of Polymer Science,* 2nd ed., vol. 6, Mark, Bikales, Overberger, Menges,and Kroschwitz, Eds., Wiley Interscience, 1986, p. 65.

least level of additives such as extrusion aids. These products are used in sheet extrusion and thermoforming, and extruded film applications for oriented food packaging.[331]

2.2.26.5 Styrene-Acrylonitrile Copolymers (SAN). Styrene-acrylonitrile polymers are copolymers prepared from styrene and acrylonitrile monomers. The polymerization can be done under emulsion, bulk, or suspension conditions.[332] The polymers generally contain between 20 to 30 percent acrylonitrile.[333] The acrylonitrile content of the polymer influences the final properties with tensile strength, elongation, and heat distortion temperature increasing as the amount of acrylonitrile in the copolymer increases.

SAN copolymers are linear, amorphous materials with improved heat resistance over pure polystyrene.[334] The polymer is transparent but may have a yellow color as the acrylonitrile content increases. The addition of a polar monomer, acrylonitrile, to the backbone gives these polymers better resistance to oils, greases, and hydrocarbons when compared to polystyrene.[335] Glass-reinforced grades of SAN are available for applications requiring higher modulus combined with lower mold shrinkage and lower coefficient of thermal expansion.[336]

As the polymer is polar, it should be dried before processing. It can be processed by injection molding into a variety of parts. SAN can also be processed by blow molding, extrusion, casting, and thermoforming.[337]

SAN competes with polystyrene, cellulose acetate, and polymethyl methacrylate. Applications for SAN include injection-molded parts for medical devices, PVC tubing connectors, dishwasher-safe products, and refrigerator shelving.[338] Other applications include packaging for the pharmaceutical and cosmetics markets, automotive equipment, and industrial uses.

2.2.26.6 Olefin-Modified SAN. SAN can be modified with olefins, resulting in a polymer that can be extruded and injection molded. The polymer has good weatherability and is often used as a capstock to provide weatherability to less expensive parts such as swimming pools, spas, and boats.[339]

2.2.26.7 Styrene-Butadiene Copolymers. Styrene-butadiene polymers are block copolymers prepared from styrene and butadiene monomers. The polymerization is performed using sequential anionic polymerization.[340] The copolymers are better known as thermoplastic elastomers, but copolymers with high styrene contents can be treated as thermoplastics. The polymers can be prepared as either a star block form or as a linear, multiblock polymer. The butadiene exists as a separate dispersed phase in a continuous matrix of polystyrene.[341] The size of the butadiene phase is controlled to be less than the wavelength of light, resulting in clear materials. The resulting amorphous polymer is tough with good flex life and low mold shrinkage. The copolymer can be ultrasonically welded, solvent welded, or vibration welded. The copolymers are available in injection-molding grades and thermoforming grades. The injection-molding grades generally contain a higher styrene content in the block copolymer. Thermoforming grades are usually mixed with pure polystyrene. Styrene-butadiene copolymers can be processed by injection molding, extrusion, thermoforming, and blow molding. The polymer does not need to be dried prior to use.[342]

Styrene-butadiene copolymers are used in toys, housewares, and medical applications.[343] Thermoformed products include disposable food packaging such as cups, bowls, "clam shells," deli containers, and lids. Blister packs and other display packaging also use styrene-butadiene copolymers. Other packaging applications include shrink wrap and vegetable wrap.[344]

2.2.27 Sulfone-Based Resins

FIGURE 2.38 General structure of a polysulfone.

Sulfone resins refer to polymers containing SO_2 groups along the backbone as depicted in Fig. 2.38. The R groups are generally aromatic. The polymers are usually yellowish, transparent, amorphous materials and are known for their high stiffness, strength, and thermal stability.[345] The polymers have low creep over a large temperature range. Sulfones can compete against some thermoset materials in performance, while their ability to be injection molded offers an advantage.

The first commercial polysulfone was Udel (Union Carbide, now Amoco), followed by Astrel 360 (Minnesota Mining and Manufacturing), which is termed a *polyarylsulfone*, and finally Victrex (ICI), a polyethersulfone.[346] The different polysulfones vary by the spacing between the aromatic groups, which in turn affects their T_gs and their heat distortion temperatures. Commercial polysulfones are linear with high T_g values in the range of 180 to 250°C, allowing for continuous use from 150 to 200°C.[347] As a result, the processing temperatures of polysulfones are above 300°C.[348] Although the polymer is polar, it still has good electrical insulating properties. Polysulfones are resistant to high thermal and ionizing radiation. They are also resistant to most aqueous acids and alkalis but may be attacked by concentrated sulfuric acid. The polymers have good hydrolytic stability and can withstand hot water and steam.[349] Polysulfones are tough materials, but they do exhibit notch sensitivity. The presence of the aromatic rings causes the polymer chain to be rigid. Polysulfones generally do not require the addition of flame retardants and usually emit low levels of smoke.

The properties of the main polysulfones are generally similar, although polyethersulfones have better creep resistance at high temperatures and higher heat distortion temperature, but more water absorption and higher density than the Udel-type materials.[350] Glass-fiber-filled grades of polysulfone are available, as are blends of polysulfone with ABS.

Polysulfones may absorb water, leading to potential processing problems such as streaks or bubbling.[351] The processing temperatures are quite high, and the melt is very viscous. Polysulfones show little change in melt viscosity with shear. Injection molding melt temperatures are in the range of 335 to 400°C, and mold temperatures are in the range of 100 to 160°C. The high viscosity necessitates the use of large cross-sectional runners and gates. Purging should be done periodically, as a layer of black, degraded polymer may build up on the cylinder wall, yielding parts with black marks. Residual stresses may be reduced by higher mold temperatures or by annealing. Extrusion and blow-molding grades of polysulfones are higher molecular weight, with blow molding melt temperatures in the range of 300 to 360°C and mold temperatures between 70 and 95°C.

The good heat resistance and electrical properties of polysulfones allows them to be used in applications such as circuit boards and TV components.[352] Chemical and heat resistance are important properties for automotive applications. Hair dryer components can also be made from polysulfones. Polysulfones find application in ignition components and structural foams.[353] Another important market for polysulfones is microwave cookware.[354]

2.2.27.1 Polyaryl Sulfone (PAS). This polymer differs from the other polysulfones in the lack of any aliphatic groups in the chain. The lack of aliphatic groups gives this polymer excellent oxidative stability, as the aliphatic groups are more susceptible to oxidative degradation.[355] Polyaryl sulfones are stiff, strong, and tough polymers with very good chemical resistance. Most fuels, lubricants, cleaning agents, and hydraulic fluids will not affect the polymer.[356] However, methylene chloride, dimethyl acetamide, and dimethyl formamide will dissolve the polymer.[357] The glass transition temperature of these polymers is about 210°C, with a heat deflection temperature of 205°C at 1.82 MPa.[358] PAS also has good hydrolytic stability. Polyarylsulfone is available in filled and reinforced grades as well as both opaque and transparent versions.[359] This polymer finds application in electrical applications for motor parts, connectors, and lamp housings.[360]

The polymer can be injection molded, provided the cylinder and nozzle are capable of reaching 425°C.[361] It may also be extruded. The polymer should be dried prior to processing. Injection molding barrel temperatures should be 270 to 360°C at the rear, 295 to 390°C in the middle, and 300 to 395°C at the front.[362]

2.2.27.2 Polyether Sulfone (PES). Polyether sulfone is a transparent polymer with high temperature resistance and self-extinguishing properties.[363] It gives off little smoke when burned. Polyether sulfone has the basic structure as shown in Fig. 2.39.

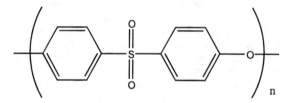

FIGURE 2.39 Structure of polyether sulfone.

Polyether sulfone has a T_g near 225°C and is dimensionally stable over a wide range of temperatures.[364] It can withstand long term use up to 200°C and can carry loads for long times up to 180°C.[365] Glass-fiber-reinforced grades are available for increased properties. It is resistant to most chemicals with the exception of polar, aromatic hydrocarbons.[366]

Polyether sulfone can be processed by injection molding, extrusion, blow molding or thermoforming.[367] It exhibits low mold shrinkage. For injection molding, barrel temperatures of 340 to 380°C with melt temperatures of 360°C are recommended.[368] Mold temperatures should be in the range of 140 to 180°C. For thin-walled molding, higher temperatures may be required. Unfilled PES can be extruded into sheets, rods, films, and profiles.

PES finds application in aircraft interior parts due to its low smoke emission.[369] Electrical applications include switches, integrated circuit carriers, and battery parts.[370] The high-temperature oil and gas resistance allow polyether sulfone to be used in the automotive markets for water pumps, fuse housings, and car heater fans. The ability of PES to endure repeated sterilization allows PES to be used in a variety of medical applications, such as parts for centrifuges and root canal drills. Other applications include membranes for kidney dialysis, chemical separation, and desalination. Consumer uses include cooking equipment and lighting fittings. PES can also be vacuum metallized for a high-gloss mirror finish.

2.2.27.3 Polysulfone (PSU).   Polysulfone is a transparent thermoplastic prepared from bisphenol A and 4,4´-dichlorodiphenylsulfone.[371] The structure is shown below in Fig. 2.40. It is self-extinguishing and has a high heat distortion temperature. The polymer has a glass transition temperature of 185°C.[372] Polysulfones have impact resistance and ductility below 0°C. Polysulfone also has good electrical properties. The electrical and mechanical properties are maintained to temperatures near 175°C. Polysulfone shows good chemical resistance to alkali, salt, and acid solutions.[373] It has resistance to oils, detergents, and alcohols, but polar organic solvents and chlorinated aliphatic solvents may attack the polymer. Glass- and mineral-filled grades are available.[374]

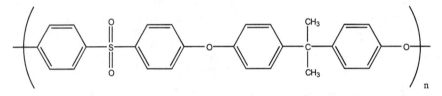

FIGURE 2.40   Structure of polysulfone.

Properties such as physical aging and solvent crazing can be improved by annealing the parts.[375] This also reduces molded-in stresses. Molded-in stresses can also be reduced by using hot molds during injection molding. As mentioned above, runners and gates should be as large as possible due to the high melt viscosity. The polymer should hit a wall or pin shortly after entering the cavity of the mold, as polysulfone has a tendency toward jetting. For thin-walled or long parts, multiple gates are recommended. For injection molding, barrel temperatures should be in the range of 310 to 400°C, with mold temperatures of 100 to 170°C.[376] In blow molding, the screw type should have a low compression ratio, 2.0:1 to 2.5:1. Higher compression ratios will generate excessive frictional heat. Mold temperatures of 70 to 95°C with blow air pressures of 0.3 to 0.5 MPa are generally used. Polysulfone can be extruded into films, pipe, or wire coatings. Extrusion melt temperatures should be from 315 to 375°C. High-compression-ratio screws should not be used for extrusion. Polysulfone shows high melt strength, allowing for good drawdown and the manufacture of thin films. Sheets of polysulfone can be thermoformed, with surface temperatures of 230 to 260°C recommended. Sheets may be bonded by heat sealing, adhesive bonding, solvent fusion, or ultrasonic welding.

Polysulfone is used in applications requiring good high-temperature resistance such as coffee carafes, piping, sterilizing equipment, and microwave oven cookware.[377] The good hydrolytic stability of polysulfone is important in these applications. Polysulfone is also used in electrical applications for connectors, switches, and circuit boards and in reverse osmosis applications as a membrane support.[378]

2.2.28 Vinyl-based Resins

2.2.28.1 Polyvinyl Chloride (PVC). Polyvinyl chloride polymers (PVC), generally referred to as vinyl resins, are prepared by the polymerization of vinyl chloride in a free radical addition polymerization reaction. Vinyl chloride monomer is prepared by reacting
ethylene with chlorine to form 1,2-dichloroethane.[379] The 1,2 dichloroethane is then
cracked to give vinyl chloride. The polymerization reaction is depicted in Fig. 2.41.

$$n\ CH_2{=}CHCl\ \rightarrow\ \text{-}(CH_2\text{-}CHCl)_n\text{-}$$

FIGURE 2.41 Synthesis of polyvinyl chloride.

The polymer can be made by suspension, emulsion, solution, or bulk polymerization
methods. Most of the PVC used in calendering, extrusion, and molding is prepared by
suspension polymerization. Emulsion polymerized vinyl resins are used in plastisols and
organisols.[380] Only a small amount of commercial PVC is prepared by solution polymerization. The microstructure of PVC is mostly atactic, but a sufficient quantity of syndiotactic portions of the chain allow for a low fraction of crystallinity (about 5 percent). The
polymers are essentially linear, but a low number of short-chain branches may exist.[381]
The monomers are predominantly arranged head to tail along the backbone of the chain.
Due to the presence of the chlorine group, PVC polymers are more polar than polyethylene. The molecular weights of commercial polymers are M_w = 100,000 to 200,000;
M_n = 45,000 to 64,000.[382] M_w/M_n = 2 for these polymers.

The polymeric PVC is insoluble in the monomer; therefore, bulk polymerization of PVC
is a heterogeneous process.[383] Suspension PVC is synthesized by suspension polymerization. These are suspended droplets approximately 10 to 100 nm in diameter of vinyl chloride monomer in water. Suspension polymerizations allow control of particle size, shape,
and size distribution by varying the dispersing agents and stirring rate. Emulsion polymerization results in much smaller particle sizes than suspension polymerized PVC, but soaps
used in the emulsion polymerization process can affect the electrical and optical properties.

The glass transition temperature of PVC varies with the polymerization method but
falls within the range of 60 to 80°C.[384] PVC is a self-extinguishing polymer and therefore
has application in the field of wire and cable. PVC's good flame resistance results from removal of HCl from the chain, releasing HCl gas.[385] Air is restricted from reaching the
flame, because HCl gas is more dense than air. Because PVC is thermally sensitive, the
thermal history of the polymer must be carefully controlled to avoid decomposition. At
temperatures above 70°C, degradation of PVC by loss of HCl can occur, resulting in the
generation of unsaturation in the backbone of the chain. This is indicated by a change in
the color of the polymer. As degradation proceeds, the polymer changes color from yellow
to brown to black, visually indicating that degradation has occurred. The loss of HCl accelerates the further degradation and is called *autocatalytic decomposition*. The degradation can be significant at processing temperatures if the material has not been heat
stabilized, so thermal stabilizers are often added at additional cost to PVC to reduce this
tendency. UV stabilizers are also added to protect the material from ultraviolet light,
which may also cause the loss of HCl.

There are two basic forms of PVC: rigid and plasticized. Rigid PVC, as its name suggests, is an unmodified polymer and exhibits high rigidity.[386] Unmodified PVC is stronger
and stiffer than PE and PP. Plasticized PVC is modified by the addition of a low-molecular-weight species (plasticizer) to flexibilize the polymer.[387] Plasticized PVC can be formulated to give products with rubbery behavior.

PVC is often compounded with additives to improve the properties. A wide variety of applications for PVC exist, because one can tailor the properties by proper selection of additives. As mentioned above, among the principal additives are stabilizers. Lead compounds are often added for this purpose, reacting with the HCl released during degradation.[388] Among the lead compounds commonly used are basic lead carbonate or white lead and tribasic lead sulphate. Other stabilizers include metal stearates, ricinoleates, palmitates, and octoates. Of particular importance are the cadmium-barium systems with synergistic behavior. Organo-tin compounds are also used as stabilizers to give clear compounds. In addition to stabilizers, other additives such as fillers, lubricants, pigments, and plasticizers are used. Fillers are often added to reduce cost and include talc, calcium carbonate, and clay.[389] These fillers may also impart additional stiffness to the compound.

The addition of plasticizers lowers the T_g of rigid PVC, making it more flexible. A wide range of products can be manufactured by using different amounts of plasticizer. As the plasticizer content increases, there is usually an increase in toughness and a decrease in the modulus and tensile strength.[390] Many different compounds can be used to plasticize PVC, but the solvent must be miscible with the polymer. A compatible plasticizer is considered a nonvolatile solvent for the polymer. The absorption of solvent may occur automatically at room temperature or may require the addition of slight heat and mixing. PVC plasticizers are divided into three groups, depending on their compatibility with the polymer: primary plasticizers, secondary plasticizers, and extenders. Primary plasticizers are compatible (have similar solubility parameters) with the polymer and should not exude. If the plasticizer and polymer have differences in their solubility parameters, they tend to be incompatible or have limited compatibility and are called *secondary plasticizers*. Secondary plasticizers are added along with the primary plasticizer to meet a secondary performance requirement (cost, low-temperature properties, permanence). The plasticizer can still be used in mixtures with a primary plasticizer, provided the mixture has a solubility parameter within the desired range. Extenders are used to lower the cost and are generally not compatible when used alone. Common plasticizers for PVC include dioctyl phthalate, di-iso-octyl phthalate, and dibutyl phthalate, among others.[391]

The plasticizer is normally added to the PVC before processing. Since the plasticizers are considered solvents for PVC, they will normally be absorbed the polymer with only a slight rise in temperature.[392] This reduces the time the PVC is exposed to high temperatures and potential degradation. In addition, the plasticizer reduces the T_g and T_m, therefore lowering the processing temperatures and thermal exposure. Plasticized PVC can be processed by methods such as extrusion and calendering into a variety of products.

Rigid PVC can be processed using most conventional processing equipment. Because HCl can be given off in small amounts during processing, corrosion of metal parts is a concern. Metal molds, tooling, and screws should be inspected regularly. Corrosion-resistant metals and coatings are available but add to the cost of manufacturing.

Rigid PVC products include house siding, extruded pipe, thermoformed, and injection-molded parts. Rigid PVC is calendered into credit cards. Plasticized PVC is used in applications such as flexible tubing, floor mats, garden hose, shrink wrap, and bottles.

PVC joints can be solvent welded rather than heated so as to fuse the two part together. This can be an advantage when heating the part is not feasible.

2.2.28.2 Chlorinated PVC. Post-chlorination of PVC was practiced during World War II.[393] Chlorinated PVC (CPVC) can be prepared by passing chlorine through a solution of PVC. The chlorine adds to the carbon that does not already have a chlorine atom present. Commercial materials have chlorine contents around 66 to 67 percent. The materials have a higher softening point and higher viscosity than PVC. They are known for good chemical resistance. Compared to PVC, chlorinated PVC has higher modulus and tensile strength. Compounding processes are similar to those for PVC but are more difficult.

Chlorinated PVC can be extruded, calendered, or injection molded.[394] Extrusion screws should be chrome plated or stainless steel. Dies should be streamlined. Injection molds should be chrome or nickel plated or stainless steel. CPVC is used for water distribution piping, industrial chemical liquid piping, outdoor skylight frames, automotive interior parts, and a variety of other applications.

2.2.28.3 Copolymers. Vinyl chloride can be copolymerized, with vinyl acetate giving a polymer with a lower softening point and better stability than pure PVC.[395] The compositions can vary from 5 to 40 percent vinyl acetate content. This material has application in areas where PVC is too rigid and the use of plasticized PVC is unacceptable. Flooring is one application for these copolymers. Copolymers with about 10 percent vinylidene chloride and copolymers with 10 to 20 percent diethyl fumarate or diethyl maleate are also available.

2.2.28.4 Dispersion PVC. If a sufficient quantity of solvent is added to PVC, it can become suspended in the solvent, giving a fluid that can be used in coating applications.[396] This form of PVC is called a *plastisol* or *oganisol*. PVC in the fluid form can be processed by methods such as spread coating, rotational casting, dipping, and spraying. The parts are then dried with heat to remove any solvent and fuse the polymer. Parts such as handles for tools and vinyl gloves are produced by this method.

The plastisol or organisols are prepared from PVC produced through emulsion polymerization.[397] The latex is then spray dried to form particles from 0.1 to 1 μm. These particles are then mixed with plasticizers to make plastisols or with plasticizers and other volatile organic liquids to make organisols. Less plasticizer is required with the organisols so that harder coatings can be produced. The polymer particles are not dissolved in the liquid but remain dispersed until the material is heated and fused. Other additives such as stabilizers and fillers may be compounded into the dispersion.

As plasticizer is added, the mixture goes through different stages as the voids between the polymer particles are filled.[398] Once all the voids between particles have been filled, the material is considered a paste. In these materials, the size of the particle is an important variable. If the particles are too large, they may settle out, so small particles are preferred. Very small particles have the disadvantage that the particles will absorb the plasticizer with time, giving a continuous increase in viscosity of the mixture. Paste polymers have particle sizes in the range of 0.2 to 1.5 μm. Particle size distribution will also affect the paste. It is usually better to have a wide particle size distribution so that particles can pack efficiently. This reduces the void space that must be filled by the plasticizer, and any additional plasticizer will act as lubricant. For a fixed particle-to-plasticizer ratio, a wide distribution will generally have lower viscosity than for a constant particle size. In some cases, very large particles are added to the paste, as they will take up volume, again reducing the amount of plasticizer required. These particles are made by suspension polymerization. With the mixture of particle sizes, these larger particles will not settle out as they would if used alone. Plastisols and organisols require the addition of heat to fuse. Temperatures in the range of 300 to 410°F are used to form the polymer.

2.2.28.5 Polyvinylidene Chloride (PVDC). Polyvinylidene chloride (PVDC) is similar to PVC except that two chlorine atoms are present on one of the carbon groups.[399] Like PVC, PVDC is also polymerized by addition polymerization methods. Both emulsion and suspension polymerization methods are used. The reaction is shown below in Fig. 2.42. The emulsion polymers are either used directly as a latex or dried for use in coatings or melt processing.

This material has excellent barrier properties and is frequently used in food packaging applications. Films made from PVDC have good cling properties, which is an advantage

$$n \ CH_2{=}CCl_2 \ \rightarrow \ (\text{-}CH_2\text{-}CCl_2\text{-})_n$$

FIGURE 2.42   Preparation of vinylidene chloride polymers.

for food wraps. Commercial polymers are all copolymers of vinylidene chloride with vinyl chloride, acrylates, or nitriles. Copolymerization of vinylidene chloride with other monomers reduces the melting point to allow easier processing. Corrosion-resistant materials should be considered for use when processing PVDC.

2.3 ADDITIVES

There is a broad range of additives for thermoplastics. Some of the more important additives include plasticizers, lubricants, anti-aging additives, colorants, flame retardants, blowing agents, cross-linking agents, and UV protectants. Fillers are also considered additives but are covered in Chap. 1.

Plasticizers are considered nonvolatile solvents.[400] They act to soften a material by separating the polymer chains, allowing them to be more flexible. As a result, the plasticized polymer is softer, with greater extensibility. Plasticizers reduce the melt viscosity and glass transition temperature of the polymer. For the plasticizer to be a "solvent" for the polymer, it is necessary for the solubility parameter of the plasticizer to be similar to the polymer. As a result, the plasticizer must be selected carefully so it is compatible with the polymer. One of the primary applications of plasticizers is for the modification of PVC. In this case, the plasticizers are divided into three classes, namely, primary and secondary plasticizers and extenders.[401] Primary plasticizers are compatible, can be used alone, and will not exude from the polymer. They should have a solubility parameter similar to that of the polymer. Secondary plasticizers have limited compatibility and are generally used with a primary plasticizer. Extenders have limited compatibility and will exude from the polymer if used alone. They are usually used along with the primary plasticizer. Plasticizers are usually in the form of high-viscosity liquids. The plasticizer should be capable of withstanding the high processing temperatures without degradation and discoloration, which would adversely affect the end product. The plasticizer should be capable of withstanding any environmental conditions that the final product will see. This might include UV exposure, fungal attack, or water. In addition, it is important that the plasticizer show low volatility and migration so that the properties of the plasticized polymer will remain relatively stable over time. There is a wide range of plasticizer types. Some typical classes include phthalic esters, phosphoric esters, fatty acid esters, fatty acid esters, polyesters, hydrocarbons, aromatic oils, and alcohols.

Lubricants are added to thermoplastics to aid in processing. High-molecular-weight thermoplastics have high viscosity. The addition of lubricants acts to reduce the melt viscosity to minimize machine wear and energy consumption.[402] Lubricants may also be added to prevent friction between molded products. Examples of these types of lubricants include graphite and molybdenum disulphide.[403] Lubricants that function by exuding from the polymer to the interface between the polymer and machine surface are termed *external lubricants*. Their presence at the interface between the polymer and metal walls acts to ease the processing. They have low compatibility with the polymer and may contain polar groups so that they have an attraction to metal. Lubricants must be selected based on the thermoplastic used. Lubricants may cause problems with clarity, ability to heat seal, and printing on the material. Examples of these lubricants include stearic acid or other carboxylic acids, paraffin oils, and certain alcohols and ketones for PVC. Low-molecular-weight materials that do not affect the solid properties, but act to enhance flow in the melt state, are termed *internal lubricants*. Internal lubricants for PVC include amine waxes, montan wax ester derivatives, and long-chain esters. Polymeric flow promoters are also examples of internal lubricants. They have solubility parameters similar to the thermoplastic, but lower viscosity at processing temperatures. They have little effect on the mechanical properties of the solid polymer. An example is the use of ethylene-vinyl acetate copolymers with PVC.

Anti-aging additives are incorporated to improve the resistance of the formulation. Examples of aging include attack by oxygen, ozone, dehydrochlorination, and UV degradation. Aging often results in changes in the structure of the polymer chain such as cross-linking, chain scission, addition of polar groups, or the addition of groups that cause discoloration. Additives are used to help prevent these reactions. Antioxidants are added to the polymer to stop the free-radical reactions that occur during oxidation. Antioxidants include compound such as phenols and amines. Phenols are often used because they have less of a tendency to stain.[404] Peroxide decomposers are also added to improve the aging properties of thermoplastics. These include mecaptans, sulfonic acids, and zinc dialkylthiophosphate. The presence of metal ions can act to increase the oxidation rate, even in the presence of antioxidants. Metal deactivators are often added to prevent this from taking place. Chelating agents are added to complex with the metal ion.

The absorption of ultraviolet light by a polymer may lead to the production of free radicals. These radicals react with oxygen resulting in what is termed *photodegradation*. This leads to the production of chemical groups that tend to absorb ultraviolet light, increasing the amount photodegradation. To reduce this effect, UV stabilizers are added. One way to accomplish UV stabilization is by the addition of UV absorbers such as benzophenones, salicylates, and carbon black.[405] They act to dissipate the energy in a harmless fashion. Quenching agents react with the activated polymer molecule. Nickel chelates and hindered amines can be used as quenching agents. Peroxide decomposers may be used to aid in UV stability.

In certain applications, flame resistance can be important. In this case, flame retarders may be added.[406] They act by one of four possible mechanisms. They may act to chemically interfere with the propagation of flame, react or decompose to absorb heat, form a fire resistant coating on the polymer, or produce gases that reduce the supply of air. Phosphates are an important class of flame retarders. Tritolyl phosphate and trixylyl phosphate are often used in PVC. Halogenated compounds such as chlorinated paraffins may also be used. Antimony oxide is often used in conjunction to obtain better results. Other flame retarders include titanium dioxide, zinc oxide, zinc borate, and red phosphorus. As with other additives, the proper selection of a flame retarder will depend on the particular thermoplastic.

Colorants are added to produce color in the polymeric part. They are separated into pigments and dyes. Pigments are insoluble in the polymer, while dyes are soluble in the polymer. The particular color desired and the type of polymer will affect the selection of the colorants.

Blowing agents are added to the polymer to produce a foam or cellular structure.[407] They may be chemical blowing agents that decompose at certain temperatures and release a gas, or they may be low boiling liquids that become volatile at the processing temperatures. Gases may be introduced into the polymer under pressure and expand when the polymer is depressurized. Mechanical whipping and the incorporation of hollow glass spheres can also be used to produce cellular materials.

Peroxides are often added to produce cross-linking in a system. Peroxides can be selected to decompose at a particular temperature for the application. Peroxides can be used to cross-link saturated polymers.

2.4 POLYMER BLENDS

There is considerable interest in polymer blends. This is driven by consideration of the difficulty in developing new polymeric materials from monomers. In many cases, it can be more cost effective to tailor the properties of a material through the blending of existing materials. One of the most basic questions in blends is whether the two polymers are miscible or exist as a single phase. In many cases, the polymers will exist as two separate phases. In this case, the morphology of the phases is of great importance. In the case of a miscible single phase blend, there is a single T_g, which is dependent on the composition of the blend.[408] Where two phases exist, the blend will exhibit two separate T_gs—one for each of the phases present. In the case where the polymers can crystallize, the crystalline portions will exhibit a melting point (T_m), even in the case where the two polymers are a miscible blend.

Although miscible blends of polymers exist, most blends of high-molecular-weight polymers exist as two-phase materials. Control of the morphology of these two-phase systems is critical to achieve the desired properties. A variety of morphologies exist, such as dispersed spheres of one polymer in another, lamellar structures, and co-continuous phases. As a result, the properties depend in a complex manner on the types of polymers in the blend, the morphology of the blend, and the effects of processing, which may orient the phases by shear.

Miscible blends of commercial importance include PPO-PS, PVC-nitrile rubber, and PBT-PET. Miscible blends show a single T_g that is dependent on the ratios of the two components in the blend and their respective T_gs. In immiscible blends, the major component has a large effect on the final properties of the blend. Immiscible blends include toughened polymers in which an elastomer is added, existing as a second phase. The addition of the elastomer phase dramatically improves the toughness of the resulting blend as a result of the crazing and shear yielding caused by the rubber phase. Examples of toughed polymers include high-impact polystyrene (HIPS), modified polypropylene, ABS, PVC, nylon, and others. In addition to toughened polymers, a variety of other two-phase blends are commercially available. Examples include PC-PBT, PVC-ABS, PC-PE, PP-EPDM, and PC-ABS.

2.5 REFERENCES

1. Carraher, C.E., *Polymer Chemistry, An Introduction,* 4th ed., Marcel Dekker, New York, 1996, p. 238.
2. Brydson, J.A., *Plastics Materials,* 6th ed., Butterworth-Heinemann, Oxford, 1995, p. 516.
3. Kroschwitz, J.I., *Concise Encyclopedia of Polymer Science and Engineering,* John Wiley and Sons, New York, 1990, p. 4.
4. Brydson, J.A., *Plastics Materials,* 6th ed., Butterworth-Heinemann, Oxford, 1995, p. 517.
5. Brydson, J.A., *Plastics Materials,* 6th ed., Butterworth-Heinemann, Oxford, 1995, p. 518.
6. Billmeyer, F.W., Jr., *Textbook of Polymer Science,* 2nd ed., John Wiley & Sons, New York, 1962, p. 439.
7. Brydson, J.A., *Plastics Materials,* 6th ed., Butterworth-Heinemann, Oxford, 1995, p. 519.
8. Berins, M.L., *Plastics Engineering Handbook of the Society of the Plastics Industry,* 5th ed., Chapman and Hall, New York, 1991, p. 61.
9. Brydson, J.A., *Plastics Materials,* 6th ed., Butterworth-Heinemann, Oxford, 1995, p. 521.
10. Brydson, J.A., *Plastics Materials,* 6th ed., Butterworth-Heinemann, Oxford, 1995, p. 523.
11. Brydson, J.A., *Plastics Materials,* 6th ed., Butterworth-Heinemann, Oxford, 1995, p. 524.
12. Berins, M.L., *Plastics Engineering Handbook of the Society of the Plastics Industry,* 5th ed., Chapman and Hall, New York, 1991, p. 62.

13. Strong, A.B., *Plastics: Materials and Processing,* Prentice-Hall, New Jersey, 1996, p. 193.
14. Brydson, J.A., *Plastics Materials,* 6th ed., Butterworth-Heinemann, Oxford, 1995, p. 525.
15. Modern Plastics, Jan. 1998, p. 76.
16. Brydson, J.A., *Plastics Materials,* 6th ed., Butterworth-Heinemann, Oxford, 1995, p. 527.
17. Carraher, C.E., *Polymer Chemistry, An Introduction,* 4th ed., Marcel Dekker, New York, 1996, p. 524.
18. Carraher, C.E., *Polymer Chemistry, An Introduction,* 4th ed., Marcel Dekker, New York, 1996, p. 524.
19. McCarthy, S.P., "Biodegradable Polymers for Packaging," in *Biotechnological Polymers,* C.G. Gebelein, Ed., Technomic Publishing, Lancaster, PA, 1993.
20. Carraher, C.E., *Polymer Chemistry, An Introduction,* 4th ed., Marcel Dekker, New York, 1996, p. 525.
21. Brydson, J.A., *Plastics Materials,* 6th ed., Butterworth-Heinemann, Oxford, 1995, p. 858.
22. Brydson, J.A., *Plastics Materials,* 6th ed., Butterworth-Heinemann, Oxford, 1995, p. 858.
23. Brydson, J.A., *Plastics Materials,* 6th ed., Butterworth-Heinemann, Oxford, 1995, p. 859.
24. McCarthy, S.P., "Biodegradable Polymers for Packaging," in *Biotechnological Polymers,* C.G. Gebelein, Ed.
25. McCarthy, S.P., "Biodegradable Polymers for Packaging," in *Biotechnological Polymers,* C.G. Gebelein, Ed.
26. Brydson, J.A., *Plastics Materials,* 6th ed., Butterworth-Heinemann, Oxford, 1995, p. 608.
27. Byrom, D., "Miscellaneous biomaterials," in *Biomaterials*, D. Byrom, Ed., Stockton Press, New York, 1991, p. 341.
28. Byrom, D., "Miscellaneous biomaterials," in *Biomaterials*, D. Byrom, Ed., Stockton Press, New York, 1991, p. 341.
29. Byrom, D., "Miscellaneous biomaterials," in *Biomaterials*, D. Byrom, Ed., Stockton Press, New York, 1991, p. 343.
30. Brydson, J.A., *Plastics Materials,* 6th ed., Butterworth-Heinemann, Oxford, 1995, p. 859.
31. Brydson, J.A., *Plastics Materials,* 6th ed., Butterworth-Heinemann, Oxford, 1995, p. 718.
32. McCarthy, S.P., "Biodegradable Polymers for Packaging," in *Biotechnological Polymers,* C.G. Gebelein, Ed.
33. Brydson, J.A., *Plastics Materials,* 6th ed., Butterworth-Heinemann, Oxford, 1995, p. 860.
34. Byrom, D., "Miscellaneous biomaterials" in *Biomaterials*, D. Byrom, Ed., Stockton Press, New York, 1991, p. 338.
35. Brydson, J.A., *Plastics Materials,* 6th ed., Butterworth-Heinemann, Oxford, 1995, p. 860.
36. Brydson, J.A., *Plastics Materials,* 6th ed., Butterworth-Heinemann, Oxford, 1995, p. 862.
37. McCarthy, S.P., "Biodegradable Polymers for Packaging," in *Biotechnological Polymers,* C.G. Gebelein, Ed.
38. Byrom, D., "Miscellaneous biomaterials" in Biomaterials, D. Byrom, Ed., Stockton Press, New York, 1991, p. 351.
39. Byrom, D., "Miscellaneous biomaterials" in *Biomaterials*, D. Byrom, Ed., Stockton Press, New York, 1991, p. 353.
40. *Encyclopedia of Polymer Science and Engineering,* 2nd ed., vol. 3, Mark, Bilkales, Overberger, Menges, Kroschwitz, Eds., Wiley Interscience, 1986, p. 60.
41. *Encyclopedia of Polymer Science and Engineering,* 2nd ed., vol. 3, Mark, Bilkales, Overberger, Menges, Kroschwitz, Eds., Wiley Interscience, 1986, p. 68
42. *Encyclopedia of Polymer Science and Engineering,* 2nd ed., vol. 3, Mark, Bilkales, Overberger, Menges, Kroschwitz, Eds., Wiley Interscience, 1986, p. 92.
43. *Encyclopedia of Polymer Science and Engineering,* 2nd ed., vol. 3, Mark, Bilkales, Overberger, Menges, Kroschwitz, Eds., Wiley Interscience, 1986, p. 182.
44. *Encyclopedia of Polymer Science and Engineering,* 2nd ed., vol. 3, Mark, Bilkales, Overberger, Menges, Kroschwitz, Eds., Wiley Interscience, 1986, p. 182.
45. *Plastics Materials,* 5th ed., J.A. Brydson, Butterworths, 1989, p. 583.
46. *Plastics Materials,* 5th ed., J.A. Brydson, Butterworths, 1989, p. 187.
47. Williams, R.W., "Cellulosics," in *Modern Plastics Encyclopedia Handbook,* McGraw-Hill, New York, 1994, p. 8.
48. Brydson, J.A., *Plastics Materials,* 6th ed., Butterworth-Heinemann, Oxford, 1995, p. 349.
49. Brydson, J.A., *Plastics Materials,* 6th ed., Butterworth-Heinemann, Oxford, 1995, p. 349.

50. Berins, M.L., *Plastics Engineering Handbook of the Society of the Plastics Industry,* 5th ed., Chapman and Hall, New York, 1991, p. 62.
51. Billmeyer, F.W., Jr., *Textbook of Polymer Science,* 2nd ed., John Wiley & Sons, New York, 1962, p. 423.
52. Berins, M.L., *Plastics Engineering Handbook of the Society of the Plastics Industry,* 5th ed., Chapman and Hall, New York, 1991, p. 63.
53. Carraher, C.E., *Polymer Chemistry, An Introduction,* 4th ed., Marcel Dekker, New York, 1996, p. 319.
54. Brydson, J.A., *Plastics Materials,* 6th ed., Butterworth-Heinemann, Oxford, 1995, p. 359.
55. Billmeyer, F.W., Jr., *Textbook of Polymer Science,* 2nd ed., John Wiley & Sons, New York, 1962, p. 426.
56. Brydson, J.A., *Plastics Materials,* 6th ed., Butterworth-Heinemann, Oxford, 1995, p. 359.
57. Brydson, J.A., *Plastics Materials,* 6th ed., Butterworth-Heinemann, Oxford, 1995, p. 359.
58. Berins, M.L., *Plastics Engineering Handbook of the Society of the Plastics Industry,* 5th ed., Chapman and Hall, New York, 1991, p. 63.
59. Berins, M.L., *Plastics Engineering Handbook of the Society of the Plastics Industry,* 5th ed., Chapman and Hall, New York, 1991, p. 63.
60. Berins, M.L., *Plastics Engineering Handbook of the Society of the Plastics Industry,* 5th ed., Chapman and Hall, New York, 1991, p. 63.
61. Brydson, J.A., *Plastics Materials,* 6th ed., Butterworth-Heinemann, Oxford, 1995, p. 360.
62. Billmeyer, F.W., Jr., *Textbook of Polymer Science,* 2nd ed., John Wiley & Sons, New York, 1962, p. 427.
63. Berins, M.L., *Plastics Engineering Handbook of the Society of the Plastics Industry,* 5th ed., Chapman and Hall, New York, 1991, p. 62.
64. Billmeyer, F.W., Jr., *Textbook of Polymer Science,* 2nd ed., John Wiley & Sons, New York, 1962, p. 428.
65. Brydson, J.A., *Plastics Materials,* 6th ed., Butterworth-Heinemann, Oxford, 1995, p. 361.
66. Berins, M.L., *Plastics Engineering Handbook of the Society of the Plastics Industry,* 5th ed., Chapman and Hall, New York, 1991, p. 62.
67. Billmeyer, F.W., Jr., *Textbook of Polymer Science,* 2nd ed., John Wiley & Sons, New York, 1962, p. 423.
68. Brydson, J.A., *Plastics Materials,* 6th ed., Butterworth-Heinemann, Oxford, 1995, p. 352.
69. Billmeyer, F.W., Jr., *Textbook of Polymer Science,* 2nd ed., John Wiley & Sons, New York, 1962, p. 424.
70. Brydson, J.A., *Plastics Materials,* 6th ed., Butterworth-Heinemann, Oxford, 1995, p. 351.
71. Billmeyer, F.W., Jr., *Textbook of Polymer Science,* 2nd ed., John Wiley & Sons, New York, 1962, p. 425.
72. Brydson, J.A., *Plastics Materials,* 6th ed., Butterworth-Heinemann, Oxford, 1995, p. 355.
73. Brydson, J.A., *Plastics Materials,* 6th ed., Butterworth-Heinemann, Oxford, 1995, p. 353.
74. Berins, M.L., *Plastics Engineering Handbook of the Society of the Plastics Industry,* 5th ed., Chapman and Hall, New York, 1991, p. 62.
75. Brydson, J.A., *Plastics Materials,* 6th ed., Butterworth-Heinemann, Oxford, 1995, p. 356.
76. Brydson, J.A., *Plastics Materials,* 6th ed., Butterworth-Heinemann, Oxford, 1995, p. 357.
77. Brydson, J.A., *Plastics Materials,* 6th ed., Butterworth-Heinemann, Oxford, 1995, p. 357.
78. Brydson, J.A., *Plastics Materials,* 6th ed., Butterworth-Heinemann, Oxford, 1995, p. 357.
79. Billmeyer, F.W., Jr., *Textbook of Polymer Science,* 2nd ed., John Wiley & Sons, New York, 1962, p. 426.
80. Billmeyer, F.W., Jr., *Textbook of Polymer Science,* 2nd ed., John Wiley & Sons, New York, 1962, p. 428.
81. Brydson, J.A., *Plastics Materials,* 6th ed., Butterworth-Heinemann, Oxford, 1995, p. 362.
82. Carraher, C.E., *Polymer Chemistry, An Introduction,* 4th ed., Marcel Dekker, New York, 1996, p. 319.
83. Brydson, J.A., *Plastics Materials,* 6th ed., Butterworth-Heinemann, Oxford, 1995, p. 363.
84. Berins, M.L., *Plastics Engineering Handbook of the Society of the Plastics Industry,* 5th ed., Chapman and Hall, New York, 1991, p. 63.
85. Berins, M.L., *Plastics Engineering Handbook of the Society of the Plastics Industry,* 5th ed., Chapman and Hall, New York, 1991, p. 63.

86. Brydson, J.A., *Plastics Materials,* 6th ed., Butterworth-Heinemann, Oxford, 1995, p. 362.
87. Billmeyer, F.W., Jr., *Textbook of Polymer Science,* 2nd ed., John Wiley & Sons, New York, 1962, p. 434.
88. *Modern Plastics,* Jan. 1998, p. 76.
89. Berins, M.L., *Plastics Engineering Handbook of the Society of the Plastics Industry,* 5th ed., Chapman and Hall, New York, 1991, p. 64.
90. Brydson, J.A., *Plastics Materials,* 6th ed., Butterworth-Heinemann, Oxford, 1995, p. 462.
91. Berins, M.L., *Plastics Engineering Handbook of the Society of the Plastics Industry,* 5th ed., Chapman and Hall, New York, 1991, p. 64.
92. Billmeyer, F.W., Jr., *Textbook of Polymer Science,* 2nd ed., John Wiley & Sons, New York, 1962, p. 433.
93. Deanin, R.D., *Polymer Structure, Properties and Applications,* Cahners, York, PA, 1972 p. 455.
94. Brydson, J.A., *Plastics Materials,* 6th ed., Butterworth-Heinemann, Oxford, 1995, p. 470.
95. Strong, A.B., *Plastics: Materials and Processing,* Prentice-Hall, New Jersey, 1996, p. 190.
96. Berins, M.L., *Plastics Engineering Handbook of the Society of the Plastics Industry,* 5th ed., Chapman and Hall, New York, 1991, p. 64.
97. Brydson, J.A., *Plastics Materials,* 6th ed., Butterworth-Heinemann, Oxford, 1995, p. 477.
98. Strong, A.B., *Plastics: Materials and Processing,* Prentice-Hall, New Jersey, 1996, p. 191.
99. Brydson, J.A., *Plastics Materials,* 6th ed., Butterworth-Heinemann, Oxford, 1995, p. 471.
100. Carraher, C.E., *Polymer Chemistry, An Introduction,* 4th ed., Marcel Dekker, New York, 1996, p. 233.
101. Brydson, J.A., *Plastics Materials,* 6th ed., Butterworth-Heinemann, Oxford, 1995, p. 472.
102. Brydson, J.A., *Plastics Materials,* 6th ed., Butterworth-Heinemann, Oxford, 1995, p. 472.
103. Galanty, P.G. and Bujtas, G.A., "Nylon," in *Modern Plastics Encyclopedia Handbook,* McGraw-Hill, New York, 1994, p. 12.
104. Brydson, J.A., *Plastics Materials,* 6th ed., Butterworth-Heinemann, Oxford, 1995, p. 473.
105. Brydson, J.A., *Plastics Materials,* 6th ed., Butterworth-Heinemann, Oxford, 1995, p. 472.
106. Brydson, J.A., *Plastics Materials,* 6th ed., Butterworth-Heinemann, Oxford, 1995, p. 473.
107. Strong, A.B., *Plastics: Materials and Processing,* Prentice-Hall, New Jersey, 1996, p. 190.
108. Brydson, J.A., *Plastics Materials,* 6th ed., Butterworth-Heinemann, Oxford, 1995, p. 484.
109. Brydson, J.A., *Plastics Materials,* 6th ed., Butterworth-Heinemann, Oxford, 1995, p. 484.
110. Berins, M.L., *Plastics Engineering Handbook of the Society of the Plastics Industry,* 5th ed., Chapman and Hall, New York, 1991, p. 64.
111. Brydson, J.A., *Plastics Materials,* 6th ed., Butterworth-Heinemann, Oxford, 1995, p. 461.
112. Billmeyer, F.W., Jr., *Textbook of Polymer Science,* 2nd ed., John Wiley & Sons, New York, 1962, p. 435.
113. Billmeyer, F.W., Jr., *Textbook of Polymer Science,* 2nd ed., John Wiley & Sons, New York, 1962, p. 436.
114. Billmeyer, F.W., Jr., *Textbook of Polymer Science,* 2nd ed., John Wiley & Sons, New York, 1962, p. 437.
115. Brydson, J.A., *Plastics Materials,* 6th ed., Butterworth-Heinemann, Oxford, 1995, p. 486.
116. Brydson, J.A., *Plastics Materials,* 6th ed., Butterworth-Heinemann, Oxford, 1995, p. 480.
117. Strong, A.B., *Plastics: Materials and Processing,* Prentice-Hall, New Jersey, 1996, p. 191.
118. Brydson, J.A., *Plastics Materials,* 6th ed., Butterworth-Heinemann, Oxford, 1995, p. 492.
119. Brydson, J.A., *Plastics Materials,* 6th ed., Butterworth-Heinemann, Oxford, 1995, p. 492.
120. Brydson, J.A., *Plastics Materials,* 6th ed., Butterworth-Heinemann, Oxford, 1995, p. 494-495.
121. Brydson, J.A., *Plastics Materials,* 6th ed., Butterworth-Heinemann, Oxford, 1995, p. 493.
122. Professor Driscoll, U. Mass. Lowell Plastics Engineering faculty, course notes, Chap. 12, "High Temperature Polyamides," p. 30, October 2005.
123. Brydson, J.A., *Plastics Materials,* 6th ed., Butterworth-Heinemann, Oxford, 1995, p. 496.
124. Brydson, J.A., *Plastics Materials,* 6th ed., Butterworth-Heinemann, Oxford, 1995, p. 497.
125. Strong, A.B., *Plastics: Materials and Processing,* Prentice-Hall, New Jersey, 1996, p. 192.
126. Brydson, J.A., *Plastics Materials,* 6th ed., Butterworth-Heinemann, Oxford, 1995, p. 400.
127. Billmeyer, F.W., Jr., *Textbook of Polymer Science,* 2nd ed., John Wiley & Sons, New York, 1962, p. 414.
128. Kroschwitz, J.I., *Concise Encyclopedia of Polymer Science and Engineering*, John Wiley and Sons, New York, 1990, p. 28.

129. Kroschwitz, J.I., *Concise Encyclopedia of Polymer Science and Engineering*, John Wiley and Sons, New York, 1990, p. 29.
130. Brydson, J.A., *Plastics Materials,* 6th ed., Butterworth-Heinemann, Oxford, 1995, p. 401.
131. Brydson, J.A., *Plastics Materials,* 6th ed., Butterworth-Heinemann, Oxford, 1995, p. 402.
132. Kroschwitz, J.I., *Concise Encyclopedia of Polymer Science and Engineering*, John Wiley and Sons, New York, 1990, p. 29.
133. Billmeyer, F.W., Jr., *Textbook of Polymer Science,* 2nd ed., John Wiley & Sons, New York, 1962, p. 413.
134. Kroschwitz, J.I., *Concise Encyclopedia of Polymer Science and Engineering*, John Wiley and Sons, New York, 1990, p. 23.
135. Brydson, J.A., *Plastics Materials,* 6th ed., Butterworth-Heinemann, Oxford, 1995, p. 507.
136. Berins, M.L., *Plastics Engineering Handbook of the Society of the Plastics Industry,* 5th ed., Chapman and Hall, New York, 1991, p. 65.
137. Carraher, C.E., *Polymer Chemistry, An Introduction,* 4th ed., Marcel Dekker, New York, 1996, p. 533.
138. Johson, S.H., "Polyamide-imide," in *Modern Plastics Encyclopedia Handbook,* McGraw-Hill, New York, 1994, p. 14.
139. Johson, S.H., "Polyamide-imide," in *Modern Plastics Encyclopedia Handbook,* McGraw-Hill, New York, 1994, p. 14.
140. Berins, M.L., *Plastics Engineering Handbook of the Society of the Plastics Industry,* 5th ed., Chapman and Hall, New York, 1991, p. 65.
141. Brydson, J.A., *Plastics Materials,* 6th ed., Butterworth-Heinemann, Oxford, 1995, p. 507.
142. Berins, M.L., *Plastics Engineering Handbook of the Society of the Plastics Industry,* 5th ed., Chapman and Hall, New York, 1991, p. 65.
143. Brydson, J.A., *Plastics Materials,* 6th ed., Butterworth-Heinemann, Oxford, 1995, p. 507.
144. Brydson, J.A., *Plastics Materials,* 6th ed., Butterworth-Heinemann, Oxford, 1995, p. 507
145. Brydson, J.A., *Plastics Materials,* 6th ed., Butterworth-Heinemann, Oxford, 1995, p. 708.
146. Berins, M.L., *Plastics Engineering Handbook of the Society of the Plastics Industry,* 5th ed., Chapman and Hall, New York, 1991, p. 66.
147. Dunkle, S.R. and Dean, B.D., "Polyarylate," in *Modern Plastics Encyclopedia Handbook,* McGraw-Hill, New York, 1994, p. 15.
148. Berins, M.L., *Plastics Engineering Handbook of the Society of the Plastics Industry,* 5th ed., Chapman and Hall, New York, 1991, p. 66.
149. Dunkle, S.R. and Dean, B.D., "Polyarylate," in *Modern Plastics Encyclopedia Handbook,* McGraw-Hill, New York, 1994, p. 16.
150. DiSano, L., "Polybenzimidazole," in *Modern Plastics Encyclopedia Handbook,* McGraw-Hill, New York, 1994, p. 16.
151. Carraher, C.E., *Polymer Chemistry, An Introduction,* 4th ed., Marcel Dekker, New York, 1996, p. 236.
152. Kroschwitz, J.I., *Concise Encyclopedia of Polymer Science and Engineering*, John Wiley and Sons, New York, 1990, p. 772.
153. DiSano, L., "Polybenzimidazole," in *Modern Plastics Encyclopedia Handbook,* McGraw-Hill, New York, 1994, p. 16.
154. DiSano, L., "Polybenzimidazole," in *Modern Plastics Encyclopedia Handbook,* McGraw-Hill, New York, 1994, p. 16.
155. Kroschwitz, J.I., *Concise Encyclopedia of Polymer Science and Engineering*, John Wiley and Sons, New York, 1990, p. 773.
156. DiSano, L., "Polybenzimidazole," in *Modern Plastics Encyclopedia Handbook,* McGraw-Hill, New York, 1994, p. 17.
157. Brydson, J.A., *Plastics Materials,* 6th ed., Butterworth-Heinemann, Oxford, 1995, p. 259.
158. Kroschwitz, J.I., *Concise Encyclopedia of Polymer Science and Engineering*, John Wiley and Sons, New York, 1990, p. 100.
159. Berins, M.L., *Plastics Engineering Handbook of the Society of the Plastics Industry,* 5th ed., Chapman and Hall, New York, 1991, p. 55.
160. Brydson, J.A., *Plastics Materials,* 6th ed., Butterworth-Heinemann, Oxford, 1995, p. 259.
161. Kroschwitz, J.I., *Concise Encyclopedia of Polymer Science and Engineering*, John Wiley and Sons, New York, 1990, p. 100.

162. Brydson, J.A., *Plastics Materials,* 6th ed., Butterworth-Heinemann, Oxford, 1995, p. 259.
163. Berins, M.L., *Plastics Engineering Handbook of the Society of the Plastics Industry,* 5th ed., Chapman and Hall, New York, 1991, p. 55.
164. Brydson, J.A., *Plastics Materials,* 6th ed., Butterworth-Heinemann, Oxford, 1995, p. 260.
165. Berins, M.L., *Plastics Engineering Handbook of the Society of the Plastics Industry,* 5th ed., Chapman and Hall, New York, 1991, p. 55.
166. 166166 Kroschwitz, J.I., *Concise Encyclopedia of Polymer Science and Engineering*, John Wiley and Sons, New York, 1990, p. 101.
167. *Plastics Materials,* 5th ed., J.A. Brydson, Butterworths, 1989, p. 525.
168. Domininghaus, H., *Plastics for Engineers, Materials, Properties, Applications,* Hanser Publishers, New York, 1988, p. 423.
169. Domininghaus, H., *Plastics for Engineers, Materials, Properties, Applications,* Hanser Publishers, New York, 1988, p. 424.
170. Domininghaus, H., *Plastics for Engineers, Materials, Properties, Applications,* Hanser Publishers, New York, 1988, p. 426.
171. Brydson, J.A., *Plastics Materials,* 6th ed., Butterworth-Heinemann, Oxford, 1995, p. 711.
172. Berins, M.L., *Plastics Engineering Handbook of the Society of the Plastics Industry, Inc.,* 5th ed., Chapman and Hall, New York, 1991, p. 67.
173. Brydson, J.A., *Plastics Materials,* 6th ed., Butterworth-Heinemann, Oxford, 1995, p. 707.
174. *Concise Polymer Handbook,* p. 477.
175. Brydson, J.A., *Plastics Materials,* 6th ed., Butterworth-Heinemann, Oxford, 1995, p. 712.
176. Brydson, J.A., *Plastics Materials,* 6th ed., Butterworth-Heinemann, Oxford, 1995, p. 713.
177. Brydson, J.A., *Plastics Materials,* 6th ed., Butterworth-Heinemann, Oxford, 1995, p. 714.
178. Brydson, J.A., *Plastics Materials,* 6th ed., Butterworth-Heinemann, Oxford, 1995, p. 712.
179. McChesney, C.E, in *Engineering Plastics,* vol. 2, Engineering Materials Handbook, ASM International, Metals Park, OH, 1988, p. 181.
180. Brydson, J.A., *Plastics Materials,* 6th ed., Butterworth-Heinemann, Oxford, 1995, p. 712.
181. McChesney, C.E., in Engineering Plastics, vol. 2, *Engineering Materials Handbook,* ASM International, Metals Park, OH, 1988, p. 181.
182. Brydson, J.A., *Plastics Materials,* 6th ed., Butterworth-Heinemann, Oxford, 1995, p. 713.
183. *Modern Plastics Encyclopedia Handbook,* McGraw-Hill, New York, 1994. p. 20.
184. McChesney, C.E, in Engineering Plastics, vol. 2, *Engineering Materials Handbook,* ASM International, Metals Park, OH, 1988, p. 181.
185. Berins, M.L., P*lastics Engineering Handbook of the Society of the Plastics Industry, Inc.,* 5th ed., Chapman and Hall, New York, 1991, p. 67.
186. *Modern Plastics,* Jan. 1999, vol. 76, no. 1, McGraw-Hill, New York, p. 65.
187. *Plastics Materials,* 5th ed., J.A. Brydson, Butterworths, Boston, 1989, p. 681.
188. *Plastics Materials,* 5th ed., J.A. Brydson, Butterworths, Boston, 1989, p. 677.
189. *Modern Plastics Encyclopedia Handbook,* McGraw-Hill, New York, 1994. p. 23.
190. *Modern Plastics Encyclopedia Handbook,* McGraw-Hill, New York, 1994. p. 23.
191. *Modern Plastics,* Jan. 1999, vol. 76, no. 1, McGraw-Hill, New York, pp. 74, 75.
192. *Principles of Polymerization,* 2nd ed., G. Odian, John Wiley & Sons, New York, 1981, p. 103.
193. *Plastics Materials,* 5th ed., J.A. Brydson, Butterworths, Boston, 1989, p. 675.
194. *Encyclopedia of Polymer Science and Engineering,* vol. 12, John Wiley & Sons, New York, 1985, p. 223.
195. *Principles of Polymerization,* 2nd ed., G. Odian, John Wiley & Sons, New York, 1981, p. 103.
196. *Encyclopedia of Polymer Science and Engineering,* vol. 12, John Wiley & Sons, New York, 1985, p. 223.
197. *Principles of Polymerization,* 2nd ed., G. Odian, John Wiley & Sons, New York, 1981, p. 105.
198. *Encyclopedia of Polymer Science and Engineering,* vol. 12, John Wiley & Sons, New York, 1985, p. 223.
199. *Encyclopedia of Polymer Science and Engineering,* vol. 12, John Wiley & Sons, New York, 1985, p. 222.
200. *Encyclopedia of Polymer Science and Engineering,* vol. 12, John Wiley & Sons, New York, 1985, p. 223.
201. *Encyclopedia of Polymer Science and Engineering,* vol. 12, John Wiley & Sons, New York, 1985, p. 195.

202. *Encyclopedia of Polymer Science and Engineering,* vol. 12, John Wiley & Sons, New York, 1985, p. 228.
203. *Encyclopedia of Polymer Science and Engineering,* vol. 12, John Wiley & Sons, New York, 1985, p. 194.
204. *Encyclopedia of Polymer Science and Engineering,* vol. 12, John Wiley & Sons, New York, 1985, pp. 195, 204-209.
205. *Encyclopedia of Polymer Science and Engineering,* vol. 12, John Wiley & Sons, New York, 1985, p. 197.
206. *Encyclopedia of Polymer Science and Engineering,* vol. 12, John Wiley & Sons, New York, 1985, p. 213.
207. Brydson, J.A., *Plastics Materials,* 7th ed., Butterworth-Heinemann, Oxford, 1995, p. 723.
208. Berins, M.L., *Plastics Engineering Handbook of the Society of the Plastics Industry,* 5th ed., Chapman and Hall, New York, 1991, p. 67.
209. Kroschwitz, J.I., *Concise Encyclopedia of Polymer Science and Engineering*, John Wiley and Sons, New York, 1990, p. 327.
210. Brydson, J.A., *Plastics Materials,* 6th ed., Butterworth-Heinemann, Oxford, 1995, p. 508.
211. Brydson, J.A., *Plastics Materials,* 6th ed., Butterworth-Heinemann, Oxford, 1995, p. 508.
212. Berins, M.L., *Plastics Engineering Handbook of the Society of the Plastics Industry,* 5th ed., Chapman and Hall, New York, 1991, p. 68.
213. Berins, M.L., *Plastics Engineering Handbook of the Society of the Plastics Industry,* 5th ed., Chapman and Hall, New York, 1991, p. 68.
214. Brydson, J.A., *Plastics Materials,* 6th ed., Butterworth-Heinemann, Oxford, 1995, p. 508.
215. Berins, M.L., *Plastics Engineering Handbook of the Society of the Plastics Industry,* 5th ed., Chapman and Hall, New York, 1991, p. 68.
216. Domininghaus, H., *Plastics for Engineers, Materials, Properties, Applications,* Hanser Publishers, New York," 1988, p. 24.
217. *Modern Plastics Encyclopedia,* 1998, vol. 74, Number 13, McGraw-Hill, p. B-4.
218. *Plastics Materials,* 5th ed., J.A. Brydson, Butterworths, 1989, p. 217.
219. Domininghaus, H., *Plastics for Engineers, Materials, Properties, Applications,* Hanser Publishers, New York, 1988, p. 55.
220. *Encyclopedia of Polymer Science and Engineering,* 2nd ed., vol. 6, Mark, Bilkales, Overberger, Menges, Kroschwitz, Eds., Wiley Interscience, 1986, p. 383.
221. *McGraw-Hill Encyclopedia of Science and Technology,* 5th ed., vol. 10, 1982, p. 647.
222. *Encyclopedia of Polymer Science and Engineering,* 2nd ed., vol. 6, Mark, Bilkales, Overberger, Menges, Kroschwitz, Eds., Wiley Interscience, 1986, p. 385.
223. *Modern Plastics Encyclopedia,* 1998, p. A-15.
224. *Encyclopedia of Polymer Science and Engineering,* 2nd ed., vol. 6, Mark, Bilkales, Overberger, Menges, Kroschwitz, Eds., Wiley Interscience, 1986, p. 486.
225. *Encyclopedia of Polymer Science and Engineering,* 2nd ed., vol. 6, Mark, Bilkales, Overberger, Menges, Kroschwitz, Eds., Wiley Interscience, 1986, p. 493.
226. *Plastics Materials,* 5th ed., J.A. Brydson, Butterworths, 1989, p. 262.
227. *Encyclopedia of Polymer Science and Engineering,* 2nd ed., vol. 6, Mark, Bilkales, Overberger, Menges, Kroschwitz, Eds., Wiley Interscience, 1986, p. 422.
228. Kung, D.M., "Ethylene-ethyl acrylate," in *Modern Plastics Encyclopedia Handbook,* McGraw-Hill, New York, 1994, p. 38.
229. Kung, D.M., "Ethylene-ethyl acrylate," in *Modern Plastics Encyclopedia Handbook,* McGraw-Hill, New York, 1994, p. 38.
230. *Plastics Materials,* 5th ed., J.A. Brydson, Butterworths, 1989, p. 262.
231. Kung, D.M., "Ethylene-ethyl acrylate," in *Modern Plastics Encyclopedia Handbook,* McGraw-Hill, New York, 1994, p. 38.
232. Baker, G., "Ethylene-methyl acrylate," in *Modern Plastics Encyclopedia Handbook,* McGraw-Hill, New York, 1994, p. 38.
233. *Encyclopedia of Polymer Science and Engineering,* 2nd ed., vol. 6, Mark, Bilkales, Overberger, Menges, Kroschwitz, Eds., Wiley Interscience, 1986, p. 422.
234. *Plastic Materials*, J.A. Brydson, 5th ed., Butterworths, London, 1989, p. 261.
235. Brydson, J.A., *Plastics Materials,* 7th ed., Butterworth-Heinemann, Oxford, 1995, p. 279.
236. *Plastic Materials,* J.A. Brydson, 5th ed., Butterworths, London, 1989, p. 229.

237. *Concise Encyclopedia of Polymer Science and Engineering*, Jacqueline Kroschwitz, Ex. Ed., Wiley-Interscience Publication, New York, 1990, p. 357.
238. Domininghaus, H., *Plastics for Engineers, Materials, Properties, Applications,* Hanser Publishers, New York, 1988, p. 65.
239. Domininghaus, H., *Plastics for Engineers, Materials, Properties, Applications,* Hanser Publishers, New York, 1988, p. 67.
240. *Plastic Materials,* J.A. Brydson, 5th ed., Butterworths, London, 1989, p. 284.
241. Domininghaus, H., *Plastics for Engineers, Materials, Properties, Applications,* Hanser Publishers, New York, 1988, p. 68.
242. Strong, A.B., *Plastics: Materials and Processing,* Prentice-Hall, New Jersey, 1996, p. 165.
243. Brydson, J.A., *Plastics Materials,* 6th ed., Butterworth-Heinemann, Oxford, 1995, p. 268.
244. Brydson, J.A., *Plastics Materials,* 6th ed., Butterworth-Heinemann, Oxford, 1995, p. 268.
245. MacKnight, W.J. and Lundberg, R.D. in *Thermoplastic Elastomers,* 2nd ed., Holden, G., Legge, N.R., Quirk, R.P., and Schroeder, H.E., Eds., Hanser Publishers, New York, 1996, p. 279.
246. Kroschwitz, J.I., *Concise Encyclopedia of Polymer Science and Engineering*, John Wiley and Sons, New York, 1990, p. 126.
247. Strong, A.B., *Plastics: Materials and Processing,* Prentice-Hall, New Jersey, 1996, p. 165.
248. Rees, R. W., "Ionomers," in *Engineering Plastics,* vol. 2, *Engineering Materials Handbook,* ASM International, Metals Park, OH, 1988, p. 120.
249. Rees, R. W., "Ionomers," in *Engineering Plastics,* vol. 2, *Engineering Materials Handbook,* ASM International, Metals Park, OH, 1988, p. 122.
250. Rees, R. W., "Ionomers," in *Engineering Plastics,* vol. 2, *Engineering Materials Handbook,* ASM International, Metals Park, OH, 1988, p. 123.
251. Strong, A.B., *Plastics: Materials and Processing,* Prentice-Hall, New Jersey, 1996, p. 165.
252. Rees, R.W., in *Thermoplastic Elastomers,* 2nd ed., Holden, G., Legge, N.R., Quirk, R.P., and Schroeder, H.E., Eds., Hanser Publishers, New York, 1996, p. 263.
253. Brydson, J.A., *Plastics Materials,* 6th ed., Butterworth-Heinemann, Oxford, 1995, p. 268.
254. Brydson, J.A., *Plastics Materials,* 6th ed., Butterworth-Heinemann, Oxford, 1995, p. 269.
255. Kroschwitz, J.I., *Concise Encyclopedia of Polymer Science and Engineering*, John Wiley and Sons, New York, 1990, p. 827.
256. Berins, M.L., *Plastics Engineering Handbook of the Society of the Plastics Industry,* 5th ed., Chapman and Hall, New York, 1991, p. 69.
257. Albermarle, "Polyimide, Thermoplastic," in *Modern Plastics Encyclopedia Handbook,* McGraw-Hill, New York, 1994, p. 43.
258. Berins, M.L., *Plastics Engineering Handbook of the Society of the Plastics Industry,* 5th ed., Chapman and Hall, New York, 1991, p. 69.
259. Kroschwitz, J.I., *Concise Encyclopedia of Polymer Science and Engineering*, John Wiley and Sons, New York, 1990, p. 827.
260. Berins, M.L., *Plastics Engineering Handbook of the Society of the Plastics Industry,* 5th ed., Chapman and Hall, New York, 1991, p. 69.
261. Brydson, J.A., *Plastics Materials,* 6th ed., Butterworth-Heinemann, Oxford, 1995, p. 504.
262. Brydson, J.A., *Plastics Materials,* 6th ed., Butterworth-Heinemann, Oxford, 1995, p. 501.
263. Berins, M.L., *Plastics Engineering Handbook of the Society of the Plastics Industry,* 5th ed., Chapman and Hall, New York, 1991, p. 69.
264. Brydson, J.A., *Plastics Materials,* 5th ed., Butterworth-Heinemann, Oxford, 1989, p. 565.
265. *Modern Plastics Encyclopedia,* 1998, mid-November 1997 issue, vol. 74, Number 13, McGraw-Hill, pp. B-162, B-163.
266. Brydson, J.A., *Plastics Materials,* 6th ed., Butterworth-Heinemann, Oxford, 1995, p. 586.
267. Brydson, J.A., *Plastics Materials,* 6th ed., Butterworth-Heinemann, Oxford, 1995, p. 564.
268. Brydson, J.A., *Plastics Materials,* 6th ed., Butterworth-Heinemann, Oxford, 1995, p. 389.
269. Brydson, J.A., *Plastics Materials,* 6th ed., Butterworth-Heinemann, Oxford, 1995, p. 391.
270. Domininghaus, H., *Plastics for Engineers, Materials, Properties, Applications,* Hanser Publishers, New York, 1988, p. 280.
271. Domininghaus, H., *Plastics for Engineers, Materials, Properties, Applications,* Hanser Publishers, New York, 1988, p. 122.
272. Brydson, J.A., *Plastics Materials,* 6th ed., Butterworth-Heinemann, Oxford, 1995, p. 261.
273. Brydson, J.A., *Plastics Materials,* 6th ed., Butterworth-Heinemann, Oxford, 1995, p. 263.

274. Brydson, J.A., *Plastics Materials*, 6th ed., Butterworth-Heinemann, Oxford, 1995, p. 567.
275. Brydson, J.A., *Plastics Materials*, 6th ed., Butterworth-Heinemann, Oxford, 1995, p. 568.
276. Brydson, J.A., *Plastics Materials*, 6th ed., Butterworth-Heinemann, Oxford, 1995, p. 570.
277. Brydson, J.A., *Plastics Materials*, 6th ed., Butterworth-Heinemann, Oxford, 1995, p. 570.
278. Domininghaus, H., *Plastics for Engineers, Materials, Properties, Applications,* Hanser Publishers, New York, 1988, p. 490.
279. Brydson, J.A., *Plastics Materials*, 6th ed., Butterworth-Heinemann, Oxford, 1995, p. 575.
280. Brydson, J.A., *Plastics Materials*, 6th ed., Butterworth-Heinemann, Oxford, 1995, p. 576.
281. Brydson, J.A., *Plastics Materials*, 6th ed., Butterworth-Heinemann, Oxford, 1995, p. 575.
282. Domininghaus, H., *Plastics for Engineers, Materials, Properties, Applications,* Hanser Publishers, New York, 1988, p. 529.
283. Harris, J.H. and Reksc, J.A., "Polyphthalamide," in *Modern Plastics Encyclopedia Handbook,* McGraw-Hill, New York, 1994, p. 47.
284. Brydson, J.A., *Plastics Materials*, 6th ed., Butterworth-Heinemann, Oxford, 1995, p. 499.
285. Harris, J.H. and Reksc, J.A., "Polyphthalamide," in *Modern Plastics Encyclopedia Handbook*, McGraw-Hill, New York, 1994, p. 47.
286. Harris, J.H. and Reksc, J.A., "Polyphthalamide," in *Modern Plastics Encyclopedia Handbook*, McGraw-Hill, New York, 1994, p. 47.
287. Brydson, J.A., *Plastics Materials*, 6th ed., Butterworth-Heinemann, Oxford, 1995, p. 499.
288. Harris, J.H. and Reksc, J.A., "Polyphthalamide," in *Modern Plastics Encyclopedia Handbook*, McGraw-Hill, New York, 1994, p. 47.
289. Brydson, J.A., *Plastics Materials*, 6th ed., Butterworth-Heinemann, Oxford, 1995, p. 499.
290. Harris, J.H. and Reksc, J.A., "Polyphthalamide," in *Modern Plastics Encyclopedia Handbook*, McGraw-Hill, New York, 1994, p. 47.
291. Harris, J.H. and Reksc, J.A., "Polyphthalamide," in *Modern Plastics Encyclopedia Handbook*, McGraw-Hill, New York, 1994, p. 48.
292. Harris, J.H. and Reksc, J.A., "Polyphthalamide," in *Modern Plastics Encyclopedia Handbook*, McGraw-Hill, New York, 1994, p. 48.
293. *Modern Plastics,* Jan. 1998, p. 58.
294. Brydson, J.A., *Plastics Materials*, 6th ed., Butterworth-Heinemann, Oxford, 1995, p. 244.
295. Cradic, G.W., "PP Homopolymer," in *Modern Plastics Encyclopedia Handbook*, McGraw-Hill, New York, 1994, p. 49.
296. Colvin, R., *Modern Plastics,* May 1997, p. 62.
297. Brydson, J.A., *Plastics Materials*, 6th ed., Butterworth-Heinemann, Oxford, 1995, p. 245.
298. Odian, G., *Principles of Polymerization,* 2nd ed., John Wiley & Sons, New York, 1981, p. 581.
299. Brydson, J.A., *Plastics Materials*, 6th ed., Butterworth-Heinemann, Oxford, 1995, p. 258.
300. Brydson, J.A., *Plastics Materials*, 6th ed., Butterworth-Heinemann, Oxford, 1995, p. 244.
301. Strong, A.B., *Plastics: Materials and Processing,* Prentice-Hall, New Jersey, 1996, p. 168.
302. Billmeyer, F.W., Jr., *Textbook of Polymer Science,* 2nd ed., John Wiley & Sons, New York, 1962, p. 388.
303. Billmeyer, F.W., Jr., *Textbook of Polymer Science,* 2nd ed., John Wiley & Sons, New York, 1962, p. 387.
304. Brydson, J.A., *Plastics Materials*, 6th ed., Butterworth-Heinemann, Oxford, 1995, p. 256.
305. Strong, A.B., *Plastics: Materials and Processing,* Prentice-Hall, New Jersey, 1996, p. 169.
306. Brydson, J.A., *Plastics Materials*, 6th ed., Butterworth-Heinemann, Oxford, 1995, p. 245.
307. Brydson, J.A., *Plastics Materials*, 6th ed., Butterworth-Heinemann, Oxford, 1995, p. 246.
308. Brydson, J.A., *Plastics Materials*, 6th ed., Butterworth-Heinemann, Oxford, 1995, p. 248.
309. Brydson, J.A., *Plastics Materials*, 6th ed., Butterworth-Heinemann, Oxford, 1995, p. 245.
310. Cradic, G.W., "PP Homopolymer," in *Modern Plastics Encyclopedia Handbook*, McGraw-Hill, New York, 1994, p. 49.
311. Brydson, J.A., *Plastics Materials*, 6th ed., Butterworth-Heinemann, Oxford, 1995, p. 253.
312. Brydson, J.A., *Plastics Materials*, 6th ed., Butterworth-Heinemann, Oxford, 1995, p. 254.
313. Brydson, J.A., *Plastics Materials*, 6th ed., Butterworth-Heinemann, Oxford, 1995, p. 255.
314. Brydson, J.A., *Plastics Materials*, 6th ed., Butterworth-Heinemann, Oxford, 1995, p. 251.
315. Brydson, J.A., *Plastics Materials*, 6th ed., Butterworth-Heinemann, Oxford, 1995, p. 257.
316. Leaversuch, R.D., *Modern Plastics,* Dec. 1996, p. 52
317. Brydson, J.A., *Plastics Materials*, 6th ed., Butterworth-Heinemann, Oxford, 1995, p. 756.

318. Brydson, J.A., *Plastics Materials,* 6th ed., Butterworth-Heinemann, Oxford, 1995, p. 767.
319. Brydson, J.A., *Plastics Materials,* 6th ed., Butterworth-Heinemann, Oxford, 1995, p. 768.
320. Sardanopoli, A.A., "Thermoplastic Polyurethanes," in *Engineering Plastics,* vol. 2, Engineering Materials Handbook, ASM International, Metals Park, OH, 1988, p. 203.
321. Sardanopoli, A.A., "Thermoplastic Polyurethanes," in *Engineering Plastics*, vol. 2, Engineering Materials Handbook, ASM International, Metals Park, OH, 1988, p. 206.
322. Sardanopoli, A.A., "Thermoplastic Polyurethanes," in *Engineering Plastics*, vol. 2, Engineering Materials Handbook, ASM International, Metals Park, OH, 1988, p. 205.
323. Sardanopoli, A.A., "Thermoplastic Polyurethanes," in *Engineering Plastics*, vol. 2, Engineering Materials Handbook, ASM International, Metals Park, OH, 1988, p. 205.
324. Sardanopoli, A.A., "Thermoplastic Polyurethanes," in *Engineering Plastics*, vol. 2, Engineering Materials Handbook, ASM International, Metals Park, OH, 1988, p. 207.
325. Brydson, J.A., *Plastics Materials,* 6th ed., Butterworth-Heinemann, Oxford, 1995, p. 427.
326. Domininghaus, H., *Plastics for Engineers, Materials, Properties, Applications,* Hanser Publishers, New York, 1988, p. 226.
327. Akane, J., "ACS," in *Modern Plastics Encyclopedia Handbook*, McGraw-Hill, New York, 1994, p. 54.
328. Akane, J., "ACS," in *Modern Plastics Encyclopedia Handbook*, McGraw-Hill, New York, 1994, p. 54.
329. Ostrowski, S., "Acrylic-styrene-acrylonitrile," in *Modern Plastics Encyclopedia Handbook*, McGraw-Hill, New York, 1994, p. 54.
330. *Principles of Polymer Engineering,* 2nd ed., McCrum, Buckley and Bucknall, Oxford Science Publications, p. 372.
331. *Encyclopedia of Polymer Science and Engineering,* 2nd ed., vol. 16, Mark, Bilkales, Overberger, Menges, Kroschwitz, Eds., Wiley Interscience, 1986, p. 65.
332. Kroschwitz, J.I., *Concise Encyclopedia of Polymer Science and Engineering*, John Wiley and Sons, New York, 1990, p. 30.
333. Brydson, J.A., *Plastics Materials,* 6th ed., Butterworth-Heinemann, Oxford, 1995, p. 426.
334. Berins, M.L., *Plastics Engineering Handbook of the Society of the Plastics Industry,* 5th ed., Chapman and Hall, New York, 1991, p. 57.
335. Brydson, J.A., *Plastics Materials,* 6th ed., Butterworth-Heinemann, Oxford, 1995, p. 426.
336. Brydson, J.A., *Plastics Materials,* 6th ed., Butterworth-Heinemann, Oxford, 1995, p. 426.
337. Kroschwitz, J.I., *Concise Encyclopedia of Polymer Science and Engineering*, John Wiley and Sons, New York, 1990, p. 30.
338. Berins, M.L., *Plastics Engineering Handbook of the Society of the Plastics Industry, Inc.,* 5th ed., Chapman and Hall, New York, 1991, p. 57.
339. Berins, M.L., *Plastics Engineering Handbook of the Society of the Plastics Industry,* 5th ed., Chapman and Hall, New York, 1991, p. 57.
340. Brydson, J.A., *Plastics Materials,* 6th ed., Butterworth-Heinemann, Oxford, 1995, p. 435.
341. Salay, J.E. and Dougherty, D.J., "Styrene-butadiene copolymers," in *Modern Plastics Encyclopedia Handbook*, McGraw-Hill, New York, 1994, p. 60.
342. Salay, J.E. and Dougherty, D.J., "Styrene-butadiene copolymers," in *Modern Plastics Encyclopedia Handbook*, McGraw-Hill, New York, 1994, p. 60.
343. Brydson, J.A., *Plastics Materials,* 6th ed., Butterworth-Heinemann, Oxford, 1995, p. 435.
344. Salay, J.E. and Dougherty, D.J., "Styrene-butadiene copolymers," in *Modern Plastics Encyclopedia Handbook*, McGraw-Hill, New York, 1994, p. 60.
345. Strong, A.B., *Plastics: Materials and Processing,* Prentice-Hall, New Jersey, 1996, p. 205.
346. Brydson, J.A., *Plastics Materials,* 6th ed., Butterworth-Heinemann, Oxford, 1995, p. 577.
347. Kroschwitz, J.I., *Concise Encyclopedia of Polymer Science and Engineering*, John Wiley and Sons, New York, 1990, p. 886.
348. Brydson, J.A., *Plastics Materials,* 6th ed., Butterworth-Heinemann, Oxford, 1995, p. 580.
349. Kroschwitz, J.I., *Concise Encyclopedia of Polymer Science and Engineering*, John Wiley and Sons, New York, 1990, p. 886.
350. Brydson, J.A., *Plastics Materials,* 6th ed., Butterworth-Heinemann, Oxford, 1995, p. 582.
351. Brydson, J.A., *Plastics Materials,* 6th ed., Butterworth-Heinemann, Oxford, 1995, p. 582.
352. Brydson, J.A., *Plastics Materials,* 6th ed., Butterworth-Heinemann, Oxford, 1995, p. 583.

353. Carraher, C.E., *Polymer Chemistry, An Introduction,* 4th ed., Marcel Dekker, New York, 1996, p. 240.
354. Kroschwitz, J.I., *Concise Encyclopedia of Polymer Science and Engineering*, John Wiley and Sons, New York, 1990, p. 888.
355. Berins, M.L., *Plastics Engineering Handbook of the Society of the Plastics Industry,* 5th ed., Chapman and Hall, New York, 1991, p. 71.
356. Berins, M.L., *Plastics Engineering Handbook of the Society of the Plastics Industry,* 5th ed., Chapman and Hall, New York, 1991, p. 71.
357. Sauers, M.E., "Polyaryl Sulfones," in *Engineering Plastics*, vol. 2, Engineered Materials Handbook, ASM International, Metals Park, OH, 1988, p. 146.
358. Sauers, M.E., "Polyaryl Sulfones," in *Engineering Plastics*, vol. 2, Engineered Materials Handbook, ASM International, Metals Park, OH, 1988, p. 145.
359. Berins, M.L., *Plastics Engineering Handbook of the Society of the Plastics Industry,* 5th ed., Chapman and Hall, New York, 1991, p. 72.
360. Berins, M.L., *Plastics Engineering Handbook of the Society of the Plastics Industry,* 5th ed., Chapman and Hall, New York, 1991, p. 72.
361. Berins, M.L., *Plastics Engineering Handbook of the Society of the Plastics Industry,* 5th ed., Chapman and Hall, New York, 1991, p. 71.
362. Sauers, M.E., "Polyaryl Sulfones," in *Engineering Plastics*, vol. 2, Engineered Materials Handbook, ASM International, Metals Park, OH, 1988, p. 146.
363. Berins, M.L., *Plastics Engineering Handbook of the Society of the Plastics Industry,* 5th ed., Chapman and Hall, New York, 1991, p. 72.
364. Watterson, E.C., "Polyether Sulfones," in *Engineering Plastics*, vol. 2, *Engineered Materials Handbook,* ASM International, Metals Park, OH, 1988, p. 161.
365. Watterson, E.C., "Polyether Sulfones," in *Engineering Plastics*, vol. 2, *Engineered Materials Handbook,* ASM International, Metals Park, OH, 1988, p. 160.
366. Berins, M.L., *Plastics Engineering Handbook of the Society of the Plastics Industry, Inc.,* 5th ed., Chapman and Hall, New York, 1991, p. 72.
367. Berins, M.L., *Plastics Engineering Handbook of the Society of the Plastics Industry, Inc.,* 5th ed., Chapman and Hall, New York, 1991, p. 72.
368. Watterson, E.C., "Polyether Sulfones," in *Engineering Plastics*, vol. 2, *Engineered Materials Handbook,* ASM International, Metals Park, OH, 1988, p. 161.
369. Berins, M.L., *Plastics Engineering Handbook of the Society of the Plastics Industry, Inc.,* 5th ed., Chapman and Hall, New York, 1991, p. 72.
370. Watterson, E.C., "Polyether Sulfones," in *Engineering Plastics*, vol. 2, Engineered Materials Handbook, ASM International, Metals Park, OH, 1988, p. 159.
371. Dunkle, S.R., "Polysulfones," in *Engineering Plastics*, vol. 2, *Engineered Materials Handbook,* ASM International, Metals Park, OH, 1988, p. 200.
372. Dunkle, S.R., "Polysulfones," in *Engineering Plastics*, vol. 2, *Engineered Materials Handbook,* ASM International, Metals Park, OH, 1988, p. 200.
373. Berins, M.L., *Plastics Engineering Handbook of the Society of the Plastics Industry,* 5th ed., Chapman and Hall, New York, 1991, p. 71.
374. Berins, M.L., *Plastics Engineering Handbook of the Society of the Plastics Industry,* 5th ed., Chapman and Hall, New York, 1991, p. 71.
375. Dunkle, S.R., "Polysulfones," in *Engineering Plastics*, vol. 2, Engineered Materials Handbook, ASM International, Metals Park, OH, 1988, p. 200.
376. Dunkle, S.R., "Polysulfones," in *Engineering Plastics*, vol. 2, Engineered Materials Handbook, ASM International, Metals Park, OH, 1988, p. 201.
377. Berins, M.L., *Plastics Engineering Handbook of the Society of the Plastics Industry, Inc.,* 5th ed., Chapman and Hall, New York, 1991, p. 71.
378. Dunkle, S.R., "Polysulfones," in *Engineering Plastics*, vol. 2, Engineered Materials Handbook, ASM International, Metals Park, OH, 1988, p. 200.
379. Brydson, J.A., *Plastics Materials,* 6th ed., Butterworth-Heinemann, Oxford, 1995, p. 301.
380. Billmeyer, F.W., Jr., *Textbook of Polymer Science,* 2nd ed., John Wiley & Sons, New York, 1962, p. 420.
381. Brydson, J.A., *Plastics Materials,* 6th ed., Butterworth-Heinemann, Oxford, 1995, p. 304.
382. Brydson, J.A., *Plastics Materials,* 6th ed., Butterworth-Heinemann, Oxford, 1995, p. 307.

383. Brydson, J.A., *Plastics Materials,* 6th ed., Butterworth-Heinemann, Oxford, 1995, p. 302-304.
384. Strong, A.B., *Plastics: Materials and Processing,* Prentice-Hall, New Jersey, 1996, p. 171.
385. Strong, A.B., *Plastics: Materials and Processing,* Prentice-Hall, New Jersey, 1996, p. 170.
386. Billmeyer, F.W., Jr., *Textbook of Polymer Science,* 2nd ed., John Wiley & Sons, New York, 1962, p. 420.
387. Strong, A.B., *Plastics: Materials and Processing,* Prentice-Hall, New Jersey, 1996, p. 172.
388. Brydson, J.A., *Plastics Materials,* 6th ed., Butterworth-Heinemann, Oxford, 1995, p. 314-316.
389. Strong, A.B., *Plastics: Materials and Processing,* Prentice-Hall, New Jersey, 1996, p. 171.
390. Strong, A.B., *Plastics: Materials and Processing,* Prentice-Hall, New Jersey, 1996, p. 172.
391. Brydson, J.A., *Plastics Materials,* 6th ed., Butterworth-Heinemann, Oxford, 1995, p. 317-319.
392. Strong, A.B., *Plastics: Materials and Processing,* Prentice-Hall, New Jersey, 1996, p. 173.
393. Brydson, J.A., *Plastics Materials,* 6th ed., Butterworth-Heinemann, Oxford, 1995, p. 346.
394. Martello, G.A., "Chlorinated PVC," in *Modern Plastics Encyclopedia Handbook,* McGraw-Hill, New York, 1994, p. 71.
395. Brydson, J.A., *Plastics Materials,* 6th ed., Butterworth-Heinemann, Oxford, 1995, p. 341.
396. Strong, A.B., *Plastics: Materials and Processing,* Prentice-Hall, New Jersey, 1996, p. 173.
397. Hurter, D., "Dispersion PVC," in *Modern Plastics Encyclopedia Handbook,* McGraw-Hill, New York, 1994, p. 72.
398. Brydson, J.A., *Plastics Materials,* 6th ed., Butterworth-Heinemann, Oxford, 1995, p. 309.
399. Brydson, J.A., *Plastics Materials,* 6th ed., Butterworth-Heinemann, Oxford, 1995, p. 450.
400. Brydson, J.A., *Plastics Materials,* 6th ed., Butterworth-Heinemann, Oxford, 1995, p. 127.
401. I.W. Sommer, "Plasticizers," in *Plastics Additives,* 2nd ed., R. Gachter and H. Muller, Eds., Hanser Publishers, New York, 1987, p. 253-255.
402. W. Brotz, "Lubricants and Related Auxiliaries for Thermoplastic Materials," in *Plastics Additives,* 2nd ed., R. Gachter and H. Muller, Eds., Hanser Publishers, New York, 1987, p. 297.
403. Brydson, J.A., *Plastics Materials,* 6th ed., Butterworth-Heinemann, Oxford, 1995, p. 129.
404. Brydson, J.A., *Plastics Materials,* 6th ed., Butterworth-Heinemann, Oxford, 1995, p. 136.
405. Brydson, J.A., *Plastics Materials,* 6th ed., Butterworth-Heinemann, Oxford, 1995, p. 130-141.
406. Brydson, J.A., *Plastics Materials,* 6th ed., Butterworth-Heinemann, Oxford, 1995, p. 141-145.
407. Brydson, J.A., *Plastics Materials,* 6th ed., Butterworth-Heinemann, Oxford, 1995, p. 146-149.
408. Kroschwitz, J.I., *Concise Encyclopedia of Polymer Science and Engineering*, John Wiley and Sons, New York, 1990, p. 830-835.
409. *Encyclopedia of Polymer Science and Engineering,* 2nd ed., vol. 6, Mark, Bilkales, Overberger, Menges, Kroschwitz, Eds., Wiley Interscience, 1986, p. 433.
410. *Encyclopedia of Polymer Science and Engineering,* 2nd ed., vol. 16, Mark, Bilkales, Overberger, Menges, Kroschwitz, Eds., Wiley Interscience, 1986, p. 65.

CHAPTER 3
THERMOSETS

Rudolph D. Deanin

University of Massachusetts
Lowell, Massachusetts

Plastics are organic polymers that can be poured or squeezed into the shape we want and then solidified into a finished product. Thermoplastics are linear polymer molecules that soften or melt when heated and solidify again when cooled. This is a reversible physical process that can be repeated many times. Thus, it is a simple low-cost process that accounts for 85 percent of the plastics industry.

Thermosetting plastics are low-molecular-weight monomers and oligomers with multiple reactive functional groups, which can be poured, melted, or squeezed into the shape we want and then solidified again by chemical reactions forming multiple primary covalent bonds that cross-link them into three-dimensional molecules of almost infinite molecular weight. These are irreversible chemical processes that cannot be repeated. They account for 15 percent of the plastics industry, they include a great variety of chemical reactions and conversion processes, and they go into a very broad range of final products.

Thus, there is a great difference between thermoplastics and thermosets, both in terms of materials chemistry and applications, and in terms of the mechanical processes used to produce finished products.

3.1 MATERIALS AND APPLICATIONS

The major thermosetting plastics, in order of decreasing market volume, are polyurethanes, phenol-formaldehyde, urea-formaldehyde, and polyesters. More specialized thermosets include melamine-formaldehyde, furans, "vinyl esters," allyls, epoxy resins, silicones, and polyimides. While they may sometimes compete with each other and with thermoplastics, for the most part, each of them has unique properties and fills unique markets and applications.

3.1.1 Polyurethanes

With a U.S. market of 6 billion pounds per year, polyurethanes are the leading family of thermosetting plastics. Of the 100 or so families of commercial plastics, they are the most versatile, finding use in rigid plastics, flexible plastics, elastomers, rigid foams, flexible foams, fibers, coatings, and adhesives. They offer unique qualities in processability, strength, abrasion resistance, energy absorption, adhesion, recyclability, and resistance to oxygen, ozone, gasoline, and motor oil. Thus, they find major use in appliances, autos, building, furniture, industrial equipment, packaging, textiles, and many other fields.

Their versatility comes from the range of liquid monomers and oligomers that can be mixed, poured, polymerized, and cured in a minute or so at room temperature. Thus, we start with a look at their basic chemistry.

3.1.1.1 Polyurethane Chemistry (Figure 3.1)

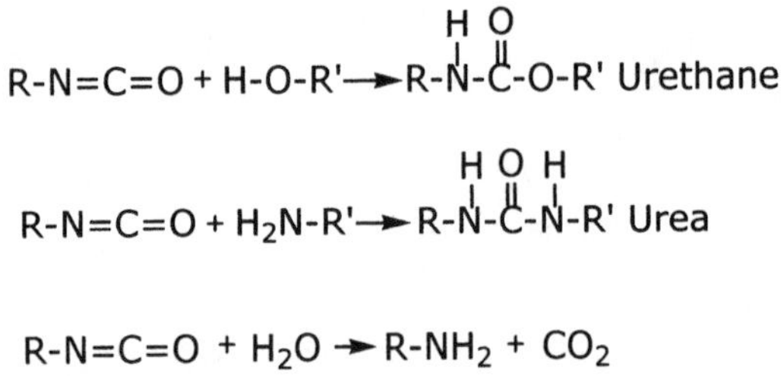

$$R\text{-}N\text{=}C\text{=}O + H\text{-}O\text{-}R' \rightarrow R\text{-}\overset{\overset{\displaystyle H}{|}}{N}\text{-}\overset{\overset{\displaystyle O}{\|}}{C}\text{-}O\text{-}R' \text{ Urethane}$$

$$R\text{-}N\text{=}C\text{=}O + H_2N\text{-}R' \rightarrow R\text{-}\overset{\overset{\displaystyle H}{|}}{N}\text{-}\overset{\overset{\displaystyle O}{\|}}{C}\text{-}\overset{\overset{\displaystyle H}{|}}{N}\text{-}R' \text{ Urea}$$

$$R\text{-}N\text{=}C\text{=}O + H_2O \rightarrow R\text{-}NH_2 + CO_2$$

FIGURE 3.1 Polyurethane chemistry.

Isocyanates and alcohols react readily to form urethanes. When the alcohols and isocyanates are multifunctional,

Polyols $R(OH)_n$

Polyisocyanates $R(NCO)_n$

they form polyurethane polymers. If they are difunctional, they form linear thermoplastic polyurethanes, which are useful in spandex fibers and thermoplastic elastomers. More often, they have higher functionality and form cross-linked thermoset polyurethanes. Most often, the polyols are trifunctional or higher, typically 3-6 OH groups. Less often, the polyisocyanates may be trifunctional or higher, typically 3-7 NCO groups. The liquid monomers are easy to mix, and the polymerization/cure reactions take a few minutes or less at room temperature. The combination of polarity, hydrogen bonding, and cross-linking in thermoset polyurethanes gives them high strength, adhesion, and chemical resistance.

Isocyanates react even more readily with amines to form ureas. So when the amines and isocyanates are multifunctional,

Polyamines $R(NH_2)_n$

Polyisocyanates $R(NCO)_n$

they form polyurea polymers. The urea groups give even stronger hydrogen bonding than the urethane groups, so they make the polymers even stronger. Many polyurethane processors use polyamines to speed the polymerization/cure reactions and to build greater strength into the finished polymer. Thus, many "polyurethanes" are actually urethane/urea copolymers, even though the manufacturers rarely mention the fact.

Isocyanates also react with water. The intermediate carbamic acid is so unstable that it decomposes immediately to form amine plus carbon dioxide. This reaction is important for two reasons: (1) carbon dioxide bubbles foam the polyurethane as it forms; this is the leading process for making foam, and (2) the amine by-product reacts to form more urea groups, which therefore strengthen the final polymer.

Isocyanates have several more reactions that are important in some more specialized applications (Fig. 3.2). Cyclotrimerization produces the isocyanurate ring, which is extremely stable, and can be used to build more heat resistance into polyurethanes. Excess isocyanate can react with the N-H group in polyurethanes to produce allophanate cross-links, which add to the cure of the polyurethane. And excess isocyanate can similarly react with the N-H groups in polyureas to produce biuret cross-links, which add to the cure of the polyurea.

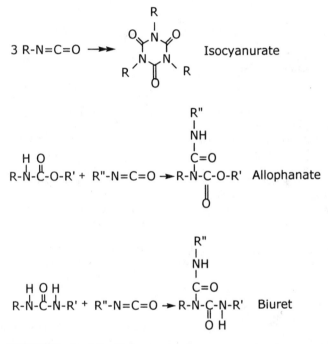

FIGURE 3.2 Specialized isocyanate reactions.

3.1.1.2 Raw Materials.

The versatility of polyurethanes is due to the variety of raw materials that can be used to build different structures into the polymers.

3.1.1.2.1 Isocyanates (Figure 3.3).

Toluene diisocyanate (TDI) is a mixture of mostly 2,4- plus some 2,6-isomer. Two commercial ratios are 80/20 and 65/35. The 4- position is more reactive; the 2- and 6- positions are sterically hindered. This gives the processor the ability to make prepolymers (oligomers) and run two-stage reactions.

Methylene diisocyanate (MDI) in the pure form gives a symmetrical structure that permits the processor to build some crystallinity, and thus greater strength, into the polymer.

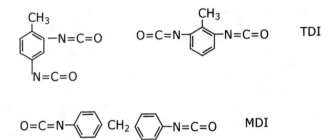

FIGURE 3.3   Isocyanates.

Polymeric MDI is a cruder mixture with 2-7 isocyanate groups, which offers lower cost and higher cross-linking for rigid products.

Hexamethylene diisocyanate (HDI) is completely aliphatic, which offers better UV stability against outdoor weathering. Because of its toxicity, it must be handled carefully in polymeric form.

Hydrogenated MDI (HMDI) is also completely aliphatic and therefore useful for UV stability against outdoor weathering.

A variety of other isocyanates are mentioned occasionally in the literature. The extent of their use is unclear.

3.1.1.2.2 Polyols (Figure 3.4).   Polyoxypropylene gives flexibility and water resistance. Since the secondary OH end group is slow to react with isocyanate, it is usually end-capped with ethylene oxide to give primary OH groups of higher reactivity.

Polyoxybutylene is more expensive but gives stronger rubbery products.

Polyesters such as poly(ethylene adipate) are more expensive and less stable toward hydrolysis but give stronger products.

These polyols build flexibility into the polymer molecule. For flexible foam and rubber, typically $n = 50$ to 60. For rigid products, n is a much lower value such as 8.

For cross-linking, there must be at least three OH groups in the polyol molecule. For flexible products, light cross-linking is introduced by a few glycerol or trimethylol propane units in the molecule. For rigid products, high cross-linking is introduced by higher polyols such as pentaerythritol or sorbitol.

Natural polyols such as castor oil are also used to some extent.

3.1.1.2.3 Catalysts (Figure 3.5).   Isocyanate + polyol reactions go quite rapidly at room temperature. Isocyanate + amine reactions go rapidly at room temperature. However, most processors add catalysts to make the polymerization/cure reactions even faster and to control the foaming process.

They generally use a combination of two synergistic catalysts: tertiary amine and organotin. Tertiary amines such as triethylene diamine promote the isocyanate-water reaction,

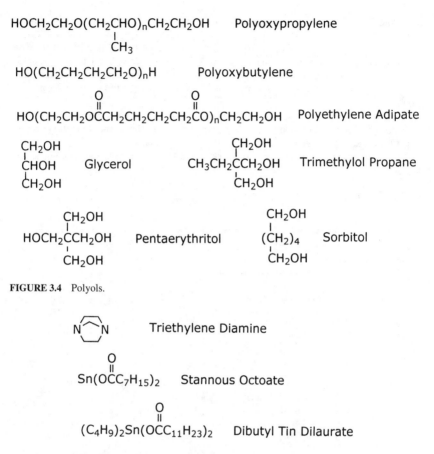

HOCH$_2$CH$_2$O(CH$_2$CHO)$_n$CH$_2$CH$_2$OH   Polyoxypropylene
              |
             CH$_3$

HO(CH$_2$CH$_2$CH$_2$CH$_2$O)$_n$H   Polyoxybutylene

HO(CH$_2$CH$_2$OCCH$_2$CH$_2$CH$_2$CH$_2$CO)$_n$CH$_2$CH$_2$OH   Polyethylene Adipate

CH$_2$OH
|
CHOH   Glycerol   CH$_3$CH$_2$CCH$_2$OH   Trimethylol Propane
|
CH$_2$OH

CH$_2$OH
|
HOCH$_2$CCH$_2$OH   Pentaerythritol   (CH$_2$)$_4$ Sorbitol
|
CH$_2$OH

FIGURE 3.4 Polyols.

Triethylene Diamine

Sn(OCC$_7$H$_{15}$)$_2$ Stannous Octoate

(C$_4$H$_9$)$_2$Sn(OCC$_{11}$H$_{23}$)$_2$ Dibutyl Tin Dilaurate

FIGURE 3.5 Polyurethane catalysts.

whereas organotin compounds such as stannous octoate or dibutyl tin dilaurate promote the isocyanate-polyol reaction. They balance these against each other to optimize the process.

For cyclotrimerization to isocyanurate, various tertiary amines, quaternary ammonium compounds, and other basic salts are mentioned in the literature.

3.1.1.2.4 Stoichiometry. Theoretically, the processor should use exactly equivalent amounts of isocyanate groups and active hydrogen groups (polyol ± amine) to favor high molecular weight. Practically, the processor varies the isocyanate/active hydrogen ratio (isocyanate index) to find the ratio that gives him the best properties. In most cases, the optimum isocyanate index is 1.05 to 1.10. There are two reasons for this: (1) ambient moisture wastes some isocyanate (see Fig. 3.1 above), and (2) excess isocyanate may give beneficial side-reactions (see Fig. 3.2 above).

3.1.1.2.5 One-Shot vs. Prepolymerization Reactions. If isocyanate and active hydrogen compounds can be mixed all at once, this "one-shot" process is simpler and more economical. In large-scale commodity production, this is usually the ultimate development.

The alternative is a two-stage process. In the first stage, polyol is mixed with excess isocyanate to form a low-molecular-weight polyurethane with isocyanate end-groups. In the second stage, the isocyanate end-groups are reacted with the stoichiometric amount of polyol to finish the polymerization reaction, or with water to link them into polyurea groups.

A more extreme two-stage process is called "quasi-prepolymer." Here, all the isocyanate is mixed with a small amount of polyol in the first stage. Then, the remaining polyol is added for the second-stage polymerization to high molecular weight.

These two-stage processes give the processor more control over the reaction and the product.

3.1.1.3 Polyurethane Products (Table 3.1)

TABLE 3.1   Polyurethane Markets

Material	%	%
Flexible foam		51
Furniture	18	
Transportation	13	
Rug underlay	11	
Bedding	5	
Other	4	
Rigid foam		26
Building insulation	14	
Home and commercial refrigeration	5	
Industrial insulation	2	
Packaging	2	
Transportation	1	
Other	2	
Reaction injection molding		6
Transportation	4	
Other	2	
Cast elastomers		2
Other (sealants, adhesives, coatings, etc.)		15
Total		100

3.1.1.3.1 Flexible Foam.   Compared to foam rubber, polyurethane is stronger and much more resistant to oxidative aging and embrittlement. Compressive stress-strain behavior can be matched to that of natural rubber, which established the preferred "feel" long ago. The largest amount of flexible foam is used for cushions in furniture, auto seating and crash-padding, rug underlay, and mattresses. Smaller amounts are used in shoe soles, winter clothing, and packaging.

Most flexible foam is manufactured by mixing 80/20 TDI with a high-molecular-weight polyether polyol, a small amount of triol for cross-linking, amine and organotin catalysts, polyol/silicone surfactant, and a measured amount of water in a one-step process, and it is then poured onto a moving belt. Foam rise takes about 1 min. Optimum soft properties depend on open (interconnecting) cells; these are produced by choice of surfactant, gas expansion while the molecular weight (melt strength) is still low, and mechanical crushing and re-expansion. This produces continuous slab stock, which is then cut into the desired individual products. About 70 percent of flexible foam is made in this way. The other 30 percent is poured into molds to make the finished products directly. This is used especially for auto and furniture seating. (See Table 3.2.)

TABLE 3.2   Flexible Polyurethane Foams: Typical Properties

Density	2 pcf
Modulus	1–10 psi
Tensile strength	28 psi
Elongation	300%
CLD/25%	0.7 psi

3.1.1.3.2 Rigid Foam.   Rigid foam is used primarily for thermal insulation. Whereas polystyrene foam must be molded and/or cut to shape before it can be used in finished products, liquid polyurethane ingredients are mixed, poured or sprayed in place, and polymerize/cure directly to the finished insulation. In addition, polyurethane foam has high adhesion to most surfaces in which it is used so, when it is poured into a sandwich structure and cured, it contributes to mechanical strength as well. Its largest use is in building, and the second largest in refrigeration. Other applications include pipes, tanks, trucks, railcars, packaging, and filling empty space in shipbuilding for flotation purposes.

Rigid foam is produced by mixing polymeric MDI with low-molecular-weight polyether polyol, high-functionality polyol for cross-linking, catalysts, and surfactants as above. Chlorofluorocarbons are technically the best foaming agents, but, because of their negative effect on the environment, they have been replaced by hydrocarbons or carbon dioxide.

Optimum insulation is achieved by low-density, small, closed cells. This foam structure is produced by choice of surfactants and by control of the temperature and the balance between rate of polymerization/cure (viscosity = melt strength) versus the rate of gas evolution. (See Table 3.3.)

TABLE 3.3   Rigid Polyurethane Foams: Typical Properties

Density	2 pcf
Flexural modulus	70 psi
Tensile strength	50 psi
Flexural strength	60 psi
Compressive strength	50 psi

Medium-density and semi-rigid foam is produced by polyols of medium molecular weight and medium functionality, and less foaming agent. These foams are used for crash padding and packaging.

3.1.1.3.3 Reaction Injection Molding (RIM). This high-speed low-cost process mixes liquid polyisocyanate and liquid polyol, injects them into a light-weight mold, and polymerizes/cures rapidly to form large, tough, durable products. It is used to make auto bumpers, front ends, and other auto parts; furniture and other imitation wood products; appliance cabinets; and shoe soles.

The process pumps liquid polyisocyanate, liquid polyol, and auxiliary ingredients including catalysts, foaming agents, and polyamine for faster cure, through an impingement mixer at 2000 to 3000 psi, at viscosity up to 3000 Cp, and injects them rapidly into a mold at 50 to 100 psi, where they polymerize/cure rapidly to structural foam or tough elastic products. (See Table 3.4.) For rubber tires on industrial equipment, the addition of glass fibers gives reinforced RIM (RRIM) with greater durability under rough conditions.

TABLE 3.4   RIM: Typical Properties

Density	60 pcf
Shore D hardness	60
Flexural modulus	25 kpsi
Tensile strength	6 kpsi
Elongation	250%

Elastomers.   Polyurethane elastomers are outstanding for their strength and for resistance to abrasion, oxygen, ozone, and gasoline. This combination of properties has proved particularly useful in shoe soles and heels, oil seals, industrial tires and wheels, chute linings, drive belts, shock absorption and vibration damping, medical products, and miscellaneous industrial applications.

They are made from long, flexible polyols with a light degree of cross-linking. They may be cast as liquids and polymerized/cured directly to solid rubber products, or they can be polymerized to linear, melt-processable rubber and then cross-linked by polyurethane chemistry or conventional rubber vulcanization chemistry. (More recently, they have also been produced as thermoplastic elastomers, in which hydrogen-bonding and/or crystallinity provide thermoplastic "cross-links," but that is another story.) This range of processability is attractive to both the thermoset plastics and rubber industries. Cast polyurethanes give the best properties (see Table 3.5).

3.1.1.3.4 Coatings.   Coatings based on polyurethanes can be applied from solution, from emulsion, or as self-curing liquid systems. The use of low-solvent or nonsolvent systems is a big help to the coatings industry in meeting the demand for better protection of the environment. In addition to simple polyurethane homopolymers, their cure reactions permit coatings technologists to copolymerize them with alkyds, epoxies, and other established coatings polymers to produce improved balance of properties.

Their adhesion, mechanical strength, flexibility, abrasion resistance, and chemical and aging resistance make them particularly useful in steel and industrial products for corrosion resistance, on wood for decoration and preservation of furniture and flooring, on ships for salt-water resistance, and on leather and textiles to upgrade their appearance and dura-

TABLE 3.5 Polyurethane Elastomers: Typical Properties

Property	Polyurethane	Natural rubber
Shore A hardness	81	71
300% modulus, psi	2000	2200
Tensile strength, psi	6300	3800
Elongation,%	600	440

bility. In aerospace, they offer resistance to rain erosion. And in cloth coating, the are superior to PVC for adhesion and freedom from plasticizers.

Adhesives and Sealants. Polyurethanes can be conveniently applied in liquid or paste form and then polymerized/cured in place without evolution of volatile by-products, a very convenient feature in making enclosed adhesive bonds. Their mechanical strength, flexibility, adhesion, and chemical resistance make them attractive in many applications. Typical applications of polyurethane sealants are in expansion joints, aerospace, architectural, electronic, and marine products.

3.1.2 Formaldehyde Copolymers

Formaldehyde reacts readily with several types of active-hydrogen monomers (phenol, urea, and melamine) to form highly cross-linked thermoset plastics. They form a family in their fundamental chemistry, and they form complementary families in terms of materials properties, markets, and practical applications.

3.1.2.1 Phenol-Formaldehyde. Phenol-formaldehyde resins were the first commercial synthetic plastics. Since their invention in 1908, they have grown and matured into the second most important family of thermoset plastics, with a U.S. market volume of 4 billion lb/yr (see Table 3.6).

TABLE 3.6 Phenolic Resin Markets

Market	%
Plywood	49
Adhesives and bonding	30
Laminates	6
Molding compounds	5
Protective coatings	1
Other	9

3.1.2.1.1 Chemistry (Figure 3.6). The phenolic hydroxyl group activates the ortho- and para-hydrogens. Formaldehyde adds readily to these positions, forming methylol

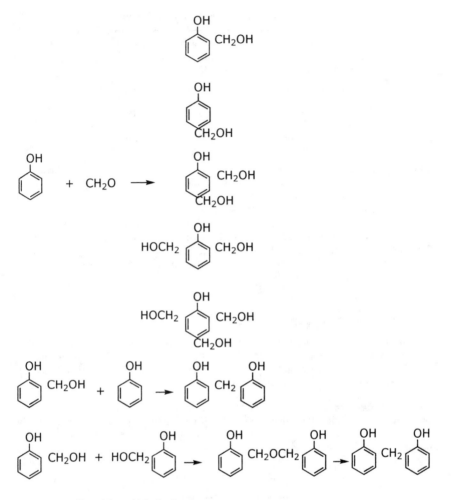

FIGURE 3.6 Phenol-formaldehyde chemistry.

groups. These are very reactive. They can condense with each other, with the ortho- or parahydrogens on other phenol molecules, or with active hydrogens in cellulose or other materials.

Since the condensation evolves volatiles and heat, it must be controlled to give useful products. The reaction is controlled by monomer ratio, pH, and temperature. It is generally run in several separate successive stages. First, it goes to low-molecular-weight "A-stage" resin, which is soluble and fusible. Then, it is compounded with fillers and additives and reacted further to moderate-molecular-weight, somewhat cross-linked "B-stage" resin, which is hard and less soluble but still fusible. Finally, the resin is formed into the shape of the desired product and thermally cured into fully cross-linked thermoset "C-stage" resin, which is rigid, insoluble, and infusible.

3.1.2.1.2 Resoles and Novolacs . There are two types of phenolic A-stage resins: resoles and novolacs (Fig. 3.7). Resoles have many methylol groups that make them water
soluble and highly reactive; novolacs are stable oligomers, which can be cross-linked by
adding more formaldehyde. Therefore, they are sometimes referred to as "one-step" and
"two-step" resins, respectively.

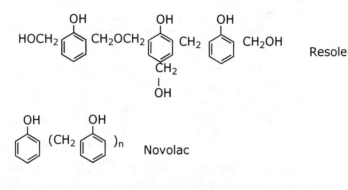

$C_6H_{12}N_4$ Hexamethylene Tetramine

FIGURE 3.7   Resoles and novolacs.

Resoles are typically prepared from 1.1 to 1.5 mols of aqueous formaldehyde + 1 mol
of phenol, with an alkaline catalyst, by heating 1 hr at 100°C and then cooling to stop the
reaction as an aqueous solution of A-stage resin. This is highly reactive, so shelf life is
usually less than 60 days. It is useful in laminating, bonding, and adhesive applications.
On heating, it is self-curing, giving off water and excess formaldehyde.

Novolacs are typically prepared from 0.8 mol of formaldehyde + 1 mol of phenol, with
(sulfuric or oxalic) acid catalyst, by refluxing 2 to 4 hr, up to 160°C to remove water of
condensation. The molten resin is poured into steel tubs or onto a concrete floor, cooled to
solidify, crushed to a powder, and blended with hexamethylene tetramine curing agent
(Fig. 3.8) for use in molding powder. This has almost infinite shelf life.

3.1.2.1.3 Adhesive and Bonding Applications.   Adhesives and bonding applications
make up 89 percent of the phenolic resin market.

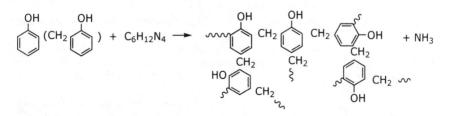

FIGURE 3.8   Novolac: cure by hexamethylene tetramine.

Plywood.   Wood is skived into thin layers of veneer. Paper is impregnated with aqueous resole resin. Alternating layers of wood and paper-phenolic are stacked to the desired thickness and pressed at 100 to 150°C and 700 to 6000 psi to make weatherproof exterior plywood for building, autos, boats, ships, trucks, and trains (Table 3.7). This uses 49 percent of the total phenolic resin market.

TABLE 3.7   Plywood: Typical Properties

Flexural modulus	1,450,0000 psi
Tensile strength	2,750 psi
Flexural strength	5,000 psi
Compressive strength	4,000 psi
Thermal expansion	$6 \times 10^{-6}/°C$

Particle board.   A mixture of 90 percent wood chips + 10 percent resole resin is pre-pressed at room temperature then hot pressed at 160 to 220°C and 290 to 590 psi. Cure is finished by hot-stacking in storage. Wafer board is made from larger chips. These boards are used for furniture core, floor underlay, prefab housing, freight cars, and ships.

Fiber board is made from wood filaments. Pressing at low pressure gives low-density boards for heat and sound insulation. Pressing at high pressure gives decorative and structural board.

These particle boards use 16 percent of the phenolic resin market.

Insulation materials.   Fiberglass wool insulation is bonded by spraying with 10 percent of aqueous resole and curing at 200°C. This is used for thermal insulation in housing and appliances. It is good up to 260°C. For higher-temperature industrial insulation—pipes, boilers, and reactors—mineral-based rock wool is used instead of glass wool; it is good up to 385°C.

Textile fiber mats are bonded by phenolic resin and used for sound insulation in autos, offices, auditoriums, and industrial plants.

These applications use 12 percent of the phenolic resin market.

"Laminates."   Kraft paper is impregnated with low-molecular-weight (300) phenolic resin, bonded with medium-molecular-weight phenolic resin, then cut and stacked to the desired thickness and pressed at 170 to 190°C and 200 psi, or wound onto a mandrel and cured to form a tube (Table 3.8). This uses 6 percent of the phenolic resin market. Such "high-pressure laminates" are used for furniture and counter tops (3 and 2 percent of the market, respectively), and electrical and mechanical applications (1 percent) such as printed circuit boards, switches, transformers, pulleys, bobbins, guide rolls for paper and textile machinery, gears, bearings, bushings, and gaskets.

Filters are made by impregnating paper with 20 to 30 percent of phenolic resin and curing in a 180°C oven. Battery separator plates are made the same way.

Synthetic fabric laminates are made by impregnating with phenolic resin and are used for helmets, aircraft interiors, and ablative nose cones for rockets. Recent research on such ablative nose cones showed that 28 percent loading with carbon nanofibers gave the lowest erosion rate at 2200°C.

Foundry moldings.   In the auto, construction, machine parts, and steel industries, molten metal is poured into sand molds to produce the shapes of the products. The sand is bonded by phenolic resin, cured at 270°C. In the *cold box* process, the binder is phenolic

TABLE 3.8 Laminated Phenolics: Typical Properties

Property	Kraft paper	Cotton fabric
Specific gravity	1.34	1.34
Tensile strength, psi	11,400	10,800
Compressive strength, psi	17,500	18,800
Impact strength, fpi	2.1	4.1

resin copolymerized with polyurethane, which cures simply at room temperature. This uses 3 percent of the phenolic resin market.

Friction materials. Brake linings and clutch facings use 2 percent of the phenolic resin market. The resin is compounded with rubber for toughness; mica, talc, and glass for friction; and powdered metal for thermal conductivity to prevent over-heating.

Abrasives. Grinding wheels (bonded abrasives) and sandpaper (coated abrasives) are made from abrasive grit bonded by phenolic resin. The abrasive grit may be alumina for cutting and polishing steel, or silicon carbide for handling glass, ceramics, and stone. This uses 1 percent of the phenolic resin market.

3.1.2.1.4 Molding Applications. Novolac resins are compounded with hexamethylene tetramine curing agent and about an equal volume of filler to produce thermosetting molding powders (Table 3.9). Wood flour is the most common filler; the short cellulose fibers are low cost, permit easy melt processing, and prevent cracking and brittleness. For higher strength, and especially impact strength, cotton flock, paper, fabric, cord, and especially glass fiber offer higher performance (Table 3.10), and fiber length is a major factor (Table 3.11). For maximum thermal, electrical, and chemical resistance, silica, clay, talc, mica, and glass are commonly used. In general, phenolic molding powders offer easy molding, low mold shrinkage, high modulus (1 to 3 million psi), superior creep resistance (Table 3.12), and good resistance to heat and chemical attack. Their main limitation is dark color, limited to dark brown to black; this may be overcome by copolymerization with melamine or soybean protein.

Compression molding is most common, because it minimizes fiber damage and warpage and gives high strength and dimensional stability. The molding powder is preheated by infrared or radio frequency, and moldings are pressed at 2 to 20 kpsi and 140 to 200°C. Transfer molding is better for thin walls and delicate inserts. Injection molding is faster, at 10 to 20 kpsi, with the melt at 104 to 116°C and the mold at 160 to 194°C. A newer method is *runnerless injection compression,* in which the melt is injected into a partially open mold (1/4 to 1/2 in), and then the mold is closed for compression; this is fast, easy venting, and gives less scrap and good dimensional stability.

Typical applications are appliances, closures, housewares, bottle caps, knobs, utensil handles, refrigerator switch boxes, sealed switches, steam irons, and sterilizable hospital equipment. High-impact grades are used for autos, industrial pulleys, electrical switch gear and switch blocks, fuse holdings, and motor housings. Electrical grades (high dielectric strength) are used for auto ignition, wiring devices, circuit breakers, commutators, brush holders, and electrical connectors. Heat-resistant grades are used for stove tops, toasters, thermostats, switch cases, terminal blocks, and many auto under-the-hood applications. This uses 5 percent of the phenolic resin market.

TABLE 3.9 Phenolic Moldings: Typical Properties

Property	Wood flour	Glass fiber
Specific gravity	1.33	1.85
Tensile strength, kpsi	7	13
Flexural strength, kpsi	11	38
Compressive strength, kpsi	28	48
Impact strength, fpi	0.4	9
Elongation, %	0.6	0.2
Heat deflection temp., °C	168	246
Thermal expansion, $10^{-6}/°C$	38	15
Linear mold shrinkage, %	0.7	0.3
Dissipation factor	0.17	0.055
Water absorption, %	0.9	0.6

TABLE 3.10 Fillers for High-Impact-Strength Phenolics

Filler	Notched Izod impact strength, fpi
None	0.25
Wood flour	0.3
Fabric	2.4
Cord	7.0

TABLE 3.11 Effect of Fiber Length on Phenolic Impact Strength

Fiber	Notched Izod impact strength, fpi
Ramie fiber, 1 in	1.0
Ramie fiber, 4 in	2.4
Sisal fiber, 1/8 to 1/4 in	3.9
Sisal fiber, 1–2 in	10.0
Tire cord, 1/2 in	10.1
Tire cord, 1 in	17.0

TABLE 3.12 Phenolics: Creep
Resistance, 200 psi, 23°C, 400 Hours

Phenolic	<0.1 %
Polycarbonate	>0.4%
Polyphenylene ether	>0.6%
Acetal	>1.4%

3.1.2.1.5 Coatings. Phenolic coatings are used on metals for heat and corrosion resistance and electrical insulation. They are good for continuous use at 145°C and short-term heat to 350°C. Often, they are blended with other coating polymers for combined properties. Typical applications are autos, heat exchangers, pipelines, boilers, reaction vessels, storage tanks, brine tanks, solvent containers, food containers, railroad cars, beer and wine tanks, beer cans, pail and drum linings, water cans, rotors, blower fans and ducts in HVAC, boats, ships, wood, and paper. These use 1 percent of the phenolic resin market.

3.1.2.1.6 Rubber Compounding. Specialty phenolic resins are used as processing aids, tackifiers, adhesives to fabric, and for reinforcement.

3.1.2.2 Urea-Formaldehyde. Urea-formaldehyde resins are one of the oldest families of commercial plastics; with a U.S. market volume of 3 billion lb/yr, they are the third largest thermosetting resin. Urea and melamine have similar polymer chemistry, so they are often discussed together as "amino resins;" but their markets and applications are quite different and are best studied separately.

3.1.2.2.1 Polymerization Chemistry. The amine groups of urea react very readily with formaldehyde, forming methylol ureas (Fig. 3.9). The A-stage reaction is controlled by the urea/formaldehyde ratio (1/1.3 to 1/2.2), an alkaline buffer at pH 7.5-8.0, and refluxing up to 8 hr, to produce a mixture of mono-, di-, and trimethylol ureas. These condense to form oligomers and finally, with acid catalysis and heat, highly cross-linked thermoset polymers.

For different applications, there are different U/F ratios and B-stage oligomers. They can be stabilized by hexamethylene tetramine to keep them alkaline, or they can be reversibly etherified with methanol or butanol to make them stable and soluble in organic solvents (Fig. 3.10). They may be compounded and processed in water or organic solution or as solid powders for different applications. For final cure, they are compounded with latent acid catalysts such as ammonium sulfamate, ammonium phenoxyacetate, ethylene sulfite, and trimethyl phosphate and generally heated to accelerate the cross-linking reaction.

3.1.2.2.2 Adhesion and Bonding. The dominating application of urea-formaldehyde resins (85 percent) is the bonding of fibrous and granulated wood for doors, furniture, and flooring. Typical process conditions are 24 hr at 200 psi and room temperature ("cold press"). Hot pressing may not need any catalyst. The resin penetrates the pores of the wood and bonds the particles together to form strong isotropic boards. Another 4 percent is used to make plywood. Since urea-formaldehyde is moisture sensitive, it is used only for indoor applications. (Phenolic resin, which is more expensive, must be used for outdoor applications.)

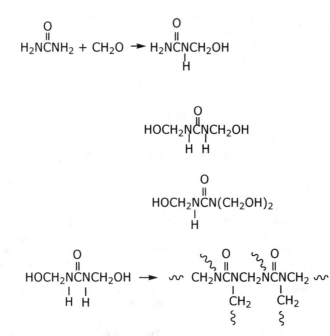

FIGURE 3.9 Urea-formaldehyde chemistry.

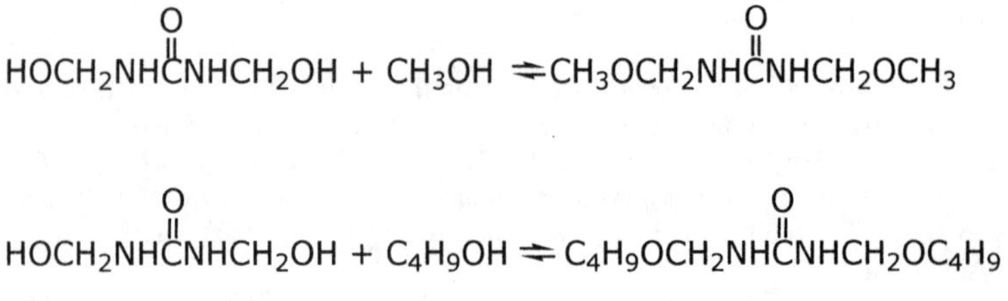

$$\underset{\substack{\|\\O}}{HOCH_2NHCNHCH_2OH} + CH_3OH \rightleftharpoons CH_3OCH_2NHCNHCH_2OCH_3$$

$$\underset{\substack{\|\\O}}{HOCH_2NHCNHCH_2OH} + C_4H_9OH \rightleftharpoons C_4H_9OCH_2NHCNHCH_2OC_4H_9$$

FIGURE 3.10 Etherification of methylol ureas.

3.1.2.2.3 Coatings. Urea-formaldehyde resins (5 percent) are used to treat paper to give it wet strength. They are also used (2 percent) to treat cotton and wool cloth to produce permanent press and increase strength, shrink resistance, and wrinkle resistance.

3.1.2.2.4 Molding Powders. Urea-formaldehyde resins are compounded with alpha-cellulose cotton fiber reinforcement to produce molding powders (4 percent) for compression, transfer, and injection molding. Typical molding conditions are 127 to 182°C and 2000 to 8000 psi. They are superior to phenolics in white color, electrical resistance, and low cost, but are limited by moisture sensitivity (Table 3.13). They are used primarily in electrical wiring devices such as wall outlets, receptacles, electric blanket controls, circuit breakers, and knob handles. Smaller amounts are used in bottle caps, housewares, buttons, and sanitary ware.

TABLE 3.13 Urea-Formaldehyde Moldings: Typical
Properties

Specific gravity	1.5
Tensile modulus	1,300,000 psi
Flexural modulus	1,450,000 psi
Tensile strength	8,250 psi
Flexural strength	13,000 psi
Impact strength	0.31 fpi
Thermal expansion	$29 \times 10^{-6}/°C$
Heat deflection temperature	133°C
Dielectric constant	6.8
Volume resistivity	10^{14} Ω-cm
Water absorption	0.6%

3.1.2.3 Melamine-Formaldehyde. Melamine-formaldehyde and urea-formaldehyde
have similar polymerization chemistry, so they are often referred to as "amino resins."
However, they differ in properties, applications, economics, and market volume, so they
are best studied independently. Melamine offers superior resistance to heat, weather, and
moisture, but it is more expensive than urea, so it is used only when its superior performance is required. The U.S. market volume is about 350 million lb/yr.

3.1.2.3.1 Polymerization Chemistry. Melamine has six amine hydrogens, all of
which can react readily with formaldehyde to produce methylol melamines (Fig. 3.11).
For different applications, the degree of methylolation is controlled by the melamine/
formaldehyde ratio, pH, temperature, and time. Polymerization reactions are buffered at
pH 8 to 10 by use of sodium carbonate or borax, and polymerization temperature 80 to
100°C. Lower pH and higher temperature produce faster reaction. Trimethylol melamine
is most common, but hexamethoxymethyl melamine (HMMM) is popular for coatings, because it is more stable and soluble in organic solvents (Fig. 3.12).

3.1.2.3.2 Coatings. The largest use of melamine-formaldehyde resins (79 percent) is
for cross-linking acrylic automotive coatings, polyester appliance coatings, and occasionally epoxy coatings as well. These polymers are designed with hydroxyl groups, and the

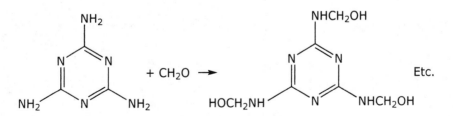

FIGURE 3.11 Melamine-formaldehyde chemistry.

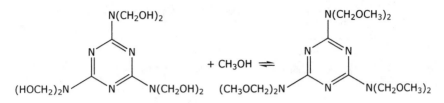

FIGURE 3.12 Hexamethoxymethyl melamine.

methylol melamine reacts with them (Fig. 3.13) to produce cross-linked thermoset coatings of excellent appearance and durability.

3.1.2.3.3 Laminates. While phenol-formaldehyde-kraft-paper laminates are used for counters, cabinets, walls, and panels in public transportation, the dark brown-black color requires a decorative overlay to make it attractive. Colored and printed paper is impregnated with melamine-formaldehyde resin and applied as the surface layer to these laminates, providing decoration along with resistance to scratches, heat, ultraviolet, water, solvents, and stains. The laminating resin is made with melamine/formaldehyde ratios of 1/2 to 3 and press-cured at 125 to 150°C. This uses 14 percent of the melamine-formaldehyde market.

3.1.2.3.4 Moldings. Melamine-formaldehyde resin (M/F = 1/2) is reinforced with alpha-cellulose cotton fiber, catalyzed with phthalic anhydride, and molded at 145 to 165°C and 4000 to 8000 psi (Table 3.14). Moldings have the highest hardness and scratch resis-

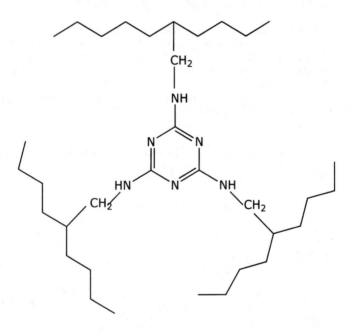

FIGURE 3.13 Melamine-formaldehyde cure of coatings.

TABLE 3.14   Melamine-Formaldehyde Moldings: Typical
Properties

Specific gravity	1.5
Rockwell hardness	M120
Tensile modulus	1,300,000 psi
Flexural modulus	1,100,000 psi
Tensile strength	10,000 psi
Flexural strength	15,000 psi
Compressive strength	39,000 psi
Notched Izod impact strength	0.3 fpi
Mold shrinkage	0.9%
Coefficient of thermal expansion	$43 \times 10^{-6}/^{\circ}C$
Heat deflection temperature	$185^{\circ}C$
Volume resistivity	10^{11} Ω-cm
Dielectric constant	7
Water absorption	0.5

tance of any plastic, along with resistance to heat and staining. Their major use is in household dinnerware (plates, bowls, cups, and glasses), where they are much lighter and more impact resistant than china. Market volume of 7 percent also includes a number of minor items such as buttons, handles, knobs, small appliances, sinks, and toilets.

3.1.2.4 Furan Resins.   Furfuryl alcohol is an agricultural by-product, which is polymerized and cured by acid catalysts, producing a very hard plastic that is very resistant to heat, flame (low smoke), water, and chemical attack (Table 3.15). Limitations are black color and brittleness.

   Initial polymerization is catalyzed by acid, and very exothermic, so it must be cooled and neutralized to produce liquid dimers and trimers (Fig. 3.14). These are then cured by strong acid such as 4 percent of p-toluene sulfonic acid. They may also be copolymerized with phenol-formaldehyde.

   Reported uses include sand molds for metal foundries and linings for chemical plant equipment such as reaction vessels, tanks, pipes, fume ducts, and sewers.

3.1.3   Vinyl Polymers

Several families of thermosetting plastics are based on cross-linking through the C=C vinyl group. The earliest, of course, were the vegetable oils used in paint for thousands of years, but their cure by atmospheric oxygen was limited to thin coatings and too slow to be useful in plastics. The leading family is the unsaturated polyesters, which form the basis of most reinforced plastics. More specialized families are the so-called "vinyl esters" and the

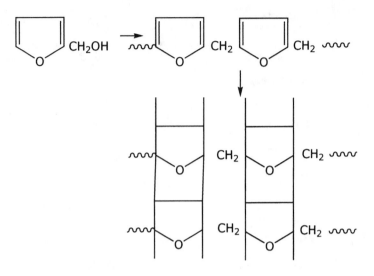

FIGURE 3.14   Furan polymerization.

TABLE 3.15   Reinforced Furan Properties

Specific gravity	1.75
Rockwell hardness	R110
Tensile modulus	1,580,000 psi
Tensile strength	3,800 psi
Flexural strength	4,800 psi
Compressive strength	11,500 psi
Water absorption	0.1%

allyl resins. They are all cross-linked (= cured) by organic peroxides, which initiate polymerization of the vinyl group.

3.1.3.1 Unsaturated Polyesters. Unsaturated polyesters are the fourth largest family of thermosetting plastics, with a U.S. market volume of 2 billion lb/yr. They are often called *thermosetting polyesters* or *alkyds*. In commercial use for 60 yr and now fairly mature, they are the largest class of reinforced plastics (Table 3.16), popularly used in building panels, chemical equipment, boats, cars, buses, trains, and planes.

3.1.3.1.1 Chemistry. Their chemistry is a fairly complex two-stage process. Typically, in the first stage, propylene glycol is mixed with maleic anhydride and phthalic anhydride (Table 3.17), and cooked 8 to 28 hr at 204 to 232°C to produce a molten prepolymer of Mn = 800 to 3000 (Fig. 3.15). This is mixed with styrene monomer to pro-

TABLE 3.16 Unsaturated Polyester Markets

Market	%
Building and construction	26
Corrosion-resistant equipment	20
Shipbuilding and marine	17
Automotive and railroad	13
Consumer products in general	8
Appliances and business machines	6
Electrical	4
Aircraft and aerospace	2
Other	4
Total	100

TABLE 3.17 Polyester Typical Recipe

Material	Mols	Pounds/pound of resin
Propylene glycol	2.7	0.2564
Maleic anhydride	1.0	0.1225
Phthalic anhydride	1.5	0.2774
Styrene		0.4000
Hydroquinone		0.0001

duce a viscous liquid (50 to 6000 cP), stabilized by hydroquinone. In the second stage, it is "catalyzed" by organic peroxide ± activators, combined with glass fiber reinforcement, shaped by a variety of mechanical processes, and cross-linked to produce the finished product.

In greater detail, during the first-stage reaction, cis-maleic ester isomerizes into trans-fumaric ester, which luckily is 40× more reactive in the second-stage cross-linking. Propylene glycol may be replaced by neopentyl glycol, trimethylpentane diol, propoxylated or hydrogenated bisphenol A to increase water and chemical resistance. Maleic/phthalic ratio may be increased to increase cure rate, hardness, and heat deflection temperature. Phthalic anhydride may be replaced by isophthalic acid to improve toughness, heat deflection temperature, and water and chemical resistance; or by tetrabromo- or tetrachloro-phthalic anhydride or chlorendic acid to increase flame retardance. And styrene may be replaced by vinyl toluene or diallyl phthalate to reduce volatility; or by triallyl cyanurate or isocyanurate to increase heat deflection temperature.

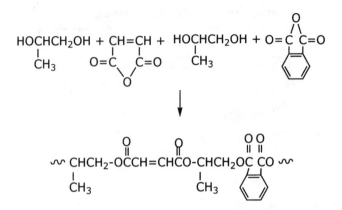

FIGURE 3.15 Unsaturated polyester chemistry.

The second-stage cross-linking (cure) reaction is initiated by organic peroxides: MEK peroxide for room-temperature cure, and benzoyl peroxide or t-butyl perbenzoate or other stabler peroxides for higher-temperature cure processes. Peroxide action may be speeded by heat and/or activators such as cobalt soaps and tertiary amines. (Nonchemists are apt to use the terms "catalyst" and "activator" rather loosely, which can be confusing or even dangerous in practice.)

3.1.3.1.2 Additives.   The most important additive is, of course, the glass fiber reinforcement, which increases modulus, strength, and impact strength (Table 3.18). In general, processes that use longer glass fiber give superior properties (Table 3.19).

TABLE 3.18   Polyester Reinforcement by Glass Fiber

1/4 in. glass fiber	Flexural modulus, kpsi	Flexural strength, psi	Notched Izod impact strength, fpi
0	550		0.3
10	1530	9,800	3.7
20	1640	16,000	6.1
30	1660	19,600	7.4
40	1720	22,300	10.7

The other, almost universal, additive is inorganic powdered fillers, used to increase viscosity, hardness, modulus, thermal conductivity, heat deflection temperature, opacity, and UV resistance, and to decrease exotherm, cure shrinkage, coefficient of thermal expansion, and cost. Calcium carbonate is the least expensive and most widely used. Clay gives higher electrical and chemical resistance. Talc gives high viscosity for gel coats and auto body repair. Alumina trihydrate gives flame retardance.

TABLE 3.19 Polyester Properties Increase with Glass Fiber Length

Process	BMC	SMC	Layup	Filament winding
Fiber length	1/4 in	1 in	Woven fabric	Continuous filament
Flexural modulus	1750	1600	2000	6000 kpsi
Flexural strength	16	26	60	175 kpsi
Impact strength	7	15	18	50 fpi

Viscosity must be controlled for most processes. For fluid processes like spraying and impregnation, it can be decreased by using lower molecular weight or higher styrene content. For many processes, it is increased by adding 0.1 to 2.0 percent of thixotropes such as silica, clay, and polyols. For leather-like tack-free BMC and SMC compounds, the polyesters are made with acid end-groups and then reacted with CaO or MgO to link them into higher-MW organometallic oligomers.

Profile is a problem when cure and shrinkage of the polymer matrix leave glass fibers at the surface, giving a rough profile. This is reduced by dissolving thermoplastic polymers such as polyvinyl acetate in the liquid system; since it does not react, it reduces the overall shrinkage of the system and thus retains a smoother profile.

Ultraviolet stability for outdoor use can be improved by opaque pigments that reflect the UV light before it can penetrate the polymer; titanium dioxide and aluminum flake are frequently used for this purpose. Ultraviolet absorbers (UVAs) such as hydroxybenzotriazoles and hydroxybenzophenones are sometimes used, and hindered amine light stabilizers (HALS) are becoming more popular.

3.1.3.1.3 Processes. Unsaturated polyesters are usually reinforced by glass fibers. This complicates conventional plastic processing, and has led to a great variety of specialized processes.

Viscosity. Different processes require different viscosity. This is most easily lowered by adding styrene monomer to the polyester oligomer (Table 3.20). It is raised by increasing polyester molecular weight and by adding thixotropic fillers. In addition, it is raised to the point of gelation by adding group II metal oxides to react with the acid end groups of the polyester, thus dramatically increasing molecular weight.

TABLE 3.20 Polyester Viscosity Is Controlled by Styrene Content

Styrene monomer	Viscosity, cP
25	5500
30	2000
35	550
40	200
43	100

Casting.   Unreinforced polyester is poured into an open mold, typically silicone rub-ber, and cured to produce bathroom sinks, counters, tubs, showers, and toilets; giftware, art objects, and "cultured marble and onyx" (Table 3.21). Appearance is controlled by choice of fillers.

TABLE 3.21   Polyester Casting Formulas

Ingredients	Cultured marble	Cultured onyx
Polyester	100	100
30-mesh $CaCO_3$	200	
80-mesh $CaCO_3$	100	
Alumina trihydrate		200
MEK peroxide	0.6	1.5
Colorants	"To suit"	"To suit"

Cast monolithic flooring is seamless, resistant to wear and chemicals, and easy to clean. "Polymer concrete" is polyester filled with aggregate.

Hand layup.   The oldest method is a purely manual operation. (1) The open female mold is first treated with mold release. (2) Optionally, a pigmented gel coat is applied 15 to 20 mils thick and partly cured. (3) Glass fiber mat, or woven or knitted cloth, is hand-laid into the mold or onto the gel coat. (4) Thixotropic polyester/styrene liquid, containing MEK peroxide and activator, is impregnated into the cloth. (5) The assembly stands and cures at room temperature. This gives a product with one good surface and somewhat ir-regular thickness.

Spray layup.   The layup process is partly mechanized by spraying. (1) Mold release is applied first. (2) Gel coat is sprayed onto the mold surface and partly cured. (3) Glass rov-ing and catalyzed liquid resin are fed through a gun, which chops the roving, mixes it with resin, and sprays the mixture into the mold.

Improvements on hand and spray layup.   A better gel coat is an acrylic sheet, which is vacuum formed to fit into the mold. This is used to make tubs, showers, spas, and toilets.

Styrene monomer emissions sometimes cause occupational and environmental health concerns. Evaporation of styrene can be reduced by adding wax to the formulation; it is immiscible and comes to the surface, forming a barrier layer. Another method of reducing volatilization is to use a less volatile monomer such as methyl styrene.

Vacuum and pressure bag molding.   A plastic film can be used to cover the layup. If a vacuum is used to pull it down onto the molding, this prevents styrene evaporation and air inhibition of the cure reaction, and it helps to compress the impregnated fiber and elim-inate empty spots. If air pressure is further applied above the film, this increases the per-formance still more.

Resin transfer molding (RTM).   This uses a closed mold. (1) Gel coat is first applied to one or both mold halves. (2) The mold is filled with reinforcing fibers and clamped shut. (3) Low-viscosity catalyzed liquid resin is pumped through a tube into the mold, impreg-nating the fibers, until the excess resin comes out of a vent in the top of the mold. Resin "transfer" can be assisted by attaching a vacuum line to the vent. (4) Cure is usually at room temperature. A low-exotherm resin is usually preferred.

Matched die molding. This uses matched male and female dies in a hydraulic press. The material may be introduced in several ways. (1) Glass fiber mat is impregnated with catalyzed resin and then pressed; this only permits small simple shaping such as cafeteria trays. (2) Glass fiber perform is made by spraying chopped fibers onto a screen prototype of the desired product then placing the impregnated perform in the mold and pressing it; this can make anything from small products up to boat hulls and auto bodies. (3) Premix is doughy molding compound of polyester, catalyst, short fibers, and fillers; this permits more versatile shaping.

Mat and perform use long glass fibers, which give better properties; premix uses shorter fibers that give easier processing. Concentrations are typically 35 to 40 glass fiber + 30 to 65 $CaCO_3$ filler.

High-temperature initiators are typically benzoyl peroxide, t-butyl peroctoate, or t-butyl perbenzoate. Molding pressures are 100 to 300 psi. Molding temperatures are 107 to 121°C for mat and perform, and 135 to 149°C for premix. Cure times are 2 min or longer.

Bulk molding compound (BMC). Polyester and low-profile resin, organic peroxide initiator, 0.25-in glass fiber, calcium carbonate filler, mold release agent, and alkaline thickener are blended into a doughy mass (Table 3.22). This is compression molded and cured. High flow permits complicated features such as ribs, bosses, and inserts. Major applications are electrical parts, dinnerware, small tools and appliances. Bulk molding compound can also be injection molded, but this requires more skill and care than simple thermoplastic injection molding.

TABLE 3.22 Bulk Molding Compound Formulation

Polyester	32.9%
Dicumyl peroxide	0.8%
Glass fiber	14.5%
$CaCO_3$	49.3%
Zn stearate	0.8%
MgO	1.7%

Sheet molding compound (SMC). (1) Polyester resin, low-profile resin, organic peroxide initiator, filler, mold release agent, and alkaline thickener form a paste (Table 3.23), which is spread onto a moving continuous carrier film. (2) Glass fiber roving is chopped (e.g. 1 in long) and spread over the paste. (3) Another layer of paste, with a top carrier film, is laid on top of the glass fiber. (4) The entire sandwich, encased between the two carrier films, is passed between pairs of rolls that knead and squeeze it to complete impregnation of the paste into the chopped fiber. (5) The sandwich is rolled up and stored until it is used. Shelf life is limited, so it is best made in house where it can be used more promptly.

Compression molding temperature is typically 135 to 149°C. Major uses are automotive body parts, business machine housings, and seating.

Pultrusion. Continuous reinforcement (fiber or fabric) is pulled continuously through liquid catalyzed resin to impregnate it and then through heated dies to shape and cure it. Properties are excellent in the machine direction but limited in the transverse direction.

Filament winding. The highest modulus and strength of any plastic material are achieved by filament winding. Continuous roving is pulled continuously through liquid

TABLE 3.23 Sheet Molding Compound Formulation

Polyester	26.2%
Low-profile resin	4.6%
t-butyl perbenzoate	0.3%
$CaCO_3$	6.2%
Zn stearate	0.9%
MgO	1.8%
Glass fiber	60.0%

resin and then wound onto a mandrel to form the shape of the desired product. The angle of winding can be calculated to maximize properties. It is then oven cured, and the mandrel is removed. This is obviously limited to hollow products of fairly simple shape, but mechanical properties can rival metals. Typical uses are tanks, pipes, boat hulls, and modular housing.

Market importance of different processes. The major processes for unsaturated polyesters may be ranked in descending order as follows:

- Sprayup
- Sheet molding compound
- Continuous laminating
- Filament winding
- Hand layup
- Bulk molding compound
- Pultrusion
- Injection molding

3.1.3.1.4 Properties. Properties of cured reinforced polyesters result from the combined effects of (1) the process technique and (2) the type of formulation used in each process (Table 3.24). Cast polyester lacks the benefits of fibrous reinforcement. Sprayup is easy but uses short fiber and achieves limited compaction. Bulk molding premix uses short fiber and high filler loading. Sheet molding compound uses longer fiber and less filler. Preform uses fairly long fiber and does not suffer shear degradation during molding. Hand layup can benefit from use of woven fabric, which contributes to higher reinforcement. Pultrusion achieves very high reinforcement in the machine direction. And filament winding packs the maximum concentration of reinforcing fiber, *and* orients it to maximize its reinforcing effect.

3.1.3.2 Vinyl Esters. The so-called "vinyl ester resins" are analogous to polyesters. They are more expensive but more resistant to hydrolysis, so they are popular for chemical-resistant equipment (e.g., tanks, pipes, ducts, scrubbers, and towers) because they are less expensive and more durable than stainless steel. Some other advantages are faster cure, higher elongation and impact strength, adhesion to glass fiber reinforcement, and resistance to heat aging.

TABLE 3.24 Reinforced Polyester Properties

Process	Flexural modulus, kpsi	Flexural strength, kpsi	Notched Izod impact strength fpi	Heat deflection temperature, °C
Cast	550	16	0.3	132
Sprayup	1000	20	10	>177
Premix/BMC	1650	16	8	>205
SMC	1600	26	15	225
Preform	1900	29	14	>205
Layup	2250	54	18	>205
Pultrusion	4500	138		>177
Filament wound	6000	175	50	>177

3.1.3.2.1 Chemistry. They are made by reaction of methacrylic acid with epoxy resins (Fig. 3.16). The reaction is catalyzed by benzyl trimethyl ammonium chloride, or oxonium or phosphonium salts (Table 3.25). Like polyesters, they are dissolved in liquid styrene monomer and stabilized by hydroquinone. And, like polyesters, they are cured by organic peroxides ± activators, at room temperature to 150°C.

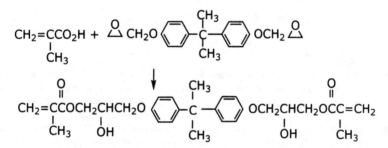

FIGURE 3.16 Vinyl ester chemistry.

3.1.3.2.2 Properties. Vinyl esters cure more easily than polyesters, because the acrylic C=C group in vinyl esters is much more reactive than the fumaric C=C group in polyesters. Vinyl esters have lower modulus, strength, and heat deflection temperature, and higher elongation and impact strength, because the bisphenol/propylene ether blocks in vinyl esters put a longer chain between cross-links, giving more molecular flexibility (Table 3.26). Vinyl esters have more adhesion to glass fiber reinforcement, because their –OH groups hydrogen-bond to the silanol surface of glass fibers. And most important, vinyl esters are more resistant to hydrolysis because (1) their ester groups are sterically hindered by the alpha-methyl groups, and (2) their polymer backbone has more C-C bonds and less ester bonds.

3.1.3.2.3 Variations. The basic vinyl ester can be modified in various ways to improve specific properties. Increasing the length of the bisphenol epoxy chain increases mo-

TABLE 3.25   Vinyl Ester Formulations

Ingredients	Standard	Acid-modified	Rubber-modified
Epoxy resin	1032	1032	609
Methacrylic acid	195	171	76
Maleic acid		32	
CTBN			228
Styrene	970	970	
Hydroquinone	0.45	0.45	0.17

TABLE 3.26   Vinyl Ester Properties

	Cast	Reinforced
Tensile modulus, kpsi	460	1590
Tensile strength, kpsi	11	24
Ultimate elongation, %	6	1
Impact strength, fpi	0.4	28
Heat deflection temperatures, °C	102	260

lecular flexibility, increasing elongation and impact strength, at the expense of modulus, strength, and heat deflection temperature. Conversely, novolac epoxy gives much higher cross-linking, increasing modulus, strength, and heat deflection temperature (148°C) at the expense of elongation and impact strength. Acrylic acid is used instead of methacrylic acid to produce vinyl esters for UV-cured coatings. Tetrabromobisphenol A builds flame-retardance into the polymer. Use of some maleic acid in place of methacrylic acid builds some acid groups onto the ends of the vinyl ester molecule; this permits MgO gelation for sheet molding compound. Use of some carboxy-terminated butadiene-nitrile oligomer (CTBN) in place of methacrylic acid builds nitrile rubber structure into the polymer, increasing impact strength. And the –OH groups of the epoxy resin can be cross-linked by diisocyanate to build some polyurethane structure and properties into the cured polymer. A number of these variations are available commercially.

3.1.3.2.4 Processing.   Reinforcement, shaping, and cure of vinyl esters is quite similar to processes described earlier for polyesters (Sec. 3.1.3.1.3).

3.1.3.3 Allyls.   The allyl group CH_2=CH-CH_2- is less reactive than conventional vinyl monomers CH_2=CH-X, which offers both advantages and disadvantages. Several allyl monomers have found unique applications in plastics: diallyl phthalate, triallyl cyanurate, and diethylene glycol bis(allyl carbonate).

3.1.3.3.1 Diallyl phthalate.   Glass-fiber-reinforced diallyl phthalate (DAP) is superior to unsaturated polyesters in shelf life and especially resistance to heat and moisture aging, leading to electrical and electronic insulation applications such as connectors for communications equipment, computers, aerospace, potentiometers, circuit boards, potting vessels, trim pots, coil forms, switches, and TV.

The liquid monomer (Fig. 3.17) is stable at room temperature, even when "catalyzed" by peroxides. Conversely, it requires high-temperature peroxides, such as t-butyl perbenzoate and dicumyl peroxide, higher temperatures (135 to 177°C), and longer times (0.5 to 4.0 min) to polymerize and cure. It is prepolymerized to solid oligomers and compounded with fillers and reinforcements to make molding powders. Cured moldings are best with glass fiber reinforcement (Table 3.27), and long glass fibers give superior strength and impact strength. They are superior to unsaturated polyester primarily in shelf life and resistance to hydrolysis and heat aging. Their limitations are cost of monomer and slow cure reactions.

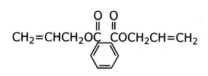

FIGURE 3.17 Diallyl phthalate.

TABLE 3.27 Diallyl Phthalate Molded Properties

Glass fiber reinforcement	Short	Long
Mold shrinkage, %	0.3	0.2
Flexural modulus, kpsi	1200	1300
Flexural strength, kpsi	12	16
Tensile strength, kpsi	7	10
Compressive strength, kpsi	25	25
Notched Izod impact strength, fpi	0.6	6.0
Heat deflection temperature, °C	204	200
Continuous heat resistance, °C	191	191
Dielectric constant	4.4	4.2
Dissipation factor	0.007	0.006
Water absorption, %	0.2	0.25

In addition to the conventional diallyl *ortho*-phthalate, diallyl *iso*-phthalate (DIAP) is also available commercially. It is more expensive but offers higher heat deflection temperature and heat aging resistance (Table 3.28).

Another use of diallyl phthalate monomer is the replacement of styrene monomer in unsaturated polyesters. DAP is superior to styrene in lower volatility and longer shelf life but is limited by higher cost and slower cure.

3.1.3.3.2 Triallyl cyanurate. The triazine ring (Fig. 3.18) is stabilized by heterocyclic resonance, giving it high heat resistance. This monomer can be used in place of styrene in unsaturated polyesters. It is less volatile than styrene. Being less reactive, it gives

TABLE 3.28   Advantages of Diallyl Iso-Phthalate

	DAP	DIAP
Heat deflection temperature, °C	204	260+
Continuous heat resistance, °C	191	232

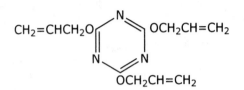

FIGURE 3.18   Triallyl cyanurate.

better shelf life but requires higher temperatures and times for the cure reaction. In the cured polyester, it gives higher heat resistance. Its use is limited by its higher price.

3.1.3.3.3 Diethylene glycol bis(allyl carbonate).   This liquid monomer can be polymerized by peroxide and heat (Fig. 3.19). When carefully cast, polymerized, and cured, it gives transparent, colorless castings of high refractive index, high hardness, and therefore scratch resistance. It is used for spectacle lenses.

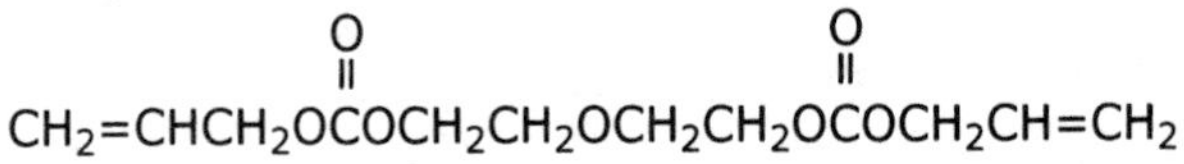

$$\underset{\text{CH}_2=\text{CHCH}_2\text{OCOCH}_2\text{CH}_2\text{OCH}_2\text{CH}_2\text{OCOCH}_2\text{CH}=\text{CH}_2}{\overset{\displaystyle \text{O} \qquad\qquad\qquad\qquad\quad \text{O}}{\overset{\displaystyle \|\qquad\qquad\qquad\qquad\quad \|}{}}}$$

FIGURE 3.19   Diethylene glycol bis(allyl carbonate).

3.1.4   Epoxy Resins

Epoxy resins enjoy a combination of fast, easy cure, high adhesion to many surfaces, and heat and chemical resistance, which leads to a U.S. market of 600 million lb/yr with a wide range of uses in plastics, coatings, and adhesives.

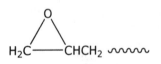

FIGURE 3.20   Epoxy or oxirane group.

The name "epoxy resins" is applied loosely both to epoxy monomers and prepolymers, and also to the cured thermoset final products. To be more precise, the epoxy (or oxirane) group (Fig. 3.20) in the monomer or prepolymer is reacted with a comonomer (curing agent or hardener) to form the cross-linked thermoset final product. The bond angles in the triangular epoxy ring are much smaller than the normal C-C and C-O bond angles, so the epoxy ring is strained and therefore very reactive, which accounts for the fast, easy cross-linking cure reactions. Furthermore, since there is no change in the number of bonds during the cure reaction, there is very little shrinkage, which permits better dimensional control than in other polymerization and cure reactions. Choice of a range of epoxy monomers and curing agents, as well as additives, leads to a wide range of final properties for different applications.

3.1.4.1 Monomers and Prepolymers.

The leading type of epoxy resin is made by reaction of bisphenol A with epichlorohydrin (Fig. 3.21). This can produce either the basic diglycidyl ether of bisphenol A (DGEBPA), or higher oligomers ($n = 1$ through 10) by increasing the BPA/ECH ratio and alkalinity (Table 3.29), producing a range from fluid liquids to soluble fusible solids (Table 3.30).

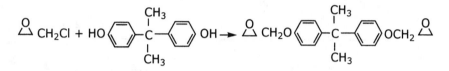

FIGURE 3.21 Reaction of bisphenol A with epichlorohydrin.

TABLE 3.29 Diglycidyl Ethers of Bisphenol A: Theory

n	Molecular weight	Epoxy equivalent weight
0	340	170
1	624	312
2	908	454
3	1192	596
4	1476	738
5	1760	880
6	2044	1022
7	2328	1164
8	2612	1306
9	2896	1448
10	3180	1590

A second type of epoxy resin is made by reaction of phenol-formaldehyde novolacs with epichlorohydrin (Fig. 3.22). Using novolacs of DP 2-6 gives solid resins and permits much higher cross-linking, giving cured products of higher heat and chemical resistance.

A third type of epoxy resin is cycloaliphatic (Fig. 3.23). These are harder to cure but offer better electrical resistance and resistance to sunlight.

A great variety of other epoxy monomers have been suggested for specialized uses (Fig. 3.24).

TABLE 3.30 Diglycidyl Ethers of Bisphenol A: Commercial

Molecular weight	Epoxy equivalent weight	Viscosity, cP	Softening point, °C
356	178	5,750	
378	189	13,000	
388	194	19,750	
980	490		70
1060	530		80
3984	1992		124

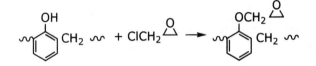

FIGURE 3.22 Reaction of novolac with epichlorohydrin.

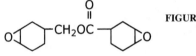

FIGURE 3.23 Cycloaliphatic epoxy resin.

3.1.4.2 Curing Agents. The epoxy ring is so strained that it opens and reacts to polymerize and cross-link very readily. It can react with a variety of basic and acidic reagents. Some of them catalyze the polymerization reaction (Fig. 3.25). Most of them are actually comonomers which then form the cross-links between the epoxy units. The epoxy resin/ curing agent ratio can be precalculated stoichiometrically but must still be adjusted experimentally to give the best balance of properties.

3.1.4.2.1 Amines. Tertiary amines R_3N are catalysts that open the epoxy ring and thus catalyze the polymerization reaction. They may be used with hydroxyl-containing molecules to catalyze homopolymerization (Fig. 3.26), but more often they are used to catalyze copolymerization of epoxy resins with amine or acid curing agents. Several more specialized amines are also mentioned as catalysts (Fig. 3.27).

Primary and secondary amines react very readily with epoxy resins (Fig. 3.28). Polyethylene polyamines $H_2N(CH_2CH_2NH)_nH$ with $n = 2$ to 4 are particularly useful, because every N-H group reacts with a different epoxy group to produce a highly cross-linked cured thermoset product. They are particularly useful for fast room-temperature cure reactions of coatings and adhesives.

They may cause problems of volatility, toxicity, shelf life, and exothermic reaction. These can be avoided in several ways. (1) Polyamine can be prereacted with part of the epoxy resin to form an adduct (Fig. 3.29), which reduces volatility and reactivity. (2) Polyamine can be blocked temporarily by prereacting with a ketone to form a ketimine, which acts as a *latent* curing agent; when this is exposed to atmospheric moisture, it hy-

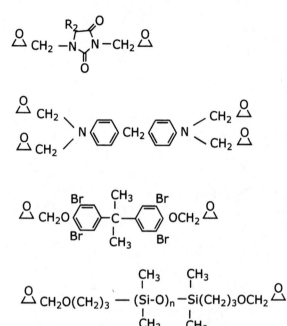

FIGURE 3.24   Miscellaneous epoxy monomers.

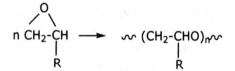

FIGURE 3.25   Homopolymerization of epoxy resins.

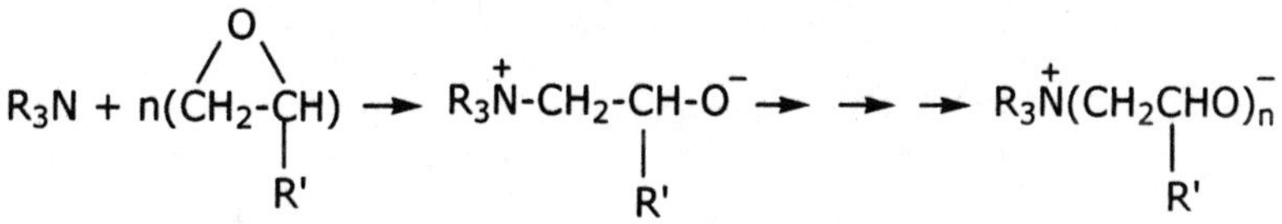

FIGURE 3.26   Tertiary amine catalysis of epoxy polymerization.

drolyzes gradually, unblocking the amine, which can then react gradually with the epoxy resin. (3) Polyamine can be reacted with fatty dibasic acid such as dimer acid, to form an amine-terminated polyamide oligomer (Fig. 3.30), which is nonvolatile, nontoxic, and less reactive. Epoxy/polyamide ratio is much less critical than the stoichiometric epoxy/

$(C_2H_5)_3N$

$(CH_3)_2NCH_2CH_2OH$

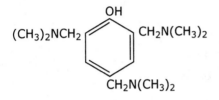

FIGURE 3.27 Amine catalysts for cure of epoxy resins.

$H_2NRNH_2 + CH_2CHR'CHCH_2 \longrightarrow \sim\sim\sim NRNCH_2CHR'CHCH_2 \sim\sim\sim$

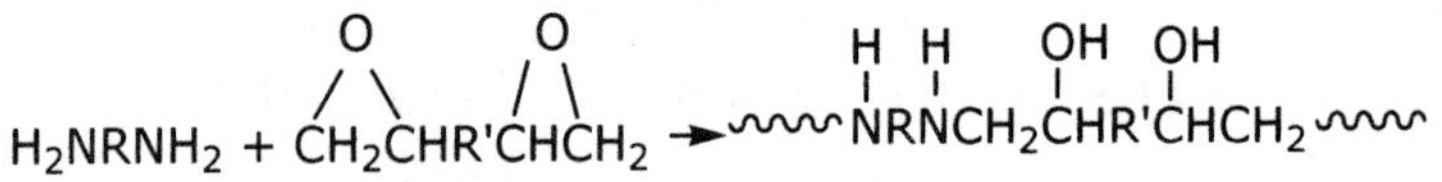

FIGURE 3.28 Polyamine cure of epoxy resins.

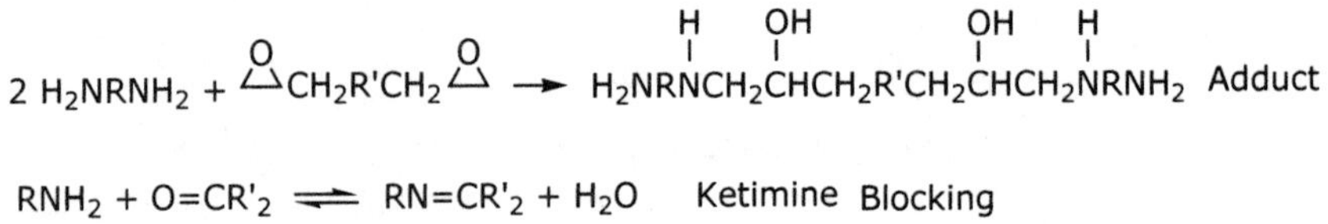

$2\ H_2NRNH_2 + CH_2R'CH_2 \longrightarrow H_2NRNCH_2CHCH_2R'CH_2CHCH_2NRNH_2$ Adduct

$RNH_2 + O{=}CR'_2 \rightleftharpoons RN{=}CR'_2 + H_2O$ Ketimine Blocking

FIGURE 3.29 Amine adducts for cure of epoxy resins.

$H_2NCH_2CH_2N(CC_{34}H_{68}C{-}NCH_2CH_2N)_nH$

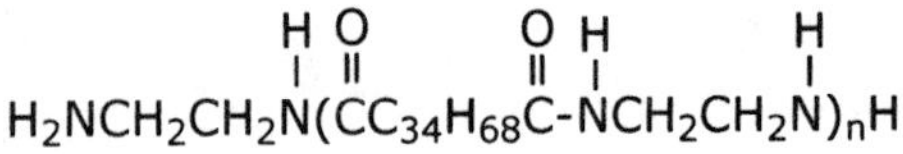

FIGURE 3.30 Amine-terminated fatty polyamides for cure of epoxy resins.

polyamine reaction. The long-chain fatty acid builds molecular flexibility into the cured epoxy resin, thus reducing its inherent brittleness. And the nonpolar hydrocarbon chains also increase moisture resistance.

Aromatic amines (Fig. 3.31) are less reactive, so they increase pot life and require heat cure. They give cured epoxies of higher heat deflection temperature and chemical resistance (Table 3.31).

Cycloaliphatic amines (Fig. 3.32) are intermediate between aliphatic and aromatic amines, in both reactivity and cured properties.

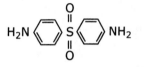

H₂N ⬡ CH₂ ⬡ NH₂

H₂N ⬡ S(=O)(=O) ⬡ NH₂

FIGURE 3.31 Aromatic amines for cure of epoxy resins.

TABLE 3.31 HDT of Amine-Cured Epoxy Resins

Polyethylene polyamines	111°C
Methylene dianiline	144°C
Metaphenylene diamine	150°C
Diamino diphenyl sulfone	190°C

$H_2NCH_2CH_2N$⬡NH N-Aminoethyl Piperazine

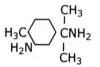 Menthane Diamine

H_2NCH_2⬡CH_2NH_2 m-Xylylene Diamine

FIGURE 3.32 Cycloaliphatic amines for cure of epoxy resins.

3.1.4.2.2 Acids. Acids can open the epoxy ring and thus produce polymerization and cure (Fig. 3.33). Some acids function primarily as catalysts, while others function as comonomers that build the cross-links into the cured epoxy resins.

Lewis acids such as $ZnCl_2$, $AlCl_3$, $FeCl_3$, and BF_3 adducts act as latent catalysts for one-part systems with good shelf life, which become active in heat cure. Phenols and organic acids also act as catalysts for cure reactions.

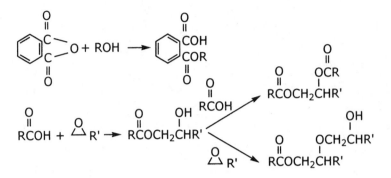

FIGURE 3.33   Acid polymerization/cure of epoxy resins.

Cyclic anhydrides are the second most important class of comonomers for cure of epoxy resins (Fig. 3.34). Whereas amine cure usually leaves linear segments and hydrophilic –OH groups, anhydride cure can also react with the –OH groups to produce many more cross-links, thus increasing molecular rigidity and water resistance. These cure reactions generally require heat and catalysis.

Conversely, the brittleness of cured epoxy resins can be ameliorated by copolymerizing with flexible curing agents. The two most popular types are carboxyl-terminated nitrile rubber and mercaptan-terminated polysulfide rubber oligomers (Fig. 3.35).

3.1.4.2.3 Relative Pot Life and Reactivity of Curing Agents.   Curing agents for different formulations and processes can be chosen according to their relative shelf-life/pot-life at room temperature and corresponding need for higher catalysis and heat to cure them (Table 3.32).

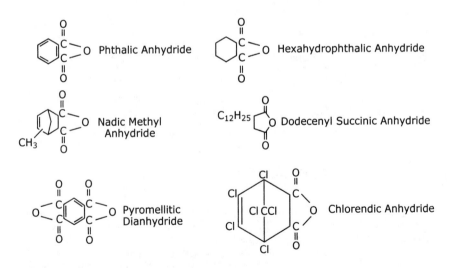

FIGURE 3.34   Cyclic anhydrides for cure of epoxy resins.

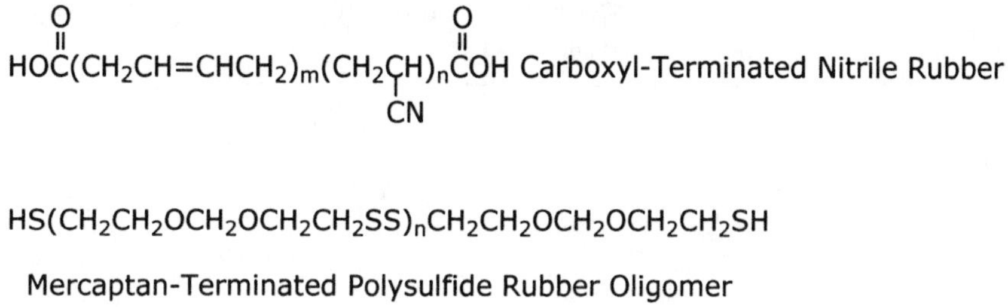

$$\overset{O}{\overset{\|}{HOC}}(CH_2CH=CHCH_2)_m(CH_2\underset{CN}{CH})_n\overset{O}{\overset{\|}{C}}OH \text{ Carboxyl-Terminated Nitrile Rubber}$$

$$HS(CH_2CH_2OCH_2OCH_2CH_2SS)_nCH_2CH_2OCH_2OCH_2CH_2SH$$

Mercaptan-Terminated Polysulfide Rubber Oligomer

FIGURE 3.35 Flexible curing agents for epoxy resins.

TABLE 3.32 Pot Life of Epoxy/Curing Agent Systems

Aliphatic amines	1 hr
Amine-terminated polyamides	3 hr
Aromatic amines	18 hr
Acid anhydrides	84 hr
Lewis acids	6 mo

3.1.4.3 Other Formulating Ingredients. A number of classes of additives are often used by individual formulators to modify or introduce new properties into the epoxy system.

3.1.4.3.1 Diluents. When epoxy resins are too viscous or too exothermic for a particular process, they can be modified by addition of low-molecular-weight aliphatic epoxides. Diepoxides can copolymerize directly into the curing process without reducing cross-linking. Monoepoxides can also copolymerize but do reduce the degree of cross-linking and thus soften properties. The literature also mentions nonreactive diluents such as plasticizers, but these would raise serious questions about degradation of properties.

3.1.4.3.2 Polymer Blends. A number of polymers are mentioned as modifiers for epoxy resins. Coal tar, phenol-formaldehyde, and polyurethane combine readily to produce intermediate properties. Silicones can add more unique properties. Polyesters and melamine-formaldehyde are also mentioned in the literature.

3.1.4.3.3 Flame Retardants. Flame retardance can be built into the epoxy resin by use of tetrabromobisphenol A or anhydride curing agents containing phosphorus or halogen. It can also be helped by nonreactive additives such as alumina trihydrate or organohalogens + antimony oxide.

3.1.4.3.4 Functional Fillers. A variety of fillers can be used to add specific properties. Metals, and beryllium and aluminum oxides, can be added to increase thermal conductivity (Table 3.33). Metals can be added to increase electrical conductivity (Table 3.34). Graphite increases lubricity and electrical conductivity. Mica increases elec-

TABLE 3.33   Thermal Conductivity of Filled
Epoxy Resins, Btu/[(ft^2-hr-°F)/ft]

Silver	240
Copper	220
Beryllium oxide	130
Aluminum	110
Steel	40
Solder	25
Aluminum oxide	20
Silver-filled epoxy	4
Aluminum-filled epoxy	2
Aluminum oxide-filled epoxy	1
Unfilled epoxy	0.1
Air	0.015

TABLE 3.34   Electrical Conductivity of Filled
Epoxy Resins, Ω-m

Silver	1.6×10^{-6}
Copper	1.8
Gold	2.3
Aluminum	2.9
Nickel	10.0
Platinum	21.5
Solder	25.0
Silver-filled epoxy	1.0×10^{-3}
Unfilled epoxy	1.0×10^{15}
Polystyrene	1.0×10^{16}
Mica	1.0×10^{16}

trical resistance. Alumina trihydrate increases arc resistance. Microballoons produce
structural foam of high compressive strength.

3.1.4.3.5 Reinforcing Fibers.   Reinforcing fibers greatly increase epoxy modulus,
strength, impact strength, heat deflection temperature, and dimensional stability
(Table 3.35).

TABLE 3.35 Properties of Reinforced Epoxy Resins

Epoxy resin	BPA			Novolac
Reinforcing fiber	None	Glass	Graphite	Graphite
Flex. modulus, kpsi	350	2500	5000	5500
Tensile str., kpsi	8.5	27.5	45	20
Flexural str., kpsi	17	60	85	40
Compressive str., kpsi	20	25	35	28
Impact str., fpi	0.6	35	18	10
Thermal exp., $10^{-6}/°C$	55	12	3	1
HDT, °C	167	288	288	260
H_2O abs., %	0.1	1.4	1.6	0.8

3.1.4.4 Markets and Applications. The largest use of epoxy resins is in coatings, comprising 53 percent of the total U.S. market (Table 3.36). They do not require solvents, so they protect the environment. They have high adhesion and chemical resistance, so they give durable protection. They are particularly useful in marine maintenance.

TABLE 3.36 Epoxy Resins, Market Analysis

Coatings		
	Can and drum lining	15%
	Plant maintenance	11
	Auto primers	7
	Pipe coating	4
	Appliances	2
	Trade sales and other	14
Printed circuit boards		12
Adhesives		8
Flooring and paving		8
Reinforced plastics		7
Tooling, casting, molding		4
Other		8
Total		100

Reinforced epoxy resins are the basis of printed circuit boards, tanks, pipes, and aerospace materials. Cast epoxies are very useful in electrical potting and encapsulation of transistors, switches, coils, integrated circuits, transformers, and switchgears.

Performance in adhesives is outstanding. Polarity, reactivity, low shrinkage, high modulus and strength, heat and chemical resistance all contribute to wide use in auto, aerospace, appliance, and mechanical construction. The total U.S. market is 600 million lb/yr, and growth rate has still not reached maturity.

3.1.5 Silicones

Silicone chemistry is a marriage of organic polymers and inorganic ceramics, which has produced synergistic benefits in abhesion, low-temperature flexibility, high-temperature stability, flame-retardance, electrical resistance, water resistance, and physiological inertness, leading to a family of elastomers and thermoset plastics with a wide variety of specialized applications.

3.1.5.1 Chemistry.   Silica sand is electrothermally reduced to silicon metal.

$$SiO_2 + C \rightarrow Si + CO_2$$

This is mixed with copper catalyst and reacted with methyl chloride at 250 to 280°C to produce a mixture of methyl chlorosilanes.

       9% CH_3SiCl_3 b.p. 66°C   designated T for trifunctional

     74% $(CH_3)_2SiCl_2$ b.p. 70°C   designated D for difunctional

      6% $(CH_3)_3SiCl$ b.p. 57°C   designated M for monofunctional

These are separated by fractional distillation.

    The chlorosilane Si-Cl bond hydrolyzes rapidly in water to form silanol Si-OH, which condenses instantly to form siloxane Si-O-Si (Fig. 3.36). Thus, $(CH_3)_2SiCl_2$ (D) produces linear silicone rubber. Introducing CH_3SiCl_3 (T) produces branching and cross-linking; at high concentrations, it produces a thermoset plastic. Conversely, introducing $(CH_3)_3SiCl$ (M) caps the ends of the growing chains and lowers the molecular weight of the rubber.

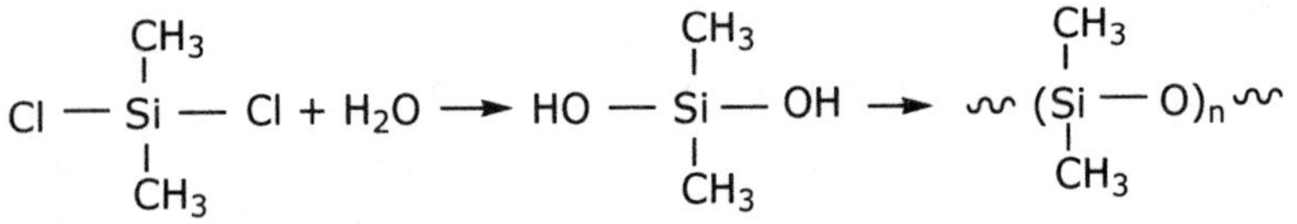

FIGURE 3.36   Silicone synthesis.

    The most common alkyl group is methyl. Introducing some phenyl groups prevents crystallization at low temperatures and thus keeps silicone rubber flexible down to lower temperatures; phenyl groups also increase heat stability at high temperatures, thus creating a wider useful temperature range for silicone rubber. $CF_3CH_2CH_2$- and $NCCH_2CH_2$- groups are used to increase resistance to fuels, oils, and organic solvents. $CH_2=CH$- groups provide reactivity for vulcanization/cure of the rubber. CH_3O- and CH_3CO_2- groups hydrolyze more slowly than Cl- and are used to provide controlled reactivity for cross-linking, coating, and adhesive bonding.

3.1.5.2 Properties.   Unlike most elastomers, silicone rubber does not contain C=C groups, so it is much more resistant to oxygen and ozone.

The Si-O and Si-C bonds in silicones are very stable, giving them high resistance to heat, electrical, and chemical attack.

The large size of the Si atom, and the oblique (150°) angle of the Si-O-Si bonds, give very little steric hindrance and very free rotation. This makes the silicone molecule very flexible and rubbery, even down to very low temperatures. On the down side, it also produces low mechanical strength and low solvent resistance.

The sheath of primary hydrogen atoms, on the methyl groups surrounding the polymer main-chain, gives low intermolecular attraction, which also contributes to rubbery behavior and low mechanical strength, and especially to low surface energy and low surface tension, which produce abhesion (nonstick) and water-repellent performance.

3.1.5.3 Rubber. Silicone rubber can be heat-cured by fairly conventional techniques. It can also be cast and cured at room temperature, producing what is called room-temperature vulcanized (RTV) rubber.

3.1.5.3.1 Heat-Cured Rubber. High-molecular-weight (500,000) linear silicone rubber is very soft and has no strength or creep resistance. It can be cross-linked by heating with peroxides (Table 3.37). The reaction of peroxide with the methyl group (Fig. 3.37) is not very efficient and levels off at 0.4 to 0.7 cross-links per 1000 Si atoms—too low to give good strength and resistance to compression set. Therefore, the rubber is usually made with a fraction of a percent of vinyl side-groups; these react readily with peroxide, giving a 90 percent yield of predicted cross-links and much better strength and compression-set resistance. If vinyl side-groups are increased up to 4 to 5 percent, silicone rubber can even be cured by conventional sulfur vulcanization.

TABLE 3.37 Peroxides for Cross-Linking Silicone Rubber

Bis(2,4-dichlorobenzoyl) peroxide	104–132°C
Benzoyl peroxide	116–138°C
Dicumyl peroxide	154–177°C
2,5-dimethyl-2,5-di(t-butylperoxy) hexane	166–182°C

Most rubber is reinforced by carbon black; silicone rubber is not. Instead, it is reinforced by fine-particle-size fumed silica. This definitely improves tensile strength, though it still cannot equal most other types of elastomers (Table 3.38). Other fillers do not increase strength but may be used to improve processability, increase hardness and reduce tack and compression set. Carbon black is used to increase electrical conductivity.

Small production runs are processed by compression or transfer molding at 800 to 3,000 psi and 104 to 188°C; mold shrinkage is 2 to 4 percent. Long production runs are more economical by injection molding at 5,000 to 20,000 psi, 188 to 252°C, and a 25 to 90 sec cycle. Extrusion requires post-cure in a 316 to 427°C hot-air oven, typically 60 ft/min; steam post-cure can run 1200 ft/min. Calendering typically runs 5 to 10 ft/min.

Specific formulations can aim at various product needs (Table 3.39). Particularly outstanding is their wide useful temperature range (Table 3.40).

3.1.5.3.2 Room-Temperature Vulcanized (RTV) Silicones. Low-molecular-weight liquid silicone oligomers, with reactive functional groups, can be poured or spread with

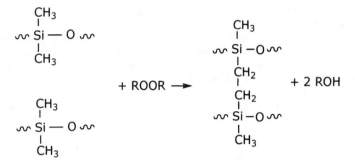

FIGURE 3.37 Peroxide cross-linking of silicone rubber.

TABLE 3.38 Fillers for Silicone Rubber

Filler	Particle size, μm	Tensile strength, psi
Fumed silica	7–10	600–1800
Precipitated silica	18–20	600–1100
Diatomaceous silica	1–5	400–800
Calcined kaolin	1–5	400–800
Calcium carbonate	1–4	400–600
Titanium dioxide	3	200–500
Iron oxide	1	200–500

little or no equipment and cross-linked (cured) at room temperature without damage to delicate electronics or other systems. They are very useful in caulking, sealants, adhesives, and arts and crafts. They are available as one- or two-part systems.

One-part systems are packaged in dry sealed cans and are perfectly stable in this state. When they are poured or spread to form products, they are activated by atmospheric moisture, and the cross-linking reaction occurs. The stable packaged oligomer has acetoxy or methoxy end-groups. When these are exposed to atmospheric moisture, they hydrolyze to hydroxyl end-groups, which condense with each other very rapidly to polymerize to high molecular weight and cross-link to thermoset rubbery products (Fig. 3.38). Acetoxy is more reactive, becoming tack-free in 1/4 to 1/2 hr and fully-cured in 12 to 24 hr; but it releases acetic acid, which can corrode copper and steel. Methoxy is slower, becoming tack-free in 2 to 4 hr and fully-cured in 24 to 72 hr; it does not cause corrosion, and it gives higher-strength products (Table 3.41). Since one-part systems depend on diffusion of atmospheric moisture, they are limited to 1/4-in thickness; thicker products require two-part systems.

Two-part systems are stable until they are mixed. The pairs are very specific chemically and must be mixed in the proper stoichiometric ratio, so the supplier specifies the procedure, and the processor simply needs to follow it. The two parts may react by con-

TABLE 3.39 Properties of Heat-Cured Silicone Rubbers

Grade	High.-temp.	High-strength Low-temp.	Solvent-resistant	Wire and cable
Shore A hardness	46	63	50	67
Tensile strength, psi	1000	1500	1000	1100
Elongation, %	430	700	220	340
Compression set, %	14	42	50	
Brittle temperature, °C	−65	−101	−68	
Oil absorption, %	6	10	1	
Volume resistivity, Ω-cm				3×10^{15}
Dielectric constant				3.3
Power Factor				0.003

TABLE 3.40 Maximum Use Temperatures of Silicone Rubbers

Temperature, °C	Time to 50% retention of elongation
121	10–20 yr
149	5–10 yr
204	2–5 yr
260	3–24 mo
316	1 wk
371	6 hr
461	10 min
518	2 min

densation or addition (Fig. 3.38). In condensation cure, the hydroxyl-terminated silicone oligomer is cross-linked by tetraethyl silicate, catalyzed by dibutyl tin dilaurate or faster by stannous octoate, and liberates alcohol, so it can be used only in an open system. In addition cure, a silicone oligomer containing vinyl $CH_2=CH-$ groups reacts with a silicone oligomer containing silane Si-H groups, catalyzed by platinum; since no volatiles are liberated, this can be done in a closed system, and it gives higher strength products (Table 3.42).

More recently, this has led to the development of liquid injection molding (LIM), in which the reactive silicone oligomer system is injection molded at 200 to 250°C and cures in a few seconds, a great advance over conventional vulcanization systems.

One-Part Systems

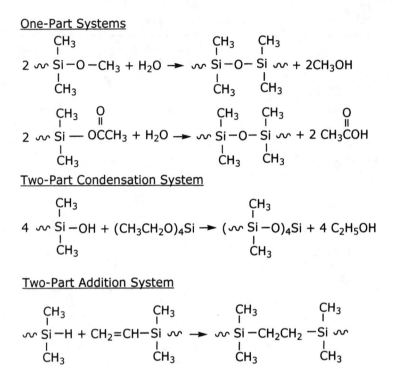

Two-Part Condensation System

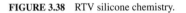

Two-Part Addition System

FIGURE 3.38 RTV silicone chemistry.

TABLE 3.41 Properties of Cured Methoxy RTV Silicone

Working time	30 min
Tack-free time	2–3 hr
Cure time (1/8 in thick)	24 hr
Shore A hardness	28
Tensile strength	150 psi
Elongation	550%
Adhesion: lap shear	100 psi
Adhesion: peel	20 lb/in
Volume resistivity	4.7×10^{14} Ω-cm
Dielectric constant	3.6
Dissipation factor	0.002

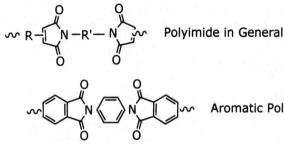

Polyimide in General

Aromatic Polyimide

FIGURE 3.39 Polyimides in general.

TABLE 3.42 Properties of Cured Two-Part RTV Silicones

Cure	Condensation cure	Addition
Shore A hardness	45	45
Tensile strength, psi	400	900
Elongation, %	120	150
Useful temperature range, °C	−115 to +204	−115 to +204
Dielectric constant	4.2	3.0
Dissipation factor	0.006	0.001

3.1.5.3.3 Silicone Resins. Hydrolysis of $(CH_3)_2SiCl_2$ produces linear flexible molecules for rubber. Hydrolysis of CH_3SiCl_3 produces highly cross-linked molecules for thermoset plastics. These are too cross-linked and brittle for most purposes. Useful thermoset plastics are prepared by copolymerizing difunctional and trifunctional monomers. In commercial practice, the ratio of difunctional to trifunctional is generally 80/20 to 40/60. For some products, methyl silicon may be partly replaced by phenyl silicon.

The mixed monomers are dissolved in organic solvent and stirred with water to produce hydrolysis and condensation to low-molecular-weight oligomers. Methyl silicon is too reactive and exothermic and must be cooled to control the A-stage reaction. Phenyl silicon is less reactive and may be heated to 70 to 75°C to promote the reaction.

The oligomer solution is then catalyzed by triethanol amine, metal octoates, or dibutyl tin diacetate and heated to increase the viscosity. At this point, it is cooled and can be stored until used. These silicone oligomers are used to make glass fabric laminates and reinforced molding powders. Phenyl silicon is compatible with epoxy, alkyd, urea, melamine, and phenolic resins and may be blended with them to increase their resistance to heat, flame, water, and weather.

Glass fabric laminates are made by dipping the glass fabric into the oligomer solution, impregnating it with 25 to 45 percent silicone resin, and evaporating the solvent. Layers of impregnated fabric are then plied to the desired thickness and press-cured. Flat sheets are cured 30 to 60 minutes at 1000 psi and 170°C. Complex shapes can be made by lower-pressure techniques such as vacuum-bag molding. These laminates are 20 to 40 percent weaker mechanically than epoxy, melamine, and phenolic but superior in electrical insulation properties, especially at high temperatures and in moist conditions (Table 3.43). They are used in electric motors, terminal boards, printed circuit boards, and transformers. They are also used for fire-resistance in aircraft firewalls and ducts.

TABLE 3.43 Electrical Properties of Silicone-Glass Cloth Laminates

Matrix resin	Phenolic	Melamine	Silicone
Power factor	0.06	0.08	0.0002
Dielectric strength, V/mil	150–200	150–200	250–300
Insulation resistance, Ω, dry	10,000	20,000	50,000
wet	10	10	10,000

Molding powders are B-stage silicone resin plus glass fiber and catalyst. They are compression molded 5 to 20 min at 1000 to 4000 psi and 160°C and then post-cured several hours to achieve optimum properties. Electrical insulation and resistance to heat and moisture are outstanding (Table 3.44). Molded parts are used in electric motors and switches.

TABLE 3.44 Silicone Resin Moldings

Specific gravity	1.65
Flexural modulus, 23°C	1,800,000 psi
200°C	900,000 psi
Flexural strength, 23°C	14,000 psi
200°C	5,000 psi
Tensile strength, 23°C	4,400 psi
200°C	1,300 psi
Dielectric constant	3.6
Power factor	0.005

3.1.5.3.4 Coatings. Silicone resin solutions are baked to produce release coatings that are resistant to heat, water, and weather. These are used in cooking and baking and for water-repellent masonry. They are also copolymerized with other thermosetting coatings to increase their heat and weather resistance.

3.1.6 Polyimides

New high-tech industries such as aerospace and electronics have created growing needs for lightweight, strong materials with increased resistance to heat, oxygen, and corrosion. Organic polymer chemists have spent the past half century developing new polymers with higher and higher performance. The guiding general principle has been the use of heterocyclic resonance to provide molecular rigidity and thermal-oxidative stability. There have been two persistent problems: (1) the syntheses are expensive, and (2) the molecular rigidity that gives heat resistance also makes processing very difficult. The most successful candidates so far have been the polyimides (Fig. 3.39).

Research has developed three synthetic routes to processability. (1) Thermoplastic polyimides contain enough single bonds in the polymer backbone to provide a certain amount of molecular flexibility and therefore processability. (2) Two-stage condensation polymerization leaves single bonds in the first stage to permit processability and then closes them to heterocyclic imide rings in the final stage of processing. (3) Second-stage addition polymerization begins with synthesis of imide-containing vinyl or acetylenic monomers in the first stage and then reacts the vinyl or acetylenic groups in the second stage to produce cross-linking cure without liberating volatile by-products.

3.1.6.1 Thermoplastic Polyimides. Several types of linear high-molecular-weight polyimides have been developed, which contain enough single bonds in the polymer backbone to make them somewhat flexible and therefore usable in conventional thermoplastic melt processing (Fig. 3.40). This does, of course, sacrifice some of the inherent thermal stability of polyimides (Table 3.45).

The best-known are General Electric Ultem poly(ether imides); these offer heat deflection temperatures of 207 to 221°C and continuous service temperatures of 170 to 180°C. Also popular are Amoco Torlon polyamide-imides, with heat deflection temperatures of 278 to 282°C. More specialized are Ciba-Geigy trimethyl phenyl indane polyimides, with heat deflection temperatures of 232 to 257°C, embrittlement times of >2000 hr at 200°C

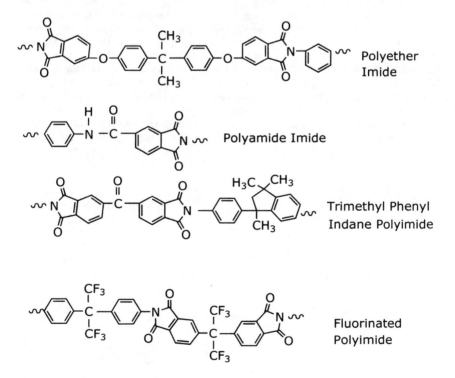

FIGURE 3.40 Thermoplastic polyimides.

TABLE 3.45 Thermoplastic Polyimide Temperature Limits

	PEI	PAI	TMPI	fpi
T_g, °C				340
HDT, °C	207–221	278–282	232–257	
Continuous service, °C	170–180			371
Embrittlement, hr/200°C			>2000	
250°C			250	
T_d, °C			450–510	

and 250 hr at 250°C, and decomposition temperatures of 450 to 510°C. And fluorinated polyimides containing the hexafluoroisopropylidene group have been reported with temperatures like $T_g = 340°C$ and continuous service temperature 371°C.

3.1.6.2 Two-Stage Condensation Polyimides.   Imides are produced by condensation re-action of amines with dibasic acids (Fig. 3.41). Diamines plus tetrabasic acids produce polyimides. When the reaction is run to completion, the highly cyclic structure is such a rigid molecule that melt processing is impossible. In fact, intramolecular cyclization com-petes with intermolecular cross-linking, so the cured polymer may actually be thermoset. However, the reaction can be run in stages by controlling temperature and time. In the first stage, it produces a polyamic acid, which still has enough single bonds in the polymer backbone to be a flexible molecule that is soluble and melt processable. When the first-stage polymer has been impregnated into reinforcing fabric and/or melt processed into the shape of the finished product, then increasing the temperature and reaction time drives the condensation cyclization reaction to the final imide structure. Since the condensation reac-tion liberates water or alcohol, special techniques are required to remove the volatiles and avoid bubbles and cracks in the solidifying polymer.

DuPont uses oxydianiline and pyromellitic dianhydride (Fig. 3.42) to produce a series of Kapton films (Table 3.46), Vespel sintered moldings (Table 3.47), and Pyralin lacquers

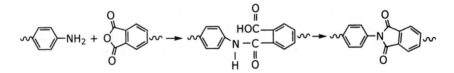

FIGURE 3.41   Two-stage condensation of polyimides.

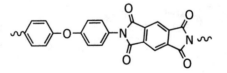

FIGURE 3.42   DuPont polyimide.

TABLE 3.46 Kapton Polyimide Films

Density	1.42
Tensile modulus, kpsi, 23°C 200°C	430 260
Tensile strength, kpsi, 23°C 200°C	25 17
Elongation, %, 23°C 200°C	70 90
Impact strength, J/mm	23
Folding endurance, cycles	10,000
Initial tear strength, g	510
Tear propagation, g	8
Volume resistivity, Ω-cm	10^{15}
Dielectric constant	3.6
Dissipation factor	0.0025
Dielectric strength, V/mil	5,400

(Table 3.48). Monsanto (Skybond) and American Cyanamid (FM-34) used m-phenylene diamine and benzophenone tetracarboxylic dianhydride (Fig. 3.43) to produce glass cloth laminates (Table 3.49). General Electric silicone polyimides (SiPI) are block copolymers of benzophenone tetracarboxylic dianhydride with methylene dianiline and bis(aminopropyl) tetramethyl disiloxane (Fig. 3.44), designed primarily for high-temperature electrical insulation (Table 3.50).

3.1.6.3 Second-Stage Addition Polymerization Cure of Polyimides.   To cure thermosetting polyimides without the problem of volatile by-products, the cross-linking reaction is based on addition polymerization instead of condensation polymerization. This again is a two-stage process. In the first stage, a low-molecular-weight oligomer is prepared containing finished imide groups; since it is low-molecular-weight, it is still easily processable, even though it contains aromatic and heterocyclic rings. Then, in the second stage, reactive groups in the oligomer are polymerized by addition reactions, building to high molecular weight and a high degree of cross-linking as well. Several types of reactive groups have been developed.

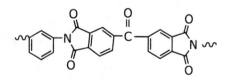

FIGURE 3.43 Monsanto and American Cyanamid polyimides.

TABLE 3.47 Vespel Polyimide Moldings

Specific gravity	1.55
Flexural modulus, 23°C, kpsi 260°C	550 305
Flexural strength, 23°C, kpsi 260°C	15 8.3
Tensile strength, 23°C, kpsi 260°C	8.8 4.6
Elongation, %, 23°C 260°C	6 4
Compressive modulus, kpsi	386
Notched impact strength, fpi	1.1
Heat deflection temperature, °C	360
Oxygen index, %	51
Volume Resistivity, Ω-cm	10^{14}
Dielectric constant	3.6
Dissipation factor	0.003
Water absorption, %	0.2

TABLE 3.48 Pyralin Lacquer Properties

Density	1.4
Tensile strength, kpsi	18
Elongation, %	18
Decomposition temperature, °C	560
Volume resistivity, Ω-cm	10^{16}
Dielectric constant	3.5
Dissipation factor	0.002
Dielectric strength, V/mil	4000

3.1.6.3.1 Bis-Maleimides. Reaction of maleic anhydride with diamines leads to two reactions. First, the amine reacts with the dianhydride groups and produces bis-maleimides (Fig. 3.45). Then, the amine adds across the double bonds ("Michael reaction"), thus lengthening the oligomer chain. These oligomers are easily impregnated into glass cloth, "catalyzed" by high-temperature peroxide such as dicumyl peroxide, stacked to the desired thickness, and press-cured or vacuum-bag cured, for example at 75 to 210 psi and

TABLE 3.49 Skybond Polyimide Laminates

Flexural modulus, kpsi	3,120
335 hr/299°C	3,120
Flexural strength, kpsi	80
30 min/407°C	53
Tensile strength, kpsi	57
335 hr/299°C	42
Volume resistivity, Ω-cm	2.47×10^{15}
Dielectric constant	4.15
Dissipation factor	0.00445
Dilectric strength, V/mil	179
Water absorption, %	0.7

TABLE 3.50 Silicone Polyimide
Electrical Properties

Bulk resistivity	10^{17} Ω-cm
Dielectric constant	3.0
Dielectric strength	5.5 MV/cm

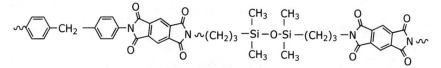

FIGURE 3.44 General Electric silicone polyimides.

200 to 250°C, followed by oven post-cure 12 to 24 hr to complete the cross-linking reaction. This produces excellent mechanical properties and heat resistance (Table 3.51).

3.1.6.3.2 Acetylene-Terminated Imide Oligomers. Oligomers containing finished imide groups can be synthesized with terminal acetylenic (ethynyl) groups (Fig. 3.46). When these are impregnated into reinforcing fabrics and heat-cured, for example 500 hr/288 to

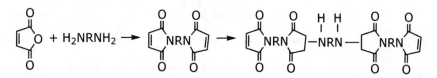

FIGURE 3.45 Bis-maleimides.

TABLE 3.51 Bis-Maleimide Cured Properties

Flexural modulus, 25°C	4000 kpsi
250°C	3200 kpsi
Aged 3000 hr/250°C	2600 kpsi
Flexural strength, 25°C	70 kpsi
250°C	50 kpsi
Aged 3000 hr/250°C	26 kpsi
Tensile strength	50 kpsi
Compressive strength	50 kpsi
Notched impact strength	13 kpsi
T_g	296°C
Volume resistivity	6×10^{14} Ω-cm
Dielectric constant	4.5
Dissipation factor	0.012
Dielectric strength	25 kV/mm

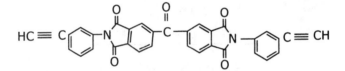

FIGURE 3.46   Acetylene-terminated imide oligomers.

316°C, they give laminates with extreme heat resistance (Table 3.52). The mechanism of the cure reaction is complex, probably producing a variety of aromatic and fused-ring structures (Fig. 3.47).

3.1.6.3.3 Nadimide-Terminated Oligomers.   Research at NASA, the U.S. Air Force, and industrial laboratories has developed a series of thermoset polyimdes that are made by impregnating the monomers into laminating fabric and then polymerizing and cross-linking them *in situ*. The body of the polyimide oligomer is made from benzophenone tetra-carboxylic acid ester or bisphenyl hexfluoropropene tetracarboxylic acid ester reacting with an aromatic diamine such as phenylene diamine or methylene dianiline (Fig. 3.48). The end-groups of the oligomer are made by end-capping with norbornene dicarboxylic acid ester. And thermosetting cross-linking cure occurs by addition polymerization of the C=C bonds in the norbornene ring. Laminate properties are very good (Table 3.53), and heat aging resistance is promising (Table 3.54). More recently, dinadimide end-capping (Fig. 3.49) has reached use temperatures of 260 to 290°C.

3.1.6.4 Polyimide Applications.   Polyimides are used where their lubricity, low coefficient of thermal expansion, heat resistance, and radiation resistance are required. Typical uses include bearings and piston rings in jet engines, appliances, office equipment, com-

TABLE 3.52   Acetylene-Terminated Polyimide Cured
Properties

T_g	386°C
Laminate flexural modulus, 23°C	4,600,000 psi
316°C	3,000,000 psi
Laminate flexural strength, 23°C	69,000 psi
316°C	45,000 psi
Shear strength, room temperature	12,000 psi
Aged 500 hr/200°C	8,700 psi
Aged 500 hr/288°C	7,400 psi
Aged 500 hr/316°C	6,000 psi
Weight loss, 1000 hr/351°C	4%
Dielectric constant, 10 MHz	5.38
12 GHz	3.12
Loss tangent, 10 MHz	0.0006
12 GHz	0.0048

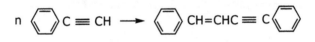

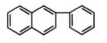

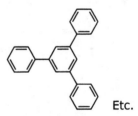

Etc.

FIGURE 3.47   Cross-linking acetylene-terminated polyimides.

pressors, and automotive transmissions; seals and insulators in nuclear applications; electric motors, wire and cable, and magnet wire; printed circuit boards; and high-temperature adhesives.

3.1.7   Miscellaneous Cross-Linking Reactions

Beyond the major thermoset plastics described above, research, development, and specialized production have explored a number of other cross-linking reactions for producing

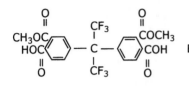

Bisphenyl Hexafluoropropene Tetracarboxylic
Acid Dimethyl Ester

 Phenylene Diamine

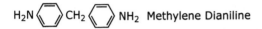 Methylene Dianiline

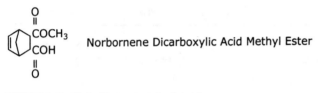 Norbornene Dicarboxylic Acid Methyl Ester

FIGURE 3.48 Nadimide-terminated polyimides.

TABLE 3.53 Nadimide-Terminated Polyimide Laminate Properties

Tensile modulus	21,700,000 psi
Tensile strength	180,000 psi
Flexural modulus	17,600,000 psi
Flexural strength	206,000 psi
Impact energy	15.2 in-lb
Coefficient of thermal expansion	0

TABLE 3.54 Nadimide-Terminated Polyimide Aging

Shear strength at 316°C before aging	7300 psi
After 400 hr	7700 psi
After 800 hr	7700 psi
After 1200 hr	7300 psi

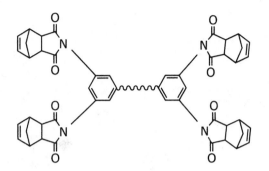

FIGURE 3.49 Dinadimide end-capped polyimides.

thermoset plastics, particularly addition reactions that do not produce volatile by-products. These may be grouped as (1) reactions of hydrocarbons, (2) triazine and other heterocyclic ring formation, and (3) polyphenylene sulfide.

3.1.7.1 Reactions of Hydrocarbons.
Several types of reactive hydrocarbon functional groups can be used to polymerize and cross-link monomers and oligomers into thermoset plastics. These include addition polymerization of acetylene-terminated molecules and ring-opening polymerization of strained carbon rings. They also include Friedel-Crafts condensation to form hydrocarbon polymers.

3.1.7.1.1 Acetylene-Terminated Monomers and Oligomers

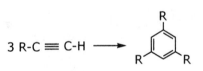

FIGURE 3.50 Addition polymerization of acetylenic monomers.

Addition polymerization of acetylene (ethynyl) groups can occur at high temperatures, for example 500 hr at 288 to 316°C followed by cure 4 to 15 hr/407 to 434°C. With monofunctional monomers, a major product is trimerization to form new aromatic rings (Fig. 3.50)— but with difunctional monomers, a great variety of cross-linked structures have been identified and/or theorized. Practically, many of these give thermoset plastics of high heat and moisture resistance, superior to epoxy resins. Since there are no volatile by-products, this offers processing advantages over many condensation-cured thermosets.

Polyimides have been cured by synthesizing acetylene-terminated oligomers containing finished imide groups, and these have shown excellent heat resistance, as discussed above (Sec. 3.1.6.3.2).

Polysulfones have been made from acetylene-terminated sulfone monomers (Fig. 3.51), and cured graphite-fiber laminates have shown T_g = 300°C and good mechanical properties at 170°C before and after heat and humid aging. Semi-interpenetrating polymer networks with linear thermoplastic polysulfones showed promise of combining the heat deflection temperature and solvent-resistance of the thermoset polymer with the impact resistance of the thermoplastic.

Polyphenylquinoxalines were cross-linked by acetylenic end-groups (Fig. 3.52), giving T_g = 321°C and good resistance to hot humid aging, but the addition of aliphatic hydrocarbon structure apparently sacrificed heat-aging resistance. Propargyl ether of bisphenol A (Fig. 3.53) was cured to a thermoset plastic with T_g = 360°C.

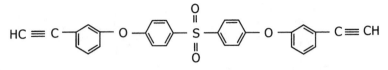

FIGURE 3.51 Acetylene-terminated sulfone.

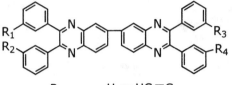

$R_{1,2,3,4}$ = H or HC≡C−

FIGURE 3.52 Acetylene-terminated polyphenylqui-noxaline.

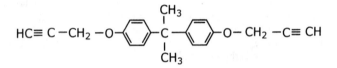

FIGURE 3.53 Propargyl ether of bisphenol A.

Phenylethynyl end-capping of polyimide oligomers (Fig. 3.54) has shown promise for high-temperature plastics and adhesives, with $T_g > 300°C$ and high adhesive strength, hot strength, and oil resistance (Table 3.55).

3.1.7.1.2 Ring-Opening Polymerization of Strained Carbon Rings. The carbon atom is tetrahedral, which means that normal C-C-C bond angles are about 109°. In small ring structures, the bond angles are much smaller than this, so they are under considerable strain, unstable, and reactive. When they break open into dienes or diradicals, they can polymerize. Several such ring-opening reactions have been suggested for cross-linking cure of thermoset plastics.

Benzocyclobutene. Polyimide oligomers with benzocyclobutene end-groups (Fig. 3.55) have been cured by electrocyclic ring-opening at 250°C. The opening of the cyclobutene ring can lead to homopolymerization, or it can copolymerize with C=C bonds in maleimides or with acetylene-terminated oligomers (Fig. 3.56), all of which lead to cross-linking and thermosetting cure. Cured samples had T_gs from 240 to 400°C or more; after 200 hr/350°C aging, they still retained 85 to 93 percent of their original weight. Similarly, a benzocyclobutene-terminated diketone (Fig. 3.57) cured to a thermoset plastic

FIGURE 3.54 Phenylethynyl end-capped polyimide oligomer.

TABLE 3.55 Phenylethynyl Cross-Linked Polyimide
Composite Properties

Shear strength, psi, room temperature	5,700
177°C	4,400
48 hr in hydraulic fluid	5,410
Flexural modulus, psi, room temperature	23,000,000
177°C	22,000,000
Flexural strength, room temperature	268,000
177°C	190,000

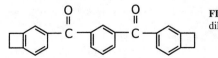

FIGURE 3.55 Benzocyclobutene-terminated polyimide oligomers.

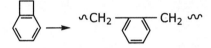

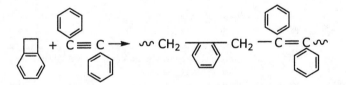

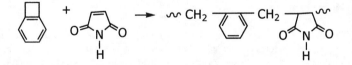

FIGURE 3.56 Benzocyclobutene cross-linking cure reactions.

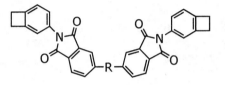

FIGURE 3.57 Benzocyclobutene-terminated diketone.

with T_g = 340°C, excellent hot/wet properties at 274°C, and useful service temperature of 260°C or above.

Paracyclophane. The strained rings of paracyclophane (Fig. 3.58) open and polymerize on heating. When paracyclophane end-groups are attached to polyimide oligomers and thermally cross-linked, cured composites have excellent heat-resistance (Table 3.56).

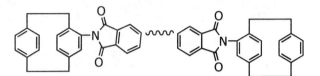

FIGURE 3.58 Paracyclophane polyimide oligomer.

TABLE 3.56 Paracyclophane-Polyimide Cured Laminates

Flexural strength, 25°C	174,000 psi
371°C	25,000 psi
Shear strength, 25°C	10,000 psi
371°C	3,700 psi
Weight loss, 500 hr/371°C	12%

FIGURE 3.59 Biphenylene.

Biphenylene. The strained ring of biphenylene (Fig. 3.59) opens at 380 to 400°C, producing a variety of cyclic and polymeric products. Organometallic catalysts work at lower temperatures, and copolymerization with acetylenic bonds looks promising.

Acenaphthylene. Acenaphthylene ring-opening (Fig. 3.60) and polymerization reactivity is comparable to bis-maleimides. Researchers have considered attaching acenaphthylene end-groups to various high-temperature oligomers (Fig. 3.61) to permit cross-linking cure reactions.

3.1.7.2 Triazine and Other Heterocyclic Ring Formation. Several types of reactions can be used to form heterocyclic rings in which multiple C-N bonds contribute high thermal stability. When these are used to cross-link heat-stable oligomers, the resulting thermoset polymers may have high thermal stability and other useful properties. These include cyanate/cyanurate, isocyanate/isocyanurate, hexaazatriphenylene trianhydride, and phthalonitrile/phthalocyanine.

3.1.7.2.1 Cyanate/Cyanurate. When aryl cyanate esters are heated to 150 to 250°C, they cyclotrimerize to cyanuric acid esters (Fig. 3.62). They can be catalyzed by organometallic and active hydrogen compounds. When the monomer is a dicyanate such as bisphenol A dicyanate (Fig. 3.63), the result is a highly cross-linked heterocyclic polymer (Table 3.57). Using a novolac polycyanate has produced T_g and useful life over 300°C.

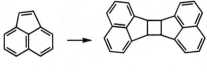

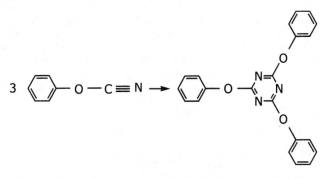

FIGURE 3.60 Acenaphthylene ring-opening reactions.

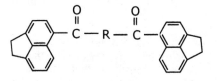

FIGURE **3.61** Acenaphthylene-terminated
high-temperature oligomers.

FIGURE 3.62 Aryl cyanate ester cyclotrimerization to cyanuric acid ester.

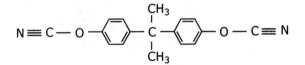

FIGURE 3.63   Bisphenol A dicyanate.

TABLE 3.57   Polycyanate Cured Properties

Density	1.19
T_g	265°C
Heat deflection temperature	230°C
Coefficient of thermal expansion	$44 \times 10^{-6}/°C$
Dielectric constant	2.7
Loss tangent	0.003
Moisture absorption at 100°C	1.3%

There is very little shrinkage (<1 percent) during cure, and the cured products have high adhesion, high heat and electrical resistance, and low moisture absorption, making them desirable for electronic applications such as printed circuit boards. There is also considerable interest in blending them with engineering thermoplastics, forming semi-interpenetrating polymer networks with polysulfones, polycarbonate, polyesters, polyarylate, and nitrile rubber, to combine optimum properties in the cured products. And there is interest in blending or copolymerizing them with other thermosetting resins such as epoxy or polyimide, and end-capping heat-resistant oligomers with aryl cyanate end-groups.

3.1.7.2.2 Isocyanate/Isocyanurate.   Isocyanates react with polyols to form rigid polyurethane foams, a major type of thermoset plastics. While these are very useful in thermal insulation, they are limited by failure at high temperature and by flammability. One way to solve these problems is to convert part of the isocyanate to isocyanurate by cyclotrimerization (Fig. 3.64). Whereas the isocyanate-polyol reaction forms polyurethane rapidly at room temperature, the cyclotrimerization of isocyanate to isocyanurate requires strong alkaline catalysis and heat to compete successfully. The resulting isocyanurate rings build considerable heat resistance (150 to 250°C, short-term ≤800°C) and flame-retardance into the polyurethane foam. They are useful for insulating pipelines and boilers.

Pure polyisocyanurate would have the highest heat and flame resistance, but the foam is too brittle to be useful. Copolymerization with polyurethane gives a more useful balance

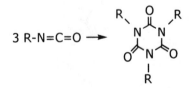

FIGURE 3.64   Cyclotrimerization of isocyanate to isocyanurate.

of properties. The preferred polyisocyanates are polymeric methylene diphenyl isocyanates with 2 to 7 isocyanate groups per molecule. Mechanical properties can be improved by glass fiber reinforcement. Copolymerization with epoxy resins to form polyoxazolidones has also been suggested (Fig. 3.65).

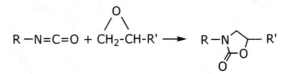

FIGURE 3.65 Copolymerization of isocyanate and epoxy to polyoxazolidone.

3.1.7.2.3 Hexaazatriphenylene Trianhydride. Researchers seeking to take polyimides to their maximum temperature limits have often concluded that the remaining weak points in their structure are the H atoms. Using hexaazatriphenylene trianhydride (Fig. 3.66) to form hydrogen-free polyimides, they have been able to produced polyimides good up to 700°C in air.

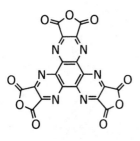

FIGURE 3.66 Hexaazatriphenylene trianhydride.

3.1.7.2.4 Phthalonitrile/Phthalocyanine. The triple bonds in phthalonitrile undergo addition polymerization, and the high heat stability of the polymers has attracted considerable research attention. Most research has focused on cyclic tetramerization to phthalocyanine (Fig. 3.67). While this simply produces a cyclic oligomer, polymerization of bis-

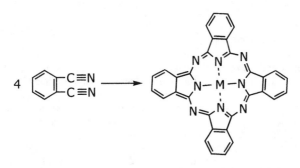

FIGURE 3.67 Cyclic tetramerization of phthalonitrile to phthalocyanine.

phthalonitriles at 250 to 350°C produces highly cross-linked thermoset heterocyclic polymers. These have high T_gs (e.g., 450°C), resist temperatures of 200 to 550°C, and have low water absorption.

Aside from simply varying the mid-section of the bis-phthalonitrile, researchers have copolymerized them with aromatic diamines. They have also inserted different metal cations into the center of the phthalocyanine ring. For example, germanium and tin give semiconductivity, and doping with iodine brings electrical conductivity up to 0.01 Ω-cm. Putting a silicon atom into the center of the ring permits them to stack layers of phthalocyanine rings around a siloxane center. This family of polymers offers a number of possibilities in terms of adhesion, heat stability, conductivity, colorability, and chemical resistance.

3.1.8 Cross-Linking of Thermoplastics

Thermoplastics are generally stable linear molecules, but we do occasionally cross-link them to improve specific processes or properties. Some of them contain reactive groups that can be cross-linked directly. Others we modify so as to make them accessible to cross-linking. Quite a number of typical improvements can be made by cross-linking:

- Foam processing
- Modulus
- Strength
- Creep-resistance
- Adhesion
- Abrasion-resistance
- Dimensional stability
- Heat deflection temperature
- Heat-shrink film and tubing
- Hot strength
- Dimensional stability
- Flame retardance
- Solvent resistance
- Water resistance
- Gelation
- Age resistance (Cage effect)

3.1.8.1 Polyethylene. Polyethylene is not very reactive, but it can be cross-linked lightly, either by chemical reaction or by high-energy radiation. Chemical cross-linking is most often accomplished by adding a high-temperature peroxide, such as dicumyl peroxide or 2,5-dimethyl-2,5-di-t-butylperoxy hexane, melt processing below the decomposition temperature of the peroxide, and then increasing the temperature to decompose the peroxide and initiate the cross-linking process (Fig. 3.68). Addition of carbon black often improves both process and properties.

A newer chemical method is to melt-blend polyethylene with vinyl trialkoxy silane and peroxide, grafting the silane as side-groups onto the polyethylene chain (Fig. 3.69). As long as it is kept very dry, the alkoxysilane groups are stable. After the grafted polyethyl-

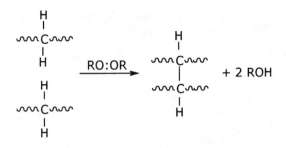

FIGURE 3.68 Peroxide cross-linking of polyethylene.

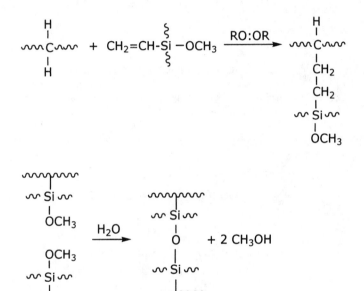

FIGURE 3.69 Silane cross-linking of polyethylene.

ene has been melt-processed into the final product, it is immersed in low-pressure steam to hydrolyze the alkoxy groups to silanol groups, which condense instantly to form siloxane cross-links.

Radiation cross-linking is often done with electron-beam equipment, using doses up to 60 Mrad. This splits off hydrogen atoms, leaving polymer radicals. The resulting radicals are trapped in the solid phase; cross-linking can be hastened by annealing, giving the radicals enough mobility to meet and combine.

Cross-linking increases melt viscosity, which improves the foaming process. It is used to produce heat-shrink film and tubing. It increases environmental stress-crack resistance, and it retains hot strength beyond the melting point of the crystalline phase. These improvements are particularly useful in wire and cable, pipe insulation, gasketing, and sports and orthopedic equipment.

3.1.8.2 Polystyrene. Polystyrene is cross-linked by copolymerization with divinyl benzene (Fig. 3.70). This is used to produce polystyrene beads for ion exchange beds such as water softeners.

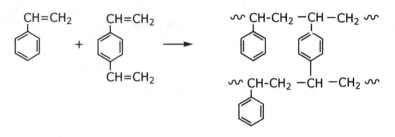

FIGURE 3.70 Styrene-divinyl benzene copolymerization.

3.1.8.3 Polyvinyl Chloride. In plastisol processing, PVC is dispersed in liquid plasticizer then fused to a sol and cooled to a gel to produce rubbery products. When the processor wants to decrease the plastisol viscosity without excessive softening of the final product, he uses a polymerizable plasticizer such as a glycol diacrylate (Fig. 3.71) plus a little peroxide. The plasticizer gives a more fluid plastisol, and then, during hot fusion, the peroxide cross-links the plasticizer to prevent excessive plasticization of the finished product.

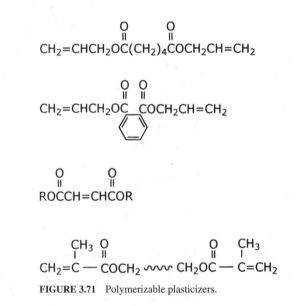

FIGURE 3.71 Polymerizable plasticizers.

3.1.8.4 Polyvinyl Alcohol and Acetals. Polyvinyl alcohol fibers are solution-spun, and then cross-linked with formaldehyde (Fig. 3.72) to make them resistant to water. Polyvinyl formal is cross-linked with phenolic resole to make baked coatings on electrical wire. Polyvinyl butyral is cross-linked with phenolic resin to make extremely tough bullet-proof helmets.

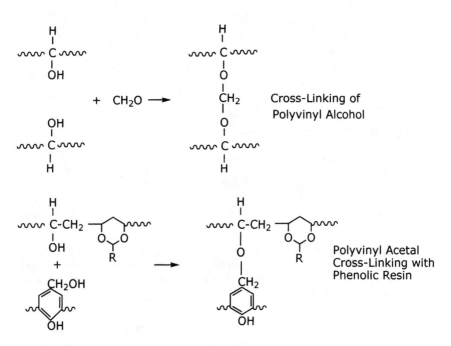

FIGURE 3.72   Polyvinyl alcohol and acetal thermosets.

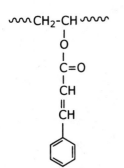

FIGURE 3.73   Polyvinyl cinnamate.

3.1.8.5 Polyvinyl Cinnamate. Polyvinyl cinnamate (Fig. 3.73) is used as a negative photoresist in photography, lithography, and production of printed circuit boards. When the polymer is partially masked and then exposed to UV light, the exposed vinyl groups cross-link and become insoluble. Then, when the unexposed polymer is dissolved away, the cross-linked portion remains to give a negative image of the mask.

3.1.8.6 Polymethyl Methacrylate. When PMMA is cast into sheets and lenses, addition of a glycol dimethacrylate comonomer produces cross-linking, which increases modulus, abrasion-resistance, heat deflection temperature, and solvent resistance.

3.1.8.7 Cellulose. Cellulose is linear but not thermoplastic. When we treat it with thermosetting polymers, it probably copolymerizes with them, cross-linking the cellulose.

The –OH groups in wood cellulose copolymerize with the methylol –CH_2OH groups in urea, melamine, and phenolic resins (Fig. 3.74) in plywood, composition board, counter-tops, cabinets, and furniture. Similarly, the –OH groups in cellulose copolymerize with the methylol groups in urea-formaldehyde to produce wet-strength paper and crease-resistant permanent-press cotton fabrics.

FIGURE 3.74 Copolymerization of cellulose with methylol resins.

3.1.8.8 Phenoxy Resin. The –OH groups in phenoxy resin (Fig. 3.75) can be cross-linked by methylol, epoxy, aldehyde, acid, and isocyanate groups for various specialty applications.

3.1.8.9 Casein. The amide groups in casein protein are reactive enough for useful cross-linking. Casein is plasticized with water and then extruded or molded. To improve properties, the products are then soaked in formaldehyde solution to cross-link the amide groups (Fig. 3.76). This is used to make products such as buttons and buckles.

3.1.9 Cross-Linking of Elastomers, Coatings, and Adhesives

Cross-linking is used not only in thermoset plastics but also in the related fields of elastomers, coatings, and adhesives. A review of these fields can broaden our view and hope-

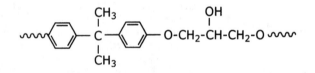

FIGURE 3.75 Phenoxy resin.

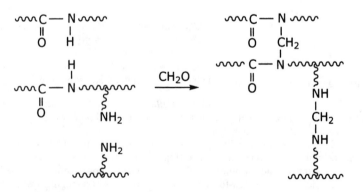

FIGURE 3.76 Casein—formaldehyde cross-linking.

fully suggest further development of cross-linking processes in thermoset plastics as well.

3.1.9.1 Elastomers. Rubber is based on long, flexible molecules. These must be bonded to each other to increase strength, elastic recovery, creep resistance, heat resistance, and chemical resistance. In thermoplastic elastomers, the bonding is based on secondary attractions such as polarity, hydrogen-bonding, and crystallinity, gathered into nano-size domains dispersed in the rubber matrix. The majority of the rubber industry, however, uses primary covalent cross-linking (vulcanization) to ensure intermolecular bonding.

Most rubber is based on polymers of isoprene or butadiene and contains many reactive C=C double bonds available for cross-linking. It is cross-linked by sulfur, aided by metal oxides and organic catalysts, producing sulfide cross-links between the polymer chains. Ethylene-propylene rubber is mostly made with several percent of diene termonomer to introduce C=C double bonds, which can then be vulcanized in the same way. Similarly, butyl rubber is made with a few percent of isoprene comonomer to introduce C=C double bonds and permit sulfur vulcanization. Even saturated elastomers are sometimes cured by sulfur, using peroxides and catalysts to activate C-H bonds, and metal oxides to attack C-Cl bonds.

Saturated elastomers are often cured by peroxide, often aided by catalysts. These include chlorinated polyethylene, fluorocarbon, acrylic ester, epichlorohydrin, polysulfide, polyurethane, and silicone. The peroxide radical abstracts an unstable hydrogen from the polymer, leaving a polymer radical, and then polymer radicals couple to produce C-C cross-links.

Halogenated elastomers are often cured by metal oxides, in combination with other ingredients. These include chlorosulfonated polyethylene, chloro- and bromo-butyl, and neoprene.

Some of the more unusual curing agents include phenolic resin, quinone dioxime, maleimide, diamine, diisocyanate, tetraethyl silicate, and triallyl isocyanurate. These are limited to very specific polymer systems.

3.1.9.2 Coatings. Although thermoplastic polymers may make good coatings, cross-linking is generally preferred to achieve maximum performance.

Alkyd resins are made from vegetable oils containing many C=C double bonds. Atmospheric oxygen attacks these bonds, causing addition polymerization and cross-linking.

Polyesters are usually produced with terminal –OH groups. These are then cured by reaction with methylol melamine or isocyanate.

Acrylic polymers are made with some hydroxyalkyl acrylate comonomer, and the –OH groups are then cured by methylol melamine or epoxy resin.

Urea-formaldehyde and melamine-formaldehyde are used to cross-link acrylic, alkyd, epoxy, and polyester coatings.

Epoxy resin coatings are cured by copolymerization with acrylic, polysulfide, polyurethane, polyamine, polyamide, amino, and phenolic oligomers.

Polyurethane coatings are cured by reaction with isocyanate or copolymerized with alkyds and then cured by atmospheric oxygen.

Phenolic resins can be cured by simple homopolymerization or by copolymerization with alkyd or epoxy resins.

Silicones are cured by hydrolysis of CH_3SiCl_3 or $(C_2H_5O)_4Si$ to form silanols, which condense with the –OH end-groups of the silicone oligomers. $CH_3Si(OH)_3$ condenses to $CH_3SiO_{1.5}$ "glass resin;" baking this at high temperatures burns out the CH_3 and produces SiO_2 ceramic coating. Silicones are also often copolymerized with alkyd, acrylic, epoxy, and polyester to upgrade coatings for resistance to heat, moisture, and weather.

3.1.9.3 Adhesives.   Ideally, an adhesive should be a low-molecular-weight fluid for easy and thorough application to a solid surface and then convert to a high-molecular-weight solid for maximum mechanical strength. This is most easily achieved by applying a low-molecular-weight reactive oligomer and then cross-linking it up to a high-molecular-weight finished adhesive bond. Additionally, if the cure reaction can react with the substrate as well, then the finished product is composed of primary covalent bonds from adherend 1 to adhesive layer to adherend 2, which should give the maximum bonding strength. Thus, it is not surprising that many of our best adhesives are thermosetting resins.

Casein and starch adhesives are cross-linked by urea-formaldehyde, melamine-fomaldehyde, or resorcinol-fomaldehyde.

Many rubber adhesives are vulcanized to increase their strength: natural rubber, SBR, neoprene, nitrile rubber, butyl rubber.

Polysulfide rubber is cured by peroxides.

Amino and phenolic resin adhesives for plywood, fiberglass insulation, laminates, friction products, and abrasive products are reactive oligomers that cure by self-condensation of methylol groups.

Epoxy adhesives are cured by amines, amides, anhydrides, polysulfides, and amino and phenolic resins.

Polyurethanes are cured by conventional polyol-isocyanate reactions.

Polyvinyl alcohol is copolymerized with cross-linkable comonomers; N-methylol acrylamide is self-curing. Carboxylic acid comonomers are cured by epoxy, urea, or phenolic resins.

Polyvinyl alcohol is cross-linked by glyoxal, epoxy, urea-formaldehyde, and phenol-formaldehyde resins. It is also cross-linked by polyvalent metal salts.

Polyvinyl formal and butyral contain –OH groups that are cured by dialdehydes, diisocyanates, anhydrides, epoxy, melamine, and phenolic resins.

Acrylics are copolymerized with a variety of functional groups that can either self-cure or react with added curing agents: amide, carboxyl, hydroxyl, epoxy, alkoxymethyl.

Anaerobic adhesives are primarily glycol dimethacrylates, catalyzed by cumene hydroperoxide, and packaged under air to inhibit premature polymerization. When they are pressed between two adherends, excluding air, they cure rapidly.

Silicone RTV sealants are HO-Silicone-OH oligomers + CH_3SiX_3 cross-linkers, where X is $-OCH_3$, $-O_2CCH_3$, or $-N=C(CH_3)_2$.

Organosilane coupling agents are used to produce adhesion between a polymer matrix and a reinforcing fiber or filler particle. They have a general formula $(RO)_3SiRX$, where RO is CH_3O- or CH_3CO_2-, and RX is designed to copolymerize with the polymer matrix during cure. For example, X would be $–CH=CH_2$ for an unsaturated polyester matrix, $-NH_2$ or epoxy for epoxy, melamine, and phenolic matrixes.

Hopefully, this comparative review will provide stimulating cross-fertilization for future developments in thermoset plastics and cross-linked polymers in general.

3.2 PROCESSES

There are a great variety of processes for manufacturing thermoset plastic products. Some of them are modifications of conventional thermoplastic processes, but most of them are uniquely designed for handling the simultaneous shaping and cross-linking that are involved. They may be classified as molding processes, reinforced plastics processes, pouring processes, and powder processes.

3.2.1 Molding Processes

Liquid B-stage resin is held in a closed cavity and heated to cross-link it to a rigid solid product. This is done in a variety of ways.

3.2.1.1 Compression Molding. The original and classic method of producing thermoset plastic products is by compression molding. A two-part steel mold is made with a cavity representing the shape of the desired product. The cavity surface is chrome-plated to give a smooth corrosion-resistant finish. The mold is mounted in a vertical compression press with two horizontal platens. The mold half with the deeper cavity is mounted on the lower platen; the other mold half is mounted on the upper platen. The mold is heated (143 to 232°C), originally by steam and more recently by electricity. The thermosetting resin is measured into the lower mold cavity, either by weight or by volume, or preferably as a cold-pressed preformed pellet. The press is closed to heat and compress the resin. If the cure reaction releases water or other volatile by-products, the press is opened briefly to release the gases and then closed again at full pressure (1,000 to 12,000 psi). Molding pressure is maintained until thermosetting cure is complete (1/2 to 5 min). Then, the press is opened, and the molded product is ejected by the help of "knock-out pins." After that, the cycle is repeated.

For faster, easier, more uniform melt flow, the performs can be preheated to 82 to 138°C by microwave or infrared heaters. This reduces abrasion of the mold and produces higher-quality molded parts.

Originally, the molding cycle was carried out manually. With progress, most molders converted to semiautomatic operation: the operator loads the resin into the mold and takes the product out of the mold, but he activates the process by simply pressing a button, and the entire molding cycle proceeds automatically. More recently, many molders have converted to fully automatic processes wherein loading, molding, and removing the molded product are all done automatically on a preset cycle.

3.2.1.2 Transfer Molding. Transfer molding is intermediate between compression and injection molding. Whereas the compression mold has just one cavity for the finished product, the transfer mold also contains a preliminary auxiliary cavity. The resin is loaded into the auxiliary cavity ("pot"), preferably as a preheated perform, where it is heated to melt processing temperature. Then, a plunger forces the molten resin into the final mold cavity, where it cures to the finished product.

This eliminates the need to "breathe" out gases. It operates at lower pressure and therefore only needs a lighter weight mold. It is more fluid, so there is less mold wear. It gives a faster cycle and a more uniform product, to very close tolerances. There is less flash, so less post-molding finishing is needed. It is particularly useful when making complicated parts, thin walls, working with fragile inserts, and encapsulation.

The original process used a single ram, both to close the mold and to transfer the resin from the pot to the mold cavity; this was a manual operation. The preferred process at present uses one ram to close the mold and another auxiliary ram to transfer the resin from the pot to the mold cavity; this is semiautomatic. A third process, which is used occasionally, uses a screw preplasticator to warm the resin and feed it to the transfer pot; this is completely automatic.

3.2.1.3 Injection Molding. Injection molding was first developed for metals and then for thermoplastics. With further progress in both materials and equipment, it has been applied successfully to thermosetting plastics. It is used mostly for phenolics and is being used more and more for polyester bulk molding compounds.

The conventional injection molding machine is modified by giving it a shorter barrel and eliminating the 3-to-5 compression ratio of the screw. Granular resin in a conical hopper is gravity-fed into the screw, which feeds it forward, heating it by both conduction and friction, to a temperature of 66 to 93°C. As molten resin builds up at the front of the screw, this pushes the screw back until it accumulates enough material to fill the mold. Then, the screw plunges forward rapidly, pumping the molten resin through the sprue and runners into the mold cavity or, very often, multiple cavities. In the mold at 177 to 204°C, the resin cures rapidly. While the resin is curing, and while the mold is opening, discharging the molded product and closing again, the screw is already plasticating another batch for the next shot. This is a faster cycle than compression or transfer molding. It is also fully automatic.

Bulk molding compound is doughy in nature, so it is fed by plunger instead of by gravity. Since it is already viscous, it does not need much preheating to make it flow. It may be injected by screw or plunger.

Since a hot runner produces considerable thermoset scrap, it may be replaced by a "cold runner" that keeps the molten resin at 66 to 99°C, still usable for the next shot. Another innovation is injection-compression molding, where the mold is left about 1/4 in open during injection and then clamped shut to finish the cycle by compression; this reduces flow orientation and improves impact strength.

3.2.1.4 *Liquid Resin Molding and Resin Transfer Molding.*

Liquid resin, such as epoxy and polyester, curing agents, and catalysts, can be mixed at room temperature and injected into a lightweight mold at low pressure (25 to 50 psi), where they are warmed and held until cured. This is useful when working with very delicate inserts and heat-sensitive components such as electronic systems. The process is called *liquid resin molding.*

It is also useful for making fabric-reinforced laminates. The multiple layers of fabric are hand-laid into the mold in a predetermined pattern to optimize strength properties. Then the mold is closed, and the liquid mixture is injected to impregnate the fabric, often using vacuum assist to replace air thoroughly by resin. The assembly is held in the press and heated until cured. The products are similar to those made by perform press molding. The process is called *resin transfer molding.*

3.2.1.5 *Reaction Injection Molding (RIM).*

In reaction injection molding, liquid polyol and polyisocyanate are metered into a mixing head, which delivers the mixed liquid into a lightweight mold, where it polymerizes and cures rapidly to a finished polyurethane molded product. This is particularly useful in the auto industry, where it is used to mold large body parts very economically.

The polyol and polyisocyanate are stored in large tanks. Catalysts, foaming agents, and other additives are generally dissolved in the polyol. Temperature must be controlled closely (30 to 38°C) to ensure proper metering and mixing. Viscosities are under 3000 Cp. The liquids are fed to a high-pressure (2000 to 3000 psi) impingement mixer, which pumps the mixed liquid rapidly into a lightweight stainless steel or aluminum mold. The polymerization reaction is very rapid and exothermic, so the mold must be cooled to keep it at 40 to 80°C. The foaming agent evaporates, creating a pressure of 30 to 100 psi in the mold.

The resulting polymers range from slightly foamed elastomers to fairly low-density rigid structural foams. These typically may have an 0.25-in foam core surrounded by a 0.0625-in solid skin, a simple approximation to sandwich structure. Typical products range from shoe soles to beer kegs to auto front ends to as large as a 600-lb boat hull.

Low capital investment, very fast cycle, and low energy requirements combine to make this process very economical. In addition to polyurethanes, the process has been used for nylon 6 monomer casting and has been suggested for epoxy and other thermosetting resins.

3.2.1.6 Rotational Molding (Rotomolding). Liquid or powdered resin is placed in a closed, heated mold and rotated on two axes so that the resin coats the walls of the mold. This produces hollow products, ranging in size from golf balls up to 22,500-gal tanks. Capital investment is low, but operating cost is high, so it is best for short production runs.

The most common equipment is a three-arm carousel, with separate stations for loading/unloading, heating, and cooling. The mold is usually cast aluminum, or electroformed nickel for fine detail.

It rotates on two perpendicular axes at speeds up to 40 RPM. It is heated in an air oven or by oil in a jacketed mold. The resin is distributed by the force of gravity—the speed is not high enough for centrifugal force. A complete cycle may be 5 to 30 min. Yields are high, because there is no waste of material.

The process is applied mainly to polyethylene but also to other thermoplastics and to a number of thermosetting resins: epoxy, phenolic, polyurethane, and silicone. It is used to produce tanks, trucks, bathrooms, boat hulls, appliance housings, and toys. It is possible to produce multiwall construction by successive loading with different resins.

3.2.2 Reinforced Plastics Processes

Addition of reinforcing fibers to plastics increases their modulus, strength (especially impact strength), dimensional stability, and heat deflection temperature. Short, chopped fibers can be dispersed in all thermoplastics and thermosets; melt processing remains fairly conventional, and property improvements are significant. Maximum improvement of properties, however, requires continuous fibers, and especially fabrics. These defy conventional melt processing. They require the fluidity of low-molecular-weight thermosetting oligomers and the development of new processing techniques, which have come to characterize reinforced plastics processing. They are often called laminated plastics, and the processes are called lamination.

3.2.2.1 Hand Layup (Contact Molding or Open Mold Process). The oldest and simplest technique is primarily manual. An open mold is waxed to lubricate it. The surface of the proposed product is applied to the mold, usually as an unreinforced gel coat or sometimes as a thermoformed plastic sheet. A layer of fabric is hand cut and hand laid into the mold. Catalyzed resin is poured over the fabric and worked into it with brush, roller, and squeegee to eliminate air bubbles. Another layer of fabric is laid over this, often oriented in a specific direction to optimize properties. Again, catalyzed resin is poured over it and worked into it. The process is repeated as many times as required to build up the desired thickness and optimum orientation of the layers. If the cure reaction is inhibited by air, a wax may be dissolved in the resin; it exudes to the surface and forms a barrier to exclude air and permit complete cure. The resin may be allowed to cure at room temperature, or the assembly may be heated to complete or hasten the cure reaction.

Polyester and epoxy resins are the most common. Capital investment is very low; labor cost is high, and skill is required. The process is ideal for small production runs of large products such as boats, tanks, flat panels, tools, and prototypes.

3.2.2.2 Sprayup. Sprayup is a mechanized version of hand layup. The spray gun is fed with (1) glass roving, (2) liquid resin, and (3) catalyst. The gun chops the glass roving into short lengths, mixes them with the catalyzed resin, and sprays them into the mold. If the gun is handled manually, skill is required, and quality will vary. If the gun is controlled automatically, less manpower is required, and quality is more uniform. As in hand layup, the sprayed glass-resin mixture must be pressed down with a roller to squeeze out air bubbles. Since hand layup can use woven fabric, it gives greater strength; sprayup delivers short, chopped fiber as a random mat, so strength is lower. The spray gun requires a modest capital investment, but the labor cost is lower. This technique is particularly common in building boat hulls.

3.2.2.3 Vacuum Bag Molding. To improve compaction of hand or spray laid assemblies, the wet product is covered with a release film such as polyvinyl alcohol, nylon, or silicone rubber. The edges of the film are clamped and sealed to the edge flanges of the mold. An internal vacuum is pumped, so atmospheric pressure presses the film down onto the product. Hand roller pressure helps the process. This squeezes out air bubbles and excess resin and results in a denser product with a better inside surface. The extra work adds to the cost, so this is used mainly in the aerospace and military fields, where the improved quality is worth the cost.

3.2.2.4 Pressure Bag Molding. A further improvement over vacuum bag molding is to place a rubber bag over the assembly, fasten it in place, and inflate it up to 50 psi air pressure. This provides additional force to squeeze out air and excess resin and produces a denser, more uniform product.

It can also be used to make hollow shapes such as tanks. The bag is covered with a preformed glass fiber mat and placed inside a hollow mold. The glass fiber mat is impregnated with catalyzed liquid resin, and air pressure inside the bag presses the assembly against the interior surface of the mold. The mold can be heated to speed the cure reaction.

Here again, labor cost and slow process cycle are economic limitations. Also, the mold must be sturdier to withstand the pressure.

3.2.2.5 Autoclave Molding. Another further improvement is to put the pressure bag assembly into a heated autoclave and fill the autoclave with up to 80 psi air pressure. Here again, equipment and labor cost are higher, but quality is higher too. Typical products are rocket nozzles, nose cones, heat shields, electronics, and aircraft parts.

3.2.2.6 Centrifugal Casting. Cylindrical products such as pipe, tubing, and tanks can be produced by centrifugal casting. Glass fiber mat is laid or sprayed onto the inner walls of a rotating cylindrical mold. Catalyzed resin is sprayed onto the mat. Speed of rotation is increased to densify the layup, and the mold is heated to speed the cure reaction. The process can also make other shapes of simple rotation such as cones and parabolas. The equipment does require some capital investment, but labor cost is low, and product quality and uniformity are high.

3.2.2.7 Rubber Plug Molding. To make a hollow product, a metal mold contains a cavity for the exterior of the product. Liquid silicone rubber is poured into the cavity and cured. The silicone rubber plug is pulled from the mold and preferably covered with a PTFE film. The glass fiber and liquid resin are built up on the surface of the silicone rubber plug. This assembly is inserted back into the mold cavity. The mold and silicone plug are heated, the silicone plug expands, and the resin cures. Then, the assembly is removed from the mold, and the silicone plug is removed from the product.

3.2.2.8 Matched Die Molding. All of the above techniques are relatively specialized processes for small production runs. For economical mass production of uniform quality products, matched die molding is the standard technique. This is used for manufacture of auto and appliance parts.

Matched steel dies are chrome plated and operated in heated compression presses. The entire premix is placed in the mold and pressed and heated 1 to 2 min to produce complete cure. Matched die molding is applied to bulk molding compound, sheet molding compound, prepregs, and performs.

3.2.2.9 Bulk Molding Compound (BMC). The entire bulk molding compound (BMC) is premixed—resin, reinforcing fiber, fillers, catalysts, mold lubricants, colorants. Optionally, a thermoplastic resin may be added to reduce cure shrinkage and give a low-profile, high-quality surface. To produce a stable doughy consistency, CaO or MgO may be added to react with acid end-groups on the resin and produce gelation. A weighed amount of doughy BMC is simply placed in the matched die mold, pressed, and cured. For good moldability, the glass fibers are generally under 1 in long.

3.2.2.10 Sheet Molding Compound (SMC). Liquid resin is spread onto a moving web of polyethylene film. Glass roving is chopped and spread over it. More resin is spread over that, and a second polyethylene film is overlaid on it. The assembly is kneaded to remove air and distribute resin and fibers more uniformly. Then, it is rolled up and stored until use. For molding, a portion of the leathery sheet is cut, placed in the matched die mold, pressed, and heated about a minute, until cured.

Glass fiber length in SMC is generally 1 to 2 in, which is less moldable than BMC but gives higher strength properties.

3.2.2.11 Prepreg Process. Reinforcing fabric is impregnated with catalyzed resin and advanced to the gelled B-stage. It has limited life, so it must be refrigerated and used up fairly rapidly.

The preimpregnated fabric is cut to the desired shape and hand laid into or onto the mold. Successive layers can be oriented to optimize properties. Then, the assembly is matched die molded or vacuum bag molded until cured.

This gives high-strength products. Manpower and skill are expensive. It is used for aerospace products such as radomes, brakes, and communication housings.

3.2.2.12 Preforms. A metal screen is made in the shape of the desired product. Glass roving is chopped and blown onto the surface of the screen. Pulling air or even vacuum on the backside of the screen helps to compact the chopped fiber onto the screen. A few percent of binder resin is sprayed onto the chopped fiber preform to stabilize it. Then, it is removed from the screen and placed in the mold. There, it is impregnated with catalyzed resin, pressed, heated, and cured. For greater strength, the fibers can be braided onto the perform to form a woven fabric reinforcement.

3.2.2.13 Foam Reservoir Molding. A sheet of open-cell polyurethane foam is impregnated with catalyzed thermosetting resin. It is sandwiched between two layers of reinforcing fabric. This is then molded at low pressure (<100 psi). This is sufficient to squeeze liquid resin from the foam out to the fabric skins. When the assembly cures, it is a sandwich structure of low density, with good flexural rigidity and impact strength. This is useful for making large fairly flat shapes such as vehicle roofs, decks, and hoods.

3.2.2.14 Continuous Lamination. Mass production of flat or corrugated sheet is accomplished by spreading liquid catalyzed resin on a web of release sheet, then chopping glass fiber onto the wet resin, covering with a second sheet of release film, kneading between pressure rollers to squeeze out air and distribute resin and fiber uniformly, passing through a heating zone to cure the resin, and cutting to the desired size. This is used for production of glazing, paneling, roofing, hoods, and ducts.

3.2.2.15 Pultrusion. This process produces long, continuous products of constant cross-section. It is a low-cost mass-production technique. Continuous filament, tape, or fabric is pulled through catalyzed liquid resin. Next, it is squeezed to remove excess resin. Then, it passes through a die to shape the cross section. Then, it passes through a heated die to cure it. Finally, it is cut to length, stacked, and packed. Strength in the lengthwise direction is outstanding. Typical products include fishing rods, flag poles, tent poles, antennas, golf club shafts, hockey sticks, bows and arrows, ski poles, fence posts, ladders, light poles, pipe supports, and tool handles.

3.2.2.16 Filament Winding. The strongest plastic products ever made, competing with or exceeding metals, are made by filament winding. Continuous filament or fabric goes continuously through a catalyzed liquid resin bath and is wound onto a mandrel in the shape of the desired product. The winding pattern is calculated and controlled to produce maximum strength. When it has been wound layer-upon-layer up to the desired thickness, the assembly is oven cured. The mandrel may remain as part of the finished product; more often, it is designed so that it can be collapsed and removed from the cured filament-wound product. Most products are simple cylindrical shapes, but more complex shapes can be produced by thoughtful design. The most common products are pressure pipes and pressure tanks. Some more specialized products include rocket motors, railroad hopper cars, turbine blades, helicopter blades, and plastic housing.

3.2.3 Pouring Processes

Liquid A-stage thermosetting resins can be poured into finished form and simply cured in place. These processes are often called *casting* and sometimes distinguished as *potting*, *encapsulation*, and *dip coating*.

3.2.3.1 Casting. Epoxy, polysulfide, polyester, polyurethane, and silicone A-stage liquids can be poured into a mold, cured, and removed from the mold as finished products. Typical products are simulated wood frames, figurines, and furniture decoration; electrical and electronic products; and solid polyurethane rubber tires for industrial equipment.

Small-scale production is manual, whereas larger production runs can be automated. For some products, rubber molds make it easy to remove the products. Where gas bubbles may disfigure the product or ruin electrical performance, vacuum degassing can prevent this problem. For penetration into fine details, vacuum and pressure impregnation are helpful.

Conversely, pouring a foamable plastic is useful for gap filling, light weight, and thermal insulation. Use of hollow glass or plastic spheres can encapsulate closed-cell bubbles in syntactic foam, which is useful for low dielectric constant and loss and for compressive resistance such as deep-sea immersion.

3.2.3.2 Potting. Electrical and electronic assemblies are often insulated and protected against mechanical abuse and environmental attack by placing them in a shell, filling the space by casting liquid A-stage thermosetting resins, and curing them in place. When the cured assembly is left in the shell, the process is called *potting*.

3.2.3.3 Encapsulation. Similarly, when the electrical/electronic assembly is cast and potted, and the shell is then removed, the process is called *encapsulation* or *embedment*.

3.2.3.4 Dip Coating. Electrical and electronic products may be insulated and protected by a conformal coating. This is produced by dipping the product into a thixotropic A-stage thermosetting resin, rotating it to ensure uniform coverage and thickness, and curing it to a finished coating which coats the entire product.

3.2.4 Powder Coating Processes

When metals are coated with polymer solutions, the solvent brings problems of flammability, toxicity, environmental pollution, and cost. As an alternative, powder coating simply applies dry powder to the metal and heats it to fuse and flow into a coating. When the powder is a thermosetting resin, heating continues to complete the cure reaction as well.

Powdered compound is prepared mainly by preblending the ingredients, fusing them in an extruder or dough-mixer, crushing, and grinding to the desired particle size. For thermosetting powders, heat history must, of course, be controlled to prevent premature cure. Powders have also been prepared by ball milling and by evaporation of solutions.

The powders are expensive, but automated coating and 100 percent utilization of powder are economical. Coatings are applied to metals for electrical insulation, corrosion resistance, and simple decoration. Thick coatings can be applied in a single pass. There are three specific powder coating techniques: fluid bed, electrostatic fluid bed, and electrostatic spray.

3.2.4.1 Fluid Bed. The 20 to 200 μm powder is placed in a tank with a porous bottom. Compressed air is blown up through the porous bottom, into the bed of resin. The bed expands and rises, and behaves much like an opaque liquid. Vibrating the bed can help.

The metal object to be coated is preheated in an oven and dipped into the fluid bed. The powder particles stick to the hot surface of the metal and flow into a continuous coat. After the object is withdrawn from the bed, it may be heated further in an oven to reflow the coating more uniformly and smoothly and to complete thermosetting cure.

Coating thickness is often 4 to 15 mils or more. Typical applications include electric motors, capacitors, resistors, bus bars, and transformers; and valves, pumps, and refinery equipment.

3.2.4.2 Electrostatic Fluid Bed. When charged air is fed up through the fluid bed, the powder particles become charged as well. When a grounded metal object is passed over the bed, the particles are attracted, fly up, and stick to it. Then, it is sent to an oven to fuse and flow the particles into a continuous film and to cure the thermosetting resin.

The leading powder is epoxy resin, applied 1 to 50 mils thick; this is used for corrosion resistance in auto underbodies, steel rebars for concrete, petroleum industry, appliances, and citrus food cans. Epoxy-polyester powders are more resistant to yellowing and are used 1 to 20 mils thick for aluminum wheels and architectural trim. Polyester-polyurethane powders are smooth, flexible, and weather resistant; 1 to 3 mil films are used for auto trim, steel wheels, lighting fixtures, patio furniture, and appliances.

3.2.4.3 Electrostatic Spray. Powder 30 to 120 μm in size is fed from a fluid bed, into a gun that puts an electric charge on it (high voltage and low amperage for safety), and compressed air fires the charged powder toward the grounded object in a hood. The powder collects on the object and clings to it; then it is heated in an oven to fuse the powder, flow it into a continuous film, and cure the thermosetting resin. This applies coatings 1 to 8 mils thick. The powder overspray is collected in the hood and recycled directly.

A less common version of this process is *friction static spray*, where the electrostatic charge is produced by friction between powder particles and the spray gun itself.

3.2.5 Post-Cure

Polymerization and cross-linking of thermosetting plastics is a gradual process. It depends on reactive groups meeting each other to complete the cure reaction. The further cross-linking proceeds, the more the molecules are frozen in place, and the harder it is for them to migrate and meet and react. Thus, complete cure may take very long.

Most thermosetting processes require heating in the process equipment. In a long cure cycle, this can tie up expensive equipment for a long time, involving high capital investment. In such cases, processors may find it more economical to use their primary process equipment to simply shape the product and then move it to a low-cost oven to complete the cure cycle more economically. Thus, such post-cure stages are often used for greater efficiency. This is particularly common after molding and extrusion processing.

3.3 CONCLUSION

Themosetting plastics involve more chemistry and higher cost than commodity thermoplastics. On the other hand, they offer many advantages in specialized processes and superior performance in final products. This is why they remain a vital and growing portion of the plastics industry. They offer many opportunities for further growth in the future.

3.4 BIBLIOGRAPHY

- General

American Plastics Council.
J. A. Brydson, *Plastics Materials,* Butterworth Scientific, 1982.
S. H. Goodman, *Handbook of Thermoset Plastics,* Noyes Publications, 1998.
R. E. Wright, Thermosets, Reinforced Plastics, and Composites, Ch. 2 in C. H. Harper, *Modern Plastics Handbook,* McGraw-Hill, 2000.
Modern Plastics Magazine.

- Polyurethanes

J. A. Brydson, *Plastics Materials,* Butterworth Scientific, 1982.
D. Klempner and K. C. Frisch, *Handbook of Polymeric Foams and Foam Technology,* Hanser, 1991.
S. Orchon, Polyurethanes, Thermoset, Ch. 42 in I. I. Rubin, *Handbook of Plastics Materials and Technology,* Wiley, 1990.
J. H. Saunders and K. C. Frisch, *Polyurethanes: Chemistry and Technology,* Interscience, 1962.

- Phenol-Formaldehyde

S. Black, *High-Performance Composites,* Nov. 2004, p. 28.
J. A. Brydson, *Plastics Materials,* Butterworth Scientific, 1982.
T. S. Carswell, *Phenoplasts,* Interscience, 1947.
C. C. Ibeh, Phenol-Formaldehyde Resins, Ch. 2 in S. H. Goodman, *Handbook of Thermoset Plastics,* Noyes Publications, 1998.

- Urea- and Melamine-Formaldehyde

J. A. Brydson, *Plastics Materials,* Butterworth Scientific, 1982.
C. C. Ibeh, Amino and Furan Resins, Ch. 3 in S. H. Goodman, *Handbook of Thermoset Plastics,* Noyes Publications, 1998.
N. E.Reyburn and A. E. Campi, Urea Melamine, Ch. 52 in I. I. Rubin, *Handbook of Plastics Materials and Technology,* Wiley, 1990.

- Furan Resins

J. A. Brydson, *Plastics Materials,* Butterworth Scientific, 1982.
C. C. Ibeh, Amino and Furan Resins, Ch. 3 in S. H. Goodman, *Handbook of Thermoset Plastics,* Noyes Publications, 1998.

- Polyesters

J. F. Dockum Jr., Ch. 14 in J. V. Milewski and H. S. Katz, *Handbook of Reinforcements for Plastics,* Van Nostrand Reinhold, 1987.
I. I. Rubin, *Handbook of Plastic Materials and Technology,* Wiley, 1990.
O. C. Zaske and S. H. Goodman, Ch. 4 in S. H. Goodman, *Handbook of Thermoset Plastics,* Noyes Publications, 1998.

- Vinyl Esters

V. B. Messick and M. N. White, Ch. 75 in I. I. Rubin, *Handbook of Plastic Materials and Technology,* Wiley, 1990.
O. C. Zaske and S. H. Goodman, Ch. 4 in S. H. Goodman, *Handbook of Thermoset Plastics,* Noyes Publications, 1998.

- Allyls

J. A. Brydson, *Plastics Materials,* Butterworth Scientific, 1982.
S. H. Goodman, *Handbook of Thermoset Plastics, Noyes Publications,* 1998, Ch. 5.
R. E. Wright and M. F. Gardner, Ch. 4 in I. I. Rubin, *Handbook of Plastic Materials and Technology,* Wiley, 1990.

- Epoxy Resins

J. A. Brydson, *Plastics Materials,* Butterworth Scientific, 1982, Ch. 26.
J. Gannon and V. Brytus, Ch. 72 in I. I. Rubin, *Handbook of Plastic Materials and Technology,* Wiley, 1990.
S. H. Goodman, *Handbook of Thermoset Plastics,* Noyes Publications, 1998, Ch. 6.
C. A. May, *Epoxy Resins: Chemistry and Technology,* Dekker, 1988.

- Silicones

J. A. Brydson, *Plastics Materials,* Butterworth Scientific, 1982, Ch. 29.
J. C. Caprino and R. F. Macander, Ch. 13 in M. Morton, *Rubber Technology,* Van Nostrand Reinhold, 1987.
R. F. Patterson, Ch. 9 in S. H. Goodman, *Handbook of Thermoset Plastics,* Noyes Publications, 1998.

- Polyimides

S. Black, *High-Performance Composites,* Nov. 2004, p. 26.
H.-G. Elias and F. Vohwinkel, *New Commercial Polymers 2,* Gordon & Breach, 1986, Ch. 11.
J. J. King and B. H. Lee, in R. B. Seymour and G. S. Kirshenbaum, *High-Performance Polymers: Their Origin and Development,* Elsevier, 1986, p. 317-330.
A. L. Landis and K. S. Y. Lau, Ch. 8 in S. H. Goodman, *Handbook of Thermoset Plastics,* Noyes Publications, 1998.

- Miscellaneous Cross-Linking Reactions

S. Black, *High-Performance Composites*, Nov. 2004, p. 27–28.
J. A. Brydson, *Plastics Materials,* Butterworths, 1982, p. 721.
H.-G. Elias and F. Vohwinkel, *New Commercial Polymers 2,* Gordon and Breach, 1986, p. 410.
R. B. Graver, in R. B. Seymour and G. S. Kirshenbaum, *High Performance Polymers: Their Origin and Development,* Elsevier, 1986, p. 309.
A. L. Landis and K. S. Y. Lau, Ch. 8 in S. H. Goodman, *Handbook of Thermoset Plastics,* Noyes Publications, 1998.
L. A. Pilato and M. J. Michno, *Advanced Composite Materials,* Springer-Verlag, 1994, Section 2.2.

- Cross-Linking of Thermoplastics

J. A. Brydson, *Plastics Materials,* Butterworths, 1982.
R. P. Patterson, Ch. 10 in S. H. Goodman, *Handbook of Thermoset Plastics,* Noyes Publications, 1998.

- Cross-Linking of Elastomers, Coatings, and Adhesives

A. Brandau, *Introduction to Coatings Technology,* Federation of Societies for Coatings Technology, 1990.
M. Morton, *Rubber Technology,* Van Nostrand Reinhold, 1987.
I. Skeist, *Handbook of Adhesives,* Van Nostrand Reinhold, 1990.
F. H. Walker, *Introduction to Polymers and Resins,* Federation of Societies for Coatings Technology, 1999.

- Processes

M. L. Berins, *Plastics Engineering Handbook,* Van Nostrand Reinhold, 1991.
C. A. Harper, *Modern Plastics Handbook,* McGraw-Hill, 2000, Ch. 6.
I. I. Rubin, *Handbook of Plastic Materials and Technology,* Wiley, 1990.

CHAPTER 4
ELASTOMERS

Aubert Y. Coran

Longboat Key, Florida

4.1 INTRODUCTION

The term *elastomer* is often used interchangeably with the term *rubber*. Elastomers (or rubbers) are amorphous polymers. Their normal-use temperatures are above their glass transition temperatures, so considerable molecular segmental motion is possible. Hard plastics normally either exist below their glass transition temperatures, or they are semicrystalline solids at room temperature.

Elastomers are different from other polymers because of their special properties. According to ASTM (D 1566), a *rubber* is a material that is capable of recovering from large deformations quickly and forcibly and can be, or already is, modified (i.e., vulcanized) to a state in which it is essentially insoluble (but can swell) in boiling solvent. A rubber, in its vulcanized state, retracts within 1 min to less than 1.5 times its original length after being stretched at room temperature to twice its length and held for 1 min before release. The term *elastomer* is often used designate polymers that have properties similar to those of a rubber. At ambient temperatures, rubbers are thus soft and deformable.

Elastomers are used in a wide variety of applications because of their unusual physical properties (flexibility, extensibility, resiliency, and durability), which are unmatched by other types of materials. Other useful properties include abrasion resistance, resistance to aqueous and other polar fluids, weathering resistance, and high frictional coefficients for traction. Elastomeric materials are frequently tailor-made for specific applications, because they can be significantly modified by compounding.

Vulcanized elastomers are thermosets (having required vulcanization). The long polymer chains are cross-linked during curing (i.e., vulcanization). Vulcanization, which generally requires some time at an elevated temperature, must take place after the elastomer is in its final shape or form (e.g., in a heated mold) because, after significant cross-linking, the polymer cannot flow (e.g., in the mold). Thermoplastic elastomers (TPEs) are elastomeric without being vulcanized. They are processed in the same way as rigid thermoplastics, e.g., polystyrene and polyethylene, without the need for time-consuming vulcanization.

The first known elastomer was natural rubber. (It was vulcanizable.) Precolumbian peoples of South and Central America used it, however, without vulcanization, to make balls, containers, and shoes and for waterproofing fabrics. Mentioned by Spanish and Portuguese writers in the 16th century, natural rubber did not attract the interest of Europeans until reports about it were made (1736 to 1751) to the French Academy of Sciences by Charles de la Condamine and François Fresneau. They called the substance by the name used by the natives, *caoutchouc*. Before 1800, natural rubber was used only for elastic bands and erasers, and these were made by cutting up pieces imported from Brazil. Joseph Priestley is credited with the discovery, c.1770, of its use as an eraser, thus the name *rubber*. In 1823, Charles Macintosh found a practical process using rubber to waterproof fabrics, and in 1839 Charles Goodyear discovered vulcanization, which revolutionized the rubber industry. In the latter half of the 19th century, the demand for rubber insulation by the electrical industry and the invention of the pneumatic tire extended the demand for rubber. Since the introduction of natural rubber, but not until fairly recently (starting in the 1930s), many synthetic vulcanizable rubbers, having various clusters of specific properties suited for different uses, have been developed.

Examples of vulcanizable elastomers include natural rubber (NR), styrene butadiene rubber (SBR), butadiene rubber (BR), ethylene-propylene-diene monomer-rubber (EPDM), butyl rubber (IIR), polychloroprene or neoprene (CR), epichlorohydrin rubber (ECO), polyacrylate rubber (ACM), millable polyurethane rubber, silicone rubber, and fluoroelastomers. Examples of thermoplastic elastomers include thermoplastic polyurethane elastomers, styrenic thermoplastic elastomers, polyolefin-based thermoplastic elastomers, thermoplastic polyether-ester (copolyester) elastomers, and thermoplastic elastomers based on polyamides.

This chapter is a perspective of the science and technology of elastomers and does not include a market analysis. Nevertheless, we must mention that the global market for these materials is large (Fig. 4.1). Global vulcanizable (conventional, vulcanizable) rubber consumption was about 20 million metric tons in 2004, whereas thermoplastic elastomer (TPE) consumption was about 1.5 million metric tons. The consumption of conventional rubbers is growing at a rate of about 3 to 4 percent, whereas the growth of TPE consumption is growing at about twice that rate.

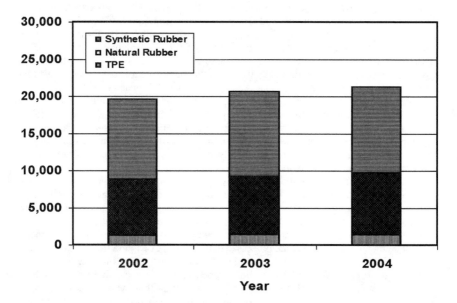

FIGURE 4.1   Estimated global consumption of rubbers and elastomers.

In this chapter, we discuss both vulcanizable elastomers and thermoplastic elastomers. There is another, albeit narrow-niche type (i.e., liquid rubbers), which we will not discuss here. Our focus will be on the various types of elastomers, their properties, applications, and processing.

4.2 DIFFERENCES AND SIMILARITIES BETWEEN ELASTOMERS AND PLASTICS

Elastomers and plastics are generally considered to be significantly different types of materials. Mechanical and functional requirements of parts made from these materials are quite different. They cannot be readily used interchangeably. Also, methods for the fabrication of plastics and conventional elastomers into usable parts are generally very different. The preponderant (vulcanizable) rubbers and elastomers are thermoset materials, requiring heat for the vulcanization process, which develops a molecular network of crosslinks between polymer backbone chains. The predominant plastics (the thermoplastic types) require heat only for flow in processing. The language and thinking of the elastomer and plastic industries are quite different. In addition, plastic materials are frequently used "out of the bag" or with relatively few ingredients added (or compounded) by the fabricator of the end-use parts. On the other hand, elastomer materials technology, the technology of making rubber parts, requires extensive compounding. A typical recipe contains elastomeric polymer, filler, antidegradants, plasticizer, processing or flow agent, vulcanizing agent, vulcanization accelerator, vulcanization activators, and so on. In addition to the selection of the polymer (or blend of polymers), the properties of the vulcanized rubber depend very strongly on the specific compound formulation. Complicated rubber recipes existed long before plastics were developed, having been first devised in the mid 1800s. Frequently, one prejudicially thinks that plastics engineering is modern technology, but that and rubber-product development is a black art.

Rigid plastics are used because they can support stress without major strain or deformation, whereas soft elastomeric materials are used because they can be easily deformed by the application of a small amounts of stress. These differences are illustrated by the stress-strain curves of Fig. 4.2.

Nevertheless, many elastomers and plastics are fundamentally very similar. Most plastics and elastomers comprise long chains of one or more types of linked monomer units. In fact, many of the same monomers are found in both thermoplastic and elastomeric polymers—e.g., styrene, acrylonitrile, ethylene, propylene, and acrylate esters. Because of the chemical similarities between elastomers and plastics, these materials are susceptible to many of the same types of chemical attack. Therefore, many of the same material-selection principles come into play for both plastics and elastomers.

Unvulcanized conventional elastomers comprise linear, long-chain molecules, exactly as do thermoplastic plastics. However, an engineering thermoplastic is a rigid solid at room temperature, whereas an uncured elastomer is a soft, pliable gum. A rigid thermoplastic has a high elastic modulus at room temperature, whereas an elastomer has a low modulus at room temperature. These differences between the moduli of elastomers and plastics are due to differences in mobility of long molecular segments. Rubber molecules, at room temperature, are far more mobile than those of rigid plastics. Segmental molecular mobility depends on the ability of main-chain atoms to rotate with respect to one another at single chemical bonds in the molecular main chains. Since the angle between adjacent bonds is less than 180° (see Fig. 4.3, wherein poly-*cis*-1,4-butadiene is used as an example), rotations between linked atoms enable bending of the chain and thus the motion of

STRESS

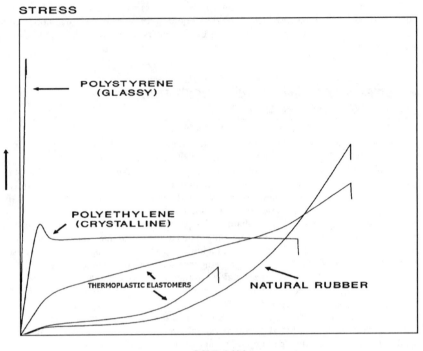

FIGURE 4.2   Stress-strain curves for rubbers and plastics.

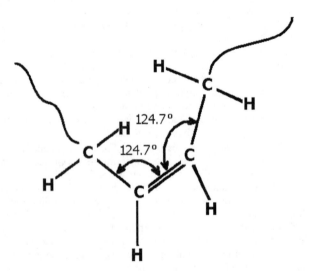

FIGURE 4.3   Backbone chain bond angles for poly-*cis*-1,4-butadiene.

segments of the linear polymer chain. Such motions, however, are suppressed by intermolecular attractions and, to some extent, by the presence of side chains and physical molecular entanglements. Effects of such factors determine the glass transition temperature (T_g) of the polymer. Below this temperature, a noncrystalline plastic or elastomeric polymer is a supercooled liquid and behaves in many ways like a rigid glass. Above this temperature, an uncross-linked noncrystalline linear polymer can flow and be processed and fabricated into final shapes and forms. Above this temperature, a cross-linked polymer has rubber-like properties unless it contains significant crystallinity. Thus, a crystalline melting point (T_m) can also be very important determinant for the behavior of polymeric materials.

The behavior of a thermoplastic material above its glass transition temperature depends on its level of crystallinity. As a noncrystalline (amorphous) polymer is slowly heated from a temperature below its T_g, it displays a large decrease in modulus as the glass transition temperature is reached. As one heats a semicrystalline plastic from a temperature below its T_g, it displays a relatively small modulus change at the glass transition temperature, followed by a plateau and then a decreasing modulus as the temperature increases and approaches the crystalline melting point.

Elastomers have glass transition temperatures well below room temperature. Rigid thermoplastics have glass transition temperatures that fall in a wide temperature range, ranging from below to above room temperature. However, if the T_g is below room temperature, the polymer must be semicrystalline to be rigid. If the T_g is well above room temperature, the noncrystalline polymer will be a rigid glassy polymer.

The presence of cross-links in a cured elastomer gives it elasticity and prevents it from becoming molten and flowing above its glass transition temperature. Elastomers can, and some do, contain small amounts of crystallinity while still being rubbery. Also, some noncrystalline elastomers can partially crystallize during stretching, and this can be a strengthening, toughening, or tear-strength-increasing mechanism.

The changes in stiffness as a function of temperature, *vis à vis* T_g and T_m are illustrated by Fig. 4.4. Table 4.1 gives values of T_g and T_m for selected polymers. Viewing Fig. 4.4 with Table 4.1 in mind, one obtains a perspective with respect to how these polymers behave as a function of temperature.

4.2.1 Differentiating Elastomers and Plastics by Measuring Dynamic Mechanical Properties

A very good way to characterize and differentiate between elastomers and rigid plastics is by the measurement of dynamic mechanical properties. A most convenient method to study dynamic mechanical properties is to impose a small, sinusoidal shear or tensile strain and measure the resulting stress. Dynamic mechanical properties are most simply determined for a small sinusoidally varying strain, for which the response is a sinusoidally varying stress. An increase in frequency of the sinusoidal deformation is equivalent to an increase in strain rate.

The shearing deformation of a sample confined between two parallel plates, as illustrated in Fig. 4.5, is described by the strain, γ. Strain is defined as the displacement of the top surface divided by the height of the sample. The stress, σ, is the tangential force per unit area producing the deformation. When a sample is subjected to oscillatory shear deformations, the strain γ varies sinusoidally with time as

$$\gamma(t) = \gamma_0 \sin \omega t \qquad (4.1)$$

where γ_0 is the strain amplitude (peak strain), ω the angular frequency, (2π times the frequency in hertz), and t the time. The stress, σ, will also oscillate sinusoidally with the an-

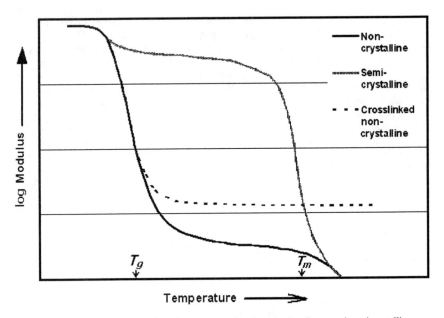

FIGURE 4.4 Modulus as a function of temperature for elastomeric, glassy, and semicrystalline materials.

gular frequency, ω, as illustrated in Fig. 4.6 (stress amplitude σ_0); however, the stress will lead the strain in time, and the phase shift is given by a phase angle, δ, resulting in

$$\sigma(t) = \sigma_0 \sin(\omega t + \delta) \qquad (4.2)$$

where ω is the angular frequency and t is time.

The stress can be decomposed into a component in phase with the strain, proportional to $\sin \omega t$, and another component 90° out of phase, proportional to $\cos \omega t$, as shown in the figure. The total stress can then be expressed as

$$\sigma(t) = \gamma_0 [G'(\omega) \sin \omega t + G''(\omega) \cos \omega t] \qquad (4.3)$$

where $G'(\omega)$ and $G''(\omega)$ are the storage and loss moduli. The storage modulus G' is a measure of energy (elastic) stored and recovered in cyclic deformation, whereas the loss modulus G'' is a measure of energy dissipated as heat, e.g., generated by a viscous process. The ratio G''/G' is tan δ, the loss tangent. This is also referred to as *hysteresis*.

One way to look at this is to consider a "spring and dashpot" model for a viscoelastic material. This model is illustrated by Fig. 4.7. The spring represents the elastic component G', whereas the dashpot represents the viscous component. The stress due to the "spring" is proportional to the strain through the proportionality constant G'. On the other hand, the stress due to the viscous-fluid behavior of the dashpot is proportional to the strain rate. Thus, elastic-component stress is proportional to the sine of ωt, and the viscous component is proportional to the cosine (derivative of sine) of ωt.

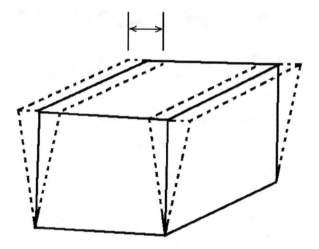

FIGURE 4.5 Dynamic shear deformation.

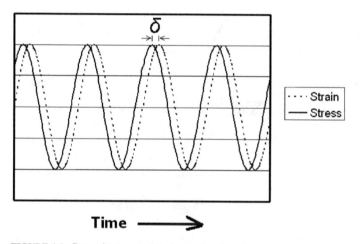

FIGURE 4.6 Dynamic stress and strain as a function of time.

The two moduli are functions of temperature and frequency. The parameters G' and tan δ, for a typical uncross-linked (unvulcanized) rubber or elastomer, are shown in Fig. 4.8 as a function of temperature (with frequency on the order of 1 Hz). The parameters are plotted against frequency, at a constant temperature, in Fig. 4.9. Thus, one observes that an increase in frequency (equivalent to an increase in strain rate) is similar to a decreased temperature.

With increasing temperature (from below T_g), tan δ increases and then decreases, with a maximum (peak) value at a temperature where the storage modulus G' decreases, with respect to temperature, at about the maximum rate. This temperature also corresponds to the glass transition. In this transition region of temperature, the material is relatively soft

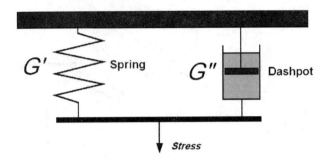

FIGURE 4.7   "Spring-and-dashpot" model for viscoelastic behavior.

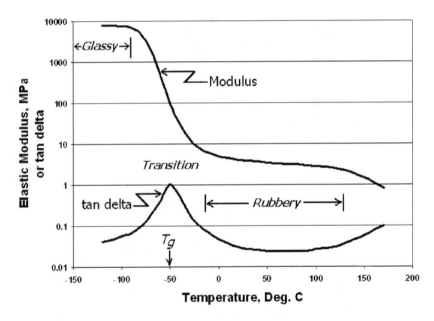

FIGURE 4.8   Dynamic storage modulus and tan δ as a function of temperature for a typical un-cross-linked elastomer.

and very sluggish and lossy (hysteretic) with a rather leathery consistency. With further heating, the material becomes rubbery.

As stated above, a major difference between an elastomer and a glassy polymer is the value of T_g. Thus, T_g is a major characteristic of a polymeric material, determining whether it is glassy or rubbery at its temperature of use.

4.2.2   The Presence of Cross-Links in Elastomers

The cross-links of a vulcanized elastomer anchor elastomer molecular segment chain-ends together at cross-link junctures such that many molecular monomer units span the distance

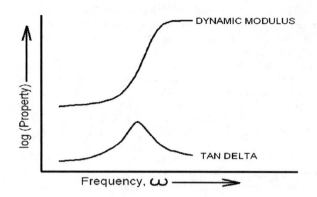

FIGURE 4.9 Dynamic storage modulus and tan δ as a function of angular frequency for a typical uncross-linked elastomer.

TABLE 4.1 Glass Transition and Melting Temperatures of Selected Plastics and Vulcanized Elastomers

	T_m, °C	T_g, °C
Elastomers		
Butadiene rubber	–	−102
Natural rubber	–	−60
Neoprene rubber	–	−43
Butyl rubber	–	−73
Silicone rubber	–	−127
Poly(butyl acrylate)	–	−55
Styrene-butadiene rubber	–	−52
Plastics		
Polyethylene	135	−78
Polypropylene	165	−13
Poly(vinyl chloride)	–	81
Poly(ethylene terephthalate)	280	69
Nylon 6	260	40
Polystyrene	–	100

between cross-links. The resistance to deformation is somewhat increased by the presence of cross-links, and T_g changes only slightly. However, the presence of the cross-links causes a rubber specimen to spring back to its original shape when the deforming stress is removed. If the deformed elastomer specimen snaps back to nearly its original shape rapidly with respect to the time scale of the experiment, it is said to be resilient. Resilience is the percentage of energy returned per cycle of *rapid* deformation upon removal of the stress. This rapid recovery from a deformation is responsible for the bouncing of a rubber ball. The opposite of resilience is *damping* or *hysteresis*, i.e., tan δ. It is a viscous effect rather than an elastic effect. The presence of cross-links reduces the viscous effect while it increases elasticity of these viscoelastic materials.

The presence of cross-links also helps in maintaining a desired geometry. Engineering thermoplastics, which contain no cross-links, are subject to creep, or cold flow, under load. They also have poor compression set resistance. Set is a permanent deformation that occurs under a load. Cross-linked elastomers can vary significantly in their set resistance. The choice of polymer, vulcanization system, and degree of cross-linking can profoundly affect set resistance (as well as many of the other properties of the vulcanizates).

4.3 TYPES OF ELASTOMERS

Until relatively recently, all elastomers were vulcanized. However, thermoplastic elastomers were first introduced in 1954 with the introduction of urethane thermoplastic elastomers. Thus, the two major types of elastomers are vulcanizable (conventional) and thermoplastic elastomers. The conventional elastomers are frequently broadly classified as natural and synthetic rubbers.

4.3.1 Conventional Vulcanizable Elastomers

A large number of chemically different conventional, vulcanizable elastomers exist. ASTM D 1418 describes many of these. Table 4.2 lists some elastomers and their properties.

- *Natural rubber* (NR) is still used in many applications. It is not one uniform product; it varies with the nature of the plant producing the sap, the weather, the locale, the care in producing the elastomer, and many other factors.
- *Isoprene rubber* (IR) is similar in chemistry to NR, but it is produced synthetically. It is a more consistent product than NR and contains no proteins that can give allergic reactions (e.g., with surgical and examination gloves).
- *Butadiene elastomers* (BRs) are usually blended with SBR or NR in tire stocks.
- *Epichlorohydrin elastomers* (CO, ECO) are flame-retardant because of the presence of chlorine. Their electrical properties are modest, but they age well and resist many chemical environments.
- *Neoprene* (CR) (*chloroprene*) was the first synthetic elastomer and is widely used in industry. It is nonflammable and has some resistance to oils.
- *Chlorosulfonated polyethylene* (CSM) is similar to CR, with some improvement in electrical properties and better heat resistance.
- *Ethylene-propylene terpolymer* (EPDM) is synthesized from ethylene, propylene, and a third monomer, a nonconjugated diene. The diene permits conventional sulfur vulcanization. The elastomer is exceptionally resistant to radiation and heat. The glass transition temperature is −60°C, and electrical properties are good.
- *Ethylene-propylene copolymer* (EPM or EPR), which was often used in wire insulation, has been largely replaced by EPDM.
- *Butyl rubber* (IIR) is highly impermeable to air and water vapor. Butyl rubber has good aging characteristics and good flexibility at low temperatures.

TABLE 4.2 Typical Properties of Vulcanized Conventional Elastomers

Elastomer type	T_g, °C	Ultimate tensile strength*	Tear resistance*	Abrasion resistance*	Ozone resistance	Compression set @ 120°C, %	Service temp., °C	Hot oil resistance*
Natural rubber, synthetic polyisoprene	−72	E	E	VG	P	70	100	P
Styrene-butadiene rubber	−50	VG	VG	VG	P		110	P
Polybutadiene rubber	−112	G	F-G	E	P		90	P
Butyl rubber	−66	G	VG	G	VG	60	150	P
Chlorinated butyl rubber	−66	G	VG	G	VG	60	150	P
Brominated butyl rubber	−66	G	VG	G	VG	60	150	P
Ethylene-propylene-diene monomer rubber	−55	G	G	G	E	50	140	P
Polychloroprene	−45	E	E	VG	P	30	125	F
Nitrile rubber, medium nitrile	−34	VG	G	E	P	50	125	E
Hydrogenated nitrile rubber	−30	VG	G	E	E	30	160	E
Polyacrylate rubber	−22 to −40	G	G	G	E	10	170	E
Epichlorohydrin rubber	−45	G	G	G	E	20	135	E
Chlorinated polyethylene	−25	VG	G	G	E		150	F
Chloro-sulfonated polyethylene	−25	VG	VG	G	E		150	F
Polyester urethane rubber	−35	E	E	E	E	70	75	VG
Polyether-urethane rubber	−55	E	G	E	E	70	75	VG
Silicone rubber	−120	F	F	F	E	3	250	VG
Fluoroelastomer	−18 to −50	VG	G	G	E	20	250	E

*.E = excellent, G = good, F = fair, P = poor, V = very.

- *Nitrile rubber* (NBR) is resistant to most oils and some other fluids.
- *Styrene butadiene rubber* (SBR) is used in tire treads and mechanical applications.
- *Fluorinated elastomers* include several types, e.g., fluorocarbons and fluorosilicones. The elastomers can be used to 315°C, do not burn, are unaffected by most chemicals, and have excellent electrical properties. In thermal stability and aging, only the silicones are better. Physical properties are good but the cost is high.

- *Silicone elastomers,* whose molecular main chains contain silicon and oxygen atom backbones, have excellent high-temperature performance (up to 315°C) and are also flexible at very low temperatures. However, their physical properties are modest.
- *Acrylic elastomer* (ACM) has heat resistance that is almost as good as that of fluorinated compounds and silicones. It also ages well and is oil resistant, but it is sensitive to water.
- *Polysulfides* (T) have good weather resistance and resistance to a variety of fluids and other chemical environments; however, their physical properties are modest.
- *Millable polyurethanes* (U) are either ester- or ether-based. Ester-based elastomers are poor in water resistance. They have good ultimate mechanical properties and abrasion resistance.

The above conventional elastomers and others are listed in Table 4.3 with their ASTM D1418 designated abbreviations.

4.3.2 Thermoplastic Elastomers

These materials have the functional requirements of elastomers (extensibility and rapid retraction) but processability similar to that of rigid thermoplastics. The principal advantages of the TPEs compared to vulcanized rubber are (1) reduction in compounding requirements, (2) easier and more efficient processing cycles, (3) scrap recycling, and (4) availability of thermoplastic processing methods. Generic classes of TPEs include

- Styrenic block copolymers
- Polyolefin blends (TPO)
- Thermoplastic vulcanizates (TPV)
- Thermoplastic polyurethanes (TPU)
- Thermoplastic copolyesters
- Thermoplastic polyamides

TPEs are processed by extrusion, injection molding, blow molding, and thermoforming, and they can be heat welded. None of these methods are applicable to thermoset-type elastomers.

4.4 PROPERTIES OF ELASTOMERS

The most frequently measured rubber properties are the mechanical properties. These include hardness, ultimate tensile strength (UTS), ultimate elongation (UE), and stress at 100, 200, and 300 percent tensile strain. Though not really moduli, these stresses are referred to, in the rubber technologists' jargon, as 100, 200, and 300 percent modulus (M100, M200, and M300). Additional properties are measured to be more or less predictive of service performance.

Hardness. Hardness is probably the least difficult and most often measured property of vulcanized elastomers. It is in almost every list of rubber specifications. As measured, hardness is the relative resistance of the rubber surface to deformation by an indentor. The hardness of rubber is measured by using a small spring-loaded gauge known as a durometer (ASTM D 2240). One measures the hardness by pressing the indentor against the rub-

TABLE 4.3 Types of Conventional Rubbers and Their Abbreviations

Description	Names	ASTM 1418 abbrev.
Chloropolyethylene rubber	Chlorinated polyethylene	CM
Chlorosulfonylpolyethylene rubber	Chlorosulfonated polyethylene, Hypolon®	CSM
Copolymers of ethylene and vinyl acetate	EVA	EAM
Terpolymer of ethylene, propylene, and a diene with the residual unsaturated portion of the diene in the side chain	Ethylene-propopylene-diene-monomer rubber	EPDM
Copolymers of ethylene and propylene	Ethylene propylene rubber, EPR	EPM
Perfluoro rubbers of the poly- methylene type having all substituent groups on the polymer chain either fluoro, perfluoroalkyl, or perfluoroalkoxy groups	Fluoroelastomer	FFKM
Fluoro rubber of the polymethylene type having substituent fluoro and perfluoroalkyl or perfluoroalkoxy groups on the polymer chain	Fluoroelastomer	FKM
Ethylene oxide (oxirane) and chloromethyl oxirane (epichlorohydrin) copolymer	Epichlorohydrin rubber, Hydrin®	ECO
Bromoisobutene-isoprene copolymer	Bromobutyl rubber	BIIR
Poly1-4-*cis*-butadiene	Butadiene rubber	BR
Chloroisobutene-isoprene copolymer	Chlorobutyl rubber	CIIR
Polychloroprene	Neoprene, poly(chlorobutadiene)	CR
Isobutene-isoprene copolymer	Butyl rubber	IIR
Polyisoprene, synthetic	Synthetic natural rubber	IR
Acrylonitrile-butadiene copolymer	Nitrile rubber, Buna N	NBR
Natural rubber	Natural polyisoprene	NR
Styrene-butadiene rubber	GRS	SBR
Carboxylic-sytrene-butadiene rubber	Carboxy SBR	XSBR
Carboxylic-acrylonitrile-butadiene rubber	Carboxy NBR	XNBR
Silicone rubber having fluorine, vinyl, and methyl substitute groups on the polymer chain	Fluorosilicone rubber	FVMQ
Silicone rubber having both methyl and vinyl substituent groups on the polymer chain	Silicone rubber, Silastic®	VMQ
Polyester urethane rubber	Millable polyurethane rubber (not thermoplastic polyurethane elastomers (TPUs)	AU
Polyether urethane rubber		EU
A rubber having either a -CH2-CH2-O-CH2-O-CH2-CH2- group or occasionally an -R- group, where R is an aliphatic hydrocarbon between the polysulfide linkages in the polymer chain	Polysulfide rubber, Thiacol® rubber	OT

ber surface and reading the scale after a designated period of time. The scale is calibrated in arbitrary units from a value of 0 to 100. There are two scales used in rubber and elastomer technology. The A scale durometer is used for most soft rubber products, whereas the D scale durometer is used for stiff rubber compounds and plastic-like materials. The chart of Fig. 4.10 shows the ranges of hardness with respect to rubbers and plastics.

At the upper end of the hardness scales (i.e., near values of 100), the hardness value is of little significance because, at these levels, the indentor moves only slightly. However, at the lower and intermittent values (i.e., 15 to 85), the hardness test is valuable as a quick and easy method to characterize the stiffness of a rubber compound. In fact, the hardness value correlates fairly well with some types of modulus measurements.

The hardness of a rubber band is about 40 on the A scale. That of a tire tread compound might be from about 55 to 65 on the A scale, whereas a shoe sole might have a hardness of about 80 A or 30 D.

Stress-Strain (Tensile) Properties.   Ultimate tensile strength (UTS), MPa (or psi); ultimate elongation (UE), percent; and stress at 100, 200, and 300 percent, MPa (or psi) strain are measured during the stretching of a dumbbell-shaped specimen (ASTM D 412). For rubber testing, the strain rate is generally 2000 percent/minute. They are normally measured at room temperature (23°C). These properties can also be measured at much higher or lower temperatures, which greatly affects them. Generally, engineering tensile strength (rather than true strength) is recorded. Engineering stress is the force to stretch is divided by the original cross-sectional area. For elastomers, true stress at break (force divided by cross-sectional area at break) is approximated by multiplying the engineering tensile strength by the extension ratio, λ_B, at break ($\lambda_B = 1 + [UE\%/100]$).

Tensile properties are important and are measured on various compositions during compound development, because they are generally part of the material specifications. Measurements of these properties are also used to, in part, control manufacturing processes. Also, resistances to various types of deterioration (i.e., by the action of heat, oil, weather, and so on) are estimated on the basis of changes in mechanical properties during exposure. For example, during aging in dry heat (in air), UE generally decreases, tensile strength can decrease or increase, and the moduli M100, M200, or M300 frequently are increased. Hardness also typically changes as a result of exposure to stressful environments.

Tearing Resistance.   Tearing resistance of rubber is measured as force per specimen thickness. There are several types of tear specimens, with limited correlation between test results obtained with the various types of specimens. Some are cut (nicked) to provide a starting point or initiation for tearing, whereas others are not. A frequently used sample is

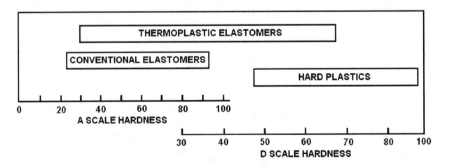

FIGURE 4.10   Hardness ranges for elastomers and plastics.

an unnicked 90° angle specimen (ASTM D 624 Die C Tear). In still another test, the test piece is a razor-nicked crescent-shaped specimen (ASTM D 624 Die A Tear). In another test, the trouser-tear specimen (so called because it resembles a pair of men's trousers) is used.

Tear-test specimens are generally pulled on a tensile tester at a cross-head rate of 8.5 mm/sec (20 in/min). The maximum force required to initiate or propagate tear is recorded as force per unit thickness. Tear strength measurements can indicate gross differences in performance in some applications. They are also useful in production control.

Abrasion Resistance.   Abrasion is the wearing of material from a rubber surface due to the action of an abrasive surface in contact with and moving with respect to the rubber surface. It is measured under a specified load, at a specified speed and type of abrasive surface. Laboratory tests may not predict service life, because the many and complex factors affecting abrasion vary greatly from application to application. However, laboratory tests are useful for quality control of rubber products intended for rough service.

The National Bureau of Standards (NBS) Abrader is a frequently used abrasion tester. Test samples are pressed under a specified load against a rotating drum covered with abrasive paper. The number of revolutions of the drum required to wear away a specified thickness of the test specimen is recorded and compared to the number of revolutions required to wear away the same thickness of a standard reference material (ASTM D 1630).

For the Pico Abrader test, a pair of tungsten carbide knives of specified geometry and controlled sharpness are rubbed over the surface of a pellet-shaped sample in a rotary fashion under controlled conditions (load, speed, and time). The volume loss of the test specimen is measured and compared to that of a reference compound (ASTM D 2228).

In the Taber Abrader (or Abraser) test, weight loss or thickness loss is measured. The specimen is placed in contact with an abrasive wheel under a load of 500 or 1000 g (ASTM D 1044). This test is generally performed on clear plastics; however, the apparatus has also been used with rubber.

Flex and Fatigue Resistance.   Flex resistance refers to the ability of a rubber part to withstand numerous flexing cycles without failure. Two common flex resistance tests are the DeMattia Flex Test (ASTM D 813) and the Ross Flex Test (ASTM D 1052). The DeMattia flex tester alternately pushes the test specimen ends together, bending it in the middle, then pulls it back to straighten it. The Ross flex tester repeatedly bends and then straightens the specimen over a metal rod. In both tests, a small cut or nick of prescribed size and shape may be made in the center of the test specimen. Data generated from these tests include flex life, which is the number of test cycles a specimen can withstand before it reaches a specified state of failure (i.e., length of grown nick or cut), and crack growth rate, which is the rate at which the cut propagates itself as the sample is flexed. These tests are particularly useful in evaluating compositions that are intended for use in products that undergo repeated flexing or bending.

Another fatigue-testing device is the (Monsanto) Fatigue to Failure Tester. This instrument measures the ultimate fatigue life as cycles to failure. Tensile-like samples are stretched at 100 cycles per minute at a preselected extension ratio. Samples can be strained over a range of 10 to 120 percent. The fatigue performance of compounds can be measured (number of cycles to failure, i.e., rupture) and compared either at constant extension ratio (strain) or at constant strain energies (work input).

Resilience.   Resilience is the stored energy that is rapidly (essentially instantaneously) returned by a vulcanized rubber object when it is suddenly released from a state of strain or deformation. Generally, the strain is also applied very rapidly. Resilience is what causes a rubber ball to bounce. It can be expressed as 100 minus the percentage of energy loss upon recovery from deformation. Either high resilience or low resilience (damping) may be required in a given application, depending on the application.

Resilience is measured by using one of several methods including the rebound pendulum test (for example, ASTM D 1054), the vertical rebound test (ASTM D 2632), or the Yerzley oscillograph test (ASTM D 945).

For a pendulum test, the pendulum is set to strike the surface of a secured stationary rubber specimen, the surface being place directly below the fulcrum of the pendulum. The pendulum is then released from a certain angle of displacement from the surface of the rubber. The percentage energy released is calculated as (1 – cos angle of rebound)/(1 – cos angle of displacement).

In the vertical rebound test, a metal plunger is allowed to fall on the test specimen, and the height to which the plunger rebounds is recorded as a percent of the release height.

Heat Buildup and Compression Fatigue.   The temperature increase in an elastomeric material during repeated deformations is due to the accumulation of thermal energy because of hysteresis or internal friction. The Goodrich Flexometer is used for heat buildup testing in the Compression Fatigue Test (ASTM D 623). A cylindrical test specimen is alternately compressed or stressed and then released in rapid cycles for a specified period of time or until it fails. The temperature rise within the test specimen is recorded. In addition, permanent set after the test is measured. The test specimen is fatigued, or weakened, by the rapidly cycling compression stress. The deterioration of the composition can actually increase the hysteresis and further increase heat buildup until, at some temperature, the sample completely degrades almost explosively. This temperature and time can also be noted.

Compression Set.   Compression set is the residual deformation of a material after removal of an applied compressive stress. For good performance in many applications, compression set values should be low. A low compression set value indicates that a material recovers much of its original height after compression and release of the compressive force.

Compression set tests are run either by applying a specified force to the test specimen (Method A of ASTM D 395) or by compressing the specimen to a specified deflection (Method B of ASTM D 395). The specimen is held in the compressed state for a specified period of time at a specified temperature, after which the compressive force is removed, and the specimen is allowed to recover at room temperature for 30 min. In Method A, compression set is the difference between the original and final thickness of specimen, expressed as a percentage of the original thickness. In Method B, compression set is expressed as the difference between the original and final thickness of the specimen as a percentage of the deflection to which the specimen was subjected. The standard specimen for a compression set test is a round pellet. O-rings of specified dimensions can also be tested. Compression set is a property that is an important consideration with seals and gaskets where the rubber part must maintain sealing force under a compressive stress or strain or in rubber hoses clamped onto nipples. Method A is a measure of compressive creep, whereas method B is a measure of compressive stress relaxation.

Compression set tests are usually run at elevated temperatures to simulate conditions or aging effects. Common test conditions are 70 hr at 70°C or 100°C, although heat-resistant materials such as fluoroelastomers may be tested for longer periods of time at temperatures up to 200°C or more. If the end product is expected to perform at low temperatures, e.g., below 0°C, compression set is measured at the expected service temperature.

Heat Resistance.   Heat resistance is the resistance to irreversible changes in properties due to exposure to elevated temperatures. Reversible changes due to elevated temperatures include expansion and softening. Permanent changes include further cross-linking, polymer chain degradation by oxidation, and stiffening due to loss of plasticizer by evaporation. In the case of nonoxidative elevated temperature aging of natural rubber, permanent softening due to a loss of cross-links (reversion) can occur.

Accelerated Aging in Hot Air. A variety of laboratory accelerated hot-air aging tests have been developed to predict the service life and aging characteristics of elastomers. For hot-air oven aging, specimens are exposed at specified elevated temperatures. Their physical properties are determined (ASTM D 573) as a function of aging time. Point changes in hardness and percentage changes in tensile strength and elongation at break are reported.

Hot-Air Test Tube Aging. In this method (ASTM D 865), specimens are heated in air but confined within individual test tubes. This prevents cross-contamination of compounds due to transfer of volatile materials (e.g., antioxidants, curatives, plasticizers, degradation products, and so on) from one sample to another. Thus, this test is free of some of the complications that can occur when numerous compounds are aged in the same enclosure. As before, hardness and tensile mechanical properties are measured before and after aging.

Deterioration by Pure Oxygen under Pressure ("Oxygen Bomb" Test). Pure oxygen under pressure is used to accelerate oxidative degradation (ASTM D 572). Test specimens are placed in a strong metal vessel under an atmosphere of pure oxygen gas pressure of up to 2070 kPa (300 psi) at a temperatures of 70 or 80°C. After the aging period, hardness and tensile properties are measured and compared with original properties.

Fluid Resistance (Chemical Resistance). Fluid resistance is the extent to which a rubber product retains its original physical characteristics and ability to function after exposure to oil, water, organic fluids, or any other liquid encountered in its use. Fluid resistance tests do not necessarily correlate with service performance, because service conditions are not easily defined. However, they are useful for screening compounds, because they give comparative assessments of expected performance.

A specimen's weight or volume and mechanical properties are measured before and after exposure to selected fluids for a specified time at a specified temperature. The effect of the fluid on the specimen is judged on the basis of the change in properties during the test. Volume, weight, hardness, tensile strength, and elongation at break of specimens are often measured. Changes in tear strength, compression set, and low-temperature properties are also frequently evaluated. ASTM D 471 describes the tests and lists the composition of various ASTM-designated fluids that have been given special ASTM designations.

Oil Resistance. This is a special case of fluid resistance. Elastomers may be swollen and weakened by oil. Standard petroleum-based ASTM test oils are used.

General-purpose rubbers (e.g., natural rubber, SBR) placed in oil absorb the fluid slowly until the oil is all imbibed or the rubber has disintegrated. Oil-resistant specialty elastomers can absorb some oil, especially at elevated temperatures, but only a limited (and sometimes negligible) amount. In most end uses requiring oil-resistant elastomers, some swelling or volume increase is acceptable. In some cases, some swelling is even desirable, e.g., to tighten up a seal. If the oil extracts a plasticizer of extender oil from a composition, a seal could even fail due to shrinkage.

Resistance to Outdoor Exposure. Elastomer compositions progressively degrade with respect to physical properties upon prolonged exposure to the elements (air, sunlight, rain, and so forth) because of chemical changes in the elastomer molecule. These changes are largely caused by the action of ozone and oxygen in the atmosphere, ultraviolet light, and in some cases water (e.g., when the polymer or other components, such as plasticizers, are subject to hydrolysis). Typical effects are surface hardening, crazing and cracking, gradual changes in tensile strength, and elongation.

To determine the actual effects of normal weathering, specimens are exposed outdoors in different locations and different climates. Changes in color, cracking, crazing, chalking, and stress-strain properties are recorded at various time intervals. (Mildew might also be observed.) Typical check periods are 1, 2, 5, 10, and 20 years exposure. Test specimens are usually stretched at low strains and mounted on wooden supporting racks for outdoor exposure. Three procedures are specified by ASTM D 518:

- Procedure A—Straight strips elongated 20 percent.
- Procedure B—Bent loops, where the elongation varies from 0 to 25 percent at different points on the loop.
- Procedure C—Tapered samples elongated to strains of 10 to 20 percent.

The specimens in their supporting racks are usually exposed facing south for maximum sunlight at an angle of 45° with the horizon. Samples may also be buried in the ground to determine resistance to moisture, soil chemicals, insects, bacteria, and oxidation effects in the absence of direct sunlight.

Simulated weathering tests are carried out by using such testing devices as the Weather-o-meter (ASTM D 750) and Fade-o-meter. In the Weather-o-meter, test specimens are continuously or intermittently exposed to water spray and artificial light produced by a carbon arc. This light is of wavelengths similar to those of natural sunlight but with increased intensity in the ultraviolet range. In the Fade-o-meter, specimens are exposed to constant controlled humidity (instead of water spray) along with carbon-arc artificial light. Specimens may be exposed unstrained or under a slight elongation

As in the other accelerated aging tests, tensile properties are measured after aging and compared with original properties. In addition, visual observations are made for cracking, crazing, or color changes.

Deterioration by the Action of Ozone. Ozone, although found in relatively small amounts in air, is a highly reactive gas. Its action on many types of rubbers can cause products to crack and fail prematurely unless they are protected by antiozonants or made from ozone-resistant elastomers.

To test for resistance to ozone attack, samples are stretched to 20 or 40 percent strain on a test rack or bent in a loop to produce a surface strain. The specimens are then placed in a chamber equipped with an ozone generator. Usual controlled test concentrations are about 0.25, 0.5, 1, or 2 ppm (parts ozone per million parts of air by volume). The test temperature is usually 40°C. The test specimens are inspected at various time intervals until initial cracking occurs.

In dynamic testing of ozone resistance, a fabric-backed vulcanized rubber specimen is continuously flexed in the ozone chamber over a roller. The fabric backing is in the form of a belt. Any protective chemical films (e.g., certain waxes and antiozonants) that might build up on the surface of the specimen in static testing are quickly broken by the continuous flexing. ASTM test method D 1149 covers static testing and D 3395 covers dynamic testing in a controlled ozone atmospheres.

Low-Temperature Properties. As elastomer compositions are gradually cooled, they gradually stiffen until the region of the glass transition, T_g, is reached where a very sharp increase in stiffness occurs as a function of decreasing temperatures. This temperature is near the brittle temperature or stiffness temperature discussed below. At temperatures below the transition temperatures, the elastomer composition becomes brittle and can crack or break on sudden impact or bending. The temperature at which this occurs becomes higher as the impact or bending becomes more rapid.

In the test for brittleness temperature (ASTM D 746), a small rectangular strip is clamped in a holder at one end and cooled to a predetermined temperature in a chilled liquid bath. The specimen is then given a sudden sharp blow by a solenoid-actuated striking arm. If the specimen does not fracture, the temperature of the liquid bath is lowered, and the process is repeated until the specimen breaks when struck.

A torsional stiffness test measures the modulus of rigidity of an elastomer composition over a broad temperature range. There are two commonly used tests: ASTM D 1043, which uses the Clash-Berg Tester, and the ASTM D 1053, which uses the Gehman Torsional Stiffness Tester. In each of these tests, a sample is chilled to a preset temperature well below the transition temperature then twisted with a known constant force. The amount of twist is measured as a function of the increasing temperature. The temperature that produces a stiffness of 69 MPa is sometimes taken as the stiffness temperature.

In the temperature retraction test (ASTM D 1329), the test specimen is stretched to a specified elongation and then frozen at –70°C. Then it is gradually warmed until it begins to retract toward its original unstretched dimensions. The temperature at which the specimen retracts by 10 percent correlates with the brittleness temperature.

Dynamic Mechanical Properties.   In an above section dynamic mechanical properties were discussed in connection with the differences between elastomers and plastics *vis à vis* the glass transition temperature, indicated by the position of the peak in the tan δ in its plot against temperature.

Dynamic mechanical properties of elastomers are important for other reasons. Rubber is often used in applications where it undergoes rapid cyclic deformations over a range of frequencies: engine mounts, tire sidewalls and treads, and so forth. In such uses, flexibility, friction, cushioning, and damping properties of the rubber are important. Such properties relate to frequency and temperature in some of the same ways as do dynamic mechanical properties. They vary as a function of the difference between the ambient temperature and the glass transition temperature, i.e., as a function of $(T - T_g)$. Also, the measurement of tan δ at a particular condition is an estimate of hysteresis, which correlates with heat buildup, and damping, this being important for such applications as engine mounts. There are even other relationships (though complex) between dynamic mechanical properties and friction (traction) and resistance to skidding on ice.

Compilation of ASTM Tests.   Table 4.4 gives a compilation of selected ASTM designated tests.

4.4.1 Rubber Elasticity

Rubber elasticity is the reason for the use of elastomers. The origin of rubbery behavior has been considered and pondered for quite a long time. On the molecular level, vulcanized and thermoplastic elastomers have something in common: their molecules have segments that are flexible at least at service or use temperatures (above T_g). The flexible segments have ends that are not free but, instead, are immobilized, either in cross-links (e.g., in vulcanized rubber) or in glassy or crystalline domains [e.g., in segmented TPEs, such as styrene-butadiene-styrene or segmented poly(butylene terephthalate) copolymers]. The rubbery molecular segments continuously change their conformations or configurations as a result of Brownian motion, which increases with temperature.

Rubber-like behavior (rubber elasticity) has been described in terms entropy (a fundamental measure of disorder). As an elastomeric material is stretched, the ends of the elastomeric molecular segments become increasingly separated from one another. The disorganized molecular chain becomes more nearly straight; that is, they become less disordered. This is an unnatural condition, which is resisted, and when the stretching stress is released (stopped), the entropy increases as the material returns to its original state of ease. Thus, a supporting molecular chain (e.g., a flexible chain between cross-links) can be viewed as an entropic spring. This idea is shown in Fig. 4.11. The arrows at the ends of some of the chain segments simply indicate that they are part of a continuous network of flexible mobile segments that terminate in junctures or cross-links. The effect of strain on the entropy of a network chain is idealized by Fig. 4.12.

The molecular origin of the recovery from large deformations, which is the essence of rubber elasticity, was not recognized until the early 1930s. Evidence for this was the fact

TABLE 4.4 Compilation of ASTM Tests Used for Elastomers

Property	Test	ASTM standard test method for:	ASTM Standards Volume
Hardness	D 2240	Rubber property—durometer hardness	9.01
Stress-strain (tensile) properties	D 412	Rubber properties in tension	9.01, 9.02
Tearing resistance	D 624	Rubber property—tear resistance	9.01
Abrasion resistance	D 1630	Rubber property—abrasion resistance (NBS Abrader)	9.01
Abrasion resistance	D 2228	Rubber property—abrasion resistance (PICO Abrader)	9.01
Abrasion resistance	D 1044	Standard test method for resistance of transparent plastics to surface abrasion	8.01
Flex and fatigue resistance	D 813	Rubber deterioration—crack growth	9.01
Flex and fatigue resistance	D 1052	Measuring rubber deterioration—cut growth using Ross Flexing Apparatus	9.01
Resilience	D 1054	Rubber property—resilience using a rebound pendulum	9.01
Resilience	D2632	Rubber property—resilience by vertical rebound	9.01
Resilience	D 945	Rubber properties in compression or shear (mechanical oscillograph)	9.01
Heat buildup and compression fatigue	D 623	Rubber property—heat generation and flexing fatigue in compression	9.01
Compression set	D 395, Method A, Method B	Rubber property—compression set	9.01
Accelerated aging in hot air	D 573	Rubber—deterioration in an air oven	9.01
Hot-air test tube aging	D 865	Rubber—deterioration by heating in air (test tube enclosure)	9.01
Deterioration by pure oxygen under pressure ("oxygen bomb" test)	D 572	Rubber—deterioration by heat and oxygen	9.01
Fluid resistance (chemical resistance)	D 471	Rubber property—effect of liquids	9.01
Resistance to outdoor exposure	D 518, procedures A, B, C	Rubber deterioration—surface cracking	9.01

TABLE 4.4 Compilation of ASTM Tests Used for Elastomers (Continued)

Property	Test	ASTM standard test method for:	ASTM Standards Volume
Resistance to outdoor exposure	D 750	Rubber deterioration in carbon-arc weathering apparatus	9.01
Deterioration by the action of ozone	D 1149	Rubber deterioration—surface ozone cracking in a chamber	9.01
Deterioration by the action of ozone	D 3395	Rubber deterioration—dynamic ozone cracking in a chamber	9.01
Low-temperature properties	D 746	Brittleness temperature of plastics and elastomers by impact	9.02
Low-temperature properties	D 1053	Rubber property—stiffening at low temperatures	9.01
Low-temperature properties	D 1043	Stiffness properties of plastics as a function of temperature by means of a torsion test	9.02
Low-temperature properties	D 1329	Evaluating rubber property—retraction at lower temperatures (TR test)	9.01

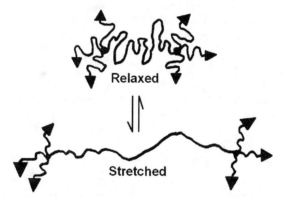

FIGURE 4.11 Entropic spring.

that the modulus of elasticity of a pure vulcanized gum (unfilled) rubber is proportional to the absolute temperature. This suggested an analogy with the kinetic theory of gases, i.e., the ideal gas law:

$$P = \frac{nRT}{V} \tag{4.4}$$

where P is pressure, R the gas constant, T the absolute temperature, V the volume, and n the number of moles of gas.

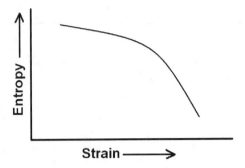

FIGURE 4.12   Entropy as a function of strain.

The statistical theory of rubber elasticity, based on estimates of the changes in entropy due to deformation gives

$$G = \frac{\nu_e RT}{V} \tag{4.5}$$

where G is the shear modulus (for elastomers 1/3 the value of Young's modulus) and ν_e is the numbers of moles effective (supportive) network chains. Equation 4.5 assumes that the natural log of number of configurations that can be assumed by a flexible segment is proportional to its configurational entropy. Shear stress is proportional to the product of absolute temperature and the derivative of this entropy with respect to shear strain. Equation 4.5 was given by Flory in the early 1930s. The relation is only a good approximation because of complexities and certain assumptions (e.g., that the chains' motion is not restricted by the presence of neighboring chains). There has been much controversy and refinement of the equation; nevertheless, all of this supports the idea of the retraction being the result of the entropic spring.

4.5 VULCANIZABLE ELASTOMERS

The first rubber to be discovered and used by industry was natural rubber. After quite a long time, synthetic rubbers were introduced. In this section, we consider natural and synthetic conventional (i.e., vulcanizable) rubbers. There is some emphasis on the science and technology of natural rubber for two reasons: (1) natural rubber has been the most studied, and (2) many of the synthetic elastomers are quite similar to natural rubber in many respects.

4.5.1   Vulcanization

Unvulcanized rubber articles are generally not strong, do not regain their shape after large deformations, and can even be sticky and have the consistency of chewing gum. The first commercial method for vulcanization has been attributed to Charles Goodyear. His process (heating natural rubber with sulfur) was first used in Springfield, Massachusetts, in 1841. Thomas Hancock used essentially the same process about a year later in England. In addition to natural rubber, many synthetic rubbers have been introduced and, in addition to sulfur, many other substances have been used in curing (vulcanization) systems. In this section, emphasis is placed on the vulcanization of general-purpose "high-diene" rubbers [natural rubber (NR), styrene-butadiene rubber (SBR), and butadiene rubber (BR)] by sulfur in the presence of organic accelerators.

The accelerated-sulfur vulcanization of these rubbers, along with the vulcanization of other rubbers that are vulcanized by closely related technology [e.g., ethylene-propylene-diene monomer rubber (EPDM), butyl rubber (IIR), halobutyl rubbers, and nitrile rubber (NBR)] comprises more than 90 percent of all vulcanization. Nevertheless, we give some consideration to vulcanization by the action of other vulcanization agents such as organic peroxides, phenolic curatives, and quinoid curatives.

Definition of Vulcanization. Vulcanization is a process generally applied to rubbery or elastomeric materials, which forcibly retract to their approximately original shape after a rather large mechanically imposed deformation. Vulcanization can be defined as a process that increases the retractile force and reduces the amount of permanent deformation remaining after removal of the deforming force. Vulcanization increases elasticity while it decreases plasticity. It is accomplished by the formation of a cross-linked molecular network (Fig. 4.13).

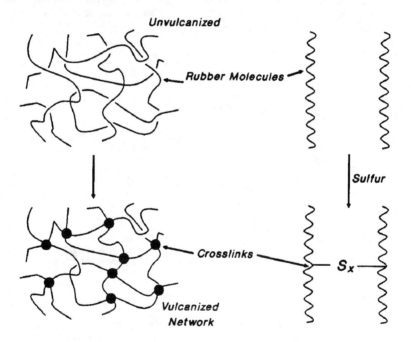

FIGURE 4.13 Network formation.

The theory of rubber elasticity (above) proposes that the retractile force to resist a deformation is proportional to the number of network supporting polymer chains per unit volume of elastomer. A supporting polymer chain is a linear polymer molecular segment between network junctures. An increase in the number of junctures (i.e., cross-links) gives an increase in the number of supporting chains. In an unvulcanized linear high polymer (above its melting point), only molecular chain entanglements act as junctures.

Vulcanization chemically produces network junctures by the insertion of cross-links between polymer chains. A cross-link may be a group of sulfur atoms in a short chain, a single sulfur atom, a carbon to carbon bond, a polyvalent organic radical, an ionic cluster, or a polyvalent metal ion. The process is usually carried out by heating the rubber, mixed with vulcanizing agents.

4.5.1.1 *Effect of Vulcanization on Elastomer Properties.* Vulcanization causes significant changes at the molecular level. The long rubber molecules (molecular weight usually between 100,000 and 500,000 daltons) become linked together with junctures (cross-links) spaced along the polymeric chains. The average distance between junctures corresponds to a molecular weight between cross-links of about 4,000 to 10,000 daltons. Because of network formation, the rubber becomes almost entirely insoluble in any solvent, and it cannot be processed by any means that requires it to flow—e.g., in a mixer, in an extruder, on a mill, on a calender, or during shaping, forming, or molding. Thus, vulcanization must occur only after the rubber article is in its final shape or form.

Major effects of vulcanization on use-related properties are indicated by Fig. 4.14. It should be noted that static modulus increases with vulcanization to a greater extent than does the dynamic modulus. (Here, static modulus is more correctly the equilibrium modulus, approximated by a low strain, slow-strain-rate modulus. Dynamic modulus is generally measured with the imposition of a sinusoidal, small strain at a frequency of 1 to 100 Hz.)

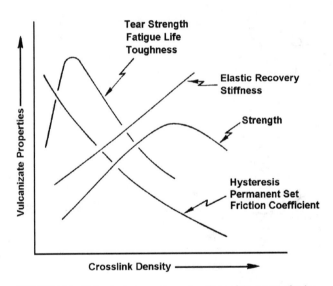

FIGURE 4.14   Vulcanizate properties as functions of the extent of vulcanization.

Hysteresis is reduced with increasing cross-link formation. Hysteresis is the ratio of the rate-dependent or viscous component to the elastic component of deformation resistance. It is also a measure of deformation energy that is not stored (or borne by the elastic network) but is converted to heat. Vulcanization then causes a trade-off of elasticity for viscous or plastic behavior. Tear strength, fatigue life, and toughness are related to the breaking energy. Values of these properties increase with small amounts of cross-linking, but they are reduced by further cross-link formation. Properties related to the energy to break increase with increases in both the number of network chains and hysteresis. Since hysteresis decreases as more network chains are developed, the energy-to-break related properties are maximized at some intermediate cross-link density.

Reversion.   Reversion is usually the loss of network structures by nonoxidative thermal aging. It is generally associated with isoprene rubbers vulcanized by sulfur. It can be the result a long vulcanization time (overcure) or of hot aging of thick sections. It is most severe at temperatures above about 155°C. It occurs in vulcanizates containing large proportions of polysulfidic cross-links. Sometimes the term "reversion" is applied to other types of nonoxidative degradation, especially with respect to rubbers not based on isoprene. For example, thermal aging of SBR (styrene-butadiene rubber), which can cause increased cross-link density and hardening, has been called *reversion* since it can be the result of overcure.

4.5.1.2 Characterization of the Vulcanization Process.

Important characteristics of the vulcanization process are the time period before cross-linking starts, the rate of cross-link formation once it starts, and the extent of cross-linking when the process ends. There must be sufficient delay or scorch resistance (resistance to premature vulcanization or induction period) to permit mixing, shaping, forming, and flowing in the mold, autoclave, or curing oven before vulcanization begins. Then cross-link formation should be rapid and controlled. The importance of heat history with respect to the vulcanization process is illustrated by Figs. 4.15 and 4.16.

Scorch resistance is frequently measured by the time at a given temperature required for the onset of cross-link formation as indicated by an abrupt increase in viscosity. The Mooney viscometer is usually used. Fully compounded, unvulcanized rubber is contained in a heated cavity. Embedded in the rubber is a rotating disc. Viscosity is continuously measured (by the torque required to keep the rotor rotating at a constant rate) as a function of time. The temperature is selected to be characteristic of rather severe processing (extrusion, calendering, and so on).

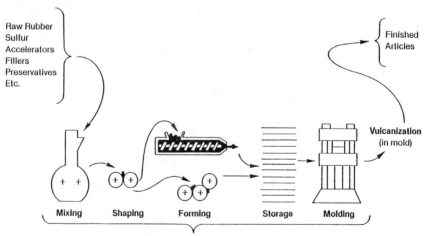

FIGURE 4.15   The effect of processing on heat history.

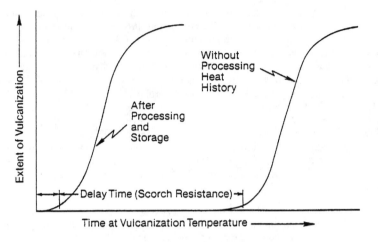

FIGURE 4.16 The effect of heat history (processing) on scorch safety.

Both the rate of vulcanization after the scorch period and the final extent of vulcanization are measured by devices called *cure meters*. The development of the oscillating disc rheometer was the beginning of modern vulcometry, which is the standard industrial practice.

During the cure meter test, a rubber is enclosed in a heated cavity (Fig. 4.17). Embedded in the rubber is a metal disc, which oscillates sinusoidally about its axis. The degree of vulcanization is measured by the increase in the torque required to maintain given amplitude (e.g., degrees of arc) of oscillation at a given temperature or during a specified temperature-time profile. The torque is proportional to a low strain elastic shear modulus and, according to the theory of rubber elasticity, proportional to cross-link density. Since this torque is measured at the elevated temperature of vulcanization, the portion of it due to viscous effects is minimal. Thus, it has been assumed that the increase in torque during vulcanization is proportional only to the number of cross-links formed per unit volume of rubber.

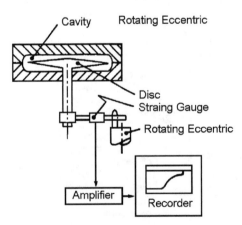

FIGURE 4.17 Oscillating disc rheometer.

There is a newer type of cure meter (e.g., Fig. 4.18). The cavity is much smaller, and there is no rotor. In this type of cure meter, one-half of the cavity (e.g., the upper half) is stationary, and the other half oscillates. These instruments are called *moving-die rheometers*. The sample is much smaller and heat transfer is faster. Also, the cavity temperature can be changed more rapidly.

In either case, the torque is plotted against time to give a so-called rheometer chart, rheograph, or cure curve (Fig. 4.19). The cure curve gives a rather complete picture of the overall kinetics of cross-link formation and even cross-link disappearance (reversion). In some cases, instead of reversion, a long plateau or marching cure can occur.

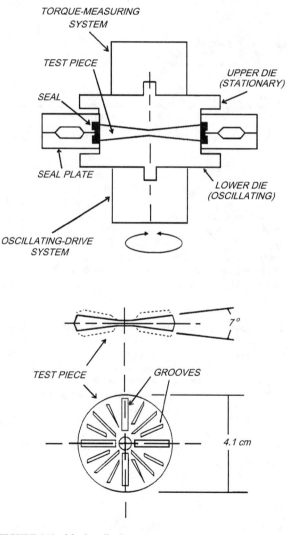

FIGURE 4.18 Moving-die rheometer.

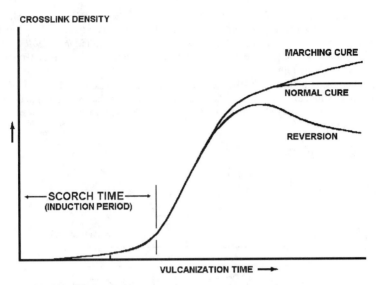

FIGURE 4.19   Rheometer cure curve.

4.5.1.3 Vulcanization by Sulfur without Accelerator.   At first, vulcanization was done by using elemental sulfur at a concentration of 8 parts per 100 parts of rubber (phr) and required 5 hr at 140°C. The addition of zinc oxide reduced the time to 3 hr. Then, the use of accelerators in concentrations as low as 0.5 phr reduced the vulcanization cycle time to as little as 1 to 3 min. As a result, elastomer vulcanization by sulfur without accelerator is no longer of much commercial significance. (An exception is the use of about 30 or more phr of sulfur to produce molded hard rubber or "ebonite" parts.) Even though unaccelerated sulfur vulcanization is not of commercial significance, its chemistry has been the object of much research and study.

4.5.1.4 Accelerated-Sulfur Vulcanization.   Organic chemical accelerators were not used until 1906 (Fig. 4.20), when the effect of aniline on sulfur vulcanization was discovered by Oenslager. Aniline, however, was too toxic for use in rubber products. Its less toxic reaction product with carbon disulfide, thiocarbanilide, was then used as an accelerator in 1907. Then, guanidine accelerators were introduced. Reaction products formed between carbon disulfide and aliphatic amines (dithiocarbamates) were first used as accelerators in 1919. These were and are still among the most active accelerators; however, most of them give little or no scorch resistance, and their use is impossible in many factory-processing situations. The somewhat delayed-action accelerators, 2-mercaptobenzothiazole (MBT) and 2-benzothiazole di-sulfide (or 2,2′-dithiobisbenzothiazole) (MBTS) were introduced in 1925. Even more delayed action and yet faster curing vulcanization was possible in 1937 with the first commercial benzothiazolesulfenamide accelerator. Still more delay became possible in 1968 with the availability of an extremely effective premature vulcanization inhibitor (PVI). This compound was N-(cyclohexylthio)phthalimide (CTP), small concentrations of which were used along with benzothiazolesulfenamide accelerators. The history of the progress toward faster vulcanization but with better control of premature vulcanization or scorch is illustrated by Figs. 4.20, 4.21, and 4.22.

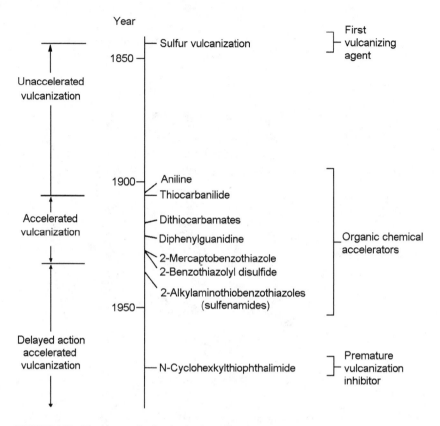

FIGURE 4.20   The history of vulcanization by sulfur.

Accelerated-sulfur vulcanization is the most widely used method. For many applications, it is the only rapid cross-linking technique that can, in a practical manner, give the delayed action required for processing, shaping, and forming before the formation of the intractable vulcanized network. It can be used to vulcanize natural rubber (NR), synthetic isoprene rubber (IR), styrene-butadiene rubber (SBR), nitrile rubber (NBR), butyl rubber (IIR), chlorobutyl rubber (CIIR), bromobutyl rubber (BIIR), and ethylene-propylene-diene-monomer rubber (EPDM). The reactive moiety for all of these elastomers can be represented by

Typically, a recipe for the vulcanization system for one of the above elastomers contains 2 to 10 phr of zinc oxide, 1 to 4 phr of fatty acid (e.g., stearic), 0.5 to 4 phr of sulfur, and 0.5 to 2 phr of accelerator. Zinc oxide and the fatty acid are vulcanization-system activators. The fatty acid, with zinc oxide, forms a salt that can form complexes with acceler-

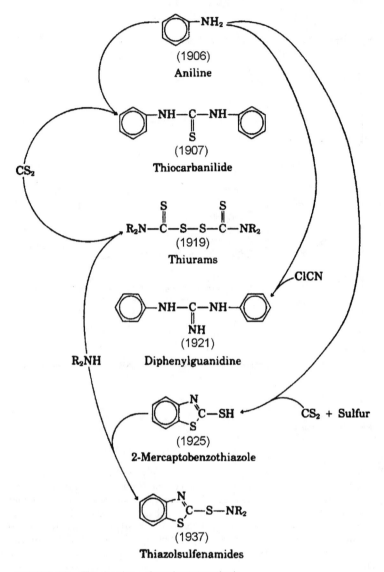

FIGURE 4.21 The chemistry of accelerator synthesis.

ators and reaction products formed between accelerators and sulfur. Accelerators are classified and illustrated in Table 4.5.

Frequently, mixtures of accelerators are used. Typically, a benzothiazole type is used with smaller amounts of a dithiocarbamate (thiuram) or an amine type. An effect of using a mixture of two different types of accelerator can be that each activates the other and better-than-expected cross-linking rates can be obtained. Mixing accelerators of the same type gives intermediate or average results.

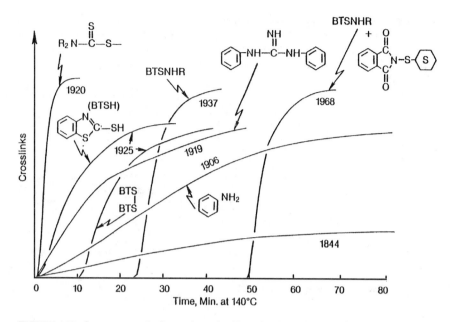

FIGURE 4.22   Improvements in the accelerated-sulfur vulcanization of natural rubber.

The use of accelerators based on secondary amines is being reduced or eliminated. This is because secondary amines can react with nitrogen oxides to form suspected carcinogenic nitrosamines. This is especially a problem with dithiocarbamate-type accelerators. Proposed accelerators that do not give carcinogenic nitrosamine derivatives include dibenzylamine derived dithiocarbamates.

Different types of accelerators impart vulcanization characteristics, which differ in both scorch resistance and cross-linking rate. Fig. 4.23 is a map of accelerator system characteristics. The effect of the addition of small concentrations of the premature vulcanization inhibitor (PVI), N-(cyclohexylthio)phthalimide, is also given by Fig. 4.23. This retarder is frequently used to independently control scorch resistance with little effect on the rate of cross-linking.

4.5.1.5 The Chemistry of Accelerated-Sulfur Vulcanization.   The general reaction path of accelerated-sulfur vulcanization is thought to be as follows: Accelerator reacts with sulfur to give monomeric polysulfides of the structure $Ac-S_x-Ac$, where Ac is an organic radical derived from the accelerator (e.g., benzothiazyl-). The monomeric polysulfides interact with rubber to form polymeric polysulfides, e.g., rubber-S_x-Ac. If a benzothiazole-derived accelerator is used, during the reaction between rubber and $Ac-S_x-Ac$, 2-mercaptobenzothiazole (MBT) is formed if the elastomer is natural rubber. (In SBR, the MBT becomes bound to the elastomer molecular chain probably as the thioether rubber-S-Ac.) When MBT itself is the accelerator in natural rubber, it first disappears then reforms with the formation of rubber-S_x-Ac. Finally, the rubber polysulfides react, either directly or through an intermediate, to give cross-links, rubber-S_x-rubber. A reaction scheme with a benzothiazolesulfenamide accelerator can be written as shown in Scheme 1. The sulfuration of rubber could proceed by the mechanism shown in Scheme 2. Cross-links could form in a number of ways, e.g., as shown in Scheme 3.

TABLE 4.5 Accelerators for Sulfur Vulcanization

Compound	Abbrev.	Structure

Benzothiazoles

| 2-Mercaptobenzthiazole | MBT | 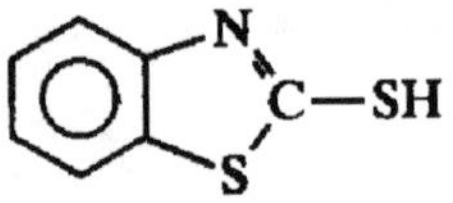 |

| 2,2′-Dithiobisbenzothiazole | MBTS | 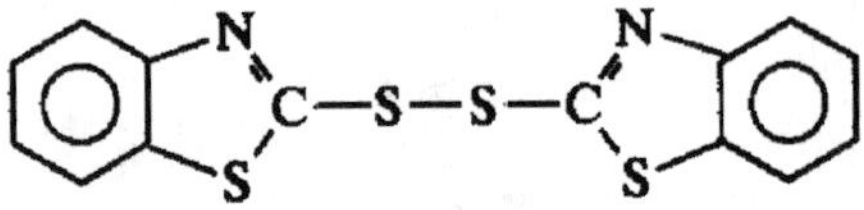 |

Benzothiazolesulfenamides

| N-Cyclohexylbenzothiazole-2-sulfenamide | CBS | 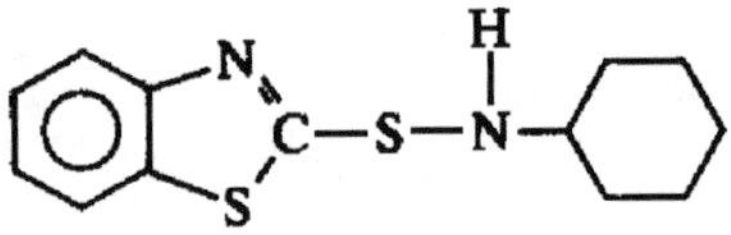 |

| N-t-butyobenzothiazole-2-sulfenamide | TBBS | 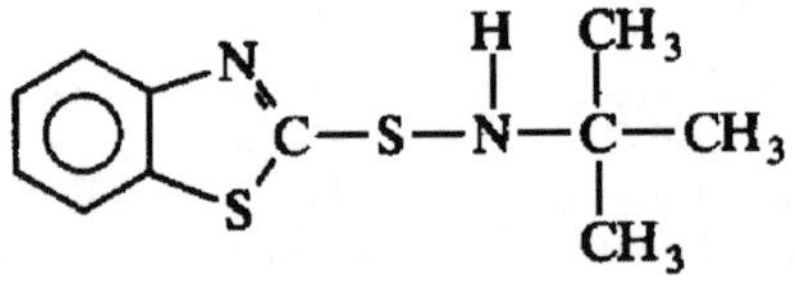 |

| 2-Morpholinothiobenzothiazole | MBS | 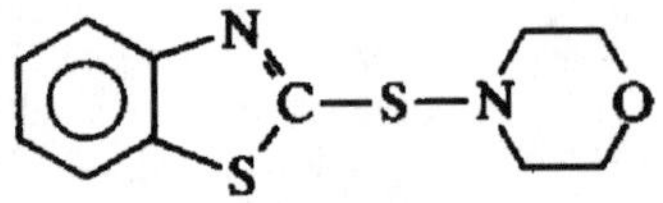 |

| N-Dicyclohexylbenzothiazole-2-sulfenamide | DCBS | 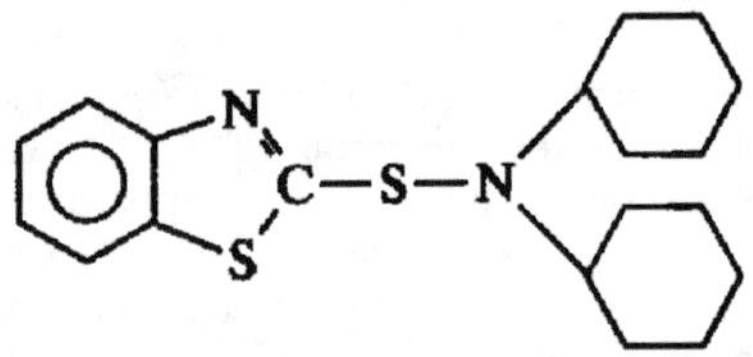 |

Dithiocarbamates

| Tetramethylthiuram monosulfide | TMTM | 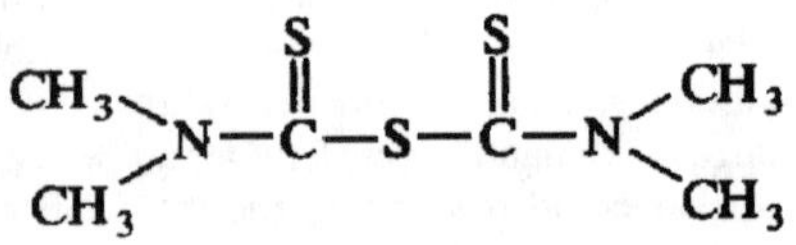 |

TABLE 4.5 Accelerators for Sulfur Vulcanization (Continued)

Compound	Abbrev.	Structure

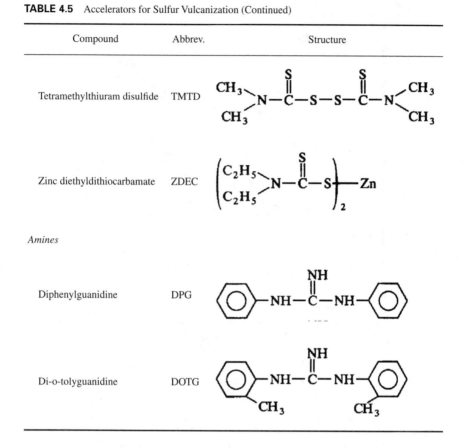

Delayed-Action Accelerated Vulcanization. If cross-link formation is by a free radical mechanism, delayed action could be the result of a quenching action by the monomeric polysulfides formed by reactions between accelerator and sulfur. If the polymeric polythiyl radicals (cross-link precursors) are rapidly quenched by an exchange reaction before they are able to form cross-links, cross-link formation would be impeded until substantial depletion of the monomeric polysulfides. This is illustrated in Scheme 4.

It has been shown that the early reaction products formed by the interaction between accelerator and sulfur (Ac-S_x-Ac) are inhibitors of cross-link formation. The very substances that give rise to the formation of the cross-link precursor (rubber-S_x-Ac) inhibit the formation of the cross-links.

Other mechanisms for delayed action have been proposed. In the case of acceleration by benzothiazolesulfenamides, the accelerator is depleted in an autocatalytic fashion with the formation of 2-mercaptobenzothiazole (MBT). The rate of this depletion is about proportional to the amount of MBT present. There is strong evidence that indicates that reactions occur in sulfenamide-accelerated systems as shown in Scheme 5.

If MBT could be taken out of the system as fast as it forms, substantial increases in processing safety would result. Such is the case when the premature vulcanization inhibitor, N-(cyclohexylthio)phthalimide (CTP) is present. This compound and others like it re-

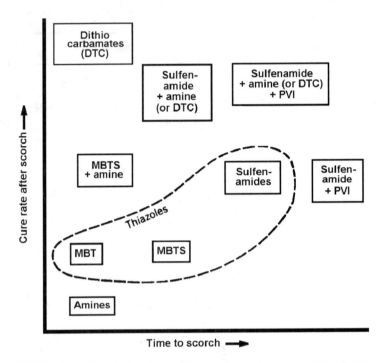

FIGURE 4.23   Vulcanization characteristics given by various accelerators and combinations.

act rapidly with MBT to form 2-(alkyldithio)benzothiazoles, R-S-S-BT, which are active accelerators but which do not interact rapidly with the sulfenamide accelerator. See Scheme 6), where L is a "leaving group" of the premature vulcanization inhibitor (e.g., N-phthalimido- for CTP).

The importance of scorch control cannot be overemphasized. Present-day tire plants could not compete without good control of scorch resistance, or processing safety as it is commonly called. Such safety is necessary to rapidly process rubber mixes at high temperatures (through extrusion, calendering, and so forth) into preforms for molding (e.g., tire components).

The Role of Zinc in Benzothiazole Accelerated Vulcanization.   An increase in the concentration of fatty acid, and hence increases in the concentration of available Zn^{++}, causes an increased overall rate in the early reactions (during the delay period), which lead to the formation of rubber-S_x-Ac. However, it gives rise to a decrease in the rate of cross-link formation but an increase in the extent of cross-linking. The increase in the rates of the early reactions has been explained by the interaction shown in Scheme 7, where the chelated form of the accelerator is more reactive than the free accelerator during the early reactions (see Scheme 8).

Here, I-S_y- is an ionized form of linear sulfur. It could be rapidly formed in a reaction between sulfur and any of a number of initiating species. It is also proposed that the presence of Zn^{++} can increase the rate of sulfurization through the formation of complexes of the type:

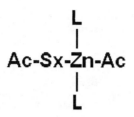

where L is a ligand such as an amine molecule.

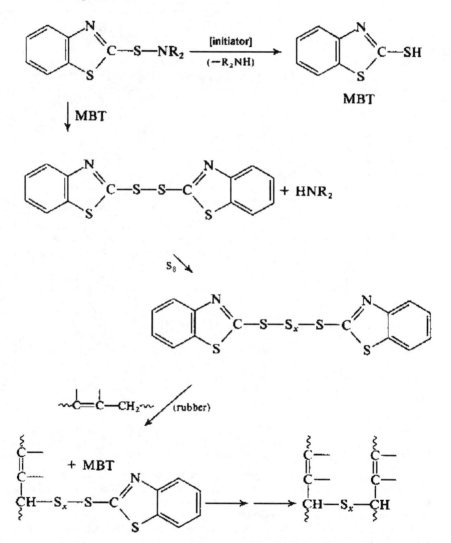

SCHEME 1

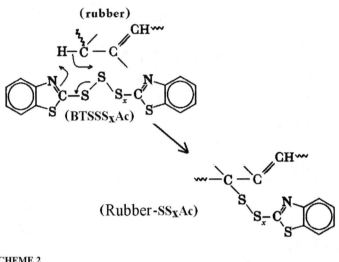

(Rubber-SS$_x$Ac)

SCHEME 2

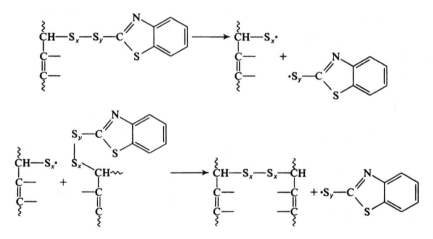

SCHEME 3

The decreased specific rate of cross-link formation, and the increased extent of cross-linking due to the presence of Zn^{++} in benzothiazole accelerated vulcanization, have been explained by the scheme shown in Scheme 9.

Zinc chelation changes the position of the S-S bond most likely to break. Since a stronger bond must break, the rate is slower. Though the rate of cross-linking is slower, the extent of cross-link formation is increased, since less sulfur is used in each cross-link. That is, the cross-links are of lower sulfidic rank.

The presence of zinc compounds can also promote the reduction the sulfur rank of cross-links during high-temperature aging of the vulcanizate, e.g., during reversion. In some cases, zinc compounds actually promote the decomposition of cross-links.

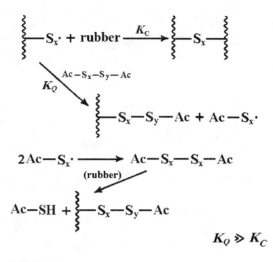

$$K_Q \gg K_C$$

SCHEME 4

BT-S-NHR + BT-SH ⟶ BT-S-S-BT + RNH₂

sulfenamide   MBT   MBTS
accelerator

SCHEME 5

L-S-R + BT-SH ⟶ BT-S-S-R + L-H

premature   MBT   2-(alkyldithiyl)-   inactive
vulcanization   benzothiazole   compound
inhibitor

SCHEME 6

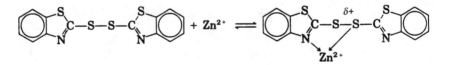

SCHEME 7

4.5.1.6 Effects of Changes in Curing System on Vulcanizate Properties. Increases in sulfur and accelerator concentrations give higher cross-link densities and, therefore, higher moduli, stiffness, hardness, and so on. However, as the ratio of the concentration of accelerator to the concentration of sulfur increases, the proportion of monosulfidic cross-links increases in natural rubber compounds. Greater amounts of accelerator (with respect to sulfur) also give an abundance of pendent groups of the type, $-S_x$-Ac, attached to and

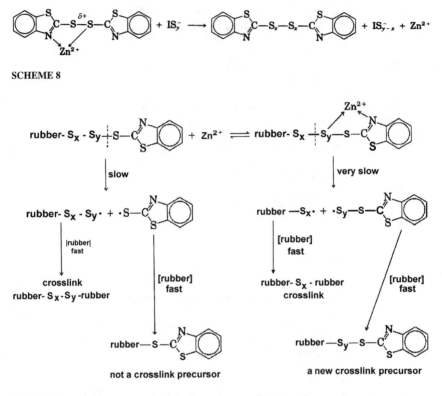

SCHEME 8

SCHEME 9

dangling from the rubber molecular chains. Higher ratios of sulfur concentration to accelerator concentration give both more polysulfide cross-links and more sulfur combined with the rubber chains, forming sulfur-containing six-membered heterocyclic rings along the rubber molecular chains. These features are indicated by Fig. 4.24. Such changes in the vulcanizate network structure are responsible for changes in vulcanizate properties due to changes in the curing-system.

Effects of changes in vulcanizate properties due to changes in accelerator and sulfur concentrations were studied with the following recipe (parts by weight): natural rubber, 100; N330 carbon black, 50; N-isopropyl-N´-phenyl-p-phenylenediamine (IPPD antidegradant) 2; zinc oxide, 5; stearic acid, 3; plasticizer, 3; sulfur, variable; N-cylohexylbenzothiazolesulfenamide (CBS), variable.

The effects of changes in the accelerator concentration on 300 percent modulus (jargon for stress at 300 percent tensile strain), thermal-oxidative aging, and fatigue life (DeMattia flex crack) are given in Fig. 4.25. The effects on 300 percent modulus are indicated by the diagonal contours of negative slope. They are parallel and show that the stress at 300 percent strain increases with an increase in either sulfur or accelerator concentration. The contours for percent retention of ultimate elongation after hot air aging (at 100°C for two days) indicate that oxidative aging, in the presence of IPPD, depends only on the concentration of sulfur. Higher concentrations of sulfur give poor aging characteristics in correlation with the higher number of points of chain sulfuration.

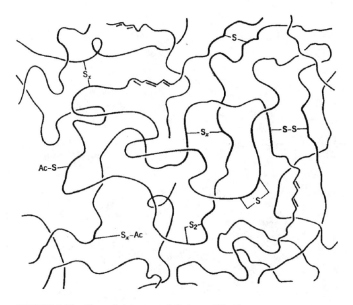

FIGURE 4.24   Cross-link types and chain modifications.

The contours for flex fatigue life are complex. The test is run such that the specimens are about equally strained; however, there is some question as to whether the tests should be run at equal strain or at equal strain energy. For some cases, where strain is restricted by fabric reinforcement, fatigue test data should be compared at equal strain amplitude. For other applications, where the strain is not limited, the tests should be run at equal strain energy. The contours as presented here can be interpreted in terms of either constant strain or constant strain energy. All points on the chart can be compared at an approximately equal strain per cycle; however, if we interpolate between the flex-life contours but only along a constant modulus contour, we can extract values corresponding to approximately equal strain energy per cycle. By choosing higher modulus contours, we are considering higher strain energies.

The low values for fatigue life at low levels of sulfur, but high levels of accelerator, have been attributed to high concentrations of accelerator-terminated appended groups and high concentrations of monosulfidic cross-links. Monosulfidic cross-links are not able to exchange, rearrange, or break to relieve stresses without the breakage of main chains.

On the other hand, polysulfidic cross-links are able to rearrange under stress. The rearrangement of a cross-link occurs in two steps: (1) breaking and (2) reforming. Recent data indicate that only the breaking of the weak polysulfide cross-links is required for the strengthening of the vulcanizate network. It is better to relieve the stress by the breaking of a cross-link than by the breaking of a polymer chain.

When even higher concentrations of sulfur are used (with the maintenance of constant modulus), flex life decreases. It is possible that this is due to the large amount of cyclic chain modification associated with high levels of sulfur. As always, there are compromises.

4.5.1.7 Accelerated-Sulfur Vulcanization of Various Unsaturated Rubbers.   Over the years, much of the research on accelerated-sulfur vulcanization was done by using natural

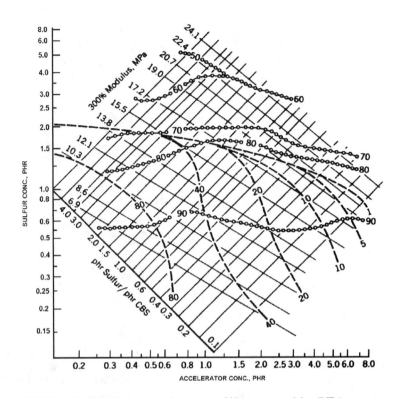

FIGURE 4.25 Vulcanizate properties, _____ 300 percent modulus (MPa); _____, De Mattia flex fatigue life (kHz × 10^{-1}); -O-O-O-O-, percent retention of ultimate elongation after two days at 100°C.

rubber as a model substrate. Natural rubber was the first elastomer and, therefore, the search for understanding of vulcanization originated with work on natural rubber. Most of what we have discussed so far *vis à vis* vulcanization has been related to natural rubber.

The chemistry of the accelerated vulcanization of BR, SBR, and EPDM appears to have much in common with the vulcanization of natural rubber. Before the formation of cross-links, the rubber is first sulfurated by accelerator-derived polysulfides (Ac-S$_x$-Ac) to give macromolecular, polysulfidic intermediates (rubber-S$_x$-Ac), which then form cross-links (rubber-S$_x$-rubber). As in the case of natural rubber, the average length of a cross-link (its sulfidic rank, the value of x in the cross-link, rubber-S$_x$-rubber) increases with the ratio of sulfur concentration to accelerator concentration (S/Ac) used in the compounded rubber mix. However, in the case of BR or SBR, the cross-link sulfidic rank is not nearly as sensitive to S/Ac as it is in the case of natural rubber. Model compound studies of the vulcanization of EPDM (e.g., wherein ethylidenenorbornane was used as a model for EPDM) indicate that the polysulfidic rank of the EPDM cross-links probably responds to changes in S/Ac in a natural rubber-like fashion.

Reversion (when defined as the loss of cross-links during nonoxidative thermal vulcanizate aging) is a problem associated mainly with natural rubber or synthetic isoprene polymers. It can occur only under severe conditions in butadiene rubber; in SBR, instead of the softening associated with the nonoxidative aging of natural rubber, one can observe hardening (the so-called marching modulus) during extensive overcure. In natural rubber and synthetic isoprene-polymer rubbers, the cross-links tend to be more polysulfidic than in the case of BR or SBR. The highly polysulfidic cross-links are more heat-labile than their lower rank cousins in BR and SBR; they are more likely to break and then form cyclic chain modifications.

The effect of zinc is much greater in the vulcanization of isoprene rubbers than it is in the vulcanization of BR and SBR. Again, the reason for the difference is not known, but a strong speculation is that this difference is also related to the presence of methyl groups only in the case of the isoprene rubbers.

Curing-System Recipes for Accelerated-Sulfur Vulcanization. Recipes for only the curing-system part of formulations are given in Table 4.6.

TABLE 4.6 Recipes for Accelerated Sulfur Vulcanization Systems[*]

			Nitrile (NBR)			
	NR	SBR	1	2	Butyl (IIR)	EPDM
Zinc oxide	5.00	5.00	3.00	2.00	3.00	5.00
Stearic acid	2.00	2.00	0.50	0.50	2.00	1.00
Sulfur	2.50	1.80	0.50	0.25	2.00	1.50
DTDM[†]	–	–	–	1.00	–	–
TBBS[†]	0.60	1.20	–	–	–	–
MBTS[†]	–	–	2.00	–	0.50	–
MBT[†]	–	–	–	–	–	0.50
TMTD[†]	–	–	1.00	1.00	1.00	1.50
Vulcanization temp., °C	148	153	140	140	153	160
Time, minutes	25	30	60	60	20	20

[*].Concentrations in phr.
[†].DTDM, 4.4′-dithiodimorpholine; TBBS, N-t-butylbenzothiazole-2-sulfenamide; MBTS, 2,2′-dithiobisbenzothiazole (2-benzothiazole disulfide); MBT, 2-mercaptobenzothiazole; TMTD tetramethylthiuram disulfide.
Note: conditions change depending on other aspects of the compositions.

4.5.1.8 Vulcanization by Phenolic Curatives, Benzoquinone Derivatives or Bismaleimides.

Diene rubbers such as natural rubber, SBR, and BR can be vulcanized by the action of phenolic compounds, which are (usually di-substituted by $-CH_2-X$ groups where X is an -OH group or a halogen atom substituent. A high-diene rubber can also be vulcanized by the action of a dinitrosobenzene which forms *in situ* by the oxidation of a quinonedioxime, which had been incorporated into the rubber along with the oxidizing agent, lead peroxide.

The attack upon rubber molecules by the vulcanization system can be visualized in a way similar to that which was postulated for the sulfurization of the rubber molecules by the action of accelerated-sulfur vulcanization systems. Reaction schemes for these two types of vulcanization can be written as shown in Schemes 10 and 11.

As shown, the chemical structural requirements for these types of vulcanization are that the elastomer molecules contain allylic hydrogen atoms. The attacking species from the vulcanization system must contain sites for proton acceptance and electron acceptance in proper steric relationship. This will then permit the rearrangement shown in Scheme 12, where A is the proton acceptor site and B is the electron acceptor site.

This is an explanation for the fact that this type of vulcanization is not enabled by double bonds *per se,* without allylic hydrogens in the elastomer molecules. (It should be pointed out that the phenolic curative can also act by a slightly different mechanism to give cross-links that contain chromane structural moieties, the allylic hydrogens still being required.)

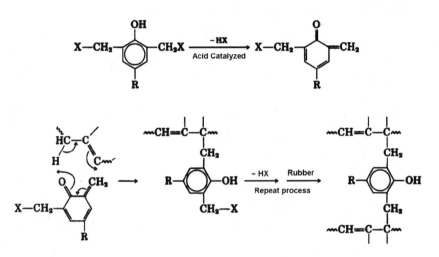

SCHEME 10 Vulcanization by phenolic curatives.

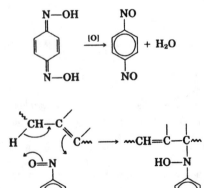

SCHEME 11 Vulcanization by benzoquinonedioxime.

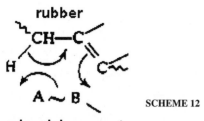

rubber

SCHEME 12

vulcanizing agent

Another vulcanizing agent for high-diene rubbers is m-phenylenebismaleimide. A catalytic free-radical source such as dicumyl peroxide or benzothiazyl disulfide (MBTS) is usually used to initiate a free-radical reaction. Although a free-radical source is frequently used with a maleimide vulcanizing agent, at high enough vulcanization temperatures, the maleimides react with the rubber without the need for a free-radical source. This could occur as shown in Scheme 13.

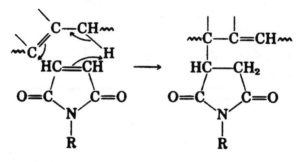

SCHEME 13

This is similar to the reaction written for the attack of rubber molecules by phenolic curatives or the *in situ* formed nitroso derivative of the quinoid (e.g., benzoquinonedioxime) vulcanization system. It is also closely related to the sulfurization scheme written for accelerated-sulfur vulcanization. Comparisons between accelerated sulfur, phenolic, quinoid, and maleimide vulcanization can then be visualized as shown in Scheme 14.

Selected recipes for vulcanization by phenolic curatives, benzoquinone-dioxime, or m-phenylenebismaleimide are given by Table 4.7. Vulcanizates based on these types of curatives are particularly useful in cases where thermal stability is required.

4.5.1.9 Vulcanization by the Action of Metal Oxides. Chlorobutadiene, i.e., chloroprene rubbers (CR), also called neoprene rubbers, are generally vulcanized by the action of metal oxides. CR can be represented by the following structure:

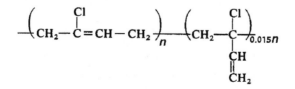

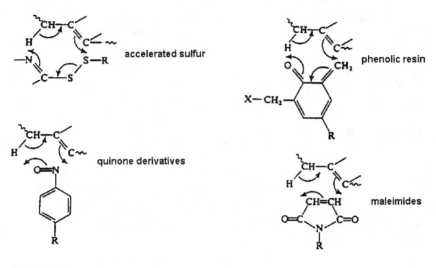

accelerated sulfur

phenolic resin

quinone derivatives

maleimides

SCHEME 14

TABLE 4.7 Recipes for Vulcanization by Phenolic Curatives, Quinone Derivatives, or Maleimides[*]

	IIR		SBR		
	1	2	1	2	NBR
Zinc oxide	5.00	5.00	–	–	–
Lead peroxide (Pb$_3$O$_4$)	–	10.00	–	–	–
Stearic acid	1.00	–	–	–	–
Phenolic curative (SP-1056)[†]	12.00	–	–	–	–
Benzoquinonedioxime (GMF)	–	2.00	–	–	–
m-Phenylenebismaleamide (HVA-2)[‡]	–	–	0.85	0.85	3.00
2-Benzothiazyl disulfide (MBTS)	–	–	2.00	–	–
Dicumyl peroxide	–	–	–	0.30	0.30
Vulcanization condition[**]					
Temperature, °C	180	153	153	153	153
Time, min	30	20	25	25	30

[*].Concentrations in phr.
[†].Schenectady Chemicals.
[‡].Du Pont.
[**].Conditions change depending on other aspects of the compositions.

Zinc oxide is the usual cross-linking agent. It is used along with magnesium oxide. The magnesium oxide is used for scorch resistance. The reaction is thought to involve the allylic chlorine atom, which is the result of the small amount of 1,2-polymerization:

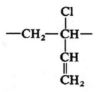

A mechanism that has been written for the vulcanization of CR by the action of zinc oxide and magnesium oxide is shown in Scheme 15.

Most accelerators for accelerated-sulfur vulcanization do not work for the metal oxide vulcanization of neoprene rubbers. An exception to this is in the use of the so-called mixed curing system for CR, in which metal oxide vulcanization is combined with accelerated-sulfur vulcanization. In this case, along with the metal oxides, accelerators such as tetramethylthiuram disulfide(TMTD) or N,N´-di-o-tolylguanidine (DOTG) are used with sulfur. This may be desirable for high resilience or for good dimensional stability.

The accelerator that has been most widely used with metal oxide cures is ethylenethiourea (ETU), N,N´-diphenylthiourea or 2-mercaptoimidazoline. The use of ETU in the vul-

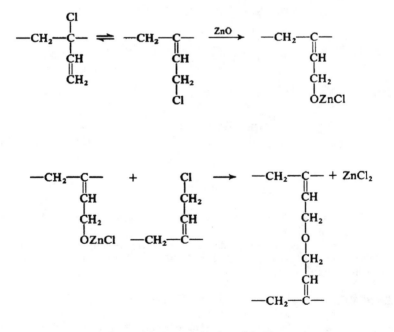

SCHEME 15

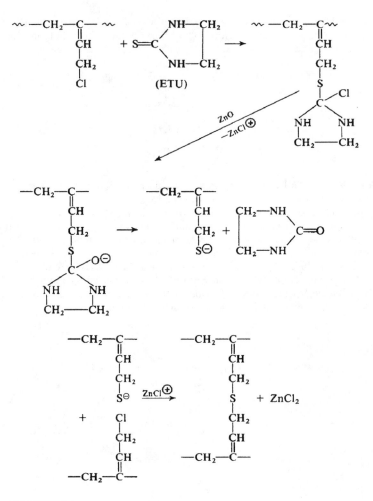

SCHEME 16

canization of CR is somewhat in doubt since it is a suspected carcinogen. A mechanism for ETU acceleration is shown in Scheme 16.

Examples of recipes for metal oxide vulcanization are given in Table 4.8. It should be noted that, in one case, calcium stearate was used instead of magnesium oxide to obtain better aging characteristics.

4.5.1.10 Vulcanization by the Action of Organic Peroxides. Peroxides are vulcanizing agents for elastomers, which contain no sites for attack by other types of vulcanizing agents. They are useful for ethylene-propylene rubber (EPR), ethylene-vinylacetate copolymers (EAM), certain millable urethane rubbers, and silicone rubbers. They are not generally useful for vulcanizing butyl rubber, poly(isobutylene-co-isoprene). Elastomers derived from isoprene and butadiene are readily cross-linked by peroxides, but many of

TABLE 4.8   Vulcanization Systems for Chloroprene Rubber[*]

ZnO	5.00	5.00	5.00
MgO	4.00	–	4.00
Calcium stearate	–	5.50	–
Stearic acid	–	–	1.00
TMTM	–	–	1.00
DOTG	–	–	1.00
ETU	0.5	0.5	–
Sulfur	–	–	1.0
Vulcanization[†]			
Temperature, °C	153	153	153
Time, min	15	15	15

[*].Concentrations in parts by weight per 100 parts of neoprene W.
[†].Conditions change depending on other aspects of the compositions.

the vulcanizate properties are inferior to those of accelerated-sulfur vulcanizates. However, peroxide vulcanizates of these diene rubbers may be desirable in applications where improved thermal ageing and compression set resistance are required.

Peroxide Vulcanization of Unsaturated Hydrocarbon Elastomers. The initiation step in peroxide-induced vulcanization is the decomposition of the peroxide to give free radicals. If the elastomer is derived from butadiene or isoprene, the next step is either the abstraction of a hydrogen atom from an allylic position on the polymer molecule or the addition of the peroxide-derived radical to a double bond of the polymer molecule. In either case, polymeric free radicals are the result (Scheme 17).

For isoprene rubber, the abstraction route predominates over radical addition. Two polymeric free radicals then unite to give a cross-link. Cross-links could also form by a chain reaction that involves the addition of polymeric free radicals to double bonds.

In this case, cross-linking occurs without the loss of a free radical, so that the process can be repeated until termination by radical coupling. Coupling can be between two

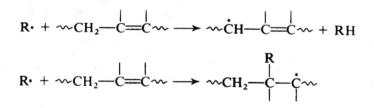

SCHEME 17

polymeric radicals to form a cross-link or by unproductive processes. A polymeric radical can unite with a radical derived from the peroxide. Also, if a polymeric radical decomposes to give a vinyl group and a new polymeric radical, a scission of the polymer chain is the result.

Few monomeric radicals are lost by coupling with polymeric radicals when dialkyl peroxides are used as the curative. Also, if the elastomer is properly chosen, the scission reaction is not excessive. For dicumyl peroxide in natural rubber, the cross-linking efficiency has been estimated at about 1.0. One "mole" of cross-links is formed for each mole of peroxide; cross-linking is mainly by the coupling of two polymeric radicals. One peroxide moiety gives two monomeric free radicals that react with rubber to give two polymeric radicals, which couple to form one cross-link.

In the case of BR or SBR, the efficiency can be much greater than 1.0, especially if all antioxidant materials are removed. A chain reaction is indicated here. One might expect that nitrile rubber would also be vulcanized with efficiencies greater than 1.0; however, though the double bonds in nitrile rubber are highly accessible, the cross-linking efficiency is somewhat less than 1.0.

Peroxide Vulcanization of Saturated Hydrocarbon Elastomers. Saturated hydrocarbon polymers are also cross-linked by the action of organic peroxides, though the efficiency is reduced by branching. Polyethylene is cross-linked by dicumyl peroxide at an efficiency of about 1.0, saturated EPR gives an efficiency of about 0.4, while butyl rubber cannot be cured at all. For polyethylene, the reaction scheme is similar to that of the unsaturated elastomers. However, branched polymers undergo other reactions. Though the peroxide is depleted, no cross-links may be formed between polymer chains, and the average molecular weight of the polymer can even been reduced by scission. Sulfur or the so-called coagents can be used to suppress scission. Examples of coagents are m-phenylenebismaleimide, high-1,2 (high-vinyl) polybutadiene, triallyl cyanurate, diallyl phthalate, ethylene diacrylate, and others.

Peroxide Vulcanization of Silicone Rubbers. Silicone rubbers (high-molecular-weigh polydimethylsiloxanes) can be represented by

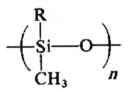

where R can be methyl, phenyl, vinyl, trifluoropropyl, or 2-cyanoethyl. Silicone rubbers that contain vinyl groups can be cured by dialkyl peroxides such as dicumyl peroxide. Saturated silicone rubbers require diacyl peroxides such as bis-(2,4-dichlorobenzoyl)peroxide. In the case of saturated siloxane rubbers, the mechanism is hydrogen atom abstraction followed by polymeric radical coupling to give cross-links. The incorporation of vinyl groups in the rubber molecule improves the cross-linking efficiency.

Vulcanization is frequently done in two steps. After a preliminary vulcanization in a mold, a high-temperature (e.g., 180°C) postcure is carried out in air. The high-temperature postcure removes acidic materials that can catalyze hydrolytic decomposition of the vulcanizate. Also, the high temperature enables the formation of additional cross-links of the following type:

Peroxide Vulcanization of Urethane Elastomers. Urethane elastomers suitable for peroxide vulcanization are typically prepared from an hydroxyl-group-terminated oligo-

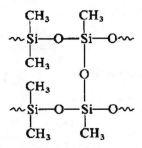

meric adipate polyester and 4,4′-methylenediphenylisocyanate (MDI). A typical structural representation is as follows:

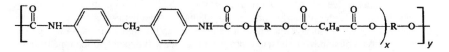

Hydrogen atoms can be abstracted from arylated methylene groups, but hydrogen atoms may also be abstracted from alpha-methylene groups of the adipate moieties. Though they are usually sufficient, vulcanization efficiencies can be increased by the incorporation of urea structures into the polymer chain.

Recipes for Peroxide Vulcanization.   Examples of starting-point recipes are given in Table 4.9. Outstanding characteristics of peroxide vulcanizates are low permanent set and thermal stability of the network.

TABLE 4.9   Recipes for Peroxide Vulcanization[*]

	NR	SBR	EPR	Silicone rubber	Millable urethane
Dicumyl peroxide	1.0	1.0	2.7	–	2
Bis(2,4-dichlorobenzoyl peroxide	–	–	–	1.0	–
Triallyl cyanurate	–	–	1.5	–	–
Vulcanization conditions[†]					
Temperature, °C	150	150	160	115, 250[‡]	153
Time, min	45	45	30	14, 1440[‡]	45

[*].Concentrations in phr.
[†].Conditions change depending on other aspects of the compositions.
[‡].Temperature and time of postcure in air.

4.5.1.11 Other Types of Vulcanization.

There are still other types of vulcanization systems based on other types of chemistry. These are applied to elastomers such as acrylates, fluoroelastomers, chlorosulfonylpolyethylene, and epichlorohydrin type elastomers. These are very specific curing systems. Some of them will be dealt with later sections of this chapter.

4.5.2 Preservation of Vulcanizates

As with other polymers, elastomers are subject to degradation due to elements of their environment. These elements include oxygen, atmospheric ozone, and ultraviolet light as from sunlight. Molecular chain scission or cross-linking can be the result, giving rise to substantial losses in performance properties (ultimate elongation, strength, flexibility, fatigue life, and so on). Thus, it is generally necessary to include additives in rubber recipes or formulations to protect elastomeric or rubber compositions from damage by the environment.

4.5.2.1 Oxidation of Polymers. The need for stabilization of organic polymers is essential, because they are exposed to oxygen throughout phases of their lifetime: the polymer production phase, the fabrication phase, and the application stage. Thus, antioxidants are added to polymers just prior to isolation, before exposure to oxygen. Such a stabilizer is expected to maintain polymer properties and to suppress gel formation or changes in viscosity. This protection is then expected to continue during storage before fabrication.

Fabrication involves shear and thermal energy, e.g., by Banbury mixing or mill mixing, extrusion, and calendering. Additional antioxidants are frequently added. The end-use product is expected to survive the environmental stresses throughout its service life. The amount and type of stabilizer chosen to protect a product will depend on the type of polymer and its use.

The oxidative degradation of the polymer proceeds by a free-radical chain reaction mechanism. Initiation usually occurs by exposure to heat, light, or mechanical stress. The process is sometimes catalyzed by certain transition metal-ion impurities. The oxidation of hydrocarbon or related polymers by oxygen is an autocatalytic process with primary products being hydroperoxides.

Autoxidation occurs in three mechanistic phases: initiation, propagation, and termination steps:

Initiation	R-R	→	2R•
	ROOH	→	RO• + HO•
	2ROOH	→	RO• + RO$_2$• +H$_2$O
Propagation	RO$_2$• + RH	→	ROOH + R•
	R• + O$_2$	→	RO$_2$•
Termination	2R•	→	R-R
	R• + RO$_2$•	→	ROOR
	2 RO$_2$•	→	Nonradical products + O$_2$

This oxidation model, however, ignores factors such as relativity differences between polymers and differences in oxygen permeability. Also, it does not account for the fact that the nature and character of the propagating radicals is not the same for all polymers. Polymers show a wide variation in susceptibility to thermal autoxidation. The ease of autoxidation depends primarily on the relative C-H bond dissociation energies for the component parts of the polymer structure. Once the free-radical process is initiated, autoxidation proceeds. The abstraction of the most labile hydrogen atom in the polymer by alkylperoxy radical predominates in the propagation step. This follows this order:

Strongest C-H bonds			Weakest C-H bonds
Primary	Secondary	Tertiary	Allylic
RCH$_2$-H	R$_2$CH$_2$-H	R$_3$C-H	RCH=CH-CH$_2$-H

Also, radicals also differ in relative lifetime.

Different radical types exist in different diene polymers; e.g., polyisoprene produces tertiary alkyl radicals, polybutadiene gives secondary alkyl types, and SBR gives allylic and benzylic types, while a nitrile polymer gives allylic and tertiary cyano types. These radicals and their corresponding peroxy radicals interact with parts of their own polymer chains, giving shielding effects, rearrangements, cleavages, and internal additions to neighboring double bonds. Also, free radicals undergo disproportionation, addition to carbon-carbon double bonds, and coupling.

Polyisoprene softens on oxidative aging, while rubber polymers, which are based on butadiene, harden on aging. The tertiary allylic radical in polyisoprene, being sterically protected, does not easily undergo cross-linking with other radicals or neighboring chain double bonds. It reacts primarily with oxygen and subsequently leads to polymer cleavage. The secondary allylic system is more reactive. It can undergo cross-chain reactions by radical addition; this can cause hardening of the composition.

4.5.2.2 Antioxidants. Factors affecting the performance of antioxidants include the intrinsic activity of the antioxidant system, the solubility of the antioxidant system in the polymer matrix, and the volatility of the antioxidant. For rubber, unfortunately, the most effective antioxidants are the staining and discoloring derivatives of aryl amines; however, the need for nondiscoloring antioxidants has been filled by phenolic nonstaining antioxidants. Amines find application as both raw polymer stabilizers and also as final vulcanizate stabilizers. On the other hand, phenolics can be used in nonblack reinforced vulcanizates where nonstaining and nondiscoloration is desired.

Amine and phenolic antioxidants are considered free-radical scavengers. They probably work by the direct abstraction of amine hydrogen by the RO• group. Increased steric hindrance at the 2 and 6 position of phenolic antioxidant has resulted in improved antioxidant performance. There may be an optimum amount of steric hindrance of the phenolic group, which should be matched to the oxidizable matrix polymer. This will allow a balanced interference of the radical chain process so that both propagating species, R• and RO•, are effectively neutralized.

Amine and phenolic type antioxidants, acting as free-radical scavengers, are illustrated below, where general structures for hindered phenol and *p*-phenylenediamine types are shown. (AH is the antioxidant, and RH is the rubber molecule.) The relative rates of free-radical quenching and free-radical-chain propagation are indicated by k_Q and k_P.

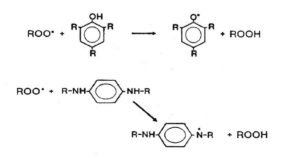

Aromatic amines (e.g., N,N′-alkyl- or aryl-substituted *p*-phenylenediamines) find application as both raw polymer stabilizers and also as final vulcanizate stabilizers. On the other hand, phenolics (i.e., 2,6-dialkyl substituted) can be used in nonblack reinforced vulcanizates where nonstaining and nondiscoloration are desired.

Examples of amine and phenolic-type antioxidants are given in Table 4.10.

In addition to the use of the above phenolic and amine type antioxidants, peroxide decomposers are used to harmlessly decompose the peroxides, which otherwise could decompose to give free radical propagating species, e.g., R-O• or H-O•. Examples of such peroxide decomposers, which act synergistically with the phenolic or amine antioxidants, are dilauryl-β,β-thiodiproprionate and tris(*p*-nonylphenyl)phosphite. These and others are also listed in Table 4.10, with their chemical structures being listed in Table 4.11. We note that some of the peroxide decomposers are also accelerators for sulfur vulcanization.

4.5.2.3 Degradation by the Action of Ozone.
The degradation of polydiene rubbers by the action of atmospheric ozone is characterized by the appearance of cracks on the surface of a finished rubber product. This degradation is caused by direct ozone attack and reaction with the double bond sites of unsaturation in a polydiene rubber.

Early work established the following:

1. Ozone absorption occurs at a linear rate.

2. The absorption of ozone is proportional to its concentration.

3. Rubber that is not strained undergoes reaction with ozone to form oxidized film, but it does not show the characteristic "ozone cracking."

This last result was a key observation and has come to be known as the *critical strain effect,* i.e., no crack growth occurs unless a specific strain for the particular polymer is exceeded.

Examples of rubbers that are thus affected include natural rubber (NR), synthetic cis-polyisoprene, styrene-butadiene rubber (SBR), polybutadiene (BR), and nitrile rubber (NBR). Rubbers with highly saturated backbones, such as ethylene-propylene-diene rubber (EPDM) or halobutyl rubber (XIIR), react very slowly with ozone and do not show this cracking phenomenon. Typically, these ozone cracks develop in the polydiene elastomers in a direction that is perpendicular to an applied stress. They are the result of rubber chain scission and lead to the formation of several oxygen-containing decomposition products. Ozone reacts with the main-chain double bonds of an elastomer molecule to give the three-oxygen-atom-containing ozonide structure, which decomposes to give chain scission products (peroxy zwitterion- and carbonyl-terminated), which can reform the ozonide, unless the zwitterion (>C$^+$-O-O-) and the carbonyl (>C=O) chain-end groups are removed from one another due to strain in the rubber composition. The reformation of the ozonide, in the absence of strain, is essential a repair of the chain breakage or scission. This is shown below.

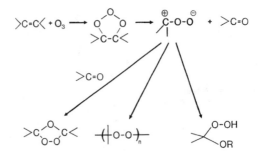

TABLE 4.10 Examples of Chemicals that Have Been Used as Antioxidants and Chemical Antiozonants

Chemical name	Abbrev.	Chemical name	Abbrev.
p-Phenylinediamine derivatives (strongly discoloring)		*Bisphenol derivatives (nondiscoloring)*	
N-Isopropyl-N´-phenyl-p-phenylenediamine	IPPD	2.2´-Methylene-bis-(4-methyl-6-tert.butylphenol)	BPH
N-(l,3-dimethylbutyl)-N´-phenyl-p-phenylenediamine	6PPD	2.2´-Methylene-bis-(4-methyl-6-cyclohexylphenol)	CPH
N-N´-Bis-(l,4-dimethylpentyl)-p-phenylenediamine	77PD	2.2´-Isobutylidene-bis-(4-methyl-6-tert.butylphenol)	IBPH
N,N´-Bis-(1-ethyl-3-methylpentyl)-p-phenylenediamine	DOPD		
N,N´-Diphenyl-p-phenylenediamine	DPPD	*Monophenol derivatives (nondiscoloring)*	
N,N´-Ditolyl-p-phenylenediamine	DTPD	2,6-Di-tert.butyl-p-cresol	BHT
N,N´-Di-β-naphthyl-p-phenylenediamine	DNPD	Alkylated phenol	APH
		Styrenated and alkylated phenol	SAPH
Dihydroqumoline derivatives (strongly discoloring)		Styrenated phenol	SPH
6-Ethoxy-2,2,4-trimethyl-1,2-dihydroquinoline	ETMQ		
2,2,4-Trimethyl-1,2-dihydroquinoline, polymerized	TMQ	*Other antidegradants (nondiscoloring)*	
		Tris-nonylphenylphosphite	TNPP
Naphthylamine derivatives (strongly discoloring)		dilauryl-β,β-thiodipropionate	DLTDP
Phenyl-α-naphthylamine	PAN	β-naphthyl disulfide	
Phenyl-β-naphthylamine	PBN	thio-β-naphthol	
		2-mercaptobenzothiazole	MBT
Diphenylamine derivatives (strongly discoloring)		benzothiazyl disulfide	MBTS
Octylated diphenylamine	ODPA		
Styrinated diphenylamine	SDPA	tris(p-nonylphenyl)phosphite	TNPP
Acetone/disphenylamine condensation product	ADPA	zinc dimethyldithiocarbamate	ZDMC
Benzimidazole derivatives (nondiscoloring)			
2-Mercaptobenzimidazole	MBI		
Zinc-2-mercaptobenzimidazole	ZMBI		
Methyl-2-mercaptobenzimidazole	MMBI		
Zinc-methylmercaptobenzimidazole	ZMMBI		

TABLE 4.11   Peroxide Decomposers

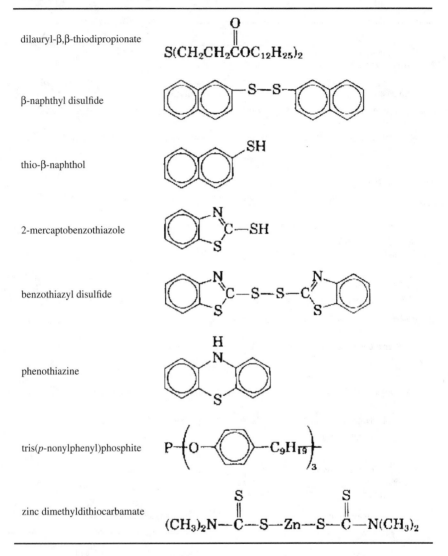

dilauryl-β,β-thiodipropionate	$S(CH_2CH_2\overset{\overset{\displaystyle O}{\|}}{C}OC_{12}H_{25})_2$
β-naphthyl disulfide	
thio-β-naphthol	
2-mercaptobenzothiazole	
benzothiazyl disulfide	
phenothiazine	
tris(p-nonylphenyl)phosphite	
zinc dimethyldithiocarbamate	

For molecular layers of the rubber polymer, this is shown schematically in Scheme 18. If there is sufficient strain in the vulcanized rubber, even if the ozonide reforms, it does so in a manor to permit a crack to form and grow.

Effective chemical antiozonants share certain common functions as follows:

1. They react directly with ozone.

2. They migrate to the surface of the rubber product to react with ozone.

3. They decrease the rate of cut crack growth.

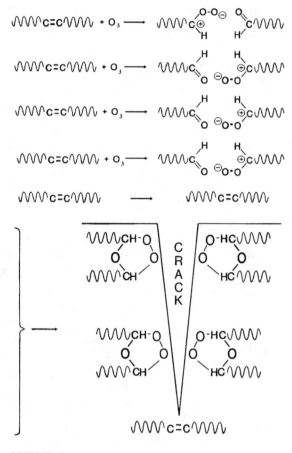

SCHEME 18

4. Antiozonants such as N,N´-dialkyl-*p*-phenylenediamines and N-alkyl-N´-aryl-*p*-phenylenediamine raise a polymer's apparent critical stress; i.e., polymers containing these materials require greater elongation for ozone cracks to occur.

4.5.2.4 *Protection against the Effects of Ozone*

Waxes. Paraffinic waxes function by blooming to the rubber surface to form a thin inert protective film. Since the wax is unreactive toward ozone, this film is a physical but not a chemical barrier to ozone. The number of carbon atoms per molecule of wax varies from 18 to 50. Microcrystalline waxes are heavier and less crystalline. They have between 37 and 70 carbon atoms per molecule. The migration rate of waxes is dependent on several factors. These include the type of rubber or blend, the amount and type of reinforcing filler, the concentration and structure of the wax, and the temperature range that the product will experience in use.

Unfortunately, waxes do not protect against ozone under dynamic conditions, e.g., for a rolling tire. Under such conditions, rupture of a barrier wax film can occur and cause fault points on the rubber surface. Instead of total surface involvement, the ozone attack occurs at these relatively few fault areas, causing rather large cracks to develop. Thus, other types of additives are needed for protection against the effects of ozone under dynamic conditions.

Chemical Antiozonants. The first effective chemical antiozonant was a dihydroquinoline type, 1,2-dihydro-6-ethoxy-2,2,4-trimethylquinoline (DETQ). However, *polymerised* quinoline derivatives provide only slight ozone protection, although they are good antioxidants. DETQ provides protection against the action of ozone, but it is highly staining and discoloring and is lost from rubber compounds because of its volatility.

p-Phenylenediamine antiozonants such as N,N′-di-*sec*-alkyl-*p*-phenylenediamines were then introduced. They surpassed the dihydroquinolines in their ability to protect rubber from ozone attack.

The success of the dialkyl-*p*-phenylenediamines led to the development of related antiozonants, i.e., alkyl/aryl-analogs, N-isopropyl-N′-pheny-*p*-phenylenediamine (IPPD) and N-cyclohexyl-N′-phenyl-*p*-phenylenediamine, and mixed N,N′-diaryl-*p*-phenylenediamine mixtures.

The longer-chain alkyl substituents served to reduce the volatility. However, the fugitive nature of the protection offered by the di-octyl-*p*-phenylenediamine was attributed to the fact that it reacts directly with oxygen (O_2) as well as ozone. Alkyl/aryl- or diaryl-*p*-phenylenediamines are less subject to depletion by reaction with oxygen and are longer-lasting in rubber compounds.

The antiozonants, N-(1,3-di-methylbutyl)-N′-phenyl-*p*-phenylenediamine (6PPD) and the C_7 and C_5 alkyl-analogs were later introduced. In addition to their use as antiozonants, *p*-phenylenediamines, primarily the more persistent mixed diaryl-derivatives, have replaced N-phenyl-β-naphthylamine as antioxidants flex crack inhibitors and synthetic polymer antioxidant stabilizers.

All of these antiozonants are staining and discoloring. This has limited their use primarily to carbon black-loaded compounds. We also note that the antiozonants and amine-based antioxidants cause a reduction in scorch resistance.

Multiple Functions. The N,N′-disubstituted-*p*-phenylenediamines (PPDs) are unique stabilizers. Many of them simultaneously are potent antioxidants as well as antiozonants (especially the sec-alkyl-*p*-phenylenediamines). They are also good flex-crack inhibitors.

In general, the best antioxidant protection is afforded at levels slightly less than 1 phr. Higher levels of use actually become detrimental, as the system can become pro-oxidative. Nevertheless, to provide antiozonant protection, antiozonant levels in excess of 2 phr or more may be necessary. The use of di-aryl-*p*-phenylenediamines with di-alkyl or alkyl/aryl derivatives is beneficial, since it reduces the pro-oxidative effect. Their low volatility and low extractability provide for long-term protection.

Differences between Polymers. The degree of required ozone protection varies with the type of rubber. Saturated elastomers need no antiozonant protection, because they have no sites for reaction with ozone. Rubbers such as EPDM, which have a low olefin concentration, need essentially no protection against the effects of ozone. Styrene-butadiene rubber (SBR) requires antiozonant, while NR and synthetic polyisoprene (IR) may require somewhat increased dosages of antiozonant. Nitrile rubber (NBR) is very difficult to protect against ozone attack.

Antiozonant activity changes with time. For short periods of aging time, the dialkyl-*p*-phenylenediamines are the most effective antiozonants, very closely followed by the alkyl/aryl analogs, with the diaryl being less effective. However, with increased aging time, the order of effectiveness is completely reversed, as oxidation and reaction with ozone occur—another reason for using mixtures of antiozonants, i.e, to provide protection for an extended period of aging.

Mechanisms for Protection against the Effects of Ozone Attack.   The mechanism of antiozonant protection is still not fully understood. However, there are several theories, which detail the mechanism of protection by chemical antiozonants: inert barrier, competitive reaction, reduced critical stress, and polymer back-bone chain repair.

The inert barrier theory says that a material that is nonreactive migrates to the surface and forms a physical barrier that prevents the ozone from reaching the reactive double bonds in the polymer. Waxes are thought to behave in this manner.

According to the competitive-reaction or "scavenger" model, the antiozonant migrates to the surface of the rubber and then selectively reacts with ozone so that the rubber is not harmed until the antiozonant is consumed. A protective-film theory suggests, that after the antiozonant has done the above and behaved as a "scavenger," the reaction products become an inert film. Any chemical antiozonant might function in both of these ways. There is much evidence to support the "scavenger" mechanism as the dominant one. There is also good support for the formation of a protective film. Surface films on rubber have been seen visually and microscopically. With partial removal of the film and reexposure to ozone, only the cleaned surface is degraded.

According to the reduced critical stress theory, certain materials migrate to or near the rubber surface and modify the internal stress of the polymer such that cracks do not appear. Although this phenomenon is poorly understood, it is easy to observe. The use of increasingly higher levels of antiozonant raises the critical stress level required for cracks to form.

The chain repair theories suggest that severed polymer chains (terminated by carboxy or aldehyde groups) can be relinked by reaction with the antiozonant or that the antiozonant reacts with the ozonide or zwitterion (carbonyl oxide) to give a low-molecular-weight, inert, self-healing film. Either way, the antiozonant would be chemically linked to the rubber. However, the chain repair or self-healing film theories do not appear to be as strongly supported as the other theories.

Ideal Antiozonants.   An ideal antiozonant should be competitively reactive with ozone in the presence of carbon-carbon double bonds in the rubber-molecule backbone. However, it should not too reactive with ozone (or even oxygen) lest it not persist to give long-term protection. It should not react with sulfur accelerators or other ingredients in the cure package. It should be nonvolatile and persist at the surface of the rubber. In addition, the ideal antiozonant should not discolor the rubber. Unfortunately, an ideally active nonstaining chemical antiozonant has not yet been found.

4.5.3   Types of Vulcanizable Elastomers and their Applications

4.5.3.1 Natural Rubber (NR).   Natural rubber, as stated above, was the first elastomer to be used in commercial applications. Although the polymer (cis-1,4-polyisoprene) occurs in over 200 plants, the rubber tree, *hevea brasiliensis,* is the source of essentially all that is used. The chemical structure of the polymer is given here:

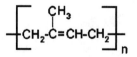

There are two possible structures for poly-1,4,polyisoprene. Natural rubber structure is of the *cis* form. The *trans* forms (the structure of guta perch or balata gum) have higher melting point and higher glass transition temperatures (see below).

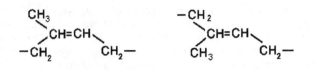

1,4- *cis* Structure   1,4- *trans* Structure

The rubber is harvested from the tree in the form of a latex, the aqueous emulsion obtained from the tree by tapping into the inner bark and collected in cups attached to the trees. The latex itself can be used for the fabrication of rubber articles, but most of the NR is used as a dry raw rubber taken from the coagulated latex. There are many types and grades of the dry rubber. However, the Malaysian rubber industry produces standard NR grades that correspond to technical specifications. Their system is being followed by other producer countries, thus the designations SMR (Standard Malaysian Rubber), SIR (Standard Indonesian Rubber), SSR (Specified Singapore Rubber), SLR (Standard Lank Rubber), TTR (Thai Tested Rubber), and NSR (Nigerian Standard Rubber). Within a national standard type, there are grades that differ with respect to color, viscosity, molecular weight, and other qualities.

NR contains small amounts of highly important nonrubber constituents, e.g., proteins, sugars, and fatty acids. Some of these nonrubber components are vulcanization activators, antidegradants, and, unfortunately, allergens.

NR can be vulcanized by using any of a number of vulcanizing systems, e.g, accelerated sulfur, peroxide, phenolic (resole), quinonedioxime, and others. However, by far, accelerated sulfur systems are the most used.

Properties. NR vulcanizates have a range of interesting properties. Individual properties of NR can be surpassed by those of synthetic rubbers, but the combination of high tensile strength, high resilience, good low-temperature flexibility, and low hysteresis and heat buildup is unique. In addition, the building tack and green strength of NR are unsurpassed by synthetic rubbers. Building tack is the ability for unvulcanized pieces of rubber to stick together during the building process, for example, for a tire, where plies and other components must adhere and "become one" before vulcanization. Green strength is the mechanical strength of the uncured polymer. It is high in the case of natural rubber because, even before vulcanization, natural crystallizes during straining. This property is likely also related to building tack, wherein there would be crystallization at the autoadhesive interface due to high local strains as one attempts to pull apart one component from the other.

NR vulcanizates can be produced in a wide hardness range (Shore A 30 to that of hard rubber or ebonite). Due to its crystallization during strain, NR has high tensile strength even without reinforcing fillers (e.g., carbon black). Also, because of strain-induced crystallization, the tear resistance of NR vulcanizates is quite high. The ultimate elongation of a NR vulcanizate is generally between about 500 and 1100 percent. Also, NR vulcanizates have very good fatigue resistance (resistance to repeated strains, each one alone less than ultimate). With respect to elastic rebound, NR vulcanizates are surpassed only by those of BR.

The heat resistance of NR is not good enough for many uses, and it is exceeded by many synthetic rubbers. It is affected by the choice of vulcanization system, vulcanization conditions, choice of protective agents, and even choice of filler. To obtain good aging resistance of NR vulcanizates, one must use protective agents in the compound and use relatively short curing cycles at relatively low temperatures.

Because of its main-chain double bonds, unstabilized NR exhibits extremely poor resistance to atmospheric ozone. Its light-colored vulcanizates have poor resistance to weathering out of doors. All of this can be improved if antidegradants, carbon black, and waxes are used. Nevertheless, NR cannot compete with essentially saturated rubbers (containing few, if any, main-chain double bonds) such as EPDM, ACM, IIR, CSM, and so on.

On the other hand, NR vulcanizates have very good low temperature flexibility. Here also, it is only surpassed by BR, which has a somewhat lower glass transition temperature, T_g.

NR vulcanizates exhibit relatively low (favorable) values of compression set at ambient and somewhat higher temperatures. At lower temperatures, compression set is less favorable, possibly due to a tendency for the rubber to crystallize. At more elevated temperatures, poor heat resistance and cross-link rearrangement can have a detrimental effect on compression set.

NR vulcanizates have very poor resistance to swelling in gasoline, mineral oils, and other nonpolar liquids, swelling as much as 1000 percent or more by volume. The resistance of NR vulcanizates to swelling by polar liquids such as water, alcohol, and others is very good.

Uses.   NR is used mostly in the form of solid rubber and to a lesser extent as latex. Minor amounts of NR are used in adhesives, rubber solutions (cements), art gum, and other products. A relatively small amount of NR is used in the production of hard rubber (a high-temperature reaction product of NR with about 30 weight percent of sulfur). Before the introduction of other rubbers or elastomers, all rubber products were produced from NR. Because of improved specific properties of many synthetic rubbers, NR has been replaced in many applications. This is especially true where resistance to heat and weathering, and oil and solvent resistance are required. There are also economic reasons for using other rubbers. The price of natural rubber can be high when demand is greater than what can be produced by the existing rubber trees. With the exceptions of tires and possibly a few other applications, NR is no longer the preferred elastomer. NR is well suited for tires because of its relatively low heat buildup, tearing resistance, low-temperature flexibility, fatigue resistance, and building tack.

In addition to the above, NR is important in the production of thin-walled products such as surgical gloves, balloons, condoms, and so forth. Here, latex dipping is the method of shaping or forming. Because of its allergen content, the human-contact products are now threatened.

Its low damping and high elasticity allow NR to be used for producing vehicle suspension elements and bumpers. An interesting suspension element is a building support to "tune" the structure to resist earthquakes. NR has also been used in supports for bridges. Early work in the 1950s on laminated rubber bearings for bridges, now used to accommodate bridge deck movements, gave rise to the development of bearings for the base isolation of whole buildings against ground-borne vibrations (e.g., underground railway systems). These, in turn, were further developed in the mid 1980s and 1990s for bearings to protect buildings against earthquakes.

4.5.3.2 Synthetic Polyisoprene (IR).

Synthetic polyisoprene is similar to natural rubber in chemical structure and properties. Although it has lower green strength, lower hot tear, and inferior aging characteristics than NR, synthetic polyisoprene exceeds the natural types in consistency of product, processing, and purity. In addition, it has better mixing, extrusion, molding, and calendering characteristics. (Processing methods for vulcanizable rubbers are discussed in a later section.)

The successful synthesis of stereoregular polyisoprene (IR) fulfilled a goal sought by polymer chemists for nearly a century. The polymer chains in the early synthetics contained mixtures of all possible molecular configurations joined together in a random fashion. Specifically, they lacked the very high cis-1,4 structure content of the natural rubber backbone that gives it the ability to undergo strain-induced crystallization.

New types of catalyst systems that could selectively join together monomer units in a well ordered fashion were discovered in the 1950s. Shortly after the discovery of the breakthrough Ziegler-Natta catalyst systems for the polymerization of ethylene, stereospecific catalysts were developed for the polymerization of isoprene. This enabled the production of a nearly pure cis-1,4-polymer, the so-called "synthetic natural rubber." In 1962, Goodyear introduced Natsyn®, a strain-crystallizable isoprene polymer with a cis-1,4-content of 98.5 percent.

Properties.   Typical raw polymer and vulcanized properties of synthetic IR are similar to those of NR. Both exhibit good inherent tack, high compounded gum tensile strength (green strength), and good vulcanizate hysteresis (low values) and tensile properties. In synthetic IR, there is minimal variance in physical properties lot to lot, and there is a low concentration of nonrubber constituents compared to natural rubber.

Because of the lower raw polymer viscosity of synthetic polyisoprene, part or the entire breakdown step normally used for natural rubber (premastication) might be eliminated.

Synthetic polyisoprene is especially well suited for injection molded compounds. Because of its uniform cure rate. Time/temperature press cycles can be established with assurance that parts will be uniformly cured.

Uses.   Synthetic polyisoprene is used in a variety of applications requiring low water swell, high gum tensile strength, good resilience, high hot tensile strength, and good tack. Gum compounds based on synthetic polyisoprene are used in rubber bands, cut thread, baby bottle nipples, and extruded hose. Black-loaded compounds find use in tires, motor mounts, pipe gaskets, shock absorber bushings, and many other molded and mechanical goods. Mineral-filled systems find applications in footwear, sponge, and sporting goods. In addition, recent concerns about allergic reactions to proteins present in natural rubber have prompted increased usage of the more pure synthetic polyisoprene in some applications. Synthetic IR, converted to a latex, is used in the production of nonallergic gloves for medical and related uses.

4.5.3.3 Butadiene Rubber (BR).   Polybutadiene rubber was originally produced by emulsion polymerization of 1,3-butadiene, generally with rather poor results. Now it is generally prepared by solution polymerization. Its general chemical structure is as follows:

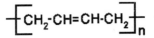

The polymer is generally prepared in the presence of a stereo-specific catalyst system that can produce a polymer that contains in excess of 92 percent cis-1,4 structure. The cis and trans are shown here:

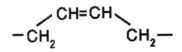

1,4- *cis* Structure   1,4- *trans* Structure

A 1,2-polybutadiene structure also occurs as shown below.

1,2-polybutadiene structure

The so-called "high cis" BRs generally contain a 93 to 97 percent *cis*-1,4 structure, 2 to 3 percent *trans*-1,4 structure, and 1 to 4 percent 1,2 structure. However, a wide variety of *cis*, *trans*, and 1,2-polymer content is possible.

Because of its poor behavior on mills and certain poor (for tire) performance properties, BR is mostly used in blends with NR or SBR or with both of them. BR can be vulcanized by using the same types of curing systems as those used for NR.

Properties.   The ultimate tensile properties of BR vulcanizates with high *cis*-1,4 content are significantly lower than those of comparable NR or SBR vulcanizates. However, its blends with NR or SBR can give excellent properties. In addition, vulcanizate properties of NR or SBR are, in some respects, improved by blending with BR. Such blends have high abrasion resistance, high resilience, and good low-temperature flexibility (due to the very low glass transition temperature of the BR, i.e., around –90°C).

Heat buildup and resistance to groove cracking in tire treads are also improved for NR and SBR in their blends with BR. Reversion resistance due to overcure and aging resistance of NR vulcanizates are improved when it is blended with BR.

The rolling resistance of tires made from NR or SBR is reduced by the presence of increasing amounts of BR. This reduces the fuel consumption of vehicles on the road. However, the presence of BR gives rise to poor wet traction, but the presence of about 40 percent (of the polymer) BR gives improvements in ice traction. In addition, BR has a great tolerance for high levels of extender oil and carbon black.

Uses.   BR was first used largely in the blend of elastomers in tire treads to give improved abrasion resistance. Because of the emergence of radial tires, BR is largely used in tire carcasses, sidewalls and bead compounds. BR is important in winter tire treads because it gives improved ice traction that it confers. Over 90 percent of the BR production is used in tires.

BR is also used in shoe soles and conveyor belts when there is a need for high abrasion resistance. BR is also used in compounds processed by injection molding because o its good flow properties.

4.5.3.4 Styrene-Butadiene Rubber (SBR).   SBR is produced by both emulsion and solution polymerization of mixtures of 1,3-butadiene and styrene. The general chemical structure of SBR polymers is as follows:

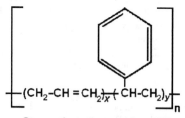

Styrene-butadiene rubber SBR

Emulsion polymerization is done hot (about 50°C) or cold (about 5°C), depending on the selection of polymerization initiator. SBR prepared by emulsion polymerization (emulsion SBR or E-SBR) generally contains about 23 percent styrene-derived units, randomly distributed in the molecular polymer chains. SBR made by solution polymerization (solution SBR or S-SBR) has about the same styrene-derived unit content. Both random and block polymers can be made by solution polymerization. Both emulsion SBR and solution SBR are available in oil-extended versions (OE-SBR). These have as much as 50 parts of extender oil per 100 parts by weight of polymer (phr).

E-SBR is available in Mooney viscosities (ML 1+4 100°C) ranging from about 30 to 120, corresponding to average molecular weights of about 250,000 to 800,000. It is supplied as dry gum, oil-extended or carbon-black-filled polymer. In some respects, the lower-viscosity grades are more easily processed, while the higher-viscosity grades have better green strength, accept higher filler and oil loadings, and tend to give less porous vulcanizates.

Cold E-SBRs (those produced at the lower temperatures) contain less long-chain branching than do the so-called hot rubbers. An effect of this is that the cold-process rubbers generally can be more easily processed than the hot-process rubbers. SBRs can be vulcanized by the same types of systems as used for NR. As with NR, accelerated sulfur curing systems are, by far, the most used.

Properties.   The mechanical properties of E-SBR vulcanizates depend on the type and level of filler in the compound. Unfilled gum vulcanizates have very poor tensile strength and ultimate elongation, because the rubber lacks self reinforcing of the type found NR rubber vulcanizates, i.e., strain-induced crystallization. This inadequacy is offset by the addition of reinforcing fillers, i.e., carbon black or chemically coupled silica. At optimum loadings with reinforcing carbon black, mechanical properties similar to those of NR can be achieved. However, NR compounds exceed SBR compounds in tear strength because of NR's strain-induced crystallization.

Emulsion SBR vulcanizates have better aging, fatigue, and heat resistance than do those of NR. Antidegradants, however, are required for this. E-SBR vulcanizates, unlike those of NR, are reversion resistant. By using reinforcing fillers, one can achieve better abrasion resistance with E-SBR than with NR. In part, for these reasons, emulsion SBRs have replaced very much of NR. However, E-SBR vulcanizates are more hysteretic than those of NR and, thus, heat buildup during heavy duty flexing is a greater problem with E-SBR than with NR.

E-SBR vulcanizates are resistant to many polar solvents, dilute acids and bases, and so on. However SBR vulcanizates swell considerably in contact with oils, fats, gasoline, kerosene, and others.

Random-distribution solution SBR vulcanizates are less hysteretic than are comparable vulcanizates of E-SBR. Also, solution polymers contain less nonrubber material. This is because there is absence of emulsifier (e.g., soap) during polymerization. During coagulation of the polymerized emulsion to obtain the rubber, fatty acids are formed. The presence of such fatty acid, in part, reduces the rate of vulcanization with respect to that of solution SBR compounds. The absence of such nonrubber components also reduces the electrical conductivity of S-SBR compounds compared to those of E-SBR. Vulcanizates of solution SBRs, having blocky monomer distributions, have very low brittleness temperatures due to the presence of relatively long polybutadiene chain segments. They have good elastic properties, low water adsorption, low electrical conductivity, and excellent abrasion resistance.

Oil-extended SBR (OE-SBR) grades contain polymer of very high molecular weight. This enables the presence of high concentrations of oil with the maintenance of viscosities similar to those of nonoil-extended SBRs for easy processing.

Uses. Emulsion SBRs are generally used in combination with BR in the production of car and light truck tires. Other applications include belting, shoe soles, hose, molded rubber goods, cable insulation, roll coverings, and so forth. Uses of various grades of E SBRs are shown in Table 4.12.

Random solution SBR is used in blends with emulsion SBR to improve processability. Blocky S-SBR is used in hard shoe soles, roll coverings, and special mechanical goods.

SBRs compete with NR, IR, and BR in many applications and as components in blends.

4.5.3.5 Butyl, Chlorobutyl, and Bromobutyl Rubbers (IIR, CIIR, and BIIR)

Butyl Rubber (IIR). This polymer is produced by cationic copolymerization of isobutylene mixed with minor amounts of isoprene. Its molecules are 97 to 99.5 mole percent derived from isobutylene, the rest being derived from isoprene. Its structure can be represented as follows:

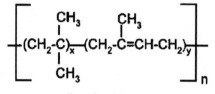

Butyl rubber IIR

It is prepared by cationic copolymerization of isobutylene with minor amount of isoprene generally in the solvent methylene chloride with aluminum chloride as a catalyst or in hexane with a dialkyl aluminum chloride as catalyst at very low temperatures (-100 to $-50°C$). Average molecular weights of IIR are generally between 300,000 and 500,000.

Properties of Butyl Rubber. Because it is a largely saturated polymer, it has good resistance against oxidation and ozonolysis. It also has a very low gas permeability. However, because of its low amount of unsaturation, butyl rubber is rather slow to vulcanize. It can be cured by sulfur, generally in the presence of fast accelerators (e.g., dithiocarbamates). It can be cured by quinone dioxime in the presence of the oxidant PbO_2. Butyl rubber is also vulcanized by using the resin curing system (phenol/formaldehyde resoles, e.g., the product SP1045) in the presence of a Lewis acid activator (e.g, $SnCl_4$), added or formed *in situ,* e.g. by the action of ZnO with a source of HCl, e.g., a halogenated polymer of a resole containing halomethylene groups (e.g., the product SP 1056). The heat resistance and resistance to weathering of IIR vulcanizates are excellent.

Uses of Butyl Rubber. Butyl rubber is used in the manufacture of inner liners of tubeless tires, inner tubes, cable insulation, pharmaceutical stoppers, curing bags, and bladders for tire manufacture. When tires are in the molds for vulcanization, the inside of the tire is filled with a butyl rubber bag or bladder of steam under enough pressure to obtain the vulcanization temperature. This is possible only because of the good resistance of butyl rubber to heat and water.

Chlorobutyl Rubber and Bromobutyl Rubbers (CIIR and BIIR). The addition of chlorine or bromine to IIR in an inert solvent (e.g., hexane) gives the facile attachment of one halogen atom per isoprene unit in the allylic position. Compared with IIR, the halogenated butyl rubbers have certain advantages. The cure reactivity is increased to give faster vulcanization rates, greater extents of vulcanization, and reduced reversion. Also, the halogenation improves the compatibility of the isoprene polymer with other types of rubber (e.g., NR) to make useful rubber-blend compositions possible.

TABLE 4.12 Uses of Emulsion SBRs

Emulsion SBR Grade	Emulsifier type	Mooney viscosity (ML 1+4 @ 100°C)	Color	Oil Grade	Oil PHR	Carbon black Grade	Carbon black PHR	Uses
1500	R	50–52	S	–	–	–	–	Passenger tire treads and mechanical rubber goods
1502	F	50–52	NS	–	–	–	–	Light-colored technical rubber goods, for blends requiring good flow properties in injection molding or calendering
1507	F	30–35	NS	–	–	–	–	
1509	F	30–35	NS	–	–	–	–	Cable and electricals
1516	F	40	NS	–	–	–	–	Injection and compression molded goods with high surface finish, brake and transmission pads, belting, adhesives
1573	R	115	NS	–	–	–	–	
1707	R	49–55	NS	NAPH	37.5	–	–	Light-colored and transparent hose, profiles, shoe soles, floor tiles
1712	F	49–55	S	HAR	37.5	–	–	Passenger tire treads, belting, dark-colored rubber products
1778	F	49–55	NS	NAPH	37.5	–	–	Same as 1707 and cable insulation
1609	R	61–68	S	HAR	5.0	N110	40	Tire retreads
1808	F	48–58	S	HAR	47.5	N330	76	Passenger tire treads, retreads, electrical products
1843	F	86	NS	NAPH	15.0	N770	100	V-belts and other dynamic applications

R = resin acid blend, F = fatty acid blend, S = staining, NS = nonstaining, NAPH = naphthenic, HAR = highly aromatic oil

Properties of Halobutyl Rubbers. Properties of vulcanizates are enhanced over those of IIR. BIIR vulcanizates have lower gas permeability, offer better ozone and weather resistance, and are faster curing than those of CIIR or IIR. The properties of CIIR are between those of BIIR and IIR.

Uses of Chlorobutyl and Bromobutyl Rubbers. CIIR and BIIR are used in inner liners of tubeless tires with improved (over IIR) covulcanization (in blends) and adhesion to other components of the tires, in inner tubes for heavy-duty applications such as in truck and bus tires, and in belts, hoses, seals, injection molded parts, and pharmaceutical stoppers.

4.5.3.6 Ethylene-Propylene Rubbers (EPR and EPDM). Polyethylene, though it is a semicrystalline solid, has a very low glass transition temperature (about − 80°C). Polypropylene is also a semicrystalline solid with at glass transition temperature of about −10°C. Copolymers of similar amounts of ethylene and propylene are noncrystalline (or only slightly crystalline) and have glass transition temperatures of roughly the weight average between that of polyethylene and polypropylene (i.e., between −40 and −60°C). They are rubbery materials. The chemical structure of ethylene-propylene rubber can be expressed as follows:

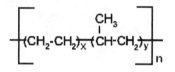

Most EPR rubbers contain 40 to 80 percent by weight of monomer repeat units derived from ethylene. The grades containing the higher concentrations of ethylene-derived monomer units generally contain some crystallinity.

Since EPR rubber molecules do not contain unsaturation, they can be vulcanized only by organic peroxide curing systems. If a third monomer is added during the polymerization, i.e., a diene monomer (wherein only one of the two double bonds takes part in the polymerization), unsaturation can be introduced into the molecule, and it can then be vulcanized by accelerated sulfur curing systems. A chemical structure for ethylene-propylene-diene-monomer (EPDM) rubbers can be expressed as follows:

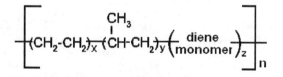

The diene monomers are nonconjugated, and the double bonds are thus located in side groups of the polymer chains. Common third monomers are dycyclopentadiene (DCP), ethylidene norbornene (ENB), and trans-1,4-hexadiene (HX). EPDM rubbers are generally vulcanized by accelerated-sulfur systems. The amount of ter-monomer can range between about 2 and 10 percent. Higher concentrations of unsaturation give faster vulcanization and can give higher cross-link densities. The grades derived from HX vulcanize less rapidly than do the others.

There are many types of commercially available EPDM rubbers. They differ with respect to ethylene/propylene monomer ratio, amount and selection of ter-monomer (unconjugated diene), molecular weight, molecular-weight distribution, viscosity, amount and type of extender oil (if present), processability, and other qualities.

The high-viscosity (high-molecular-weight) EPDMs are generally sold with added extender oil. The amount of oil can be 100, 200, or more parts per 100 parts of rubber (phr).

Properties of Ethylene-Propylene Rubbers. Elastic properties of EPR or EPDM are better than those of many synthetic rubbers, but hysteresis is not as good (low) as in the case of NR and BR. Resistance to compression set for EPR and EPDM vulcanizates is excellent. For EPDM, this is especially true in grades containing larger amounts of termonomer units derived from ENB. Fatigue resistance for EPDM vulcanizates is very good, comparable to that of SBR vulcanizates. The resistance to heat and aging of EP rubber vulcanizates is better that for NR SBR and NBR vulcanizates. Peroxide vulcanizates are notably resistant to oxidative heat aging. Also, these largely saturated polymers are very resistant to attack by ozone. Low-temperature flexibility and oil swelling properties of these elastomers are similar to those of NR.

Electrical properties (insulating, dielectric breakdown, corona resistance, and so on) of ethylene-propylene rubbers are excellent. This is especially true for peroxide-cured EPR.

Uses of Ethylene-Propylene Rubbers. EPDM and EPR vulcanizates are used in extruded profiles, cable insulation and jacketing, and roofing membranes. There are many automotive uses: radiator hose, door and trunk seals, insulation, jacketing, and others. These elastomers are also used in applications such as window and architectural profiles, dock fenders, and washing-machine hoses. In short, their applications are extensive and diverse. Ethylene-propylene rubbers may be the most versatile of general-purpose rubbers. In addition, EP rubbers are added to polyolefin plastics as impact modifiers and as components of certain thermoplastic elastomer compositions (e.g., thermoplastic vulcanizates, which are discussed later in this chapter).

4.5.3.7 Chloroprene or Neoprene Rubber (CR). Chloroprene rubber (poly-2-chlorobutadiene) is produced by the emulsion polymerization of 2-chlorobutadiene, in the presence of a free-radical initiator. Its general chemical structure can be represented as follows:

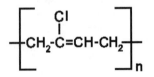

There are a number of commercially available grades of CR. They differ, for example, with respect to processability, mercaptan modification, polymerization temperature (effects tendency to crystallize), viscosity (processability), stabilizer, copolymerization with other monomers (crystallizability), gel content (effects processing), reactive groups (important for lattices), and so on. With increasing polymerization temperatures, there is less uniformity in chain structure due to incorporation of 1,2- and 3,4-structural units and different isomers in the monomer sequences. This reduces the rate of crystallization of the resulting polymers. If the structure is too regular, too much crystallization occurs for the production of rubber products, because they tend to harden very rapidly with loss of elasticity. Nevertheless, the extensively crystallizable CRs are useful in adhesive compositions. For general-purpose rubber applications, there are three grades: G, W, and T, with selected features to offer a range of processing, curing, and performance properties.

CR compounds are vulcanized by metal oxides, i.e., ZnO. MgO is also present in the compound to somewhat retard the action of ZnO. Thioureas are used as accelerators. For low-water-absorption vulcanizates, PbO or Pb_2O_4 can be used as the vulcanizing agent.

Accelerated sulfur vulcanization systems can be employed, but generally as components of mixed vulcanization systems containing both metal oxide and accelerated sulfur.

Properties of Chloroprene Rubber Vulcanizates. Unfilled (nonreinforced) gum vulcanizates are much stronger for CR than for other synthetic rubbers (except for synthetic polyisoprene) because of the tendency for strain-induced crystallization (similar to the case of isoprene rubbers). Similarly, resistance to tearing is very good for CR vulcanizates.

Because of the chlorine content, CRs are more flame resistant than other elastomers. However, when chlorinated polymers burn, they liberate great amount s of corrosive, toxic vapors. They are not used in areas where there is a strong need for safety.

The stiffening of CR vulcanizates with cooling can be due to crystallization. Fillers have little effect on this, but ester-type plasticizers can greatly reduce the crystallization temperature.

Because of the increased polarity of CR over other (i.e., hydrocarbon) rubbers, CR vulcanizates are sufficiently oil resistant for many uses. Also, CR vulcanizates are fairly resistant to chemicals such as concentrated alkalis, dilute acids, and aqueous salt solutions.

CR is superior to NR or SBR as a barrier against gas permeation, but it does not exhibit as low permeability as do butyl rubbers, epichlorohydrin rubbers, or nitrile rubbers.

Uses of CR. CR vulcanizates are used in many rubber products that are flame resistant, resistant to fats and oils, and resistant to weathering and ozone (when compounded with proper antidegradants). Products include moldings, extrusions, seals, hoses, rolls, belts, shoe soles, bearings, rubberized fabrics, linings, and cable jackets. However, in some areas, the use of CRs has been reduced by competition form nitrile rubbers (for better oil resistance) and EPDM compositions because of price.

In contrast to the low crystallinity required for rubber products, crystallization is a great benefit for adhesives. In solution, crystallization doesn't occur but, after drying, the adhesive film hardens rapidly due to crystallization.

4.5.3.8 Nitrile Rubber (NBR). Nitrile rubbers are copolymers of butadiene and acrylonitrile produced by emulsion polymerization. There are "hot" and "cold" polymerized types. The hot-polymerized types generally have the higher green strengths but are somewhat more difficult to process. The acrylonitrile monomer repeat units impart resistance to oil swelling. There are grades containing 18 to 50 percent acrylonitrile-derived backbone units. Glass transition temperatures and oil resistance increase with the nitrile content, the glass transition temperatures ranging from -38 to $-2°C$. An example of a structure for nitrile rubber can be given as follows:

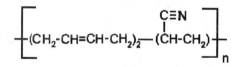

Relatively small amounts of acrylic acid can be used in the monomer polymerization mix to give carboxylated nitrile rubber (X-NBR), whose polymer chains have carboxylic-acid side groups.

NBR can be partially or even completely hydrogenated (to eliminate carbon-carbon double bonds) in nonaqueous solution by using suitable catalysts (e.g., cobalt, rhodium, ruthenium, iridium, or palladium complexes) to give hydrogenated nitrile rubbers (H-NBRs). Completely saturated H-NBR grades are cross-linked with peroxides.

Properties of Nitrile Rubbers.   Used with reinforcing fillers, NBR vulcanizates of excellent mechanical properties are obtained. With proper compounding, a wide range of hardness grades are possible, with good resistance to compression set. Elastic properties of unplasticized NBRs are somewhat less favorable than those of NR or SBR. The use of ester-type plasticizers and non- or semireinforcing fillers can nevertheless give rise to compositions of good elasticity. The presence of reinforcing fillers can give abrasion resistance that is considerably better than that of comparable NR or SBR vulcanizates, and X-NBR vulcanizates have extremely good abrasion resistance. The heat resistance of NBR vulcanizates is generally better than for vulcanizates of NR of SBR. With reduced amounts of oxygen, as in an oil environment, the heat aging resistance is even more impressive. The weather and ozone resistance of NBR vulcanizates is similar to that of NR vulcanizates, but antiozonants are somewhat less effective with NBR. Low-temperature flexibility improves with decreasing amounts of nitrile content but improves with increasing concentrations of ester-type plasticizers, e.g., adipate esters.

Because nitrile rubbers are polar elastomers, their vulcanizates are very resistant to swelling in hot oil, gasoline, grease, and other nonpolar substances. The resistance to swelling in nonpolar oils, solvents, and so on, is improved greatly with increasing nitrile content. Here we see the need for compromises: increasing nitrile content improves oil resistance but reduces elasticity and set resistance and most severely reduces low-temperature flexibility. Plasticizers are very necessary to surmount this problem.

NBR vulcanizates are less permeable to gasses than are NR and SBR, with permeability decreasing with increasing nitrile content. Permeability by gasoline vapors and the like, however, is very high.

Nitrile rubber vulcanizates have a considerably higher electrical conductivity than do those of nonpolar elastomers and are, thus, not generally used in parts that require low electrical conductivities.

Completely saturated HNBR vulcanizates have the excellent resistance to hot air and hot oils, and they have high resistance to oxidative and ozone attack. They have good resistance to sulfur-containing oils, sulfur and nitrogen-containing oil additives, and industrial chemicals in general. The fully saturated H-NBRs have very high tensile strength, good low-temperature flexibility, and good abrasion resistance. Unfortunately, H-NBRs are costly.

Uses of Nitrile Rubbers.   NBR vulcanizates are used where, in addition to good mechanical properties, good resistance to swelling in oils and resistance to abrasion are required. Typical uses are in seals, O-rings, packings for crank shafts and valves, membranes, bellows for coupling, hose, high-pressure hose, and others. NBRs are also used in oil-rig applications, roll coverings, conveyor belts, linings, containers, work boots, shoe soling, and so on. NBRs are also used in products for the food industry. Liquid grades of NBR, whose molecules contain reactive groups, are used to make liquid-cast elastomeric parts and used as impact modifiers for epoxy resins. NBR gloves are prepared by the latex dipping process.

4.5.3.9 Chlorinated Polyethylene (CM).   Commercial grades of CM are produced by random chlorination of high-density polyethylene in aqueous suspension. Uniform chlorination of the polymer requires elevated temperatures for the reaction. The general chemical structure of CM can be represented as follows:

$$\left[(CH_2\text{-}CH_2)_x (CH_2\text{-}CH)_y \atop {} \right]_n$$
with Cl substituent on the CH.

The crystallinity of the polymer is greatly reduced by chlorination. The amount of crystallinity remaining depends on the extent of chlorination. Grades of CM with 25 percent chlorine are somewhat crystalline, because their backbones contain long polyethylene sequences and are harder than the noncrystalline grades. Grades containing 34 percent chlorine or more are not crystalline at all. Grades containing 35 percent chlorine have the lowest brittleness temperatures, in the range of about –40°C. Chlorination increases the polarity of the polymer and gives elastomers that are resistant to swelling in nonpolar fats, oils, solvents, and so forth. Because these elastomers contain no molecular double bonds, they are vulcanized by peroxides. As with other chlorine-containing elastomers, CM should be protected against dehydrohalogenation. This is done by the addition of MgO, lead compounds, epoxidized oils or epoxy compounds, or combinations of them.

Properties of Chlorinated Polyethylene Vulcanizates. CM vulcanizates have good mechanical properties; low compression set; low brittleness temperatures; very good dynamic fatigue; excellent aging, weathering, and ozone resistance; fairly good hot oil resistance; good chemical resistance; flame resistance; and good color stability.

Uses of Chlorinated Polyethylene Vulcanizates. CM is used in applications where aging resistance in hot air, oils, or chemicals is required and where good ozone, weathering, and flame resistance are required. Many such applications are in the wire and cable industry.

4.5.3.10 Chlorosulfonated Polyethylene (CSM). CSM is produced by ultraviolet radiation of low-density polyethylene in an inert chlorinated solvent at 70 to 75°C in the presence of chlorine and sulfur dioxide. Its chemical structure can be represented as follows:

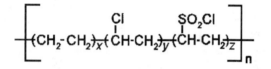

Commercial grades contain 25 to 43 percent chlorine and 0.8 to 1.5 percent sulfur randomly distributed along the polymer chains. Cure rates increase with increases in chlorosulfonation. Grades with low chlorine content (25 percent) are best for heat resistance and optimum electrical resistivity. Flame resistance increases with chlorine content. CSM was commercialized by Du Pont as Hypalon® rubber.

CSM is more easily vulcanized than is CM. Polyvalent metal oxides, such as those of lead and magnesium, react in the presence of small amounts of acids (such as stearic) or sulfur vulcanization accelerators, e.g., TMTD or MBT.

Properties of Chlorosulfonated Rubber Vulcanizates. Properties of CSM vulcanizates are very similar to those of CM vulcanizates. It has a combination of toughness, resistance against dry heat and weathering, ozone resistance, flame resistance, resistance to hot oils, corrosive chemicals, and so forth.

Uses of Chlorosulfonated Rubber. CSM is used in applications similar to those of CM. It is used in automotive hoses, tubes, gaskets, electrical wire insulation (even in high-voltage power stations), industrial hoses, tank linings, coated fabrics, conveyor belting, and other applications.

4.5.3.11 Polysulfide Rubbers (TM). Polysulfide rubbers are produced in aqueous solution at 60°C by the polycondensation of aliphatic dihalides (e.g., ethylene dichloride, di-2-chloroethyl formal, or di-2-chloroethyl ether) and alkali polysulfides, e.g., sodium tetrasulfide. The polymer precipitates in the form of small particles that are washed, coagulated, and dried. By varying the organic dihalide and the alkali polysulfide, a wide variety of products can be obtained. A general chemical structure for these rubbers can be given as follows:

where R is a divalent organic radical. There has been some question as to whether the sulfur atoms are connected one to another, in a row, as:

$$-S\text{-}S\text{-}S\text{-}S-$$

or in some other conformation such as:

$$\begin{array}{cc} S & S \\ \| & \| \\ -S\text{-} & S- \end{array}$$

where the sulfide rank (i.e., the subscript x) is about 4.

The polysulfide rubbers have been commercial products for a longer time than any other synthetic rubber (since 1929). The products are sold as Thiokol® rubbers in the following grades:

- Thiokol A, where the dihalide is ethylene dichloride
- Thiokol ST, where the dihalide is di-2-chloroethyl formal
- Thiokol FA, where the dihalide is a mixture of di-2-chloroethyl formal and ethylene dichloride

Type A was the first commercial grade but has been largely superseded by type FA. Type ST is prepared with a small percentage of 1,2,3 trichloropropane to provide a branch point for improving the cure state, thus reducing compression set. It has a lower molecular weight than Thiokol A or FA, and its molecules contain mercaptan, -SH, groups.

There is another type, Thiokol LP, grades of which are liquid polymers used in sealant and mastic applications. They are formed by breaking down a high-molecular-weight polymer in a controlled way. These grades also have mercaptan end groups.

Both Thiokol A and FA require peptization (e.g., in the presence of MBT or DPG) to enable easy processing.

Types A and FA can be cured by the addition of ZnO alone at about 10 phr. Additions of small amounts (up to 1 phr) of sulfur accelerate the curing.

Properties of Polysulfide Rubbers.   Vulcanizates of polysulfide elastomers are better than all other elastomers with respect to resistance to aromatic and chlorinated hydrocarbons, and to ketones. They have good weather and ozone resistance, and they are relatively inexpensive. However, they have somewhat lower level of mechanical properties and particularly poor compression set.

Uses of Polysulfide Rubbers.   Polysulfide elastomers are used in roller covering applications, hose liners, and solvent- and oil-resistant molded goods. The sealants are used in construction and in aerospace industries.

4.5.3.12 Epichlorohydrin Elastomers (CO, ECO, and ETER). Amorphous polymers, which have the structure of polyethylene ether with chloromethyl side groups, are obtained by a ring-opening polymerization of epichlorohydrin. These elastomers, having relatively high glass transition temperatures, are designated CO and have the general chemical structure given here:

$$\begin{array}{c} CH_2Cl \\ | \\ {-}(CH_2{-}CH{-}O{\,})_n \end{array}$$

By copolymerization with ethylene oxide, copolymers, designated ECO and having lower glass transition temperatures, of the following general structure are obtained:

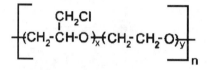

Also, terpolymers, designated ETER and having sulfur-curable functionality, have been prepared.

The ring-opening polymerization, catalyzed by Al(alkyl)$_3$/water, can be carried out in aliphatic, aromatic, or chlorinated hydrocarbon solvents or in ethers at somewhat elevated temperatures.

CO and ECO are vulcanized without sulfur. Rather, they are generally cured by the action of thioureas or triazines in the presence of acid acceptors such as MgO or dibasic lead phosphite. The terpolymers can be cured by accelerated sulfur or peroxide curing systems as well as by the action of thioureas, and so on.

Properties of Epichlorohydrin Rubbers. The homopolymer CO has the highest polarity, the highest vulcanization rate, and best resistance to heat and oil swelling, but the poorest low-temperature flexibility. Gas permeability is low, and flame resistance is very good. ECO, or terpolymers, having fewer chloromethyl groups, compromise the good properties of CO in exchange for improvements in low-temperature flexibility. Unvulcanized epichlorohydrin rubbers tend to stick to mill rolls and are difficult to process unless processing aids are added to their compounds.

Because the backbones of the molecules of these rubbers are saturated, ozone and oxidation resistance are very good.

The average molecular weight of a CO is about 500,000 or more, corresponding to Mooney viscosities (ML1+4, 100°C) in the range of 45 to 70.

ECO and ETER vulcanizates exhibit damping characteristics similar to those of NR vulcanizates but better high-temperature resistance than NR.

Uses of Epichlorohydrin Rubbers. Because of their properties and moderate price, epichlorohydrin rubbers are used in automotive applications such as seals, gaskets, hoses, and tubing. They are also used in coated fabrics and roll covers.

4.5.3.13 Acrylic Rubbers (ACMs). Polar elastomers are obtained by the copolymerization of acrylate esters with monomers, which contain reactive sites for cross-linking reactions to take place during vulcanizations. The general chemical structure for ACM rubbers is as follows:

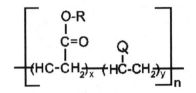

where R is ethyl or butyl (or mixtures thereof), and Q is a reactive moiety for cross-linking. The subscript x is much larger than the subscript y (e.g., 30 times larger).

ACM is generally produced by emulsion polymerization in the presence of a free-radical initiator. The polymerization is initiated by organic peroxides or azo compounds. Potassium persulfate or redox systems are also used.

The incorporation a comonomer gives the reactive site Q. Examples of reactive groups for vulcanization are as follows:

$$-O-CH_2-CH_2Cl$$

$$-O-CO-CH_2Cl$$

$$-CO-O-CH_2Cl$$

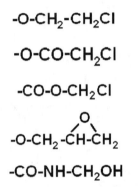

$$-CO-NH-CH_2OH$$

corresponding to the comonomers 2-chloroethyl vinyl ether, vinyl chloroacetate, chloromethyl acrylate, allyl glycidyl ether, and N-methylolacrylamide, respectively. The later monomer is used in acrylic rubbers for latex applications.

The choice of acrylate ester determines the glass transition temperature. It also determines the oil swelling and heat resistance of the vulcanizates. Poly(ethyl acrylate) is more polar than is poly(butyl acrylate), and it is more oil resistant than is the poly(butyl acrylate), but it has a glass transition temperature, T_g, of –21°C versus –49°C for poly(butyl acrylate). Mixed ethyl and butyl ester polymerizates are elastomers of intermediate properties.

The type and level of cure-site bearing comonomer, because it is used at low levels, has little influence on the oil swelling, heat resistance, and low-temperature flexibility. Instead, it determines the cure behavior and cross-link density. Thus, it has effects on mechanical properties, elasticity, and permanent set. Older grades of ACM are being replaced because of their poor vulcanization characteristics. New grades, prepared with undisclosed commoners, are much more rapidly vulcanized and give excellent properties without postcure (e.g., in an oven, as required for earlier grades).

The selection of the reactive group determines the choice of vulcanization system. Chlorine groups require amines, sulfur, and accelerators or combinations of metal soaps and sulfur. Glycidyl groups are cured with ammonium benzoate or dicarboxylic acids. ACMs with methylolacrylamide groups are mixed with those containing unmethylolated acrlylamide groups. With heat, the mixtures spontaneously cross-link by splitting off water to form methylene-bisamide bridges between polymer chains. This system is used in latex applications.

ACM compounds are difficult to process and processing aids must be used, e.g., stearic acid, zinc soaps, fatty alcohol residues, octadecylamine, or pentaerythritol tetrastearate. This later processing aid does not affect vulcanization characteristics, whereas the others retard or accelerate vulcanization.

Properties of Acrylic Rubbers. The ultimate tensile properties of ACM vulcanizates are not as good as those of NR or NBR, but the tensile properties are sufficient for their applications. ACM grades can be used under certain conditions for 1000 hr at 160 to 170°C. ACM vulcanizates can withstand exposures of 1000 hr in oil at 150°C. In addition, ACM vulcanizates are very resistant to degradation by the action of ozone.

ACM vulcanizates are very resistant to swelling in animal, vegetable, and mineral oils, but not motor fuels.

ACM grades based on ethyl acrylate, without plasticizers, have a brittleness temperature of −18°C. The addition of plasticizers or the use of the butyl acrylate-based elastomers (or both) can give a brittleness temperature of −40°C.

Uses of Acrylic Rubbers. The main uses of ACM are in automotive and engineered products. Applications include seals and O-rings (for crankshafts, automatic and differential transmissions, valves, and so on) and oil hose. ACM vulcanizates are also used for roll coverings, tank linings, and fabric covering.

4.5.3.14 Urethane Elastomers (AU and EU). One can prepare polymers with a variety of chemical structures and physical properties by reacting a great variety of low-molecular-weight compounds or oligomers with diisocyanates. The introduction of polyurethanes made it possible to produce a range of materials from hard plastics to soft rubbers and polymers with properties between these extremes. Before the introduction of polyurethanes, tough, useful polymers with properties between those of rubbers and plastics were not known. Polyurethane chemistry enables the tailor-making of materials of specific properties by changing the chemical starting materials, their concentrations, and the processing conditions.

Liquid and viscous reactants permit processing techniques such as reaction injection molding, which have led to many new applications that will not be discussed. Here, we are concerned with polyurethane-technology polymers, which can be processed and cured by using conventional mixing, shaping, and vulcanization processes of the general rubber industry. These have been called *millable polyurethane elastomers.* Thermoplastic polyurethane elastomers, which will be considered in a later section, have been designated TPUs.

To produce the millable polyurethane elastomers, one can bring diisocyanate molecules into reaction with oligomeric hydroxyl-terminated polyesters or polyethers and lower-molecular-weight diols (e.g., 1,4-butanediol). The ester-derived elastomers are designated AU, while the ether-derived materials are designated EU. An illustrative chemical structure is given here, where toluene diisocyanate was used:

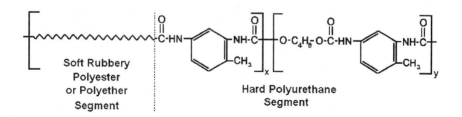

The molecules are segmented block copolymers. The hardness, toughness, and abrasion resistance increase with increases in the length or concentration of hard polyurethane segment.

Cross-link sites are urethane active hydrogen atoms, active methylene groups (e.g., from the use of a diisocyanate such as methylenebis(4-phenylisocyanate), or allylic hydrogen atoms from double bonds incorporated into for vulcanization by sulfur.

Isocyanates and peroxides are the usual vulcanizing agents for AU elastomers. Sulfur vulcanization, which can be done with some grades, is far less prevalent. Antidegradants are not generally required for AUs. Polyurethane elastomers can be degraded by the action of water. EU grades are generally more resistant to hydrolysis than are the AU grades. Thus, hydrolysis inhibitors are important, especially for the AU grades. Polycarbodiimides are use for this purpose.

AU grades are frequently compounded with filler. A small amount of a reinforcing black or fumed silica can considerably improve tear strength and increase hardness.

Properties of Polyurethane Vulcanizates. Polyurethane vulcanizates can exhibit ultimate tensile strengths as high as 40 MPa. Hardnesses are generally high, ranging from 70 to 99 Shore A. Vulcanizates have high degrees of elasticity, even at the higher levels of hardness. Abrasion resistance is better than that of other rubber vulcanizates. However, compression set is relative high when measured at elevated temperatures.

The vulcanizates have relatively good resistance to heat at temperatures up to about 100°C or a bit higher. However, they are attacked by water, steam, acids, and bases, as well as by some lubricants. Except for in tropical climates, weathering resistance is good. Ozone resistance of the vulcanizates is excellent.

The brittleness temperatures of polyurethane rubber vulcanizates generally range between –22 and –35°C. The vulcanizates swell very little in aliphatic solvents and motor fuels. Their swelling resistance to polar liquids (e.g., chlorinated hydrocarbon, esters, and ketones), however, is not as good. The harder grades generally exhibit the better resistance to swelling.

The gas permeability of polyurethane elastomer vulcanizates is very low, on the order of that of butyl rubbers.

Uses of Polyurethane Vulcanizates. Polyurethane elastomers are used in the automotive and engineered products industries for seals, shock absorbers, power-transmission flexible joints, and suspension and support members. Polyurethane vulcanizates are also used in solid tires, elastic thread, footwear, and others. Problems with hydrolytic stability and resistance to heat must be considered when one uses these vulcanizates.

4.5.3.15 Polysiloxane or Silicone Rubber (Q). The polymer backbones of silicone rubbers contain no carbon atoms, but they contain alternating silicon and oxygen atoms. The predominate structural unit is generally dimethylsiloxane, as follows:

Polydimethylsiloxane is designated MQ.

Another repeat unit is the phenylmethylsiloxane diradical (PMQ):

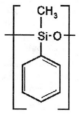

which can be used along with the dimethylsiloxane repeat units. A third type of repeat unit, which contains a vinyl group (VMQ), can be used to improve curing characteristics:

The starting material for the production of polydimethylsiloxane (MQ) is dimethyldichlorosilane, which, in the presence of acid, is hydrolyzed and subsequently condenses with the elimination of water to give a mixture of straight-chain and cyclic oligo-dimethylsiloxanes. A second step is needed to convert the reaction mixture to high-molecular-weight polymer.

High-temperature-vulcanizable grades of MQ have average molecular weights in the range of about 300,000 to 700,000. (The room-temperature-vulcanizable grades used for calking and adhesives have much lower molecular weights, in the range of about 10,000 to 100,000. These polymers are not discussed here.)

The high-temperature-vulcanizable grades have broad molecular weight distributions and high viscosity-average molecular weights. Yet, they have very low viscosities in the unvulcanized state.

Properties of Silicone Rubbers.   Gum vulcanizates have essentially no tensile strength. Fillers are therefore essential. Reinforcing silicas are frequently used. Even then, the tensile properties and abrasion resistance of silicone rubber vulcanizates are poor in comparison with of other types of elastomer. However, the properties change very little with increasing temperatures. These polymers excel in high-temperature applications.

Silicone rubber vulcanizates withstand long-term exposure to hot air at temperatures as high as 180 to 250°C. They remain elastic for 1000 hr at such temperatures. However, high-temperature (120 to 140°C) steam attacks silicone rubbers. The vulcanizates are very resistant to weathering and ozone attack and be used in hoses to convey ozone gas. These materials are also resistant to high-energy radiation.

MQ and VMQ elastomers stiffen only below –50°C, but some PVMQ vulcanizates are flexible even at temperatures below –100°C, and this is without plasticizers.

The oil swelling resistance of silicone elastomer vulcanizates is similar to that of chloroprene rubber vulcanizates. Silicone elastomers are resistant to certain heat transfer fluids but not to motor fuels, chlorinated hydrocarbons, esters, ketones, and ethers.

Silicone elastomers are very permeable to gases—generally about 100 times more so than are butyl or nitrile rubbers.

Silicone rubbers are good electrical insulators and remain so at temperatures up to 180°C.

These polymers resist burning and ignite in air only at temperatures of about 400°C or more. When they do burn, relatively sturdy silica structures form, which are good insulators and are functional in wire insulation, even for a short time, in fire.

Uses of Silicone Rubbers. Because of their high cost, silicone elastomers are used only in applications where other elastomers fail. They are mainly used in applications requiring high heat resistance and extreme low-temperature flexibility. These elastomers are used in the electrical, electronic, aerospace, automotive, mechanical equipment, lighting, cable, and textile industries. They are also used in pharmaceutical and medical applications for components in contact with food. They are now used in high-temperature cooking utensils (e.g., scrapers and spatulas).

Silicone elastomers are used in the automotive industry in such applications as ignition cables, coolant and heater hoses, O-rings, and seals. In aircraft applications, they are used in seals, connectors, cushions, and hoses. In home applications, they are used in O-rings, seals, and gaskets. There are also uses in naval and other applications.

4.5.3.16 Fluoroelastomers (FKMs). Fluoroelastomers can be prepared by co- or terpolymerization of the following monomers:

$$CF_2=CF-CF_3$$

Hexafluoropropylene

$$CF_2=CF_2$$

Tetrafluoroethylene

$$CHF=CF-CF_3$$

1-Hydropentafluoropropylene

$$CF_2=CF-O-CF_3$$

Perfluoro(methyl vinyl ether)

An example of such a copolymer is given by the chemical structure shown here:

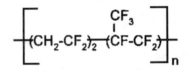

Other comonomers such as vinylidenefluoride and chlorotrifluoroethylene are used, generally in smaller amounts. In addition, some fluoroelastomers incorporate bromine-containing curing-site monomers and can be vulcanized with peroxides.

Fluoroelastomers are prepared by emulsion polymerization at elevated temperatures in the presence of peroxides as initiators.

Different grades (prepared from different monomer mixtures) require different types of vulcanization chemistry, with different curing systems. Examples of curing agents are (1) hexamethylenediamine with MgO, CaO red lead, or ZnO/dibasic lead phosphite as acid acceptors, (2) bisphenol A/organophosphonium salt with MgO/calcium hydroxide as acid acceptor, and (3) organic peroxides for certain grades, again with acid acceptors.

FKM elastomers are available in a range of different viscosities.

Properties of Fluoroelastomers. FKM elastomers can have reasonably low glass transition temperatures, in the range of –18 to –40°C. The tensile properties of fluoroelastomers vulcanizates are fairly good but can decrease considerably with increasing temperatures.

FKM vulcanizates have excellent heat resistance, giving continuous service for 1000 hr at 220°C. Useful service is even possible at 250°C. These elastomers are also highly resistant to weathering and ozone attack. FKM vulcanizates are resistant to swelling in hot oils and aliphatic compounds. They also are resistant to aromatics, chlorinated hydrocarbons, and motor fuels. In addition, they are very resistant to most mineral acids. The gas-permeability resistance of FKM vulcanizates even exceeds that of butyl rubber vulcanizates.

Uses of Fluoroelastomers. FKM elastomers are expensive, but their demand is high because of their unusual stability in very severe environments. They are used in specialty products, e.g., shaft seals of internal combustion engines, and components in aircraft and rockets. Products include seals, gaskets, liners, hoses, protective fabric coatings, diaphragms, roll covers, and cable jacketing.

4.5.4   Fillers, Plasticizers, and Other Compounding Ingredients

In addition to polymers, vulcanization-system ingredients, and antidegradants, other compounding ingredients include fillers, pigments, plasticizers, reinforcing resins, processing aids, flame retarders, and others.

4.5.4.1   Fillers.
Fillers for elastomers are not generally used just to fill space and cheapen the compositions. They are very important to modify the properties of rubber compositions in very positive ways. This is especially true for the so-called reinforcing fillers. Their presence in the compound can improve the strength- and durability-related properties of vulcanizates and can strongly enhance processing characteristics. The choice and amount of filler can have a profound effect on vulcanizate properties. These effects depend on several factors: level of use (concentration), primary particle size, surface area (inverse function of primary particle size), and structure (shape factor, e.g., spherical, chain or rod-like, plate-like, and so forth).

One can produce soft natural rubber vulcanizates or chloroprene rubber vulcanizates with good strength-related properties because of their tendency to crystallize during deformation. This is not true for most of the synthetic elastomers (e.g., SBR, BR, IIR, EPDM, and others), which do not significantly crystallize due to strain. It is only with the use of reinforcing fillers that serviceable vulcanizates can be made from such elastomers. Reinforcing fillers even improve many properties of the crystallizing rubbers.

In addition to increasing the stiffness, hardness, and modulus of a vulcanizate, the presence of a reinforcing filler increases properties such as tensile strength, abrasion resistance, and tear resistance. (Other properties, such as rebound or resilience, can be reduced.)

The major types of fillers (in the approximate descending order of their amounts used in the market) are carbon black, silica, kaolin clay, and calcium carbonate. These probably account for 95 percent of the filler used in rubber vulcanizates.

Reinforcement. There is little agreement on the mechanism for reinforcement of elastomers by fillers in elastomer vulcanizates. Some investigators believe that chemical bonding between rubber molecules and filler particles is necessary. Others believe that no chemical bonding is necessary. Others feel that the truth is somewhere in between these two views. It is possible that weaker-than-primary-chemical bonding occurs. If the elastomer molecules are bonded in such a fashion that bonds are easily broken during deformation but rapidly reform during the same process, significant energy dissipation will occur (and heat will evolve) with an increase of the energy to rupture the rubber. The restraint of the polymer chains due to even the weak bonding would cause increases in stiffness and modulus, over and above that due to the presence of a significant volume of nondeformable material. The bonding of rubber to filler explains the fact that reinforcing fillers give steeper stress-strain curves at elongations greater than about 100 to 200 percent. In the absence of any bonding, elastomer would be peeled away from the filler particle surfaces in a dewetting process. One measure of reinforcement is the ratio of stress at 300 percent tensile strain to that at 100 percent tensile strain, the ratio being higher for well reinforced elastomers. The measurements are made during the standard stress-strain test.

The changes that occur during strain give rise to what is known as the *Mullins effect* or *Mullins softening*. When a filler-reinforced vulcanizate is prestretched and then relaxed, the force to stretch it again is less than for the unprestretched sample, but only up the strain used in the prestretch. At higher strains, the stress-strain curve resumes that of the unprestretched specimen. This implies breakage or movement of rubber-filler bonds during stretching. After aging, the prestretched sample reverts to the nonprestretched state.

Particle Size, Surface Area, and Structure. Fillers with primary particle sizes greater than 10 µm act as flaws, which can initiate rupture during flexing, bending, or stretching. Fillers with primary particle sizes between 1 and 10 µm are diluents, usually having only small effects on vulcanizate properties. Semireinforcing fillers, with primary particle sizes ranging between 0.1 and 1 µm (between 100 and 1000 nm), can improve the strength of vulcanizates and increase modulus and hardness. Fillers of primary particle sizes ranging between 10 and 100 nm greatly improve strength, tearing resistance, wearing resistance, and other qualities. The reason for using the phrase, "primary particle size," is that, in many cases (e.g., silica or carbon black fillers), during manufacture, first essentially spherical primary particles are formed, which coalesce into aggregates. The aggregates are in the form of compact structures, chains, or branched chains of high shape factor, i.e., high structure. Various types of fillers are classified according to type and primary particle size in Fig. 4.26. Types of structures formed from aggregated primary particles are illustrated by Fig. 4.27.

Fillers that have very small primary particle sizes have high surface area per unit weight of filler. Fillers that have high surface area have larger amounts of contact area available for interaction and bonding with the elastomeric matrix polymer.

The average particle diameter of a filler sample can be directly determined by electron microscopy where, for example, 1000 to 1500 single particles are measured under magnifications from 50 to 75,000. Particle-size distributions are then determined.

From the average primary particle diameter, a theoretical total surface area can be calculated, assuming spherical particle shapes. This does not account for the aggregate structure or porosity. Another method to determine surface area (per 100 g of filler) is by measuring the absorption of a gas of small atoms, e.g., nitrogen. The gas penetrates the finest crevices. According to the nitrogen adsorption method developed by Brunauer, Emmett and Teller, one obtains the so-called BET-value of surface area, expressed as m^2/g.

Structure can be thought of as degree of difference from a spherical shape. It is similar to shape factor. High structure aggregates are in the form of chains, branched chains, and so on. We note that some of this aggregate structure can break down during processing due to the development of high stress on the structures.

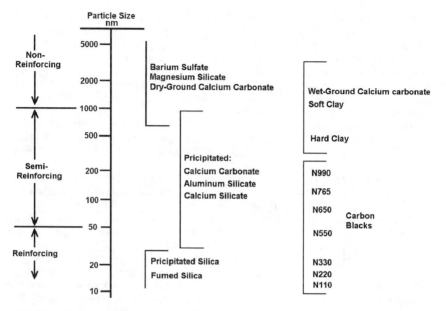

FIGURE 4.26 Types of fillers.

Large Particles, Low Surface Area, High Structure (High DBP Absorption)

Small Particles, High Surface Area High Structure (High DBP Absorption)

Large Particles, Low Surface Area, Low Structure (Low DBP Absorption)

Small Particles, High Surface Area Low Structure (Low DBP Absorption)

FIGURE 4.27 Particle size and structure.

High-structure fillers give rise to reduced elasticity in the uncured state. Unfilled elastomers, when extruded in the uncured state (e.g., during processing) expand or swell when they leave the extruder die (have memory or nerve). Along with this die swell is a shortening of the extruded profile. It is called *extrusion shrinkage*. Extrusion shrinkage is greatly reduced by fillers, especially those of high structure. Also, as the structure of the filler increases, the viscosity of the uncured composition or the stiffness of the vulcanizate increases. This is because the higher-structure fillers immobilize more of the elastomer during its straining in either the cured or uncured state.

The amount of structure is measured by using the dibutyl phthalate (DBP) absorption method. Small amounts of DBP (a nonvolatile liquid) are added to dry filler until a non-crumbling paste is obtained. The DBP absorption is expressed in ml of DBP per 100 g filler.

Filler Surface Activity. A filler can have high surface area and high structure and still give poor reinforcement if its surface does not interact at all with the elastomeric matrix. For example, carbon black, which is a highly effective reinforcing filler, loses much of its reinforcing effect if it is graphitized. During the graphitization processing (high-temperature heating in the absence of reactive gases such as air), most of the reactive chemical functional groups are removed from the particulate surfaces.

A way to infer the activity of a filler toward an elastomer is to measure so called "bound rubber." When an uncured elastomer-filler mixture is extracted with a solvent (e.g., toluene), then the gel-like elastomer, which is bound to filler, cannot be dissolved, whereas the rest of the elastomer is soluble and is extracted away from the gel-like mixture. The more the bound rubber, the more active the filler is assumed to be.

In the case of carbon black, chemical functional groups on the filler that may have some relation to reinforcement include carboxyl, lactone, quinone, hydroxyl, and so forth. These are located at the edges of graphitic planes.

4.5.4.2 Carbon Black. Carbon black has been used in rubber compounds for well over a 100 years. First, there was lamp black, produced by the deposition from oil flames onto china plates. It was used as a black pigment. Then, channel blacks (formed by exposing an iron plate to a natural gas flame and collecting the deposited soot) were used as reinforcing fillers in 1910. More recently, furnace black (produced industrially from petroleum oil in a furnace by incomplete combustion in an adjustable and controllable process) was introduced. Thermal carbon blacks are generally produced from natural gas in preheated chambers without air. They are essentially nonreinforcing fillers that improve tensile strength only slightly. However, they give only moderate hardness, even at high loadings, and their compounds are easily processed. Furnace blacks are the main types used today. ASTM designations, the older nomenclature, particle size, surface area, and structure of some blacks are given in Table 4.13.

The first letter of the ASTM classification indicates the expected type of cure rate for the compound as below:

- N for normal cure rate (indicates that the compounds will cure at a normal rate)
- S for slow cure rate

The letters N and S correspond, respectively, to the furnace blacks and channel types.
The first digit indicates particle size ranges as follows:

- 1 for 10 to 19 nm
- 2 for 20 to 25 nm
- 3 for 26 to 30 nm

TABLE 4.13 Colloidal Properties of Rubber-Grade Carbon Blacks

	ASTM classification	Abbrev.	Common name	Particle size, nm	DBP absorption, mil/100 g
Furnace blacks	N110	SAF	Super abrasion furnace	21	113
	N220	ISAF	Intermediate abrasion furnace	23	115
	N326	HAF-LS	High abrasion furnace, low structure	28	72
	N330	HAF	High abrasion furnace	29	101
	N550	FEF	Fine extrusion furnace	50	120
	N660	GPF	General-purpose furnace	62	91
	N770	SRF	Semireinforcing furnace	66	75
Thermal blacks	N880	FT	Fine thermal	150	52
	N990	MT	Medium thermal	400	40
Channel blacks	S301	MPC	Medium processing channel	27	72
	S300	EPC	Easy processing channel	32	75

- 4 for 31 to 39 nm
- 5 for 40 to 48 nm
- 6 for 49 to 60 nm
- 7 for 61 to 100 nm
- 8 for 101 to 200 nm
- 9 for 201 to 500 nm

The second and third digits are arbitrary.

Carbon blacks of the smaller, and mid-sized primary particles are extremely good reinforcing fillers. They are the most used. At optimum loading, the finer the particle size (the higher the surface area per gram of carbon black), the higher the tensile strength, the higher the tear strength, and the higher abrasion resistance—however, the greater the difficulty of dispersion and the higher the cost of the carbon black.

Carbon blacks are typically used at levels of about 50 parts by weight per 100 parts of the rubber and extender and plasticizers combined. That is, for a recipe containing 100 parts of elastomer and 30 parts of extender oil, 65 parts of carbon black could typically be used. Adjustment changes in hardness (i.e., to meet specific specifications) are easily made by, for example, increasing the carbon black level or reducing the extender oil level to increase hardness. A rough idea of how vulcanizate properties change with carbon black loading is given by Fig. 4.28.

4.5.4.3 Silica. Silicas are highly active, light-colored fillers. The most important silicas for the rubber industry are prepared by precipitation, wherein alkali silicate solutions are acidified under controlled conditions. The precipitated silica is washed and dried. Colloidal silicas of very high surface area (small primary particles) are produced by this method.

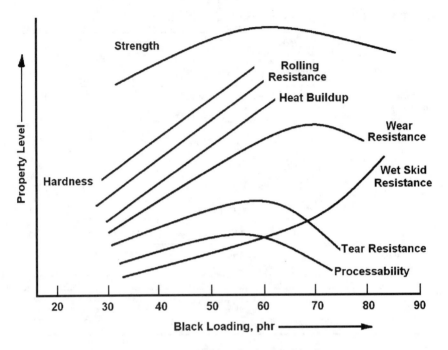

FIGURE 4.28 Vulcanizate properties as a function of carbon black loading.

Silicates (e.g., calcium or aluminum silicates) are not as active as fillers as are the silicas. Colloidal silicas can also be prepared by the so called pyrogenic process, wherein silicon tetrachloride is hydrolyzed at high temperatures as follows:

$$SiCl_4 + 2\,H_2O \rightarrow SiO_2 + 4\,HCl$$

This process produces very finely divided silicas, important as fillers for silicone rubbers.

All precipitated silicas and silicate fillers contain some water. Since the water content can influence processing and vulcanizate properties, it is necessary to control the amount of water present during processing and packaging.

As with carbon blacks, silica fillers are characterized on the basis of primary particle size and specific area. The smallest observable single filler particles (primary) have diameters of about 15 nm. The surface forces of the primary filler particles are so high that thousands of them agglomerate to form extremely robust secondary particles that cannot be broken apart. These secondary particles further agglomerate to form chain-like tertiary structures, many of which can be more or less degraded by shear forces. Determination of surface areas is done using the BET nitrogen absorption method.

As with carbon blacks, precipitated silicas are classified with respect to structure by the degree of oil absorption. Typical values of oil absorption for various silicas are as follows:

- For very high structure silica, >200 ml/100g
- For high structure silica, 175 to 200 ml/g
- For medium structure silica, 125 to 175 ml/g

- For low structure silica, 75 to 125 ml/g
- For very low structure silica, <75 ml/g

Silicas have strongly polar surface characteristics. This is because of the many hydroxyl groups occupying the silica surfaces. This causes the silica particles to bond to one another, as apposed to bonding to or being wetted by the rubber molecules. This can cause problems *vis à vis* the dispersion of silica into the rubber matrix during mixing. It can also interfere with the general processability of the uncured rubber compound.

Because silicas are acidic, they retard the cure during accelerated-sulfur vulcanization. Also, because of their polarity, they can adsorb such rubber chemicals as vulcanization accelerators and reduce the efficiency of curing. It may be necessary to add additional amounts of accelerator (e.g., DPG or DOTG) to compensate for the effects on the curing system. Polyols and polyethers have been used as additives to compete for the silica polar groups, reducing the amount of curing-system ingredients that are adsorbed by the silica.

To improve the bonding of silica to rubber molecules rather than to one another, silane coupling agents are used. One of these is the commercially available bis-(triethoxysilyl-propyl)tetrasulfide. An ethyoxy group of this molecule can react with a silica -OH group to give ethanol and a linkage to a silica particle, whereas the tetrasulfide part of the coupling molecule can interact with rubber, the overall result being a rubber-to-silica linkage: *silica*-O-Si($[O$-$C_2H_5]_2$)CH_2-CH_2-CH_2-S_x-*rubber*. This coupling-agent additive also can be used to reduce the reversion in natural vulcanizates. It is a slow curative that slowly cross-links the natural rubber, compensating for the loss of cross-links during reversion. For coupling silica particles to rubber molecules during high-temperature mixing, care must be taken that the curing reaction is not so extensive so as to cause premature vulcanization (scorch).

There is much interest in using silica fillers in tires, because it is possible to obtain abrasion-resistant treads of lower hysteresis (thus better fuel economy) than that of carbon-black-filled treads. However, there have been problems with the processing of the silica-filled compounds. Also silica, being nonelectrically conductive, gives vulcanizates that can hold static electrical charges due to rolling on the road. Efforts to get around these and other problems have led to the introduction of hybrid silica-carbon fillers.

4.5.4.4 Clays. Kaolin clay fillers are generally used to reduce cost while improving certain physical and processing characteristics. There are two basic types of rubber filler clays: (1) "hard clays," having median particle sizes of 250 to 500 nm, and (2) "soft clays," having median particle sizes of 1000 to 2000 nm. The hard clays give vulcanizates of higher tensile strength, stiffness, and abrasion resistance than do the soft clays. They are semireinforcing. Soft clays can be used with higher loadings than can hard clays. Also, faster extrusion rates are obtained with the soft clays.

More hard clays than soft clays are used in rubber compounds, because they are semireinforcing fillers. Aminosilane and mercaptosilane treatment of hard clays enhances reinforcement. Sometimes, hard clay is used with other fillers, for example, to improve the tensile strength and increase the modulus of calcium carbonate-filled vulcanizates. Clay is sometimes used to replace a portion of the more expensive carbon black or silica, with little loss of performance.

Airfloat clay, the type most used in rubber compounds, is dry-ground hydrous kaolin that has been air-separated to reduce impurities and control particle-size distribution. However, some water-washed clay (slurried in water and centrifuged or hydrocycloned to remove impurities) is used, because it contains a lower level of impurities and gives compounds that are more colorable. The water-washed clay also causes less die wear during extrusion.

Calcined clay (produced by heating a fine natural china clay to high temperatures in a kiln) is used mostly in wire and cable coverings because of excellent water resistance and electrical properties of its vulcanizates. Delaminated clays are also used in rubber compounds. They are made by attrition milling the coarse clay fraction from the water-washing of soft clay. This breaks down the kaolinite stacks into thin, wide individual plates, improving brightness, opacity, and barrier properties. Such clays impart very high stiffness and low die swell because of their high shape factors.

4.5.4.5 Calcium Carbonate (Chalk, Whiting).   Two general types of calcium carbonates are used in the rubber industry: (1) wet or dry ground natural limestone, having particle sizes between 700 and 5000 nm, and (2) precipitated calcium carbonate with fine and ultra-fine products having average particle sizes as low as 40 nm. The ground products have particles of low anisotropy (low structure or shape factor), low surface area, and low surface activity. They are widely used only because of their low cost, and they can be used at very high concentrations. Ground-calcium-carbonate vulcanizates have poor abrasion and tear resistance. The dry-ground is the least expensive filler, and it can be used at the highest of levels. Precipitated calcium carbonates have much higher surface areas because of their smaller particle size. Ultra-fine calcium carbonates, having particle sizes less than 100 nm, can have specific surface areas similar to those of hard clays.

Both the ground and precipitated calcium carbonates can by treated with stearic acid to control water absorption, improve dispersability, and promote better wetting of the filler by rubber. Silane treatment of these fillers is not effective. However, there is an ultra-fine grade coated with carboxylated polybutadiene, which reactively links to the particle surfaces. Such treated ultra-fine products can give reinforcement of about the same level of the semireinforcing thermal carbon blacks.

4.5.4.5.1 Other Fillers

Talc.   Talc is little used in rubber applications. Platy talcs are hydrophobic, white, alkaline, and of high particulate asymmetry. They are readily treated with silanes and other coupling agents. Unfortunately, particles of talc are generally too large for effective elastomer reinforcement. Nevertheless, talcs can be micronized to reduce median particle sizes to 1000 to 2000 nm. Such products are used but compete with less expensive clays.

Aluminum Oxyhydrate.   This material is used for its ability to give off water at high temperatures as a flame retardant.

Barite.   Barite, ground barium sulfate, is used in acid-resistant vulcanizates, because it is resistant to even strong acids that would attack other mineral fillers. It is also used where high-density products are desired. It has little effect on cure, stiffness, or vulcanizate stability.

Mica.   Because of its high aspect ratio and platyness, this material is sometimes used as a semireinforcing filler. The platyness can also reduce swelling of compounds in oils, solvents, and others.

Diatomite (Kieselguhr).   Diatomaceous earth (as it is also called) is chemically inert, but it has high adsorptive power. This can account for adsorption of curing ingredients that interfere with accelerated-sulfur vulcanization. However, diatomite is used as a filler in silicone rubber. Because of its high adsorptive capacity, it is used as a process aid in high-oil rubber compounds.

4.5.4.6 Reinforcing Resins.   The main types of reinforcing resins used in rubber compounds are high-styrene resins and phenolic resins. The high-styrene resins are copolymers of styrene and butadiene wherein 50 to 85 percent of the polymer is derived from styrene. They are used to stiffen NR and SBR rubber compounds, for example, in shoe soles. Phenolic resins are used for reinforcing NBR compounds. The phenolic resin is cross-linked during vulcanization, and its presence can give rise to increased hardness, tensile strength, tear strength, and abrasion resistance. Before curing, the phenolics act as processing aids.

4.5.4.7 Pigments. Both organic and inorganic pigments are used in colored rubber compounds. Pigments are insoluble in rubber and rubber solvents. They must be easily dispersed in rubber compounds and insensitive to vulcanization conditions, vulcanizing agents, and other additives. They must be light fast and insensitive to conditions encountered in product use (e.g. acid or base). They are generally free of strong pro-oxidants such as copper and manganese compounds.

White Pigments. Various types of titanium dioxide are probably the most important white pigments for rubber. Although they are fairly expensive, they are economically used because of their great whitening power, and only small amounts are required. They also have minimum effects on vulcanizate properties unless concentrations of about 20 phr or more are used. Lithopone (a white pigment consisting of a mixture of zinc sulfide, zinc oxide, and barium sulfate) has relatively low whitening power; thus, large amounts must be used. This can degrade the vulcanizate properties. For this reason, titanium dioxide is preferred.

There are two forms of titanium dioxide used in rubber: anatas and rutile types that differ in crystalline structure. An anatase-type titanium oxide pigmented vulcanizate can have an outstanding (bluish) white color, while most rutile titanium dioxides give a cream-colored white rubber vulcanizate. However, rutile types have 20 percent more covering power than do anatas types. Also, rutile types give the more light- and weather-resistant vulcanizates. Nevertheless, anatas types are used where a more nearly pure white material is required.

Inorganic Colored Pigments. Inorganic pigments do not have the brilliance of some of the organic ones, but they have the better weathering properties and good chemical resistance. Also, they can be low in cost. They are used in low concentrations lest they unfavorably influence the performance properties of the vulcanizates.

Iron oxide pigments are used to obtain reddish, brown, beige, and yellow hues. Iron oxide pigments should be free of such pro-oxidants as manganese impurities. Chromium oxide pigments are used for greenish and yellowish green hues. Cadmium-containing pigments are used for brilliant yellow, orange, and red colors. However, cadmium compounds are restricted in some countries for toxicological reasons. Ultramarines are used for blue colors.

Organic Colored Pigments. The organic pigments are more efficient than the inorganic ones. They give brilliant colors but are not as resistant to light and weather, and they have less covering ability. They are also generally more expensive. Suitable materials include azo dyes, for example, from the diazo coupling of *o*-chloroaniline with *p*-nitrophenyl-3-methyl-5-pyrazole to produce an orange pigment. Other examples are alizarine dyes, and for blues and greens, the phthalocyanine dies. These pigments are available as pure powders or in paste form.

4.5.4.8 Other Compounding Ingredients

Softeners, Tackifiers, and Processing Aids. Softeners (e.g., extender oils, process aids, and tackifiers) are added to (1) improve processing characteristics of the compound, (2) to modify the final compound properties (e.g., hardness), (3) to reduce the cost of the compound (i.e., an extender oil, being inexpensive and enabling greater levels of inexpensive filler), and (4) to reduce the power consumption during processing.

Differences among softeners, tackifier resins, and softeners are blurred, and many are dual-purpose ingredients of rubber compounds. Plasticizers also act as softeners and processing aids but will be considered separately. Unlike petroleum oils, the term *plasticizer* will be generally applied to synthetic ingredients, which are frequently added to lower the T_g of the composition (i.e., to impart low-temperature flexibility).

Petroleum Oils.   Petroleum oils are generally mixtures of paraffinic, naphthenic, and aromatic hydrocarnbons. The relative amounts of these components determine the compatibility of a particular oil with a particular rubber. The paraffinic oils are more compatible with EPDM and IIR. The more aromatic oils are more compatible with the more polar rubbers (e.g. CR, NBR, and CSM). Most petroleum extender oils are compatible with NR, IR, BR, and SBR. The effects of adding extender oil are to lower viscosities of uncured compounds and allow the use of greater amounts of filler, and with respect to the vulcanizates, to reduce hardness, reduce modulus, and somewhat reduce tensile strength.

The viscosity and volatility of the oil are important. Generally, low-viscosity oils give vulcanizates of lower glass transition temperatures. The lower-molecular-weight paraffinic oils generally have lower viscosities, but they are also more volatile and thus somewhat fugitive, especially at elevated temperatures.

As well as acting as plasticizers, the extender oils are considered to be process aids because of the reduced viscosities of the rubber compounds wherein they are used. This allows easier processing, especially with rubber stocks that are highly loaded with filler.

Process Aids.   Fatty acids, their metal salts (soaps), fatty acid esters, fatty alcohols, and other substances are used to improve processing characteristics of rubber compounds. Many such additives are available. They can have strong influences on processability. They act as lubricants for flow during extrusion, molding, and so forth, allowing easy slippage between the rubber stock and the metal surfaces. They can also improve the dispersion of fillers, and so forth. In addition to aiding in flow during molding and extrusion, the presence of lubricating process aids reduces the temperature of mixing in internal mixers.

Fatty acids are used in small amounts, with zinc oxide, as vulcanization activators. In addition to their activating effect in the vulcanization process, the acid and its *in-situ-*formed zinc soap do act as lubricants as well as activators.

In addition to the fatty acids and their metal soaps, fatty acid esters and fatty alcohols are used, because they give outstanding processing improvements but without other types of action—for example, cure and activation or breakdown enhancement during the mastication of NR or IR. Pentaerythritol tetrastearate is a example of an ester-type process aid with a broad range of applications. It does not bloom or exert unwanted effects.

Tackifiers.   Pine tar, coumarone-indene resins, zylol-formaldehyde, and other resins are used to increase the tack of rubber compounds. *Tack*, here, means stickiness of the uncured rubber stock to itself, rather than to other things, such as metal surfaces. Tack has also been called *autoadhesion*. It is extremely important for building up structures such as tires. Natural rubber inherently has good natural tack, but most synthetic rubbers do not.

Synthetic Plasticizers.   The most important types of synthetic plasticizers are esters. Phthalate esters are used to improve elasticity and low-temperature flexibility, especially in NBR and CR vulcanizates. Common examples are dibutyl phthalate (DBP), di(2-ethylhexyl) phthalate (DOP), diisooctyl phthalate (DIOP), and diisononyl phthalate (DINP). They are generally used at levels of 5 to 30 phr.

Adipate and sebacate esters are used, in particular when low-temperature flexibility is especially desired. Examples are di-2-ethylhexyl adipate (DOA) and di-2-ethylhexyl sebacate (DOS). Azelaic acid esters are also used. Trimellitates [e.g., triisooctyl trimellitate (TIOTM)] are plasticizers with extremely low volatility. Phosphate esters are use to give softness when flame retardance is also required.

Other ester plasticizers include polyesters of adipic and sebacic acids and 1,2-propyleneglycol. These are used where nonvolatile and nonmigrating plasticizers are needed. Other types of esters are also used, such as citrates, ricinoleates, and octyl-iso-butyrate.

Chlorinated hydrocarbons are used as plasticizers in rubber articles (e.g., at a level of 20 phr) to lower flammability (e.g., chlorinated paraffins in combination with antimony trioxide).

Flame Retardants. Hydrocarbon elastomers are flammable and thus require flame retardants if their service conditions include the possibility of fire. Alumina trihydrate, magnesium hydroxide, and zinc borate are used, because they give off blanketing vapors at high temperatures. Also, typical flame-retardant systems include chlorinated paraffins or brominated aromatic resins in combination with antimony trioxide.

Blowing Agents. Blowing agents are used to produce cellular rubber (e.g., sponge rubber). These additives give off gas during vulcanization to form bubbles in the vulcanizate. Usually, highly plasticized compounds are used.

At one time, sodium bicarbonate (e.g., in combination with oleic acid) was used to give off carbon dioxide during curing. However, it was difficult to disperse very finely and uniformly to give a uniform fine cellular structure.

Organic blowing agents that liberate nitrogen are more commonly used. They are dispersed more easily and give greater processing safety and regularity of the foam. Common examples are sulfonyl hydrazides, certain N-nitroso compounds (e.g., dinitrosopentamethylenetetramine), and azo dicarbonamides.

Peptizers. Certain elastomers such as NR must be broken down (reduced in molecular weight) by mastication, for example in an internal mixer or (less commonly) on an open two-roll mill. With NR, this can be done purely by mechanical means but, as the temperature rises due to mixing, the viscosity drops, and the mechanochemical action is greatly reduced (because there is not enough shear stress). Certain additives can facilitate the breakdown. They are called *peptizers* and are used in small concentration (0.05 to 0.15 phr) for breaking down the elastomer (generally NR) before adding the general compounding ingredients. An appropriate peptizer is zinc pentachlorothiophenate, with or without a zinc soap activator. The activator increases the temperature range for the peptization process. The soap also reduced the effective viscosity and lowers the mastication temperature, possibly because of its lubricant activity.

4.5.5 Processing of Vulcanizable Elastomers

Many of the production methods used for rubbers are similar to those used for plastics. However, rubber processing technology is also different in certain respects. Processing rubber into finished goods consists of compounding, mixing, shaping, generally molding, and vulcanizing. Rubber is always compounded with additives: vulcanization chemicals, and usually fillers, antidegradants, oils or plasticizers, and so on. It is through compounding that the specific rubber vulcanizate obtains its characteristics (properties, cost, and processability) to satisfy a given application.

4.5.5.1 Mixing

Mastication. The first step in rubber compounding and mixing is mastication (breakdown of the polymer). This is especially essential for natural rubber. During the mixing of the rubber polymer or polymers with other ingredients, the rubber must be more plastic than elastic so as to accept the additives during mixing. Some rubbers have molecular weights that are large enough to permit entanglements that act as cross-links during the deformation motion of the material in the internal mixer on a two-roll mill. Working the rubber, especially in the presence of peptizers, reduces the molecular weight sufficiently to permit good mixing.

In early times, rubber breakdown and subsequent compounding was done on open roll mills. A schematic representation of such a mill is represented by Fig. 4.29. The rolls ro-

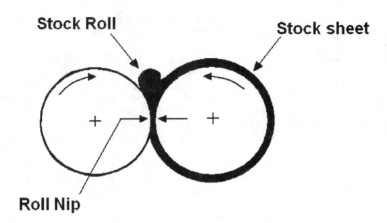

Stock Roll

Stock sheet

Roll Nip

FIGURE 4.29 Schematic of a two-roll mill.

tate in opposite directions, each turning toward the nip. Normally, the gears are such that one roll turns, typically, about 20 percent faster than the other to give a "friction ratio" (drive to driven) of 1.2. This ratio can vary. The nip distance is adjusted to give the desired amount of working of the rubber. Somewhat more material is on the mill than to just give a sheet, and the excess forms the roll of rubber over the nip.

Now, mastication is predominately performed in an internal mixer. Schematic diagrams of two types of internal mixers are given in Fig. 4.30. The two rotors rotate toward one another. In the case of the tangential-type mixer, the rotors are generally operated at different speeds, whereas, in the case of the intermeshing mixer, the rotational speeds must be the same. The intermeshing-rotor mixers may be able to give faster dispersive mixing with the better cooling efficiency, but the payload is greater with the tangential mixers. The cavity of the mixer is fed from a loading chute through which the rubber and, in later steps, the fillers and other compounding ingredients can be added. Such mixers can be very large, handling payloads as great as 500 kg or more. The temperature is partly controlled by the fluid jacketing, which can contain cold water, warm water, or steam. Importantly, the temperature is largely dependant on the work put into the rubber mass during its mixing.

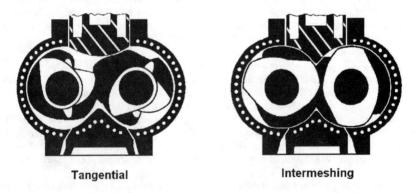

Tangential

Intermeshing

FIGURE 4.30 Schematic of internal mixers with tangential or intermeshing rotors.

For NR, it is necessary to first achieve a temperature above about 60 to 70°C to melt out any crystallinity. Then, if the temperature becomes too high, the viscosity will drop too much, preventing the development of sufficient stresses for the mechanochemical breakdown of the polymer.

Stage-One Mixing (Masterbatching).   The additives must be thoroughly mixed with the rubber polymer (or polymers) to achieve uniform dispersion of ingredients. Uncured rubbers have high viscosity and, therefore, working of the rubber during mixing can increase its temperature up to 150°C or more. If vulcanizing agents are present from the start of mixing, premature vulcanization (scorch) might be the result. To avoid premature vulcanization, a two-stage mixing process can be employed. In the first stage, nonvulcanizing ingredients (filler, antidegradant, softeners or oils, wax, processing aids, and so forth) are combined with the raw rubber in, for example, an internal mixer (although, for small quantities, an open roll mill can be used). The cure activators can be added in this first stage, but not sulfur and frequently not accelerator. The objectives of masterbatching are to achieve good homogeneous blending of the polymer with chemical additives and good dispersive mixing to achieve deagglomeration as well as distribution of the filler. The product of the first stage is generally called a *masterbatch*. Because considerable heat is generated the first stage of mixing, the masterbatch can be cooled by milling on a cooled open two-roll mill. It may also be necessary to do some milling of the masterbatch to improve the dispersive mixing and to make it homogeneous.

Stage-Two Mixing (Finish Mixing).   After the first-stage mixing has been completed and the masterbatch has been allowed to cool, stage two mixing is carried out, during which the vulcanizing agents, such as sulfur, and accelerator are added. This second-stage mixing, which finishes the mixing process, has been carried out on an open mill, but it is more frequently done in a carefully temperature-controlled internal mixer. The maximum temperature allowed is incorporated into mixing procedures as the controlled dump temperature.

4.5.5.2 Shaping and Forming.   Shaping processes for rubber products can be divided into the basic categories of extrusion, calendering, and molding. Vulcanization usually happens in a heated mold, but it can also occur in a steam autoclave, salt bath, or a hot air oven after extrusion. Some products require assembly work as well as shaping or forming. This is required for built-up products such as tires.

Extrusion.   Screw extruders are generally used for extrusion of uncured rubber into shaped rubber sections for later use (e.g., treads and side walls of tires) or for essentially forming the shape of the final product (e.g., hoses, vehicle seals, and so on). The length/diameter (L/D) ratio of the extruder barrel is less than for thermoplastics, typically in the range 10 to 15, to reduce the risk of premature cross-linking due to heat build-up in the barrel. The earlier "hot-feed" extruders were shorter, and the rubber was fed as a heated strip. Die swell occurs in rubber extrudates, since the rubber polymer is highly elastic (due to entanglements of its very long molecules) and exhibits die swell or "memory."

Either hot or cooled finished mix is fed into the extruder (as a hot strip or crumb fed to a hot-feed extruder, or a cooled strip or pellets into a cold-feed extruder). The extruder can form extruded profiles suitable as components of built-up products, such as side walls or treads of tires. The extruder can also form profiles that are vulcanized "on the run" by passing the "endless" extrudates through a heated salt bath or hot oven, either of which is long enough for vulcanization to sufficiently occur before the profile exits the oven or hot bath. Profiles of unvulcanized products, such as hoses, can be vulcanized after extrusion in a steam autoclave or hot oven.

Calendering.   Rubber calenders consist of at least three rolls, which can be adjusted for gap, speed, and temperature. Calendering can be used for forming sheets of uncured rubber that will be later used as components of built-up products such as tires. Textile fabric for reinforcement can be embedded during the process. Calendering is a process for producing sheets of uncured rubber (for later vulcanization), such as for roofing membranes.

The uncured rubber stock is passed through a series of gaps of decreasing size made by a stand of rotating rolls, and the final roll gap determines the sheet thickness (Fig. 4.31). A variant of this is the use of a calendering process for the production of coated fabric for such applications as carcass plies in tires (Fig. 4.32).

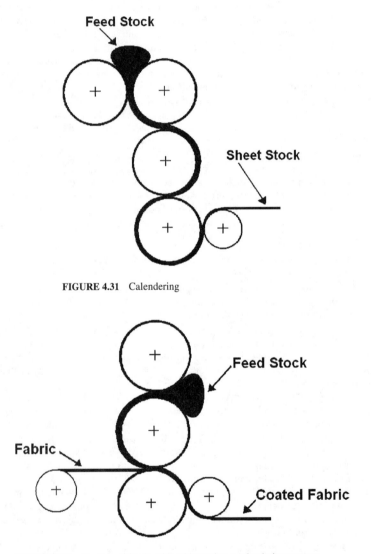

FIGURE 4.31   Calendering

FIGURE 4.32   Coating of fabric with rubber using a calendering process.

4.5.5.3 Molding. A molding process is used for the production of many types of products, including shoe soles and heals, gaskets and seals, suction cups, bottle stops, tires, and others. There are basically three types of molding: compression molding, transfer molding (Fig. 4.33), and injection molding (Fig. 4.34). Vulcanization is accomplished in the heated mold in all three processes.

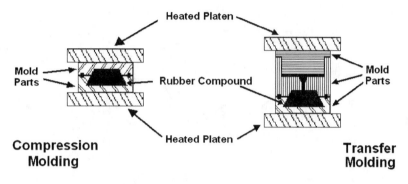

FIGURE 4.33   Schematic of press molding.

Compression Molding.   Hydraulic presses are frequently used for compression molding. These presses consist of two or more press platens, most commonly heated by steam. The presses are connected to hydraulic systems used to open and close the presses. The preforms to be vulcanized are placed into the hot, closable, two-part (Fig. 4.33) molds, which are placed between the press platens. The press is closed, the molds being held under pressure (35 to 100 bar) for the period of time required for sufficient vulcanization. (If the press is opened too soon, there will be insufficient vulcanization to prevent the formation of bubbles.) After sufficient curing in the mold, the press is opened, and articles are removed from the mold. Articles sufficiently cured to avoid bubbles or blow can be allowed to finish curing as they slowly cool down. Routinely, each mold contains a multitude of cavities for producing many parts with each molding cycle. Also, the two-piece molds can be stacked, each pair being hinged.

There is a variation of compression molding wherein the so-called *toggle press* is used. Instead of using a hydraulic press, one uses an electrically operated press, which is opened and closed mechanically via toggles. The molds are built into the specialized press. The most important press of his type is used for molding tires (wherein a pressurized steam-filled bladder or bag is used as a collapsible core).

Transfer Molding.   Transfer molding is a variation or refinement of compression molding. It is somewhat related to injection molding. In its simplest form, transfer molding uses a mold having three parts (Fig. 4.33). The upper and lower parts are attached to the platens of a hydraulic press, whereas the middle part is removable. The upper part of the mold is generally a piston, and the middle part contains a cylindrical cavity that receives the rubber compound to be molded. The middle part also contains nozzle openings in the bottom of its cavity. The bottom part of the mold contains a cavity that will contain the vulcanized part after the process is completed. As the press is closed, the piston of the top part of the mold forces uncured rubber stock through the nozzle openings into the product-mold cavity in the bottom part of the mold. Thus, the rubber compound is "transferred" during the closure of the three-part mold.

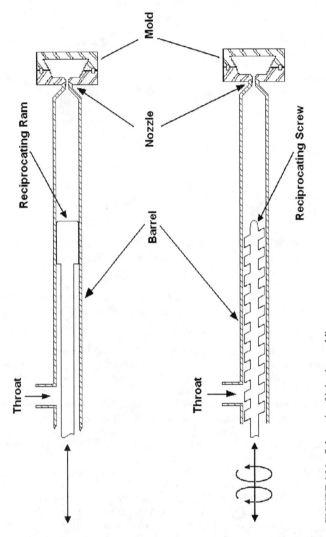

FIGURE 4.34 Schematic of injection molding.

There are certain disadvantages with the use of such multiple-part molds. Much time can be lost when one removes the mold, and the time required to heat the mold can be considerable. Also, the separation of all three parts of the mold with the extraction of the part can be difficult. On the other hand, transfer molding simplifies the loading of molds, in comparison to compression molding.

Injection Molding. The injection molding process has become a mainstay of rubber part manufacture. It is generally very rapid in comparison to the other types of molding processes. Schematics for two types of injection molding processes are given by Fig. 4.34. Injection molding in the thermoset rubber industry is different from what it is in the thermoplastic plastics industry. The rubber is only heated to a processing temperature for flow, but the mold, rather than being cooler than the polymer, is much hotter, at a vulcanization temperature. The ram, or plunger, injection molding process is a descendent of transfer molding. In the modern machine, rubber is plasticized and heated by a screw in a separate plastication cylinder (not shown in the schematic) and transported to the injection cylinder (shown in the schematic) and then transported to the injection cylinder through a nonreturn valve (e.g., in the throat), with the ram in the retracted position. When the required amount has been accumulated in the cylinder before the ram, the feed screw stops, and the rubber is injected by means of an injection plunger or ram.

In reciprocating-screw injection molding, the rubber compound is heated by the retractable (reciprocating) screw, in a position where its front end is near the nozzle. Then, the screw retracts as it "winds" itself ("unscrews") out, away from the nozzle and the warm plasticated rubber, which is now up against the nozzle. The screw is then moved forward (toward the mold) and, in so doing, it injects the rubber into the mold.

The higher the temperature of the rubber in the barrel before it is injected, the faster the rubber can be injected and the faster it will be heated up to vulcanization temperature in the hot mold. However, if the rubber stock temperature is too high before injection, the rubber might begin to cure prematurely (scorch). Care must be taken to avoid this.

4.5.6 Example Recipes of Selected Rubber Compounds

Table 4.14 gives example recipes of compounds of various types of elastomer. The recipes should not be used as a formulary. They are just to give a flavor of the types of compounds developed by rubber compounders. Since many end-use products have different specifications, and different rubber-product manufacturing facilities have different types of equipment, compounds must be developed specifically with respect to both the requirement for the end-use applications and the manufacturing equipment that is available.

4.6 THERMOPLASTIC ELASTOMERS

A thermoplastic elastomer (TPE) is a rubbery material with properties and functional performance very similar to those of a conventional thermoset rubber, yet it can be fabricated in the molten state as a thermoplastic. ASTM D 1566 defines TPEs as "a diverse family of rubber-like materials that, unlike conventional vulcanized rubbers, can be processed and recycled like thermoplastic materials." Many TPEs meet the standard ASTM definition of a rubber, since they recover quickly and forcibly from large deformations, they can be elongated by more than 100 percent, their tension set is less than 50 percent, and they are sometimes insoluble in boiling organic solvents. Figure 4.35 indicates hardness ranges for various types of TPEs and conventional elastomers.

TABLE 4.14   Example Rubber Compound Recipes

Natural rubber extrusion compound		EPDM vacuum tube	
SMR-5	100	EPDM (Keltan 5508)	100
Stearic acid	0.5	N-650 black	170
Zinc oxide	5		
Octylated diphenylamines (anti-degradant)	2	Paraffinic process oil (Sunpar 2280)	100
Processing aid	2	Wax (A-C Polyethylene 617)	5
Extender oil	15	Zinc oxide	5
Vulcanized soybean oil (Neophax A)	30	Stearic acid	1
Carbon black N-550	32	MBT (accelerator)	1
Soft clay	40	TMTM (accelerator)	1.6
Sulfur	2.75	Sulfur	1.5
N-*t*-Butylbenzothiazolesulfena-mide (accelerator)	1		
Tetramethylthiuram disulfide (accelerator)	0.2		
Cure 10 min at 153°C		Cure 20 min at 160°C	

NR-BR-SBR tread		Chloroprene v-belt	
SBR 1204	50	Neoprene GRT	100
BR 1203	20	N550 carbon black	30
NR-SMR5	30	Reinforcing silica (Hi-Sil 233)	15
Carbon black N-339	40	Octylated diphenylamines (anti-degradant)	2
Hi-Sil 210 (treated silica)	20	MBTS (accelerator)	0.5
Extender oil	17	Novalak resin (SRF 1501) (adhesion promoter)	2
Zinc oxide	3	Methylated melamine formaldehyde resin (Cyrez 963) (adhesion promoter)	6
Stearic acid	1	Stearic acid	2
6PPD (antidegradant)	1.5	Magnesium oxide (Maglite D)	5
Sulfur	1.5	Zinc oxide	5
t-Butylbenzothiazolesulfena-mide (accelerator)	2		
Cure 15 min at 150°C		Cure 40 min at 155°C	

TABLE 4.14   Example Rubber Compound Recipes (Continued)

Silica-filled butyl rubber compound		NBR injection molding compound	
IIR (Butyl 365)	100	NBR (Chemigum N685B)	100
Hi-Sil 233	40	N774 (SRF-NS carbon black)	65
Naphthenic oil	5	Zinc oxide	4
Tris(2-butoxyethyl)phosphate (plasticizer)	3	Mixed diaryl-*p*-phenylenedi-amines (Wingstay 100 anti-degradant)	2
Stearic acid	1	Plasticizer	15
Zinc oxide	5	Stearic acid	0.5
Polyethylene glycol PEG 4000	3	Tetramethylthiuram disulfide (TMTD accelerator)	1.7
Sulfur	1	N-Oxydiethylenebenzothiazole-2-sulfenamide (OBTS accelerator)	1.7
60% tetramethylthiuram disulfide/40% tetraethylthiuram disulfide	2	Sulfur	0.6
Dipentamethylenethiuram hexasulfide	1		
Cure 30 min at 160°C		Cure 15 min at 160°C	

Epichlorohydrin seal compound		Acrylic elastomer seal compound	
Epichlorohydrin rubber (Hydrin C2000)	100	Soap/sulfur-curable ACM	100
N-550	45	FEF Black (N-550)	55
Di-(butoxyethoxyethyl) adipate	15	Stearic acid	2
Zinc 2-mercaptobenzimidazole (antidegradant)	1	Agerite Superflex (diphenyl)-amine-acetone reaction product) antidegradant	2
Stearic acid	1	Sodium stearate	3
Calcium carbonate	5	Potassium stearate	0.4
Magnesium oxide	3	Sulfur	0.3
2,4,6-Trimercapto-S-triazine (Zisnet F-PT)	0.8		
Cure 15 min at 200°C		Cure 15 min at 175°C and post-cure 8 hr a 175°C	

TABLE 4.14 Example Rubber Compound Recipes (Continued)

Silicone rubber compound		Fluoroelastomer compound	
VMQ gum	100	FKM (peroxide curable—Viton® GLT)	100
Precipitated silica (Hi-Sil 915)	40	Zinc oxide	3
Silicone fluid (processing aid)	4	N990 carbon black	30
2,5-Bis(t-butylperoxy)-2,3-dime-thylhexane (Luperox 101)	0.8	Triallyl isocyanurate (co-agent)	3
		2,5-Bis(t-butylperoxy)-2,5-dime-thylhexane (Luperox 101 XL)	3
Cure 10 min at 170°C and post-cure 1 hr at 250°C		Cure 7 min at 177°C and postcure 2 hr at 232°C	

Urethane rubber compound	
Millable polyurethane elastomer (Millathane 76)	100
Stearic acid	0.5
N-550 black	30
Dibutoxyethyl adipate (TP-95)	20
Adaphax 758 (castor oil polymer)	5
Di-Cup 40C (40% dicumyl per-oxide)	2
Cure 9 min at 160°C	

4.6.1 Comparisons of TPEs with Thermoset Rubbers

TPEs have replaced thermoset rubber in a wide range of parts. This is because of the favorable balance between the advantages and disadvantages of TPEs in comparison with thermoset rubbers.

Practical advantages offered by TPEs over thermoset rubbers include the following:

1. Processing is simpler and requires fewer steps. Figure 4.36 contrasts the simple thermoplastics processing used to make TPE parts with the multistep process required for conventional thermoset rubber parts. Each processing step adds cost to the finished part and, in the case of thermoset rubbers, may generate significant amounts of scrap.

2. Processing time cycles are much shorter for TPEs. These times are on the order of seconds, compared to minutes for thermoset rubber parts, which must be held in the mold while vulcanization takes place.

3. TPEs usually require little or no compounding with other materials. They are available fully compounded and ready for a wide range of uses. Their compositional consistency is higher than that of thermoset rubbers, which must be mixed with curatives, stabilizers, processing aids, and specialty additives such as flame retardants.

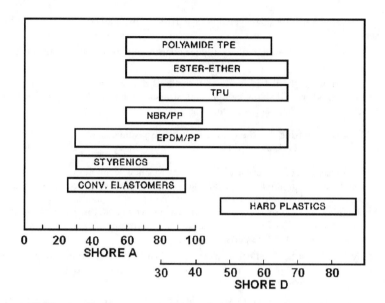

FIGURE 4.35   Hardness ranges for thermoplastic elastomers.

Part Fabrication with Conventional Rubbers

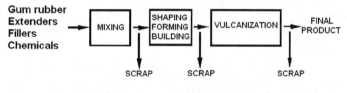

Part Fabrication with Thermoplastic Elastomers

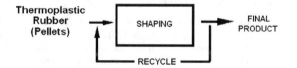

FIGURE 4.36   TPE part fabrication in a single step vs. three or more steps for conventional elastomers.

4. TPE scrap (regrind) may be recycled. Such scrap is generated, for example, in runners and sprues from injection molding and during startup and shutdown of any processing unit. Thermoset rubber scrap is often discarded, causing an added cost and a load on the environment. Most TPEs will tolerate several regrind-recycle steps without significant change in properties.

5. TPE parts can readily be recycled after they have given a normal, useful lifetime of service.

6. Thermoplastics processing allows tighter control on part dimensions than thermoset rubber processing.

7. TPEs permit thermoplastic fabrication methods not feasible for thermoset rubbers. These methods include blow molding, coextrusion with rigid thermoplastics, thermoforming, heat welding, and film blowing.

It should not be surprising that there are offsetting disadvantages to the use of TPEs compared to thermoset rubbers, as listed below:

1. To thermoset rubber processors, TPEs belong to a new technology requiring unfamiliar processing equipment and techniques. Thermoplastics processors are familiar with this technology and have the necessary equipment, although they are generally not familiar with the markets for rubber articles. The capital investment for thermoplastics equipment is often a major hurdle for a thermoset rubber processor to participate in the market for TPE parts.

2. Many TPEs must be dried before processing. While this is a familiar step to thermoplastics processors, it is not necessary for thermoset rubbers. Drying equipment is usually not available in a rubber shop.

3. A TPE becomes molten at a specific elevated temperature, above which a part will not maintain its structural integrity. Cross-linked thermoset rubbers do not display such melting behavior and are limited in upper service temperature only by chemical degradation such as oxidation.

4. TPEs require moderately high production volume for good processing economics. Thermoplastics tooling costs are generally higher than those for thermoset rubber parts, some of which are compression molded in volumes of only a few hundred per year.

A compounded rubber stock is often less costly on a volume basis than a competitive TPE. However, lower processing costs can more than compensate for the material cost difference. The needed equipment investment and production volumes must be weighed against the fabrication savings and material cost differences.

4.6.2   General Characteristics of TPEs

A TPE generally comprises two polymeric phases: a hard thermoplastic phase and a soft elastomeric phase. The properties of the resulting TPE depend, at least in part, on the properties of each of the two phases and their mutual interactions. The two phases may result from simply mixing two different polymers, as in a blend of a hard thermoplastic such as polypropylene (PP) with a soft elastomer such as ethylene-propylene terpolymer (EPDM rubber), to give a thermoplastic elastomeric olefin (TPO). Dynamic vulcanization (under conditions of high shear and temperature) of the elastomer phase of such a blend gives rise to a thermoplastic vulcanizate (TPV), with properties close to those of a conventional thermoset rubber. The two phases of a TPE may also be present as hard and soft segments along a common polymer backbone. This is the case for block copolymers, the basis for many commercially important TPEs. Table 4.15 compares the performance characteristics of six different generic classes of TPEs.

The performance characteristics of a TPE depend on the T_m (or T_g if the hard phase is glassy rather than crystalline) of the hard thermoplastic phase and the glass transition tem-

TABLE 4.15 Characteristics of Generic Classes of TPEs*

Property	Styrenic	TPO	TPV	Copolyester	Polyurethane	Polyamide
Specific gravity range	0.9–1.20	0.89–1.00	0.94–1.00	1.10–1.40	1.10–1.30	1.00–1.20
Shore hardness	20A–60D	60A–65D	35A–50D	35D–72D	60A–55D~	60A–65D
Low-temperature limit, °C	–70	–60	–60	–65	–50	–40
High-temperature limit, °C	70–100	70–100	135	125	120	170
Compression set resistance, 100°C	P	P	G/E	F	F/G	F/G
Resistance to aqueous fluids	G/E	G/E	G/E	P/G	F/G	F/G
Resistance to hydrocarbon fluids	P	P	F/G	G/E	G/E	G/E

*P = poor, F = fair, G = good, E = excellent.
Does not include grades containing a special flame-retardant package, which generally raises the specific gravity by 20 to 30 percent.

perature (T_g) of the soft elastomeric phase. The useful temperature range for a TPE is between T_m or T_g of the hard phase and T_g of the soft, rubbery phase. Within this range, the TPE displays its desirable elastomeric properties. At temperatures above T_g and T_m of the hard thermoplastic phase, the TPE is a molten fluid and can be processed by usual thermoplastics techniques. Below the rubbery-phase T_g, the TPE is brittle, without its useful elastomeric properties. This is indicated by Fig. 4.37. Types of phase morphologies are indicated by Fig. 4.38. The upper left quadrant of the figure represents the nanoscale phase domain morphology of the molecularly segment triblock styrenic TPEs, the molecular segments being derived from styrene-butadiene-styrene (S-B-S), styrene-isoprene-styrene (S-I-S), and styrene-hydrogenated butadiene-styrene (S-EB-S for styrene ethylene/butylenes-styrene). The styrenic segments become involved is glassy nano-domains at temperatures below the T_g of the styrenic phase. (Above this T_g, the TPE can flow.)

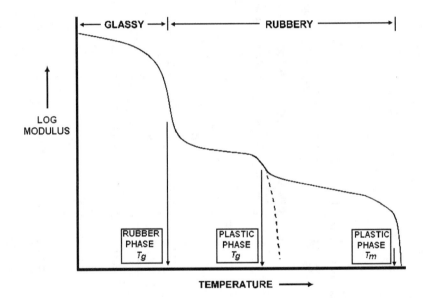

FIGURE 4.37 Modulus as a function of temperature for a thermoplastic elastomer.

The upper right quadrant represents segmented block polymers, wherein the hard-phase domains crystallize at temperatures below the melting temperature. Above the melting temperature, the polymer can flow.

The lower left quadrant represents the simple blends. One type of TPE blend is the TPO, which is a simple polymer blend of a rigid thermoplastic polymer with a technologically compatible (not necessarily thermodynamically compatible) elastomeric polymer. Each of the polymeric components exists in its own phase (with its own T_g or T_m). The properties of the blend are partly predictable by the properties of the components; however, the phase morphology is also extremely important. The hard phase must be continuous for the blend to be thermoplastic. Commercial TPO blends include those of EPDM rubber with PP or polyethylene (PE). Another type of simple polymer blend is that of poly(vinyl chloride) and nitrile rubber (NBR). This may have been the first useful rubber-plastic blend.

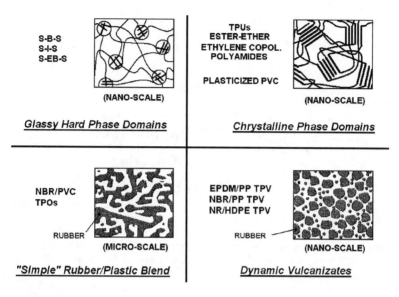

FIGURE 4.38   Phase morphology of different types of thermoplastic elastomers.

A thermoplastic vulcanizate (TPV), represented by the lower right quadrant of Fig. 4.38, is a TPE produced by dynamic vulcanization, the process vulcanizing a vulcanizable elastomer during its intimate mixing with a thermoplastic polymer in the molten state. A TPV comprises finely divided particles of highly cross-linked rubber in a continuous matrix of rigid thermoplastic.

Figure 4.39 shows a schematic cost-performance comparison for generic classes of conventional thermoset rubbers and for TPEs. Very approximately, the properties and performance of a given TPE class are somewhat comparable to those of the thermoset rubber at the same position on the cost-performance chart. Thus, the styrenics are candidates to replace NR and styrene-butadiene rubber (SBR), and the TPVs logically replace EPDM and chloroprene (neoprene) rubber.

4.6.3   Types of Thermoplastic Elastomers

4.6.3.1 Styrenic Block Copolymers.   Styrenic TPEs are copolymers whose molecules have the S-D-S structure, where S is a hard segment of polymerized styrene or styrene derivative, and D is a soft central segment of polymerized diene or hydrogenated diene units. Polybutadiene (B), polyisoprene (I), and polyethylenebutylene (EB) are the most commonly used rubbery segments (D). Structures for these triblock copolymers are represented as follows:

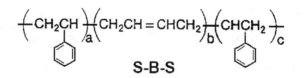

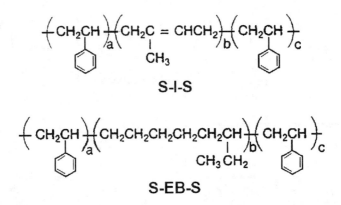

S-I-S

S-EB-S

The polystyrenic segments, with molecular weights of 5000 to 8000, form glassy aggregates on cooling with many of the properties of polystyrene itself. The upper glass transition temperature of such TPEs based on styrene is thus close to that for polystyrene (100 to 110°C). On cooling from the molten state to temperatures below the hard-segment T_g, the styrene-derived segments coalesce to form domains that act both as reinforcing filler and as cross-links since the ends of the rubbery chains (D) are bound therein. Diblock (S-S) copolymers are not TPEs, because the D-segments cannot be supporting chains, each having only one of its ends restrained in the glassy domain.

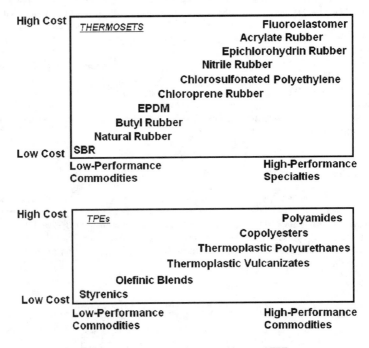

FIGURE 4.39 Cost/performance for thermoset rubbers and TPEs.

The characteristics of these TPEs depend on the relative proportions of the polymerized styrene and diene units as well as the chemical nature of the monomers. At low styrene levels, the TPEs are soft and rubbery. With increasing styrene content, the TPE progressively becomes stiffer at room temperature, and it becomes a glossy, hard material similar to an impact-modified polystyrene (HIPS). Removal of the rubbery-block unsaturation by hydrogenation (to give S-EB-S materials) makes the styrenic TPEs much more resistant to oxidation and ozone attack. Commercially available styrenic TPEs range in hardness from 20 Shore A to 60 Shore D.

Styrenics have a broad service temperature range, from about –70 to about 70 to 100°C, but the upper limit is sufficiently low to restrict their use to lower-temperature applications. They are very resistant to water and other polar fluids but lose much of their effectiveness if exposed to oils, fuels, and other nonpolar organic solvents. Their compression and tension set (resistance to plastic deformation under stress) is good (that is, low) at ambient temperatures but becomes progressively poorer with increasing temperature. Above 70°C, their compression set is poor (high). The common styrenic TPEs with butadiene- or isoprene-derived segments are among the lowest-cost, lowest-performance classes of TPEs. Their performance, however, is adequate for a wide variety of nondemanding rubber applications.

Uses of Styrenic TPEs.   The huge market for styrenic TPEs includes applications as fully compounded materials that can function as replacements for thermoset rubber, leading to uses in such articles as shoe soles, sporting goods, and a variety of other applications not requiring service above 70°C, and not requiring hydrocarbon fluid resistance or exceptionally good physical properties. The styrenics are the generic class of TPEs most widely used as compounding ingredients in other useful compositions. They are major constituents in adhesives, resins, sealants, caulking, motor vehicle lubricants, and thermoset automotive body parts.

4.6.3.2 Ester-Ether Block Copolymers [Copolyester TPEs (COPs)].   Ester-ether block copolymer TPEs, also called copolyester TPEs, are segmented block copolymers with the -A-B-A-B- structure, where A and B are alternating hard and soft polymeric segments connected by ester linkages. A general chemical structure can be expressed as follows:

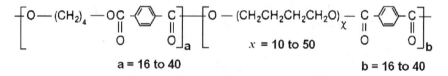

a = 16 to 40   b = 16 to 40

Hard Segment   **Soft Segment**

These block copolymers have an excellent combination of properties and are priced higher than styrenics, TPOs, or TPVs.

The morphology of COPs is that shown in the upper right-hand quadrant of Fig. 4.38. These materials perform as TPEs if the structures of A and B are chosen to give rubbery properties to the copolymer over a useful temperature range. The glass transition point of the soft segment should be low enough to prevent brittleness at the lowest temperature to which the working material will be exposed. The melting point of the hard segment should be high enough to allow the material to maintain a fabricated shape at its highest service temperature.

The ester-ether copolymers have a material cost higher than that of most thermoset rubber compounds. This higher cost can be more than offset by the high strength and modulus of the COP, which permits thinner, lighter parts and markedly lower part weights. The efficiency of thermoplastics processing can combine with the lower part weight to give pronounced cost savings relative to a thermoset rubber.

The tensile strengths of these TPEs are high, and they have hardnesses ranging from 40 to 63 Shore D (generally harder than conventional thermoset rubber compounds). As hardness increases, these copolymers become more like hard plastics and less like rubber vulcanizates.

COPs have a useful service temperature range from –40 to 150°C. The lower limit is set by the soft segment T_g. Retention of physical properties measured at elevated temperatures is quite good. They have very good resistance to a wide range of fluids. However, the ester and ether linkages in the polymer backbone render them susceptible to hydrolysis in the presence of both acids and bases.

COPs are resilient, with low hysteresis and heat buildup for uses requiring rapid, repeated flexing. In their elastic, low-strain region, COPs have very good resistance to flex fatigue and to tensile and compressive creep.

4.6.3.3 Thermoplastic Polyurethanes (TPUs). The first commercial TPEs were the TPUs, which have the same block copolymer morphology as do the COPs. Their general structure is -A-B-A-B-, where A represents a hard crystalline block derived by chain extension of a diisocyanate with a glycol. The soft block, represented by B, can be derived from either a polyester or a polyether. Typical TPU structures, both polyester and polyether types, are represented here:

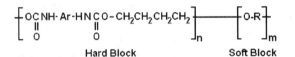

Hard Block   Soft Block

Ar = divalent aromatic radical

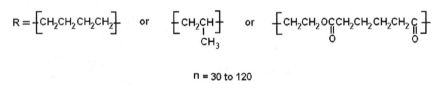

n = 30 to 120

m = 8 to 50

As with other block copolymers, the nature of the soft segments determines the elastic behavior and low-temperature performance. TPUs based on polyester soft blocks have excellent resistance to nonpolar fluids and high tear strength and abrasion resistance. Those based on polyether soft blocks have excellent resilience (low heat buildup, or hysteresis), thermal stability, and hydrolytic stability.

TPUs can be made much softer than can the copolyester TPEs—down to a Shore A hardness of 50. The properties of TPU TPEs are largely determined by the ratio of the amounts of hard to soft phases, the length and length distribution of the segments, and the crystallinity of the hard segments.

TPUs are noted for their outstanding abrasion resistance and low coefficient of friction on other surfaces. However, TPUs deteriorate slowly but noticeably between 130 and

170°C. Melting of the hard phase causes morphological changes, which are reversible, while oxidative degradation is slow and irreversible. Both processes become progressively more rapid with increasing temperature. Polyether soft blocks give TPUs having greater resistance to thermal and oxidative attack than do polyester blocks.

TPUs are polar materials and are therefore resistant to nonpolar organic fluids such as oils, fuels, and greases, but they are readily attacked and even dissolved by polar organic fluids such as dimethylformamide and dimethylsulfoxide. TPUs behave like copolyester TPEs toward water and aqueous solutions, being resistant to these media except at very high or low pH. Polyether TPUs are more resistant to such hydrolytic degradation than are the polyester TPUs.

The premium cost of TPUs can be justified for those applications requiring high levels of abrasion resistance and toughness or low coefficients of friction. These applications include caster wheels, shoe soles, automotive fascia, and heavy-duty hose and tubing.

4.6.3.4 Polyamide Thermoplastic Elastomers. The newest and highest-performance class of TPEs are segmented block copolymeric polyamides. The soft segments may have either aliphatic polyester or polyether structures.

$$HO \left[\overset{O}{\underset{\|}{C}} - PA - \overset{O}{\underset{\|}{C}} - O - PE - O \right]_n H$$

PA = polyamide segment
PE = polyether or polyester segment

The morphology is that of typical block copolymers, as shown in the upper right quadrant of Fig. 4.38. The polyamide TPEs typically have higher temperature and chemical resistance than do TPUs or copolyesters, and their cost is greater.

The soft segments may consist of polyester, polyether, or polyetherester chains. Polyether chains give better low-temperature properties and resistance to hydrolysis, while polyester chains in the soft segment give better fluid resistance and resistance to oxidation at elevated temperatures. As in other block copolymer TPEs, the nature of the hard segments determines the melting point of a polyamide TPE and its performance at elevated temperatures.

These TPEs have a wide hardness range, from a high of 65 Shore D down to 60 Shore A. They have useful tensile properties at ambient temperatures and excellent retention of these properties at higher temperatures. For example, a 90 Shore A polyamide TPE can retain more than 50 percent of its tensile strength and modulus at 100°C. Annealing at a can result in significant increases in tensile strength, modulus, and ultimate elongation. The polyamide TPEs are second only to TPUs in abrasion resistance and show excellent fatigue resistance and tear strength.

With service temperatures ranging from –40 up to 170°C, these materials give the highest performance of any generic class of TPE. They are also the most expensive. Polyester-based polyamide block copolymers give excellent properties retention after aging at 175°C for five days. These TPEs have good resistance toward hydrocarbon fuels, oils, and greases. Their good resistance to water and aqueous solutions decreases as the temperature is raised. Polyester block copolyamides are more sensitive to hydrolysis in humid air at higher temperatures than are polyether block copolyamides.

Processing temperatures (220 to 290°C) are higher than those for other TPEs because of their higher melting points. As with the TPUs, the polyamides require thorough drying before processing.

Their physical properties, chemical stability, and resistance to temperatures above 135°C qualify these materials as competitors for silicone rubber and fluoroelastomers. Applications include hose, tubing, gaskets, and protective covers for use in high-temperature environments.

4.6.3.5 Thermoplastic Elastomeric Olefins (TPOs).

EPDM rubber and PP are the constituents of the most common TPOs. An EPDM/PP TPO has a T_m near that of the hard PP phase and a rubber-phase T_g close to that for the soft EPDM (plus additives) phase. These TPOs thus melt in the range of 150 to 165°C and can be processed above these temperatures. They show excellent low-temperature performance with brittle points often below –60°C. The T_m clearly determines the upper theoretical service temperature limit of these TPOs. The maximum long-term service temperature is usually 25 to 50°C below the T_m, depending in part on the resistance of the polymers to oxidative attack.

EPDM/PP TPOs compete directly with styrenic TPEs as low-cost, low-specific-gravity (0.9 to 1.0) materials with fair to good mechanical performance and environmental resistance. They range in hardness (Table 4.15) from 60 Shore A up to 65 Shore D, with the harder products being more commonly found in commercial applications. The harder TPOs are essentially impact-modified thermoplastics and not true rubbers. The softer TPOs are rubbery at room temperature, but these characteristics are rapidly lost at elevated temperatures. EPDM/PP TPOs are therefore generally useful only below 70 to 80°C.

At ambient temperatures (0 to 40°C), TPOs have quite rubber-like properties. As the temperature is raised, however, these properties deteriorate quite sharply. On the other hand, the absence of unsaturation in the polymer backbones of both PP and EPDM makes these polymers and the TPOs derived from them very resistant to degradation by oxidation or ozone attack. The nonpolar nature of EPDM/PP TPOs makes them highly resistant to water, aqueous solutions, and other polar fluids such as alcohols and glycols, but they swell extensively with loss of properties when exposed to halocarbons and oils and fuels.

EPDM/PP TPOs have good electrical properties such as high resistivity, dielectric strength, and low power factor, allowing their use as primary electrical insulation where temperature and fluid resistance are not critical.

TPOs (olefinic blends) comprise a lower-performance, lower-cost class of TPEs (Fig. 4.39). Their performance and properties are generally inferior to those of thermoset rubbers. Yet, they are suitable for uses where (1) the maximum service temperature is modest (below 80°C), (2) nonpolar fluid resistance is not needed, and (3) a high level of creep and set can be tolerated. Thus, TPOs are marketed more on the basis of cost rather than performance, competing directly with the lower-cost general-purpose rubbers (NR, SBR, and the like). TPOs are associated with the traditional practice of rubber compounding and mixing. They can be prepared by the same techniques and equipment as for thermoset rubber; however, they need to be processed at temperatures above the T_m of the thermoplastic hard phase. The amounts of elastomer, rigid thermoplastic, plasticizer, and other ingredients can be varied to achieve specific properties in much the same manner as with rubber compounds.

Having been first commercialized in 1972, EPDM/PP TPOs are used mainly in external automotive and electrical applications up to 80°C. Automotive uses include exterior trim such as bumpers, fascia, and nonsealing moldings; however, under-the-hood uses are generally excluded because of temperature and fluid-resistance requirements in the engine compartment.

4.6.3.6 Thermoplastic Vulcanizates (TPVs).

TPVs differ from TPOs in that the rubber phase is highly vulcanized (cross-linked). This phase of a TPO has little or no cross-linking. As a result, the properties and performance of a TPV are much closer to those of a conventional thermoset rubber. Key properties for distinguishing between a TPV and a TPO for a given elastomer-thermoplastic system are as follows:

1. Reduced permanent set
2. Improved ultimate mechanical properties
3. Improved fatigue resistance
4. Greater resistance to attack by fluids, e.g., hot oils
5. Improved high-temperature utility
6. Greater stability of phase morphology in the melt
7. Greater melt strength
8. More reliable thermoplastic fabricability

In earlier literature, these TPEs have been called *elastomeric alloys*. Most workers in the TPE field have come to prefer the term *TPV*, since it conveys more clearly the specific nature of these materials.

TPVs are prepared by dynamic vulcanization, the process of vulcanizing an elastomer during its intimate molten-state mixing with a nonvulcanizing thermoplastic polymer. Small elastomer droplets are vulcanized to give a particulate vulcanized elastomer phase of stable domain morphology during melt processing and subsequently. The process generates a TPE with properties closer to those of a thermoset rubber than those of a comparable unvulcanized composition if there is a high concentration of rubbery phase.

On melt mixing of the thermoplastic and rubbery polymers under high stress, the less viscous thermoplastic tends to become the continuous phase with the more viscous rubber dispersed in it. The dispersed rubber particles are then vulcanized and form a three-dimensional polymer network within each particle; they cannot recombine or agglomerate into larger particles. The most common polymer system in TPVs is PP/EPDM rubber; however, a number of other polymer systems have been used commercially. These include NBR/PP, IIR/PP, and NR/PP.

The morphology of a TPV is best understood as a dispersion of very small, highly cross-linked elastomer particles in a continuous phase of hard thermoplastic. The size of the elastomer phase particles is one key to the performance of the TPV. If the particles have diameters as small as 1 to 5 μm, mechanical properties of the TPV are surprisingly good, almost reaching those of a corresponding thermoset rubber and vastly exceeding those of a TPO from the same polymers. TPVs are the TPEs, which are closest to conventional thermoset rubbers with respect to their behavior. A high degree of cross-linking of the dispersed elastomer phase is necessary for the rubber-like performance of a TPV.

TPVs have mechanical properties such as modulus (compression or tensile), tear strength, abrasion resistance, and compression set resistance that make them suited for a broad range of rubber applications. Also, their resistance to fatigue is superior to that of a thermoset rubber of comparable hardness. The useful long-term (weeks, years) service temperature range for TPVs is between the T_g of the soft elastomer phase and the temperature at which oxidative degradation of the TPV becomes significant or the material melts. Because there is no unsaturation in the polymer backbone of EPDM rubber, TPVs based on EPDM have a higher service temperature limit (approaching T_m) than those based on unsaturated elastomers such as NBR and NR. For EPDM/PP TPVs, this range is –60 to 135°C; for NBR/PP TPVs, it is –40 to 125°C; and it is –40 to 100°C for NR/PP TPVs. The lower temperature of each of these ranges is determined by the brittle point or T_g of the elastomer. The upper temperature is based on properties retention after continuous aging in hot air for 1000 hr.

Just as with thermoset rubbers, nonpolar fluids such as oils or fuels cause varying degrees of swelling, fluid absorption, and loss of properties by the TPV. Resistance to such fluids is similar to the fluid resistance of the elastomeric component of the TPV. Thus, NBR/PP TPVs were developed as TPEs of improved oil resistance.

4.6.3.7 Other TPEs. An emerging group of TPOs are produced by the copolymerization of ethylene and alpha olefins, in the presence of metallocene catalysts, to give a polymers with a narrow molecular-weight distributions, controlled levels of long-chain branching, and homogeneous comonomer distribution. The polymer structure is similar to that of a block copolymer. Produced in a single reactor, these TPEs have a major cost advantage over the older TPOs and TPVs. They likely will compete most directly with the TPOs. It is highly unlikely that their performance can be brought up to the level of the TPVs.

Blends of PVC and NBR (possibly the oldest of commercial polymer blends) have found some use in areas where the service temperature is at or near ambient and oil resistance is needed, such as in the hose and electrical wire and cable markets. They are normally custom compounded for a specific use.

Comparison of Properties. Properties of examples of the above TPEs are given in Table 4.16.

TABLE 4.16 Properties of Selected Thermoplastic Elastomers[*]

	PU	Styr.	TPO	Copolyester	TPVs	Polyamides
Hardness, Shore A or D	70A	72A	71A	40D	72A	75A
UTS, MPa	24	26	6.4	26	8.3	29
Stress at 100% strain, MPa	3.5	2	2.2	6.4	3.2	4.3
Ultimate elongation, %	500	560	800	450	375	350
Dry heat performance	F	P	P	G	G	VG
Hot oil resistance	E	VP	VP	E	G	E
Elevated-temp. set resistance	P	VP	VP	G	E	F–G

[*].P = poor, F = fair, G = good, E = excellent.

4.6.4 Processing of TPEs

4.6.4.1 Need for Drying TPEs. Many TPEs can absorb sufficient moisture in a short time to cause difficulties such as poor surface appearance (or *splay*) on injection molded parts and rough surfaces and porosity in extrusions. A moisture pickup of 0.10 percent or greater can give highly significant problems, even in the case of nonpolar hydrocarbon TPEs. An exposure of less than one day can be unacceptable.

While some TPEs have been developed to specifically avoid the need for drying, most should be dried immediately before use at a temperature of 70 to 100°C for two to six hours, depending on the specific TPE. Because of high surface areas, regrind material (for recycling to the process) should be dried an extra one to two hours.

4.6.4.2 Extrusion. Simple extrusion can be used to manufacture a variety of TPE shapes such as tubing, hose, sheet, and complex profiles. A thermoplastic extruder with a screw length-to-diameter (L/D) ratio of at least 20:1, and preferably 24:1 to 32:1, is rec-ommended for most TPEs. The extruder should be capable of operating in the melt tem-perature range of 170 to 250°C. A variety of screw designs (polyolefin type, flighted barrier, and pin mixing) have been used. Screw cooling is generally not required. Typical polyolefin-type metering screws with compression ratios in the range of 2:1 to 4:1 are nor-mally used, although other thermoplastics screws have also been used. Screen packs of 20 to 60 mesh are used to provide a clean melt stream with even flow. Finer mesh screens can be used depending on the viscosity of the particular TPE and its sensitivity to the higher temperatures caused by the finer screen. The polymer melt temperature in the extrusion process should be about 30 to 70°C above the melting point of the TPE.

Thermoplastic materials generally exhibit a die swell on exiting from an extrusion die. TPEs can give die swells significantly lower than those of typical rigid thermoplastic poly-mers.

TPE sheet can be produced (e.g., from TPVs) by extrusion processing. Sheet extrusion of TPEs can produce thicknesses from about 0.2 to 4 mm and widths of 2 m or more. Very thin TPE sheeting can be produced by the blown film process used for many thermoplas-tics. This technique allows hard TPE sheet to be made down to a thickness as low as 0.05 mm, and even soft TPEs can yield sheets 0.2 mm thick.

4.6.4.3 Injection Molding.   Injection molding is the most widely used process for fabri-cating TPE parts. Short molding cycles and elimination of scrap by regrinding quite often outweigh the generally higher material cost of a TPE compared to that of a thermoset. Sprues and runners from injection molding can be recycled.

TPEs can be molded in the same type of equipment used for injection molding thermo-plastics such as PP and PE. Good part definition and integrity are obtained by adequate mold packing, which gives strong weld lines and minimizes shrinkage. This is achieved by mold designs with a balanced layout. Runners should be as short as feasible, and small gates are recommended to provide high shear rates for uniformity of the material as it en-ters the cavity. Operating conditions should be selected to give melt temperatures in the range of 20 to 50°C above the melting point of the TPE so as to allow adequate mold pack-ing.

4.6.4.4 Blow Molding.   TPEs can be blow molded (Fig. 4.40) to produce hollow shapes in the same manner and equipment used for rigid thermoplastics, either via injection or ex-trusion blow molding techniques. Extrusion blow molding is the simpler process in which a hollow molten parison is extruded vertically downward into a mold cavity. As the mold is clamped around the parison, blowing takes place, forcing the molten TPE against the water-cooled mold. The part is then cooled to give it sufficient structural integrity for re-moval from the mold, normally by gravity. Injection blow molding is similar except that very close control of the parison dimensions is obtained by injection molding the parison. While still hot, the molded parison is transferred into a blow mold for final part shaping. The need for two sets of tooling makes this process costly but allows for precise wall-thickness control. Injection blow molding is very capital intensive and thus requires large production volumes for favorable economics.

Blow molding cannot be used to produce hollow rubber parts from conventional ther-moset rubbers. These must be injection molded over a manually removable collapsible core to form thin-walled rubber shapes such as boots, bellows, and covers. Blow molding is uniquely suited to TPEs as a material for hollow, thin-walled rubber articles.

4.6.4.5 Other Fabrication Methods.   Another thermoplastics processing technique that is suited to some TPEs, but not to thermoset rubbers, is thermoforming. A sheet of TPE is heated to 10 to 40°C above the melting temperature of the hard crystalline phase (or glass

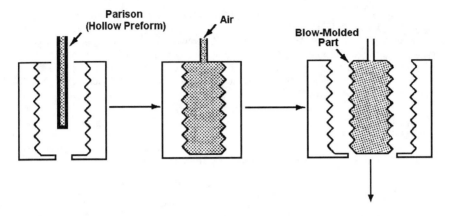

FIGURE 4.40 Schematic of blow molding.

transition temperature, if there is no crystallinity), then pressure or vacuum is used to stretch the softened sheet over or into a mold or form. Draw ratios up to 3:1 (thickness of original sheet to final part thickness) can be attained with some TPEs. TPVs are particularly suited for both this technique and blow molding.

Welding is a fast and simple method for bonding a TPE to itself or to another compatible thermoplastic material. The surfaces to be joined must be heated above the melting point and held together under slight pressure (preferably after some flow has occurred) until the joint cools and solidifies. Surface heating can be by direct contact with hot air or a hot surface, radiation heating, or other methods.

TPEs can readily be via foamed extrusion to lower their macroscopic hardness and density. Closed-cell foamed tubing, sheeting, and profiles with a thin, solid skin are now routinely fabricated by extruding TPEs with a chemical or physical blowing agent.

4.6.5 Uses for TPEs

TPEs find uses in many applications where thermoset rubbers are used, but not in pneumatic vehicle tires, which account for slightly more than one half of the worldwide rubber consumption. There are, however, a number of TPEs that are used for solid, nonpneumatic tires, and caster wheels for carts, lawn equipment, toys, and so on.

Nontire automotive uses where the service requirements are not too demanding (temperature below 70°C and little or no fuel and oil resistance required) have provided numerous markets for both the styrenic and TPO materials. These uses include weather stripping, bumpers, fascia, dashboard trim, plugs, and grommets, to mention a few.

The higher-performance COPs, TPUs, and TPVs are used in those areas where the service-temperature range, mechanical abuse, and fluid resistance demand a higher level of performance. Such applications include seals and gaskets, convoluted grease-filled boots for steering and front-wheel drive, and assemblies and covers for safety air bags.

The growth of TPE usage in nonautomotive applications has also been rapid. Specific parts are as diverse as dishwasher sump boots, architectural window glazing, flashlight housings, hand-tool gaskets, typewriter and printer rollers, and household plumbing seals.

Low dielectric constant, high electrical resistance, high dielectric strength, and low power factor make hydrogenated styrenic block copolymers and EPDM/PP TPVs especially useful in electrical applications. Flame-retardant TPEs are available for electrical uses requiring rubber that will not support combustion.

A growing number of uses for TPEs in food-processing equipment include beverage-dispenser pumps and food-container seals. In the medical-use area, uses include peristaltic pump tubing, syringe stoppers, catheters, and hospital tubing and sheeting.

4.7 ADDITIONAL READING

1. J. E. Mark, E. Burak, and F. R. Eirich, Eds., *The Science and Technology of Rubber,* 3rd ed., Elsevier, New York, 2005.
2. K. Bhowmick, M. M. Hall, and H. A. Benarey, Eds., *Rubber Products Manufacturing Technology,* Marcel Dekker, New York, 1994.
3. W. Hofmann, *Rubber Technology Handbook,* Hanser, New York, 1989.
4. J. D. Ferry, *Viscoelastic Properties of Polymers,* 3rd ed., Wiley, New York, 1980.
5. J. L. White, *Rubber Processing Technology, Materials, and Principles,* Hanser, New York, 1995.
6. P. A. Ciullo and N. Hewitt, *The Rubber Formulary,* Noyes, New York, 1999.
7. K. Bhowmick and H. L. Stephens, Eds., *Handbook of Elastomer,* 2nd ed., Marcel Dekker, New York, 2001.
8. G. Holden, N. R. Legge, R. P. Quirk, and H. E. Schroeder, Eds., *Thermoplastic Elastomers,* 2nd ed., Hanser, New York, 1996.
9. C. P. Rader, in *Handbook of Plastics, Elastomers, and Composites,* C. A. Harper, Ed., Ch. 5., McGraw-Hill, New York, 1996.
10. J. A. Kuczkowski, in *Oxidation Inhibition in Organic Materials,* vol. 2, J. Pospísil and P. P. Klemchuk, Eds., Ch. 8., CRC Press, Boca Raton, Florida, 1990.

CHAPTER 5
PLASTICS ADDITIVES

Rudolph D. Deanin

University of Massachusetts
Lowell, Massachusetts

There are about 100 families of commercial polymers in use in the plastics industry. Most of these are available in a range of molecular weights, and many of them are also available in a range of copolymers. This certainly provides a wide range of properties for use in different products.

But the many thousands of different plastic products require a much greater variety of properties and balance of properties. It might be possible to develop a new polymer to meet each of these needs, but the time and cost would be prohibitive. In most cases, it is much more convenient to use the polymers that are commercially available and to improve individual properties, and balance of properties, by the use of additives.

Thus, polymers are hardly ever used in pure form; they are almost always optimized by the use of additives. In most cases, a few additives are sufficient. But in some polymer families, compounders escalate to much larger numbers of additives so as to produce the processability and final properties they need. The most extreme examples are in vinyls, rubber, coatings, and adhesives, where the number of additives in a compound can easily reach 10 to 20 or even more!

The total tonnage of additives is conventionally about 18 percent of plastics production. If we include polymers added in making polymer blends, the total increases to about 25 percent. The major families of additives may be listed in order of total tonnage as follows:

- Fillers and reinforcements
- Polymers for polymer blends
- Plasticizers
- Flame Retardants
- Colorants
- Stabilizers (antioxidants, antiozonants, vinyl heat stabilizers, UV, and biostabilizers)
- Lubricants
- Organic peroxides
- Foaming agents
- Antistats

5.1 STABILIZERS

The primary covalent bonds that hold polymer molecules together are generally quite stable, permitting their wide use in plastics, rubber, fiber, coatings, and adhesives. They are not perfectly stable, however (Table 5.1). To a slight extent, they are in equilibrium with radical and ionic states,

$$A{:}B \rightarrow A^{\cdot} \cdot B \text{ or } A^{+} {}^{-}B$$

which would be unstable and reactive. Input of energy via the following processes:

- Mechanical (especially shear and dynamic)
- Thermal (in processing and in use)
- Electrical (especially high voltage and high frequency)
- Optical (especially UV and IR)
- Chemical (especially oxygen, ozone, and moisture)
- Biological (especially microorganisms)

increases this dissociation, resulting in cleavage or cross-linking, structural changes, and reactions with atmospheric oxygen, moisture, and chemical environments in general. These reactions can be used to modify polymers and perhaps to recycle them but, more often, they cause degradation and loss of properties in aging.

TABLE 5.1 Stability of Chemical Bonds

Chemical bond	Bond energy, kcal/mol	Chemical bond	Bond energy, kcal/mol
C-F	116	$(CH_3)_2CH$-H	95
C_6H_5-N	110	C-O	93
C_6H_5-O	107	CH_3COCH_2-H	92
Si-O	106	$(CH_3)_3C$-H	91
CH_3-H	104	$CH_2=CHCH_2$-H	85
C_6H_5-H	104	C-N	82
C-C	101	N-N	37
C_6H_5-C_6H_5	100	O-O	34

We can modify polymer structure to make it stabler and more resistant to degradation. Generally, perfluorination, aromatic and heterocyclic resonance, and polysiloxane backbone produce great stability, most dramatically in our ultra-high-temperature plastics, but involve high cost and difficult processability. Conversely, groups that can cause instability include tertiary branch points, C=C, aliphatic nitrogen and oxygen atoms, and hydrolyzable groups such as ester, urethane, and amide; so avoiding such groups can improve stability.

More commonly, however, we use a small amount of additive in a fairly stable polymer to make it more stable and useful. Individual polymers have individual stability problems, for which we use different types of additives. It is a mistake to think of stabilizers in general producing stability in general. We can distinguish five major types of stabilizers to solve five types of stability problems:

1. *Antioxidants* are used to protect against combined attack of atmospheric oxygen and heat during processing and during use above room temperature.
2. *Antiozonants* are used to protect the C=C double bonds in rubber against attack by atmospheric ozone.
3. *PVC heat stabilizers* are used to protect polyvinyl chloride against thermal dehydrochlorination during processing and during use above room temperature. Similar additives are sometimes useful in other halogenated systems as well.
4. *Ultraviolet stabilizers* are used to protect polymers in outdoor use against the high-energy ultraviolet radiation from the sun, and occasionally indoors when exposed to fluorescent lights.
5. *Biostabilizers* are used to protect polymers against biological attack, primarily by microorganisms.

5.1.1 Antioxidants

The attack of atmospheric oxygen on polymers is a free-radical reaction accelerated by heat, particularly during processing, and also during long-term use above room temperature. Typical autoxidation reactions such as

$$R{:}H \rightarrow R{\cdot} + {\cdot}H$$

$$R{\cdot} + O_2 \rightarrow RO_2{\cdot}$$

$$RO_2{\cdot} + R'H \rightarrow RO_2R' + H{\cdot} \text{ or } RO_2H + R'{\cdot}$$

$$H{\cdot} + O_2 \rightarrow HO_2{\cdot}$$

$$HO_2{\cdot} + RH \rightarrow HO_2R + H{\cdot} \text{ or } HO_2H + R{\cdot}$$

proceed rapidly forming peroxides and hydroperoxides. These are unstable and decompose into free radicals

$$RO{:}OR \rightarrow RO{\cdot} + {\cdot}OR$$

$$RO{:}OH \rightarrow RO{\cdot} + {\cdot}OH$$

which decompose further forming carbonyl, carboxyl, cross-linking, cleavage, disproportionation, and/or depolymerization products. Change in molecular weight affects most properties. Change in unsaturation causes discoloration. Change in polarity ruins electrical resistance. And oxidized species are more susceptible to biological attack.

Antioxidants are additives to prevent or at least delay the attack of oxygen. There are three main types, as described in the following sections.

5.1.1.1 Free-Radical Chain-Breakers. These are most commonly and most widely used and are therefore referred to as *primary antioxidants*. They react with a decomposing free radical and convert it into a stable product,

$$R^{\cdot} + AH \rightarrow RH + A^{\cdot}$$

The antioxidant radical A$^{\cdot}$ is a relatively inactive species, stabilized by aromatic resonance, and disposes of the excess energy harmlessly. Two major classes of compounds are used as free-radical chain-breakers: hindered phenols and diaryl amines.

5.1.1.1.1 Hindered Phenols. The aromatic hydroxyl group in phenols

$$ArO{:}H$$

is very effective for terminating free-radical chain-reactions. In simple phenols, it tends to form quinoid structures, which cause discoloration. Substitution of the ortho and para positions on the benzene ring prevents the formation of quinoid structures and produces non-discoloring antioxidants (Fig. 5.1). The most common and widely used is named di-t-

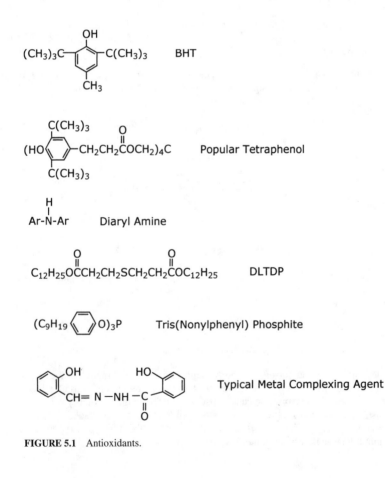

FIGURE 5.1 Antioxidants.

butyl-p-cresol (TBC) or butylated hydroxy toluene (BHT). This is perfectly satisfactory in polymers that can be processed at relatively low temperatures, such as LDPE. In higher-temperature processing, it is too volatile and can be lost too rapidly to be useful. Then, we resort to polyphenols of higher molecular weight and therefore lower volatility. Di- and tetra-phenols are commonly used for such higher-temperature processing, giving much better permanence (Tables 5.2 through 5.5).

TABLE 5.2 HDPE Stabilization: Days in 125°C Oven to Reach Embrittlement

Hindered phenol alone		Hindered phenol + 0.25 percent DLTDP
0	10 days	10 days
0.025 percent	80	170
0.05	140	250
0.1	190	310

TABLE 5.3 Polypropylene Stabilization: Days in 135°C Oven to Reach Embrittlement

Hindered phenol, %	Days
0	5
0.03	70
0.09	120
0.15	150
0.2	160
0.4	200
0.5	220

5.1.1.1.2 Diaryl Amines. The most active free-radical chain-breakers are diaryl amines. These generally discolor badly due to formation of quinoid structures, so they are not often used in plastics. In rubber, where the many C=C bonds create a great need for stabilization, and where carbon black generally masks any discoloration, diaryl amines are almost universally used. In plastics containing carbon black, they may also be used very effectively.

5.1.1.2 Nonradical Peroxide-Decomposers. Free-radical chain breakers alone may not completely prevent the formation of peroxides. A second additive is then included to provide a second mechanism for protection. This is an additive to decompose the peroxide that did form, but decompose it by a nonradical process, so that it does not kick off the

TABLE 5.4   Impact Styrene Stabilization: ASTM D-1925
Yellowness Index After 80°C Oven Aging

Hours	Unstabilized	0.1 percent hindered phenol
0	2	2
250	35	2
500		5
750		10
1000		17

TABLE 5.5   Nylon 6 Stabilization: Hours
to 20 Percent Loss of Tensile Strength in
160°C Oven

Hindered Phenol, %	Hours
0	5
0.2	60
0.4	115
0.6	170

free-radical degradation reaction. Two types of compounds are commonly used as nonradical peroxide-decomposers: aliphatic sulfides and organic phosphites.

5.1.1.2.1 Aliphatic Sulfides.   The most popular one is dilauryl thio dipropionate (DLTDP), nicknamed dillydip. For lower polarity and volatility, the lauryl groups are extended to stearyl (C_{18}) (DSTDP).

5.1.1.2.2 Organic Phosphites.   These are also often used. When they are used to supplement the hindered phenol primary antioxidant, they are considered nonradical peroxide decomposers. On the other hand, in many polymers, they may be used as the primary antioxidant, presumably both to prevent the formation of peroxide and to decompose any peroxide that does form. The most popular is tris(nonylphenyl) phosphite, but many others are also used.

5.1.1.3 Complexing Agents: Metal Deactivators.   Transition metals (elements with more than one valence state) catalyze oxidation reactions by complex redox (reduction-activation) mechanisms,

$$RO:OH \leftrightarrow RO^{\cdot} + {}^{\cdot}OR$$

$$RO^{\cdot} + M^{++} \rightarrow RO^{-} + M^{+++}$$

When traces of transition metal compounds are present in polymers, either as catalyst residues (Ziegler-Natta, metallocene, single-site, polyesterification) or as impurities in fillers (iron particularly), or from corrosion of process equipment, or as adjacent surfaces on final products (insulation on copper wire), they may aggravate the attack of atmospheric oxygen and the resulting degradation of the polymer. One way to remove the metal from the system is to tie it up in an inactive complex, in which form it is no longer able to catalyze the oxidation reaction. These complexing agents are usually organo nitrogen compounds or polyols. They are not used alone but are added as synergists to a system that already contains primary antioxidants.

5.1.1.4 Acid Scavengers. Oxidation of polymers produces organic acids. Chlorine and bromine, from catalyst residues and flame-retardants, produce stronger acids. These can cause hydrolysis of polymers and corrosion of process equipment. Therefore, it is fairly common practice to add acid scavengers to neutralize them. These are mildly alkaline substances such as calcium and zinc stearates, hydrotalcite, hydrocalumite, and zinc oxide.

5.1.1.5 Use in Commercial Plastics. LDPE is usually stabilized by 0.005 to 0.05 percent BHT. DLTDP and nonylphenyl phosphite may be added as well. For wire and cable insulation, metal deactivator is also needed.

LLDPE and HDPE use higher-molecular-weight phenols and higher concentrations. Cross-linked polyethylene, containing carbon black, permits use of thiodiphenols and diaryl amines, since their discoloration is masked by the carbon black. For wire and cable, hydrazides and triazines are common metal deactivators to protect against copper catalysis of oxidation.

Polypropylene contains less-stable tertiary hydrogens and processes at higher temperatures, so it requires higher concentrations (0.25 to 1.0 percent) of higher-molecular-weight phenols and more vigorous use of aliphatic sulfides and aromatic phosphites. Poly-1-butene is similar.

ABS contains 10 to 30 percent of butadiene rubber, whose C=C bonds are very sensitive to oxidation, producing embrittlement and discoloration. Triaryl phosphites are used as primary antioxidants, in concentrations up to 2.5 percent, producing excellent stabilization.

"Crystal" polystyrene is resistant to oxidation, but most "polystyrene" is actually impact styrene containing 2 to 10 percent of butadiene rubber. Like ABS, it requires similar stabilization, but lower concentrations are sufficient.

Acetal resins are sensitive to oxidation and are generally stabilized by high-molecular-weight phenols. Polyesters and polyurethanes are commonly stabilized by phosphites. Polyamides are stabilized by phosphites and also (surprisingly) by copper and manganese salts, presumably through complex formation with the amide groups themselves.

5.1.1.6 Market Analysis

See Table 5.6 for an analysis of worldwide consumption of antioxidants.

5.1.2 Antiozonants

The C=C in most rubber molecules, and in many high-impact plastics, is very sensitive to traces of natural and man-made ozone in the atmosphere. Ozone adds to the double bonds, forming ozonides that break down into various oxidized species, causing severe embrittlement. This requires vigorous protection to give products with useful lifetimes. Two types of additives are used: physical and chemical.

5.1.2.1 Physical Antiozonants. Saturated waxes are added during rubber compounding. Being immiscible, they migrate to the surface (bloom), forming a barrier coating that

TABLE 5.6 World Consumption of Antioxidants

Type	Percent
BHT	14
Higher phenols	42
Phosphites	31
Sulfides	9
Other	4
Thousand metric tons	207

keeps ozone from reaching the rubber. Paraffin waxes bloom rapidly but are too brittle. Microcrystalline waxes bloom more slowly but are less brittle. Mixture of the two types gives broader protection. These are adequate for static performance but are too brittle for dynamic stretching and flexing.

A saturated rubber can be coated on the surface to provide a barrier against ozone. Ethylene/propylene, plasticized PVC, and polyurethane are typical coatings. However, these involve problems of adhesion, elasticity, and cost, so they are not commonly used.

5.1.2.2 Chemical Antiozonants. These are mostly secondary alkyl aryl amines R-NH-Ar and related compounds. They give excellent protection. Most of them discolor badly, but several are recommended for nonstaining applications.

Most compounders use a combination of physical and chemical antiozonants and achieve excellent protection in this way. For more severe ozone-resistance problems, there are, of course, a number of specialty elastomers that are saturated and therefore completely ozone-resistant: ethylene/propylene rubber, chlorinated and chlorosulfonated polyethylene, ethylene/vinyl acetate, ethylene/acrylic esters, butyl rubber, SEBS, plasticized PVC, butyl acrylate copolymers, polyepichlorohydrin and copolymers, polyetherester block copolymer, polyurethane, and silicone.

5.1.3 PVC Heat Stabilizers

PVC is very heat sensitive. When it is heated during processing, or even during use, it loses HCl, which is toxic and corrosive; forms C=C bonds which cause discoloration; and cross-links, causing clogging of process equipment and embrittlement of products (Fig. 5.2). The problem is caused by an occasional unstable Cl atom that is destabilized by being adjacent to a branch point, a C=C group, a C=O group, or an oxygen atom. It requires strong and precise stabilization for practical use. There are three major classes of heat stabilizers for PVC, as described below.

5.1.3.1 Lead Compounds. These were the earliest in commercial practice. "Normal" lead salts included sulfate, silicate, carbonate, phosphite, stearate, maleate, and phthalate. "Basic" lead salts combined these with lead oxide, giving greater stability. They were low-cost, efficient, and gave excellent electrical resistance. Disadvantages were opacity, sulfur-staining, and toxicity. Due to worries about toxicity, their use has been restricted to electrical wire and cable insulation.

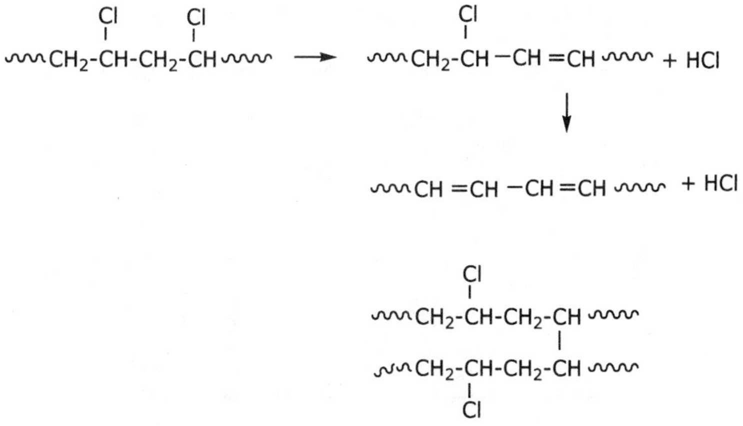

$$\underset{\underset{\text{Cl}}{|}}{\text{Cl}}\quad\underset{\underset{\text{Cl}}{|}}{\text{Cl}}$$

$$\text{ммCH}_2\text{-CH-CH}_2\text{-CH}\text{ммм} \longrightarrow \text{ммCH}_2\text{-CH} - \text{CH} = \text{CH}\text{ммм} + \text{HCl}$$

$$\downarrow$$

$$\text{ммCH} = \text{CH} - \text{CH} = \text{CH}\text{ммм} + \text{HCl}$$

$$\underset{\underset{\text{Cl}}{|}}{}$$

$$\text{ммCH}_2\text{-CH-CH}_2\text{-CH}\text{ммм}$$

$$\text{ммCH}_2\text{-CH-CH}_2\text{-CH}\text{ммм}$$

$$\underset{\text{Cl}}{|}$$

FIGURE 5.2   Thermal degradation of PVC.

5.1.3.2 Ba/Ca Soap + Cd/Zn Soap + Epoxidized Fatty Ester + Organic Phosphite. This synergistic combination has always been unnecessarily secretive, sold under vague names such as "mixed metal," "synergistic," and so on. It is universally used for plasticized PVC, because it is soluble, economical, and effective. The metal soap may be phenate, octoate, neodecanoate, naphthenate, benzoate, laurate, myristate, palmitate, or stearate.

The Group IIB metal soap (Cd or Zn) is the primary stabilizer. It replaces an unstable Cl atom by a stable ester group,

$$\text{Polymer-Cl} + \text{M(O}_2\text{CR)}_2 \rightarrow \text{Polymer-O}_2\text{CR} + \text{MCl}_2$$

Cd is more reliable, but worries about toxicity have practically eliminated its use. Zn is more powerful but tricky, so compounders have had to learn how to handle it very carefully.

The Group IIA metal soap (Ba or Ca) is a reservoir to regenerate the essential Group IIB metal soap:

$$\text{Ba(O}_2\text{CR)}_2 + \text{ZnCl}_2 \rightarrow \text{BaCl}_2 + \text{Zn(O}_2\text{CR)}_2$$

Ba works best, but there is some worry about toxicity. Ca is less effective but completely nontoxic, so it is used when there is worry about toxicity.

The epoxidized fatty ester may be epoxidized soybean oil for compatibility and nontoxicity, or epoxidized tall oil esters for low cost and low-temperature flexibility. It is generally believed to function by neutralizing HCl. It may also replace unstable Cl on the polymer or complex $ZnCl_2$ to keep it from degrading the PVC.

The organic phosphite is generally believed to function by complexing $ZnCl_2$ to keep it from degrading the PVC.

The synergistic effect is clearly seen by comparing the individual ingredients with the total system (Table 5.7). Typical concentrations are about 2 percent metal soap, 5 percent epoxidized fatty ester, and 1 percent organic phosphite.

5.1.3.3 Organotin Salts. The most powerful and expensive stabilizers for PVC are organotin compounds, most generally of the type R_2SnX_2. The R group is most often butyl, but

TABLE 5.7 Synergistic Stabilization of PVC: Gardner Color After Aging in 150°C Oven

Aging time, minutes	0	50	200
1 percent barium laurate	1	13	14
1 percent cadmium laurate	1	3	3
1 percent zinc laurate	1	18	18
5 percent epoxidized soybean oil	2	10	13
1 percent alkyl diaryl phosphite	1	17	18
All five together	1	1	2

sometimes is octyl for food packaging or methyl for higher efficiency. The most powerful X group is $-SCH_2CO_2C_8H_{17}$, which is called *isooctyl thioglycollate* or *isooctyl mercaptoacetate*. For greater lubricity or UV stability, the X group may be maleate or laurate (Table 5.8). The relative amounts of R and X are sometimes varied for subtle reasons. In rigid PVC, where high melting point and high viscosity cause the most serious instability problems, organotin is always used. Concentrations range from 2 to 3 percent down to one tenth as much, depending on the equipment and process.

TABLE 5.8 Organotin Stabilization of PVC: Gardner Color After Aging in 175°C Oven

Aging time, minutes	30	60
Unstabilized	13	15
3% dibutyl tin dilaurate	2	5
3% dibutyl tin maleate	2	3
3% dioctyl tin bis-octylthioacetate	1	2
3% dibutyl tin bis-octylthioacetate	1	2
3% dimethyl tin bis-octylthioacetate	1	1

5.1.3.4 Miscellaneous Stabilizers. A variety of other stabilizers are vaguely mentioned in the literature, mainly by vendors. Polyols and organo-nitrogen compounds may be added to complex iron impurities in fillers and keep them from catalyzing degradation of PVC. Other additives are more secretive and their benefits less clear. Bisphenol is added to wire and cable insulation to stabilize the plasticizer rather than the PVC. UV stabilizers may be added for outdoor use, and biostabilizers are important to protect the plasticizer.

5.1.3.5 Other Organohalogens. Thermal instability is also a problem in other polymers such as chlorinated polyethylene, chlorinated PVC, polyvinylidene chloride, chlorinated rubber, and chlorinated and brominated flame-retardants. PVC heat stabilizers may help here, too, but require careful adjustment for optimum performance in each system.

5.1.3.6 Market Volume. The total market for PVC heat stabilizers may be about 100 million pounds in the United States and 1 billion pounds worldwide, half for organotin and half for metal soap-epoxidized fatty ester-organic phosphite systems.

5.1.4 Ultraviolet Light Stabilizers

Five percent of the sunlight that penetrates the ozone layer and reaches the Earth is high-energy short-wavelength ultraviolet (UV) radiation, 290 to 400 nm. When polymers are used out of doors, absorption of this UV energy raises the electrons of primary covalent bonds from their low, stable energy level up to higher unstable energy levels that lead to degradation (Tables 5.9 and 5.1). Polymer structures that can absorb UV include benzene rings, C=C, C=O, OH, ROOH (Table 5.10), and especially conjugated groups of such structures. Even polymers that do not contain such groups may still degrade, and the blame is then placed on impurities or complex-formation. UV degradation can lead to cleavage to lower molecular weight or cross-linking to higher molecular weight, unsaturation, photooxidation, and photohydrolysis, all of which result in weathering deterioration.

TABLE 5.9 UV Wavelengths and Energy Levels

UV wavelength, nm	Energy level, kcal
259	111
272	105
290	100
300	95
320	90
340	84
350	81
400	71

TABLE 5.10 UV Absorption by Functional Groups in Polymers

Benzene rings	<350 nm
C=C	<250 nm
C=O	<360 nm
O-H	<320 nm
ROOH	<300 nm

There are a number of ways to protect plastic products for use outdoors, as described below.

5.1.4.1 UV Reflectors. If a UV-resistant material will reflect UV light away from the polymer, this can increase its lifetime tremendously. A metallized surface can give such protection, and, if it is made extremely thin, it may be able to combine UV stability and visible transparency. Pigmented fluoropolymer and acrylic coatings can be applied to the polymer, either by coextrusion of capstock or by post-coating, and provide such stability. More simply, dispersion of TiO_2 and especially aluminum flake in the polymer can reflect away most of the UV before it reaches more than a few surface molecules of the polymer, and this technique has been very popular.

5.1.4.2 UV Absorbers. Certain classes of additives absorb UV so efficiently that there is very little UV left to attack the polymer. They also have the little-understood ability to dispose of the excess energy harmlessly. o-hydroxy benzophenones, and especially o-hydroxyphenyl benzotriazoles, are quite successful, even in concentrations below 1 percent (Fig. 5.3, Tables 5.11 through 5.13). Salicylic esters are less effective at lower cost. Carbon black is the most effective additive for stabilizing against UV degradation (Table 5.14), but, of course, it limits color to opaque black; also, it may generate so much heat that it can cause thermal degradation. Zinc oxide is the most efficient inorganic UV

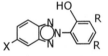 o-Hydroxy Benzophenone

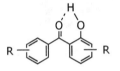 o-Hydroxyphenyl Benzotriazole

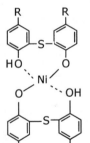 Typical Quencher

 General Type of HALS

FIGURE 5.3 Ultraviolet light stabilizers.

TABLE 5.11 Polypropylene UV Stabilization: Laboratory-Accelerated UV to 50 Percent Loss of Tensile Strength

Stabilizer	Hours
None	350
0.50 percent UV absorbers	800–2000
0.25 percent HALS	4000
0.50 percent HALS	6800

TABLE 5.12 ABS UV Stabilization: Retention of $20 \ kg/m^2$ Impact Strength After Lab-Accelerated UV Aging

Stabilizer	Hours
None	225
1 percent UV absorber	500
1 percent HALS	1225
0.5 percent UVA + 0.5 percent HALS	2000

TABLE 5.13 Polycarbonate UV Stabilization: Laboratory-Accelerated UV Aging to Yellowness Index +5

Unstabilized	700 hr
0.25 percent UV absorbers	2800 hr

TABLE 5.14 ABS Stabilization by Carbon Black Impact Strength Retained After Five Years Outdoor Weathering

Natural color	40%
Black	82%

absorber and is useful especially when combined with organic synergists, typically in HDPE and polypropylene.

5.1.4.3 Quenchers. When a polymer absorbs UV energy, it may be able to dispose of it harmlessly by intermolecular transfer to certain additives that can then carry the energy away and dispose of it harmlessly. These additives are referred to as energy *quenchers*. Organo-nickel compounds are often useful as quenchers. Carbon black probably functions partly as a quencher.

5.1.4.4 Hindered-Amine Light Stabilizers (HALS). Most UV degradation is actually photooxidation—UV-accelerated free-radical attack by atmospheric oxygen. The most recent and most popular way of stabilizing polymers against it is by addition of hindered amines to interfere with the free-radical chain reaction.

$$R_2NH + O_2 \rightarrow R_2NO^{\cdot}$$

This nitroxide radical reacts with a degrading polymer radical $R^{\cdot}$

$$R_2NO^{\cdot} + \cdot R' \rightarrow R_2NOR'$$

This reacts with another degrading polymer radical $R''^{\cdot}$

$$R_2NOR' + \cdot R'' \rightarrow R_2NO^{\cdot} + R'R''$$

This produces stable polymer $R'R''$ and regenerates the nitroxide radical to continue its work (Tables 5.11 and 5.12).

Since UV absorbers and HALS operate by different mechanisms, combined use of the two types of stabilizers offers beneficial synergism (Table 5.12).

5.1.4.5 Market Volume. Table 5.15 provides market volume information for some leading stabilizers.

TABLE 5.15   Leading UV Stabilizers

HALS	46 percent
Benzotriazoles	27 percent
Benzophenones	20 percent
Others	7 percent
Total	24,800 tons

Use in polymers	
Polypropylene	45 percent
Polyethylene	29 percent
PVC	9 percent
Engineering plastics	7 percent
Styrenics	5 percent
Others	5 percent

5.1.4.6 Prodegradants. When plastics accumulate in solid waste, it might be desirable to accelerate their UV degradation. This has been accomplished semicommercially by incorporating enough C=O groups to absorb UV energy and initiate photodegradation processes. It has also been demonstrated experimentally by adding transition metal compounds such as ferrous laurate to catalyze photooxidation of the polymer (Table 5.16).

These techniques do not destroy the polymer, but they embrittle it enough to crumble, and oxidize it enough to promote biodegradation later (Table 5.17).

TABLE 5.16 Accelerated UV Embrittlement of Polypropylene

	Time to embrittlement	
Ferrous laurate, %	Unstabilized, hr	Heat-stabilized, hr
0	118	384
0.01	0	167
0.1	0	167
1.0	0	95
2.0	0	47

TABLE 5.17 Fungus Growth[*] on Molded Plastics: Effect of UV Degradation

UV degradation before fungus test	None	4 months
High-density polyethylene	Trace	Heavy
Polystyrene	Trace	Trace
90 percent PS + 10 percent styrene/vinyl ether copolymer	Trace	Slight
50 percent PS + 50 percent styrene/vinyl ether copolymer	Trace	Moderate

*.Trace = barely noticeable, slight = 10–30% of surface, moderate = 30–60% of surface, heavy = 60–90% of surface.

5.1.5 Biostabilizers

Microorganisms such as bacteria, actinomycetes, and fungus can attack plastics, producing discoloration and degradation of mechanical and electrical properties. They thrive primarily at 20 to 30°C and high humidity, whenever they can find a source of food. Natural polymers such as cellulose and protein are a good source of food. Animal fats and vegetable oils are a good source of food; when they are used in paints, alkyds, and urethanes, these polymers are biodegradable. Synthetic polymers that contain aliphatic hydroxyl and ester groups may be a good source of food; these include polycaprolactone, polyester urethanes, and the new purposely biodegradable polylactic acid, polyhydroxybutyrate, and polyhydroxyvalerate. Fairly sensitive polymers include polyvinyl acetate, polyvinyl alcohol, and ethylene/vinyl acetate. Most other polymers are not inherently biodegradable. However, monomeric additives are often an excellent source of food and primary focus of biological attack: ester plasticizers, epoxy ester stabilizers, and natural esters used in polyurethanes and fatty ester lubricants are the most common problems. (Starch fillers have actually been used to incorporate biodegradability in plastics.) A variety of chemicals can be used to stabilize plastics against biological attack.

Testing usually begins by placing plastics samples in Petri dishes, injecting microorganisms, and observing whether they grow. Further testing may include humidity, soil burial, and other natural exposures. A major problem is that species of microorganisms vary from one geographic region to another, so it is hard to design reliable broad-spectrum laboratory tests and to recommend successful additives from one region to another.

The greatest problem is differential toxicity. Any chemical that is toxic to microorganisms will probably be toxic to macroorganisms such as ourselves. Thus, it is necessary to distinguish those additives that offer maximum toxicity toward microorganisms along with minimum toxicity toward macroorganisms, and to define the critical balance for different plastic products.

5.1.5.1 10,10´-oxy-bis(phenoxarsine) (OBPA). This (Fig. 5.4-I) is the leading commercial antimicrobial. It is very efficient, so it can be used at very low concentration (0.04 percent) and can be synergized by bis(trichloromethyl) sulfone.

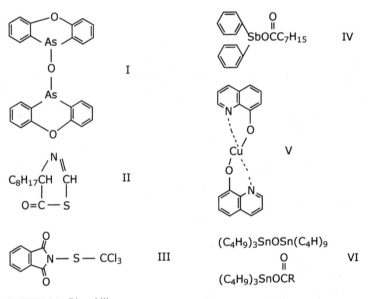

FIGURE 5.4 Biostabilizers.

5.1.5.2 2-n-octyl-4-isothiazoline-3-one. This (Fig. 5.4–II) is a newer antimicrobial that is nontoxic to humans and is used at 3 percent in vinyls and paints.

5.1.5.3 Trichloromethyl Thio Phthalimide. This (Fig. 5.4–III) is harmless to humans and is useful at 0.25 to 0.50 percent to control actinomycetes, which cause pink staining of plasticized vinyls.

5.1.5.4 Diphenyl Antimony 2-Ethylhexoate. This (Fig. 5.4–IV) is approved for use in vinyl shower curtains, wallpaper, upholstery, and rug underlay.

5.1.5.5 Copper Quinolinolate. This (Fig. 5.4–V) is relatively harmless to humans. Used at 0.5 percent, it controls mildew. Because of its deep yellow-green color, it is used mainly for military purposes.

5.1.5.6 Tributyl Tin Oxides. These (Fig. 5.4–VI) have been useful in vinyls, polyure-thanes, and marine paints. Use is decreasing because of worry about toxicity.

5.1.5.7 Copper Powder. At high loading (70 percent), copper powder has been recommended for control of fouling in marine paints.

5.1.5.8 Alkyl Amines. Alkyl amines have been grafted onto polymer surfaces in recent research to make them bactericidal.

5.1.5.9 Use in Commercial Plastics. The major use is in plasticized PVC to protect the ester plasticizers. Other wide uses are in polyester urethanes and in oil paints. Typical products include shower curtains, wall and floor coverings, carpet underlay, marine uphol-stery, awnings, refrigerator gasketing, weatherstripping, swimming pool liners, water beds, and hospital sheeting.

5.2 FILLERS AND REINFORCEMENTS

When large amounts of solid materials are finely dispersed in a polymer matrix, we call these materials *fillers* or *reinforcements*. In terms of total tonnage, these are the leading type of additives in plastics. Some of their effects are quite general. Many of their specific effects are so different that it is best to study them in four distinct classes.

1. Extender fillers
2. Reinforcing fillers
3. Reinforcing fibers
4. Specialty, or "functional" fillers

5.2.1 General Effects

Most fillers and fibers are inorganic materials of high density, polarity, modulus, melting point, refractive index, and solvent resistance, so incorporating them into organic poly-mers produces major changes in properties.

5.2.1.1 Packing. Many of these properties are proportional to the volume fraction of fillers or fibers added. Maximum packing fraction can be calculated geometrically and confirmed experimentally. For spherical particles, maximum packing fraction can go as high as 85 percent. For conventional fibers, it can go as high as 91 percent. Man-made fi-bers with rectangular or hexagonal cross sections are easy to make and theoretically can be packed neatly to approach 100 percent!

5.2.1.2 Processability. Dispersion of polar fillers and fibers in the molten polymer re-quires special care to produce interfacial wetting and shear mixing to produce dispersion. Fillers and fibers rubbing against screws and channels produce frictional heating, and they add thermal conductivity; both effects can speed the processing cycle. They do increase viscosity considerably; which makes processing more difficult, and they are so hard that abrasion of process equipment requires more frequent replacement.

5.2.1.3 Mathematical Modeling. Mathematical modeling can attempt to predict and ra-tionalize effects on properties but requires so many assumptions that it leaves quite a gap between theory and practice.

5.2.1.4 Modulus. Modulus is increased greatly, because, when flexible polymer molecules bump against the hard surface of inorganic particles, they lose much of their inherent flexibility. The effect is most pronounced for fibers in the axial direction, because, even if the polymer is willing to respond, the high-modulus fibers absolutely refuse to respond at all. *Creep Resistance* correlates with modulus, both theoretically and practically. This can bring performance of plastics much closer metals and ceramics.

5.2.1.5 Breaking Strength. Breaking strength is increased greatly by continuous fibers; when the polymer is ready to fail, the high-strength fibers absolutely are not. Short fibers may or may not increase strength somewhat, depending on stress-transfer across the fiber/polymer interface; they may actually decrease it, because the fiber ends act as stress concentrators, causing premature failure. Particulate fillers usually decrease strength due to stress concentration at sharp edges and corners of the filler particles.

5.2.1.6 Impact Strength. This is increased tremendously by continuous fibers (Fig. 5.5); they seem to distribute the shock over the entire length of the fiber so that the stress at any one point is very small. Short fibers are unpredictable; they may increase impact strength moderately or not at all, or even decrease it, their ends acting as stress concentrators. Particulate fillers almost always decrease impact strength, again due to stress concentration at their sharp edges and corners. Impact strength theory is seriously handicapped by the assumption that the same failure mechanisms operate at both low speed and high speed; it would be much better to recognize that high-speed impact failure is a completely different phenomenon that deserves its own theoretical analysis.

5.2.1.7 Friction and Abrasion Resistance. These qualities are increased by the sharp edges of filler particles and the sharp ends of fibers that protrude from the surface of the polymer matrix.

FIGURE 5.5 Unbreakable plastics.

5.2.1.8 Coefficient of Thermal Expansion (CTE). CTE is inverse to the attractive forces holding the molecules together. The weak secondary attractions between polymer molecules permit a high rate of thermal expansion, whereas the strong primary forces in inorganic materials restrict them to a much lower rate of thermal expansion. For simple extender fillers, the expansion rates of polymer and filler are simply additive, so the CTE simply decreases in proportion to volume fraction of simple extender fillers (Fig. 5.6). Reinforcing fillers are more effective, and reinforcing fibers are most effective in reducing thermal expansion, because they restrict the molecular motion of the polymer molecules. This brings plastics closer to the performance of metals and ceramics.

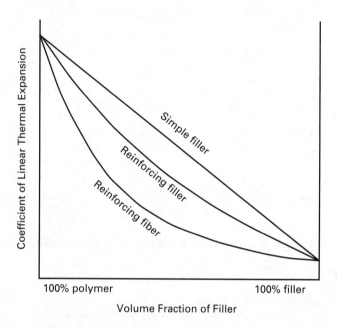

FIGURE 5.6   Effect of fillers on coefficient of thermal expansion.

5.2.1.9 Heat Deflection Temperature.   This is increased slightly in amorphous polymers, because the fillers or fibers reduce the mobility of the polymer molecules. It may be increased tremendously in crystalline polymers, because fillers and especially fibers raise the plateau of the modulus versus temperature curve just enough to extend the pass/fail limit of the standard test by hundreds of degrees (Fig. 5.7, Table 5.18). The practical significance of this obviously depends on the judgment of the product designer.

5.2.1.10 Thermal Conductivity.   The thermal conductivity of inorganic fillers and fibers is higher than organic polymers, so adding them does increase conductivity in proportion to volume fraction (Sec. 5.2.5.2).

5.2.1.11 Flame Retardance.   Flame retardance is increased somewhat, because fillers and fibers increase both viscosity and thermal conductivity (Secs. 5.2.5.3 and 5.7).

5.2.1.12 Dielectric Constant and Loss.   These are much higher in highly polar inorganic materials, so fillers and fibers generally increase them proportionally in plastics.

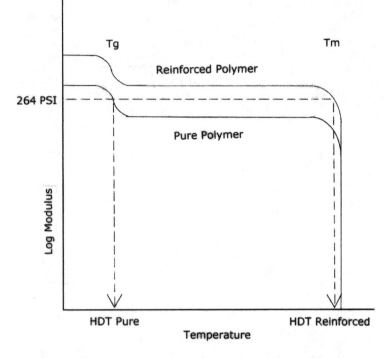

FIGURE 5.7 Effect of fillers on heat deflection temperature of crystalline polymers.

TABLE 5.18 Effect of Fillers on Heat Deflection Temperature

Polymer	Unfilled HDT, °C	Glass fiber, %	Filled HDT, °C
Acetal	110	10	160
Nylon 6	60	14	205
Nylon 66	71	13	243
PBT	67	10	200
PET	70	15	210

5.2.1.13 Opacity. Opacity results from the fact that inorganic fillers and fibers are denser than organic polymers, so the speed of light is slower, so their refractive index is higher, so light waves are scattered and dispersed as they pass through the interface. Fillers are often used to produce opacity. Conversely, to seek transparency in filled and reinforced polymers, one must either match the refractive indices of the two phases or reduce the particle size below the wavelength of visible light; both of these approaches are very difficult.

5.2.1.14 Swelling and Permeation. These are reduced, because fillers and fibers restrict free volume and mobility of the polymer matrix, making it harder for small molecules of liquids and gases to dissolve and diffuse through the polymer, and because the small molecules must permeate around the impervious particles—a "tortuous" route that further impedes permeability. On the other hand, the high polarity of most fillers and fibers may attract moisture to penetrate along their interface with the polymer, weakening stress transfer across the interface, and often plasticizing and even hydrolyzing the polymer; this is particularly noticeable in outdoor weathering.

5.2.1.15 Cost. The cost of simple extender fillers may be lower than polymers on a weight basis, but their higher density, more difficult processability, and decrease in strength properties may eliminate any overall economy. Fillers should be chosen primarily for their beneficial effects on technical properties; if they also decrease cost, this is simply an added benefit. Reinforcing fibers increase the cost of commodity plastics, but they may actually reduce the cost of some high-end engineering thermoplastics.

5.2.2 Extender Fillers

Simple inorganic particles are generally added to plastics to increase modulus, friction, and opacity, and to reduce raw material cost.

5.2.2.1 Glass Microspheres. Glass microspheres range in size from 5000 down to 4 μm and may be solid or hollow down to one tenth of solid density. Solid spheres improve melt processability and give smooth surfaces, high modulus, compressive strength, and dimensional stability. Hollow spheres are added to epoxy and other thermoset resins to produce syntactic foams for deep-sea and low-dielectric applications. For lower cost and lower performance, coal-fire fly-ash is sometimes recommended in place of costly glass spheres.

5.2.2.2 Calcium Carbonate. This is the most widely-used economic extender for polymers. Benefits commonly reported include processability, hardness, dimensional stability, whiteness, opacity, gloss, and mar resistance. Particle sizes range from 0.125 in for ground mineral grades down to submicron sizes for chemically precipitated grades that may even reinforce strength and impact strength; price is generally inverse to particle size. Calcium stearate surface treatment improves most of these properties.

5.2.2.3 Titanium Dioxide. This is the leading white pigment in coatings and is also widely used in paper and plastics. (Relative market volumes are coatings 50 percent, paper 25 percent, and plastics 25 percent.) Its high refractive index produces opacity, and its chemical and UV stability produce weather resistance. (It is important to use the rutile grade for weather resistance; the less-stable anatase grade is strictly for paints that erode gradually, producing a chalky surface that is self-cleaning, washing away easily to shed dirt and mold.)

5.2.2.4 Clays. Clays such as kaolin are finer than calcium carbonate, typically 0.2 to 10 μm, providing more reinforcement. They are used to increase the viscosity of polyester bulk molding and sheet molding compounds; to give hardness, opacity, and whiteness in vinyl flooring; and to increase heat and electrical resistance in wire and cable insulation. They are improved by calcining and by silane surface treatment. New delamination treatments to produce extremely small particle size are the basis of current developments in nanotechnology (Sec. 5.2.3.6).

5.2.2.5 Silica. Silica is a naturally occurring mineral that is ground down to particle sizes of 2 to 10 μm and used as a low-cost, stable, white filler.

5.2.2.6 Talc. Talc is a magnesium silicate mineral, often used in polypropylene to improve processing, rigidity, creep resistance, and heat deflection temperature.

5.2.3 Reinforcing Fillers

Fibers increase strength but make melt processing much more difficult. Reinforcing fillers are fine particles that permit fairly normal melt processing but do increase strength; they are often referred to as *mineral reinforcement.* When they are examined under a microscope, they are generally plate-like or fiber-like in appearance. Theoretically, the strength of reinforced plastics depends on the force required to pull a fiber out of the polymer matrix: if the fiber is embedded far enough into the matrix, it must break before pulling out. Model calculations often conclude that, when the aspect (L/D) ratio is greater than 20/1, the fiber will not pull out before it breaks. Many reinforcing fillers appear to have more than the critical aspect ratio of 20/1.

5.2.3.1 Wood Flour. Wood flour is made by controlled attrition of wood and contains microscopic cellulose fibers. It was first used in phenolic plastics to increase their strength, and it remains the basis of general-purpose phenolic moldings. It was occasionally used in other plastics and is currently gaining popularity in vinyl and other thermoplastic wood/plastic composites for processability and durability superior to wood alone. At high loadings in HDPE and PVC, these "wood/plastics composites" look like wood but are more durable, and they are finding growing use as "plastic lumber" in outdoor construction and furniture.

5.2.3.2 Wollastonite. This is a calcium silicate mineral, acicular, with aspect ratios of 3 to 20/1. It is of interest for reinforcement of strength and as a safe replacement for asbestos.

5.2.3.3 Franklin Fiber. This is a calcium sulfate crystal with aspect ratios of 60/2 μm. It is an easy-processing reinforcement but suffers from water sensitivity.

5.2.3.4 Mica. Mica is a potassium aluminum silicate mineral that occurs as flakes with aspect ratio up to 50/1. The best grades offer good processability, reinforcement, and impermeability.

5.2.3.5 Asbestos. Asbestos is a low-cost magnesium silicate mineral that occurs as very short, fine fibers of high modulus, strength, and thermal and chemical resistance. It was a popular filler until it was noticed that it collected in the lungs and caused serious health problems. Its use has been discontinued except in critical applications such as brake linings. Since then, a number of other promising short, fine fibers have been abandoned for fear that they may cause similar problems.

5.2.3.6 Nanofillers. Nanofillers are extremely fine particles, under a micron in size. The most successful ones have been made by intercalating quaternary ammonium surfactants between the layers of montmorillonite clay, followed by fluid polymer, to exfoliate them down to 1-nm platelets with aspect ratio of 1000/1. When these are dispersed in nylon at low concentrations of 2 to 10 percent, the tremendous numbers of plate-like particles can produce easy processing, high modulus and strength, heat deflection temperature, transparency, and impermeability. Typical studies report flexural modulus increased 126 percent, flexural strength increased 60 percent, HDT increased 87°C, and impermeability increased fourfold. Since they are smaller than the wavelength of visible light, they do not reduce transparency. The technology is being extended into commercial practice, including a variety of other fillers and polymers.

5.2.3.7 Carbon Black. Carbon black is made by cracking organic oils in a high-temperature furnace, producing particle sizes in the 10 to 100 nm range. The best grades, seen under an electron microscope, are clusters of particles, referred to as *high-structure*. The aromatic carbon rings are attracted to the C=C bonds in rubber and may graft to them during vulcanization. They give such high-strength reinforcement of rubber that their use is almost universal. For some reason, they do not reinforce the strength of plastics but are very useful for UV stabilization and electrical semiconductivity.

5.2.3.8 Fumed Aerosil Silica. This is produced by mixing $SiCl_4$ with steam. Here again, the particle size is in the nanometer range, and high-structure clusters give good reinforcement to silicone rubber. They also give extreme viscosity and thixotropy to liquid systems such as vinyl plastisols and epoxy resins.

5.2.4 Reinforcing Fibers

Fibers have much higher modulus and strength, and much lower thermal expansion, than bulk polymers, so dispersing them in a polymer matrix can produce an excellent increase in modulus, strength, and dimensional stability.

5.2.4.1 Glass. Continuous glass fibers are typically calcium/aluminum/boron/magnesium silicate, melt spun at 2400°F (1316°C), and 9 to 18 μm in diameter. When they are incorporated into plastics, they produce the highest modulus, strength, and impact strength ever achieved (Table 5.19). However, processing of continuous fiber is limited to specialized techniques such as filament winding, pultrusion, and compression molding. For broader application, glass fibers are chopped 1 to 2 in (~25 to 50 mm) long for sheet molding compound, 0.5 to 1 in (~13 to 25 mm) for bulk molding compound, and 0.125 to 0.5 in (~3 to 13 mm) for thermoplastic molding and extrusion. This does permit fairly conventional melt processing, but it certainly sacrifices a good portion of the potential properties, whether processors admit it or not (Table 5.20). Optimum performance depends on stress transfer between polymer and fiber, and fiber ends act negatively as stress concentrators. Furthermore, fiber breakage during melt flow severely reduces the final length of the fibers, reducing properties even further. Nevertheless, it is still possible to improve thermoplastic properties considerably by adding glass fibers, so the technique is very popular (Table 5.21).

5.2.4.2 Mineral Wool. Mineral wool is a low-cost silicate fiber spun from molten slag in steel refineries. It is widely used as thermal insulation in housing and appliances. Since its composition and structure are not well controlled, it is not comparable with chopped glass fibers; however, it is sometimes used as a partial replacement for them. Jim Walters Processed Mineral Fiber (PMF) in particular has been reported for such applications.

5.2.4.3 Specialty Fibers. Specialty fibers offer benefits, but, because they are expensive, they are only used in special high-performance products. Carbon fibers are made by pyrolyzing polyacrylonitrile, producing amorphous carbon reinforced by crystalline graphite fibrils; they offer high strength, lubricity, and electrical conductivity. Aramide fibers are aromatic polyamides; they offer low density, impact strength, vibration damping, and wear resistance.

TABLE 5.19 Maximum Properties of Reinforced Plastics

Epoxy/glass	Modulus to 5,500,000 psi
	Flexural strength to 70,000 psi
	Impact strength to 10 fpi
	HDT to 600°F
Thermoset polyester/glass	Modulus to 3,000,000 psi
	Flexural strength to 80,000 psi
	Impact strength to 30 fpi
	HDT to 500°F

TABLE 5.20 Impact Strength of Reinforced Nylon 6,6

Fiber length, inches	fpi
0.500	7.5
0.375	6.5
0.250	5.5
0.125	3.0

TABLE 5.21 Properties of Reinforced Nylon 6,6

Glass content, %	0	10	20	30	40	50	60
Flexural modulus, kpsi	410	650	850	1300	1600	2200	2800
Flexural strength, kpsi	15	20	29	38	42	47	50
Tensile strength, kpsi	12	14	19	27	31	32	33
Impact strength, fpi	0.9	0.8	1.2	2.0	2.6	2.6	2.6
HDT, °F/264 psi	150	485	485	490	500	500	500
Thermal expansion, $10^{-5}/°F$	4.5	2.7	2.3	1.8	1.4	1.0	0.9

5.2.5 Specialty, or "Functional" Fillers

Aside from their use for economics or mechanical reinforcement, a number of fillers are used to improve a variety of specific properties, as described below.

5.2.5.1 Lubricity and Abrasion Resistance. In plastic gears and bearings, these qualities are improved by adding solid powders such as brass, molybdenum sulfide, graphite, polyethylene, and especially polytetrafluoroethylene.

5.2.5.2 Thermal Conductivity. Thermal conductivity can be increased to shorten molding cycles and to avoid overheating of electrical equipment. Silver, copper, and aluminum have conductivities 1000 times that of unfilled plastics; loading them into plastics can increase conductivity considerably, in proportion to their volume fraction (Table 5.22). Beryllium oxide, boron nitride, aluminum oxide, aluminum nitride, and graphite are also quite effective.

TABLE 5.22 Thermal Conductivity of Filled Plastics

Material	Thermal conductivity, Btu/[(ft^2-hr-°F)/ft]
Silver	240
Copper	220
Aluminum	110
Steel	40
Al_2O_3	20
Epoxy + silver	4
Epoxy + aluminum	2
Epoxy + Al_2O_3	1
Epoxy alone	0.1
Air	0.01

5.2.5.3 Flame Retardance. Flame retardance is commonly produced by adding solid powders of organo-bromine, organo-chlorine, antimony oxide, and inorganic hydrates (Sec. 5.7). It is also reported that fillers and reinforcements in general can contribute to flame retardance by increasing melt viscosity and heat transfer.

5.2.5.4 Electrical Conductivity. This quality is important to bleed off static charge and to avoid electromagnetic interference (EMI) (Sec. 5.9). It can be produced by adding carbon black, graphite, and especially metallic fillers (Table 5.23). This requires particle-toparticle contact, so flakes are more efficient than simple powders, and fibers are most efficient of all (Table 5.24).

5.2.5.5 Magnetism. Magnetism can be produced by magnetic fillers such as barium ferrite. This produces moldable magnets that are nonconductive and rust resistant.

5.2.5.6 Color and Opacity. These features are, of course, produced by fillers, both inorganic and organic (Sec. 5.8), in much lower concentrations than are normally considered "fillers."

5.2.5.7 Ultraviolet Light Stabilization. UV light stabilization is produced by fillers that reflect UV, particularly aluminum flake and TiO_2, and by fillers that absorb UV radiation and reduce it to harmless wavelengths, particularly carbon black and zinc oxide (Sec. 5.1.4).

TABLE 5.23   Electrical Conductivity of Filled Polymers

Material	Log volume resistivity (Ω-cm)
Unfilled polymers	15 to 16
Graphite-filled coatings	1 to 2
Nickel-filled epoxy	0 to –2
Graphite	–3
Silver-filled epoxy	–4
Nickel	–5
Aluminum, copper, silver	–6

TABLE 5.24   Electrical Conductivity of Reinforced Plastics: 40 Percent by Weight of Fiber

Fiber	Log volume resistivity (Ω-cm)
Carbon	0
Aluminum	–1
Brass	–2
Copper	–4

5.2.5.8 Impermeability.   Impermeability (barrier performance) is produced by plate-like flakes, which increase the tortuous path that permeating molecules must seek.

5.2.5.9 Controlled Degradability.   This has been produced by use of biodegradable fillers such as starch powder. Once the filler has disappeared, the polymer crumbles, and the high surface area accelerates oxidative and biodegradation.

5.2.5.10 Carbon Nanotubes.   These are tiny hollow fibers made up of carbon atoms arranged in a hexagonal pattern, in flat sheets that roll up into seamless tubes. Diameters range from 1 to 200 nm and aspect ratios up to 10,000! Their modulus, strength, and thermal and electrical conductivity are superior to graphite and carbon fiber. Used at 1 to 5 percent in plastics, they provide very high modulus, strength, and thermal and electrical conductivity. Processing is difficult, and cost is extremely high, but researchers are optimistic about their future.

5.2.6   Technical Summary

The relative effects of fillers and reinforcements on plastics may be clarified by summarizing them in tabular form (Table 5.25). In the table, (+) means an increase in the property, (++) means a great increase, (–) means a decrease, (– –) means a great decrease, and (±) means the effect varies depending on the specific filler, fiber, polymer, or test.

TABLE 5.25 Summary of Fillers and Reinforcements

Property	Extender Fillers	Reinforcing Fillers	Short Fibers	Continuous Fibers
Melt processability	±	±	−	− −
Modulus	+	+	++	++
Creep resistance	+	+	++	++
Strength	−	+	+	++
Impact strength	−	±	±	++
Friction	±	+	+	+
Abrasion resistance	+	+	+	+
Thermal conductivity	+	+	+	+
Coeff. of thermal exp.	−	−	− −	− −
Heat deflection temp.	+	+	+	+
Flame retardance	+	+	+	+
Transparency	−	−	−	−
Cost	±	+	+	++

5.2.7 Markets

Approximate U.S. tonnage of fillers and reinforcements for plastics is about 16 billion pounds per year (Tables 5.26 and 5.27).

TABLE 5.26 Leading Fillers and Reinforcements

Substance	Millions of lb
Calcium carbonate	8500
Glass Fiber	4000
Alumina trihydrate	520
Clay	510
Titanium dioxide	490
Wollastonite	370
Talc	320
Silica	250

TABLE 5.27 Markets for Reinforced Plastics

Market	Percent
Transportation	25
Construction	21
Corrosion resistance	16
Marine	13
Electrical/electronic	9
Consumer products	7
Appliances and business machines	5
Aerospace	2
Other	2

5.3 COUPLING AGENTS

5.3.1 Polymer/Filler Interface

When inorganic fillers and fibers are dispersed in an organic polymer matrix, the interface is weakened by sharp differences in modulus, thermal expansion, polarity, chemical attraction, and chemical reactivity. This gives rise to many practical problems as listed below:

- *Dispersion* of solid fillers in the liquid matrix is difficult, slow, and incomplete.
- *Strength* is limited by premature failure at the weak interface.
- *Impact strength* is critically lowered by the weakness of the interface.
- *Thermal cycling* produces mismatch at the interface, resulting in premature failure.
- *Pigmentation* with colored fillers is inefficient and expensive.
- *Humidity* attacks the interface preferentially, causing hydrolysis and premature failure.

These problems are often solved by using coupling agents to strengthen the interface.

5.3.2 Commercial Coupling Agents

Some commercial coupling agents are illustrated in Fig. 5.8.

5.3.2.1 Dispersants. Dispersants for blending pigments into liquid systems include a wide range of anionic, nonionic, and cationic surfactants that help to wet the pigment particles and disperse them in the liquid system.

5.3.2.2 Stearic Acid. This is frequently used to coat filler particles. In calcium carbonate, it is the preferred coupling agent. Presumably the $-CO_2H$ group orients toward the filler particle, and probably reacts with it, while the $-C_{17}H_{35}$ chain penetrates into the polymer matrix.

Methacrylato Chromic Chloride

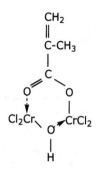

Organosilanes
Vinyl
Chloropropyl

Epoxy

Methacrylate
Primary Amine
Diamine
Mercapto
Cationic Styryl

$CH_2=CHSi(OCH_3)_3$
$ClCH_2CH_2CH_2Si)OCH_3)_3$

$\overset{O}{\triangle}CH_2OCH_2CH_2CH_2Si(OCH_3)_3$

CH_3
$CH_2=CCO_2CH_2CH_2CH_2Si(OCH_3)_3$
$H_2NCH_2CH_2CH_2Si(OCH_3)_3$
$H_2NCH_2CH_2NHCH_2CH_2CH_2Si(OCH_3)_3$
$HSCH_2CH_2CH_2Si(OCH_3)_3$
$CH_2=CHC_6H_4CH_2NHCH_2CH_2NHCH_2CH_2CH_2Si(OCH_3)_3$

Organotitanates
$(RO)_{1-4}Ti(OXR'Y)_{2-5}$

FIGURE 5.8 Commercial coupling agents.

5.3.2.3 Methacrylato Chromic Chloride. This was one of the earliest coupling agents for glass fibers in thermosetting polyesters. Presumably, the Cr-Cl groups react with the Si-OH groups on the glass surface to create Si-O-Cr bonds, while the methacrylate groups copolymerize with the styrene and unsaturated polyester during cure, producing true primary covalent bonding from glass to coupling agent to polymer matrix.

5.3.2.4 Organosilanes. These have been the leading class of coupling agents for many years. The general type structure $(RO)_3SiR'X$ contains three R = methoxy, ethoxy, or acetoxy groups that react with the Si-OH surface of glass fibers or mineral fillers to produce Si-O-SiR'X bonds to the coupling agent. In thermosetting polymers, the X group is chosen to copolymerize with them during cure, producing true covalent bonding from filler or fiber, through coupling agent, to the polymer matrix. In thermoplastics, the R'X group is chosen for similar polarity and/or hydrogen-bonding to give optimum secondary attraction to the polymer matrix.

5.3.2.5 Organotitanates. Organotitanates appear analogous to organosilanes at lower cost. Experimental results vary, but their main service appears to be as dispersing agents. More recently, zircoaluminates and other organometallics have been added to this broad family.

5.3.2.6 Fatty Esters and Amides. These have been offered as low-cost coupling agents. Their main service appears to be as dispersing agents.

5.3.2.7 Polypropylene/Acid Grafts. Grafts have been made with maleic anhydride or acrylic acid. These are useful for example in coupling talc and mica into polypropylene.

5.3.3 Application Techniques

Most often, the filler and coupling agent are slurried in water, the coupling agent hydrolyzes Si-OR + $H_2O \rightarrow$ Si-OH and reacts with the Si-OH on the filler surface to form Si-O-Si filler-to-coupling agent bonds. Alternatively, the coupling agent can be dry-blended with the filler by tumbling at controlled humidity, but this requires more skill. *In-situ* treatment is based on adding both filler and coupling agent to the molten polymer; this wastes some coupling agent, but it eliminates the cost of a separate pretreatment step. Perhaps most promising is vapor-phase application of the coupling agent, as is done in the manufacture of glass fiber, where the coupling agent acts first to protect the glass fibers and later to bond them to the polymer matrix.

5.3.4 Coupling Agent Theory

There are a variety of theories to explain the action of coupling agents. *Primary covalent bonding* is quite probable when organosilanes are used in thermosetting plastics. *Secondary attractions* are more likely when coupling agents are used in thermoplastics. *Interpenetrating polymer networks (IPNs)* may be postulated when the organosilane extends into the polymer matrix. This concept may be broadened to consider the formation of a gradually *modulated interphase* rather than a sharp monomolecular interface. *Morphology* of the coupling agent layer has been studied by electron microscopy, and some researchers believe the coupling agent accumulates in tiny hills on the glass fiber surface, and these hills act like the pins in a mechanical assembly, preventing the fiber from pulling out of the polymer matrix. Coupling agent may create *friction* between the fiber and the polymer matrix, increasing the stress needed to pull the fibers out of the matrix.

5.3.5 Practical Benefits of Coupling Agents

Coupling agent theory and salesmanship are often more optimistic than practical results. It is important to be realistic about practical benefits outlined below.

5.3.5.1 Protection of Glass Fibers. This is definitely produced by vapor phase treatment with organosilanes. They coat the glass fibers and keep them from scratching and weakening each other. Since they provide coupling later, this is a double benefit.

5.3.5.2 Dispersion. Dispersion of fillers in liquid systems is faster and more complete. This optimizes mechanical and optical properties. It is most evident in coatings, but it is also important in plastics.

5.3.5.3 Lower Viscosity. Lower viscosity is often reported in liquid systems. Again, this is most evident in coatings, but it may also be important in plastics.

5.3.5.4 Mold Wear. This is a distinct problem, especially in glass-fiber-reinforced thermoplastics. Compounders often claim that their proprietary coupling agents reduce mold wear, but molders remain rather skeptical.

5.3.5.5 Strength. The strength of filled and reinforced plastics, resistance to thermal cycling, and efficiency of flame retardants may be improved by coupling agents, but commercial secrecy and salesmanship have obscured any objective benefits.

5.3.5.6 Wet Aging Resistance. This may be the most demonstrable benefit in reinforced plastics. When glass-fiber-reinforced thermoset polyesters are immersed in water, particularly boiling water, they lose strength rapidly. When they are properly prepared with organosilane coupling agents, their strength retention is markedly improved. The coupling agent strengthens the interfacial bond between glass fiber and polyester matrix, and it provides hydrophobicity to repel moisture and keep it from intruding into the interface.

5.4 PLASTICIZERS

Plasticizers are most commonly liquid esters of low volatility, which are blended into rigid thermoplastic polymers to make them soft and flexible. Most are esters of phthalic, phosphoric, and adipic acids. Major use is in polyvinyl chloride (PVC) elastoplastics. Another major use, rarely mentioned in the literature, is the addition of hydrocarbon oils to rubber to improve processability. Plasticizers are also used to improve melt processability and toughness of rigid plastics such as cellulose esters and ethers, and they are used in a variety of specialized applications. In some cases, they perform dual functions such as thermal stabilization or flame retardance. This gives the individual processor the ability to tailor properties for each product.

5.4.1 Compatibility

The first requirement of a plasticizer is that it should be compatible with the polymer; that is, it should be completely miscible and remain permanently in the polymer. In general, this requires that polymer and plasticizer should have solubility parameters within one to two units of each other. Strong mutual hydrogen-bonding is a second factor favoring compatibility. And low molecular weight also favors miscibility.

When a plasticizer meets these requirements, it is actually a solvent for the polymer and can speed melt processing. When solubility parameters are a little farther apart, the plasticizer must be heated to dissolve the polymer; on cooling to room temperature, it forms a gel, which may favor optimum balance of flexibility and strength. When solubility parameters are still farther apart, it is incompatible unless used in combination with a compatible "primary" plasticizer and is referred to as a "secondary" plasticizer.

5.4.2 Efficiency

There are 600 commercial plasticizers. Some are very efficient in softening the polymer. Others are much less efficient and are used for other reasons. Efficiency is measured by plotting modulus versus plasticizer concentration and comparing the plots for different plasticizers. It is reported either (1) as the amount of plasticizer required to reach a standard modulus or (2) as the modulus produced by a standard amount of plasticizer.

Three factors determine plasticizer efficiency. (1) A flexible plasticizer molecule, containing long $(CH_2)_n$ chains, is more efficient in flexibilizing the polymer; rigid units such as benzene rings are much less efficient. (2) Low polarity and low hydrogen-bonding provide less attraction between polymer and plasticizer (borderline compatibility), permitting easier bonding-debonding equilibrium, which favors molecular mobility and flexibility. (3) Lower-molecular-weight plasticizers have more mobility and favor greater flexibility.

5.4.3 Effects on Other Properties

5.4.3.1 Processability. Plasticizers are often used to increase melt flow during processing, particularly in rubber, cellulosic plastics, and coatings. In vinyl plastisol technology, powdered resin is slurried in nonsolvent plasticizer at room temperature for easy processing in low-cost equipment and then heated to dissolve and cooled to gel into the final product.

5.4.3.2 Mechanical Properties. Addition of plasticizer lowers hardness and strength and increases elongation and creep. It increases adhesion and friction.

5.4.3.3 Acoustic Properties. Addition of plasticizer produces damping of vibration (auditory frequencies) and thus reduces the resulting noise.

5.4.3.4 Thermal Properties. Addition of plasticizer slides the modulus-versus-temperature curve horizontally toward lower and lower temperatures, in proportion to the amount of plasticizer added. This lowers the stiffening temperature of flexible plastics. It also lowers the softening temperature of rigid plastics, which limits maximum use temperature but improves melt processability.

When epoxidized fatty esters are used to synergize thermal stabilization of PVC (Sec. 5.1.3.2), these are liquids which therefore also act as secondary plasticizers.

Flammability is also a function of the plasticizer used. When phthalates are added to PVC, they reduce the total chlorine content, which reduces flame retardance. On the other hand, when organic phosphate esters are used as plasticizers, they actually increase flame-retardance (Sec. 5.7). This makes them particularly important in cellulosic plastics.

5.4.3.5 Electrical Properties. Addition of plasticizer increases mobility of ionic impurities, which lowers electrical resistance (increases conductivity). Plasticizer increases the mobility of polar groups in the polymer, which increases dielectric constant and shifts the loss peak to higher frequencies.

5.4.3.6 Chemical Properties. Plasticizer increases free volume and molecular mobility, which increases permeability of small molecules of liquids and gases. Toxicity of some plasticizers to small laboratory animals has led to considerable environmental and political concern, but they have never caused any trouble to human beings; on the contrary, the inventor of plasticized PVC lived healthily to the age of 101.

5.4.4 Permanence

Plasticizer failure can shorten the useful life of the plastic product. Failure can be either loss of plasticizer (it is "fugitive") or degradation of the plasticizer ("aging").

5.4.4.1 Fugitivity. Plasticizer may escape for a number of reasons. *Volatility* is gradual evaporation at higher temperatures, such as fogging of auto windshields in hot weather, and shrinkage and embrittlement of electrical wire and cable insulation. *Extraction* of plasticizer by gasoline, motor oil, solvents, or even soapy water causes stiffening of jackets and gloves in dry cleaning and of shower curtains and baby pants in soapy water. *Exu-*

dation of oily droplets on the surface is generally evidence of poor compatibility. *Migration* of plasticizer into an adjacent unplasticized surface can ruin the surface. *Humidity* can reduce compatibility in borderline cases.

Whenever a plasticizer is too fugitive for the application, the problem is easily solved by using a higher-molecular-weight plasticizer. Thus, changing from dioctyl to didecyl phthalate is enough to prevent windshield fogging or to get a higher temperature rating from Underwriters Laboratories. Going to ditridecyl phthalate achieves an even higher temperature rating. Going from dialkyl phthalate to trialkyl trimellitate is even better. In serious problems, the best solution is to use polymeric plasticizers—linear polyesters of MW 1000 to 3000. Ultimately, high-molecular-weight polar elastomers such as ethylene/vinyl acetate, chlorinated polyethylene, butadiene/acrylonitrile rubber, and polyurethane can provide absolute permanence.

5.4.4.2 Aging. Plasticizers have less age resistance than polymers such as PVC. Thermal oxidation occurs in electrical wire and cable insulation and is commonly retarded by addition of antioxidants. Microbiological attack on plasticizers is commonly retarded by addition of biostabilizers (Sec. 5.1.5).

5.4.5 Special Effects

5.4.5.1 Air Molecules. Air molecules have lower molecular weight and higher mobility than any liquid plasticizer. Consequently, foaming a polymer (Sec. 5.6) can produce much greater softness than conventional plasticization.

5.4.5.2 Water Absorption. Water absorption in polar hydrogen-bonding polymers can provide powerful but unreliable plasticization, depending on humidity and immersion. The effect is most often noticed in nylon. It also occurs in cellulosic plastics.

5.4.5.3 Antiplasticization. This is an unexpected stiffening effect observed when small amounts of plasticizer are added to PVC, acrylic, and polycarbonate plastics. Probably, these stiff polymer molecules do not pack neatly on cooling from the melt, and the residual free volume leaves some molecular mobility. Adding a little plasticizer may fill the free volume or may give the polymer molecules enough mobility to pack more neatly, eliminating free volume and thus increasing their rigidity.

5.4.5.4 Internal Plasticization. This is produced by building the plasticizing structure right into the polymer molecule. In *homopolymers* such as polyolefins and vinyl and acrylic esters, lengthening the side-chains up to 6 to 12 CH_2 groups pushes the polymer main-chains farther apart ("tent-pole effect"), creating free volume that increases the mobility of the polymer molecules, changing rigid plastics into flexible, rubbery, and even tacky polymers.

In some polymer families, *copolymerization* with more flexible comonomer units is very effective in producing the amount of flexibility desired. Major commercial examples are ethylene/propylene rubber, styrene/butadiene plastics and latex paint, vinyl chloride/vinyl acetate plastics, vinyl acetate/acrylic ester latex paints, and methyl methacrylate/acrylic ester plastics and latex paints.

In all these cases, the resulting plasticization is absolutely permanent. However, there are two drawbacks. (1) A plasticizing structure bound into the polymer molecule is less efficient than an independent monomeric structure. (2) The product can be made by the polymer manufacturer only on a large scale, not by the independent processor on a daily basis for individual customer needs.

5.4.6 Commercial Practice

5.4.6.1 Polyvinyl Chloride. Polyvinyl chloride is about 85 percent amorphous regions that can absorb plasticizer and produce flexibility, and 15 percent crystalline regions that remain firmly bonded and retain strength and creep resistance. This is an ideal balance, which accounts for the fact that PVC uses about 80 percent of the total plasticizer market.

Most of this is phthalic esters. Di(2-ethylhexyl) phthalate was the first and remains the leading material, and it is called "general-purpose." For higher permanence, diisodecyl and ditridecyl phthalates are common; trimellitates are more permanent but more expensive. Surprisingly, linear alkyl phthalates are superior to branched (isoalkyl) phthalates for both efficiency and permanence. For still greater permanence, liquid polyesters of MW 1000 to 3000 are used, but they are less efficient and more expensive. For flooring, butyl benzyl phthalate is used for fast fusion and stain resistance. For low-temperature flexibility, dioctyl adipate is most common; azelate and sebacate esters are more efficient but more expensive. For greater flame-retardance, tricresyl phosphate is common, and, for efficiency as well, trialkyl phosphates are excellent but more expensive. And, of course, for thermal stabilization synergists, epoxidized soybean and linseed oils and tall oil esters are standard practice (Sec. 5.1.3.2).

5.4.6.2 Rubber. Rubber is commonly plasticized by hydrocarbon oils from the petrochemical industry, particularly to improve melt flow in calendering and extrusion. This practice rarely finds its way into the technical literature or market analyses.

5.4.6.3 Cellulose Esters. These and ethyl cellulose are rigid plastics that are lightly plasticized to improve melt processability and to impart some impact resistance and flexibility in sheet form. Lower alkyl phthalates are most common, while phosphates are used for flame retardance.

5.4.6.4 Ultra-High-Molecular-Weight Polyethylene. This is extremely difficult to process. Addition of paraffin wax or low-molecular-weight polyethylene wax permits conventional melt processing.

5.4.6.5 Polystyrene. Polystyrene melt processability is improved by addition of a small amount of mineral oil or butyl oleate to increase melt flow.

5.4.6.6 Nylon. Nylon's melting point is reduced by addition of toluene sulfonamide for use as a laminating adhesive in multilayer textile construction.

5.4.6.7 Market Analysis. See Table 5.28 for a plasticizer market analysis.

TABLE 5.28 Plasticizer Market Analysis

Plasticizer	%	Plasticizer	%
Phthalates	69	Phosphates	3
Epoxy esters	9	Trimellitates	2
Glutarates	4	Azelates	1
Adipates	4	Others	5
Polyesters	3		

5.5 LUBRICANTS AND PROCESSING AIDS

The term *lubricants* is used by a variety of specialists to cover a range of chemicals that are added to plastics to improve a variety of performance characteristics in processing or in final properties. When they are used to improve processing, the field may expand to include other types of processing aids. When they are used to improve final properties, the field may expand to include other properties beyond lubrication.

Most of these characteristics are difficult to quantify, difficult to correlate between laboratory and commercial practice, and depend very much on the individual processor, processing equipment, process, and type of product. Theoretical understanding is limited, problems are solved by trial and error, and secrecy is dominant, making this one of the most obscure and difficult areas in the entire field of plastics additives.

5.5.1 Lubricants in Processing

Lubricants may improve processing in a number of different ways. These are discussed below.

When pellets or powder fail to flow smoothly down the feed hopper, they tend to bridge or arc. This may require mechanical tapping to keep them flowing, and coating the pellets or powder particles with a small amount of lubricant may help them to flow more freely.

Screw feed produces frictional heating, which speeds fusion of pellets or powder into a fluid mass. When processors want to delay fusion, usually in processing rigid vinyl, lubricants that reduce friction between the resin and the steel screw and channel may be useful to reduce frictional heating and thus delay fusion.

Lubricants that are liquid and miscible in the molten polymer can act as plasticizers to decrease melt viscosity, decrease power requirements, and give faster flow.

Shear degradation occurs when polymer molecules are very stiff and/or very entangled, and friction from the screw forces them to flow anyway. Then, the molecules may simply break and degrade to lower molecular weight. Lubricants can prevent shear degradation, either by plasticizing the disentanglement of the polymer molecules or by reducing the friction between the polymer and the steel screw.

Steel molds, extrusion dies, and rolls have high surface energy; molten polymers have much lower surface energy. This makes the polymer coat and stick to the steel equipment. Lubricants that interface between polymer and steel can help to release the plastic product from the mold, die, or roll. This can speed processing and avoid distortion of the product when it is pulled away from the steel.

Die swell following extrusion can be reduced by addition of lubricants. They may function by plasticizing the polymer, permitting molecules to disentangle and accept the shape of the die, or they may function by reducing adhesion between the polymer extrudate and the steel die, thus releasing the product without distortion.

Gloss may be hurt by friction between polymer and extrusion die, producing microscopic roughening of the surface. A lubricant that reduces adhesion and friction between polymer and steel can thus reduce roughening and increase surface gloss.

More serious adhesion and friction between polymer and extrusion die can cause melt fracture in the form of sharkskin, matte, or orange peel. All of these can be reduced by a lubricant that reduces adhesion and friction at the interface.

"Slip agents" are added to polymers for film production. They exude to the surface of the film and lubricate it, keeping the films from sticking to each other and making it easier for them to slide over each other and to separate from each other. This is helpful in processing, both for speeding production and to avoid distortion.

5.5.2 Lubricants in Final Products

Lubricants may be added to a polymer to produce lubricity in the final product. The effects are discussed below.

Antiblocking agents ("flatting agents") may be added to polymers for film production. They roughen the surface of the films, reducing surface contact and thus reducing adhesion between them, making it easier to separate them in use. This is particularly important in packaging film and bags.

Gears and bearings made from metals have high polar attraction, which produces high friction and requires excessive use of contaminating lubricants such as motor oil and grease. By comparison, organic plastics have much lower attraction and may require much less use of contaminating lubricants. Even here, however, they may not give perfect self-lubrication. Thus, it may be very helpful to add lubricants that come to the surface of the gears and bearings and make them completely self-lubricating.

5.5.3 Lubricant Chemistry

It is common to distinguish between "internal" and "external" lubricants. The worst definition is based on *who* adds the lubricant: if the supplier added it, it comes to the processor as an "internal" lubricant; if the processor must add it himself, it is an "external" lubricant. A semimeaningful distinction: an *internal* lubricant is compounded into the polymer; an *external* lubricant is applied to the surface of the mold. The proper distinction is based on miscibility/immiscibility of the additive in the polymer: an *internal* lubricant is miscible with the polymer and acts like a small amount of plasticizer to increase molecular flexibility and mobility, and to help the disentanglement and flow of the polymer molecules in the melt. An *external* lubricant is immiscible in the polymer and tends to exude to the surface of the plastic and form an abhesive (nonstick) interface between the polymer and the steel process equipment, or it tends to come to the surface of the finished plastic product and give it continual lubricity in the use of the product.

Some lubricants are purely internal; some are purely external. Many exhibit a balance of internal/external activity, which depends both on the lubricant and on the polymer in which it is used. This depends on the relative polarity of the polymer and the lubricant. If they have very similar polarity, the lubricant is completely soluble in the polymer and acts as an internal lubricant. If they have very different polarity, the lubricant is insoluble in the polymer and exudes to the surface, where it acts as an external lubricant. If the lubricant molecule contains a nonpolar segment, -----, and a polar segment, *****, the relative length or strength of these segments

$$\text{-------}***$$

$$\text{-----}*****$$

$$\text{---}*******$$

can be adjusted to balance the miscibility of the lubricant in the polymer and thus adjust its relative internal versus external performance.

5.5.4 Major Market Classes of Lubricants

The major market classes of lubricants are as follows:

- Fatty esters (mono, di, and triglycerides and straight-chain esters)
- Fatty amide waxes (erucic, oleic, ethylene-bis-stearamide)

- Metal stearates (Ca, Zn, Mg)
- Paraffin waxes (MW 300-600)
- Polyethylene waxes (MW 1,000-10,000)
- Fatty alcohols and acids
- Lecithin
- Organic phosphate esters
- Fluoropolymers
- Silicones
- Antiblocking agents (diatomaceous earth, talc, calcium carbonate, silica, silicates)

5.5.5 Major Use in Polymers

5.5.5.1 Market Distribution. A breakdown of lubricant market distribution is provided in Table 5.29.

TABLE 5.29 Use of Lubricants in Plastics

Lubricant	%
Polyvinyl chloride	44
Styrenics	12
Polyolefins	7
Other thermoplastics	4
Thermosets	33
Total lubricants/total plastics	0.19

5.5.5.2 Polyvinyl Chloride. Average lubricant concentration 0.6 percent. Paraffin and polyethylene waxes; fatty alcohols, acids, esters, and amides; calcium stearate; diatomaceous earth or fumed silica for antiblocking.

5.5.5.3 Polystyrene and ABS. Average lubricant concentration 0.35 percent. One to 3 percent mineral oil, butyl oleate or stearate as internal lubricants for melt flow; 0.1 to 0.5 percent stearic acid, Zn stearate, or fatty amides for mold release. Polyethylene wax, montan wax, ethylene-bis-stearamide, or glycerol monostearate for ABS.

5.5.5.4 Polyolefins. Average lubricant concentration 0.09 percent. Two to 5 percent polyethylene or polypropylene wax as internal lubricant; 0.25 to 1.0 percent ethylene-bis-stearamide as internal/external lubricant; 0.1 percent erucamide and oleamide as slip agents; 1 to 3 percent glycerol monostearate or hydrolyzed montan wax as external lubricant/slip agent/antifog coating; diatomaceous earth, talc, calcium carbonate, silica, and silicates as anti-blocking agents; Ca and Zn stearates; 0.05 to 0.1 percent fluoroelastomer (VDF/HFP) in LLDPE extrusion to prevent melt fracture, die buildup, and gel particles; also high-MW silicone for the same reasons.

5.5.5.5 Polycarbonate.   0.1 to 1.2 percent Ca stearate, stearyl stearate, or montan ester as mold release.

5.5.5.6 Acetal.   0.5 percent amide wax.

5.5.5.7 Polyethylene Terephthalate.   Silica, talc, zeolites, and calcium carbonate for antiblocking.

5.5.5.8 Unsaturated Polyester.   Butcher wax, Ca and Zn stearates, fatty amides, lecithin, vegetable oil, or organic phosphates as mold release agents.

5.5.5.9 Polyurethanes.   0.5 to 2.0 percent polar ester such as montan for mold release. 0.2 to 1.0 percent for calendaring.

5.5.5.10 Nylon.   Silica, talc, zeolites, and calcium carbonate for antiblocking.

5.5.5.11 Phenolics.   Ca and Zn stearates or fatty amides for mold release.

5.5.5.12 Common Mold Release Agents.   Fatty esters and amides, wax, polyethylene, oxidized polyolefins, PTFE powder, silicones, metal stearates, talc, mica.

5.5.6   Other Processing Aids

5.5.6.1 Acrylics for Rigid PVC.   Melt processability of rigid PVC is greatly improved by addition of several percent of an acrylic processing aid, generally a graft of PMMA onto a polyacrylic ester such as ethyl or butyl acrylate. Improvements reported include increased friction producing faster fusion, increased melt viscosity, reduced jetting in injection molding, increased shear producing faster mixing, increased melt strength and elongation for blow molding and thermoforming, die swell for blow molding, and increased surface gloss. Typical concentrations are 0.5-3.0 percent for injection and blow molding; 1-5 percent for extrusion, calendering, and thermoforming; and 5-12 percent for foaming.

5.5.6.2 Branched Polypropylene.   This is added to regular polypropylene to produce high melt strength for blow molding and foam production.

5.5.6.3 Viscosity Depressants.   PVC plastisols are made more fluid by 0.5 to 2.0 percent of ethoxylated fatty esters. Filled plastisols use silanes and titanates.

5.5.6.4 Thixotropes.   Latex, plastisols, epoxies, and polyesters are thickened by precipitated or fumed silica as well as clays. Latexes are also thickened by cellulose ethers, Na CMC, and Na polyacrylate.

5.5.6.5 Surfactants.   Foaming of plastisols and polyurethanes is improved and stabilized by liquid silicones.

5.5.6.6 Emulsifiers.   Emulsifiers are used for emulsion polymerization and for compounding additives into polymer latexes. They may be anionics such as sodium soaps, sulfonates, sulfates, and phosphated organic molecules; cationics such as quaternary ammonium organics; or nonioncs, generally polyethoxylated organic acids and alcohols. Anionics and nonionics are often used together, anionics for more powerful emulsification and nonionics for resistance to acid and hard water.

5.5.6.7 Dispersants.   Dispersants are used to separate pigment particles from each other, distribute them uniformly in the liquid polymer system, and stabilize the dispersed state.

5.5.6.8 Bonding Agents.   Phenolic resins are added to plastisols for adhesion to metals. Isocyanates are added to plastisols for adhesion to fabrics.

5.5.6.9 Nucleating Agents.   Crystallization of polymers from the melt occurs in two successive stages: *nucleation* (formation of tiny incipient crystal lattices), and *growth* (addition of layers of polymer molecules to the surface of the nucleus or growing crystal. Low temperature (near T_g) promotes the formation of nuclei, but high temperature (near T_m) favors growth. To solve this dichotomy and promote formation of small uniform crystallites, compounders add nucleating agents, which then permit processors to crystallize finely, uniformly, and rapidly at higher temperatures.

Nucleating agents give faster crystallization, finer crystal size, and higher percent crystallinity, hardness, modulus, strength, elongation, impact strength, heat deflection temperature, and transparency.

Two types of materials nucleate crystallization: tiny filler particles with sharp edges and corners, and high-MP polymers that crystallize first and then nucleate crystallization of the lower-MP major polymer.

5.5.6.9.1 Polyethylene.   Crystallization is fast but may be coarse. One percent of K stearate reduces spherulite size.

5.5.6.9.2 Polypropylene.   Coarse crystallinity makes it opaque. To reduce crystal size and increase transparency, 0.1 to 0.3 percent of substituted sorbitols are added as "clarifiers." Dibenzylidene sorbitol, and especially its methyl derivatives, are the best. One percent of Na benzoate is used for injection molded food and drug packaging. Other metal benzoates, t-butyl benzoate and naphthoate, are also mentioned.

5.5.6.9.3 Polyethylene Terephthalate.   Crystallization is difficult and slow. Particularly for injection molding, 0.5 percent of 3-μm metal oxide, metal salt, pigment, or other minerals, and ionomer are mentioned. For thermoformed food trays, 1 to 3 percent of low-MW polyolefin.

5.5.6.9.4 Nylon.   Mentioned most often is 0.1 percent of silica. Others mentioned include Na benzoate, minerals, MoS_2, FeS, TiO_2, talc, Na phenyl phosphinate, and higher-MP polymers.

5.5.7   Effects of Lubricants on Other Final Properties

Fogging is caused by condensation of moisture on clear film, forming droplets that scatter light. It is a problem in food packaging for refrigerator and freezer display and in agriculture for greenhouses. It is worst on hydrophobic films of low surface energy, particularly polyolefins. Some lubricants also serve as antifogging agents. Glycerol monoesters, sorbitol esters, and polyethoxylates of fatty acids and alcohols are designed for semimiscibility, to exude to the surface of the film. Once there, they provide a hydrophilic surface of higher surface energy. Moisture that condenses on the surface is able to wet and spread across the surface in a uniform film, which remains transparent.

As mentioned earlier, when lubricants smooth the surface of extruded products, this can increase gloss.

One negative effect is often encountered in the use of silicones. While they are very abhesive and prevent polymer from sticking to steel process equipment, they also prevent later postprocessing such as printing, decorating, and adhesive bonding. It has been suggested that this is a problem mainly in dimethyl silicones and may be avoided by using mixed methyl/aryl silicones.

5.6 FOAMING AGENTS

5.6.1 General Foaming Process

There are a great variety of foaming processes. They all involve these three basic steps:

1. *Liquid state.* This can be a polymer melt, low-MW thermosetting prepolymer, vinyl plastisol, or latex.
2. *Blow bubbles.* This can be done with a permanent gas, hollow microspheres, a volatile liquid, or a chemical reaction that produces a gas.
3. *Solidify.* A thermoplastic melt can be cooled, a thermosetting prepolymer can be cured, a plastisol can be gelled, and a latex can be coagulated and/or evaporated.

5.6.2 Physical Foaming Agents

5.6.2.1 Permanent Gases. *Air* can be whipped into liquid plastisol or used to froth liquid polyurethane formulations. In hot processing, where air could cause oxidative degradation, *nitrogen* alone is used; typically, it is compressed into molten polymer during molding or extrusion to make thermoplastic structural foam.

5.6.2.2 Hollow Microspheres. These may be glass or plastic (polyolefin, VDC/AN, phenolic). They may contain air or a volatile liquid. They can be stirred into a liquid system, such as epoxy, which is then cured. This is interesting for deep-sea and low-dielectric products.

5.6.2.3 Volatile Liquids. These are chilled and mixed into the liquid polymer system at room temperature. They provide temporary plasticization. Then, the heat of processing volatilizes them, plasticization ceases, and the expanding gas bubbles produce the foam.

5.6.2.3.1 Pentane. Pentane (BP 34°C) is 10 percent soluble in polystyrene. It is low in cost and nontoxic but somewhat flammable. Isopentanes, butanes, and hexanes, and mixtures of these, are often used.

5.6.2.3.2 Fluorocarbons. Fluorocarbons were used in rigid polyurethane foam and sometimes in flexible polyurethane and polystyrene as well. They are nonflammable and nontoxic and give excellent foaming and thermal insulation. Unfortunately, it was discovered that they accumulate in the stratosphere, destroy the ozone layer, which protects us from UV radiation, and contribute to global warming. By international agreement, they are being phased out in stages until they have been completely eliminated. The industry is still searching for suitable substitutes.

5.6.2.3.3 Supercritical CO_2. This is liquefied at high pressure, forced into the molten polymer, and then allowed to expand and form gas bubbles. This process has been successful in thermoplastic structural foams.

5.6.3 Chemical Foaming Agents

5.6.3.1 Sodium Bicarbonate + Citric Aid. The mixed powders are dry-coated onto foamable polystyrene beads. When the beads are steamed to foam them, the powders dissolve and react to form CO_2. The CO_2 is not the primary foaming agent; it is a nucleating agent for the volatilizing pentane, serving to produce smaller and more uniform pentane bubbles and resulting foam.

5.6.3.2 Isocyanate + Water. Flexible polyurethane foam is made primarily by the reaction of excess isocyanate with a stoichiometric amount of water during the polymerization/cure reaction. This releases CO_2 and foams the polymer as it forms.

5.6.3.3 Azo Compounds. Commercial chemical foaming agents are all organic nitrogen compounds of the general formula R-N-N-R. During the heat of melt processing, they decompose, liberating nitrogen and other gases. Their two most critical properties are (1) *decomposition temperature* and (2) *gas yield* in ml from 1 g of solid foaming agent. The leading commercial materials may be arranged by decomposition temperature (shown in Table 5.30).

TABLE 5.30 Chemical Foaming Agents

Abbreviation	Decomposition temperature, °C, in Plastics and **Pure**	Gas yield in ml/g at STP
TSH	**105**–132	115
OBSH	**149**–190	125
AZDN (ABFA)	110–**200**–235	220
TSSC	193–**235**	146
5PT	**232**–288	190

5.6.3.3.1 TSH: p-Toluene Sulfonyl Hydrazide. This is used in low-temperature processing such as plastisol and epoxy.

5.6.3.3.2 OBSH: Oxy-bis(Benzene Sulfonyl Hydrazide). This is used in fairly low-temperature processing such as LDPE, EVA, PVC, epoxy; about 5 percent of the market.

5.6.3.3.3 AZDN [Azo-Di(Carbonamide)] or ABFA [Azo-bis(Formamide)]. Catalysts permit use at lower temperatures; coarse particle size permits use at higher temperatures. It is widely used in commodity thermoplastics and even some engineering thermoplastics; about 90 percent of the market.

5.6.3.3.4 TSSC: p-Toluene Sulfonyl Semicarbazide. Somewhat better for engineering thermoplastics.

5.6.3.3.5 5PT: 5-Phenyl Tetrazole. For engineering thermoplastics.

5.6.3.3.6 Polyphenylene Sulfoxide (300 to 400°C). This has been recommended for high-temperature thermoplastics such as fluoropolymers, polyphenylene ether, polyphenylene sulfide, polyether ketones, liquid crystal polymers, nylons, and polyetherimide.

5.6.3.3.7 Why So Many Azo Compounds?   Theoretically, the gas must be released at a critical melt viscosity that permits cell expansion but is strong enough to prevent bursting of cell walls. Each polymer and each process reaches this critical viscosity at a different temperature.

5.6.3.3.8 Concentrations Used in Processing.   The amount of chemical foaming agent used depends mainly on the process (see Table 5.31).

TABLE 5.31   Concentrations of Foaming Agents Used in Processing

Process	%
Soft plastisol	2–4
Extrusion of structural foam	0.2–1.0
Injection molding of structural foam	0.3–0.5
Elimination of sink marks in injection molding	0.05–0.1

5.6.4   Combinations of Foaming Agents

Processors have always had some interest in using combinations of foaming agents, such as combinations of volatile hydrocarbons or combinations of fluorocarbons. The effort to replace fluorocarbons has led to much greater study of such combinations, between classes as well as within classes of foaming agents. Typical combinations are as follows:

- Air + isocyanate/water
- Hydrocarbon + carbon dioxide
- Volatile liquid + chemical foaming agent

5.7 FLAME RETARDANTS

Conventional materials (wood, fabric, paper, paint) burn. In a fire, steel melts, and concrete crumbles. This causes much damage to property and loss of life. People have had to cope with this for thousands of years, and they simply accept it. When new synthetic polymers burn in a fire, people are very upset, and they demand that we remedy the problem.

Aside from simple burning, fires produce several related problems that may be even more serious. When fires produce smoke, people caught in the fire may be unable to see the escape route. Indoor fires in a limited air supply produce carbon monoxide, and fires involving halogens will produce halogen and hydrogen halide gases, whose toxicity may rob people of the will to escape. Sophisticated electronic equipment can be corroded by acid combustion gases, resulting in huge replacement costs. Thermoplastic ceilings may melt and drip hot molten burning plastic on firemen trying to fight the fire.

5.7.1   Testing

Testing is a major problem. Many experts consider all laboratory-scale tests to be unrealistic and irrelevant. Large-scale tests require larger batches of material than are available in research and development. Real fires occur under such a variety of conditions that they are very difficult to correlate with laboratory data. And legal problems have arisen from laymen's misunderstanding of technical terminology, greatly limiting our ability to record and communicate test data.

Laboratory testing has gone through a series of fads: burning/nonburning, rate of burning, oxygen index, radiant panel, char formation, smoke generation, toxicity of combustion gases, calorimetry, and thermal degradation. Practical testing has been governed by Underwriters Laboratory pass/fail, building industry tunnel tests, insurance company room-corner tests, and review of actual fires.

5.7.2 Flammability of Plastics

For simple laboratory evaluation, *oxygen index* is the percentage of oxygen in air that is just sufficient to support burning (see Table 5.32).

TABLE 5.32 Oxygen Index of Common Polymers

Polymer	Oxygen index, %	Polymer	Oxygen index, %
Polyoxymethylene	15	Nylon	22
Polyethylene	17	(Wood	24)
Polypropylene	17	Polyphenylene ether	26
Polymethyl methacrylate	17	Polycarbonate	26
Cellulose acetate	17	Polysulfone	30
Polystyrene	18	Silicone	30
ABS	18	Polyvinylidene fluoride	44
Phenol-formaldehyde	18	Polyvinyl chloride	45
Epoxy	18	Polyvinylidene chloride	60
Polyesters	20	Polytetrafluoroethylene	95

Normal air is 21 percent oxygen. Polymers with an oxygen index below 21 will burn. Polymers above 21 can be burned in a strong flame but may self-extinguish if left alone.

Thus, it is possible to choose a polymer to meet most flame-retardant requirements. However, in terms of other properties, especially economics, it is more common to use additives to increase the flame retardance of the commonly used polymer.

5.7.3 Flame Retardants

There are certain ingredients that definitely increase flame retardance: organic phosphorus, organic bromine, organic chlorine, antimony oxide as synergist for the halogens, and water of hydration.

5.7.3.1 Organic Phosphorus. This is the most flame-retardant element in the periodic table. It is generally used as phosphate ester plasticizers or built into polyols for polyesters and polyurethanes. It is fairly expensive but reduces flammability very efficiently, without creating smoke or toxic corrosive gases. (Elemental red phosphorus may be used as a flame retardant, but its inherent flammability makes it tricky.)

5.7.3.2 Organic Bromine. This is an efficient flame retardant that can be greatly synergized by addition of antimony trioxide. Since aliphatic bromine is too unstable for plastic processing, preferred compounds are polybrominated diphenyl ethers for thermoplastics, and tetrabromo bisphenol A and tetrabromophthalic anhydride for epoxies and polyesters. In a fire, it does produce smoke and toxic corrosive gases, so this must be considered in specific applications. In Europe, environmental concerns may limit the use of bromine.

5.7.3.3 Organic Chlorine. This is a good flame retardant, less expensive but less efficient than bromine. It too is greatly synergized by antimony trioxide. A particularly popular compound is Dechlorane Plus, the adduct of hexachlorocyclopentadiene with cyclooctadiene. Chlorinated paraffins are also widely used. Like bromine, chlorine may be limited by problems with smoke, toxic corrosive gases, and environmental concerns.

5.7.3.4 Water of Hydration. A fireman uses water to put out a fire. A plastics chemist can do the same. Alumina trihydrate $Al(OH)_3$ is a low-cost filler that contains 35 percent water, which it loses above 205°C. In a fire, evaporation of water removes so much heat that it cools the plastic below the temperature at which it can burn and then surrounds it with water vapor, which excludes atmospheric oxygen and thus chokes out the fire without creating any smoke or toxic corrosive gases. Its main problem is that it loses water during high-temperature plastic processing, so it is useful mainly in low-temperature processes such as polyethylene, vinyl plastisols, epoxies, and thermosetting polyesters.

Magnesium hydroxide $Mg(OH)_2$ contains 31 percent water, which it loses at above 320°C. This is stable enough for many plastic processes and provides just as good flame retardance as alumina trihydrate, so it has become very popular in applications such as polypropylene.

Both of these compounds must be used at very high concentrations to provide enough water to put out a fire (typically 30 to 75 percent), at which they cause brittleness, so compounders often add rubber to counteract the brittleness and then add more flame retardant to protect the rubber.

5.7.3.5 Other Flame Retardants. Other retardants frequently mentioned in the literature include zinc borate, ammonium molybdate, other metal oxides, and melamine derivatives. One particularly intriguing approach is intumescence, a mixture of a "carbonific" such as pentaerythritol, an acid catalyst such as an amine phosphate, and a "spumific" such as an amine or melamine, which react to form a foamed char that insulates the plastic against the heat of the fire and also acts as a barrier to exclude atmospheric oxygen. Recent research interest also includes nanoclay fillers, which are believed to trap polymer molecules between their platelets, preventing degradation and thus inhibiting burning.

5.7.3.6 Method of Use. Compounders add flame retardants to the finished polymer; these "additive" flame-retardants are 87 percent of the market. The other 13 percent of "reactive" flame-retardants are built into the polymer molecule, either during polymerization or in the curing agents added to thermosetting plastics.

5.7.4 Markets

The total world consumption of flame retardants is over 2 billion pounds per year. Half is alumina trihydrate, one-quarter is organic bromine, with organic phosphorus, organic chlorine, and antimony oxide in smaller amounts.

The largest use is in thermosets, second in PVC, and third in styrenics, with smaller amounts in polyolefins and engineering thermoplastics. In terms of applications, electrical and electronics, transportation, and construction use the largest amounts.

5.8 COLORANTS

One-third of all plastics are used in natural uncolored form, and the other two-thirds are colored, using on the average about 1 percent of colorant.

5.8.1 Color Theory

When white light shines on a plastic material, it may be absorbed, reflected, and/or transmitted. Absorption removes certain wavelengths; those that remain are the color we see. Opaque pigments, insoluble in the polymer, reflect an opaque color to the eye. Transparent organic dyes, soluble in the polymer, transmit a "see-through" color to the eye.

Solar radiation is a normal distribution of wavelengths, from shortwave ultraviolet, through the visible spectrum (rainbow) from blue (4000 A = 400 nm) to green to yellow to red (7000 A = 700 nm) and on into the infrared. Colored materials absorb some of these wavelengths and leave the others that we see; each colored material has a characteristic absorption spectrum that defines it in this way.

Color scientists often use a color sphere to characterize a particular color (Fig. 5.9). The horizontal circumference of the sphere is called the *hue*, and it represents the rainbow spectrum from purple to blue to green to yellow to orange to red around the rim of the circle. A vertical axis through the sphere is called the *lightness* or *brightness*, and it ranges from black at the south pole, through dark to light grays through the sphere, to white at the north pole. A radius from the lightness/brightness axis, out to the hue circumference, is called the *saturation* or *purity* or *chroma*, and it ranges from pure bright colors near the circumference to pastel or grayish colors toward the center of the sphere. Thus, any color may be specified by its location within this three-dimensional color sphere.

More recently, color scientists have developed the much more complex CIE chromaticity diagram (Fig. 5.10) and computerized it so they can calculate precisely the matching of an experimental color with a desired model color. More practically, color technologists have assembled thousands of shades of color chips, arranging and organizing them and giving them code numbers for convenient reference. One of the best known is the Munsell Book of Color.

5.8.2 Chemical Classification

Colorants are generally classified as inorganic or organic. Generally, inorganic colors are coarser particles of lower coloring efficiency, stabler to heat, light, and chemical environment, and less expensive per pound. But many of the brightest colors are made from metals whose toxicity is causing health concerns, both in manufacturing and in the ultimate environment. Organic colors are finer particles of higher coloring efficiency, less stable, and more expensive per pound (although not necessarily in the low concentrations required to produce the desired color); they are generally nontoxic, but they must be monitored for possible impurities that could cause problems.

Colorants may also be classified as *dyes* or *pigments*. Dyes are either soluble in the polymer, or particles so fine (<2 μm) that they do not scatter and reflect light, so they give transparent colors. They are mainly organic compounds and very efficient colorants, but they tend to be unstable and extractable. Pigments are either insoluble particles or, less often, dyes bonded onto the surface of insoluble particles, producing opaque colors; they

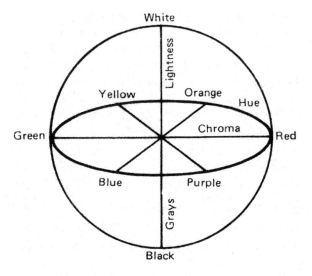

FIGURE 5.9   Color sphere.

may be inorganic or organic. In the plastics industry, pigments are commonly used to produce opaque colors; dyes are used only as specialties where a transparent color is particularly desired.

5.8.3   Major Classes of Colorants in Plastics

5.8.3.1 Inorganics

- Titanium dioxide is used in tremendous tonnage (270 million pounds per year) both as opaque pigment and as white colorant in plastics. This is 90 percent of the inorganic pigment usage in the plastics industry.
- Iron oxides (10 million pounds per year) produce a broad range of reds, yellows, browns, and blacks, which are stable, nontoxic, and low in cost but rather dull in appearance.
- Cadmiums offer a range of yellow to red colors that are bright, fairly expensive, and raise health problems.
- Chromiums and molybdenums offer a range of yellow, orange, and green colors that are attractive but raise health problems.
- Iron blues are attractive and inexpensive.
- Cobalt blues are attractive and expensive.

5.8.3.2 Organic

Carbon black is often listed as the leading "organic" colorant (85 million pounds per year), but most of this is used for reinforcement, stabilization, conductivity, and other purposes. Actually, nigrosine dye is a much more effective black colorant.

Generally, the conjugated unsaturation of aromatic carbonyl and nitrogen compounds produces a wide range of useful colors (Fig. 5.11). For example, carbazoles, perylenes,

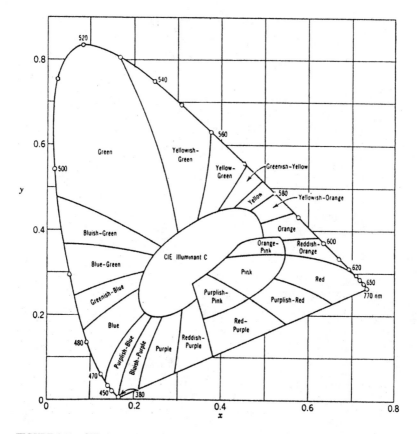

FIGURE 5.10   CIE chromaticity diagram.

and quinacridones produce reds; disazos, isoindolines, and isoindolinones produce reds to yellows; anilines produce orange; monoazos produce orange to yellows; anthraquinones, diarylides, and nickel azos produce yellows; and phthalocyanines produce greens to blues.

Overall, there is a trend to give up inorganics of suspected toxicity and replace them by organics, but the organics must be chosen carefully to retain the heat and light stability required in processing and using plastics.

5.8.4   Criteria in Choosing Colorants

A typical checklist includes dispersability, rheology, plate-out, thermal stability, appearance, light fastness, weathering, migration, and toxicity in both processing and use, particularly in leaching from solid waste.

5.8.5   Market Analysis

In worldwide tonnage, titanium dioxide is about 8 billion pounds. In plastics, inorganic white pigments were 72 percent of the total market, inorganic colored pigments 8 percent,

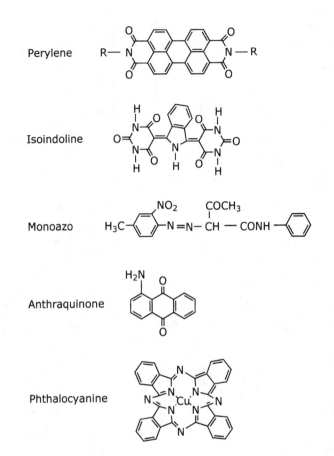

Perylene

Isoindoline

Monoazo

Anthraquinone

Phthalocyanine

FIGURE 5.11   Organic colorants.

carbon black 13 percent, organic pigments 5 percent, and dyes 2 percent. But recent trends have probably favored the replacement of inorganic by organic colorants.

Prices for inorganic colorants are mostly $1 to $3 per pound, and $3 to $30 for organic colorants.

5.8.6   Compounding Techniques

5.8.6.1 Powdered Color Pigments.   The primary particles are individual crystals. These are firmly bonded into tight clumps called *aggregates*. These are further bonded into loose clumps called *agglomerates*. It takes skill and energy to disperse these into molten plastics, and this is best done by experts. The average compounder/processor may waste much time looking for the optimum technique.

5.8.6.2 Colored Compound.   Processors can buy the plastic compound already precolored. This is commonly done with specialty plastics. It is expensive and leaves the processor with inventory problems.

5.8.6.3 Color Concentrate (Masterbatch). Expert compounders disperse colorants at 20 to 60 percent concentration in a carrier polymer, using high shear to break down agglomerates and produce uniform dispersion of maximum coloring efficiency. This color concentrate is used by processors, simply blending it with virgin (natural color) resin ("letdown with natural"). Typical ratios of concentrate/natural are 1/20 to 1/100. This technique is low in cost, does not create inventory problems, and is most commonly used with commodity resins.

5.8.6.4 Liquid Color. The colorant is predispersed in a liquid carrier, hopefully compatible with the resin. It is metered into the base of the hopper or the beginning of the screw in extrusion or injection molding and blends uniformly with the resin by the time it reaches the exit from the screw. Although originally billed as a universal technique, it has rather found applications in certain processes where it is the optimum technique.

5.8.6.5 Color Infusion. This immerses the finished plastic product in a hot aqueous dispersion of colorant + dispersant. In several minutes, the color diffuses into the plastic product, giving it permanent coloration. The length of time determines the depth of the color.

5.8.7 Special Colorants

Fluorescent colors are used to produce brighter reds and yellows. Phosphorescent colors are used to produce brighter yellows-greens-blues. Pearlescent colors combine internal and external reflections; they are made by techniques such as coating titanium dioxide on mica.

Metallic flakes are added to colorants to give them a metallic sheen. Aluminum flakes give a silvery sheen and also improve UV stability and impermeability. Bronze flakes can be formulated into a range of colors from green to red to gold.

5.8.8 Fluorescent Whiteners

Most polymers tend to form conjugated unsaturation during aging, absorbing blue light from the visible spectrum and therefore turning somewhat yellow. One way to mask this is to add fluorescent whiteners.

These are primarily bis-benzoxazoles, triazines and triazoles of phenyl coumarins, and bis-styryl biphenyls (Fig. 5.12). They absorb invisible UV light, dispose of part of the en-

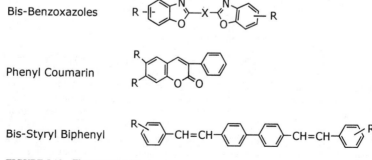

FIGURE 5.12 Fluorescent whiteners.

ergy, and re-emit the rest as visible light at the blue-violet end of the visible spectrum. This neutralizes the yellowness in the polymer, and emits a brilliant white. They are frequently used in polyolefins, polystyrene, ABS, PVC, polycarbonate, and polyurethanes. Concentrations are typically 0.01 to 0.1 percent.

5.8.9 Reference Texts

The major texts in the field are T. C. Patton's *Pigment Handbook* (John Wiley & Sons, now out of print) and F. W. Billmeyer's *Textbook of Color Science* (Wiley-Interscience).

5.9 ANTISTATS

When two materials are in contact with each other, electrons migrate across the interface. When they are separated, some electrons may be caught on the wrong side, producing a static charge.

Conventional structural materials are conductive enough to bleed off the charge to ground. Organic polymers are nonconductors, and may hold the charge for a long time.

The charge on plastics may develop during separation from the mold or roll, from friction during manufacture or use, or simply from evaporation of water from the surface. This sometimes causes problems in processing, particularly in handling thin films and fibers. It causes a much greater range of problems in the use of the product: collection of dust; unsightly packaging; cling and discomfort of clothing and upholstery; shock; occasional dust explosions; oxygen hazard in hospitals; "noise" in sound recordings and photography and magnetic tapes and discs, computer chips, military electronics; and electromagnetic interference (EMI) of electronic equipment in general.

These are arranged more or less in order of increasing need for static dissipation. They are generally classified in terms of electrical resistance. For example, over 10^{12} Ω-cm is nonconductive insulation, 10^{10-12} is antistatic, 10^{6-10} is statically dissipative, 10^{2-6} is slightly conductive, 10^{1} is EMI shielding, $10^{0 \text{ to} -3}$ is semiconductive, and $10^{-3 \text{ to} -5}$ is metallic conductivity.

Various techniques are used to minimize these problems. In manufacturing, it is possible to ionize the air and thus neutralize static charges. In textile manufacturing, it is common to humidify the air to make fiber surfaces more conductive. Organic additives can make plastics fairly conductive to dissipate a static charge. In more extreme cases, high loading with carbon black makes rubber and plasticized PVC fairly conductive. And loading with carbon fibers and metallic fillers (particularly aluminum flakes and fibers) makes plastics conductive for EMI shielding.

5.9.1 Mechanisms of Antistatic Action

When organic antistats are used to reduce static charge on plastics, several theories are offered to explain their action. Most commonly, it is assumed that the additive is polar enough to exude to the surface of the plastic, where it absorbs moisture from the air, permitting ionic impurities to conduct current electrolytically. The most effective antistats actually contain ionic groups that are free to migrate and conduct. Some theorists believe that simple passage of water vapor over the surface of the plastic may be enough to carry away the static charge. From a different point of view, static charge is created by friction; the antistat acts as a surface lubricant, reducing friction and therefore reducing the buildup of a static charge.

5.9.2 Commercial Antistats

Quaternary ammonium soaps, $R_4N^+ X^-$, have the most powerful antistatic action. Unfortunately, they tend to decompose in high-temperature processing, so they are sometimes post-applied as a 1 to 2 percent solution. They also encounter objections from the FDA.

Ethoxylated amines, $RNH(CH_2CH_2O)_nH$, approach quaternary ammonium soaps, both in effectiveness and in problems.

Ethoxylated esters, $RCO_2(CH_2CH_2)_nOH$, are the most widely used class. By balancing the organic acid portion (R) against the polyoxyethylene portion, it is possible to control polarity and therefore semicompatibility and rate of migration to the surface of the plastic, thus making it self-renewable over the lifetime of the product. They adsorb water to the surface, making it conductive and lubricating it to reduce friction. They are usually non-toxic and stable enough for melt processing. Ethoxylated alcohols, $RO(CH_2CH_2O)_nH$, are also used.

Glycerol mono- and di-esters perform fairly similarly to ethoxylated esters and are used for this reason. Being derived from natural products, they are easily acceptable to FDA.

Organic phosphate esters are also reported in similar use.

More recently, alkali sulfonates have been reported in PS and PVC.

5.9.3 Use in Commercial Plastics

LDPE typically uses 0.05 to 1.0 percent, HDPE 0.2 to 0.3, PP 0.5, and PS 2 to 4 percent. Rigid PVC uses 1 to 2 percent, and plasticized PVC 2 to 5 percent.

5.9.4 Test Methods

5.9.4.1 Dust Attraction. Dust attraction is the oldest and crudest method. The technician rubs the plastic sample against his clothing, and then lowers it toward a dish of dust, and notes the height at which the dust jumps up to the charged plastic. A more sophisticated test uses a sooty flame to deposit soot on the plastic, and then measures the amount of soot collected.

5.9.4.2 Surface Conductivity. Determining the surface conductivity of the plastic sample is a popular, simple measurement that is often assumed to correlate with antistatic behavior. Practical proof would be more reassuring.

5.9.4.3 Electrostatic Decay. A high static charge is applied to the sample electrically. Then the rate of decay is measured instrumentally.

In all these methods, relative humidity is the most treacherous variable that must be considered. This can produce a 10^4 range in electrical resistivity over the normal range of ambient humidity.

5.9.5 Market Analysis

Ethoxylated fatty amines are 48 percent of the market, aliphatic sulfonates 25 percent, fatty acid esters 16 percent, quaternary ammonium compounds 2 percent, others 9 percent.

For use in individual plastic materials, styrenics used 39 percent of the market, LDPE/LLDPE 20 percent, HDPE 13 percent, PVC 12 percent, PP 11 percent, and others 5 percent.

5.10 ORGANIC PEROXIDES

The O:O bond in peroxides is quite unstable.

$$RO:OR \rightarrow RO^{\cdot} + {}^{\cdot}OR$$

While they are difficult to make, ship, store, and handle, the radicals they produce are very useful in vinyl free-radical polymerization, cure of unsaturated polyesters, cross-linking of thermoplastics, grafting, and compatibilization of polymer blends.

Stability/reactivity is generally measured by the temperature at which the half-life of the peroxide is 10 hr, called "the ten-hour half-life temperature." It is controlled by choice of the R groups and accelerated by raising the temperature, radiation, catalysis by cobalt soaps, amines, or redox reaction with reducing agents.

5.10.1 Major Classes of Peroxides

Major classes of peroxides are shown in Fig. 5.13.

5.10.1.1 Acyl Peroxides

- Benzoyl peroxide is the longest-established and most widely used. With 10-hr half-life at 71°C, it is used to polymerize styrene and other vinyl polymers, for medium-temperature cure of unsaturated polyesters, and for a variety of grafting and compatibilization reactions.
- Lauroyl peroxide (61°C) is used for somewhat higher reactivity. Its aliphatic structure also gives lighter color in polymers than can be obtained with the aromatic benzoyl peroxide.
- Decanoyl peroxide is used to a lesser extent.

5.10.1.2 Ketone Peroxides.
MEK peroxide is used for room-temperature cure of unsaturated polyesters. Typical concentrations are 0.5 to 2.0 percent. It may be catalyzed by 0.05 to 0.3 percent of cobalt naphthenate and also further catalyzed by amines.

5.10.1.3 Peroxy Esters.
These cover a wide range of reactivities and uses.

- t-butyl peroxy pivalate is a typical low-temperature peroxide.
- t-butyl peroctoate (70°C) is a typical medium-temperature peroxide.
- t-butyl perbenzoate (101°C) is a typical high-temperature peroxide, useful in polymerizing styrene and in cure of BMC and SMC unsaturated polyesters.

5.10.1.4 Dialkyl Peroxides.
These are typically high-temperature materials.

- Dicumyl peroxide (dicup or DCP) (104°C) is useful in cross-linking LDPE, EVA, EPR, and EPDM.
- Di-t-butyl peroxide (125°C) is useful for the high-temperature finish of styrene polymerization to reduce residual styrene monomer content and thus improve modulus, HDT, taste, and odor.

5.10.1.5 Hydroperoxides.
Hydroperoxides such as cumene hydroperoxide are used primarily for low-temperature emulsion polymerization of butadiene to make "cold rubber."

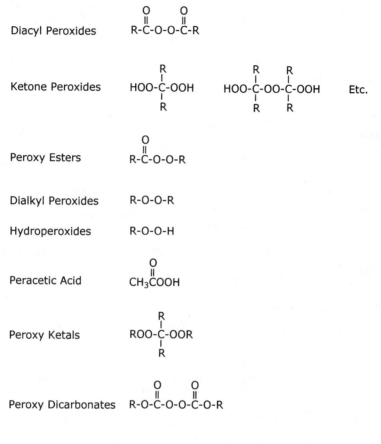

Diacyl Peroxides	$R-\overset{\overset{O}{\|\|}}{C}-O-O-\overset{\overset{O}{\|\|}}{C}-R$
Ketone Peroxides	$HOO-\overset{\overset{R}{\|}}{\underset{R}{C}}-OOH \qquad HOO-\overset{\overset{R}{\|}}{\underset{R}{C}}-OO-\overset{\overset{R}{\|}}{\underset{R}{C}}-OOH \qquad$ Etc.
Peroxy Esters	$R-\overset{\overset{O}{\|\|}}{C}-O-O-R$
Dialkyl Peroxides	$R-O-O-R$
Hydroperoxides	$R-O-O-H$
Peracetic Acid	$CH_3\overset{\overset{O}{\|\|}}{C}OOH$
Peroxy Ketals	$ROO-\overset{\overset{R}{\|}}{\underset{R}{C}}-OOR$
Peroxy Dicarbonates	$R-O-\overset{\overset{O}{\|\|}}{C}-O-O-\overset{\overset{O}{\|\|}}{C}-O-R$
Dimethyl Di-t-Butyl Peroxy Hexyne	$CH_3-\overset{\overset{CH_3}{\|}}{\underset{C_4H_9}{C}}-C\equiv C-\overset{\overset{CH_3}{\|}}{\underset{C_4H_9}{C}}-CH_3$

FIGURE 5.13 Peroxides.

They are catalyzed by redox systems consisting of reducing sugars, iron soaps, and phosphates.

5.10.1.6 Peracetic Acid (CH₃CO₃H).   This is used mainly in epoxidizing olefins such as soybean oil for vinyl stabilizers and in synthesis of aliphatic epoxy resins.

5.10.1.7 Peroxyketals.   These are particularly popular for cure of BMC and SMC unsaturated polyesters.

5.10.1.8 Peroxydicarbonates.   These are the least stable class, often too unstable for shipment, in which case they must be synthesized where they are going to be used. They have become the leading initiator for vinyl chloride polymerization.

5.10.1.9 2,5-Dimethyl-2,5-di-t-Butyl Peroxy Hexyne-3. This, having a 10-hr half-life temperature 135°C, was developed specifically for the higher temperature processing required in the cross-linking of HDPE.

5.10.2 Safety Precautions

- Peroxides are often sold and handled in dilute form to reduce the danger of explosive reaction.
- Heat should be avoided in shipping, storing, and handling. Some must be kept refrigerated.
- Friction should be avoided, both in packaging (no tight-fitting or screwed lids) and in handling and processing.
- Organic impurities should not be allowed to contaminate peroxides, as the attack on them would be exothermic and kick off the entire batch.
- Peroxides and catalysts/promoters should never be mixed together in the pure state. The batch of polymer should be divided in half. Then, the peroxide is added to one half and the catalysts/promoters to the other half. In this diluted form, the two halves can then be mixed to start the reaction.

5.10.3 U.S. Market Analysis

Table 5.33 provides an analysis of peroxides used in plastics in the United States.

TABLE 5.33 Peroxides Used in Plastics

Type	Amount, millions of pounds
MEK peroxide	11
Benzoyl peroxide	10
Peroxy esters	9
Dialkyl peroxides	5
Others	5

5.11 POLYMER BLENDS

Polymer properties may be improved by adding a second polymer. This is not a general rule, but a number of polymer blends have offered so much improvement that 40 percent of commercial plastics are now based on polymer blends.

5.11.1 Miscibility

If the two polymers are completely miscible down to the molecular and even segmental level, they form a single homogeneous phase, and properties are generally proportional to the ratio of the two polymers in the blend. Miscibility depends on equal polarity or mutual attraction such as hydrogen bonding or cocrystallization. This is not very common, but there are several important examples of such completely miscible blends. It gives the compounder simple straightforward control over balance of properties.

5.11.2   Practical Compatibility

Most polymer pairs are too dissimilar for complete miscibility. They reject each other and separate into two or more phases. Generally, the major polymer forms a continuous matrix phase and retains most of its original properties. The minor polymer separates into dispersed "domains" and may affect certain specific properties. When the domains are extremely fine, sensitive properties may detect the phase separation, but many practical properties may resemble homogeneous single-phase blends. When the domains are larger in size, they will have distinct effects on certain specific properties; when these effects are beneficial, the blend is described as theoretically *immiscible* but practically *compatible*. When the domains are too coarse, most properties will suffer, and the blend is described as *incompatible*.

5.11.3   Interface/Interphase

In multiphase polyblends, a critical factor is the interface between the phases. If the two polymers reject each other and separate into phases, they are likely to reject each other at the interface as well. Such a weak interface will fail under stress, and most properties will suffer. Thus, most polymer blends are practically incompatible. Yet, most successful commercial polyblends are multiphase systems. This means that there must be a mechanism to strengthen the interface.

In some cases, the two polymers have some partial miscibility, so the interface is not a sharp separation of one polymer from the other but, rather, a modulating solution of the two polymers in each other, offering a gradual *interphase* rather than a sharp *interface*. Such an interphase can modulate properties gradually from one phase to the other and thus reduce the stress.

5.11.4   Compatibilizers

In most cases, it is necessary to add a compatibilizing agent to strengthen the interface. In basic research, the preferred compatibilizing agent is a diblock copolymer, with one block soluble in one phase and the other block soluble in the other phase. The block copolymer tends to locate at the interface. This creates primary covalent bonds across the interface and thus strengthens it. In commercial practice, the compatibilizing agent is usually a graft copolymer, with a backbone soluble in one phase and side-chains soluble in the other phase; this is not as theoretically satisfying, but it is usually easier and more economical to make and appears to work perfectly well in practice. In some cases, the graft copolymer is made separately and then added to the polyblend during compounding; in other cases, it may be formed directly during compounding by reactive processing.

5.11.5   Effect of Polyblend Ratio on Polyblend Properties

When two polymers are blended in ratios from 100/0 to 0/100, and the effect on properties is measured, we may observe one of four types of behavior (Fig. 5.14).

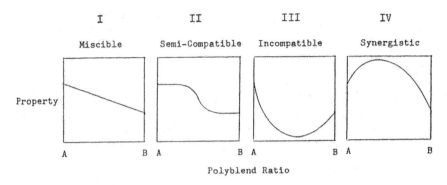

FIGURE 5.14 Properties vs. polymer/polymer ratio in a polyblend.

5.11.5.1 Type I. If the two polymers are completely miscible down to the molecular level and form a single homogeneous phase, properties are generally proportional to the ratio of the two polymers in the blend. Even if the two polymers are immiscible and form fine phase separation, many property tests are relatively insensitive to fine-phase separation and may still show such "homogeneous behavior." Practically, this is useful to compounders who want the ability to produce a spectrum of balance of properties at low cost.

5.11.5.2 Type II. When two polymers are immiscible and form two separate phases, the major polymer will form the continuous matrix phase and retain most of its original properties, while the minor polymer will form finely dispersed domains and contribute certain specific properties. Thus, high A/B ratios will have properties similar to poly-A, and high B/A ratios will have properties similar to poly-B. Obviously, at fairly equal ratios of A and B, there will be a phase inversion with a rapid change of properties from one plateau to the other.

This explains the two leading uses of polymer blends. (1) When rigid plastics suffer from brittleness, dispersion of fine rubbery domains in the rigid matrix can add great impact strength with little sacrifice of rigidity. (2) Rubber molecules must be tied together to give them strength, creep resistance, and insolubility; while this is usually done by thermoset vulcanization, it can also be done by dispersion of fine rigid thermoplastic domains, either glassy or crystalline, to form thermoplastic elastomers.

5.11.5.3 Type III. When two polymers are immiscible and separate into two phases, there may be so little attraction between them that the interface between the phases is extremely weak and will fail under stress. This is most often seen in ultimate tensile strength and ultimate elongation. In most products, this would be labeled "incompatibility." However, there are occasional examples where such behavior is actually beneficial. For example, adding an immiscible polymer may decrease melt viscosity and thus improve melt processing. Or it may decrease breaking strength, producing a package that is easier to open and therefore more customer friendly. Thus, it is safer to label Type III behavior "U-shaped" or "trough-shaped," rather than simply incompatible.

5.11.5.4 Type IV. Once in a while, the polymer blend may exhibit properties greater than either of the individual polymers, a major synergistic improvement in practical utility. The leading example of this phenomenon is the use of finely dispersed rubbery domains to increase the impact strength of a brittle glassy matrix polymer. Commodity examples are

high-impact polystyrene, ABS, rigid PVC, and high-impact polypropylene; more specialized examples are toughened epoxy resins and super-tough nylon.

5.11.5.5 Modulus vs. Temperature.   When a rigid polymer and a rubbery polymer are
completely miscible, blending them in rigid/rubbery ratios from 100/0 to 0/100 simply
shifts the *log modulus versus temperature* curve horizontally along the temperature axis
(Fig. 5.15, Type I). This makes it easy for the processor to adjust balance of properties to
suit the individual customers' needs.

On the other hand, when the two polymers separate into separate phases, this adds an
intermediate plateau to the original curves (Fig. 5.15, Type II). Here, the height of the in-

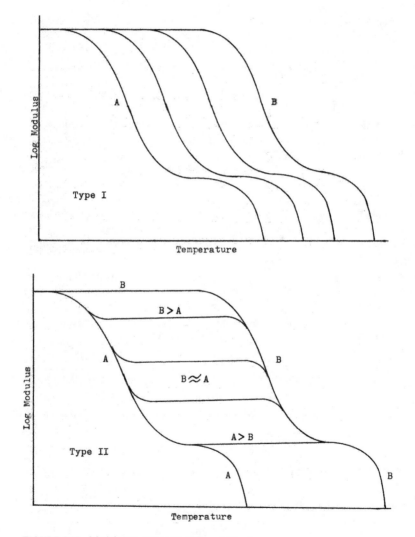

FIGURE 5.15   Modulus vs. temperature for polyblends.

termediate modulus plateau is proportional to the ratio of the two polymers in the blend, and the useful temperature range extends from the glass transition of the rubber to the glass transition or melting point of the rigid polymer. This explains the successful use of immiscible, compatible polymer blends to make both high-impact rigid plastics and thermoplastic elastomers.

5.11.6 Major Commercial Polyblends

- *Low-density polyethylene* is added to linear-low-density polyethylene to produce non-Newtonian shear sensitivity needed for blown-film production.
- *EPDM rubber* is added to high-density polyethylene and polypropylene to provide environmental stress-crack resistance. It is grafted into polypropylene to increase low-temperature impact strength. It is grafted with maleic anhydride (maleated) and then grafted onto nylon to increase its impact strength. Recent news releases suggest that this technique is also being applied to other engineering thermoplastics.
- *Polybutadiene* is grafted into polystyrene and SAN to produce high-impact polystyrene and ABS.
- *Nitrile rubber* is used to increase impact strength of epoxy resins. It is made as a low-molecular-weight liquid oligomer with carboxy end-groups (CTBN) and used as a curing agent for the epoxy resin.
- *Polystyrene* is added to polyphenylene ether to improve melt processability and decrease cost. When impact styrene is used, it also increases impact strength.
- *ABS* is added to rigid polyvinyl chloride to increase melt processability and impact strength. It is added to polycarbonate to increase melt processability and environmental stress-crack resistance and to decrease notch sensitivity and cost.
- *Polytetrafluoroethylene* is added to acetal, polycarbonate, and nylon to decrease friction and increase abrasion resistance.
- *Polyvinyl acetate* is added as low-profile resin in unsaturated thermosetting polyester to decrease shrinkage and prevent reinforcing fibers from protruding. It thus gives improved surface.
- *Polyethyl acrylate* is added to rigid polyvinyl chloride to improve melt processability. It is compatibilized by grafting polymethyl methacrylate onto it.
- *Polybutyl acrylate* is added to rigid polyvinyl chloride to increase impact strength. It is compatibilized by grafting polymethyl methacrylate onto it.
- *Polyethylene terephthalate and polybutylene terephthalate* are added to polycarbonate to provide environmental stress-crack resistance.
- *Nylon* is added to high-density polyethylene to make it impermeable for use in gasoline tanks.

5.11.7 Other Uses of Polyblending

- *Melt flow* may be increased by adding a more fluid polymer, or sometimes an immiscible polymer.
- *Melt strength and elasticity* may be increased by adding a high-MW polymer or one with long-chain branching.

- *Modulus* of rubber may be increased by adding a miscible rigid plastic. Modulus of a rigid plastic may be decreased by adding a miscible rubber to act as a polymeric plasticizer.
- *Strength* of rubber can be increased by adding a rigid plastic.
- *Abrasion resistance* can be increased by adding PTFE powder or by blending with polyurethane rubber.
- *Heat deflection temperature* can be increased by blending with a miscible polymer of higher HDT.
- *Flame retardance* can be increased by blending with a halogenated polymer such as PVC.

5.12 MISCELLANEOUS ADDITIVES

5.12.1 Polymerization, Cross-Linking, and Curing Agents

These additives create reactive (unstable) systems, so they are usually added by the processor just before final plastic processing. In most cases, the chemistry of these systems is very precise and sophisticated, so these additives are best specified by the polymer producer, not casually chosen and used by the average compounder/processor. They are best described according to the polymer system in which they are used. (See also Chap. 3.)

5.12.1.1 Cross-Linking of Thermoplastics (Polyethylene, Saturated Elastomers, Acrylic Ester Polymers). These are most often cross-linked by peroxides, choosing the peroxide according to the processing temperature. Peroxide forms free radicals, which abstract less-stable hydrogen atoms from the polymer, leaving polymer radicals. When two polymer radicals join, this forms a cross-link. The degree of cross-linking is low—not enough to cause rigidity, but enough to improve strength, creep-resistance, hot strength, and insolubility. Thus, polyethylene is cross-linked by 2 to 10 percent of peroxide during reactive extrusion to form piping and wire and cable insulation. Conversely, polypropylene radicals tend to cleave rather than cross-link, so peroxide is used to decrease melt viscosity for easier processability.

Several companies have experimented with the use of vinyl organosilanes as cross-linking agents. For example, 2 percent of vinyl trimethoxy silane is first activated by 0.1 percent of dicumyl peroxide and grafted onto the thermoplastic polymer. The system is kept dry to stabilize the methoxy groups. After melt processing, the solid product is exposed to moisture to hydrolyze the methoxy groups, which then condense with each other to form cross-links.

5.12.1.2 Cure of Epoxy Resins. The reactivity of the epoxy ring permits polymerization and cross-linking reactions with many types of additives. Polyamines are most commonly used, particularly polyethylene polyamines. Since these are often too volatile, allergenic, and reactive, they are usually reacted into epoxy adducts or polyamides of somewhat higher molecular weight, lower volatility, greater safety, and more controlled reactivity. Polybutadiene oligomers with acid end-groups, and polysulfide oligomers with mercaptan end-groups, are curing agents that are used to reduce the inherent brittleness of cured epoxy resins. Polyanhydrides are used for higher heat resistance, because they can react further with the hydroxyl groups formed during the polymerization of the epoxy resin. For solid molding compounds and impregnated tapes, aromatic amines, and high-temperature catalysts like BF_3:amine adducts and dicyandiamide are often precompounded, giving systems that are fairly stable at room temperature. For flame retardance, halogenated anhydrides are often used in place of normal anhydrides.

5.12.1.3 Unsaturated Polyesters. Copolymerization and cure of the fumarate esters with styrene monomer is initiated by peroxides, choosing the peroxide appropriate to the processing temperature chosen. For room-temperature cure, MEK peroxide is generally used; it can be accelerated by cobalt naphthenate and further catalyzed by tertiary amines. For higher-temperature cure reactions such as BMC and SMC, higher-temperature peroxides are chosen (Sec. 5.10).

5.12.1.4 Polyurethanes. The two major types of catalysts, for the polyol-polyisocyanate reaction to form polyurethanes, are tertiary amines and/or organotin compounds such as dibutyl tin dilaurate. For delayed reactions, the amines and/or the isocyanates can be temporarily blocked by adducts, which are removed and liberated during the cure reaction.

5.12.1.5 Furfuryl Alcohol Resins. Polymerization and cure, and copolymerization/cure with urea-formaldehyde and phenol-formaldehyde, are generally catalyzed by acids such as p-toluene sulfonic acid and zinc chloride.

5.12.2 Surface Properties

A variety of additives are used primarily to modify surface properties, either during processing or during use of the finished product. They are collected here for this general purpose.

- *Hyperdispersants* are low-molecular-weight block copolymers designed to separate filler and pigment particles from each other, disperse them more readily in liquid systems, and stabilize these dispersions for more efficient use of the solid particles. One block is designed to be attracted to the surface of the solid particle, the other block to be attracted into the plasticizer, polyolefin, or other matrix being used for the masterbatch, and also into the final matrix polymer in the finished product. The exact nature of these block copolymers is still a secret of their developers and producers.

- *Corrosion inhibitors* are commonly included in coatings on steel. These include phosphates of iron, manganese, and zinc; chromates of zinc and strontium; soaps of calcium, lead, sodium, and zinc; lead oxide, carbonate, and sulfate; ferric complexes; and zinc dust. Some newer types include organic phosphates and sulfonates. Due to worries about toxicity and the environment, this entire field is in a state of change.

- *Prebonding etch* is needed to activate the perfluoro surface of PTFE before it can be bonded with adhesives. This is typically a solution of sodium in naphthalene, which is extremely alkaline. It pulls some fluorine atoms off the surface, or even carbonizes it, leaving a surface which is more ready to accept adhesives such as epoxy resins.

- *Antiblocking agents* are often fine filler particles that roughen film surfaces enough to prevent them from coming into good contact with each other, and thus reduce adhesion between them. These are typically 0.1 percent of chalk or 1 percent of amide wax.

- *Antislip agents* are sometimes needed to overcome excessive lubrication. For example, lubricants are added to films to keep them from sticking together during handling. If the films are converted into bags, filled with heavy solids, and stacked on a pallet, they may be so slippery that they slide off the stack and fall all over the floor. In such cases, an antislip agent may be added to the formulation to create enough friction/adhesion to prevent such sliding. Antislip agents for polyolefin films are often materials such as oleamide in PE or erucamide in PP, used at about 0.05 percent.

- *Antifog agents* are added to films to keep moisture condensation from clouding them and preventing the passage of light. Typical uses are in polyolefins, polystyrene, PVC, and polyester films, as an aid in marketing refrigerated and frozen foods, and in plastic film for greenhouses. These are hydrophilic organic compounds such as mono- and diglycerides, higher-polyol partial esters, and ethoxylated phenols and fatty alcohols and acids, used at 0.5 to 4.0 percent, at a cost of $0.80 to 4.00 per pound. They adsorb moisture and spread it into a continuous transparent surface film rather than the opacifying droplets that normally form on a low-polarity plastic surface.

- *Water repellants* are sometimes used in surface treatments such as isobutyl trimethoxy silane.

5.12.3 Degassing Agents

Compounding and mixing of liquid systems often traps air, or volatile liquids that volatilize during processing, forming bubbles and other flaws in the finished products. Several types of additives are used to remove these volatiles before they cause trouble.

- *Humidity eliminators* are used to absorb moisture from PVC plastisols and other liquid systems, to prevent blistering, bubbles, and craters. A typical system would be a combination of calcium oxide plus a wetting agent.

- *Air-release agents* are added to liquid epoxy, polyester, and polyurethane systems to remove air bubbles before cure. These function by their surface activity. They are mostly proprietary compositions.

- *Antifoam agents* are added to latexes to prevent air from producing foam that would ruin the dried final coatings. Typical antifoams are octyl alcohol and liquid silicones. They must be chosen with care to avoid negative effects on adhesion and decoration of the finished products.

5.12.4 Oxygen Scavengers

These may be added to PET bottle resins to protect the contents against attack by atmospheric oxygen. They are mostly proprietary.

5.12.5 Epoxy Diluents

When an epoxy resin formulation is too viscous for the intended process, it may be thinned by adding low-molecular-weight liquids. These are preferably mono- or diepoxy monomers, smaller than conventional bisphenol epoxy molecules, which can react right into the finished resin without sacrifice of properties. Some formulators may use nonreactive diluents, but these may detract from finished properties.

5.12.6 Fragrances

Chemical fragrances are added to products either to produce a desired odor ("decorative") or to mask an undesirable odor ("functional") in a material or an environment. They are generally perfume oils that are masterbatched into thermoplastics, often combined with colorant and sold in powder or pelletized form to processors who blend the masterbatch with natural resin in conventional molding, extrusion, and foam processes.

They may be used at concentrations from 10 ppm (0.001 percent) (food packaging film) up to 40 percent (air fresheners). The concentrates are sold at $2 to $8 per pound.

Lifetime of the odor is controlled by vapor pressure of the odorant, surface/volume ratio of the product, temperature, and controlled air flow. Lifetimes may range from several days to many years. Shelf life, properly packaged, is "almost infinite."

Major developed uses are in garbage bags, films in general, room and auto air fresheners, toys, and housewares. Developing markets include textiles, hospital supplies, consumer packaging to stimulate sales, wall tile, air conditioning, and enhancement of work efficiency. In most of these uses, the odor is perfectly apparent. However, in some uses, it is kept at a subliminal level for subtle psychological effect.

5.12.7 Masterbatches

Compounders are always free to buy individual additives and combine them to their specific needs. In some cases, particularly when the additive is difficult to compound, it is best masterbatched by specialists and then sold to the processor who simply "lets it down with natural." Many processors prefer not to become involved in the chemical details and buy combined masterbatches containing several additives all together in the same masterbatch. This is particularly common in stabilizer packages, surface-treated fillers and fibers, lubricant packages, colorants + odorants, polyblends, and antifog/antistat/lubricant additives. While this saves the processor a lot of detail formulating effort, it makes it much more difficult for him to identify and solve problems when the compound does not perform properly.

5.13 GENERAL REFERENCES

1. Jesse Edenbaum, *Plastics Additives and Modifiers Handbook,* Van Nostrand Reinhold, 1992.
2. J. T. Lutz and R. F. Grossman, *Polymer Modifiers and Additives,* Dekker, 2000.
3. Hans Zweifel, *Plastics Additives Handbook,* Hanser, 2001.

5.14 SPECIALIZED REFERENCES

1. H. S. Katz and J. V. Milewski, *Handbook of Fillers for Plastics,* Van Nostrand Reinhold, 1987.
2. J. W. Lyons, *The Chemistry and Uses of Fire Retardants,* Wiley-Interscience 1970.
3. J. V. Milewski and H. S. Katz, *Handbook of Reinforcements for Plastics,* Van Nostrand Reinhold, 1987.
4. D. R. Paul and C. B. Bucknall, *Polymer Blends,* John Wiley & Sons, 2000.
5. E. P. Plueddemann, *Silane Coupling Agents,* Plenum, 1982.
6. J. K. Sears and J. R. Darby, *The Technology of Plasticizers,* John Wiley & Sons, 1982.

CHAPTER 6
NANOMANUFACTURING WITH POLYMERS

Daniel Schmidt, Joey Mead, Carol Barry
Department of Plastics Engineering
University of Massachusetts
Lowell, Massachusetts

Julie Chen
Department of Mechanical Engineering
University of Massachusetts
Lowell, Massachusetts

6.1 INTRODUCTION

Nanotechnology offers the promise of unique and wonderful products as a result of the dramatic changes that occur when structures approach nanoscale dimensions. At these sizes, on the molecular level, the surface or interfacial properties play a more significant role. These effects are used to produce more effective drug delivery systems, higher-performance plastic parts, and faster and lighter-weight memory devices—products that affect our everyday life. Many of the current nanotechnology products and reports are focused on semiconductor processing and ceramic-based materials; however, these materials have limitations due to high density and rigidity. Polymer-based products offer an advantage because of their lighter weight, flexibility, and biological compatibility. In addition, polymers also provide the benefit of ease of fabrication using high-rate and continuous processing. As a result, it is anticipated that polymeric materials will play a more important role in the future of the nanotechnology revolution.

One of the critical issues in advancing the field of nanotechnology is the need to develop economic and robust manufacturing methods. Since polymers can be fabricated in a wide array of shapes and forms, their manufacturing approaches can be utilized to develop numerous products. Some of the potential new products include extruded multicomponent thin films for conformable, high-density data storage or displays, injection-molded low-cost calibration standards, and electrospun nanotextiles for energy storage. The issue for developing these products is the need to develop commercially viable nanomanufacturing methods. Nanomanufacturing of polymers is likely to look quite different from current macroscale processes; however, we can make modifications to existing equipment to produce a host of new products. Although it is anticipated that many new products will emerge, we can explore four basic geometries: nanocomposites, nanofibers, nanolayered films, and nanofeatured polymers. This chapter briefly covers these current technologies, some already appearing in commercial products.

6.2 NANOCOMPOSITES

Nanocomposites consist of a nanometer-scale phase in combination with another phase. While this section focuses on polymer nanocomposites, it is worth noting that other important materials can also be classed as nanocomposites—super-alloy turbine blades, for instance, and many sandwich structures in microelectronics. Dimensionality is one of the most basic classifications of a (nano)composite (Fig. 6.1). A nanoparticle-reinforced system exemplifies a zero-dimensional nanocomposite, while macroscopic particles produce a traditional filled polymer. Nanofibers or nanowhiskers in a matrix constitute a one-dimensional nanocomposite, while large fibers give us the usual fiber composites. The two-dimensional case is based on individual layers of nanoscopic thickness embedded in a matrix, with larger layers giving rise to conventional flake-filled composites. Finally, an interpenetrating network is an example of a three-dimensional nanocomposite, while co-continuous polymer blends serve as an example of a macroscale counterpart.

Well before the term *nanocomposite* was coined, researchers, especially those in the rubber industry, were working toward the use of nanoscopic filler particles (carbon black, fumed

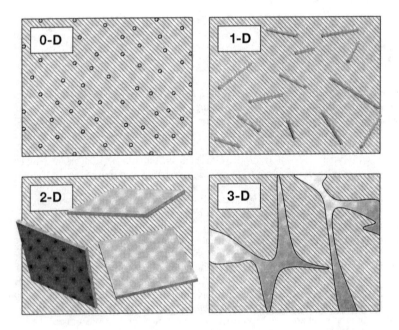

FIGURE 6.1 A dimensionality-based classification system for nanocomposites, covering nanoparticles (0-D), nanorods/nanofibers (1-D), nanolayers/nanodiscs/nanoplatelets (2-D), and interpenetrating networks (3-D).

silica, and so on). In the prenanometer age, their size was measured in either angstroms or "millimicrons," but these were some of the first true nanocomposite systems. Even then, some of the basic truths of nanocomposite research had already been explicitly stated:

> Fine particle size does not necessarily lead to good reinforcement. In practice, the situation is complicated by the fact that very finely divided fillers tend to agglomerate and are extremely difficult to disperse.... The use of organic or other coatings in filler surfaces sometimes promotes dispersion, and increases the effective use of fillers of very fine particle sizes.[1]

Why were nanometer dimensions so important? Again, the same questions raised nowadays by nanocomposite research were beginning to be answered far earlier, in the silicone industry:

> The factor common to all reinforcing fillers is high specific surface area, though whether this is the only—or even the principal—requirement has not yet been demonstrated with certainty."[2]

This issue of specific surface area hints at how one might change the nature of reinforcement. In typical micro- and macrocomposites, the properties are dictated by the bulk properties of both the matrix and the filler. This relationship between the properties of the composite and the properties of the filler is what leads to the stiffening and degraded elongation mentioned earlier. In the case of nanocomposites, the properties of the material are instead tied to the interface. Terms like *bound polymer, bound rubber,* and *interphase* have been used to describe the polymer at or near the interface, where significant deviations from bulk structure and properties are known to occur (Fig. 6.2).

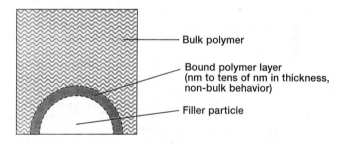

Bulk polymer

Bound polymer layer
(nm to tens of nm in thickness,
non-bulk behavior)

Filler particle

FIGURE 6.2 Bound polymer. The thickness of this boundary layer is typically described as being in the range of nanometers to tens of nanometers.

In polymers filled with fillers with high specific surface areas (that is, hundreds of meters squared per gram), most of the polymer present is near an interface (and thus *bound polymer*), even with only a small weight fraction of filler. Such fillers are necessarily nanoscopic, as this is the only way to achieve such a high specific surface area. If the interaction at the interface is a strong one, or if the structure of the interfacial polymer is very different from the bulk, one can expect to see markedly different properties in the material as a whole. These changes have a fundamentally different origin from those found in micro- and macrocomposites, where the volume of the interphase is only a small fraction of the overall volume of the material. Therefore, nanocomposites are often referred to as being "different" from other reinforced systems. Some specific examples of these differences follow.[*]

6.2.1 1-D Nanocomposites

The most readily cited examples of this class of nanocomposite are systems based on carbon nanotubes. Carbon nanotubes have many interesting properties, including exceptionally high mechanical strength and remarkably versatile electronic properties. They occur in two distinct forms: single-walled nanotubes (SWNTs) and multiwalled nanotubes (MWNTs). Compared with multiwalled nanotubes, single-walled nanotubes are expensive and difficult to obtain, but they have been of great interest due to their superior electronic, mechanical, and gas adsorption properties.

The use of carbon nanotubes in polymer composites has attracted much attention, due not only to their interesting mechanical and electronic properties but also to their very high aspect ratio. These properties make them ideal reinforcing fibers in nanocomposites. However, carbon nanotubes do not spontaneously disperse in polymers, making filler dispersion a major issue. Nanotubes aggregate easily to form bundles that are very difficult to disrupt. In addition, such bundles or ropes are often heavily entangled with one another. With high shear, these ropes can be untangled, but dispersion at the single-tube level is difficult to achieve, since the attractive forces are large, and the percolation threshold is low. Likewise, high concentrations of carbon nanotubes are difficult to work with due to viscosity issues. Because of these difficulties, the potential advantages of these fillers have been difficult to realize in practice.

Generally, polymer/nanotube nanocomposites have been fabricated by direct mixing or *in-situ* polymerization. *In-situ* polymerization is more effective in producing homogenous dispersions, because the nanotubes more readily disperse monomers than in the polymers. Direct melt mixing has many advantages over *in-situ* polymerization when it comes to practical application. However, methods aimed at improving dispersion in such systems include surface functionalization, acid treatment, and the use of special surfactants. In addition to these methods, which often involve harsh conditions that may degrade the properties of the carbon nanotubes, techniques of specific interest due to milder conditions and the potential for application in polymer nanocomposites include vapor-phase amination to give good solvent compatibility[3] and the use of highly charged nanoparticles to allow for the dispersion down to the single tube level in water.[4] Epoxies,[5-7] poly(vinyl alcohol),[8] and poly(methyl methacrylate)[9-11] have all been used in the production of carbon nanotube nanocomposites, to give just a few examples. Much of the work, however, has focused on conducting polymers related to poly(phenylene vinylene),[12-16] due not only to better polymer/nanotube interactions but also the possibility for interesting electronic properties in the nanocomposites thus formed.

6.2.2 2-D Nanocomposites

6.2.2.1 Introduction. From the field of fiber composites, it is known that increasing the aspect ratio of the inorganic filler results in an increased modulus. As previously alluded to, however, the field of particle-reinforced rubbers has given rise to another school of thought, which holds that increasing the amount of specific surface area (and thus, the

---

[*]The 0-D case is readily produced with traditional (nano)fillers such as carbon black and fumed silica; as such fillers have been in use for over 50 years at the time of writing, this case will not be described here.

amount of interfacial or "bound" polymer) especially, will result in the most effective reinforcement.[1,2] In 2-D nanocomposites, these effects are combined. On the one hand, nanoscopic layered fillers allow for the retention of aspect ratios in excess of 1000 during
normal polymer processing, something that is exceedingly difficult to achieve with conventional fillers. On the other hand, such systems also benefit from the extremely large
amount of interface and interphase polymer produced due to full dispersion of nanometer
thick silicate layers in the polymer matrix. This combination is possible only in the nanometer regime and leads to properties not found in micro- and macrocomposites.

The nanofillers of choice in the polymer nanocomposites described here are layered
silicates, both naturally occurring and synthetic. A variety of layered silicates have been
used, including naturally occurring ion-exchangeable clay minerals (smectites such as
montmorillonite, hectorite, bentonite, and so on) and synthetically prepared mineral analogs (smectite analogs fluorohectorite and laponite, synthetic fluoromica, layered double
hydroxides, and so forth) to mesogen-modified sol-gel derived layered structures of amorphous silica.[17] General characteristics of most of these materials include layer thicknesses
of ~1 nm and lateral dimensions ranging from ~25 nm to ~5 μm, and cation-exchange capacities between 0.65 and 1.50 meq/g. Naturally occurring silicates are known to contain
quartz and other particulate impurities, which, because of their size (submicron to micron
diameter) and lack of interactions, can significantly degrade impact toughness, even in
small amounts. It is also known that significant variations in chemistry can occur from
clay deposit to clay deposit. Likewise, the synthesis of smectite analogs can produce non-
lamellar impurities as well as significant variations in layer chemistry.[18]

From many reported observations of nanocomposites via transmission electron microscopy (TEM), as well as molecular dynamics simulations,[19] it is known that these layers are highly flexible. In fact, they have been described by one investigator as behaving
"like wet tissue paper," showing significant flexibility and surface adhesion[20] (although
with much higher strength, thankfully). The mechanical properties of a single silicate
layer are often assumed to be similar to materials like glass and mica, and elastic moduli
around 170 GPa are frequently mentioned in the literature. However, ultrasonic measurements of related clay minerals give an elastic modulus of 6.29 GPa,[21] indicating that
even this most basic assumption is likely a significant overestimation, and that we have
much to learn about the mechanical behavior of such materials. Likewise, another molecular dynamics study validates idea that the mechanical properties of such layers are
anisotropic,[22] further complicating matters. Whether an elastic modulus can even be defined for something less than 10 atoms thick is another question entirely. All of these issues become important when attempting to model or describe the mechanical properties
of these systems.

6.2.2.2 Nanolayer Dispersion.

Three general states of dispersion are often described as
existing (or coexisting) in a layered silicate nanocomposite (Fig. 6.3). They are the immiscible (or phase- separated) state, the intercalated state, and the exfoliated (or dispersed)
state. Intermediate states of dispersion have been described in the literature,[23–25] but, as a
first approximation, these three are sufficient.

The immiscible state corresponds to the situation in which the individual silicate layers
do not separate at all and are confined to multilayer stacks, which may even be agglomer-
ated, as expected when the silicates are incompatible with the polymer matrix. The intercalated state is typically described as a situation in which the silicate layers expand to
accommodate some small number of polymer molecules, but where the layers retain a relatively small interlayer spacing and an ordered layered structure, analogous to the swelling of a multilayer stack. Finally, the exfoliated state refers to the situation in which the
silicate layers are individually dispersed.

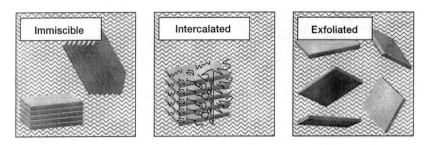

FIGURE 6.3 Basic dispersion states in layered silicate nanocomposites. From left to right: immiscible (effectively a 0-D micro-/macrocomposite), intercalated, and exfoliated. Lines represent polymer chains.

This is a gross oversimplification, of course, as geometric concerns alone indicate that, for such high-aspect ratio particles, true exfoliation cannot take place except at extremely low inorganic loading levels, even lower than those typically used in nanocomposites, thanks to the ease with which the percolation threshold is reached in such systems. However, significant dispersion, beyond intercalation, can be made to take place even in samples containing 10 wt% silicate, resulting in a larger separation of the silicate layers and the degradation of loss of the lamellar ordering found in the immiscible and intercalated cases. The multilayer stacks break up, and, while the polymer chains may still be confined with respect to their random coil dimensions, and the silicate layers may still be within the influence of one another and even engage in cooperative motion, the situation is distinctly different from the intercalated case. It is very likely that multiple states of nanoscale dispersion will coexist within the same nanocomposite; this is of clear importance and must be taken into consideration when studying both properties and nanostructure. With effective mixing, however, the greater the compatibility between the polymer and the silicate, and the closer the system should come to complete exfoliation.

In practice, intercalation is relatively easy to achieve, while something close to true exfoliation is much more difficult. It is worth noting that enhancement of properties may be seen in both cases. While mechanical properties are most improved through exfoliation, improved thermal, fire, and barrier properties and decreased thermal expansion coefficients may be expected without complete exfoliation based solely on arguments of silicate layer properties and increased tortuosity with respect to the path of diffusing species.

6.2.2.3 Compatibilization: The Vehicle of Dispersion. In general, these layered silicates are hydrophilic. Therefore, it is necessary that they be organically modified to improve compatibility with all but the most polar polymers. Without compatibilization, such materials are effectively particle-reinforced micro-/macrocomposites, the particles being larger agglomerates of multilayer silicate stacks. To access the unique properties of nanocomposites, the dispersion must be significantly better.

Many phyllosilicates and their synthetic analogs are capable of cation exchange, while layered double hydroxides are capable of anion exchange. In these materials, regions between adjacent layers contain highly mobile ions (Na^+, Li^+, NO_3^-, and others) that are readily extracted via exchange with other ions. If an ionic surfactant is used in this reaction, the products arer layers organically modified by ionically bound surfactant molecules. It is possible to produce a huge variety of such materials based on combinations of different layers (variations in size and exchange capacity) and surfactants (variations in chemistry). The structures of some typical surfactant modifiers for cation-exchangeable layered silicates are shown in Fig. 6.4. With the proper choice of silicate and surfactant, a thermodynamically favorable nanocomposite structure can be formed; surfactant tails are displaced from the silicate layer surface (where they are found in undisturbed organosilicates) and are replaced by polymer chains that interact more strongly with the silicate, preferably without interacting too negatively with the surfacant. Alternatively, the clays may be forced apart by *in-situ* polymerization or even ion exchanged with a reactive species of some sort so as to enhance the polymerization rate in the interlayer regions and better separate the layers.

Alternatives to ion exchange include silane treatment of exposed hydroxyls, generally located on the edge of the silicate layers, to produce covalently bound organic functionalities, or the use of copolymers as compatibilizers in otherwise incompatible systems. Specifically, copolymers containing at least one segment with an affinity to the (organo)silicate (polar or hydrophyllic polymers) typically may be added as compatibilizers by themselves or in addition to others already found in the system.

However it is obtained, the achievement of a nanostructure with significant silicate dispersion has been the goal of the majority of polymer/layered silicate nanocomposite research. As a result, a number of thermodynamic descriptions of such systems have been attempted.[26-28] As with any additive, however, the presence of unbound compatibilizing agents will impact the properties of the blend as a whole, potentially giving rise to problems like plasticization, thermal degradation, and outgassing, to name just a few. Ironically, the silicate layers themselves, like many oxide surfaces, can catalyze the degradation of organic compounds, including the polymers they are meant to reinforce.[29-31]

6.2.2.4 Nanocomposite Production.

Methods for nanocomposite production can be classified into two general (and sometimes overlapping) categories. One approach involves enhancing the mobility of the polymer of interest, in the presence of silicate layers, and with thermodynamic compatibility between the two, so as to achieve silicate layer dispersion. This best describes the techniques of melt blending and solvent casting. The other approach attempts to literally force the layers apart through the preferential insertion of material in the interlayer galleries, regardless of thermodynamic compatibility between the silicate and the polymer. This best describes a number of different *in-situ* polymerization techniques. Both have been used successfully to produce polymer/layered silicate nanocomposites, although, not surprisingly, there are strong indications that we ignore thermodynamic compatibility at our peril.[32]

These techniques are present in many variations and, in some instances, they are even combined. *In-situ* polymerization, for example, may be used to produce heavily compatibilized layered silicates, to act as concentrated master batches, and melt-blended with pure

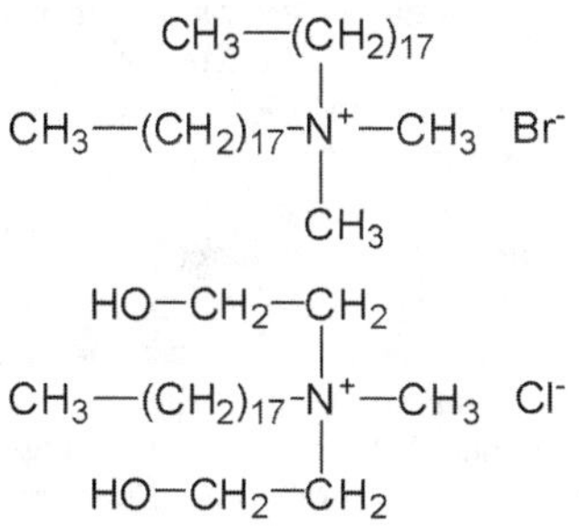

FIGURE 6.4 Typical silicate modifiers. These alkylammonium surfactants are used to modify layered silicates capable of cation exchange, either in aqueous solution or using a surfactant melt.

polymer to produce the final nanocomposite material. The preparation of epoxy nanocomposites might also be classified as a combination technique, as it involves the intercalation of epoxy prepolymers into the silicate layers, followed by a cross-linking reaction (which can be construed as a form of polymerization) to form a network. Here, the silicate-bound head-groups of the alkylammonium surfactants used to compatibilize the system perform double duty as catalysts, enhancing the interlayer reaction rate and resulting in the production of more material between the layers than without, forcing them apart.

Solvent casting is one of the simplest techniques by which nanocomposites are produced. A polymer, a silicate, and a solvent are combined and thoroughly mixed, and the solvent is then allowed to evaporate, leaving the nanocomposite behind, typically as a thin film. The solvent imparts the enhanced mobility the polymer needs to intercalate between the silicate layers, while thermodynamic compatibility and physical mixing give rise to a dispersed system. A solvent should be chosen that completely dissolves the polymer and completely disperses the silicate.

Melt blending is akin to solvent casting in that, here too, the polymer is given the enhanced mobility it needs, in combination with physical mixing, to disperse compatibilized layered silicates on the nanoscale. There is (typically) no solvent used in this technique, and, as the name implies, the enhanced polymer mobility comes simply from thermal energy. Using an extruder or heated mixing chamber of some sort, the molten polymer and the compatibilized layered silicate are physically mixed, and a nanocomposite is obtained. Again, the mixing can be performed via something akin to stirring (i.e., the use of mixing screws in an extruder setup) as well as by using ultrasonics, where a probe is immersed in a polymer melt. The two mixing techniques may again be combined, as long as the polymer viscosity is low enough to allow for effective sonication. The possibility of ultrasonic degradation of the polymer, however, is something that must be considered.

The lack of solvent in this technique solves many problems associated with solvent casting. No impurities or residues are introduced into the sample, settling is generally not an issue (due to the high viscosities of most polymer melts), and there are no concerns about complete evaporation or the retention of a small amount of a plasticizing solvent. For a given set of mixing conditions, the polymer microstructure is affected only by the composition of the nanocomposite, and pieces of highly varying geometry and size may be produced without problems. As in the case of solvent casting, however, there is still an alignment effect. The silicate layer orientation may be randomized during melt blending, but extrusion or compression molding will result in layer alignment along the flow direction or perpendicular to any applied compressive stresses.

In addition to these concerns, new problems are created when heat is used to enhance the polymer mobility. To achieve good mixing, a long mixing time is preferable, but this must be balanced against the normally undesirable thermal degradation of the polymer and the silicate modifier. Finding an optimum can be difficult, especially for polymers that thermally degrade readily (e.g., some biodegradable polyesters) or for those systems where the processing temperature must be high to achieve polymer flow and effective mixing (many polyamides and polycarbonates fall into this category). In the latter case, it is the organic silicate compatibilizers that degrade. Intuitively, this loss of compatibilizer will result in the loss of thermodynamic compatibility and thus nanoscale dispersion, although, in at least one example, degradation of the silicate compatibilizer seems to have improved mixing.[33] Be that as it may, the risks of modifier degradation usually outweigh the benefits, and processing conditions are chosen to minimize the degradation of both the polymer and whatever silicate compatibilizers are present. This can make processing all the more challenging and has spurred the search for more thermally stable compatibilizers than the traditionally used alkylammonium salts.[34–36]

In-situ polymerization covers any process in which the nanocomposite is made by performing some sort of polymerization reaction in the presence of a layered silicate. There are many variations on this technique, all stemming from the need to disperse the silicate layers. The simplest involves mixing a monomer with a layered silicate and polymerizing from there. For this to succeed, the silicates must be compatibilized to allow the monomers to intercalate, at which point the polymerization reaction can take place in the interlayer galleries. If the silicate surface layer or some silicate-bound functionality catalyzes the polymerization reaction and enhances its rate with respect to material outside of the interlayer galleries, dispersion may be strongly enhanced, up to the gel point at least. Such a situation can be encouraged through the use of silicate-bound monomers or initiators. Alternatively, emulsion polymerization may be performed by dispersing an unmodified silicate in a water/monomer emulsion.

In any case, dispersion obtained via *in-situ* polymerization may improve the barrier and thermal properties based on a tortuosity argument alone but may or may not improve the mechanical performance, depending on the level of polymer/silicate interactions. Without thermodynamic compatibility, the silicate layers may even collapse to form multilayer stacks if the nanocomposite is heated.[32] Finally, all of these techniques become somewhat complicated by any changes the silicate layers cause in terms of the polymer's molecular weight distribution as compared to silicate-free controls.

6.2.2.5 Property Enhancements: The Result of Dispersion. The work that inspired many to look at polymer/layered silicate nanocomposites in the first place was performed by the Toyota Corporation and showed that, in nylon-6, as little as a few weight percent of silicate could afford significant simultaneous enhancements in stiffness, tensile strength, and heat distortion temperature without sacrificing impact strength.[37–40] Since then, many attempts have been made at nanocomposite reinforcement and property enhancement in a number of other systems. Numerous epoxies have also been studied,[41–44] again with the potential for simultaneous improvements of a variety of mechanical properties in unique combinations and with improved toughness.[45–47] Poly(vinylidene fluoride) nanocomposites have shown dramatic enhancements in elongational toughness,[48,49] while some polyurethanes have also shown promise.[50,51] Improvements have even been seen in carefully modified/processed polypropylene based systems,[52,53] despite generally poor PP/silicate compatibility. Work relating abrasion resistance and surface adhesion/friction in polymers to the number of cross-links in the system supports the claim that surface characteristics of polymer nanocomposites (where strong polymer/silicate interactions act as physical cross-links) are enhanced.[54,55] Improvements in barrier properties have been reported,[56–60] in addition to enhancements in thermal stability* and self-extinguishing characteristics attributed to char formation.[61–64]

Further studies on layered silicate nanocomposites have shown enhancement in polymer crystallization rates[65–68] and decreases in thermal expansion coefficient,[69] which may allow for decreased cycle times during processing and significant economic benefits in addition to the properties enhancements already mentioned. Variations in nanocomposite formulation can allow for precise control of biodegradability,[70–72] and there is every reason to believe that they should be readily recyclable as well. Even the weatherability and resistance to aging these materials display can surpass the analogous homopolymers.[73,74] In short; nanocomposites offer the promise of polymers with minimal filler content (thus having low density) whose properties are either on par with or superior to

---

*While clays cometimes catalyze polymer degradation, giving rise to a decrease in the temperature at which degradation begins, the enhanced barrier properties of the nanocomposites hinder transport of the degradation by-products, decreasing the rate of degradation and increasing the temperature at which the degradation rate is at a maximum.

the pure polymer as well as conventionally reinforced systems, and that can be produced more rapidly and at a lower cost. Although this will not occur in all systems under all conditions, all of these things have contributed to the growing interest in nanocomposite materials.[23,75-79]

6.2.2.6 *Explaining Behavior with Structure.*

To answer the question of why these materials behave so differently, we must revisit the issue of the polymer interphase (or bound polymer). These nanocomposite systems can be prepared such that nearly the entire polymer in the system is interphase polymer, due to the large specific surface area of the layered silicates (when properly dispersed) and the large amount of interface thus created. Coupled with the many descriptions of "bound polymer" reported in composite literature, molecular dynamics simulations of nanocomposite systems also indicate that local density variations occur in proximity to organically modified layered silicates, both in the absence and presence of intercalated polymer chains.[80-82] In fact, such interlayer density variations (albeit with small molecules rather than polymers) have been reported for many years in field of clay science.[83]

Additional changes may arise if the polymer has a significant crystalline component. Changes in the crystalline phase present have been reported in some of the most well studied nanocomposites, those based on nylon-6. In these systems, the layered silicate appears to stabilize the otherwise metastable gamma phase, whose crystal structure is more densely packed than that of the normally most-favored alpha phase. Whether this is due to hydrogen bonding of the polymer to the silicate surface or issues of confinement or local density variations, this change is of great importance, as the properties of gamma nylon-6 are not the same as those of alpha nylon-6, and the prevalence of the former will greatly affect the properties (mechanical, barrier, and others) of the materials as a whole. Other polymers may show similar behavior as long as the polymer/silicate interactions are strong enough and the polymer has multiple crystalline forms—as observed in the case of poly(vinylidene fluoride) nanocomposites, for example.[48,49]

The rate and degree of crystallization may also be affected. Any interface, any surface, any defect in any system may act as a nucleation point, and the addition of more of these nucleation points will necessarily produce systems with finer-grained crystalline microstructures. In nanocomposites, the material is full of nucleation sites; as long as silicate dispersion is favorable, materials containing finer-grained polymer crystallites (versus the pure polymer) will be produced.[66,84-86] Physical partitioning may limit the crystallite size, while the anisotropic nature of the silicate layers may induce anisotropic crystallization. Even if the favored crystalline phase remains unchanged, crystallization rates may be affected, and the overall crystallinity of the sample may also change.

In addition to its structure, the dynamics of the interphase are extremely important. Studies on poly(styrene) have shown not only that there exists significant polymer mobility below T_g,[87] but that these nanocomposites display a large range of mobilities both above and below T_g, indicating a large range of distinctly different polymer environments.[88] This, in turn, points to fundamentally different matrix mobility near the silicate surface. Similar results have also been found in poly(methylphenylsiloxane) nanocomposites,[89] while a number of other studies on a variety of systems have reached similar conclusions.[90,91]

The arrangement of the layers themselves also has an impact. The self-extinguishing properties of these materials, for instance, have been described qualitatively. A protective char layer forms and acts as a diffusion barrier to further combustion. Likewise, before the advents of nanocomposites, models of the barrier properties of glass-ribbon reinforced composites[92] foreshadowed the increased tortuosity arguments often heard with regard to nanocomposite barrier properties.[93,94] Improvements on these first approximations of bar-

rier properties are being developed in the form of attempts to take into account the significant changes in the polymer matrix itself, especially near internal interfaces in
nanostructured materials in general[95] and in layered silicate nanocomposites in particular.[96,97] All of these investigations into behavior at various levels, from the macro- to the
nano- and even the molecular, are extremely important. As a better understanding of the
fundamentals of these systems is developed, this hierarchy should begin to fall into place.

6.3 NANOFIBERS

6.3.1 Fiber Formation/Fiber Spinning

Polymer fibers have a broad range of applications from carpets to ballistic protection. Fibers are typically fabricated through extrusion, with common process variations designated as wet spinning, dry spinning, melt spinning, and gel spinning. After extrusion, the
fibers can be drawn in an intermediate or solidified state to achieve the desired final diameter and crystallinity.

Thermoplastic fibers are typically fabricated via a melt spinning process wherein the
melt is extruded through the many holes in a spinneret and cooled to harden. Other fiber
materials can be dissolved in a solvent and then extruded, with the solid fibers precipitat-
ing from solution either due to a chemical reaction in a liquid bath or due to solvent evaporation in a gas. The latter method, dry spinning, is closest to the focus of this section—
electrospinning of nanofibers.

The use of spinnerets becomes impractical as the desired fiber diameter approaches the
nanoscale because of the high pressures needed at much smaller hole diameters and the effect of dimensions approaching the radius of gyration of individual molecular chains.

One particular method of submicron diameter fiber formation is the electrospinning
process, first patented by Formhals[98] in 1934. Despite this early presence in the literature,
it has only been in the last decade that electrospinning has garnered significant attention.
In this process, a polymer solution in a small tube (pipette or syringe) is subjected to an
electric field as seen in the schematic in Fig. 6.5. As the field is increased, the solution at
the tip of the tube becomes elongated, forming what is termed a *Taylor cone*.[99] Once the
electric force is sufficient to overcome the surface tension of the liquid, a jet of charged solution is ejected toward a grounded target.[100] Along the way to the target, solvent is evaporated, and the fiber undergoes a whipping process. This whipping process is thought to be
caused by a combination of electro- and hydrodynamic instabilities.[101,102] The fibers are
typically collected on a grounded target in a nonwoven mat.

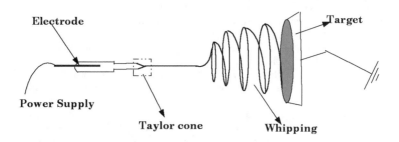

FIGURE 6.5 Schematic of basic electrospinning setup.

6.3.2 History, Applications, and Challenges of Electrospinning

In his patent, Formhals[98] claimed of an apparatus designed to collect yarns. The apparatus, shown in Fig. 6.6, has a channel carrying polymer solution (cellulose acetate in ethylene glycol) connected to the negative electrode of a high-voltage power source. The positive electrode is connected to pointed metal strips on a moving belt. On applying the voltage, polymer fibers from the nozzles are attracted to the electrodes on the moving belt. Due to the charge concentration on the tip, a large amount of fibers would be attached on the tip. A rotating wheel at one end of the belt was designed to collect the fibers from the tip of the electrodes. A high voltage of 56 kV was used for this process. The method and the apparatus had some limitations in regard to the collection of the fibers. The fibers would adhere to the moving belt, drum, and other parts of the apparatus, thereby making it difficult to collect fibers in the form of yarns.

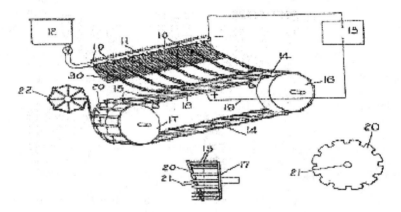

FIGURE 6.6   Apparatus used by Formhals[98] in his patent for electrospinning.

During the decades between Formhals' patent and the recent upsurge in electrospinning activity, the primary record of work on electrospinning was in patents focused on innovative methods for collection, including patterned targets. In 1966, Simons[103] invented a process for forming a thin, lightweight, nonwoven fabric with a pattern simulating woven fabric. He used segmented targets for the process with selective charge on different sections. Fibers were formed from a solution of polyurethane in methyl ethyl ketone. Figure 6.7 shows some of the target geometries used in the research.

Simons also conducted experiments to determine the effect of solution viscosity, dielectric constant, conductivity, and solvent volatility on the fiber forming process. Simm et al.[104,105] invented a process by which they modified the electrode, using an annular electrode spinning at a rate of 30 rpm. The polymer used was a solution of polystyrene in methylene chloride, and the applied potential difference between the electrodes was 120 kV. The sprayed solution was collected in the form of a thick, dry, porous fiber mat. Finn et al.[106] used a similar process in which he employed the combined action of centrifugal and electrostatic forces to form a fiber mat. He used a polyurethane solution in an open cup held at a high potential and spun about its vertical axis. The solution is propelled out of the cup onto the target, which is mounted above the cup and driven around by rollers.

Berry,[107] Ho,[108] and Bornat,[109] in their patents, modified the electric field acting on the fibers to form tubular fibrous structures. Figure 6.8 shows the setup used. Berry used

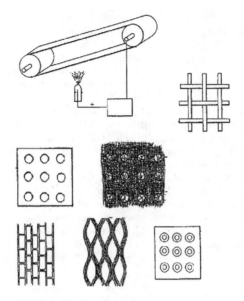

FIGURE 6.7 Experimental set-up and patterned targets used by Simons.[103]

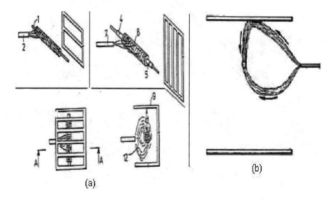

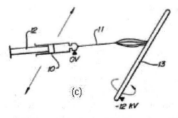

FIGURE 6.8 Experimental set-up used by (a) Berry,[107] (b) Bornat,[109] and (c) How.[108]

horizontal plates with voltage potential on the path between the source and target for guiding the fibers on the rotating mandrel. The distance of the mandrel from either of the plates could be varied to have fibers with different diameters deposited on the mandrel. He claimed that fibers taking a longer path to the target resulted in randomly oriented fibers with smaller diameters. The larger-diameter fibers embedded circumferentially along with the smaller diameter fibers provided for a better void control. The mandrel was rotated at a speed of 5000 rpm. He also varied the voltages applied to the grid (horizontal plate) and the rotating mandrel. How[108] also designed an apparatus for collecting fibers on a rotating mandrel to make synthetic vascular grafts. He applied a high voltage of –12 kV on the rotating mandrel. The speed of rotation (1500 to 9000 rpm) was controlled to achieve a desired degree of anisotropy in the product. The source containing the polymer solution was traversed along the length of the tube.

Bornat, in his apparatus, used a voltage on a grid surrounding the target to attract the polymer fibers from the source. He used different configurations of grids to obtain fiber deposition on the rotating mandrel. He also increased the number of source points to increase the deposition of the fibers. More recently, Scardino[110] used electrospun fibers in combination with strengthening fibers/filaments to form two-dimensional and three-dimensional fabrics by weaving, braiding, or knitting. He used an air vortex to prevent the sticking of fibers to surfaces and to guide them on the filaments. The apparatus is shown in Fig. 6.9.

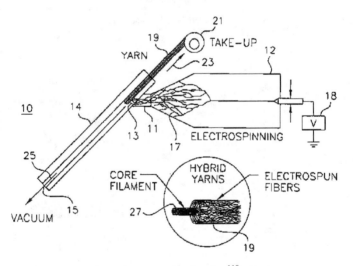

FIGURE 6.9 Experimental set-up used by Scardino[110] with air vortex.

In the mid to late 1990s, the growing ability to image and generally characterize objects at the nanoscale was fueling an explosion of activity in nanotechnology. At the same time, Reneker and coworkers[100,111,112] were carrying out detailed experiments on the electrospinning process and were sharing their findings openly. Reneker's encouragement and dissemination of both problem statements and problem solutions significantly increased the rate at which interest and activity in electrospinning expanded.

These earlier studies in the mid to late 1990s focused on electrospinning with different kinds of polymers. More than 20 polymers have been electrospun by Reneker's group.

Some experimental work was also carried out to study the effect of process parameters on structure and morphology of the fibers.[111–119] Gibson et al.[120–122] studied the transport properties, water vapor diffusion, and gas convection properties of electrospun fiber mats for use in chemical protective clothing. Although there has been research effort into electrospinning, electrospun fibers have not been fully utilized to make commercial products due to the large number of parameters affecting the process and difficulties associated with collecting the fiber themselves.

While most of this activity was taking place in academia, a few companies were actively pursuing commercial applications. While most were startup companies, at the other extreme was Donaldson Company, Inc. (www.donaldson.com), a global company founded in 1915 and based in Minneapolis, MN. As a worldwide leader in filtration products, Donaldson had been using the electrospinning process (well before its recent spike in popularity) to create nanofiber webs on fabric substrates to obtain improved performance of air filters for machinery and many other industrial filtration applications. To generate the surface area coverage necessary for their commercial manufacturing rates, Donaldson utilized multiple nozzles on a production line to achieve coverage of tens of thousands of square meters per day.

Another great early supporter of the technology was the Army Research Laboratory. With the in-house efforts at the Natick Army Soldier Systems Center (http://www.natick.army.mil/) and ARL, as well as collaborations with academia and industry, Schreuder-Gibson and her colleagues also did much to advance both the fundamental knowledge and the potential for applications of interest to the military, such as flame, chemical, biological, and environmental protection in clothing for soldiers.

Many of the perceived commercial applications are in filtration and similar areas, but applications are also growing rapidly in biomedical applications such as wound healing, tissue scaffolds, and chemical protective clothing.

There are a number of current challenges to the electrospinning process. These include

- Uniformity of fiber diameter
- Continuous fiber collection
- Multiple nozzles (high-volume production)
- Patterned, oriented fibers (versus random mat)
- 3-D, thick structures

6.3.3 Process Parameters and Observed Morphologies

A broad range of polymers have been electrospun,[112,114,123–125] including polyamides,[126] biopolymers,[127] polyacrylonitrile, polyethylene, polyethylene terephthalate, polypropylene,[128] polybenzimidazole],[129] and polycaprolactone. In general, the desired baseline fiber morphology from electrospinning is a continuous fiber of uniform diameter. Observed deviations from this baseline include beaded fibers[130] or fibers with sections of much larger diameter, droplets, and sheath/core fibers.

Parameters affecting the electrospun fibers can be classified as material system and process parameters. Material system parameters include viscosity, conductivity, and surface tension of the polymer solution, along with the molecular weight and molecular weight distribution of the polymer.[100,131,132] The process parameters include the applied voltage, flow rate, polymer concentration, source to target distance, target geometry, temperature, humidity, and the air velocity in the chamber.

Table 6.1 provides a list of typical process parameters studied for PEO. The electrospinning process has been studied by varying the polymer used, solution concentration,

TABLE 6.1 List of Process Input Variables and Typical Ranges[133]

Parameter studied	Range							
Polymer	PEO in ethyl alcohol and water							
Deposition duration (min)	5 to 10							
Applied voltage (kV)	(<5) Droplets at syringe	5	10	15	20	25	(>20) Unstable jet	
Flow rate (ml/min)	(<0.01) Inconsistent	0.01	0.03	0.06	0.10	0.12	0.15	(>0.15) Droplets at syringe
Source to target distance (cm)	Polymer deposition on target	5	10	15	16	20	Inconsistent—loss of fibers	
Humidity range (%)	—	15–25		35–45		>50% Droplet deposition on target		
Polymer concentration (wt%)	No fiber formation	4.5	7.5	9.0	12.5	Highly viscous		

Base case (shaded in table): 9 wt% PEO in ethyl alcohol and water at a source-to-target distance of 16 cm, polymer flow rate = 0.1 ml/min.

applied flow rate, and source to target distance. The voltage range used was from 5 to 25 kV, depending on the source to target distance (typically 5 to 25 cm). Polymer flow rate selected is based on the polymer and the solution concentration to form fibers.

Effects of many of the process parameters have been identified; for example, the concentration of the polymer in the solution should be sufficient to cause polymer entanglements (fiber formation), but higher concentrations will increase the viscosity resistance to the electrical force. Because of its ease of spinning and utilization of a relatively harmless solvent (water and alcohol), polyethylene oxide (PEO) has been the most studied polymer. Doshi et al.[100,111] used PEO of MW of 145×10^3 g/mol, surface tension of 61 dynes/cm, conductivity of 400 μS/cm, and solution viscosity of 400 to 800 centipoise for their process optimization studies. Deitzel et al.[116] and Bunyan[133] also studied the effect of polymer feed rate, applied voltage, and the properties of the solution (surface tension, viscosity, concentration) on the morphology of the PEO fibers produced. Deitzel's solution was a concentration in the range from 4 to 10 percent, viscosity in the range of 1 to 20 poise, and surface tension of 55 to 35 dynes/cm. In both studies, fiber diameter increased with increasing solution concentration (Fig. 6.10).

Figure 6.11 shows Bhowmick's[134] results on the effect of varying concentration on the fiber morphology (for a fixed electric field strength). The 8-wt% PCL polymer solution is characterized by bead formation in the electrospun nanofiber web. The number of beads significantly decreased at the higher concentration of 13 wt%. When the PCL concentration reached 15 wt%, no noticeable beads were observed. By increasing the PCL concentration, the electrospun fiber diameter increased because the viscosity increases limited the attenuation. A mixture of fine and coarse fibers is observed at 20 wt%. When the PCL concentration is greater than 20 wt%, the solution precipitated. Therefore, for a PCL/acetone solution, only concentrations between 15 and 20 wt% can be used to form continuous, uni-

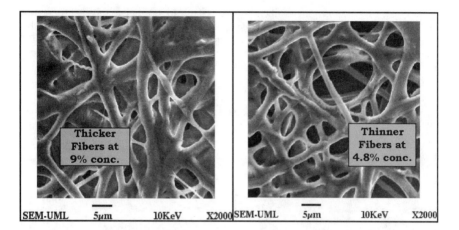

FIGURE 6.10   SEM micrographs of PEO fibers showing increasing fiber diameter with increased polymer concentration.[133]

FIGURE 6.11   SEM micrographs of PCL electrospun scaffolds for different polymer concentrations: (a) 8 wt%; (b) 13 wt%; (c) 15 wt% and (d) 20 wt%.[134]

form diameter fibers. At high solution viscosities, drying of the solution at the tip may cause difficulty in forming fibers.

Fong and Reneker[115] concluded that, with an increase in polymer concentration, the required spinning voltage for initiation also increased. The paper also made note of the increase in the jet diameter with a decrease in the distance between source and target. This shape change corresponds to the decrease in the stability of the initiating jet (on increasing the electric field strength) leading to an increase in the number of bead defeats observed on the electrospun fibers. Fong and Reneker were able to produce fibers with diameters in the range of 50 to 80 nm by electrospinning aqueous solutions of calf thymus Na-DNA (biopolymer).

In Fig. 6.10, some merging of fibers is observed due to incomplete solvent evaporation prior to deposition on the target. The effect of humidity on solvent evaporation is further seen in Fig. 6.12, where less evaporation at higher humidity leads to greater merging of fibers deposited on the target. Figure 6.13 is an example of a process chart for PEO, indicating how combinations of parameters can affect the fiber formation.

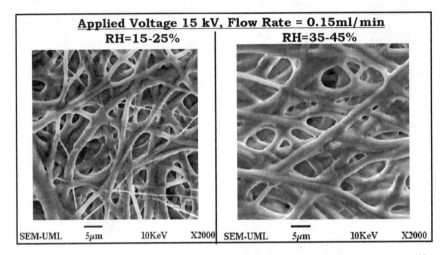

FIGURE 6.12   SEM micrographs showing higher humidity leading to greater merging of fibers.[133]

Many similar studies have been published in the literature for different materials. Optimization of process parameters is fairly well understood for these single nozzle studies, albeit mostly based on empirical studies. Some efforts have been made in process modeling; for example, Hohman et al.[135,136] and Spivak et al.[137] Hohman et al. concentrated on explaining the reasons for instability in the electrically forced fluid jets and on predicting the conditions during the process. Spivak et al. considered the motion of a viscous jet accelerated by an external electric field for modeling purposes. They also took into account the inertial, hydrostatic, viscous, electric, and surface tension forces. Additional modeling and experimental methods to better control the fabrication of nanofibers and nanofiber structures via electrospinning are currently under investigation.

6.4 NANOLAYERED FILMS

Most commercial products use a combination of materials as a single material cannot meet the range of service requirements.[138] Some conventional techniques to combine polymers include blending, lamination, and coextrusion. In the area of films, multilayering approaches offer the potential to combine different polymer types to obtain desirable properties. Each layer is designed to offer different properties; for example, one can combine a scratch-resistant outer layer with a barrier layer in the middle. These layered systems have shown a number of advantages, such as enhanced fracture toughness.[139,140] Layering approaches have a number of applications, such as for enhanced food protection, high-

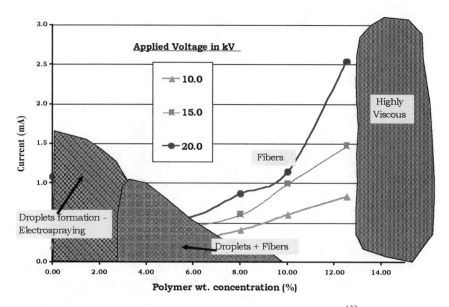

FIGURE 6.13 Example of process parameter chart for PEO fiber morphology.[133]

strength materials, and anticorrosion coatings.[141] Optical applications are another area of interest; for example, nanolayers allow an octopus to focus underwater.[142] The general structure of a layered system is shown in Fig. 6.14.

FIGURE 6.14 Multilayer structure.

There are a number of approaches to develop these nanolayered films. One technique to produce multilayered films is through chaotic mixing processes.[143] A second technique is called layer-by-layer assembly. The approach consists of preparing positively and negatively charged layers. The surface to be coated is dipped into the first solution—for example, negatively charged. It is rinsed and dried, then dipped into the second solution, which in this case would be positively charged. In this way, many layers can be built up and held together with electrostatic forces as depicted in Fig. 6.15. Nanoparticles can also be attached in this technique.[144] Limitations of this process include the requirement for solutions and the need for charged polymers.

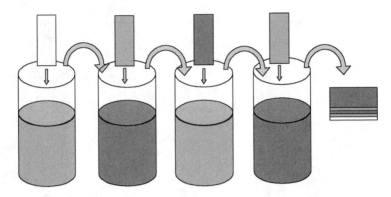

FIGURE 6.15   Layer-by-layer assembly technique.

Another approach to manufacturing these types of structures is a modification of the coextrusion process or multilayer coextrusion.[145,146] In this way, the need for solvents is eliminated, and a continuous manufacturing process is developed. The pioneering research in multilayer coextrusion was conducted by Alfrey and Schrenk in the late 1960s.[147] They extended their initial research in feedblock design and modified it to produce multilayer blown films and sheets.[148] They demonstrated that this technique could be used to prepare multilayer films having nanometer-thick layers. Commercial viability of the procedure was demonstrated when the process was used to produce designed birefringence films.[149] Much of the early research was extended by researchers at Dow[150] during the years from 1969 to 1990, studying both coextrusion and multilayer coextrusion. In the late 1990s, this work was extended by researchers at Case Western Reserve University. They used materials supplied by Dow to investigate the mechanical properties and fracture mechanics of multilayer films. Developing new die designs, they have looked more closely at diffusion mechanics[151] and the formation of unique blends,[152] and they have recently produced multilayer films having layers 25 nm thick.[153] Rheological properties of micrometer-thick multilayer films have been investigated by researchers from the University of Minnesota.[154]

These layered films can be produced using modifications to the feedblock. For example, coextrusion of materials with viscosity ratios up to 40:1 and temperature differences up to 80°C has been reported.[155] Current research is investigating the limits when the layer size approaches the nanoscale. In this case, the limits of viscosity ratio, temperature, and material type remain largely unknown, although such limits are an area of intense research. One of the major roadblocks toward producing nanometer-thick layers is the breakup of the layers into droplets].[156] The basic technique to produce the multilayer films is performed using what are termed *layer multiplying elements*. These elements divide the coextruded melt stream, stack them, and recombine them. This process is repeated many times, increasing the number of layers with each division. This process is shown in Fig. 6.16.

This coextrusion technique has been applied to the fabrication of breathable films, using polyethylene oxide and polyolefin filled with $CaCO_3$.[157] As the layer thickness decreased, the water vapor transmission rate changed dramatically due to changes from a continuous to discontinuous layer structure. Microlayer extrusion of a number of polymers has been accomplished, including combinations of high-density polyethylene (HDPE) and linear low-density polyethylene (LLDPE),[158] ethylene-styrene copolymers and low-density polyethylene (LDPE),[159] polypropylene with both metallocene and Ziegler-Natta

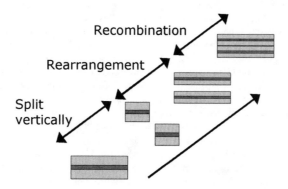

FIGURE 6.16 Schematic of layer multiplying process.[156]

polyethylenes,[160] polycarbonate (PC) and styrene-acrylonitrile (SAN) polymers,[161] polystyrene and polyethylene,[139] and polystyrene and polypropylene.[139,162] The microlayered systems have also been combined with the injection molding process to produce structures with "platelets" of one polymer in another.[163] This resulted in materials with enhanced barrier properties.

Particulate fillers can be incorporated into one of the layers, producing novel structures. Nazarenko et al.[164] incorporated nickel, TiO_2, and talc in LLDPE/low-density polyethylene (LDPE) combinations. Novel structures can also be prepared by combining filled and unfilled layers of the same material, such as filled and unfilled polypropylene (PP).[161] Microlayered structures using poly(ethylene terephthalate)(PET) and talc filled PET have also been prepared.[165]

When considering the properties of the layered structures, adhesion between the layers becomes a critical issue. If the materials are highly incompatible, the adhesion will be weak at the interface, leading to delamination. To circumvent this problem, tie layers can be used to increase the interfacial toughness. In the case of polypropylene (PP) and polyamide coextrusion, maleated PP has been used to increase the interfacial toughness of the structure.[166]

6.5 MOLDING OF NANOSCALE FEATURES

Although many processes are available or have been proposed for fabricating polymer-based micro- and nanoanalytical devices, including microtransfer,[167] microcontact printing,[168] dip pen nanolithography,[169–171] nanoimprint lithography,[172,173] hot embossing,[174–176] injection compression molding,[177] and injection molding.[178–181] The latter two processes offer the best potential for high-rate, high-volume manufacturing of a wide range of "disposable" nanodevices, such as protein assays and tissue scaffolding,[182] and can match the relatively short lifetimes of biomedical products. In addition, molded circuits would significantly reduce the infrastructure investment, manufacturing costs, and environmental and worker safety impact associated with manufacturing of electronic devices.[183]

The current product containing nanoscale features is the digital versatile disk (DVD). This 0.6-mm thick disk contains depressions or pits that are 320 nm wide and 120 nm deep. While DVDs are typically injection or injection compression molded in less than

5 s,[184–186] the machines, tooling, materials, and process were optimized for this one application. These disks must be easily demolded,[187] retain minimal stress (usually measured as birefringence) so as to prevent distortion of the laser signal,[188] and meet a flatness specification. Complete replication of the tooling projections, however, is not required since the "pits" and "lands" must only reflect light differently enough that an opto-electronic detector and electronics can translate these feature differences into binary data. Materials must have optical clarity, good scratch resistance, low melt viscosity, good heat resistance, and low moisture absorption. Therefore, only a limited number of materials have been tested for compact disk and DVD molding, with bisphenol A polycarbonate being the leading polymer used for DVDs.

Parts with nanoscale features have been injection molded to determine the effect of processing conditions and material properties on feature replication.[189,190] Since the tooling and mold parts were characterized using atomic force microscopy, two measurements, the depth ratio (relative depth of the molded features to the tooling depth) and the mean surface roughness, were developed to quantify feature replication. While injection velocity and processing pressures had little effect on replication, the melt and mold temperatures affected the ability to replicate the nanoscale features from the nickel tooling. Higher melt and mold temperatures usually produced greater depth ratios and smoother part surfaces, thereby linking feature replication to bulk melt viscosity and cooling rates. These results were consistent with findings for micromolded parts. Unexpected reductions in replication quality suggested that (1) viscosity increases near the mold wall reduced filling rates, (2) melt adhered to the tooling surface when the mold temperature exceeded $T_g - 20°C$, (3) melt degraded at the tooling surface, and (4) surface energies play an important role in surface roughness of nanoscale replication. Polymethylmethacrylate yielded the best replication and, as expected, semicrystalline materials were more difficult to handle than amorphous materials.

Tooling development is also required for the injection molding process of nanoscale features. Available tooling technologies are listed in Table 6.2. Conventional injection

TABLE 6.2 Comparison of High-Resolution Tooling Techniques

Method	Max. resolution	Aspect ratio	Material
CNC machining	100 μm		
Micromilling	50–100 μm	N/A	Steel
Microwire EDM	1–50 μm		
LIGA	0.5 μm	≤500	Nickel
Laser ablation	1–50 μm	≤600	Metals, polymers
Electroforming	1 μm	2.5	Nickel alloys
Lithography—UV	157 nm	Typically	Silicon or other
Lithography—EUV	13 nm	low, but up	semiconductors
Lithography—e-beam	<10 nm	to 30[203]	

molds are machined from steel, aluminum, and beryllium-copper alloys,[191] and CNC machining has a resolution of 100 μm.[192] To attain the feature sizes needed for micromolds, toolmakers have turned to high-precision electro wire discharge machining (WEDM), lithographic and etching techniques traditionally used in semiconductor fabrication, Lithographie Galvanoformung Abformung (LIGA), and laser ablation.[193] A very sophisticated version of micro-wire EDM uses 20-μm diameter wire to cut through a conductive workpiece, but it is not a sufficiently reliable technique for nanoscale resolution.[194] Pulsed laser ablation provides a means of depositing thin coatings under vacuum and in the presence of lower pressures of background gas (He, Ar, H_2, N_2), of a wide range of target materials (e.g., graphite, CVD diamond, copper and aluminium, ZnO and LiF, and various polymeric materials), on a wide range of substrates, at room temperature,[195] and it can provide a resolution of 1 μm for some materials.[196] With LIGA, a high-energy x-ray source creates high aspect-ratio features as small as 0.5 μm in prepolymer-coated silicon; the cured and cleaned pattern is later electroplated, usually with nickel. Although LIGA is a leading method for manufacturing of miniature metal parts, there are several drawbacks, including the relatively soft nickel surface (which is subject to abrasion and wear), slow processing due to multiple steps including electroplating, and an expensive synchrotron light source.[197] Due to the drawbacks, only a few products, such as a spectrometer, aspirator, and miniature motor, have been commercialized since the first publication of LIGA in the mid 1980s.[198,199] DVD tooling, disks containing depressions or pits that are 100 to 400 nm wide and 140 nm deep,[200] are prepared using photolithography and electroforming of nickel. Finally, ultraviolet (UV) and electron beam lithography of silicon wafers provide current resolution limits of 157 nm[201] and less than 10 nm,[202] respectively, but the primary substrate material for these processes is silicon rather than steel.

As a serious candidate for future tooling material, silicon has several attractive factors, including high hardness, low linear thermal expansion, high thermal conductivity, and a flatter surface than steel.[203,204] Such tooling has been employed for hot embossing, which is performed at temperatures near the softening temperature of a polymer using relatively low pressures and heating times of 5 to 10 min.[205,206] Becker and Heim[204] used silicon wafers for hot embossing 8 μm features, and Heyderman et al.[206,207] were able to hot emboss features with a minimum size of 50 nm using silicon and chromium-germanium coated silicon as the tooling. More recently, Grewell et al.[208,209] were able to reduce the heating time of hot embossing down to less than 10 s by using infrared (IR), ultrasonic, and laser heating sources focused on a small area (i.e., 7×30 mm for IR and 4×25 mm for laser). Currently, 1 to 3 min of cycle time is available by commercial hot embossing equipment.[210] With UV nanoimprint lithography, a UV-transmittable quartz mold is pressed into a thin film of low-viscosity UV-curable prepolymer (i.e., photoresist), spin coated onto a substrate (usually silicon). After curing the photoresist, the mold is separated from the substrate, and residual prepolymer is removed using high anisotropic reactive ion etching. Nanoimprint lithography produces a polymeric pattern at room temperature and has a cycle time about 2 min.[211] In contrast, Shah et al.[212] introduced wet-etched silicon wafers with a minimum feature size of 40 μm as a direct tooling material for injection molding, and Yu et al.[213] employed silicon wafers as a substrate for making a nickel mold insert through UV-lithography and electrodeposition. Using silicon wafer with V-shaped grooves of 1 μm opening width and aspect ratio of 0.707, D'Amore et al.[214] were able to evaluate the moldability and achieved filling aspect ratio of 0.5 by using the injection-compression molding process. With some modifications to the molding parameters and insert mounting, Yoon et al.[215] found that silicon wafers could survive molding pressures of 50 MPa, but gallium arsenide fractured prematurely at pressures of 5 MPa. This silicon tooling survived more than 3000 injection molding cycles without damage to the silicon surface.[216] A comparison of these techniques is given in Table 6.3.

Direct use of silicon and gallium arsenide as mold inserts for injection molding showed that tooling surface quality, geometric effects, and heat transfer were critical for replication. Passive heating was effective for replication, but, unlike the other micromolding methods, longer cooling times were not always beneficial for replication quality. The replication quality tended to decreases with the aspect ratio (depth-to-width) of the tooling due to heat transfer issues as well as deteriorating quality of the tooling.

TABLE 6.3 Comparison of Manufacturing Techniques for Nanoscale

	Hot embossing	Nanoimprint	Injection molding
Materials	Thermoplastics	Photoresist (PMMA, epoxy)	Thermoplastics or thermosets
Pressure	>10 MPa	<0.1 MPa	>50 MPa
Temperature	$>T_g$	~25°C	150–400°C
Cycle time	1–10 min	2 min	3–14 s

6.6 SUMMARY

The development of modified and completely new high-rate processes for controlling nanoscale reinforcements, fibers, layers, and features represents an area of significant promise and growth. While much of the past and current efforts in nanoscale manufacturing have resided in the semiconductor and microelectronics industry, processing advances for "soft" materials—i.e., polymeric and biological materials—are still at an early stage. Commercial products are already on the market but are typically limited to slight modifications of existing equipment to incorporate nanoscale particles and fillers. Near-term advances are likely to continue in this vein, since this approach creates products that readily solve the nano-to-micro-to-macro integration issue; for example, nanofeatures molded onto a microscale component or nanolayers in a micro or macro film. Even for existing manufacturing processes, however, necessary modifications to equipment and processes require a better fundamental understanding of nanoscale issues. Barriers such as dispersion, orientation control, layer stability, and replication quality at high rates, over long lengths and high volumes, and for a broad range of materials, still remain. As these challenges to high-rate processing are overcome, these materials will find application in not only the biomedical field but also in electronics, sensing, displays, packaging, and many other product areas.

6.7 REFERENCES

1. Meals, R. N. and Lewis, F. M. *Silicones*. Reinhold Publishing Co., New York (1959).
2. Fordham, S. *Silicones*. Philosophical Library Inc., New York (1961).
3. Basiuk, E. V., Monroy-Peláez, M., Puente-Lee, I., and Basiuk, V. A. *Nano Lett.* 4, 863 (2004).
4. Zhu, J., Yudasaka, M., Zhang, M., and Iijima, S. *J. Phys. Chem.* B 108, 11317 (2004).
5. Ajayan, P. M., Stephan, O., Colliex, C., and Trauth, D. *Science* 265 1212 (1994).
6. Sandler, J., Shaffer, M. S. P., Prasse, T., Bauhofer, W., Schulte, K., and Windle, A. H. *Polymer* 40, 5967 (1999).
7. Gong, X. Y., Liu, J., Baskaran, S., Voise, R. D., and Young, J. S. *Chem. Mater.* 12, 1049 (2000).
8. Shaffer, M. S. P. and Windle, A. H. *Adv. Mater.* 11, 937 (1999).
9. Jia, Z., Wang, Z., Xu, C., Liang, J., Wei, B., Wu, D., and Zhu, S. W. *Mater. Sci. Eng.* A 271, 395 (1999).
10. Stéphan, C., Nguyen, T. P., Lamy de la Chapelle, M., Lefrant, S., Journet, C., and Bernier, P. *Synth. Met.* 108, 139 (2000).
11. Haggenmueller, R., Gommans, H. H., Rinzler, A. G., Fischer, J. E., and Winey, K. I. *Chem. Phys. Lett.* 330, 219 (2000).

12. Curran, S. A., Ajayan, P. M., Blau, W. J., Carroll, D. L., Coleman, J. N., Dalton, A. B., Davey, A. P., Drury, A., McCarthy, B., Maier, S., and Strevens, A. *Adv. Mater.* 10, 1091 (1998).

13. Ago, H., Petritsch, K., Shaffer, M. S. P., Windle, A. H., and Friend, R. H. *Adv. Mater.* 11, 1281 (1999).

14. Dalton, A. B., Stephan, C., Coleman, J. N., McCarthy, B., Ajayan, P. M., Lefrant, S., Bernier, P., Blau, W. J., and Byrne, H. J. *J. Phys. Chem.* B 104, 10012 (2000).

15. McCarthy, B., Coleman, J. L., Curran, S. A., Dalton, A. B., Davey, A. P., Z. Konya, Z., Fonseca, A., Nagy, J. B., and Blau, W. J. *J. Mater. Sci. Lett.* 19, 2239 (2000).

16. Dalton, A. B., Blau, W. J., Chambers, G., Coleman, J. N., Henderson, K., Lefrant, S., McCarthy, B., Stephan, C., and Byrne, H. J. *Synth. Met.* 121, 1217 (2001).

17. Schmidt, D. F., Qian, G., and Giannelis, E. P. *Polym. Mater. Sci. Eng.* 82, 215 (2000).

18. Breu, J., Seidl, W., Stoll, A. J., Lange, K. G., and Probst, T. U. *Chem. Mater.* 13, 4213 (2001).

19. Sato, H., Yamagishi, A., and Kawamura, K. *J. Phys. Chem.* B 105, 7990 (2001).

20. Piner, R. (Northwestern University, Evanston, IL), personal communications 6/02–11/02.

21. Prasad, M., Kopycinska, M., Rabe, U., and Arnold, W. *Geophys. Res. Lett.* 29, 13–1 (2002).

22. Manevitch, O. L., and Rutledge, G. C. *J. Phys. Chem.* B 108, 1428 (2004).

23. Giannelis, E. P., Krishnamoorti, R., and Manias, E. *Adv. Poly. Sci.* 138, 107 (1999).

24. Morgan, A. B., Gilman, J. W., and Jackson, C. L. *Polym. Mat. Sci. Eng.* 82, 270 (2000).

25. Manias, E., "Origins of the Materials Properties Enhancements in Polymer/Clay Nanocomposites," *Proc. Nanocomposites 2001: Delivering New Value to Plastics,* A. Golovoy, Ed., Executive Conference Management, Chicago, IL (2001).

26. Vaia, R. A., and Giannelis, E. P. *Macromol.* 30, 7990 (1997).

27. Balazs, A. C., Singh, C., and Zhulina, E. *Macromol.* 31, 8370 (1998).

28. Guenthner, A. J., and Kyu, T. *J. Poly. Sci.* B Poly. Phys. 38, 2366 (2000).

29. Griessbach, E. F. C., and Lehmann, R. G. *Chemosphere* 38 1461 (1999).

30. Singh, U. B., Gupta, S. C., Flerchinger, G. N., Moncrief, J. F., Lehmann, R. G., Fendinger, N. J., Traina, S. J., and Logan, T. *J. Environ. Sci. Technol.* 34, 266 (2000).

31. Davis, R. D., Gilman, J. W., and VanderHart, D. L. *Polym. Degrad. Stab.* 79 111 (2003).

32. Alexandre, M., Dubois, P., Sun, T., Garces, J. M., and Jérôme, R. *Polymer* 43, 2123 (2002).

33. VanderHart, D. L., Asano, A., and Gilman, J. W. *Chem. Mater.* 13, 3796 (2001).

34. Zhu, J., Morgan, A. B., Lamelas, F. J., and Wilkie, C. A. *Chem. Mater.* 13, 3774 (2001).

35. Gilman, J. W., Awad, W. H., Davis, R. D., Shields, J., Harris (Jr.), R. H., Davis, C., Morgan, A. B., Sutto, T. E., Callahan, J., Trulove, P. C., and DeLong, H. C. *Chem. Mater.* 14, 3776 (2002).

36. Xie, W., Xie, R., Pan, W.-P., Hunter, D., Koene, B., Tan, L.-S., and Vaia, R. *Chem. Mater.* 14, 4837 (2002).

37. Okada, A., Kawasumi, M., Kurauchi, T., and Kamigaito, O. *Polym. Prepr.* 28, 447 (1987).

38. Okada, A., Kawasumi, M., Usuki, A., Kojima, Y., Kurauchi, T., and Kamigaito, O. in *Polymer Based Molecular Composites (Mater. Res. Soc. Symp. Proc. 171),* 45, D. W. Schaefer and J. E. Mark, Eds., Materials Research Society, Pittsburgh, PA (1990).

39. Kojima, Y., Usuki, A., Kawasumi, M., Okada, A., Kurauchi, T., and Kamigaito, O. *J. Poly. Sci.* A Poly. Chem. 31, 983 (1993).

40. Kojima, Y., Usuki, A., Kawasumi, M., Okada, A., Fukushima, Y., Kurauchi, T., and Kamigaito, O. *J. Mater. Res.* 8, 1185 (1993).

41. Messersmith, P. B., and Giannelis, E. P. *Chem. Mater.* 6, 1719 (1994).

42. Lan, T., and Pinnavaia, T. J. *Chem. Mater.* 6, 2216 (1994).

43. Lan, T., Kaviratna, P. D., and Pinnavaia, T. J. *Polym. Mater. Sci. Eng.* 71, 527 (1994).

44. Pinnavaia, T. J., Lan, T., Wang, Z., Shi, H., and Kaviratna, P. D. in *Nanotechnology (Am. Chem. Soc. Symp. Ser. 622),* 250, G.-M. Chow and K. E. Gonsalves, Eds., American Chemical Society, Washington, DC (1996).

45. Zilg, C., Mülhaupt, R., and Finter, J. *Macromol. Chem. Phys.* 200, 661 (1999).

46. Liu, W., Hoa, S. V., and Pugh, M. *Polym. Eng. Sci.* 44, 1178 (2004).

47. Lu, H., Liang, G-Z., Ma, X., Zhang, B., and Chen, X. *Polym. Int.* 53, 1545 (2004).

48. Shah, D., Maiti, P., Gunn, E., Schmidt, D. F., Jiang, D. D., Batt, C. A., and Giannelis, E. P. *Adv. Mater.* 16, 1173 (2004).

49. Shah, D., Maiti, P., Jiang, D. D., Batt, C. A., and Giannelis, E. P. *Adv. Mater.* 17, 525 (2005).

50. Wang, Z. and Pinnavaia, T. J. *Chem. Mater.* 10, 3769 (1998).

51. Plummer, C. J. G., Garamszegi, L., Leterrier, Y., Rodlert, M., and Månson, J-A. E. *Chem. Mater.* 14, 486 (2002).
52. Recihert, P., Nitz, H., Klinke, S., Brandsch, R., Thomann, R., Mülhaupt, R. *Macromol. Mater. Eng.* 275, 8 (2000).
53. Manias, E., Touny, A., Wu, L., Strawhecker, K., Lu, B., and Chung, T. C. *Chem. Mater.* 13, 3516 (2001).
54. Tervoort, T. A., Visjager, J., and Smith, P. *Macromol.* 35, 8467 (2002).
55. Maeda, N., Chen, N., Tirrell, M., and Israelachvili, J. N. *Science* 297, 379 (2002).
56. Yano, K., Usuki, A., Okada, A., Kurauchi, T., and Kamigaito, O. *J. Poly. Sci. A Poly. Chem.* 31, 2493 (1993).
57. Messersmith, P. B., and Giannelis, E. P. *J. Poly. Sci. A Poly. Chem.* 33, 1047 (1995).
58. Wang, Z., and Pinnavaia, T. J. *Chem. Mater.* 10, 3769 (1998).
59. Usuki, A., Tukigase, A., and Kato, M. *Polymer* 43, 2185 (2002).
60. Xu, R., Manias, E., Snyder, A. J., and Runt, J. *J. Biomed. Mater. Res.* 64A, 114 (2003).
61. Lee, J., Takekoshi, T., and Giannelis, E. P. in *Nanophase and Nanocomposite Materials II (Mater. Res. Soc. Symp. Proc. 457),* 513, S. Komarneni, J. C. Parker, and H. J. Wollenberger, Eds., Materials Research Society, Pittsburgh, PA (1997).
62. Gilman, J. W., Kashiwagi, T., and Lichtenhan, J. D. *SAMPE J.* 33, 40 (1997).
63. Lee, J., and Giannelis, E. P. *Polym. Prepr.* 38, 688 (1997).
64. Gilman, J. W., Kashiwagi, T., Giannelis, E. P., Manias, E., Lomakin, S., Lichtenhan, J. D., and Jones, P. in *Fire Retardancy of Polymers: The Use of Intumescence* (Spec. Pub. Roy. Soc. Chem. 224), 203, M. Le Bras, G. Camino, S. Bourbigot, and R. Delobel, Eds., Royal Society of Chemistry Information Services, Cambridge, UK (1998).
65. Ke, Y., Long, C., and Qi, Z. *J. Appl. Poly. Sci.* 71, 1139 (1999).
66. Kuchta, F.-D., Lemstra, P. J., Keller, A., Batenburg, L. F., and Fischer, H. R. in *Organic/Inorganic Hybrid Materials II (Mater. Res. Soc. Symp. Proc. 576),* 363, L. C. Klein et al., Eds., Materials Research Society, Pittsburgh, PA (1999).
67. Tseng, C.-R., Lee, H.-Y., and Chang, F.-C. *J. Poly. Sci. B Poly. Phys.* 39, 2097 (2001).
68. Wu, T.-M., Chen, E.-C., and Liao, C.-S. *Poly. Eng. Sci.* 42, 1141 (2002).
69. Hsu, S. L.-C., and Chang, K.-C. *Polymer* 43, 4097 (2002).
70. Bandyopadhyay, S., Chen, R., and Giannelis, E. P. *Polym. Mater. Sci. Eng.* 81, 159 (1999).
71. Ratto, J. A., Steeves, D. M., Welsh, E. A., and Powell, B. E. in *ANTEC 1999 Plastics (Proc. 57th Ann. Tech. Conf. Soc. Plast. Eng.)* 2, 1628, Society of Plastics Engineers, New York (1999).
72. Lee, S.-R., Park, H.-M., Lim, H., Kang, T., Li, X., Cho, W.-J., and Ha, C.-S. *Polymer* 43, 2495 (2002).
73. Goldman, A. Y., Montes, J. A., Barajas, A., Beall, G., and Eisenhour, D. D. in *ANTEC 1998 Plastics (Proc. 56th Ann. Tech. Conf. Soc. Plast. Eng.)* 2, 2415, Society of Plastics Engineers, New York (1998).
74. Fischer, S., De Vlieger, J., Koch, T., Batenburg, L., and Fischer, H. in *Filled and Nanocomposite Polymer Materials (Mater. Res. Soc. Symp. Proc. 661),* KK2. 2/1, A. I. Nakatani et al., Eds., Materials Research Society, Pittsburgh, PA (2001).
75. Giannelis, E. P. in *Biomimetic Materials Chemistry,* 337, S. Mann, Ed., Wiley VCH, New York (1996).
76. LeBaron, P. C., Wang, Z., and Pinnavaia, T. J. *Appl. Clay Sci.* 15, 11 (1999).
77. Alexandre, M., and Dubois, P. *Mat. Sci. Eng. R. Rep.* 28, 1 (2000).
78. Vaia, R. A., and Giannelis, E. P. *MRS Bull.* 26, 394 (2001).
79. Schmidt, D., Shah, D. and Giannelis, E. P. *Curr. Op. Sol. St. Mat. Sci.* 6, 205 (2002).
80. Hackett, E., Manias, E., and Giannelis, E. P. *Chem. Mater.* 12, 2161 (2000).
81. Hackett, E. Computer Simulations of Confined Organic Molecules in 2:1 Layer Silicates. Ph.D. dissertation, Dept. of Materials Science & Engineering, Cornell University, Ithaca, NY (2001).
82. Manias, E., and Kuppa, V. in *Polymer Nanocomposites (Am. Chem. Soc. Symp. Ser. 804),* 193, R. Krishnamoorti and R. A. Vaia, Eds., American Chemical Society, Washington, DC (2002).
83. Moore, D. M., and Reynolds, R. C. *X-Ray Diffraction and the Identification and Analysis of Clay Minerals,* 2nd ed. Oxford University Press, New York (1997).
84. Kuchta, F.-D., Lemstra, P. J., Keller, A., Batenburg, L. F., and Fischer, H. R. in *Organic/Inorganic Hybrid Materials (Mater. Res. Soc. Symp. Proc. 628),* CC11. 12. 1, R. Laine et al., Eds., Materials Research Society, Pittsburgh, PA (2001).

85. Saujanya, C., Imai, Y., and Tateyama, H. *Poly. Bull.* 49, 69 (2002).
86. Svoboda, P., Zeng, C., Wang, H., Lee, L. J., and Tomasko, D. L. *J. Appl. Poly. Sci.* 85, 1562 (2002).
87. Zax, D. B., Yang, D.-K., Santos, R. A., Hegemann, H., Giannelis, E. P., and Manias, E. *J. Chem. Phys.* 112, 2945 (2000).
88. Manias, E., Kuppa, V., Yang, D.-K., and Zax, D. B. *Coll. Surf. A Physicochem. Eng. Asp.* 187–188, 509 (2001).
89. Anastasiadis, S. H., Karatasos, K., Vlachos, G., Manias, E., and Giannelis, E. P. *Phys. Rev. Lett.* 84, 915 (2000).
90. Krishnamoorti, R., Vaia, R. A., and Giannelis, E. P. *Chem. Mater.* 8, 1728 (1996).
91. Kwiatkowski, J., and Whittaker, A. *J. Poly. Sci. B Poly. Phys.* 39, 1678 (2001).
92. Mehta, B. S., DiBenedetto, A. T., and Kardos, J. L., *Int. J. Poly. Mater.* 3, 269 (1975).
93. Fredrickson, G. H., and Bicerano, J. *J. Chem. Phys.* 110, 2181 (1999).
94. Bharadwaj, R. K. *Macromol.* 34, 9189 (2001).
95. Leddy, J. *Langmuir* 15, 710 (1999).
96. Beall, G. W., and Tsipursky, S. J. in *Chemistry and Technology of Polymer Additives,* Ch. 15, 266, S. Al-Malaika, A. Golovoy, and C. A. Wilkie, Eds., Blackwell Science, Malden, MA (1999).
97. Beall, G. W. in *ANTEC 2001 Plastics (Proc. 59th Ann. Tech. Conf. Soc. Plast. Eng.)* 2, 2195, Society of Plastics Engineers, New York (2001).
98. Formhals, A., Method and Apparatus for Production of Fibers, U.S. Patent No. 1,975,504 (filed 1934).
99. Taylor, G. I., *Proc. Roy. Soc. London,* A313, 453 (1969).
100. Doshi, J., The Electrospinning Process and Applications of Electrospun Fibers, Ph.D. thesis, University of Akron, Akron, OH (1994).
101. Reneker, D. H., Yarin, A. L., Fong, H., and Koombhongse, S., *J. Appl. Phys.* 87, 4531 (2000).
102. Shin, Y. M., Hohman, M. M., Brenner, M. P., and Rutledge, G. C., "Electrospinning: A whipping fluid jet generates submicron polymer fibers," *Appl. Phys. Lett.* 78, 1149 (2001).
103. Simons, H. L., Process and Apparatus for Producing Non-Woven Fabric, U.S. Patent no. 3,280,229 (1966).
104. Simm, W., et al., Fiber made of electrostatically spun fibers, U.S. Patent no. 4,069,026, Jan. 17 (1978).
105. Simm, W., et al., Fiber fleece of electrostatically spun fibers and methods of making same, U.S. Patent no. 4,143,196, Mar. 6 (1976).
106. Finn, J., De Tora, S. A., Method of producing fibrous structure, U.S. Patent No. 4,223,101, Sep. 16 (1980).
107. Berry, J. P., Method and Apparatus for Manufacturing Electrostatically Spun Structure, U.S. Patent no. 5,024,789, June (1991).
108. How, T. V., Synthetic vascular grafts and methods of manufacturing such grafts, U.S. Patent no. 4,552,707, November (1985).
109. Bornat, A., Production of Electrostatically Spun Products, U.S. Patent no. 4,689,186, June (1987).
110. Scardino, F. L. and Richard. J. B., Fibrous Structures Containing NanoFibrils and Other Textile Fibers, U.S. Patent no. 6,308,509, Oct. (2001).
111. Doshi, J., Reneker, D. H., *J. Electrostat.* 35, 151–160 (1995).
112. Reneker, D. H. and Chun, I., *Nanotechnology* 7, 216–223 (1996).
113. Warner, S. B., Buer, A., Ugbolue, S. C., Rutledge, G. C., Shin, M. Y., National Textile Center Annual Report (www.ntcresearch.org) (1998).
114. Srinivasan, G., Reneker, D. H., *Polymer Intl.* 36(2) (1995).
115. Fong, H., Chun, I., Reneker, D. H., *Polymer* 40, 4585–4592 (1999).
116. Deitzel, J. M., Kleinmeyer, J. D., Harris, D., Beck Tan, N. C., *Polymer* 42, 261–272 (2001).
117. Kim, J. S., Reneker, D. H., *Polymer Engg. and Science* 39(5) (1999).
118. Bognitzki, M., Frese, T., Steinhart, M., Greiner, A., Wendorff, J. H., *Polymer Engg. and Science* 41(6) (2001).
119. Fridrikh, S. V., Yu, J. H., Brenner, M. P., Rutledge, G. C., *Phys. Review Lett.* 90 (14), 11 (2003).
120. Gibson, P., et al., *Textile Res. J.* 69 (5), 311–317 (1999).
121. Gibson, P. W., Gibson, H. L., Rivin, D., *AIChE J* 45, 1, 190–195 (1999).

122. Gibson, P. W., Schreuder-Gibson, H. L., Rivin, D., *Colloids and Surfaces, A: Physicochemical and Engineering Aspects,* 469–481 (2001).
123. Drew, C., Wang, X., Senecal, K., Schreuder-Gibson, H., He, J., Tripathy, S., and Samuelson, L., *SPE Technical Papers* 46(2), 1477 (2000).
124. Buer, A., Ugbolue, S. C., and Warner, S., *Textile Res. J.* 71(4), 323 (2001).
125. Jaeger, R., Schönherr, H., and Vancso, G. J., *Macromolecules* 29, 7643 (1996).
126. Bergshoef. M. and Vansco, G. J., "Transparent Nanocomposites with Ultrathin Electrospun Nylon 4,6 Fiber Reinforcement," *Adv. Mat.* 11, 1362 (1999).
127. Buchko, C. J., Chen, L. C., Shen, Y., Martin, D. C., "Processing and microstructural characterization of porous biocompatible protein polymer thin films," *Polymer* 40, 7397 (1999).
128. Reneker, D. H. and Chun, I. "Nanometre diameter fibres of polymer, produced by electrospinning," *Nanotechnology* 7, 216 (1996).
129. Kim, J.-S. and Reneker, D. H., "Polybenzimidazole Nanofiber Produced by Electrospinning," *Polym. Eng. & Sci.* 39, 849 (1999).
130. Fong, H., Chun, I., and Reneker, D. H., "Beaded nanofibers formed during electrospinning," *Polymer* 40, 4585 (1999).
131. Larrondo, L. and Manley, R., *J. Polym. Sci B: Polym. Phys.* 19, 909 (1981).
132. Doshi, J. and Reneker, D. H., *J. Elec.* 35, 151 (1995).
133. Bunyan, N., Control of Deposition and Orientation of Electrospun Fibers, MS thesis, Department of Mechanical Engineering, University of Massachusetts Lowell, Lowell, MA (2003).
134. Bhowmick, S., University of Massachusetts Dartmouth (2005).
135. Hohman, M. M., Shin, M., Brenner, M. P., Rutledge, G. C., *Phys. Fluids* 13, 2201–2221(2001).
136. Hohman, M. M., Shin, M., Brenner, M. P., Rutledge, G. C., *Phys. Fluids* 13, 2221–2236 (2001).
137. Spivak, A. F., Dzenis, Y. A., Reneker, D. H., *Mechanics Research Comm.* 27(1), 37–42 (2000).
138. Utracki, L. A., Part 1: *Introduction to Polymer Alloys and Blend, Polymer Alloys and Blends-Thermodynamics and Rheology,* 1990, Carl Hanser Verlag, New York, pp. 5.
139. Schrenk, W. J., and Alfrey, Jr., T., "Some Physical Properties of Multilayered Films," *Polym. Eng. & Sci.* 9, 393 (1969)
140. Baer, E., Jarus, D., and Hiltner, A., "Microlayer Coextrusion Technology," *SPE Technical Papers,* 1999, p. 3947.
141. http://www.sciencenews.org/articles/20030809/bob9.asp, accessed November 29, 2005.
142. http://www.nature.com/news/2004/041206/pf/041206-5_pf.html, accessed November 29, 2005.
143. Zumbrunnen, D. A., and Inamdar, S. "Novel sub-micron highly multi-layered polymer films formed by continuous flow chaotic mixing," *Chem. Eng. Sci.* 56, 3893 (2001).
144. http://www.nomadics.com/_media/documents/pdf/Nanotechnology_Fact_Sheet_Layer_by_Layer_Assembly.pdf, accessed November 29, 2005.
145. Schrenk, W. J, Alfrey, T. J., Paul, D. R. (Ed.), Newman, S. (Ed.), *Coextruded Multilayer Polymer Films and Sheets, Polymer Blends,* Vol. II, 1st ed., Academic Press, London, 1979, pp. 129.
146. Pan, S. J., Im, J., Hill, M. J., Keller, A., Hiltner, A., Baer, E., "Structure of Ultrathin Polyethylene Layers in Polyethylene Multilayer Films," *J. Poly. Sci. Part B: Poly. Phys.* 28, 1990, pp. 1105–1119.
147. Schrenk, W. J., Alfrey, T. J, "Some Physical Properties of Multilayered Films," *Poly. Engg. and Sci.* 9, 6, 1969, pp. 393.
148. Schrenk, W. J., "Some Contributions of Turner Alfrey to Fabrication of Polymers," *J. Poly. Sci.: Poly. Symp.* 72, 1985, pp. 307–319.
149. Gurne, E. F., Schrenk, W. J., Alfrey, T. J., "Physical Optics of Multi-layered Plastic Films," *Poly. Eng. and Sci.* 9, 6, 1969, pp. 400.
150. Ramanathan, R., Shanker, R., Rehg, T., Jons, S., "'Wave' Pattern Instability in Multilayer Coextrusion: an Experimental Investigation," *SPE*—54th Ann. Tech. Conf., Indianapolis, Vol. I, 1996, pp. 226.
151. Shin, E., Hiltner, A., Baer, E., "The Damage Zone in Microlayer Composition of Poly carbonate and Styrene-acrylonitrile," *J. Appl. Poly. Sci.,* 47, 1993, pp. 245–267.
152. Mueller, C. D., Nazarenko, S., Ebeling, T., Schuman, T. L., Hiltner, A., Baer, E., "Novel Structures by Microlayer Coextrusion—Talc filled PP, PC/SAN, and HDPE/ LLDPE," *Poly. Eng. and Sci.* 37, 2, pp. 355.
153. Liu, R. Y. F., Jin, Y., Hiltner, A., Baer, E., "Probing Nanoscale Polymer Interactions by Forced Assembly," *Macr. Rapid Comm.* 24, 16, 2003, pp. 943–948.

154. Zhao, R., Macosko, C. W., "Slip at Polymer-Polymer Interfaces: Rheological Measurements on Coextruded Multilayers," *J. Rheo.* 46, 1, 2002, 145–167.
155. The Cloeren Company, *Modern Plastics* (1985), Vol. 62, Dec., p. 28.
156. Ho, K., Lee, J. S., Viriyabanthron, N., Sung, C., Barry, C. M. F., Mead, J. L., "Investigation of Interfacial Instabilities in Nanolayer Extrusion," SPE-62nd Ann. Tech. Conf., Chicago, Vol. I, 376 (2004).
157. Mueller, C., Topolkaraev, V., Soerens, D., Hiltner, A., and Baer, E., "Breathable Polymer Films Produced by the Microlayer Coextrusion Process," *J. Appl. Polym. Sci.* 78, 816 (2000).
158. Schuman, T., Nazarenko, S., Stepanov, E. V., Magonov, S. N., Hiltner, A., and Baer, E., "Solid state structure and melting behavior of interdiffused polyethylenes in microlayers," *Polymer* 40, 7373 (1999).
159. Ronesi, V., Cheung, W., Chum, S. P., Hiltner, A., and Baer, E., "Adhesion of Ethylene-Styrene Copolymers to Polyethylene in Microlayers," *SPE Technical Papers* 1844 (2002).
160. Poon, B., Chum, S. P., Hiltner, A., and Baer, E., "Adhesion of Polypropylene to Metallocene/Ziegler-Natta Polyethylene Blends in Microlayers," *SPE Technical Papers* 1849 (2002).
161. Meuller, C., Nazarenko, S., Ebeling, T.,Schuman, T., Hiltner, A., and Baer, E., "Novel Structures by Microlayer Coextrusion—Talc-Filled PP, PC/SAN, and HDPE/LLDPE," *Polym. Eng. & Sci.* 37, 355 (1997).
162. Cole, P. J., and Macosko, C. W., "Polymer-Polymer Adhesion in Melt-Processed Layered Structures," *J. Plastic Film and Sheeting* 16, 213 (2000).
163. Jarus, D., Hiltner, A., and Baer, E., "Barrier properties of polypropylene/polyamide blends produced by microlayer coextrusion," *Polymer* 43, 2401 (2002).
164. Nazarenko, S., Dennison, M., Schuman, T., Stepanov, E. V., Hiltner, A., and Baer, E., "Creating Layers of Concentrated Inorganic Particles by Interdiffusion of Polyethylenes in Microlayers," *J. Appl. Polym. Sci.* 73, 2877 (1999).
165. Sekelik, R. J., Stepanov, E. V., Nazarenko, S., Schiraldi, D., Hiltner, A., and Baer, E., "Oxygen Barrier Properties of Crystallized and Talc-Filled Poly(ethylene terephthalate)," *J. Polym. Sci. Part B: Polymer Physics* 37, 847 (1999).
166. Ebeling, T., Norek, S., Hasan, A., Hiltner, A., and Baer, E., *J. Appl. Polym. Sci.* 71, 1461 (1999).
167. Zhao, X.-M., Xia, Y., Whitesides, G. M., *Advanced Materials* 8 (10), 837–840 (1996).
168. Reyes, D. R., Iossifidis, D., Auroux, P.-A., Manz, A., *Anal. Chem.* 74, 2623–2636 (2002).
169. Wang, X., Bullen, D., Zou, J., Ryu, K., Liu, C., Chung, S.-W., Mirkin, C. A., in Linear probe arrays for dip-pen nanolithography, International Conference on Micro & Nano Systems, Kunming, China, 2002; Kunming, China, 2002.
170. Zhang, H., Chung, S.-W., Mirkin, C. A., *Nanoletters* 3 (1), 43–45 (2003).
171. Yu, M., Ivanisevic, A., In Nanoscale surface patterning, *Materials Research Society Symposium Proceedings*, pp. Q8.19.1–Q8.19.5 (2003).
172. Li, H., Huck, S., *Curr. Op. Solid State and Mat. Sci.* 6 (1), 3–8 (2002).
173. McAlpine, M. C., Friedman, R. S., Lieber, C. M., *Nano Lett.* 3 (4), 443–445 (2003).
174. Grass, B., Neyer, A., Johnck, M., Siepe, D., Eisenbeiss, F., Weber, G., Hergenroder, R., *Sensors and Actuators B: Chemical* 72 (3), 249–258 (2001).
175. Jaszewski, R. W., Gobrecht, J., and Smith, P., *Microelectronic Eng.* 41–42, 575–578 (1998).
176. Schift, H., David, C., Gabriel, M., Gobrecht, J., Heyderman, L. J., Kaiser, W., Koppel, S., and Scandella, L., *Microelectronic Eng.* 53 (1–4), 171–174 (2000).
177. Liu, Y., Ganser, D., Schneider, A., Liu, R., Grodzinski, P., Kroutchinina, N., *Anal. Chem.* 73, 4196–4201 (2001).
178. McCormick, R. M., Nelson, R. J., Alonso-Amigo, M. G., Benvegnu, D. J., and Hooper, H. H., *Anal. Chem.* 69, 2626–2630 (1997).
179. Edwards, T. L., Mohanty, S. K., Edwards, R. K.,Thomas, C., Frazier, A. B., in Rapid tooling using SU-8 for injection molding microfluidic components, *Proc. SPIE* (2000).
180. Hulme, J. P., Fielden, P. R., Goddard, N. J., *Anal. Chem.* 76, 238–243 (2004).
181. Kang, S., *Japanese J. App. Phys.* 43 (8B), 5706–5716 (2004).
182. Lai, S., Lee, L. J., Yu, L., Koelling, K. W., Madou, M. J., in Micro- and nano-fabrication of polymer based microfluidic platforms for bioMEMS applications, *Materials Research Society Symposium Proceedings,* 17–27 (2002).
183. Yao, D., and Kim, B., in Thin-wall injection molding using rapidly heated molds, Annual Technical Conference, 2003; Society of Plastics Engineering, 521–525 (2003).

184. Schift, H., David, C., Gabriel, M., Gobrecht, J., Heyderman, L. J., Kaiser, W., Koppel, S., and Scandella, L., *Microelectronic Eng.* 53 (1–4), 171–174 (2000).
185. Oshiro, T., Goto, T., Ishibashi, J., Annual Technical Conference 55 (1), 409–414 (1997).
186. Sancoucy, M., private conversation (2005).
187. Ebert, W., Kaufmann, R., Schwemler, C., Haese, W., Fischer, P., Dobler, M., Genz, J., Polycarbonate molding compounds with good demolding properties and moldings therefrom (2001).
188. Park, S. J., Han, J. H., Ryim, W. G., Chang, S. K., Kim, J. H., Kang, T. G., Heo, B. S., and Kwon, T. H., Annual Technical Conference 2 (56), 1756–1760 (1998).
189. Srirojpinyo, C., Yoon, S., Lee, J. S., Mead, J. L., and Barry, C. M. F., *Proc. Annual Technical Conference of Society of Plastics Engineers* 51 (2005).
190. Srirojpinyo, C., Yoon, S., Lee, J., Sung, C., Mead, J. L., and Barry, C. M. F., "Effects of Materials When Injection Molding Nano-Scale Features," *Proc. Annual Technical Conference of Society of Plastics Engineers* 50, 743 (2004).
191. Malloy, R. A., *Plastics Part Design for Injection Molding,* Hanser Gardner Publications (1994).
192. Becker, H., and Heim, U., *Sensors and Actuators* 83, 130–135 (2000).
193. Weber, L., and Ehrfeld, W., *Proc. Annual Technical Conference of Society of Plastics Engineers,* 1998, 930.
194. Pham, D. T., Dimov, S. S., Bigot, S., Ivanov, A., and Popov, K., *J. Materials Processing Technology* 149, 50–57 (2004).
195. http://www.chm.bris.ac.uk/pt/laser/ashfold/ablation.htm, last accessed November 22, 2005.
196. Kim, J. T., Kim, B. C., Jeong, M. Y., and Lee, M. S., *J. Materials Processing Technology* 146, 163–166 (2004).
197. Hirata, Y., *Nuclear Instruments and Methods in Physics Research Section B: Beam Interactions with Materials and Atoms,* 208, 21–26 (2003).
198. Hruby, J., Sandia National Laboratories (2000).
199. Hormes, J., Göttert, J., Lian, K., Desta, Y., and Jian, L., *Nuclear Instruments and Methods in Physics Research Section B: Beam Interactions with Materials and Atoms,* 199, 332–341 (2002).
200. Sharpless, G., *CD and DVD Disc Manufacturing* (2002).
201. Fay, B., *Microelectronic Eng.* 61–62, 11–24 (2002).
202. Lehmann, F., Richter, G., Borzenko, T., Hock, V., Schmidt, G., and Molenkamp, L. W., *Microelectronic Eng.* 65, 327–333 (2003).
203. Ansari, K., Kan, J. A. v., Bettiol, A. A., and Watt, F., *App. Phys. Lett.* 85, 476–478 (2004).
204. Becker, H., and Heim, U. IEEE International Conference on Micro Electro Mechanical Systems, Orlando, 228–231 (1999).
205. Lee, W., Jin, M. K., Yoo, W. C., and Lee, J. K., *Langmuir* 20, 7665–7669 (2004).
206. Heyderman, L. J., Schift, H., David, C., Ketterer, B., Maurand, M. A. d., Gobrecht, J., *Microelectronic Engineering,* 57–58, 375–380 (2001).
207. Chou, S. Y., Krauss, P. R., and Renstrom, P. J., *App. Phys. Lett.* 67, 3114–3116 (1995).
208. Grewell, D., Mokhtarzadeh, A., Benatar, A., Lu, C., and Lee, L. J., *Proc. Annual Technical Conference of Society of Plastics Engineers,* 2003, 1094–1098 (2003).
209. Grewell, D., Lu, C., Leeand, L. J., Benatar, A., *Proc. Annual Technical Conference of Society of Plastics Engineers,* 2004, pp. 1231–1235.
210. http://www.jo-mt.de/downloads/Jenoptik_Froehling.pdf, last accessed November 22, 2005.
211. Becker, H., and Locascio, L. E., *Talanta* 56, 267–287 (2002).
212. Shah, J., Su, Y.-C., and Lin, L., *Proc. Micro-Electro-Mechanical Systems,* Nashville, TN, 295–302 (1999).
213. Yu, L., Koh, C. G., Koelling, K. W., Lee, L. J., and Madou, M. J., *Proc. Annual Technical Conference of Society of Plastics Engineers,* 2001, 85–789 (2001).
214. D'Amore, A., Gabriel, M., Haese, W., Schiftand, H., Kaiser, W., in *Kunststoffe plast europe,* 4–7 (2004).
215. Yoon, S., Srirojpinyo, C., Lee, J. S., Sung, C., Mead, J. L., and Barry, C. M. F. *Proc. Annual Technical Conference of Society of Plastics Engineers,* 738–742 (2004).
216. Yoon, S., Srirojpinyo, C., Lee, J. S., Mead, J. L., Matsui, S., and Barry, C. M. F., SPIE International Symposia, Smart Structures & Materials/NDE, San Diego, March 7–9, 107–116 (2005).

CHAPTER 7

PLASTICS JOINING

Edward M. Petrie

Industry Consultant
Cary, North Carolina

7.1 INTRODUCTION

This chapter provides practical information and guidance on an important plastic process—joining—that occurs only *after* the part is formed. An essential ingredient in most manufacturing operations is the knowledge of how to join parts made from plastics to themselves and to other substrates. This chapter provides the basic knowledge necessary for one to become skilled at selecting the optimal joining materials and cost-effective processes.

The reader should consider the individual plastic supplier as the primary source of information on joining processes for specific types of plastic materials. Generally, this knowledge is readily available, because the plastic resin producers benefit by providing the most complete and up-to-date information on how their materials can reliably and economically produce commercial products.

The joining processes that can be used will be dependent on the type of polymeric resins used to manufacture the parts to be assembled. However, there are also certain assembly opportunities for the designer because the material is a plastic material. Generally, plastics can be joined by more processes than other materials. The following processes are normally used for joining plastic materials:

- Mechanical fastening
- Adhesive bonding
- Thermal welding
- Solvent cementing

Solvent cementing and thermal welding use the resin in the part itself as the "fastener" to hold the assembly together. Adhesive bonding and mechanical fastening use another substance as the "fastener."

Each design engineer must determine the joining method that best suits the purpose. The choice will often depend on the type of plastic, the service environment, economic and time constraints, and production parameters. The designer should not force an assembly method on a plastic product originally designed for another assembly method. Usually, parts must be designed to employ a specific joining process. In fact, certain plastic materials are purposely chosen for an application because of their capability for high-volume assembly. For instance, in the automotive industry, certain plastics are often chosen because they can be assembled conveniently with very fast processes such as ultrasonic welding.

It is important that the designer realize the unique opportunities and problems posed by each method of assembly. To do this well, he or she must have an understanding of materials science, chemistry, surface science, mechanics, and industrial engineering. All of these disciplines will come into play. Even with this background, final selection of the most desirable assembly method involves some trial and error that can become costly and time consuming. The purpose of the next sections of this chapter is to give the designer a foundation to find the right assembly system for any particular combination of plastic material, part design, service environment, or production constraint.

7.2 GENERAL PLASTIC MATERIAL CONSIDERATIONS

The joining of plastics is generally more difficult than the joining of other substrates because of the polymeric material's low surface energy, poor wettability, and possible presence of mold-release agents and other contaminants that can create a weak interface.

The relative differences in thermal expansion coefficient and elastic modulus also make joining of plastics to nonplastic materials difficult. They may also cause very high loads or loose-fitting fasteners in parts assembled with mechanical fasteners.

The nature of the polymeric material could also change with the service environment. Parts may swell in solvent, become brittle when exposed to UV, eliminate plasticizer on aging, gain a plasticizer (water) during exposure to humidity, and go through many other changes. All of these will have an effect on the joint, whether it is bonded with adhesives or mechanically fastened.

It should also be noted that plastics can be significantly modified by mixing or combining different types of polymers and by adding fillers or modifiers. These modifications can heavily influence the bulk properties as well as the surface properties of the plastic, and, as a result, they influence the assembly characteristics. Thus, plastic materials must be carefully specified, and the specification requirements should reflect critical assembly criteria.

With plastic materials, the designer also has a greater choice of bonding techniques than with many other materials. Thermosets must be adhesively bonded or mechanically joined, but most thermoplastics can also be joined by solvent or heat welding. Additionally, plastic parts can be designed for assembly by means of molded-in, snap-fit, press-fit, pop-on, and threaded fasteners so that no additional adhesives, solvents, or special equipment is required.

Table 7.1 describes various joining methods for plastics and the advantages and disadvantages of each. Table 7.2 indicates which joining methods are appropriate for various common plastic substrates. The plastic manufacturer is generally the leading source of information on the proper methods of joining a particular plastic.

7.3 METHODS OF MECHANICAL JOINING

There are basically two methods of mechanical assembly for plastic parts. The first uses fasteners, such as screws or bolts; the second uses interference fit, such as press-fit or snap-fit and is primarily used in thermoplastic applications. This latter method of fastening is also called *design for assembly* or *self-fastening*.

TABLE 7.1 Bonding or Joining Plastics: What Techniques Are Available, and What Do They Offer[1]

Technique	Description	Advantages	Limitations	Processing considerations
Solvent cementing	Solvent softens the surface of an amorphous thermoplastic; mating takes place when the solvent has completely evaporated. Bodied cement with small percentage of parent material can give more workable cement, fill in voids in bond area. Cannot be used for polyolefins and acetal homopolymers.	Strength, up to 100% of parent materials, easily and economically obtained with minimum equipment requirements.	Long evaporation times required; solvent may be hazardous; may cause crazing in some resins.	Equipment ranges from hypodermic needle or just a wiping media to tanks for dip and soak. Clamping devices are necessary, and air dryer is usually required. Solvent-recovery apparatus may be necessary or required. Processing speeds are relatively slow because of drying times. Equipment costs are low to medium.
		Thermal Welding		
Ultrasonics	High-frequency sound vibrations transmitted by a metal horn generate friction at the bond area of a thermoplastic part, melting plastics just enough to permit a bond. Materials most readily weldable are acetal, ABS, acrylic, nylon, PC, polyimide, PS, SAN, phenoxy.	Strong bonds for most thermoplastics; fast, often less than 1 s. Strong bonds obtainable in most thermal techniques if complete fusion is obtained.	Size and shape limited. Limited applications to PVCs, polyolefins.	Converter to change 20 kHz electrical into 20 kHz mechanical energy is required along with stand and horn to transmit energy to part. Rotary tables and high-speed feeder can be incorporated.
Hot-plate and hot-tool welding	Mating surfaces are heated against a hot surface, allowed to soften sufficiently to produce a good bond, then clamped together while bond sets. Applicable to rigid thermoplastics.	Can be very fast, for example, 4 to 10 s in some cases; strong.	Stresses may occur in bond area.	Uses simple soldering guns and hot irons, relatively simple hot plates attached to heating elements up to semiautomatic hot-plate equipment. Clamps needed in all cases.
Hot-gas welding	Welding rod of the same material being joined (largest application is vinyl) is softened by hot air or nitrogen as it is fed through a gun that is softening part surface simultaneously. Rod fills in joint area and cools to effect a bond.	Strong bonds, especially for large structural shapes.	Relatively slow; not an "appearance" weld.	Requires a hand gun, special welding tips, an air source, and welding rod. Regular hand-gun speeds run 6 in/min; high-speed hand-held tool boosts this to 48 to 60 in/min.

TABLE 7.1 Bonding or Joining Plastics: What Techniques Are Available, and What Do They Offer[1] *(continued)*

Technique	Description	Advantages	Limitations	Processing considerations
Spin welding	Parts to be bonded are spun at high speed, developing friction at the bond area; when spinning stops, parts cool in fixture under pressure to set bond. Applicable to most rigid thermoplastics.	Very fast (as low as 1 to 2 s); strong bonds.	Bond area must be circular.	Basic apparatus is a spinning device, but sophisticated feeding and handling devices are generally incorporated to take advantage of high-speed operation.
Dielectric	High-frequency voltage applied to film or sheet causes material to melt at bonding surfaces. Material cools rapidly to effect a bond. Most widely used with vinyls.	Fast seal with minimum heat applied.	Only for film and sheet.	Requires rf generator, dies, and press. Operation can range from hand-fed to semiautomatic with speeds depending on thickness and type of product being handled. Units of 3 to 25 kW are most common.
Induction	A metal insert or screen is placed between the parts to be welded, and energized with an electromagnetic field. As the insert heats up, the parts around it melt, and when cooled form a bond. For most thermoplastics.	Provides rapid heating of solid sections to reduce chance of degradation.	Since metal is embedded in plastic, stress may be caused at bond.	High-frequency generator, heating coil, and inserts (generally 0.02 to 0.04 in thick). Hooked up to automated devices, speeds are high. Work coils, water cooling for electronics, automatic timers, multiple-position stations may also be required.
		Adhesives*		
Liquids solvent, water-based anaerobics	Solvent- and water-based liquid adhesives, available in a wide number of bases—for example, polyester, vinyl—in one- or two-part form fill bonding needs ranging from high-speed lamination to one-of-a-kind joining of dissimilar plastics parts. Solvents provide more bite, but cost much more than similar base water-type adhesive. Anaerobics are a group of adhesives that cure in the absence of air.	Easy to apply; adhesives available to fit most applications.	Shelf and pot life often limited. Solvents may cause pollution problems; water-based not as strong; anaerobics toxic.	Application techniques range from simply brushing on to spraying and roller coating-lamination for very high production. Adhesive application techniques, often similar to decorating equipment, from hundreds to thousands of dollars with sophisticated laminating equipment costing in the tens of thousands of dollars. Anaerobics are generally applied a drop at a time from a special bottle or dispenser.

Type	Description			Application
Pastes, mastics	Highly viscous single- or two-component materials which cure to a very hard or flexible joint depending on adhesive type.	Does not run when applied.	Shelf and pot life often limited.	Often applied via a trowel, knife, or gun-type dispenser; one-component systems can be applied directly from a tube. Various types of roller coaters are also used. Metering-type dispensing equipment in the $2500 range has been used to some extent.
Hot melts	100% solids adhesives that become flowable when heat is applied. Often used to bond continuous flat surfaces.	Fast application; clean operation.	Virtually no structural hot melts for plastics.	Hot melts are applied at high speeds via heating the adhesive, then extruding (actually squirting) it onto a substrate, roller coating, using a special dispenser or roll to apply dots or simply dipping.
Film	Available in several forms including hot melts, these are sheets of solid adhesive. Mostly used to bond film or sheet to a substrate.	Clean, efficient.	High cost.	Film adhesive is reactivated by a heat source; production costs are in the medium to high range depending on heat source used.
Pressure-sensitive	Tacky adhesives used in a variety of commercial applications. Often used with polyolefins.	Flexible.	Bonds not very strong.	Generally applied by spray with bonding effected by light pressure.
Mechanical fasteners (staples, screws, molded-in inserts, snap-fits, and variety of proprietary fasteners)	Typical mechanical fasteners are listed on the left. Devices are made of metal or plastic. Type selected will depend on how strong the end product must be, appearance factors. Often used to join dissimilar plastics or plastics to nonplastics.	Adaptable to many materials; low to medium costs; can be used for parts that must be disassembled.	Some have limited pull-out strength; molded-in inserts may result in stresses.	Nails and staples are applied by simply hammering or stapling. Other fasteners may be inserted by drill press, ultrasonics, air or electric gun, hand tool. Special molding—that is, molded-in-hole—may be required.

*Typical adhesives in each class are: Liquids: 1. Solvent—polyester, vinyl, phenolics acrylics, rubbers, epoxies, polyamide; 2. Water—acrylics, rubber-casein; 3. Anaerobics—cyanoacrylate; 4. Mastics—rubbers, rubbers, epoxies; 5. Hot melts—polyamides, PE, PS, PVA; 6. Film—epoxies, polyamide, phenolics; 7. Pressure-sensitive—rubbers.

TABLE 7.2 Assembly Methods for Plastics[2]

Plastic	Adhesives	Dielectric welding	Induction bonding	Mechanical fastening	Solvent welding	Spin welding	Direct heat welding	Ultrasonic welding
Thermoplastics:								
ABS	X		X	X	X	X	X	X
Acetals	X		X	X	X	X	X	X
Acrylics	X		X	X	X	X		X
Cellulosics	X				X	X		
Chlorinated polyether	X	X		X			X	
Ethylene copolymers		X						
Fluoroplastics	X	X					X	
Ionomer								X
Methylpentene								X
Nylons	X		X	X	X		X	X
Phenylene oxide–based materials	X			X	X	X	X	X
Polyesters	X			X	X	X		X
Polyamide-imide	X			X				X
Polyaryl ether	X	X		X				X
Polyaryl sulfone	X			X				X
Polybutylene	X			X				X
Polycarbonate	X	X	X	X	X	X	X	X
Polycarbonate/ABS	X	X	X	X	X	X	X	X
Polyethylenes	X	X	X	X	X	X	X	X
Polyimide	X			X				
Polyphenylene sulfide	X	X	X	X	X		X	X
Polypropylenes	X	X	X	X	X	X	X	X
Polystyrenes	X		X	X			X	X
Polysulfone	X	X	X	X			X	X
Propylene copolymers	X	X	X	X	X	X		
PVC/acrylic alloy	X	X		X			X	X
PVC/ABS alloys	X	X	X	X	X	X	X	X
Styrene acrylonitrile	X		X	X	X	X	X	X
Vinyls	X	X	X	X	X		X	

Thermosets:

Alkyds	X	:	:	X	:	:	:	:	:
Allyl diglycol carbonate	X	:	:	:	:	:	:	:	:
Diallyl phthalate	X	:	:	X	:	:	:	:	:
Epoxies	X	:	:	X	:	:	:	:	:
Melamines	X	:	:	X	:	:	:	:	:
Phenolics	X	:	:	X	:	:	:	:	:
Polybutadienes	X	:	:	:	:	:	:	:	:
Polyesters	X	:	:	X	:	:	:	:	:
Silicones	X	:	:	X	:	:	:	:	:
Ureas	X	:	:	:	:	:	:	:	:
Urethanes	X	:	:	X	:	:	:	:	:

If possible, the designer should try to design the entire product as a one-part molding, because it will eliminate the need for a secondary assembly operation. However, mechanical limitations often will make it necessary to join one part to another using a fastening device. Fortunately, there are a number of mechanical fasteners designed for metals that are also generally suitable with plastics, and there are many other fasteners specifically designed for plastics. Typical of these are thread-forming screws, rivets, threaded inserts, and spring clips.

As in other fabrication and finishing operations, special considerations must be given to mechanical fastening because of the nature of the plastic material. Care must be taken to avoid overstressing the parts. Mechanical creep can result in loss of preload in poorly designed systems. Reliable mechanically fastened plastic joints require:

- A firm strong connection

- Materials that are stable in the environment

- Stable geometry

- Appropriate stresses in the parts including a correct clamping force

In addition to joint strength, mechanically fastened joints should prevent slip, separation, vibration, misalignment, and wear of parts. Well designed joints provide the above without being excessively large or heavy or burdening assemblers with bulky tools. The design of plastic parts for mechanical fastening will depend primarily on the particular plastic being joined and the functional requirements of the application.

Besides the fundamental strength of the plastic part, important properties that must be considered when using mechanical fasteners are creep, relaxation, craze resistance, notch sensitivity, and stiffness. When these properties are disregarded, premature failure of the assembly is possible.

The main advantages of mechanical fasteners are that they are relatively simple to use, and they can provide for repeated assembly and disassembly of the product. The primary disadvantage is that mechanical fasteners are relatively heavy and bulky. They do not provide a thin joint geometry or attractive aesthetic appearance that is required for many plastic products.

Mechanical fasteners and the design of parts to accommodate them are covered in detail in later sections of this chapter. The following section describes the use of press-fit and snap-fit designs that are integrated into the molded part to achieve assembly.

7.3.1 Design for Self-Fastening

It is often possible and desirable to incorporate fastening mechanisms into the design of the molded part itself. The two most common methods of doing this are by interference fit (including press-fit or shrink- fit) and by snap-fit. Whether these methods can be used will depend heavily on the nature of the plastic material and the freedom one has in part design.

Self-fastening joint designs generally produce very high stresses in the plastic part during the assembly operation. With brittle plastics, such as thermosets, press-fit assembly may cause the plastic to crack if conditions are not carefully controlled. In addition, certain plastics, especially thermoplastics, are subject to cold flow under stress. Under continued stress, which is the fundamental "adhesive" that holds self-fastened parts together, the plastic may relax, causing the joint to fail.

Another important point to consider when designing joints for self-fastening is the thermal coefficient of expansion. For example, most metals have a thermal expansion coefficient that is about ten times smaller than plastics. When dissimilar materials are joined, changes in ambient temperatures can cause the joint to relieve stress and fail. Thus, self-assembled joints must be designed for anticipated operating conditions including the environment and stress levels.

7.3.1.1 Press Fit. In press or interference fits, a shaft of one material is joined with the hub of another material by a dimensional interference between the shaft's outside diameter and the hub's inside diameter. Press-fit joints can be made by simple application of force or by heating or cooling one part relative to the other. This simple, fast assembly method provides joints with high strength and low cost.

Press fitting is applicable to parts that must be joined to themselves or to other plastic and nonplastic parts. The advisability of its use will depend on the relative properties of the two materials being assembled. When two different materials are being assembled, the harder material should be forced into the softer. For example, a metal shaft can be press-fitted into plastic hubs.

Where press fits are used, the designer generally seeks the maximum pullout force using the greatest allowable interference between parts that is consistent with the strength of the plastic. General equations for interference fits (when the hub and shaft are made of the same materials and for when they are a metal shaft and a plastic hub) are given in Fig. 7.1. Safety factors of 1.5 to 2.0 are used in most applications.

Figures 7.2 and 7.3 show calculated interference limits at room temperature for press fitted shafts and hubs of Delrin® acetal resin and Zytel® nylon resin. These represent the maximum allowable interference based on yield point and elastic modulus data.

The following precautions must be observed to avoid over-stressed conditions in press-fit joints:

- Design the press-fit equal to or less than the creep limit of the plastic part.
- Use smooth, rounded inserts because of possible stress concentration.
- Avoid locating the mold knot line on the area to be inserted.
- Remove all incompatible chemicals from the insert.

For a press-fit joint, the effect of thermal cycling, stress relaxation, and environmental conditioning must be carefully evaluated. Testing of the factory assembled parts under expected temperature cycles, or under any condition that can cause changes to the dimensions or modulus of the parts, is obviously indicated. Differences in coefficient of thermal expansion can result in reduced interference due either to one material shrinking or expanding away from the other, or it can cause thermal stresses as the temperature changes.

Since plastic materials will creep or stress-relieve under continued loading, loosening of the press fit, at least to some extent, can be expected during service. To counteract this, the designer can knurl or groove the parts. The plastic will then tend to flow into the grooves and retain the holding power of the joint.

7.3.1.2 Snap Fit. Snap-fit joints may be the most widely used way of joining and assembling plastics. Easily identifiable examples of common annular snap-fit joints are child-resistant caps on pill bottles and the ballpoint pens with snap-on caps. A cross section of an annular snap-fit joint is illustrated in Figure 7.4. But snap-fit joints are also used in nonannular part assembly such as the joining of a plastic housing to a base. Many designs and configurations can be used with a snap-fit configuration. The individual plastic resin suppliers are suggested for design rules and guidance on specific applications.

In all types of snap-fit joints, a protruding part of one component, such as a hook, stud, or bead, is briefly deflected during the joining operation, and it is made to catch in a de-

General equation for interference

$$I = \frac{S_d D_s}{W}\left[\frac{W\mu_h}{E_h} + \frac{1-\mu_s}{E_s}\right]$$

in which

$$W = \frac{1+\left(\dfrac{D_s}{D_h}\right)^2}{1-\left(\dfrac{D_s}{D_h}\right)^2}$$

$I\ =$ Diametral interference, mm (in.)

$S_d =$ Design stress limit or yield strength of the polymer, generally in the hub, MPa (psi) (A typical design limit for an interference fit with thermoplastics is 0.5% strain at 73ºC.)

$D_h\ =$ Outside diameter of hub, mm (in.)

$D_s\ =$ Diameter of shaft, mm (in.)

$E_h\ =$ Modulus of elasticity of hub, MPa (psi)

$E_s\ =$ Elasticity of shaft, MPa (psi)

$\mu_h\ =$ Poisson's ratio of hub material

$\mu_s\ =$ Poisson's ratio of shaft material

$W\ =$ Geometric factor

If the shaft and hub are of the same material, $E_h = E_s$ and $\mu_h = \mu_s$. The above equation simplifies to:

Shaft and hub of same material

$$I = \frac{S_d D_s}{W} \times \frac{W+1}{E_h}$$

If the shaft is a high modulus metal or other material, with $E_s > 34.4 \times 10^3$ MPa, the last term in the general interference equation is negligible, and the equation simplifies to:

Metal shaft, plastic hub

$$I = \frac{S_d D_s}{W} \times \frac{W+\mu_h}{E_h}$$

FIGURE 7.1   General calculation of interference fit between a shaft and a hub.[3]

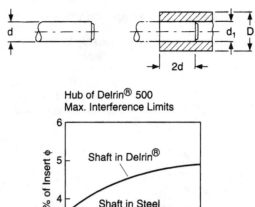

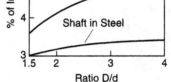

FIGURE 7.2 Maximum interference limits for Delrin acetal.[4]

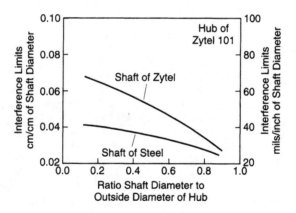

FIGURE 7.3 Theoretical interference limits for Zytel nylon. (Based on yield point and elastic modulus at room temperature and average moisture conditions.)[4]

pression (undercut) in the mating component. This method of assembly is uniquely suited to thermoplastic materials due to their flexibility, high elongation, and ability to be molded into complex shapes.

The two most common types of snap-fits are those with flexible cantilevered lugs (Fig. 7.5) and those with a full cylindrical undercut and mating lip (e.g., annular snap-fit as shown in Figure 7.4). Cylindrical snap fits are generally stronger but require deformation for removal from the mold.

Annular snap fit — snap groove close to end

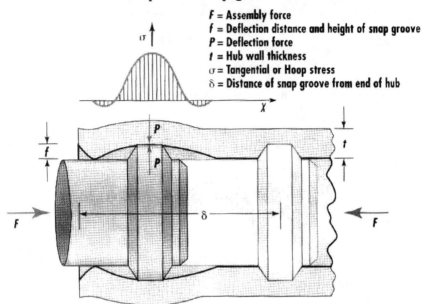

F = Assembly force
f = Deflection distance and height of snap groove
P = Deflection force
t = Hub wall thickness
σ = Tangential or Hoop stress
δ = Distance of snap groove from end of hub

FIGURE 7.4 Cross-section view of an annular snap-fit showing the stress distribution during the joining operation.[5]

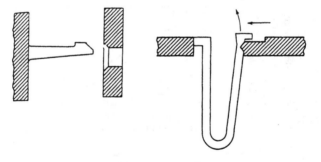

FIGURE 7.5 Snap-fitting cantilevered arms.[6]

Snap-fit assemblies are usually employed to attach lids or covers that are meant to be disassembled or that will be lightly loaded. The design should be such that, after the assembly, the joint will return to a stress-free condition. Materials with good recovery characteristics are required.

The maximum permissible deformation of the plastic hub in an annular snap-fit joint is limited by the maximum permissible strain or proportional limit of the material. This limit is typically 50 percent of the strain at break for most reinforced plastics. It can be upwards of 60 to 70 percent of strain at break for more elastic polymers.

The geometry of the ridge determines the assembly force needed to engage the snap joint and whether the snap joint will be detachable or permanent. The lead angle is generally less than 30°. A shallow return angle easily separates, and a 90° angle is permanent. A 45° return angle is generally designed for products that must be disassembled.

To obtain satisfactory results, the undercut design must fulfill certain requirements:

- The wall thickness should be kept uniform.
- The snap fit must be placed in an area where the undercut section can expand freely.
- The ideal geometric shape is circular.
- Ejection of an undercut core from the mold is assisted by the fact that the resin is still at relatively high temperatures.
- Weld lines should be avoided in the area of the undercut.

In the cantilevered snap-fit design, the retaining force is essentially a function of the bending stiffness of the resin. Cantilevered lugs should be designed in a way so as not to exceed allowable stresses during assembly. Cantilevered snap fits should be dimensioned to develop constant stress distribution over their length. This can be achieved by providing a slightly tapered section or by adding a rib. Special care must be taken to avoid sharp corners and other possible stress concentrations.

7.3.2 Mechanical Fasteners

A large variety of mechanical fasteners can be used for joining plastic parts to themselves and to other materials. These include machine screws, self-tapping screws, rivets, spring clips, and nuts. In general, when repeated disassembly of the product is anticipated, mechanical fasteners are used.

Mechanical fasteners provide an economical and reliable means of joining plastic parts to themselves and to other materials. However, mechanical fasteners are generally not employed in high-volume, fast-production operations because of the added inventory requirements. Also, the time required to join parts with mechanical fasteners relative to other processes is substantial, especially for larger parts that require multiple fasteners.

Metal fasteners of high strength can overstress plastic parts, so torque-controlled tightening or special design provisions are required. Where torque cannot be controlled, various types of washers can be used to spread the compression force over larger areas.

7.3.2.1 Machine Screws, Bolts. Parts molded of thermoplastic resin are sometimes assembled with machine screws or with bolts, nuts, and washers (Fig. 7.6), especially if it is a very strong plastic. The threads used to engage the mechanical fastener may either be molded into the part (molded-in threads or molded-in inserts) or attached after the molding operation (post-molded inserts).

Machine screws are generally used with threaded inserts, nuts, and clips. They rarely are used in pretapped holes. Particular attention should be paid to the head of the fastener. Conical heads produce undesirable tensile stresses and should not be used. Bolt or screw heads with a flat underside, such as pan heads, round heads, and so forth (Fig. 7.7), are preferred, because the stress produced is more compressive.

Flat washers are also suggested and should be used under both the nut and the fastener head. Sufficient diametrical clearance for the body of the fastener should always be provided in the parts to be joined. This clearance can nominally be 0.25 mm (0.010 in).

When the application involves infrequent disassembly, molded-in threads can be used. Molded-in inserts provide very high-strength assemblies and relatively low unit cost.

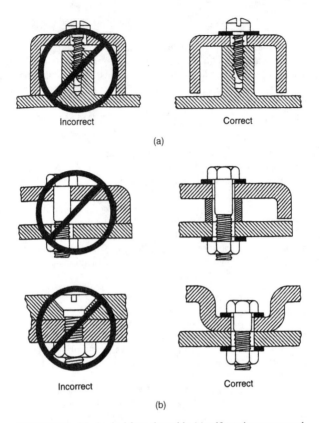

FIGURE 7.6   Mechanical fastening with: (a) self-tapping screws and (b) bolts, nuts, and washers.[6]

However, molded-in inserts could increase cycle time while the inserts are manually placed in the mold.

Coarse threads can be molded into most materials. Threads of 32 or finer pitch should be avoided, along with tapered threads, such as pipe threads. These types of threads can cause excessive stress. Other factors that should be considered when designing molded-in threads are:

- If the mating part is metal, over-torque will result in failure.
- Feather edges on thread runouts should be avoided to prevent cross threading or thread damage.
- The roots and crests of threads should be rounded to reduce stress concentrations as well as to help mold filling.

Internal threads can be formed by collapsible cores or unscrewing core inserts. External threads can be formed by split cores or by unscrewing devices. All of these designs increase mold costs.

Post-molded inserts come in four types: press-in, expansion, self-tapping, and thread forming, and inserts that are installed by some method of heating (e.g., ultrasonic). Metal

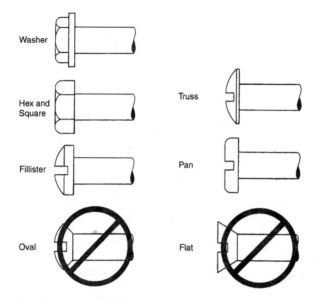

FIGURE 7.7 Common head styles of screws and bolts. Flat underside of head is preferred.[6]

inserts are available in a wide range of shapes and sizes for permanent installation. Inserts are typically installed in molded bosses, designed with holes to suit the insert to be used. Some inserts are pressed into place, and others are installed by methods designed to limit the stress and increase strength. Generally, the outside of the insert is provided with projections of various configurations that penetrate the plastic and prevent movement under normal forces exerted during assembly.

7.3.2.2 Self-Threading Screws. Self-threading or self-tapping screws provide an economical means for joining plastics. These screws cut or form a thread as they are inserted into the part, eliminating the need for molding an internal thread or conducting a separate tapping operation.

Self-threading screws can either be thread-cutting or thread-forming. To select the correct screw, the designer must know which plastic will be used and its modulus of elasticity. The advantages of using these types of screws are that they are generally off-the-shelf items and low in cost, high production rates are possible, and they require minimum tooling investment.

The principal disadvantage is limited reuse; after repeated disassembly and assembly, these screws will cut or form new threads in the hole, eventually destroying the integrity of the assembly. Self-threading screws should not be used when repeated disassembly may be required.

Molded plastic parts acquire internal stresses during the molding process, and these increase when fasteners are installed. As a result, bulk material is generally added to area of the joint in the form of a boss. The added material of a boss is intended to prevent cracking of the plastic when the fastener is installed and tightened. The boss outside diameter should also provide enough strength to withstand possible hoop stresses developed by screw insertion.

Bosses normally extend at 90° angles from the cross section of the component and can be rib supported or attached to a side wall for extra support when required. As a general rule, the outside diameter of the boss should be 2.5 to 3 times the hole diameter. The larger the boss diameter, the greater the area over which the fastening stresses can be spread. When boss diameter must be small, screws for plastics with an extra-wide thread spacing are preferred.

When planning for self-threading screws, the receiving hole diameter should be equal to the screw pitch diameter. A length of at least twice the screw's major diameter should be provided in the boss for thread engagement. The amount of torque that can be placed on a screw depends on both the cross sectional area of the boss and the total number of threads. Since sufficient threads can usually be provided, the allowable torque is most often dependent on the boss' cross-sectional area.

Thread-forming screws are usually less expensive than thread-cutting screws and have the highest resistance to back-out. The threads are formed in the plastic part by forcing the plastic to deform around the metal screw and displacing the plastic through the threading operation. Therefore, thread-forming screws are generally applied only to materials with a modulus of below 1400 MPa (200,000 psi). With certain exceptions (e.g., some acetals and nylon 6,12), they can also be used on materials with modulus in the range of 200,000 to 400,000 psi.

Assembly strengths using thread-forming screws can be increased by reducing hole diameter in the more ductile plastics, by increasing screw thread engagement, or by going to a larger-diameter screw when space permits. The most common problem encountered with these types of screws is boss cracking. This can be minimized or eliminated by increasing the size of the boss, increasing the diameter of the hole, decreasing the size of the screw, changing the thread configuration of the screw, or changing the part to a more ductile plastic.

A number of thread-forming fasteners are especially designed for use with plastics (Figure 7.8). The type used will generally depend on the modulus of the plastic and the specific requirements of the assembly operation. Characteristics of several common and specialty thread-forming screws are summarized in Table 7.3.

Thread-cutting screws are used in harder, less ductile plastics with a flexural modulus of 400,000 to 10^6 psi. They are also often employed on glass-filled resins. Thread-cutting screws remove material as they are installed, thereby avoiding high stress. These screws typically have higher thread engagement and can withstand higher clamp loads. They also produce lower residual hoop stress in the part than thread-forming screws.

TABLE 7.3   Characteristics of Various Types of Thread-Forming Screws

Type of screw	Characteristics
Type AB and B	Fast-driving, spaced-thread screws; AB has a 45° angle, B has a blunt point.
Type BP	Similar to the B type except with a 40° angle.
Type U	Intended for permanent fastening; blunt point, multiple threads.
Trilobe	Designed to reduce radial pressure.
Hi-Lo	Dual-height thread design provides holding power by increasing the amount of plastic held between threads.
Sharp thread	Thread angles are less than 60°. Sharper threads can be forced into ductile plastics more readily.
Type T	Fine threads; recommended for plastics with flexural moduli above 10^6 psi.

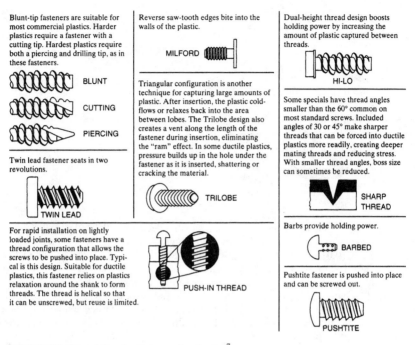

Blunt-tip fasteners are suitable for most commercial plastics. Harder plastics require a fastener with a cutting tip. Hardest plastics require both a piercing and drilling tip, as in these fasteners.

BLUNT

CUTTING

PIERCING

Twin lead fastener seats in two revolutions.

TWIN LEAD

For rapid installation on lightly loaded joints, some fasteners have a thread configuration that allows the screws to be pushed into place. Typical is this design. Suitable for ductile plastics, this fastener relies on plastics relaxation around the shank to form threads. The thread is helical so that it can be unscrewed, but reuse is limited.

Reverse saw-tooth edges bite into the walls of the plastic.

MILFORD

Triangular configuration is another technique for capturing large amounts of plastic. After insertion, the plastic cold-flows or relaxes back into the area between lobes. The Trilobe design also creates a vent along the length of the fastener during insertion, eliminating the "ram" effect. In some ductile plastics, pressure builds up in the hole under the fastener as it is inserted, shattering or cracking the material.

TRILOBE

PUSH-IN THREAD

Dual-height thread design boosts holding power by increasing the amount of plastic captured between threads.

HI-LO

Some specials have thread angles smaller than the 60° common on most standard screws. Included angles of 30 or 45° make sharper threads that can be forced into ductile plastics more readily, creating deeper mating threads and reducing stress. With smaller thread angles, boss size can sometimes be reduced.

SHARP THREAD

Barbs provide holding power.

BARBED

Pushtite fastener is pushed into place and can be screwed out.

PUSHTITE

FIGURE 7.8   Thread-forming fasteners for plastics.[7]

Similar to thread-forming screws, thread-cutting screws should also not be installed and removed repeatedly. If frequent removal and reinstalling of the screw must occur, the boss should be large enough to allow replacement of the screw with the next larger size.

7.3.2.3 Rivets.   Rivets provide permanent assembly at very low cost. The assembly costs are generally lower than for threaded fasteners, and high-volume, high production speeds can be achieved. Disadvantages of rivet assembly include lower tensile and fatigue strength and poorer precision when compared to other methods of mechanical fastening. Clamp load must be limited to low levels to prevent distortion of the part.

Rivets are metal or plastic pins manufactured with a head on one side. They are inserted into a preformed hole through the parts to be joined, and the straight end is flared to lock the rivet in place. To distribute the load, rivets with large heads should be used with washers under the flared end of the rivet. The heads should be three times the shank diameter. Standard rivet heads are shown in Figure 7.9.

Rivets are often used with high-strength plastic and composite parts, especially when they are to be joined to metal substrates. Riveted composite joints should be designed to avoid loading the rivet in tension. Generally, a hole 1/64 in larger than the rivet shank is satisfactory for composite joints. A number of patented rivet designs are commercially available for joining aircraft or aerospace structural composites.

7.3.2.4 Other Mechanical Fasteners.   *Push-on spring steel fasteners* can be used for holding light loads. They are used when low-cost, rapid mechanical assembly is required. Spring steel fasteners are also often referred to as speed nuts and speed clips.

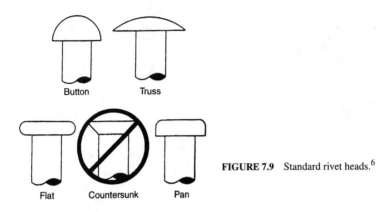

Button   Truss

Flat   Countersunk   Pan

FIGURE 7.9   Standard rivet heads.[6]

Spring steel fasteners are simply pushed over a molded stud with hand pressure. A simple push-on spring steel fastener is illustrated in Figure 7.10. The stud should have a minimum 0.38 mm (0.015 in) radius at its base. Too large a radius could create a thick section, resulting in sinks or voids in the plastic molding.

Staking is an assembly operation similar to riveting. In staking operations, the fastener is generally a deformable plastic or metal stud. A head is formed on the stud during the fastening operation via cold flow or melting. The stud protrudes through a hole in the parts being joined, and the head(s) are formed on the end(s) of the stud to lock the parts in place. The stud can either be (1) a separate fastener item such as a section of rod so that heads need to be formed on both ends, or (2) it can be molded into one of the plastic parts so that only one end needs to be formed into a head.

Staking can be produced by either cold or hot methods. In cold-staking, high pressures (generally greater than 6000 psi) are used to induce cold flow of the plastic stake. Only malleable thermoplastic materials are suitable for cold-staking. Soft or very brittle materi-

Push-on spring-steel   Molded stud
   fastener

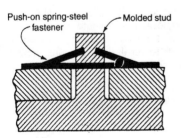

FIGURE 7.10   Push-on spring steel fasteners.[6]

als are difficult to cold-stake. Heat-staking is generally only used with thermoplastic staking material. Heat-staking can be accomplished using a heated mold so that a head can be formed. Ultrasonic horns have also been used for fast, heat-staking of certain thermoplastics. (See section on Ultrasonic Welding.)

7.3.3 Special Consideration for Composites

The structural efficiency of a composite structure is established, with very few exceptions, by its joints, not by its basic structure. The possible joining method for composites is as broad as with metals, namely riveting, bolting, pinning, and bonding. The only metal joining processes that are not suitable for thermoset composites are welding and brazing. However, thermoplastic and metallic matrix composites can be joined by welding or brazing.

Composites are mechanically fastened in a manner similar to metals. Parts are drilled, countersunk, and joined with a fastener. Rivets, pins, two-piece bolts, and blind fasteners made of titanium, stainless steel, and aluminum are all used for composites. Several factors should be considered:

1. Differential expansion of the fastener in the composite
2. The effect of drilling on the structural integrity of the composite, as well as delamination caused by the fastener under load
3. Water intrusion between the fastener and composite
4. Electrical continuity of the composite and electrical conductivity between fasteners
5. Possible galvanic corrosion at the composite joint
6. Weight of the fastening system
7. Environmental resistance of the fastening system

Table 7.4 provides a performance comparison for some of the mechanical fasteners most commonly used to join composite materials.

Aluminum and stainless fasteners expand and contract when exposed to temperature extremes, as in aircraft applications. In carbon-fiber composites, contraction and expansion of such fasteners can cause changes in clamping load. Pressure within the joint is often critical.

Drilling and machining can damage composites. Several techniques exist for producing quality holes in composites. Carbon, aramid, and boron fiber reinforced materials each require different drilling methods and tools. When composites are cut, fibers are exposed. These fibers can absorb water, which weakens the material. Sealants can be used to prevent moisture absorption in the clearance hole. Sleeved fasteners can also provide fits that reduce water absorption as well as provide tightness.

Fastener holes should be straight and round within the limits specified. Normal hole tolerance is 0.075 mm (0.003 in). Interference fits may cause delamination of the composite. Holes should be drilled perpendicular to the sheet within one degree. Special sleeved fasteners can limit the chances of damage in the clearance hole and still provide an interference fit. Fasteners can also be bonded in place with adhesives to reduce fretting.

Galvanic corrosion may occur in carbon fiber composites if aluminum fasteners are used, due to the chemical reaction of the aluminum with the carbon. Coating the fastener guards may prevent corrosion but adds cost and time to the assembly process. As a result, aluminum fasteners are often replaced by more expensive titanium and stainless steel fasteners in carbon fiber composite joints.

When joining composites with mechanical fasteners, special consideration must be given to creep. There are two kinds of creep: creep of the fastener hole and long term material compression. The greater the material modulus, the lower the creep. There are me-

TABLE 7.4   Fasteners for Advanced Composites[8]

Fastener type	Fastener material	Surface coating	Suggested application			
			Epoxy/ graphite composite	Kevlar	Fiberglass	Honeycomb
Blind rivets[a]	5056 Al	None	NR	E[h]	E[h]	}[e]
	Monel	None	G[h]	E	E	
	A-286	Passivated	G	E	E	
Blind bolts[b]	A-286	Passivated	E[h]	E[h]	E	}
	Alloy steel	Cadmium	NR	E	E	
Pull-type lockbolts	Titanium	None	E[d]	E[c]	E[c]	G or NR[f]
Stump-type lockbolts	Titanium	None	E[d]	E[c]	E[c]	G or NR[f]
Asp fasteners	Alloy steel	Cad/Nickel	G[g]	E	E	E
Pull-type lockbolts	7075 Al	Anodize	NR	E	E	NR

[a]Blind rivets with controlled shank expansion.
[b]Blind bolts are not shank expanding.
[c]Fasteners can be used with flanged titanium collars or standard aluminum collars.
[d]Use flanged titanium collar.
[e]Performance in honeycomb should be substantiated by installation testing.
[f]Depending on fastener design. Check with manufacturer.
[g]Nickel plated Asp only.
[h]Metallic structure on backside.
Note: NR = not recommended; E = excellent; G = good.

chanical ways to reinforce the hole or distribute the load so that the creep problem is minimized. For fasteners that rely on inserts, the ability of the composite to retain the fastener must be considered.

Like mechanically fastened metal structures, composites exhibit failure modes in tension, shear and bearing but, because of the complex failure mechanisms of composites, two further modes are possible, namely cleavage and pullout. Environmental degradation of a bolted joint, after exposure to hot, wet environment is most likely to occur in the shear and bearing strength properties. The evidence shows that for fiber reinforced epoxies, temperature has a more significant effect than moisture, but in the presence of both at 127°C, a strength loss of up to 40 percent is possible.

Evidence suggests that the failure behavior of thermoplastics is much the same as for thermoset composites. High joint efficiencies can be obtained with suitable consideration to the joint design, fastener type, and environmental factors. In addition to mechanical and adhesive joining, thermoplastic composites can also be heat welded depending on the concentration and type of fillers within the composite.

7.4 ADHESIVE BONDING

An *adhesive* is any substance capable of holding items together by surface attachment. The bonds formed may be permanent, such as those used to make almost everything from shoes to airplanes, or temporary such as those used to make Post-It® notes and Band-Aids®.

There are many types and forms of adhesives that can be used with plastics to provide strong structural bonds. Plastics generally have lower tensile strength than other materials, such as metals, and an adhesive strength can often be achieved that is greater than the strength of the substrate itself.

Adhesive bonding presents several distinct advantages over other methods of fastening plastic substrates. These are summarized in Table 7.5. Adhesive bonding is often preferred when different types of substrates (e.g., metals to plastics) need to be joined, when high-volume production is necessary, or when the design of the finished part prohibits the use of mechanical fasteners.

TABLE 7.5   Advantages and Disadvantages of Adhesive Bonding

Advantages	Disadvantages
1. Provides large stress-bearing area.	1. Surfaces must be carefully cleaned.
2. Provides excellent fatigue strength.	2. Long cure times may be needed.
3. Damps vibration and absorbs shock.	3. Limitation on upper continuous operating temperature (generally 350°F).
4. Minimizes or prevents galvanic corrosion between dissimilar metals.	
5. Joins all shapes and thicknesses.	4. Heat and pressure may be required.
6. Provides smooth contours.	5. Jigs and fixtures may be needed.
7. Seals joints.	6. Rigid process control usually necessary.
8. Joins any combination of similar or dissimilar materials.	7. Inspection of finished joint difficult.
9. Often less expensive and faster than mechanical fastening.	8. Useful life depends on environment.
10. Heat, if required, is too low to affect metal parts.	9. Environmental, health, and safety considerations are necessary.
11. Provides attractive strength-to-weight ratio.	10. Special training sometimes required.

Although there are various ways of joining plastics to themselves or to other materials, adhesive bonding has often proved to be the most effective assembly method. In many applications, the use of adhesives rather than metal fasteners reduces product cost and the weight of the assembly while, in some cases, providing longer service life. Adhesive bonding can also be used very effectively in prototypes and with large or intricate assemblies that, for economic or design reasons, cannot be molded or processed as a single part.

However, the joining of plastics with adhesives can be made difficult because of their low surface energy, poor wetting, and presence of contaminants such as mold-release agents, low-molecular-weight internal components (e.g., flexibilizers, UV inhibitors, processing aids), and possible susceptibility to moisture and other environmental factors. Fortunately, numerous adhesives and processing methods are available for the joining of plastic materials and have been successfully used in many applications. Many of these products and applications are described in articles and handbooks on the subject. The plastic resin manufacturer is generally the leading source of information on the proper methods of joining a particular plastic.

There are many types and ways to classify adhesives (Table 7.6). However, all adhesives share several important common characteristics:

TABLE 7.6 Examples of Common Methods of Classifying Adhesives

Function
- Structural—adhesives that are permanent and have very high strength
- Nonstructural—adhesives that are soft and relatively weak; often used for temporary fastening

Chemical composition
- Animal hide or bone—early adhesives, often referred to as "glues"
- Epoxy—modern polymeric adhesives that cure by chemical reaction
- Polyvinyl acetate—polymeric adhesives dispersed in water, often used as "wood glues"
- Acrylic—sticky or "high tack" polymeric adhesives that are applied in several ways, often used in pressure-sensitive tapes and labels
- Starch—made from plant roots, often used in school paste
- Milk curd—made from skim milk, often used in bonding envelopes, cardboard boxes, etc.
- Inorganic—cements

Application and setting
- Drying—evaporation of solvent or water (e.g., polyvinyl acetate, starch, milk curd, animal types)
- Reactive—liquid components chemically react with each other to form a solid (two-component epoxy adhesives, cyanoacrylates or "super glues" react with water, some adhesives react on exposure to light or radiation)
- Hot melt—when heated, the adhesive flows and wets the substrates, then on cooling, it gels to a hard, strong material (polyvinyl acetate and many other polymers)

Physical form
- Liquids and pastes (without water or solvent)
- Solvent solutions
- Water dispersions
- Solids (rods, tape, film, powder)
- Pressure-sensitive (adhesive generally applied to a backing material such as a tape or label)

- At some point, they must act as a liquid to flow over and *wet* the surface of the part they are bonding. The liquid form allows the adhesive to fill gaps, cavities, and spaces at the joint or bond *interface*. Wetting allows the adhesive to make very close contact with the surface—so close that the molecules of the adhesive and surface material interact, causing a strong bond.

- The adhesive must harden or *cure* from a liquid to a solid. This hardening can occur by drying or evaporation of solvent, by a chemical reaction, or by cooling from a melt form.

- Once solidified, the adhesive must be capable of holding the parts together under normal forces. It also must be capable of resisting all service environment conditions (moisture, temperature, and so on) and last the life of the product being assembled.

Most adhesives are made from naturally occurring or man-made organic polymeric molecules. These are like other molecules, except they are extremely long. These adhesives provide good bond strength as well as resistance to chemicals, impact, and other common environments. Some adhesives are made from inorganic chemicals rather than polymers. These are generally referred to as *cements*.

7.4.1 Theories of Adhesion

Various theories attempt to describe the phenomena of adhesion. There is no general consensus among experts regarding a universal theory that explains why adhesives stick. Instead, experts believe that several theories may be applicable, depending on the nature of the adhesive bond and the substrate. The adhesion theories that are applicable to plastic substrates are the following.

7.4.1.1 Mechanical Theory.

The surface of a solid material is never truly smooth but consists of a maze of microscopic peaks and valleys. According to the mechanical theory of adhesion, the adhesive must penetrate the cavities on the surface and displace the trapped air at the interface. Some mechanical anchoring appears to be a prime factor in bonding many rough or porous substrates. Adhesives also frequently bond better to abraded surfaces than to natural surfaces (although this is sometimes not true of certain low-surface-energy plastics). Mechanical abrasion is a popular surface preparation step prior to adhesive bonding. The beneficial effects of surface roughening may be due to:

- Mechanical interlocking
- Formation of a clean surface
- Formation of a more reactive surface
- Formation of a larger surface area.

7.4.1.2 Adsorption Theory.

The adsorption theory states that adhesion results from molecular contact between two materials and the surface forces that develop. The process of establishing intimate contact between an adhesive and the adherend is known as *wetting*. Figure 7.11 shows examples of good and poor wetting of a liquid adhesive spreading over

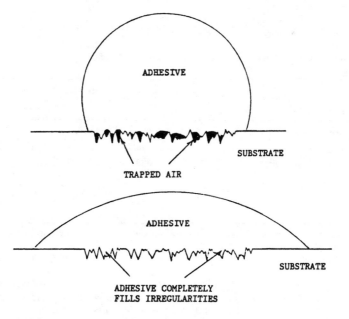

FIGURE 7.11 Illustration of poor (top) and good (bottom) wetting by an adhesive spreading over a substrate surface.

a substrate surface. For an adhesive to adequately wet a solid surface, the adhesive should have a lower surface tension than the solid's critical surface tension:

adhesive's surface tension << substrate's critical surface tension

Tables 7.7 and 7.8 list surface tensions of common adherends and liquids used as adhesives. Most liquid adhesives easily wet metallic solids because of the high surface tension of most metals. But many solid organic substrates have surface tensions less than those of common adhesive.

TABLE 7.7 Critical Surface Tensions of Common Plastics and Metals

Materials	Critical surface tension, dyne/cm
Acetal	47
Acrylonitrile-butadiene-styrene	35
Cellulose	45
Epoxy	47
Fluoroethylene propylene	16
Polyamide	46
Polycarbonate	46
Polyethylene	31
Polyethylene terephthalate	43
Polyimide	40
Polymethylmethacrylate	39
Polyphenylene sulfide	38
Polystyrene	33
Polysulfone	41
Polytetrafluoroethylene	18
Polyvinyl chloride	39
Silicone	24
Aluminum	~500
Copper	~1000

TABLE 7.8 Surface Tension of Common Adhesives and Liquids

Material	Surface tension, dyne/cm
Epoxy resin	47
Fluorinated epoxy resin*	33
Glycerol	63
Petroleum lubricating oil	29
Silicone oils	21
Water	73

*Experimental resin; developed to wet low-energy surfaces. (Note low surface tension relative to most plastics.)

From Tables 7.7 and 7.8, it can be forecast that epoxy adhesives will wet clean aluminum or copper surfaces. However, epoxy resin will not wet a substrate having a critical surface tension significantly less than 47 dynes/cm. Epoxies will not, for example, wet either a metal surface contaminated with silicone oil or a clean polyethylene substrate.

For wetting to occur, the substrate surface has to be chemically or physically altered by some mechanism to raise its surface energy. This is why there are so many prebond surface treatments for plastic substrates.

After intimate contact is achieved between adhesive and adherend through wetting, adhesion results primarily through forces of molecular attraction. The adhesion between adhesive and adherend is believed to be primarily due to van der Walls forces of attraction.

7.4.1.3 Electrostatic and Diffusion Theories. The electrostatic theory states that electrostatic forces in the form of an electrical double layer are present at the adhesive–adherend interface. These forces account for resistance to separation. The electrostatic theory of adhesion is not generally applicable for common production assembly, but it does apply to the adhesion of particulates (e.g., dust) on plastic film.

The fundamental concept of the diffusion theory is that adhesion occurs through the interdiffusion of molecules in the adhesive and adherend. The diffusion theory is primarily applicable when both the adhesive and adherend are polymeric. For example, bonds formed by solvent or heat welding of thermoplastics result from the diffusion of molecules.

7.4.1.4 Weak Boundary Layer Theory. According to the weak boundary layer theory, when bond failure seems to be at the interface, usually a cohesive break of a weak boundary layer is the real event. Weak boundary layers can originate from the adhesive, the adherend, the environment, or a combination of any of the three. When bond failure occurs, it is the weak boundary layer that fails, although failure seems to occur at the adhesive–adherend interface.

Figure 7.12 shows examples of certain possible weak boundary layers for a plastic substrate. There are many opportunities for weak boundary layers to occur on a plastic part. These can include mold release, plasticizer migration, and moisture migrating to the interface. Certain weak boundary layers can be removed or strengthened by various surface treatments.

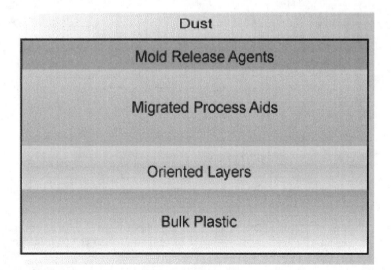

FIGURE 7.12   Schematic representation of surface regions for a plastic substrate.

7.4.2   Requirements for a Good Bond

The basic requirements for a good bond to plastic substrates are surface cleanliness, wetting of the surface by the adhesive, solidification of the adhesive, and proper selection of adhesive and joint design. These requirements are briefly defined here and then more thoroughly discussed in following sections.

To achieve an effective adhesive bond, one must start with a clean surface. Foreign materials such as dirt, oil, mold release agents, and so forth must be removed from the surface, or else the adhesive will bond to these weak boundary layers rather than to the actual substrate. To have acceptable joint strength, many plastic materials require surface treatment prior to bonding to provide one or a combination of the following: surface roughness, raised surface energy, removal of contaminants or weak boundary layers, strengthening of boundary layers. At minimum, the surface must be made free of all contaminants such as oil, grease, mold-release agents, water, and polishing compounds. Solvents or detergents are generally used to clean the plastic parts prior to bonding.

While it is in the liquid state, the adhesive must wet the substrate. The result of good wetting is greater contact area between adherend and adhesive over which the forces of adhesion may act. If the surface is one of low energy, wetting cannot be obtained without some sort of surface modification. For example, polytetrafluoroethylene, polyethylene, and certain other polymeric materials are completely unsuitable for adhesive bonding in their natural state. For these surfaces, mechanical abrasion will only provide unwettable air pockets at the interface, resulting in lower bond strength. Thus, for surfaces where wetting is difficult, mechanical abrasion is not recommended as a surface preparation. Plastics with lower surface energy may need to be chemically or physically treated prior to bonding to raise their surface energy.

The liquid adhesive, once applied, must be capable of conversion into a solid. The process of solidifying can be completed in different ways. Adhesives are generally solidified by the following processes:

- Chemical reaction by any combination of heat pressure and curing agent
- Cooling from a molten liquid to a solid state
- Drying due to solvent evaporation

The main areas of concern when selecting an adhesive are the materials to be bonded, service requirements, production requirements, and overall cost.

The adhesive joint should be designed to optimize the forces acting on and within the joint. Although adequate adhesive bonded assemblies have been made from joints designed for mechanical fastening, maximum benefit can be obtained only in assemblies specifically designed for adhesive bonding.

7.4.3   Basic Adhesive Materials

Adhesives may be classified by various methods such as function, chemical composition, mode of application and setting, physical form, cost, and end use. The most common methods of adhesive classification are summarized in Table 7.6 (p. 7.22).

7.4.3.1 Function.   The functional classification defines adhesives as being structural or nonstructural. Structural adhesives are materials of high strength and permanence. They are generally represented by thermosetting adhesives with shear strengths greater than 1000 psi. Their primary function is to hold structures together and be capable of resisting high loads.

Nonstructural adhesives are not required to support substantial loads but merely hold lightweight materials in place. Nonstructural adhesives are sometimes referred to as *holding adhesives.* They are generally represented by pressure-sensitive, contact, and hot-melt adhesives. Sealants usually have a nonstructural function. They are principally intended to fill a gap between adherends to provide a seal without having high degrees of adhesive strength.

7.4.3.2 Chemical Composition. The composition classification describes synthetic adhesives as thermosetting, thermoplastic, elastomeric, or combinations of these. They are described generally in Table 7.9.

Thermosetting adhesives are materials that cannot be heated and softened repeatedly after the initial cure. Adhesives of this sort cure by chemical reaction at room or elevated temperatures, depending on the type of adhesive. Substantial pressure may also be required with some thermosetting adhesives, and others are capable of providing strong bonds with only contact pressure. Thermosetting adhesives are sometimes provided in a solvent medium to facilitate application. However, they are also commonly available as solventless liquids, pastes, and solids.

Thermosetting adhesives may be sold as multiple- and single-part systems. Generally the single-part adhesives require elevated-temperature cure, and they have a limited shelf life. Multiple-part adhesives have longer shelf lives, and some can be cured at room temperature or more rapidly at elevated temperatures. But they require metering and mixing before application. Once the adhesive is mixed, the working life is limited. Because molecules of thermosetting resins are densely cross-linked, their resistance to heat and solvents is good, and they show little elastic deformation under load at elevated temperatures.

Thermoplastic adhesives do not cross-link during cure, so they can be resoftened with heat. They are single-component systems that harden upon cooling from a melt or by evaporation of a solvent or water vehicle. Hot-melt adhesives commonly used in packaging are examples of a solid thermoplastic material that is applied in a molten state, and adhesion develops as the melt solidifies during cooling. Wood glues are thermoplastic emulsions that are a common household item. They harden by evaporation of water from the emulsion.

Thermoplastic adhesives have a more limited temperature range than thermosetting types. It is not suggested to use thermoplastic adhesives over 65°C. Their physical properties vary over a wide range, because many polymers are used in a single-adhesive formulation.

Elastomeric-type adhesives are based on polymers having great toughness and elongation. These adhesives may be supplied as solvent solutions, latex cements, dispersions, pressure-sensitive tapes, and single- or multiple-part solventless liquids or pastes. The curing requirements vary with the type and form of elastomeric adhesive. These adhesives can be formulated for a wide variety of applications, but they are generally noted for their high degree of flexibility and good peel strength.

Adhesive alloys or hybrids are made by combining thermosetting, thermoplastic, and elastomeric adhesives. They utilize the most useful properties of each material. However, the adhesive alloy is usually never better than its weakest constituent. For example, higher peel strengths are generally provided to thermosetting resins by the addition of thermoplastic or elastomeric materials, although usually at the sacrifice of temperature resistance. Adhesive alloys are commonly available in solvent solutions and as supported or unsupported film.

7.4.3.3 Mode of Application and Setting. Adhesives are often classified by their mode of application. Depending on viscosity, adhesives are sprayable, brushable, or trowelable. Heavy-bodied adhesive pastes and mastics are considered extrudable; they are applied by syringe, caulking gun, or pneumatic pumping equipment.

TABLE 7.9 Adhesives Classified by Chemical Composition[9]

Classification	Thermoplastic	Thermosetting	Elastomeric	Alloys
Types within group	Cellulose acetate, cellulose acetate butyrate, cellulose nitrate, polyvinyl acetate, vinyl vinylidene, polyvinyl acetals, polyvinyl alcohol, polyamide, acrylic, phenoxy	Cyanoacrylate, polyester, urea formaldehyde, melamine formaldehyde, resorcinol and phenol-resorcinol formaldehyde, epoxy, polyimide, polybenzimidazole, acrylic, acrylate acid diester	Natural rubber, reclaimed rubber, butyl, polyisobutylene, nitrile, styrene-butadiene, polyurethane, polysulfide, silicone, neoprene	Epoxy-phenolic, epoxy-polysulfide, epoxy-nylon, nitrile-phenolic, neoprene-phenolic, vinyl-phenolic
Most used form	Liquid, some dry film	Liquid, but all forms common	Liquid, some film	Liquid, paste, film
Common further classifications	By vehicle (most are solvent dispersions or water emulsions)	By cure requirements (heat and/or pressure most common but some are catalyst types)	By cure requirements (all are common); also by vehicle (most are solvent dispersions or water emulsions)	By cure requirements (usually heat and pressure except some epoxy types); by vehicle (most are solvent dispersions or 100% solids); and by type of adherends or end-service conditions
Bond characteristics	Good to 150 to 200°F; poor creep strength; fair peel strength	Good to 200 to 500°F; good creep strength; fair peel strength	Good to 150 to 400°F; never melt completely; low strength; high flexibility	Balanced combination of properties of other chemical groups depending on formulation; generally higher strength over wider temp range
Major type of use	Unstressed joints; designs with caps, overlaps, stiffeners	Stressed joints at slightly elevated temp	Unstressed joints on lightweight materials; joints in flexure	Where highest and strictest end-service conditions must be met; sometimes regardless of cost, as military uses
Materials most commonly bonded	Formulation range covers all materials, but emphasis on nonmetallics—especially wood, leather, cork, paper, etc.	For structural uses of most materials	Few used "straight" for rubber, fabric, foil, paper, leather, plastics films; also as tapes. Most modified with synthetic resins	Metals, ceramics, glass, thermosetting plastics; nature of adherends often not as vital as design or end-service conditions (that is, high strength and temperature)

Another distinction between adhesives is the manner in which they flow or solidify. Some adhesives solidify simply by losing solvent, while others harden as a result of heat activation or chemical reaction. Pressure-sensitive systems flow under pressure and are stable when pressure is absent.

7.4.3.4 Physical Form. Another method of distinguishing between adhesives is by physical form. The physical state of the adhesive generally determines how it is to be applied. Liquid adhesives lend themselves to easy handling via mechanical spreaders or spray and brush. Paste adhesives have high viscosities to allow application on vertical surfaces with little danger of sag or drip. These bodied adhesives also serve as gap fillers between two poorly mated substrates.

Tape and film adhesives are poor gap fillers but offer a uniformly thick bond line, no need for metering, and easy dispensing. Adhesive films are available as a pure sheet of adhesive or with cloth or paper reinforcement. Another form of adhesive is powder or granules that must be heated or solvent-activated to be made liquid and applicable.

7.4.3.5 Cost. The cost of fastening with adhesives must include the material cost of the adhesive, the cost of labor, the cost of equipment, the time required to cure the adhesive, and the economic loss due to rejects of defective joints. Often, the actual material cost of the adhesive is rather minor compared to the total assembly cost per unit.

Adhesive material cost should be calculated on a cost per bonded area basis. Since many adhesives are sold as dilute solutions, a cost per unit weight or volume basis may lead to erroneous comparisons.

Adhesive price is dependent on development costs and volume requirements. Adhesives that have been specifically designed to be resistant to adverse environments are more expensive than general-purpose adhesives. Adhesive prices range from pennies a pound for inorganic and animal-based systems to hundreds of dollars per pound for certain heat-resistant synthetic adhesives. Adhesives in film or powder form require more processing than liquid or paste types and are more expensive.

7.4.4 Joint Design

7.4.4.1 Types of Stress. To effectively design joints for adhesive bonding, it is necessary to understand the types of stress that are common to bonded structures. Four basic loading stresses are common to adhesive joints: tensile, shear, cleavage, and peel. Any combination of these stresses, illustrated in Fig. 7.13, may be encountered in an adhesive application.

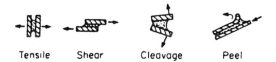

Tensile   Shear   Cleavage   Peel

FIGURE 7.13   Four basic types of adhesive stress.

Tensile stress develops when forces acting perpendicular to the plane of the joint are distributed uniformly over the entire bonded area. Adhesive joints show good resistance to tensile loading, because all of the adhesive contributes to the strength of the joint. In practical applications, unfortunately, loads are hardly ever purely axial, and cleavage or peel stresses tend to develop due to eccentricity in loading. Since adhesives have poor resistance to cleavage and peel, joints designed to load the adhesive in tension should have physical restraints to ensure purely axial loading.

Shear stresses result when forces acting in the plane of the adhesive try to separate the adherends. Joints dependent on the adhesive's shear strength are relatively easy to make and are commonly used. Adhesives are generally strongest when stressed in shear, because all the bonded area contributes to the strength of the joint.

Cleavage and peel stresses are undesirable. Cleavage occurs when forces at one end of a rigid bonded assembly act to split the adherends apart. Peel stress is similar to cleavage but applies to a joint where one or both of the adherends are flexible. Joints loaded in peel or cleavage provide much lower strength than joints loaded in shear, because the stress is concentrated on only a very small area of the total bond. Peel stress particularly should be avoided where possible, since the stress is confined to a very thin line at the edge of the bond (Figure 7.14). The remainder of the bonded area makes no contribution to the strength of the joint.

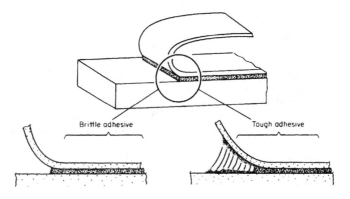

Brittle adhesive   Tough adhesive

FIGURE 7.14   Tough, flexible adhesives distribute peel stress over a larger area.[10]

7.4.4.2 Maximizing Joint Efficiency.   Although adhesives have often been used successfully on joints designed for mechanical fastening, the maximum efficiency of bonded joints can be obtained only by designing the joint specifically for adhesive bonding. To avoid concentration of stress, the joint designer should take into consideration the following rules:

1. Keep the stress on the bond line to a minimum.
2. Design the joint so that the operating loads will stress the adhesive in shear.
3. Peel and cleavage stresses should be minimized.
4. Distribute the stress as uniformly as possible over the entire bonded area.
5. Adhesive strength is directly proportional to bond width. Increasing width will always increase bond strength; increasing the depth does not always increase strength.
6. Generally, rigid adhesives are better in shear, and flexible adhesives are better in peel.

Brittle adhesives are particularly weak in peel, because the stress is localized at only a thin line, as shown in Figure 7.14. Tough, flexible adhesives distribute the peeling stress over a wider bond area and show greater resistance to peel.

For a given adhesive and adherend, the strength of a joint stressed in shear depends primarily on the width and depth of the overlap and the thickness of the adherend. Adhesive shear strength is directly proportional to the width of the joint. Strength can sometimes be increased by increasing the overlap depth, but the relationship is not linear. Since the ends of the bonded joint carry a higher proportion of the load than the interior area, the most efficient way of increasing joint strength is by increasing the width of the bonded area.

In a shear joint made from thin, relatively flexible adherends, there is a tendency for the bonded area to distort because of eccentricity of the applied load. This distortion, illustrated in Figure 7.15, causes cleavage stress on the ends of the joint, and the joint strength may be considerably impaired. Thicker adherends are more rigid, and the distortion is not as much a problem as with thin-gauge adherends. Since the stress distribution across the bonded area is not uniform and depends on joint geometry, the failure load of one specimen cannot be used to predict the failure load of another specimen with different joint geometry.

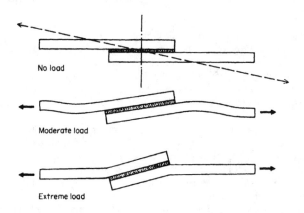

No load

Moderate load

Extreme load

FIGURE 7.15 Distortion caused by loading can introduce cleavage stresses and must be considered in the joint design.[9]

The strength of an adhesive joint also depends on the thickness of the adhesive. Thin adhesive films offer the highest shear strength, provided that the bonded area does not have "starved" areas where all the adhesive has been forced out. Excessively heavy adhesive-film thickness causes greater internal stresses during cure and concentration of stress under load at the ends of a joint. Optimum adhesive thickness for maximum shear strength is generally between 2 and 10 mils. Strength does not vary significantly with bond-line thickness in this range.

7.4.4.3 Joint Geometry. The ideal adhesive-bonded joint is one in which, under all practical loading conditions, the adhesive is stressed in the direction in which it most resists failure. A favorable stress can be applied to the bond by using proper joint design. Some joint designs may be impractical, expensive to make, or hard to align. The design engineer will often have to weigh these factors against optimum adhesive performance.

7.4.4.3.1 Joints for Flat Adherends. The simplest joint to make is the plain butt joint. Butt joints cannot withstand bending forces, because the adhesive would experience cleavage stress. If the adherends are too thick to design simple overlap-type joints, the butt joint can be improved by redesigning in a number of ways, as shown in Figure 7.16. All the modified butt joints reduce the cleavage effect caused by side loading. Tongue-and-groove joints also have an advantage in that they are self-aligning and act as a reservoir for the adhesive. The scarf joint keeps the axis of loading in line with the joint and does not require a major machining operation.

Lap joints are the most commonly used adhesive joint, because they are simple to make, are applicable to thin adherends, and stress the adhesive to be stressed in shear. However, the simple lap joint causes the adhesive to be stressed in shear. In this design, the adherends are offset, and the shear forces are not in-line, as was illustrated in Figure 7.15. This factor results in cleavage stress at the ends of the joint, which seriously impairs its efficiency. Modifications of lap-joint design (Figure 7.17) include:

1. Redesigning the joint to bring the load on the adherends in line
2. Making the adherends more rigid (thicker) near the bond area
3. Making the edges of the bonded area more flexible for better conformance, thus minimizing peel

The joggle-lap-joint design is the easiest method of bringing loads into alignment. The joggle lap can be made by simply bending the adherends. It also provides a surface to which it is easy to apply pressure. The double-lap joint has a balanced construction, which is subjected to bending only if loads on the double side of the lap are not balanced. The beveled lap joint is also more efficient than the plain lap joint. The beveled edges allow conformance of the adherends during loading thereby reducing cleavage stress on the ends of the joint.

Strap joints keep the operating loads aligned and are generally used where overlap joints are impractical because of adherend thickness. Strap-joint designs are shown in Fig. 7.18. Like the lap joint, the single strap is subjected to cleavage stress under bending forces. The double strap joint is more desirable when bending stresses are encountered. The beveled double strap and recessed double strap are the best joint designs to resist bending forces. Unfortunately, they both require expensive machining.

When thin members are bonded to thicker sheets, operating loads generally tend to peel the thin member from its base, as shown in Figure 7.19. The subsequent illustrations show what can be done to decrease peeling tendencies in simple joints. Often, thin sheets of a material are made more rigid by bonding stiffening members to the sheet. Resistance to bending forces is also increased by extending the bond area and increasing the stiffness of the base sheet.

7.4.4.3.2 Cylindrical-Joint Design. Several recommended designs for rod and tube joints are illustrated in Figure 7.20. These designs should be used instead of the simpler butt joint. Their resistance to bending forces and subsequent cleavage is much better, and the bonded area is larger. Unfortunately, most of these joint designs require a machining operation.

7.4.4.3.3 Angle- and Corner-Joint Designs. A butt joint is the simplest method of bonding two surfaces that meet at an odd angle. Although the butt joint has good resistance to pure tension and compression, its bending strength is very poor. Dado, L, and T angle joints, shown in Figure 7.21, offer greatly improved properties. The T design is the preferable angle joint because of its large bonding area and good strength in all directions.

Corner joints for relatively flexible adherends such as sheet metal should be designed with reinforcements for support. Various corner-joint designs are shown in Figure 7.22. With very thin adherends, angle joints offer low strengths because of high peel concentrations. A design consisting of right-angle corner plates or slip joints offers the most satis-

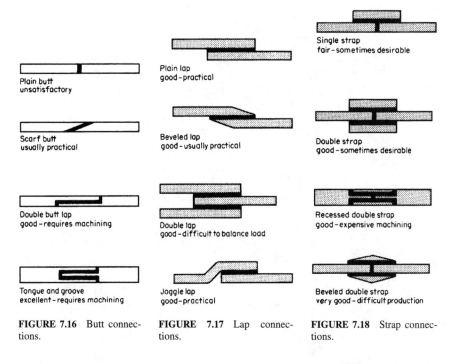

Plain butt
unsatisfactory

Scarf butt
usually practical

Double butt lap
good – requires machining

Tongue and groove
excellent – requires machining

Plain lap
good – practical

Beveled lap
good – usually practical

Double lap
good – difficult to balance load

Joggle lap
good – practical

Single strap
fair – sometimes desirable

Double strap
good – sometimes desirable

Recessed double strap
good – expensive machining

Beveled double strap
very good – difficult production

FIGURE 7.16 Butt connections.

FIGURE 7.17 Lap connections.

FIGURE 7.18 Strap connections.

factory performance. Thick, rigid members such as rectangular bars and wood may be bonded with an end lap joint, but greater strengths can be obtained with mortise and tenon. Hollow members such as extrusions fasten together best with mitered joints and inner splines.

7.4.4.3.4 Flexible Plastics and Elastomers. Thin or flexible polymeric substrates may be joined using a simple or modified lap joint. The double strap joint is best but also the most time-consuming to fabricate. The strap material should be made out of the same material as the parts to be joined, or at least have approximately equivalent strength, flexibility, and thickness. The adhesive should have the same degree of flexibility as the adherends.

If the sections to be bonded are relatively thick, a scarf joint is acceptable. The length of the scarf should be at least four times the thickness; sometimes larger scarves may be needed.

When bonding elastic material, forces on the elastomer during cure of the adhesive should be carefully controlled, since excess pressure will cause residual stresses at the bond interface. Stress concentrations may also be minimized in rubber-to-metal joints by elimination of sharp corners and using metal thick enough to prevent peel stresses that may arise with thinner-gauge metals.

As with all joint designs, polymeric joints should avoid peel stress. Figure 7.23 illustrates methods of bonding flexible substrates so that the adhesive will be stressed in its strongest direction.

7.4.4.3.5 Rigid Plastic Composites. Reinforced plastics are often anisotropic materials. This means their strength properties are directional. Joints made from anisotropic sub-

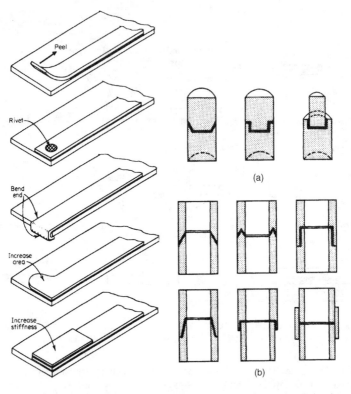

FIGURE 7.19 Minimizing peel in adhesive joints.[11]

FIGURE 7.20 Recommended designs for rod and tube joints: (a) round bars and (b) cylinders or tubes.[12]

strates should be designed to stress both the adhesive and adherend in the direction of greatest strength. Laminates, for example, should be stressed parallel to the laminations. Stresses normal to the laminate may cause the substrate to delaminate.

Single and joggle lap joints are more likely to cause delamination than scarf or beveled lap joints. The strap-joint variations are useful when bending loads may be imposed on the joint.

7.4.5 Surface Preparation

Many plastics and plastic composites can be treated prior to adhesive bonding by simple mechanical abrasion or alkaline cleaning to remove surface contaminants. In some cases it is necessary that the polymeric surface be physically or chemically modified to achieve acceptable bonding. This applies particularly to crystalline thermoplastics such as the polyolefins, linear polyesters, and fluorocarbons. Methods used to improve the bonding characteristics of these surfaces include:

1. Oxidation via chemical treatment or flame treatment
2. Electrical discharge to leave a more reactive surface

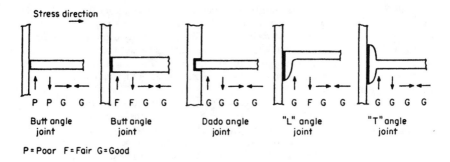

P = Poor   F = Fair   G = Good

FIGURE 7.21   Types of angle joints and methods of reducing cleavage.[11]

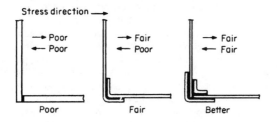

FIGURE 7.22   Reinforcement of bonded corner joints.

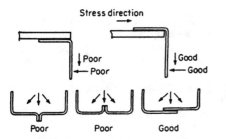

FIGURE 7.23   Methods of joining flexible rubber or plastic.[11]

3. Plasma treatment (exposing the surface to ionized inert gas)

4. Metal-ion treatment (e.g., sodium naphthalene process for fluorocarbons)

Surface preparation is most important for plastic parts that will be bonded with adhesives. Solvent and heat welding do not generally require chemical alteration of the surface; however, they do require cleaning. Welding procedures are discussed in another section of this chapter.

As with metallic substrates, the effects of plastic surface treatments decrease with time. It is necessary to prime or bond soon after the surfaces are treated. Some surface treatments, such as plasma, have a long effective shelf life (days to weeks) between treatment and bonding. However, some treating processes, such as electrical discharge and flame treating, will become less effective the longer the time between surface preparation and bonding.

Table 7.10 lists common recommended surface treatments for plastic adherends. These treatments are necessary when plastics are to be joined with adhesives. Specific surface treatments for certain plastics and their effect on surface property characteristics are discussed in Sec. 7.6. Details regarding the surface treatment process parameters may also be found in ASTM D-2093 and various texts on adhesive bonding of plastics. An excellent source of information regarding prebond surface treatments is the supplier of the plastic resin that is being joined.

Chemical or physical surface treatments are especially required for structural bonding of low-surface-energy plastics. Low-surface-energy plastics include polyethylene, polypropylene, TPO, and fluorinated polymers. These surface treatments are designed to increase the critical surface tension and improve wetting and adhesion. In addition to increasing the critical surface tension, surface treatments are designed to remove contaminants or "weak boundary layers," such as a mold release.

Abrasion and solvent cleaning are generally recommended as a surface treatment for high-surface-energy thermoplastics and for thermosetting plastics. Frequently, a mold-release agent is present and must be removed before adhesive bonding. Mold-release agents are usually removed by a detergent wash, solvent wash, or solvent wipe.

Common solvents used to clean plastic surfaces for adhesive bonding are acetone, toluene, trichloroethylene, methyl ethyl ketone (MEK), low-boiling petroleum ether, and isopropanol. A solvent should be selected that does not affect the plastic surface but is sufficiently strong to remove organic contamination. Safety and environmental factors must be considered when choosing a solvent. Solvent cleaning alone can be used for high-surface-energy plastics that do not require the maximum joint strength.

The compatibility of cleaning solvents with plastic substrates is extremely important. Solvents can affect polymeric surfaces and provide unacceptable part appearance or even degradation of properties. Solvents that are recommended for cleaning plastics are shown in Table 7.11. Suppliers of mold release agents are the best source for information on solvents that will remove their materials.

Abrasive treatments consist of scouring, machining, hand sanding, and dry and wet abrasive blasting. The abrasive medium can be fine sandpaper, carborundum or alumina abrasives, metal wools, or abrasive shot. Mechanical abrasion is usually preceded and followed by solvent cleaning. The choice is generally determined by available production facilities and cost.

Laminates can be prepared by either abrasion or the tear-ply technique. In the tear-ply design, the laminate is manufactured so that one ply of heavy fabric, such as Dacron, glass, or the equivalent, is attached at the bonding surface. Just prior to bonding, the tear-ply is stripped away, and a fresh, clean, bondable surface is exposed.

Chemical surface treatments vary with the type of plastic being bonded. These processes can involve the use of corrosive and hazardous materials. The most common processes are sulfuric acid–sodium dichromate etch (polyolefins) and sodium-naphthalene etch (fluorocarbons). Both of these processes are described in ASTM D-2093.

Flame, hot air, electrical discharge, and plasma treatments change the surface of the polymer both physically and chemically. The plasma treating process has been found to be very successful on most low-energy surface plastics. Table 7.12 shows that plasma treatment results in improved plastic joint strength with common epoxy adhesive. Plasma treatment requires vacuum and special batch processing equipment.

Most optimized surface treatment processes require prolonged production time and provide safety and environmental concerns. One should be careful not to overspecify the surface treatment required. Only the minimal process necessary to accomplish the functional objectives of the application is required.

Several new surface treatments and modifications of older, conventional surface treatments have been introduced over the last few years to provide alternatives to the common

TABLE 7.10   Surface Preparation Methods for Plastics

Adherend	Degreasing solvent	Method of treatment	Remarks

Acetal (copolymer)	Acetone	1. Abrasion. Grit or vapor blast, or medium-grit emery cloth followed by solvent degreasing
		2. Etch in the following acid solution:
		                    Parts by wt.
		Potassium dichromate 75
		Distilled water 120
		Concentrated sulfuric acid
		  (96%, sp. gr. 1.84) 1500
		  for 10 s at 25°C. Rinse in distilled water, and dry in air at RT
			For general-purpose bonding
			For maximum bond strength. ASTM D 2093

Acetal (homopolymer)	Acetone	1. Abrasion. Sand with 280A-grit emery cloth followed by solvent degreasing 2. "Satinizing" technique. Immerse the part in *Parts by wt.* Perchloroethylene 96.85 1,4-Dioxane 3.00 *p*-Toluenesulfonic acid 0.05 Cab-o-Sil (Cabot Corp.) 0.10 for 5–30 s at 80–120°C. Transfer the part immediately to an oven at 120°C for 1 min. Wash in hot water. Dry in air at 120°C	For general-purpose bonding For maximum bond strength. Recommended by DuPont
Acrylonitrile butadiene styrene	Acetone	1. Abrasion. Grit or vapor blast, or 220-grit emery cloth, followed by solvent degreasing 2. Etch in chromic acid solution for 20 min at 60°	 Recipe 2 for methyl pentane
Cellulosics: Cellulose, cellulose acetate, cellulose acetate butyrate, cellulose nitrate, cellulose propionate, ethyl cellulose	Methanol, isopropanol	1. Abrasion. Grit or vapor blast, or 220-grit emery cloth, followed by solvent degreasing 2. After procedure 1, dry the plastic at 100°C for 1 h, and apply adhesive before the plastic cools to room temperature	For general bonding purposes
Diallyl phthalate, diallyl isophthalate	Acetone, methyl ethyl ketone	Abrasion. Grit or vapor blast, or 100-grit emery cloth, followed by solvent degreasing	Steel wool may be used for abrasion

TABLE 7.10   Surface Preparation Methods for Plastics   *(Continued)*

Adherend	Degreasing solvent	Method of treatment	Remarks
Epoxy resins	Acetone, methyl ethyl ketone	Abrasion. Grit or vapor blast, or 100-grit emery cloth, followed by solvent degreasing	Sand or steel shot are suitable abrasives
Ethylene vinyl acetate	Methanol	Prime with epoxy adhesive and fuse into the surface by heating for 30 min at 100°C	
Furane	Acetone, methyl ethyl ketone	Abrasion. Grit or vapor blast, or 100-grit emery cloth, followed by solvent degreasing	
Ionomer	Acetone, methyl ethyl ketone	Abrasion. Grit or vapor blast, or 100-grit emery cloth, followed by solvent degreasing	Alumina (180-grit) is a suitable abrasive
Melamine resins	Acetone, methyl ethyl ketone	Abrasion. Grit or vapor blast, or 100-grit emery cloth, followed by solvent degreasing	For general-purpose bonding
Methyl pentene	Acetone	1. Abrasion. Grit or vapor blast, or 100-grit emery cloth, followed by solvent degreasing 2. Immerse for 1 h at 60°C in <div align="right">*Parts by wt.*</div>Sulfuric acid (96% sp. gr. 1.84) 26 Potassium chromate 3 Water 11 Rinse in water and distilled water. Dry in warm air 3. Immerse for 5–10 min at 90°C in potassium permanganate (saturated solution), acidified with sulfuric acid (96%, sp. gr. 1.84). Rinse in water and distilled water. Dry in warm air 4. Prime surface with lacquer based on urea-formaldehyde resin diluted with carbon tetrachloride	Coatings (dried) offer excellent bonding surfaces without further pretreatment
Phenolic resins phenolic melamine resins	Acetone, methyl ethyl ketone detergent	1. Abrasion. Grit or vapor blast, or abrade with 100-grit emery cloth, followed by solvent degreasing 2. Removal of surface layer of one ply of fabric previously placed on surface before curing. Expose fresh bonding surface by tearing off the ply prior to bonding	Steel wool may be used for abrasion. Sand or steel shot are suitable abrasives. Glass-fabric decorative laminates may be degreased with detergent solution

TABLE 7.10   Surface Preparation Methods for Plastics *(Continued)*

Adherend	Degreasing solvent	Method of treatment	Remarks
Polyamide (nylon)	Acetone, methyl ethyl ketone, detergent	1. Abrasion. Grit or vapor blast, or abrade with 100-grit emery cloth, followed by solvent degreasing	Sand or steel shot are suitable abrasives
		2. Prime with a spreading dough based on the type of rubber to be bonded in admixture with isocyanate	Suitable for bonding polyamide textiles to natural and synthetic rubbers
		3. Prime with resorcinol-formaldehyde adhesive	Good adhesion to primer coat with epoxy adhesives in metal-plastic joints
Polycarbonate, allyl diglycol carbonate	Methanol, isopropanol, detergent	Abrasion. Grit or vapor blast, or 100-grit emery cloth, followed by solvent degreasing	Sand or steel shot are suitable abrasives
Fluorocarbons: Polychloro-trifluoroe-thylene, polytetra-fluoro-ethylene, polyvinyl fluoride, polymono-chlorotri-fluoro-ethylene	Trichloro-ethylene	1. Wipe with solvent and treat with the following for 15 min at RT: Naphthalene (128 g) dissolved in tetrahydrofuran (1 l) to which is added sodium (23 g) during a stirring period of 2 h. Rinse in deionized water, and dry in water air	Sodium-treated surfaces must not be abraded before use. Hazardous etching solutions requiring skillful handling. Proprietary etching solutions are commercially available (see 2). PTFE available in etched tape. ASTM D 2093
		2. Wipe with solvent and treat as recommended in one of the following commercial etchants: Fluoroetch (Action Associates) Tetraetch (W. L. Gore Associates)	
		3. Prime with epoxy adhesive, and fuse into the surface by heating for 10 min at 370°C followed by 5 min at 400°C	
		4. Expose to one of the following gases activated by corona discharge: Air (dry) for 5 min Air (wet) for 5 min Nitrous oxide for 10 min Nitrogen for 5 min	Bond within 15 min of pretreatment
		5. Expose to electric discharge from a tesla coil (50,000 V ac) for 4 min	Bond within 15 min of pretreatment

TABLE 7.10   Surface Preparation Methods for Plastics *(Continued)*

Adherend	Degreasing solvent	Method of treatment	Remarks
Polyesters, polyethylene terephthalate (Mylar)	Detergent, acetone, methyl ethyl ketone	1. Abrasion. Grit or vapor blast, or 100-grit emery cloth, followed by solvent degreasing 2. Immerse for 10 min at 70–95°C in <table><tr><td></td><td>*Parts by wt.*</td></tr><tr><td>Sodium hydroxide</td><td>2</td></tr><tr><td>Water</td><td>8</td></tr></table>Rinse in hot water and dry in hot air	For general-purpose bonding For maximum bond strength. Suitable for linear polyester films (Mylar)
Chlorinated polyether	Acetone, methyl ethyl ketone	Etch for 5–10 min at 66–71°C in <table><tr><td></td><td>*Parts by wt.*</td></tr><tr><td>Sodium dichromate</td><td>5</td></tr><tr><td>Water</td><td>8</td></tr><tr><td>Sulfuric acid (96%, sp. gr. 1.84)</td><td>100</td></tr></table>Rinse in water and distilled water Dry in air	Suitable for film materials such as Penton. ASTM D 2093
Polyethylene, polyethylene (chlorinated), polyethylene terephthalate (see polyesters), polypropylene, polyformaldehyde	Acetone, methyl ethyl ketone	1. Solvent degreasing 2. Expose surface to gas-burner flame (or oxyacetylene oxidizing flame) until the substrate is glossy 3. Etch in the following: <table><tr><td></td><td>*Parts by wt.*</td></tr><tr><td>Sodium dichromate</td><td>5</td></tr><tr><td>Water</td><td>8</td></tr><tr><td>Sulfuric acid (96% sp. gr. 1.84)</td><td>100</td></tr></table>Polyethylene and polypropylene 60 min at 25°C or 1 min at 71°C Polyformaldehyde 10 s at 25°C 4. Expose to following gases activated by corona discharge: Air (dry) for 15 min Air (wet) for 5 min Nitrous oxide for 10 min Nitrogen for 15 min 5. Expose to electric discharge from a tesla coil (50,000 V ac) for 1 min	Low-bond-strength applications For maximum bond strength. ASTM D 2093 Bond within 15 min of pretreatment. Suitable for polyolefins. Bond within 15 min of pretreatment. Suitable for polyolefins.
Polymethyl methacrylate, methacrylate butadiene styrene	Acetone, methyl ethyl ketone, detergent, methanol, trichloroethylene, isopropanol	Abrasion. Grit or vapor blast, or 100-grit emery cloth, followed by solvent degreasing	For maximum strength relieve stresses by heating plastic for 5 h at 100°C

TABLE 7.10　Surface Preparation Methods for Plastics *(Continued)*

Adherend	Degreasing solvent	Method of treatment	Remarks
Poly-phenylene	Trichloro-ethylene	Abrasion. Grit or vapor blast, or 100-grit emery cloth, followed by solvent degreasing	
Poly-phenylene oxide	Methanol	Solvent degrease	Plastic is soluble in xylene and may be primed with adhesive in xylene solvent
Polystyrene	Methanol, isopropanol, detergent	Abrasion. Grit or vapor blast, or 100-grit emery cloth, followed by solvent degreasing	Suitable for rigid plastic
Polysulfone	Methanol	Vapor degrease	
Polyurethane	Acetone, methyl ethyl ketone	Abrade with 100-grit emery cloth and solvent degrease	
Polyvinyl chloride, polyvinylidene chloride, polyvinyl fluoride	Trichloro-ethylene, methyl ethyl ketone	1. Abrasion. Grit or vapor blast, or 100-grit emery cloth followed by solvent degreasing	Suitable for rigid plastic. For maximum strength, prime with nitrile-phenolic adhesive
		2. Solvent wipe with ketone	Suitable for plasticized material
Styrene acrylonitrile	Trichloro-ethylene	Solvent degrease	
Urea for-maldehyde	Acetone, methyl ethyl ketone	Abrasion. Grit or vapor blast, or 100-grit emery cloth, followed by solvent degreasing	

SOURCE: Based on the following: N. J. DeLolis, *Adhesives for Metals Theory and Technology*, Industrial Press, New York, 1970; C. V. Cagle, *Adhesive Bonding Techniques and Applications*, McGraw-Hill, New York, 1968; W. H. Guttmann, *Concise Guide to Structural Adhesives*, Reinhold, New York, 1961; "Preparing the Surface for Adhesive Bonding," Bulletin G1-600, Hysol Division, Dexter Corporation; A. H. Landrock, *Adhesive Technology Handbook*, Noyes Publications, Park Ridge, N. J., 1985; and J. Schields, *Adhesive Handbook*, CRC Press, Boca Raton, Fla., 1970.

TABLE 7.11   Common Degreasing Solvents for Polymeric Surfaces[13]

Adherend	Solvent
Acetal (copolymer)	Ketone
Acetal (homopolymer)	Ketone
Acrylonitrile-butadiene-styrene	Ketone
Cellulose, cellulose acetate, cellulose acetate butyrate, cellulose nitrate	Alcohol
Fluorocarbons	Chlorinated alcohol or ketone
Polyamide (nylon)	Ketone
Polycarbonate	Alcohol
Polyolefins	Ketone
Polyethylene terephthalate, PET (Mylar)	Ketone
Polyimide	Ketone
Polymethylmethacrylate, methacrylate butadiene	Ketone or alcohol
Polyphenylene oxide	Alcohol
Polyphenylene sulfide	Ketone, chlorinated solvents
Polystyrene	Alcohol
Polyvinyl chloride, polyvinyl fluoride	Ketone, chlorinated solvents
Thermoplastic polyester	Ketone
Thermoset plastics	Ketone

TABLE 7.12   Typical Adhesive Strength Improvement with Plasma Treatment: Aluminum-to-Plastic Shear Specimen Bonded with a Conventional Epoxy Adhesive[15]

Plastic	Strength of bond, psi	
	Control	After plasma treatment
Polyamide	846	>3956
Polyethylene	315	>3125
Polyethylene terephthalate	530	1660
Polypropylene	370	3080
Polystyrene	566	>4015
Polytetrafluoroethylene	75	750
Polyvinyl fluoride	278	>1280

processes noted above. The driving factors for these developments have primarily been related to environment and safety. Harsh chemicals and elevated-temperature processing associated with conventional chemical and flame treatment methods have inhibited many from using such processes.

In addition to providing safer and environmentally friendly processes, these newer surface treatments have also been shown to provide for easier and faster processing. They promise a potentially tremendous positive impact on both manufacturing cost and performance properties. The reduced cost impact can be in the form of equipment costs, implementation costs, operational costs, rework costs and storage/waste removal costs.

7.4.6 Adhesives Selection

Factors most likely to influence adhesive selection are listed in Table 7.13. However, thermosetting adhesives such as epoxies, polyurethanes, or acrylics are commonly used for structural application. The adhesive formulations are generally tough, flexible compounds that can cure at room temperature. The reasons that these adhesives have gained most popularity in bonding of plastics are summarized in this section.

The physical and chemical properties of both the solidified adhesive and the plastic substrate affect the quality of the bonded joint. Major elements of concern in selecting an adhesive for plastic parts are the thermal expansion coefficient and glass transition temperature of the substrate relative to the adhesive. Special consideration is also required of polymeric surfaces that can change during normal aging or exposure to operating environments.

Significant differences in thermal expansion coefficient between substrates and the adhesive can cause serious stress at the plastic's joint interface. These stresses are compounded by thermal cycling and low-temperature service requirements. Selection of a resilient adhesive or adjustments in the adhesive's thermal expansion coefficient via filler or additives can reduce such stress.

TABLE 7.13 Factors Influencing Adhesive Selection

Stress	Tension
	Shear
	Impact
	Peel
	Cleavage
	Fatigue
Chemical factors	External (service-related)
	Internal (effect of adherend on adhesives)
Exposure	Weathering
	Light
	Oxidation
	Moisture
	Salt spray
Temperature	High
	Low
	Cycling
Biological factors	Bacteria or mold
	Rodents or vermin
Working properties	Application
	Bonding time and temperature range
	Tackiness
	Curing rate
	Storage stability
	Coverage

Structural adhesives must have a glass transition temperature higher than the operating temperature to avoid a cohesively weak bond and possible creep problems. Modern engineering plastics, such as polyimide or polyphenylene sulfides, have very high glass transition temperatures. Most common adhesives have a relatively low glass transition temperature so that the weakest thermal link in the joint may often be the adhesive.

Use of an adhesive too far below its glass transition temperature could result in low peel or cleavage strength. Brittleness of the adhesive at very low temperatures could also manifest itself in poor impact strength.

Generally, the best adhesive is one that will wet the substrate and, when cured, has a modulus and thermal expansion coefficient similar to the substrate or else has necessary toughness and elongation to accommodate stresses caused by thermal movements. Differences in flexibility or thermal expansion between the adherends or between the adhesive and adherend can introduce internal stresses into the bond line. Such stresses can lead to premature failure of a bond. Thus, rigid, heavily filled adhesives are often chosen for bonding metals.

Flexible adhesives are often chosen for bonding plastics and elastomers. Lower-modulus adhesives generally have the flexibility to bond well to plastic substrates. However, these are generally weaker in shear than more rigid adhesives. Fortunately, exceptionally high shear strength is often not required for an adhesive for plastic, since the plastic substrate itself is relatively weak.

For many high-surface-energy thermosetting plastics, such as epoxies, polyesters, and phenolics, adhesive bonding is generally easy and can be accomplished with many of the same adhesives that are used on metal substrates. For thermoplastics, the surface energy is generally lower, the reactivity is greater, and the thermal expansion is higher than for thermosets. Therefore, when bonding thermoplastics, consideration must be given to the surface energy of the adhesive and the substrate, the compatibility of the adhesive with the substrate, and thermal expansion coefficients.

There are numerous families of adhesives within the structural and nonstructural types. The most common chemical families of structural and nonstructural adhesive families for bonding plastics are identified in Table 7.14.

TABLE 7.14 Common Families of Structural and Nonstructural Adhesives for Bonding Plastics

---

Structural adhesives for bonding plastics
- Cyanoacrylate
- Epoxy
- Polyurethane
- Reactive acrylic
- Light curing adhesive (acrylic and cyanoacrylate)

Nonstructural adhesives for bonding plastics
- Synthetic and natural elastomers
- Thermoplastic hot melts
- Resin latex adhesives
- Silicone

Structural adhesives are those having bond shear strength on the order of 1000 psi or greater. This is often sufficient to cause failure of the plastic substrate when the bond is tested. Structural adhesives are generally intended for applications where chemical and temperature resistance are requirements, as well as high strength and toughness.

Nonstructural adhesives are those having bond strength that is less than 1000 psi but sufficient for applications such as pressure sensitive tapes, labels, laminates, and so on. Nonstructural adhesives are usually employed where production speed, convenience, and high peel strength are required. They generally have sufficient permanence for the applications mentioned.

In selecting an adhesive system for a plastic, it is important to remember that the adhesive must retain its initial strength during the life of the product. Often plastics substrates can chemically and/or physically change during service aging. Therefore, the choice of the adhesive must be adequate for resisting initial as well as long-term stress conditions.

Plastic substrates could be chemically active, even when isolated from the operating environment. Many polymeric surfaces slowly undergo chemical and physical change. The plastic surface, at the time of bonding, may be well suited to the adhesive process. However, after aging, undesirable surface conditions may present themselves at the interface, displace the adhesive, and result in bond failure. These weak boundary layers may come from the environment or within the plastic substrate itself.

Moisture, solvent, plasticizers, and various gases and ions can compete with the cured adhesive for bonding sites. The process by which a weak boundary layer preferentially displaces the adhesive at the interface is called *desorption*. Moisture is the most common desorbing substance, being present both in the environment and within many polymeric substrates.

Solutions to the desorption problem consist of eliminating the source of the weak boundary layer or selecting an adhesive that is compatible with the desorbing material. Excessive moisture can be eliminated from a plastic part by postcuring or drying the part before bonding. Additives that can migrate to the surface can possibly be eliminated by reformulating the plastic resin. Also, certain adhesives are more compatible with oils and plasticizer than others. For example, the migration of plasticizer from flexible polyvinyl chloride can be counteracted by using nitrile-based adhesives. Nitrile adhesives resins are capable of absorbing the plasticizer without degradation.

7.4.7 The Adhesives Bonding Processes

After the adhesive is applied, the assembly must be mated as quickly as possible to prevent contamination of the adhesive surface. The substrates are held together under pressure and heated if necessary until cure is achieved. The equipment required to perform these functions must provide adequate heat and pressure, maintain constant pressure during the entire cure cycle, and distribute pressure uniformly over the bond area. Of course, many adhesives cure with simple contact pressure at room temperature, and extensive bonding equipment is not necessary.

Pressure devices should be designed to maintain constant pressure on the bond during the entire cure cycle. They must compensate for thickness reduction from adhesive flow-out or thermal expansion of assembly parts. Thus, screw-actuated devices like C-clamps and bolted fixtures are not acceptable when constant pressure is important. Spring pressure can often be used to supplement clamps and compensate for thickness variations. Dead-weight loading may be applied in many instances; however, this method is sometimes impractical, especially when heat cure is necessary.

Pneumatic and hydraulic presses are excellent tools for applying constant pressure. Steam or electrically heated platen presses with hydraulic rams are often used for adhesive bonding. Some units have multiple platens, thereby permitting the bonding of several assemblies at one time.

Many structural adhesives require heat as well as pressure. Most often the strongest bonds are achieved by an elevated-temperature cure. With many adhesives, trade-offs between cure times and temperature are permissible. But generally, the manufacturer will recommend a certain curing schedule for optimum properties.

However, often the temperature required to cure the adhesive will adversely affect heat-sensitive plastic parts. Also, heat-curing adhesives are generally more rigid than those that cure at room temperature, and the resulting modulus is too high for many plastic-bonding applications. As a result, most adhesives recommended for bonding plastic substrates cure at room temperature.

It is highly desirable to have a uniformly thin (2- to 10-mil) adhesive bond line. Starved adhesive joints, however, will yield exceptionally poor properties. Three basic methods are used to control adhesive thickness. The first method is to use mechanical shims or stops, which can be removed after the curing operation. Sometimes it is possible to design stops into the joint.

The second method is to employ a film adhesive that becomes highly viscous during the cure cycle, preventing excessive flow-out. With supported films, the adhesive carrier itself can act as the "shims." Generally, the cured bond-line thickness will be determined by the original thickness of the adhesive film. The third method of controlling adhesive thickness is to use trial and error to determine the correct pressure-adhesive viscosity factors that will yield the desired bond thickness.

7.4.8 Quality Control

Quality control systems and procedures are a requirement in almost every bonding application. Quality control should cover all phases of the bonding cycle from inspection of incoming material to the inspection of the completed assembly. In fact, good quality control will start even before receipt of materials. Quality control will encompass, at a minimum:

- Specification and inspection of incoming adherends as well as adhesives
- Control over the surface preparation process
- Control over the bond fabrication process (equipment, temperature, pressure, time, and so on)
- Inspection of the final part (destructively or nondestructively)
- Training of personnel in all aspects of adhesive bonding as well as safety and health requirements

The human element enters the adhesive-bonding process more than in other fabrication techniques. An extremely high percentage of defects can be traced to poor workmanship. This generally prevails in the surface-preparation steps but may also arise in any of the other steps necessary to achieve a bonded assembly. This problem can be largely overcome by proper motivation and education. All employees from design engineer to laborer to quality-control inspector should be somewhat familiar with adhesive bonding technology and be aware of the circumstances that can lead to poor joints. A great many defects can also be traced to poor design engineering.

The plant's bonding area should be as clean as possible prior to receipt of materials. The basic approach to keeping the assembly area clean is to segregate it from the other manufacturing operations either in a corner of the plant or in isolated rooms. The air should be dry and filtered to prevent moisture or other contaminants from gathering at a possible interface. The cleaning and bonding operations should be separated from each other. If mold release is used to prevent adhesive flash from sticking to bonding equipment, it is advisable that great care be taken to assure that the release does not contaminate the adhesive or the adherends. Spray mold releases, especially silicone release agents, have a tendency to migrate to undesirable areas.

Acceptance tests on adhesives should be directed toward assurance that incoming materials are identical from lot to lot. The tests should be those that can quickly and accurately detect deficiencies in the adhesive's physical or chemical properties. A number of standard tests for adhesive bonds and for adhesive acceptance have been specified by the American Society for Testing and Materials (ASTM). The properties usually reported by adhesive suppliers are ASTM tensile-shear (ASTM D-1002) and peel strength (ASTM D-903, D-1876, and D-3167).

Actual test specimens should also be made to verify strength of the adhesive. These specimens should be stressed in directions that are representative of the forces that the bond will see in service, i.e., shear, peel, tension, or cleavage. If possible, the specimens should be prepared and cured in the same manner as actual production assemblies. If time permits, specimens should also be tested in simulated service environments, e.g., high temperature and humidity.

Surface preparations must be carefully controlled for reliable production of adhesive-bonded parts. If a chemical surface treatment is required, the process must be monitored for proper sequence, bath temperature, solution concentration, and contaminants. If sand or grit blasting is employed, the abrasive must be changed regularly. An adequate supply of clean wiping cloths for solvent cleaning is also mandatory. Checks should be made to determine if cloths or solvent containers may have become contaminated.

The adhesive metering and mixing operation should be monitored by periodically sampling the mixed adhesive and testing it for adhesive properties. A visual inspection can also be made for air entrapment and degree of mixing. The quality-control engineer should be sure that the oldest adhesive is used first and that the specified shelf life has not been exceeded.

During the actual assembly operation, the cleanliness of the shop and tools should be verified. The shop atmosphere should be controlled as closely as possible. Temperature is in the range of 18 to 32°C and relative humidity from 20 to 65 percent is best for almost all bonding operations.

The amount of the applied adhesive and the final bond-line thickness must also be monitored, because they can have a significant effect on joint strength. Curing conditions should be monitored for heat-up rate, maximum and minimum temperature during cure, time at the required temperature, and cool-down rate.

After the adhesive is cured, the joint area can be inspected to detect gross flaws or defects. This inspection procedure can be either destructive or nondestructive in nature. Destructive testing generally involves placing samples of the production run in simulated or accelerated service and determining if it has similar properties to a specimen that is known to have a good bond and adequate service performance. The causes and remedies for faults revealed by such mechanical tests are described in Table 7.15.

Nondestructive testing (NDT) is far more economical, and every assembly can be tested if desired. However, there is no single nondestructive test or technique that will provide the user with a quantitative estimate of bond strength. There are several ultrasonic test methods that provide qualitative values. However, a trained eye can detect a surprising number of faulty joints by close inspection of the adhesive around the bonded area. Table 7.16 lists the characteristics of faulty joints that can be detected visually. The most difficult defects to be found by any method are those related to improper curing and surface treatments. Therefore, great care and control must be given to surface-preparation procedures and shop cleanliness.

TABLE 7.15 Faults Revealed by Mechanical Tests

Fault	Cause	Remedy
Thick, uneven glue line	Clamping pressure too low	Increase pressure. Check that clamps are seating properly
	No follow-up pressure	Modify clamps or check for freedom of moving parts
	Curing temperature too low	Use higher curing temperature. Check that temperature is above the minimum specified throughout the curing cycle
	Adhesive exceeded its shelf life, resulting in increased viscosity	Use fresh adhesive
Adhesive residue has spongy appearance or contains bubbles	Excess air stirred into adhesive	Vacuum-degas adhesive before application
	Solvents not completely dried out before bonding	Increase drying time or temperature. Make sure drying area is properly ventilated
	Adhesive material contains volatile constituents	Seek advice from manufacturers
	A low-boiling constituent boiled away	Curing temperature is too high
Voids in bond (that is, areas that are not bonded), clean bare metal exposed, adhesive failure at interface	Joint surfaces not properly treated	Check treating procedure; use clean solvents and wiping rags. Wiping rags must not be made from synthetic fiber. Make sure cleaned parts are not touched before bonding. Cover stored parts to prevent dust from settling on them
	Resin may be contaminated	Replace resin. Check solids content. Clean resin tank
	Uneven clamping pressure	Check clamps for distortion
	Substrates distorted	Check for distortion; correct or discard distorted components. If distorted components must be used, try adhesive with better gap-filling ability
Adhesive can be softened by heating or wiping with solvent	Adhesive not properly cured	Use higher curing temperature or extend curing time. Temperature and time must be above the minimum specified throughout the curing cycle. Check mixing ratios and thoroughness of mixing. Large parts act as a heat sink, necessitating larger cure times

TABLE 7.16   Visual Inspection for Faulty Bonds

Fault	Cause	Remedy
No appearance of adhesive around edges of joint or adhesive bond line too thick	Clamping pressure too low Starved joint Curing temperature too low	Increase pressure. Check that clamps are seating properly Apply more adhesive Use higher curing temperature. Check that temperature is above the minimum specified
Adhesive bond line too thin	Clamping pressure too high Curing temperature too high Starved joint	Lessen pressure Use lower curing temperature Apply more adhesive
Adhesive flash breaks easily away from substrate	Improper surface treatment	Check treating procedures; clean solvents and wiping rags. Make sure cleaned parts are not touched before bonding
Adhesive flash is excessively porous	Excess air stirred into adhesive Solvent not completely dried out before bonding Adhesive material contains volatile constituent	Vacuum-degas adhesive before application Increase drying time or temperature Seek advice from manufacturers
Adhesive flash can be softened by heating or wiping with solvent	Adhesive not properly cured	Use higher curing temperature or extend curing time. Temperature and time must be above minimum specified. Check mixing

7.5 WELDING

Certain thermoplastic substrates may be joined by methods other than mechanical fastening or adhesive bonding. Welding is particularly attractive for thermoplastics, because joining times are often very short, enabling high throughput. Also, the various welding processes typically provide strong joints, tolerate contaminated surfaces, and successfully join such difficult to bond substrates as polyethylene, polypropylene, and nylon.

Welding processes are of two main types: thermal and solvent. By careful application of heat or solvent to a thermoplastic substrate, one may liquefy the surface resin and use it to form the bond. The bond strength is determined by diffusion of polymer from one surface into another instead of by the wetting and adsorption of an adhesive layer. It is possible to weld plastics of different types. However, for both thermal and solvent welding, the success of the process will be heavily determined by the compatibility of the polymers being joined.

With thermal or solvent welding, surface preparation is not as critical as with adhesive bonding. However, some form of surface pretreatment may still be necessary, although difficult chemical or physical treatments to increase the surface energy are not required. Certainly, the parts should be clean, and all mold release and contaminants must be removed by standard cleaning procedures.

It may also be necessary to dry certain polymeric parts, such as nylon and polycarbonate, before welding so that the inherent moisture in the part will not affect the overall quality of the bond. It may also be necessary to thermally anneal parts, such as acrylic, before solvent welding to remove or lessen internal stresses caused by molding. Without annealing, the stressed surface may crack or craze when in contact with solvent.

7.5.1 Thermal Welding

Welding by application of heat, or "thermal welding," provides an advantageous method of joining many thermoplastics that do not degrade rapidly at their melt temperature. It is a method of providing fast, relatively easy, and economical bonds that are generally 80 to 100 percent of the strength of the parent plastic. In all thermal welding processes, the substrate surface is heated by some method until it is at a melt or flowable state. The melted surfaces are then pressed together (forged), which results in interdiffusion of the molecular chains. On cooling to a solid state, a strong and permanent joint is created.

Thermal welding process can be of two kinds: direct and indirect. Each kind of thermal welding may be further classified, as shown below, according to the method used to provide heat.

Direct Thermal Welding Processes

- Heated tool welding
- Hot gas welding
- Other (infrared radiation, laser, and others)

Indirect Thermal Welding Processes

- Friction or spin welding
- Induction welding
- Ultrasonic and vibration welding
- Dielectric welding

With direct welding, the heat is applied directly to the substrate in the form of either a heated tool or hot gas. Indirect heating occurs when some form of energy other than thermal is applied to the joint. The applied energy, which causes heating at the interface or in the plastic as a whole, is generally in the form of friction, high-frequency electrical fields, electromagnetic fields, or ultrasonic vibration. Because the heating is localized at the bonding surface, indirect heating processes are very energy efficient, generally resulting in bonds that are stress free and of higher strength than those made by direct welding methods.

7.5.1.1 Heated Tool Welding. With this method, the surfaces to be fused are heated by holding them against a hot metal surface (232 to 371°C); then the parts are brought into contact and allowed to harden under slight pressure (5 to 15 psi). Electric strip heaters, soldering irons, hot plates, and resistance blades are common methods of providing heat. Heated platens are generally employed to create a molten or plasticized region. Thus, this form of welding is often called *hot-plate welding.*

One production technique involves butting flat plastic sheets on a table next to an electrical resistance heated blade that runs the length of the sheet. Once the plastic adjacent to the blade begins to soften, the blade is raised, and the sheets are pressed together and held under pressure while they cool. The heated metal surfaces are usually coated with a high-temperature release coating such as polytetrafluoroethylene to discourage sticking to the molten plastic.

Successful heated tool welding depends on the temperature of the heated tool surface, the amount of time the plastic adherends are in contact with the hot tool, the time lapse before joining the substrates, and the amount and uniformity of pressure that is held during cooling. Heated tool welding can be used for structural plastic parts, and heat sealing can be used for plastic films. With heat sealing, the hot surface is usually hot rollers or a heated rotating metal band commonly used to seal plastic bags. Table 7.17 offers heat welding temperatures for a number of common plastics and films.

TABLE 7.17   Hot-Plate Temperatures to Weld Plastics and Plastic Films[17]

Plastic	Temperature, °F	Film	Temperature, °F
ABS	450	Coated cellophane	200–350
Acetal	500	Cellulose acetate	400–500
Phenoxy	550	Coated polyester	490
Polyethylene			
LD	360	Poly(chlorotrifluoroethylene)	415–450
HD	390	Polyethylene	250–375
Polycarbonate	650	Polystyrene (oriented)	220–300
PPO	650	Poly(vinyl alcohol)	300–400
Noryl*	525	Poly(vinyl chloride) and	
		copolymers (nonrigid)	200–400
Polypropylene	400	Poly(vinyl chloride) and	
		copolymers (rigid)	260–400
Polystyrene	420	Poly(vinyl chloride)—	
		nitrile rubber blend	220–350
SAN	450	Poly(vinylidene chloride)	285
Nylon 6, 6	475	Rubber hydrochloride	225–350
PVC	450	Fluorinated ethylene-	
		propylene copolymer	600–750

*Trademark of General Electric Company.

Resistance wire or implant welding is also a type of heated tool welding. This method generally employs an electrical resistance heating wire laid between mating substrates to generate the heat of fusion. When energized, the wire undergoes resistance heating and causes a melt area to form around the adjacent polymer. Pressure on the parts during this process causes the molten material to flow and act as a hot-melt adhesive for the joint. After the bond has been made and the joint solidifies, the resistance wire material is generally cut off and removed.

Resistance wire welding can be used on any plastic that can be joined effectively by heated tool welding. The process is typically applied to relatively large structures. Contacting the plastic resin manufacturer for details concerning the specific parameters of this process is recommended.

Radio-frequency energy has also been used to heat an implant that is placed at the joint interface. Current passing through the conductive implant generates the heat in this process. Once the joint is made, the implant can be reheated via radio-frequency heating and the parts can be disassembled. Thus, this welding process is popular for applications where recovery or efficient disassembly of parts is critical.

The resistive element can be any material that conducts current, including metal wires and braids and carbon-based compounds. Implant materials should be compatible with the intended application, since they will remain in the bondline.

7.5.1.2 Hot Gas welding.   In hot gas welding, the weld joint is filled with a partially or fully molten polymer. This process is often used for long bond lines and for outdoor applications where it is difficult to control conditions. Common applications are the bonding of pond liners, repair of large thermoplastic tanks, assembly of large air ducts, and the joining of pipe.

In the most common hot gas welding process, an electrically or gas heated welding gun with an orifice temperature of 218 to 371°C can be used to bond many thermoplastic materials. The pieces to be bonded are beveled and positioned to form a V-shaped joint as shown in Figure 7.24. A welding rod, made of the same plastic that is being bonded, is laid into the joint, and the heat from the gun is directed at the interface of the substrates and the rod. The molten product from the welding rod then fills the gap. A strong fillet must be formed, the design of which is of considerable importance.

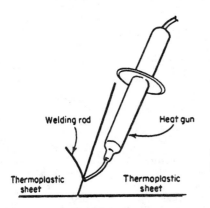

FIGURE 7.24   Hot gas welding apparatus.

A large difference between the plastic melting temperature and the decomposition temperature of the plastic is necessary for consistent, reliable hot gas welding results. Usually, the hot gas can be common air. However, for polyolefins and other easily oxidized plastics, the heated gas must be inert or nitrogen, since air will oxidize the surface of the plastic.

After welding, the joint should not be stressed for several hours. This is particularly true for polyolefins and nylons. Hot gas welding is not recommended for filled materials or substrates of less than 1/16 in thickness. Applications are usually large structural assemblies. The weld is not cosmetically attractive, but tensile strengths that are 85 percent of the parent materials are easily obtained.

A similar type of welding to hot gas welding is *extrusion welding*. In extrusion welding, a fully molten polymer is injected into the weld joint. The molten polymer is generated inside the welding tool or extrusion equipment and then pumped into the weld joint as the tool is moved along the weld.

7.5.1.3 Friction or Spin Welding.   Spin welding uses the heat of friction to cause fusion at the interface. One substrate is rotated very rapidly while in touch with the other stationary substrate so that the surfaces melt without damaging the part. Sufficient pressure is applied during the process to force out excess air bubbles. The rotation is then stopped, and pressure is maintained until the weld sets. Rotation speed and pressure are dependent on the thermoplastics being joined.

The main process parameters are the spin of rotation, weld or axial pressure, and weld time. The equipment necessary depends on production requirement, but spin welding can be adapted to standard shop machinery such as drill presses or lathes. In commercial spin welding machines, rotational speeds can range from 200 to 14,000 rpm. Welding times (heating and cooling) can range from less than 1 to 20 s, with typical times being several seconds.

A wide variety of joints can be made by spin welding. Since the outer edges of the rotating substrate move considerably faster than the center, joints are generally designed to concentrate pressure at the center. Hollow sections with thin walls are the best joint designs for this welding method, since the differential generation of heat could result in high weld zone stresses. A shallow tongue-and-groove type of joint design is also useful to index the opposite parts and provide a uniform bearing surface.

Because of its high weld quality, simplicity, speed, and reproducibility, spin welding is a popular method of joining large-volume products, packaging, and toys. Common applications are the manufacture of floats, aerosol bottles, and attachment of studs to plastic parts.

7.5.1.4 Induction Heating.

An electromagnetic induction field can be used to heat a metal grid or an insert placed between mating thermoplastic substrates (see implant welding, above). When the joint is positioned between energized induction coils, the hot insert material responds to the high-frequency AC source, causing the plastic surrounding it to melt and fuse together. Slight pressure is maintained as the induction field is turned off and the joint hardens.

In addition to metal inserts, electromagnetic adhesives can be used to form the joint. Electromagnetic adhesives are made from metal or ferromagnetic particle-filled thermoplastics. These adhesives can be shaped into gaskets or film that can easily be applied and will melt in an induction field.[18] The advantage of this method is that stresses caused by large metal inserts are avoided.

Induction welding is less dependent than other welding methods on the properties of the materials being welded. It can be used on nearly all thermoplastics. In welding different materials, the thermoplastic resin enclosing the metal particles in the electromagnetic adhesive is made of a blend of the materials being bonded. Table 7.18 shows compatible combinations for electromagnetic adhesives. Reinforced plastics with filler levels over 50 percent have been successfully electromagnetically welded.

TABLE 7.18 Compatible Plastic Combinations for Bonding with Electromagnetic Adhesives[19]

	ABS	Acetal	Acrylic	Nylon	PC	PE	PP	PS	PVC	SAN
ABS	X		X				X	X		
Acetal		X								
Acrylic	X		X				X	X		
Nylon				X						
Polycarbonate					X					
Polyethylene						X				
Polypropylene							X			
Polystyrene	X		X					X	X	
Polyvinyl chloride	X		X					X	X	
SAN										X

X = compatible combinations.

Strong and clean structural, hermetic, and high-pressure seals can be obtained from this process. Important determinants of bond quality in induction welding are the joint design and induction coil design. With automatic equipment, welds can be made in less than 1 s.

7.5.1.5 Ultrasonic and Vibration Welding. Ultrasonic welding is a well accepted method for joining high-volume, relatively small plastic parts. Energy for vertical oscillations produces intense frictional heating between two substrates. This frictional heating produces sufficient thermal energy to rapidly generate a molten weld zone.

During ultrasonic welding, a high-frequency (20 to 40 kHz) electrodynamic field is generated that resonates a metal horn. The horn is in contact with one of the plastic parts, and the other part is fixed firmly. The horn and the part to which it is in contact vibrate sufficiently fast to cause great heat at the interface of the parts being bonded. With pressure and subsequent cooling, a strong bond can be obtained with many thermoplastics.

The basic variables in ultrasonic bonding are amplitude, air pressure, weld time, and hold time. The desired joint strength can be achieved by altering these variables. Increasing weld time generally results in increasing bond strength up to a point. After that point, additional weld time does not improve the joint and can even degrade it. Average processing times, including welding and cooling, are less than several seconds.

Typical ultrasonic joint designs are shown in Figure 7.25. Usually, an energy director, or small triangular tip in one of the parts, is necessary. All of the ultrasonic energy is concentrated on the tip of the energy director, and this is the area of the joint that then heats, melts, and provides the material for the bond. Ultrasonic welding is considered a faster means of bonding than direct heat welding.

Ultrasonic welding of parts fabricated from ABS, acetals, nylon, PPO, polycarbonate, polysulfone, and thermoplastic polyesters should be considered as early in the design of the part as possible. Very often, minor modifications in part design will make ultrasonic welding more convenient. The plastic resin manufacturer or ultrasonic equipment supplier can recommend best joint design and ultrasonic horn design.

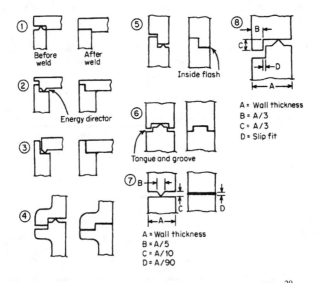

FIGURE 7.25 Typical joint designs used in ultrasonic welding.[20]

Like resin materials (such as ABS-to-ABS) are the easier to weld ultrasonically, but some unlike resins may be bonded provided they have similar melt temperatures, chemical composition, and modulus of elasticity. Generally, amorphous resins (ABS, PPO, and polycarbonate) are also easier to weld ultrasonically than crystalline resins (nylon, acetal, thermoplastic polyester).

Ultrasonic equipment can also be used for mechanical fastening operations. Ultrasonic energy can be used to apply threaded inserts to molded plastic parts and to heat stake plastic studs.

Vibration welding is similar to ultrasonic welding except that it uses a lower frequency (120 to 240 Hz) of vibration. In this way, very large parts can be bonded. The two thermoplastic parts are rubbed together, under pressure and at a suitable frequency and amplitude, until enough heat is generated at the joint interface to melt and diffuse the polymer surfaces.

The process parameters affecting the strength of the resulting weld are the amplitude and frequency of motion, weld pressure, and weld time. There are two types of vibration welding: linear, in which friction is generated by a linear motion of the parts, and orbital, in which one part is vibrated using circular motion in all directions. Examples of these processes are illustrated in Figure 7.26.

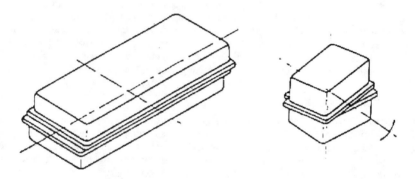

FIGURE 7.26   The vibration welding process. Linear vibration (left) is employed where the length to width ratio precludes the use of axial welding (right) where the axial shift is still within the width of the welded edge.

Vibration welding has been used on large thermoplastic parts such as canisters, pipe sections, and other parts that are too large to be excited with an ultrasonic generator. An advantage of vibration welding over ultrasonic welding is that it can provide gas-tight joints in structures having long bond lines. Ultrasonic welding is basically a spot weld technique limited by the size of the horn.

Total process time for vibration joining is generally between 5 and 15 s. This is longer than spin or ultrasonic welding but much shorter than direct heat welding, solvent cementing, or adhesive bonding. Vibration welding is applicable to most polymers, amorphous as well as semicrystalline. It is particularly useful for the semicrystalline polymers that cannot be easily welded using ultrasonic or solvent welding.

Vibration welding can be applied to ABS, acetal, nylon, PPO, thermoplastic polyesters, and polycarbonate. For vibration welding, hydroscopic resins, such as nylon, do not have to be dried as is necessary with ultrasonic welding. Joint designs do not require an energy director as in ultrasonic joint designs, but the joint area must be strong enough to resist the forces of operation without deformation. This often requires thickening the bond area or designing stiffeners into the part near the joint areas.

7.5.1.6 Other Welding Methods. *Dielectric welding* can be used on most thermoplastics except those that are relatively transparent to high-frequency electric fields. It is used mostly to seal vinyl sheeting such as automobile upholstery, swimming pool liners, and rainwear. An alternating electric field is imposed on the joint, which causes rapid reorientation of polar molecules, and heat is generated by molecular friction. The field is removed, and pressure is then applied and held until the weld cools.

Variable in the bonding operation are the frequency generated, dielectric loss of the plastic, the power applied, pressure, and time. The frequency of the field being generated can be from radio frequency up to microwave frequency. Dielectric heating can also be used to generate the heat necessary for curing polar, thermosetting adhesives, or it can be used to quickly evaporate water from water based adhesives—a common application in the furniture industry.

Other thermal welding processes that are less common than those described above but still used in industry are *infrared welding* and *laser welding*. These are generally used in specialty processes or with applications that require unique methods of heating because of the joint design or nature of the final product.

Infrared radiation (IR) is generally supplied by high-intensity quartz heat lamps. IR can penetrate into a polymer without contact with the surface of the part. The depth of penetration depends on many factors, and it varies strongly with polymer formulation. Laser welding is also a noncontact process for welding thermoplastics. Usually, a defocused laser beam must be employed to avoid burning the polymer. High-speed (50 m/min) laser welding of polyethylene films is one of several applications being explored.

7.5.2 Solvent Welding

Solvent welding or *cementing* is the simplest and most economical method of joining many noncrystalline thermoplastics. Solvent welding is accomplished by applying solvent to the bonding area between the two mating surfaces. The solvent dissolves the interface area, resulting in molecular entanglement and diffusion between the mating surfaces. As the solvent dissipates, the entanglement is effectively frozen in place.

Solvent cemented joints are less sensitive to thermal cycling than joints bonded with adhesives, and they are as resistant to degrading environments as their parent plastic. Bond strength equaling 85 to 100 percent of the parent plastic can be obtained. Solvent cements also offers aesthetically pleasing bonds with reasonable assembly time and cost.

The major disadvantage of solvent cementing is the possibility of stress cracking or crazing of the part and the possible hazards of using low-vapor-point solvents. When two dissimilar plastics are to be joined, adhesive bonding is generally desirable because of solvent and polymer compatibility problems.

Solvent cements should be chosen with approximately the same solubility parameter as the plastic to be bonded. Table 7.19 lists typical solvents used to bond major plastics. The solvent cement can be bodied to 25 percent by weight with the parent plastic to fill gaps and reduce shrinkage and internal stress during cure.

It is common to use a mixture of fast-drying solvent with a less volatile solvent to prevent crazing and to adjust drying times. With methylene chloride based solvents, glacial acetic acid (3 percent) or diacetone alcohol (10 percent) can be added to retard the solvent drying rate and reduce the effects of moisture during bonding.

Moisture is often an issue when solvent welding. Moisture absorbed by the solvent can reduce its effectiveness during bonding. Solvent containers should be sealed to minimize moisture absorption from the environment.

TABLE 7.19 Typical Solvents for Solvent Cementing of Plastics[1]

Plastic	Solvent
ABS	Methyl ethyl ketone, methyl isobutyl ketone, tetrahydrofuran, methylene chloride
Acetate	Methylene chloride, acetone, methyl ethyl ketone, ethyl acetate
Acrylic	Methylene chloride, ethylene dichloride
Cellulosics	Methyl ethyl ketone, acetone
Nylon	Aqueous phenol, solutions of resorcinol in alcohol, solutions of calcium chloride in alcohol
PPO	Trichloroethylene, ethylene dichloride, methylene chloride
PVC	Cyclohexane, tetrahydrofuran, dichlorobenzene
Polycarbonate	Methylene chloride, ethylene dichloride
Polystyrene	Methylene chloride, ethylene dichloride, trichloroethylene, toluene, xylene
Polysulfone	Methylene chloride

Note: These are solvents recommended by the various resin suppliers. A key to the selection of solvents is how fast they evaporate: a fast-evaporating product may not last long enough for some assemblies; too slow evaporation could hold up production.

The parts to be bonded should be unstressed and, if necessary, annealed. The surfaces should be smooth, mate well, and have tight tolerances. Although bodied cements will provide gap filling properties, they should not be considered a final solution for poorly matched parts.

A V-joint or rounded butt joint are generally preferred for making a solvent butt joint. Scarf joints and flat butt joints are difficult to position and to apply pressure to during the solvent evaporation phase of the process. Surfaces must also be clean. Any residual mold release agents or mold polishing compounds can inhibit solvent bonding.

The solvent cement is generally applied to the substrate with a syringe or brush. In some cases, the surface can be immersed in the solvent. After the area to be bonded softens, the parts are mated and held under pressure until dry. Pressure should be low and uniform so that the finished joint will not be stressed. After the joint hardens, the pressure is released, and an elevated temperature cure may be necessary, depending on the plastic and desired joint strength. The bonded part should not be packaged or stressed until the solvent has adequate time to escape from the joint.

Proper ventilation and exhaust of the work area where solvents are used are critical for obvious health, safety, and environmental concerns. Solvent vapors can also cause part crazing and should be removed from the assembly area.

7.6 RECOMMENDED ASSEMBLY PROCESSES FOR COMMON PLASTICS

When decisions are to be made relative to assembly methods (mechanical fastening, adhesive bonding, thermal welding, or solvent cementing), special considerations must be taken because of the nature of the substrate and possible interactions with the adhesive or the environment. The following sections identify some of these considerations and offer an assembly guide to the various methods of assemblies that have been found appropriate for specific plastics.

7.6.1 Acetal Homopolymer and Acetal Copolymer

Parts made of acetal homopolymer and copolymer are generally strong and tough, with a surface finish that is the mirror image of the mold surface. Acetal parts are generally ready for end-use or further assembling with little or no post-mold finishing.

Press fitting has been found to provide joints of high strength at minimum cost. Acetal copolymer can be used to provide snap-fit parts. Use of self-tapping screws may provide substantial cost savings by simplifying machined parts and reducing assembly costs.

Epoxies, isocyanate cured polyester, and cyanoacrylates are used to bond acetal copolymer. Generally, the surface is treated with a sulfuric-chromic acid treatment. Epoxies have shown 150 to 500 psi shear strength on sanded surfaces and 500 to 1000 psi on chemically treated surfaces. Plasma treatment has also been shown to be effective on acetal substrates. Acetal homopolymer surfaces should be chemically treated prior to bonding. This is accomplished with a sulfuric-chromic acid treatment followed by a solvent wipe. Epoxies, nitrile, and nitrile-phenolics can be used as adhesives.

Thermal welding and solvent cementing are commonly used for bonding this material to itself. Heated tool welding produces exceptionally strong joints with acetal homopolymers and copolymers. With the homopolymer, a temperature of the heated surface near 288°C and a contact time of 2 to 10 s are recommended. The copolymer can be hot plate welded from 221 to 293°C. Annealing acetal copolymer joints is claimed to strengthen them further. Annealing can be accomplished by immersing the part in 175°C oil. Acetal resin can be bonded by hot wire welding. Pressure on the joint, duration of the current, and wire type and size must be varied to achieve optimum results. Shear strength on the order of 150 to 300 lb/in or more have been obtained with both varieties, depending on the wire size, energizing times (wire temperature), and clamping force.

Hot gas welding is used effectively on heavy acetal sections. Joints with 50 percent of the tensile strength of the acetal resin have been obtained. Conditions of joint design and rod placement are similar to those presented for ABS. A nitrogen blanket is suggested to avoid oxidation. The outlet temperature of the welding gun should be approximately 332°C for the homopolymer and 293°C for the copolymer. For maximum joint strength, both the welding rod and parts to be welded must be heated so that all surfaces are melted.

Acetal components can easily be joined by spin welding, which is a fast and generally economical method to obtain joints of good strength. Spin welded acetal joints can have straight 90° mating surfaces, or surfaces can be angles, molded in a V-shape, or flanged.

Although not common practice, acetal copolymer can be solvent welded at room temperature with full-strength hexafluoroacetone sesquihydrate (Allied Chemical Corp., Morristown, NJ). The cement has been found to be an effective bonding agent for adhering to itself, nylon, or ABS. Bond strengths in shear are greater than 850 psi using "as molded" surfaces. Hexafluoroacetone sesquihydrate is a severe eye and skin irritant. Specific handling recommendations and information on toxicity should be requested from Allied Chemical Corp. Because of its high solvent resistance, acetal homopolymer cannot be solvent cemented.

7.6.2 Acrylonitrile-Butadiene-Styrene (ABS)

ABS parts can be designed for snap-fit assembly using a general guideline of 5 percent allowable strain during the interference phase of the assembly. Thread-cutting screws are frequently recommended for nonfoamed ABS, and thread-forming screws for foamed grades. Depending on the application, the use of bosses and boss caps may be advantageous.

The best adhesives for ABS are epoxies, urethanes, thermosetting acrylics, nitrile-phenolics, and cyanoacrylates. These adhesives have shown joint strength greater than that of the ABS substrates being bonded. ABS substrates do not require special surface treatments other than simple cleaning and removal of possible contaminants.

ABS can also be bonded to itself and to certain other thermoplastics by either solvent cementing or any of the heat welding methods. For bonding acrylonitrile butadiene styrene (ABS) to itself, it is recommended that the hot plate temperatures be between 221 to 287°C. Lower temperatures will result in sticking of the materials to the heated platens, while temperatures above 287°C will increase the possibility of thermal degradation of the surface. In joining ABS, the surfaces should be in contact with the heated tool until they are molten, then brought carefully and quickly together and held with minimum pressure. If too much pressure is applied, the molten material will be forced from the weld and result in poor appearance and reduced weld strength. Normally, if a weld flash greater than 1/8 in occurs, too much joining pressure has been used.

Hot gas welding has been used to join ABS thermoplastic with much success. Joints with over 50 percent the strength of the parent material have been obtained. The ABS welding rod should be held approximately at a 90° angle to the base material; the gun should be held at a 45° angle with the nozzle 1/4 to 1/2 in from the rod. ABS parts to be hot gas welded should be bonded at 60° angles. The welding gun, capable of heating the gas to 260 to 315°C, must be moved continually in a fanning motion to heat both the welding rod and bed. Slight pressure must be maintained on the rod to insure good adhesion.

Spin welded ABS joints can have straight 90° mating surfaces, or surface can be angled, molded in a V-shape, or flanged. The most important factor in the quality of the weld is the joint design. The area of the spinning part should be as large as possible, but the difference in linear velocity between the maximum and minimum radii should be as small as feasible.

One of the fastest methods of bonding ABS and acetal thermoplastics is induction welding. This process usually takes 3 to 10 s but can be done in as little as 1 s. During welding, a constant pressure of at least 100 psi should be applied on the joint to minimize the development of bubbles; this pressure should be maintained until the joint has sufficiently cooled. When used, metal inserts should be 0.02 to 0.04 in thick. Joints should be designed to enclose completely the metal insert. Inserts made of carbon steel require less power for heating, although other types of metal can be used. The insert should be located as close as possible to the electromagnetic generator coil and centered within the coil to assure uniform heating.

The solvents recommended for ABS are methyl ethyl ketone, methyl isobutyl ketone, tetrahydrofuran, and methylene chloride. The solvent used should be quick drying to prevent moisture absorption yet slow enough to allow assembly of the parts. The recommended cure time is 12 to 24 hr at room temperature. The time can be reduced by curing at 55 to 65°C. A cement can be made by dissolving ABS resin in solvent up to 25 percent solids. This type of cement is very effective in joining parts that have irregular surfaces or areas that are not readily accessible. Because of the rapid softening actions of the solvent, the pressure and amount of solvent applied should be minimal.

7.6.3 Cellulosics

Cellulosic materials (e.g., cellulose acetate, cellulose acetate butyrate, cellulose nitrate, ethyl cellulose) can be mechanically fastened by a number of methods. However, their rigidity and propensity to have internal molding stresses must be carefully considered.

Adhesives commonly used are epoxies, urethanes, isocyanate cured polyesters, nitrilephenolic, and cyanoacrylate. Only cleaning is required prior to applying the adhesive. A recommended surface cleaner is isopropyl alcohol. Cellulosic plastics may contain plasticizers. The extent of plasticizer migration and the compatibility with the adhesive must be evaluated. Cellulosics can be stress cracked by uncured cyanoacrylate and acrylic adhesives. Any excess adhesive should be removed from the surface immediately.

Cellulosic materials can also be solvent cemented. Where stress crazing is a problem, adhesives are a preferred method of assembly.

7.6.4 Fluorocarbons

Because of the lower ductility of the fluorocarbon materials (e.g., TFE, CTFE, FEP), snap-fit and press-fit joints are seldom used. Rivets or studs can be used in forming permanent mechanical joints. These can be provided with thermal techniques on the melt-processable grades. Self-tapping screws and threaded inserts are used for many mechanical joining operations. In bolted connections, some stress relaxation may occur the first day after installation. In such cases, mechanical fasteners should be tightened; thereafter, stress relaxation is negligible.

The combination of properties that makes fluorocarbons highly desirable engineering plastics also makes them nearly impossible to heat or solvent weld and very difficult to bond with adhesives without proper surface treatment. Fluorocarbons such as polytetrafluoroethylene (TFE), polyfluoroethylene propylene (FEP), polychlorotrifluoroethylene (CTFE), and polymonochlorotrifluoroethylene (Kel-F) are notoriously difficult to bond because of their low surface tension and chemical resistance. However, epoxy and polyurethane adhesives offer moderate strength if the fluorocarbon is treated prior to bonding.

The fluorocarbon surface may be made more "wettable" by exposing it for a brief moment to a hot flame to oxidize the surface. The most satisfactory surface treatment is achieved by immersing the plastic in a sodium-naphthalene dispersion in tetrahydrofuran. This process is believed to remove fluorine atoms, leaving a carbonized surface that can be wet easily. Fluorocarbon films treated for adhesive bonding are available from most suppliers. A formulation and description of the sodium-naphthalene process may be found in Table 7.10 (p. 7.37). Commercial chemical products for etching fluorocarbons are also listed.

Another process for treating fluorocarbons as well as some other hard-to-bond plastics (notably polyolefins) is plasma treating. Plasma surface treatment has been shown to increase the tensile shear strength of Teflon® bonded with epoxy adhesive from 50 to 1000 psi. The major disadvantage of plasma treating is that it is a batch process, which involves large capital equipment expense, and part size is often limited because of available plasma treating vessel volume. Epoxies and polyurethanes are commonly used for bonding treated fluorocarbon surfaces.

Melt-processable fluorocarbon parts have been successfully heat welded, and certain grades have been spin welded and hermetically sealed with induction heating. However, because of the extremely high temperatures involved and the resulting weak bonds, these processes are seldom used for structural applications. Fluorocarbon parts cannot be solvent welded because of their great resistance to all solvents.

7.6.5 Polyamide (Nylon)

Because of their toughness, abrasion resistance, and generally good chemical resistance, parts made from polyamide or resin (or nylon) are generally more difficult to finish and assemble than other plastic parts. However, nylons are used virtually in every industry and market. The number of chemical types and formulations of nylon available also provides difficulty in selecting fabrication and finishing processes.

Nylon parts can be mechanically fastened by most of the methods described in this chapter. Mechanical fastening is usually the preferred method of assembly, because adhesives bonding and welding often show variable results, mainly due to the high internal moisture levels in nylon. Nylon parts can contain a high percentage of absorbed water. This water can create a weak boundary layer under certain conditions. Generally parts are dried to less than 0.5 percent water before bonding.

Some epoxy-, resorcinol formaldehyde-, phenol resorcinol-, and rubber-based adhesives have been found to produce satisfactory joints between nylon and metal, wood, glass, and leather. The adhesive tensile shear strength is about 250 to 1000 psi. Adhesive bonding is usually considered inferior to heat welding or solvent cementing. However, priming of nylon adherends with compositions based on resorcinol formaldehyde, isocyanate modified rubber, and cationic surfactant have been reported to provide improved joint strength. Elastomeric (nitrile, urethane), hot melt (polyamide, polyester), and reactive (epoxy, urethane, acrylic, and cyanoacrylate) adhesives have been used for bonding nylon.

Induction welding has also been used for nylon and polycarbonate parts. Because of the variety of formulations available and their direct effect on heat welding parameters, the reader is referred to the resin manufacturer for starting parameters for use in these welding methods. Both nylon and polycarbonate resins should be predried before induction welding.

Recommended solvent systems for bonding nylon to nylon are aqueous phenol, solutions of resorcinol in alcohol, and solutions of calcium chloride in alcohol. These solvents are sometimes bodied by adding nylon resin.

7.6.6 Polycarbonate

Polycarbonate parts lend themselves to all mechanical assembly methods. Polycarbonate parts can be easily joined by solvents or thermal welding methods; they can also be joined by adhesives. However, polycarbonate is soluble in selected chlorinated hydrocarbons. It also exhibits crazing in acetone and is attacked by bases.

When adhesives are used, epoxies, urethanes, and cyanoacrylates are chosen. Adhesive bond strengths with polycarbonate are generally 1000 to 2000 psi. Cyanoacrylates, however, are claimed to provide over 3000 psi when bonding polycarbonate to itself. No special surface preparation is required of polycarbonate other than sanding and cleaning. Polycarbonates can stress crack in the presence of certain solvents. When cementing polycarbonate parts to metal parts, a room-temperature curing adhesive is suggested to avoid stress in the interface caused by differences in thermal expansion.

Polycarbonate film is effectively heat sealed in the packaging industry. The sealing temperature is approximately 218°C. For maximum strength, the film should be dried at 120°C to remove moisture before bonding. The drying time varies with the thickness of the film or sheet. A period of approximately 20 min is suggested for a 20-mil-thick film and 6 hr for a 0.25-in-thick sheet. Predried films and sheets should be sealed within 2 hr after drying. Hot plate welding of thick sheets of polycarbonate is accomplished at about 343°C. The faces of the substrates should be butted against the heating element for 2 to 5 s or until molten. The surfaces are then immediately pressed together and held for several seconds to make the weld. Excessive pressure can cause localized strain and reduce the strength of the bond. Pressure during cooling should not be greater than 100 psi.

Polycarbonate parts having thickness of at least 40 mils can be successfully hot gas welded. Bond strengths in excess of 70 percent of the parent resin have been achieved. Equipment should be used capable of providing gas temperature of 315 to 648°C. As prescribed for the heated tool process, it is important to adequately predry (120°C) both the polycarbonate parts and welding rods. The bonding process should occur within minutes of removing the parts from the predrying oven.

For spin welding, tip speeds of 30 to 50 ft/min create the most favorable conditions to get polycarbonate resin surfaces to their sealing temperature of 223°C. Contact times as short as 0.5 s are sufficient for small parts. Pressures of 300 to 400 psi are generally adequate. Parts my be heat treated for stress relief at 120°C for several hours after welding. However, this stress relief step is often unnecessary and may lead to degraded impact properties of the parent plastics.

Methylene chloride is a very fast solvent cement for polycarbonate. This solvent is recommended only for temperature climate zones and on small areas. A mixture of 60 percent methylene chloride and 40 percent ethylene chloride is slower drying and the most common solvent cement used. Ethylene chloride is recommended in very hot climates. These solvents can be bodied with 1 to 5 percent polycarbonate resin where gap filling properties are important. A pressure of 200 psi is recommended.

7.6.7   Polyethylene, Polypropylene, and Polymethyl Pentene

Because of their ductility, polyolefin parts must be carefully assembled using mechanical fasteners. These assembly methods are normally used on the materials having higher modulus such as high molecular weights of polyethylene and polypropylene.

Polyolefin materials can be effectively bonded only if the surface is first oxidized. Polyethylene and polypropylene can be prepared for bonding by holding the flame of an oxyacetylene torch over the plastic until it becomes glossy, or else by heating the surface momentarily with a blast of hot air. It is important not to overheat the plastic, thereby causing deformation. The treated plastic must be bonded as quickly as possible after surface preparation.

Polyolefins, such as polyethylene, polypropylene and polymethyl pentene, as well as polyformaldehyde and polyether, may be more effectively treated with a sodium dichromate-sulfuric acid solution. This treatment oxidizes the surface, allowing better wetting by the adhesive. Flame treatment and corona discharge have also been used. Table 7.20 shows the relative joint strength of bonded polyethylene and other plastic substrates pretreated by these various methods.

Another process, plasma treatment, has been developed for treating hard-to-bond plastics such as polyolefins. This process works in various ways depending on the type of plastic being treated. For most polyolefins, plasma treatment cross-links the polymeric surface by exposing it to an electrically activated inert gas such as neon or helium. This forms a tough, cross-linked surface that wets easily and is adequate for printing and painting as well as bonding. Shear strengths in excess of 3000 psi have been reported on polyethylene treated for 10 min in an oxygen plasma and bonded with an epoxy adhesive.

TABLE 7.20   Effect of Surface Treatments on Polyethylene and Other Plastic Substrates[21]

| Polymer | Relative bond strength* | | | |
	Control	Plasma	Abrasion	Chemical
High-density polyethylene	1.0	12.1		5.1
Polypropylene	1.0	221		649
Valox 310 polyester (General Electric Company)	1.0	18.9	2.9	1.0
Silicone rubber	1.0	>20	4.7	

*Results normalized to the control for each material.
SOURCE: Branson International Plasma Corporation.

Epoxy and nitrile-phenolic adhesives have been used to bond polyolefin plastics after surface preparation. Polyolefins can be thermally welded by almost any technique. However, they cannot be solvent welded, because of their resistance to most solvents.

7.6.8 Polyethylene Terephthalate and Polybutylene Terephthalate

These materials can be joined by mechanical self-fastening methods or by mechanical fasteners. However, polyethylene terephthalate (PET) and polybutylene terephthalate (PBT) parts are generally joined by adhesives.

Surface treatments recommended specifically for PBT include abrasion and solvent cleaning with toluene. A medium-strength bond can be obtained with polyethylene terephthalate plastics and films by abrasion and solvent cleaning. Gas plasma surface treatments and chemical etch have been used where maximum strength is necessary. Solvent cleaning of PET surfaces is recommended.

The linear film of polyethylene terephthalate (Mylar ®) surface can be pretreated by alkaline etching or plasma for maximum adhesion, but a special treatment often is unnecessary. A strong bond can be achieved by immersing the surface in a warm solution of sodium hydroxide or in an alkaline cleaning solution for 2 to 10 min. Commonly used adhesives for both PBT and PET substrates are isocyanate cured polyesters, epoxies, and urethanes. Polyethylene terephthalate cannot be solvent cemented or heat welded.

Ultrasonic welding is the most common thermal assembly process used with polybutylene terephthalate parts. However, heated tool welding and other welding methods have proven satisfactory joints when bonding PET and PBT to itself and to dissimilar materials. Solvent cementing is generally not used to assemble PET or PBT parts because of their solvent resistance.

7.6.9 Polyetherimide (PEI), Polyamide-imide, Polyetheretherketone (PEEK), Polyaryl Sulfone, and Polyethersulfone (PES)

These high-temperature thermoplastic materials are generally joined mechanically or with adhesives. The high modulus, low creep strength, and superior fatigue resistance make these materials ideal for snap-fit joints.

They are easily bonded with epoxy or urethane adhesives; however, the temperature resistance of the adhesives do not match the temperature resistance of the plastic part. No special surface treatment is required other than abrasion and solvent cleaning. Polyetherimide (ULTEM®), polyamide-imide (TORLON®), and polyethersulfone can be solvent cemented, and ultrasonic welding is possible.

These plastics can also be welded using vibration and ultrasonic thermal processes. Solvent welding is also possible with selected solvents and processing conditions.

7.6.10 Polyimide

Polyimide parts can be joined with mechanical fasteners. Self-tapping screws must be strong enough to withstand distortion when they are inserted into the polyimide resin, which is very hard. Polyimide parts can be bonded with epoxy adhesives. Only abrasion and solvent cleaning is necessary to treat the substrate prior to bonding. The plastic part will usually have higher thermal rating than the adhesive. Thermosetting polyimides cannot be heat welded or solvent cemented.

7.6.11 Polymethylmethacrylate (Acrylic)

Acrylics are commonly solvent cemented or heat welded. Because acrylics are a non-crystalline material, they can be welded with greater ease than semicrystalline parts. Ultrasonic welding is the most popular process for welding acrylic parts. However, because they are relatively brittle materials, mechanical fastening processes must be carefully chosen.

Epoxies, urethanes, cyanoacrylates, and thermosetting acrylics will result in bond strengths greater than the strength of the acrylic part. The surface needs only to be clean of contamination. Molded parts may stress crack when in contact with an adhesive containing solvent or monomer. If this is a problem, an anneal (slightly below the heat distortion temperature) is recommended prior to bonding.

7.6.12 Polyphenylene Oxide (PPO)

Polystyrene modified polyphenylene oxide can be joined with almost all techniques described in this chapter. Snap-fit and press-fit assemblies can be easily made with this material. Maximum strain limit of 8 percent is commonly used in the flexing member of PPO parts. Metal screw and bolts are commonly used to assemble PPO parts or for attaching various components.

Epoxy, polyester, polyurethane, and thermosetting acrylic have been used to bond modified PPO to itself and other materials. Bond strengths are approximately 600 to 1500 psi on sanded surfaces and 1000 to 2200 psi on chromic acid etched surfaces.

Polystyrene modified polyphenylene oxide (PPO) or Noryl® can be hot plate welded at 260 to 288°C and 20 to 30 s contact time. Unmodified PPO can be welded at hot plate temperatures of 343°C. Excellent spin welded bonds are possible with modified polyphenylene oxide (PPO), because the low thermal conductivity of the resin prevents heat dissipation from the bonding surfaces. Typical spin welding conditions are rotational speed of 40 to 50 ft/min and a pressure of 300 to 400 psi. Spin time should be sufficient to ensure molten surfaces.

Polyphenylene oxide joints must mate almost perfectly; otherwise, solvent welding provides a weak bond. Very little solvent cement is needed. Best results are obtained by applying the solvent cement to only one substrate. Optimum holding time has been found to be 4 min at approximately 400 psi. A mixture of 95 percent chloroform and 5 percent carbon tetrachloride is the best solvent system for general-purpose bonding, but very good ventilation is necessary. Ethylene dichloride offers a slower rate of evaporation for large structures or hot climates.

7.6.13 Polyphenylene Sulfide (PPS)

Being a semicrystalline thermoplastic, PPS is not ideally suited to ultrasonic welding. Because of its excellent solvent resistance, PPS cannot be solvent cemented. PPS assemblies can be made by a variety of mechanical fastening methods as well as by adhesives bonding.

Adhesives recommended for polyphenylene sulfide include epoxies and urethanes. Joint strengths in excess of 1000 psi have been reported for abraded and solvent-cleaned surfaces. Somewhat better adhesion has been reported for machined surfaces over as-molded surfaces. The high heat and chemical resistance of polyphenylene sulfide plastics make them inappropriate for solvent cementing or heat welding.

Polyimide and polyphenylene sulfide (PPS) resins present a problem in that their high-temperature resistance generally requires that the adhesive have similar thermal properties. Thus, high-temperature epoxies adhesives are most often used with polyimide and PPS parts. Joint strength is superior (greater than 1000 psi), but thermal resistance is not better than the best epoxy systems (150 to 204°C continuous).

7.6.14 Polystyrene

Polystyrene parts are conventionally solvent cemented or heat welded. However, urethanes, epoxies, unsaturated polyesters, and cyanoacrylates will provide good adhesion to abraded and solvent cleaned surfaces. Hot-melt adhesives are used in the furniture industry. Polystyrene foams will collapse when in contact with certain solvents. For polystyrene foams, a 100 percent solids adhesive or a water-based contact adhesive is recommended.

Polystyrene can be joined by either thermal or solvent welding techniques. Preference is generally given to ultrasonic methods because of its speed and simplicity. However, heated tool welding and spin welding are also commonly used.

7.6.15 Polysulfone

Polysulfone parts can be joined with all the processes described in this chapter. Because of their inherent dimensional stability and creep resistance, polysulfone parts can be press fitted with ease. Generally, the amount of interference will be less than that required for other thermoplastics. Self-tapping screws and threaded inserts have also been used.

Urethane and epoxy adhesives are recommended for bonding polysulfone substrates. No special surface treatment is necessary. Polysulfones can also be easily joined by solvent cementing or thermal welding methods.

Direct thermal welding of polysulfone requires a heated tool capable of attaining 372°C. Contact time should be approximately 10 s, and then the parts must be joined immediately. Polysulfone parts should be dried 3 to 6 hr at 121°C before attempting to heat seal. Polysulfone can also be joined to metal, since polysulfone resins have good adhesive characteristics. Bonding to aluminum requires 372°C. With cold rolled steel, the surface of the metal first must be primed with a 5 to 10 percent solution of polysulfone and baked for 10 min at 260°C. The primed piece then can be heat welded to the polysulfone part at 260 to 315°C.

A special tool has been developed for hot gas welding of polysulfone. The welding process is similar to standard hot gas welding methods but requires greater elevated temperature control. At the welding temperature, great care must be taken to avoid excessive application of heat, which will result in degradation of the polysulfone resin.

For polysulfone, a 5 percent solution of polysulfone resin in methylene chloride is recommended as a solvent cement. A minimum amount of cement should be used. The assembled pieces should be held for 5 min under 500 psi. The strength of the joint will improve over a period of several weeks as the residual solvent evaporates.

7.6.16 Polyvinyl Chloride (PVC)

Rigid polyvinyl chloride can be easily bonded with epoxies, urethanes, cyanoacrylates, and thermosetting acrylics. Flexible polyvinyl chloride parts present a problem because of plasticizer migration over time. Nitrile adhesives are recommended for bonding flexible polyvinyl chloride because of compatibility with the plasticizers used. Adhesives that are found to be compatible with one particular polyvinyl chloride plasticizer may not work with another formulation. Solvent cementing and thermal welding methods are also commonly used to bond both rigid and flexible polyvinyl chloride parts.

7.6.17   Thermoplastic Polyesters

These materials may be bonded with epoxy, thermosetting acrylic, urethane, and nitrile-phenolic adhesives. Special surface treatment is not necessary for adequate bonds. However, plasma treatment has been reported to provide enhanced adhesion. Solvent cementing and certain thermal welding methods can also be used with thermoplastic polyester.

Thermoplastic polyester resin can be solvent cemented using hexafluoroisopropanol or hexafluoroacetone sesquihydrate. The solvent should be applied to both surfaces and the parts assembled as quickly as possible. Moderate pressure should be applied as soon as the parts are assembled. Pressure should be maintained for at least one to 2 min; maximum bond strength will not develop until at least 18 hr at room temperature. Bond strengths of thermoplastic polyester bonded to itself will be in the 800 to 1500 psi range.

7.6.18   Thermosetting Plastics

Thermosetting plastics (e.g., epoxies; diallyl phthalate; polyesters; melamine, phenol and urea formaldehyde; and polyurethanes) are joined either mechanically or by adhesives. Their thermosetting nature prohibits the use of solvent or thermal welding processes; however, they are easily bonded with many adhesives.

Abrasion and solvent cleaning are generally recommended as the surface treatment. Surface preparation is generally necessary to remove contaminant, mold release, and gloss from the part surface. Simple solvent washing and abrasion is a satisfactory surface treatment for bonds approaching the strength of the parent plastic. An adhesive should be selected that has a similar coefficient of expansion and modulus as the part being bonded. Rigid parts are best bonded with rigid adhesives based on epoxy formulations. More flexible parts should be bonded with adhesives that are flexible in nature after curing. Epoxies, thermosetting acrylics, and urethanes are the best adhesives for the purpose.

7.6.19   Elastomeric Adherends

Elastomeric polymers can cause problems during adhesive bonding due to their elongation and swelling characteristics in certain chemicals. Elastomeric parts are also often contaminated with mold release and plasticizers or extenders that can migrate to the surface.

Solvent washing and abrading are common treatments for most elastomers, but chemical treatment is required for maximum properties. Many synthetic and natural rubbers require "cyclizing" with concentrated sulfuric acid until hairline fractures are evident on the surface.

Fluorosilicone and silicone rubbers must be primed before bonding. The primer acts as an intermediate interface, providing good adhesion to the rubber and a more wettable surface for the adhesive.

7.7 REFERENCES

1. Trauernicht, J. O., Bonding and Joining, Weigh the Alternatives, Part 1, Solvent Cements, Thermal Welding, *Plast. Technol.*, Aug. 1970.
2. Engineer's Guide to Plastics, *Mater. Eng.*, May 1972.
3. Chapter 14, Mechanical Fastening, *Handbook of Plastics Joining*, Plastics Design Library, Norwich, NY, 1997.
4. *Dupont Design Guide.*

5. Hoffman, J. M., Fundamentals of Annular Snap-Fit Joints, *Machine Design*, January, 6, 2005, p. 84.
6. Engineering Plastics, *Engineered Materials Handbook*, Vol. 2, ASM International, Materials Park, OH, 1988.
7. *Machine Design*, November 17, 1988.
8. Fastening, Joining, and Assembling Reference Issue, *Machine Design*, November 17, 1988.
9. Merriam, J. C., Adhesive Bonding, *Materials Design Engineering*, Sept. 1959.
10. Rider, D. K., Which Adhesives for Bonded Metal Assembly, *Prod. Eng.,* May 25, 1964.
11. Koehn, G. W., "Design Manual on Adhesives," *Machine Design*, April, 1954.
12. *Adhesive Bonding Alcoa Aluminum*, Aluminum Company of America, 1967.
13. Dillard, J. G., "Adhesives and Sealants," *Engineering Materials Handbook*, Vol. 3, H. F. Brinson, Ed., ASM International, Materials Park, OH, 1990.
14. Bersin, R.L., "How to Obtain Strong Adhesive Bonds Via Plasma Treatment," *Adhesives Age*, 1972, p. 37.
15. *Using Low Temperature Plasmas for Surface Treatment of Polymers*, Product Bulletin 2402, International Plasma Corporation, Hayward, CA.
16. Petrie, E. M., Trends in Plastic Surface Modification to Improve Adhesion, SpecialChem4Adhesives.com, October, 20, 2004.
17. Gentle, D. F., Bonding Systems for Plastics, in D. J. Almer, Ed., *Aspects of Adhesion*, Vol. 5, University of London Press, London, 1969.
18. EMABOND Process and Adhesives, Ashland Specialty Chemical Company.
19. Electromagnetic Bonding—It's Fast, Clean, and Simple, *Plast. World*, July 1970.
20. How to Fasten and Join Plastics, *Materials Engineer*, March, 1971.
21. Surface Preparation of Plastics, in Adhesives and Sealants, Vol. 3, *Engineered Materials Handbook*, H. F. Binson, Ed., ASM International, Materials Park, OH, 1990.

CHAPTER 8
PLASTICS RECYCLING AND BIODEGRADABLE PLASTICS

Susan E. Selke

Michigan State University
East Lansing, Michigan

8.1 INTRODUCTION

Plastics recycling continues to grow around the world, although it has declined to some extent in the United States in the last several years. Interest in and availability of biodegradable plastics has increased substantially in the last decade. As concern about high oil prices and global climate change increases, sustainability of products and processes is becoming a concern, fueling increased interest in biobased products of various kinds, including plastics. These biobased plastics may or may not be biodegradable, just as biodegradable plastics may or may not be biobased.

This chapter begins with an overview of municipal solid waste and the contribution that plastics of various types make to it. Recycling of plastics is discussed, with a concentration on postconsumer plastics—those that have served their intended use. Routine reprocessing and use of process scrap is not covered, as it is generally considered part of normal plastic manufacturing rather than categorized as recycling. Energy recovery, whether through incineration or through thermal depolymerization or pyrolysis, is also not covered. Biodegradable plastics are covered in detail, with some discussion of biobased plastics that may not be biodegradable.

The main focus of this chapter is on the United States, but some comparisons between the United States, Canada, and selected countries in Europe and Asia are included, along with discussion of important developments affecting recycling and use of biodegradables that are taking place outside the United States.

8.1.1 Municipal Solid Waste

Concerns about municipal solid waste (MSW), defined by the U.S. Environmental Protection Agency as including residential, institutional, office, and commercial waste, but not including construction and demolition debris, wastewater treatment sludge, and industrial waste,[1] stem from three main considerations. One is the amount of space occupied by the waste—space that therefore is not available for other uses. Another is the resources that are "wasted" when materials are sent to disposal. The third concern is the health and environmental impacts associated with emissions from the waste, either during or after disposal.

The United States produces a very large amount of MSW, over 236 million tons in 2003, compared to 205 million tons in 1990. Most of this growth can be attributed to population increase, as the per capita generation rate has remained relatively stable since 1990, at about 4.5 lb per person per day (Fig. 8.1).[2] The overall recycling rate reached a historic high of 23.5 percent that year, but, as can be seen in Fig. 8.2, growth in the recycling rate has been slow since the mid 1990s. Nonetheless, the per capita discard rate, the amount of MSW going to landfill or incineration, declined to 3.09 lb per person per day in 2003 (Fig. 8.1).[1]

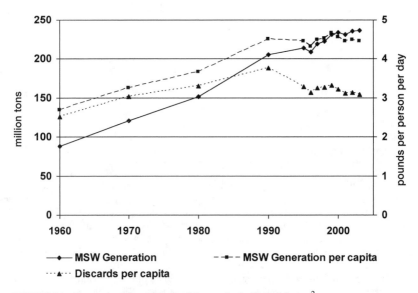

FIGURE 8.1   Generation of municipal solid waste in the United States.[2]

Landfill remains the primary method for handling MSW, although its prevalence has decreased with time (Fig. 8.3). The declining number of landfills in the United States, down to 1767 in 2002 from 7924 in 1988,[1] has been compensated by new landfills being much larger, on average, than those that have shut down. Composting, which was insignificant before the late 1980s, has grown substantially. Incineration has stayed relatively stable over the last decade.

Definitions of municipal solid waste vary from country to country, making comparisons difficult. In fact, even within the United States, definitions vary from state to state. Furthermore, there are often considerable uncertainties in estimates of waste quantities. *BioCycle Magazine,* the other major source of information on the U.S. MSW stream, compiles information from surveys of state agencies. Because of differences in definitions and methodology, *BioCycle's* reports of MSW amounts routinely exceed those from EPA. In the most recent survey, *BioCycle* reported generation of municipal solid waste totaled 369.4 million tons in 2002 (after adjusting state-reported values to adjust for items such as imported and exported waste, and construction and demolition and industrial wastes),[3] compared to 235.5 million tons reported by EPA that year.[1]

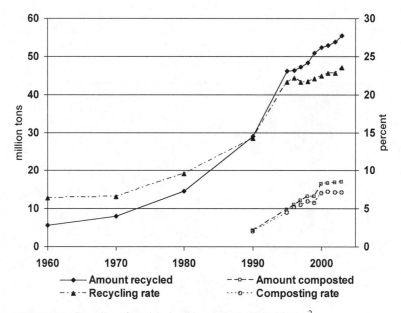

FIGURE 8.2 Recycling of municipal solid waste in the United States.[2]

8.1.2 Plastics in Municipal Solid Waste

Plastics account for only about 11.3 percent of materials in the U.S. MSW stream by weight (Fig. 8.4), but this proportion continues to increase. The proportion by volume is very difficult to determine accurately, but is certainly larger than the percent by weight, as the density of plastics is less than the average of MSW. EPA in the past has estimated proportion by density only for the portion of the waste stream going to disposal (landfill and incineration), reporting that plastics accounted for 26.1 percent of discarded MSW by volume in 1997, when their proportion was 13.0 percent by weight.[4]

Much of the plastic in MSW originates in packaging. Figure 8.5 shows the proportions of plastics, paper, metal, glass, and rubber and leather in EPA's major product categories of durable goods, nondurables, and containers and packaging. (EPA's remaining product categories are food scraps, yard trimmings, and "other.") As can be seen, the containers and packaging category has the largest share of plastics in MSW.

8.1.3 Plastics Recycling Rates

The amount of plastics entering the U.S. municipal waste stream in 2003 was estimated at 26.65 million tons. The amount in the containers and packaging segment was the largest, at 11.91 million tons. About 8.39 million tons came from durable goods and 4.6 million tons from nondurables. The recycling rates for plastics in the United States are, in general, considerably lower than for many other materials. The rates differ for plastics in different categories of goods and for different plastic resins within the same category, as will be discussed in more detail. As can be seen in Fig. 8.6, the containers and packaging category has both the highest overall recycling rate, 38.8 percent, and the highest recycling rate for

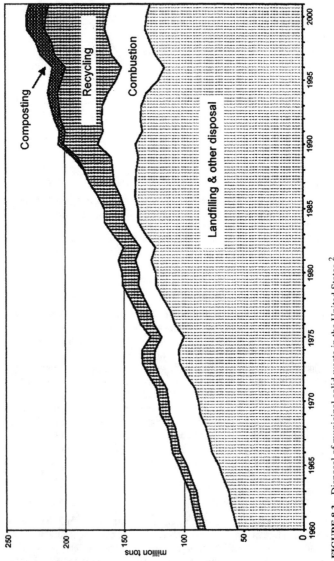

FIGURE 8.3 Disposal of municipal solid waste in the United States.[2]

8.4

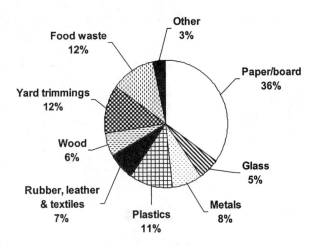

FIGURE 8.4   Materials in U.S. municipal solid waste, 2003.[1]

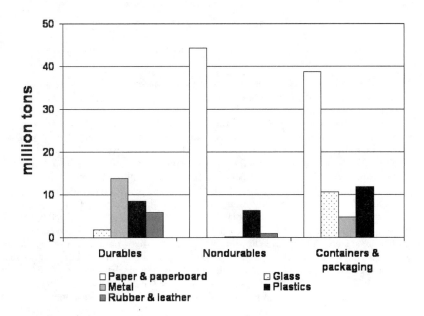

FIGURE 8.5   Materials in U.S. municipal solid waste product categories, 2003.[1]

plastics, 8.9 percent. Plastics in durable goods are recycled at a rate of 3.9 percent compared to 18.1 percent for the category as a whole. Nondurable goods have an overall recycling rate of 31.0 percent, but there is negligible recycling of plastics in this category.[1]

The amount of plastics recycled in 2003 was 330,000 tons from durable goods, and 1,060,000 tons from containers and packaging, for a total of 1.39 million tons.[1] *Plastics News* reported that sales of North American plastics recyclers and brokers totaled $2.02

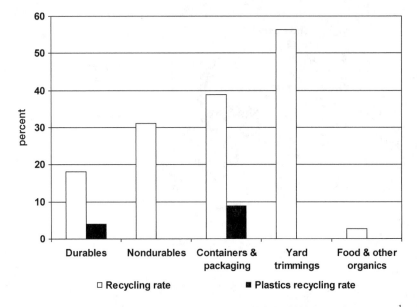

FIGURE 8.6   Overall and plastics recycling rates in the U.S. for 2003, by product category.[1]

billion in 2003, with 29 percent of that postconsumer plastics, 43 percent postindustrial plastics, and 28 percent brokered plastics. There were 219 processing plants in the United States, 21 in Canada, and 9 in Mexico.[5]

The overall recycling rate for plastics in MSW in 2003 was only 5.2 percent—much less than the rates for most other materials (Table 8.1).[1] Growth in recycling of plastics was very rapid in the 1980s and early 1990s but stagnated and even declined in more recent years (Fig. 8.5). Recycling rates for plastics differ considerably by resin and by product type, as we will discuss in more detail in this chapter. The falling recycling rates for plastics have been the focus of a great deal of concern. The plastics recycling industry sees stagnating collection and increasing competition from Asia, particularly China, as a threat to their survival. Demand for recycled plastics continues to exceed supply. Plastics processors are faced with high costs for bales of recyclable plastic, due to healthy overseas markets, but are unable to proportionally increase prices for their products, which compete with off-spec and virgin resin. The Association of Postconsumer Plastics Recyclers (APR) has warned that more plastics recyclers will close their doors if the situation continues. In fact, economic problems have beset APR and NAPCOR, the National Association for PET Container Resources, as well. Both have had to lay off staff and decrease their activities due to funding cuts.[6–9]

Historical trends in plastics recycling amounts and rates in the United States and in Europe are shown in Figs. 8.7 and 8.8. These values include feedstock recycling, but do not include energy recovery. While both are limited to recovery from postconsumer materials only, there are some differences in the types of materials included, so care should be taken in comparing the values. For example, recovery of agricultural and automotive plastics is included in the European figures but not, for the most part, in those for the United States. Figure 8.9 shows the breakdown by category for plastics recycled in Europe. As can be seen, only 39 percent of the total amount of recycled plastics comes from what Association of Plastics Manufacturers in Europe (APME) defines as municipal solid waste.[10]

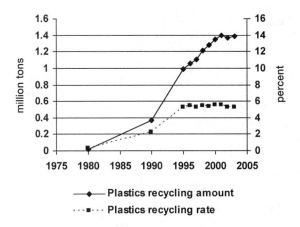

FIGURE 8.7 Amounts recycled and recycling rates for plastics in U.S. municipal solid waste.[2]

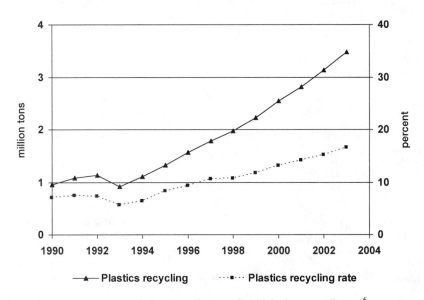

FIGURE 8.8 Amounts recycled and recycling rates for plastics in western Europe.[6]

However, straightforward comparisons between amounts of plastic recycled from "municipal solid waste" cannot be made, either, since definitions differ. Some, but not all, of the material identified in Europe as belonging to the distribution and industry, agriculture, and electrical and electronics sectors, for example, would be classified as part of municipal solid waste in the United States.

Countries within western Europe differ considerably in recycling rates for plastics. APME reports that, in 2003, Germany had the highest rate at 27.1 percent, while the rate in Greece was only 2.2 percent (Fig. 8.10). Recycling rates for packaging plastics are gen-

TABLE 8.1 Recovery of Selected Materials in U.S. Municipal Solid Waste. (Recovery includes both recycling and composting.)[1]

Material	Amount recovered (million tons)		Recovery rate (%)	
	2000	2003	2000	2003
Paper and paperboard	37.5	40.0	42.8	48.1
Glass	2.66	2.35	21.1	18.8
Steel and other ferrous metals	4.61	5.09	34.2	36.4
Aluminum	0.86	0.69	27.4	21.4
Other nonferrous metals	1.03	1.06	67.9	66.7
Plastics	1.35	1.39	5.5	5.2
Rubber and leather	0.82	1.10	12.6	16.1
Textiles	1.29	1.52	13.7	14.4
Wood	1.24	1.28	9.6	9.4
Food wastes	0.68	0.75	2.6	2.7
Yard trimmings	15.8	16.1	56.9	56.3

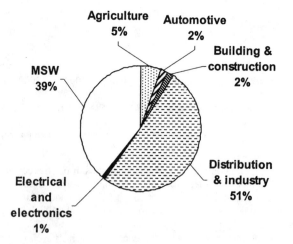

FIGURE 8.9 Plastics recycling in western Europe by category, 2003.[6]

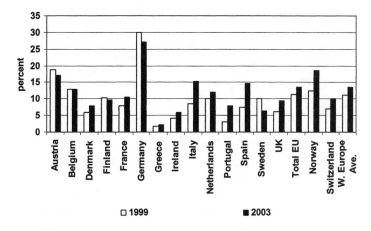

FIGURE 8.10   Plastics recycling in western Europe by country, 1999 and 2003.[6]

erally higher than the rates for plastics as a whole, exceeding 20 percent for mechanical recycling alone in Austria, Germany, Norway, Belgium, Italy, Netherlands, and Spain. France, the UK, and Switzerland have mechanical recycling rates between 15 and 20 percent; Denmark, Finland, Portugal, and Sweden have rates between 10 and 15 percent. Ireland has a rate between 5 and 10 percent, and Greece between 0 and 5 percent.[10]

In Australia, a total of 189,385 tonnes of plastics were recovered for recycling in 2003, for a recycling rate of 12.4 percent. Of this, 69 percent was processed domestically, and the remainder was exported, mostly to Asia. Plastics recycling has increased significantly in the last decade, although it did decline in 2001 and 2002. There was a major rebound in 2003, with the total amount of plastics recycled more than twice that of 1997 and 20 percent higher than in 2002 (Fig. 8.11). This material was nearly evenly divided between municipal waste (49.3 percent) and commercial and industrial waste (49.6 percent); the remaining 1.1 percent was building, construction, and demolition waste. Plastic packaging totaled 71.2 percent of the total wastes recovered, so, obviously, a significant fraction of the commercial and industrial waste would have been defined as municipal waste in the United States. The overall plastic packaging recycling rate was 20.5 percent.[11]

Polyethylene terephthalate (PET) had the highest recycling rate of all plastics in U.S. MSW in 2003, 14.3 percent, followed by high-density polyethylene (HDPE) at 9.1 percent, as can be seen in Fig. 8.12. High-density polyethylene is recovered in the greatest total amount, followed by PET. The most prevalent plastic present in the MSW stream is low and linear low-density polyethylene (LDPE/LLDPE), followed by HDPE.[1] Table 8.2 shows the amounts of the major plastic resins in U.S. MSW in 2000 and 2003, and the amounts recovered for recycling. It is readily apparent that overall plastics recycling rates have fallen during this time period. Only the HDPE rate has increased.

In Australia, PET also had the highest recovery rate in 2003, 31.5 percent, followed by HDPE at 23.1 percent (Fig 8.13). The total amount of HDPE recovered was also highest, with PET in second place.[11]

Calculation of recycling rates is controversial. There have been charges in the past that surveys that ask recyclers for data produce inflated figures and thus inflate recycling rates. Surveying organizations take steps to minimize this problem but cannot totally eliminate it. The reverse problem is the omission of organizations that do recycling, thus underestimating recycling rates.

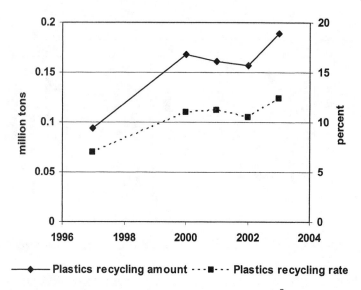

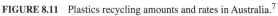

FIGURE 8.11   Plastics recycling amounts and rates in Australia.[7]

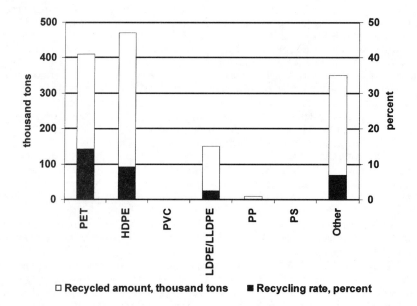

FIGURE 8.12   U.S. plastics recycling rates by resin, 2003.[1]

TABLE 8.2   Recovery of plastics in U.S. MSW by resin.[2]

	2000		2003	
Resin	Amount recycled (thousand tons)	Recycling rate (percent)	Amount recycled (thousand tons)	Recycling rate (percent)
PET	430	17.3	410	14.3
HDPE	420	8.7	470	9.1
PVC	negligible	0	negligible	0
LDPE/LLDPE	150	2.6	150	2.4
PP	10	0.3	10	0.3
PS	negligible	0	negligible	0
Other	330	7.1	350	6.9
Total plastics	1340	5.4	1390	5.2

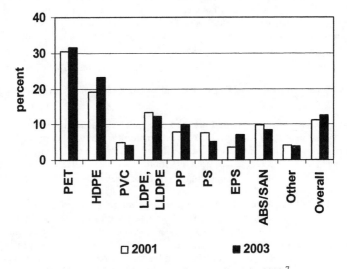

FIGURE 8.13   Australian plastics recycling rates by resin, 2003.[7]

A more fundamental problem than data accuracy is the matter of definition—what should count as recycled? The two most common options are (1) determining the amount of material collected for recycling and (2) determining the amount of material delivered for reuse. Since, typically, 5 to 15 percent of collected material is lost during processing (mostly because it is some type of contaminant such as a paper label, product residue, unwanted variety of plastic, or other material), recycling rates calculated using these two methods can differ substantially.

In the United States., the American Plastics Council (APC) is a major source of information about plastics recycling rates. In 1997, APC switched, in determining recycling rates, from using the amount of cleaned material ready for use to using the amount of material collected for processing. They justified this decision by claiming it is more in keeping with the way recycling rates are calculated for other materials—a claim that is true for some materials, such as paper, but not true for others, such as aluminum. This change brought considerable criticism, exacerbated by the fact that it occurred at a time when recycling rates, calculated in the same fashion, were declining. APC was accused of trying to mask the extent of the decline by the change in methodology. For example, the PET bottle recycling rate in 1997 was 27.1 percent based on material collected but only 22.7 percent based on clean material ready for reuse.[12] APC drew additional criticism by deleting polystyrene food service items from the definition of plastic packaging, beginning in 1995, which also had the effect of increasing the reported recycling rate for packaging plastics. The Environmental Defense Fund (EDF) even issued a report, titled "Something to Hide: The Sorry State of Plastics Recycling," in which they highlighted the difference the change in method of calculation made in the reported recycling numbers.[13] Now, however, this change in methodology has been generally accepted.

A related issue is how to deal with imports and exports of recyclable materials. Generally, imported goods that enter the waste stream are added to those produced domestically, and exported goods are subtracted, in calculating the denominator of the recycling rate—materials available for recycling. Most countries count collected recyclables that are exported for recycling as part of the recycled stream, since they do represent materials diverted from disposal. The issue of how to count collected recyclables being imported for recycling is more controversial. This has not been an issue for the plastics industry, but the U.S. aluminum industry has been criticized for including imported scrap cans in calculating the recycling rate for U.S. beverage cans.[14]

8.1.4 Environmental Benefits of Recycling and Use of Biodegradable Plastics

An obvious benefit of recycling and use of biodegradable plastics is that both reduce the requirement for landfill or incineration of waste materials. Items that are recycled are, by definition, diverted from the waste stream. Biodegradable plastics can be managed by composting, generally perceived as more environmentally beneficial than landfill or incineration. In fact, advocates of composting often refer to it as *natural* or *biological recycling.*

Often, although not always, another benefit of recycling is cost reduction. For example, use of regrind became routine because of the monetary savings it provided. Similarly, certain plastics industries for years have relied on a combination of off-spec and recycled plastics because of their lower price. The desire to benefit from consumer preferences for recycled material coupled, in some cases, with legislative pressures have led, on occasion, to the anomalous situation of recycled plastic being worth more per pound than virgin resin, but these situations are usually short lived. Recent increases in the cost of oil and natural gas, with consequent increases in prices for virgin resins, provide more opportunity for recycled plastics to be economically competitive.

Biodegradable plastics are still generally more expensive than the synthetic plastics they compete with, although the price differential is decreasing. If these biodegradable plastics are also biobased, increases in price of oil and natural gas may make them more competitive.

Additional benefits from recycling of plastics result from the fact that use of recycled resin displaces use of virgin materials and thus reduces depletion of natural resources. Recycling processes generally produce fewer environmental effluents than do processes that produce virgin resin, so the use of recycled plastics usually results in a decrease in air and water pollution. Biobased plastics use renewable materials as a feedstock, so they also can reduce resource depletion.

A factor that is certain to become increasingly important in the next decade is that the use of recycled plastics often results in significant energy savings, compared to the use of virgin resin. For example, Fenton[15] calculated the total energy requirement for a low-density polyethylene grocery bag to be 1400 kJ, while a bag with 50 percent recycled content required only 1164 kJ, for a savings of nearly 17 percent. A DOE report concluded that recycling PET products such as soft drink and ketchup bottles requires only about a third of the energy needed to produce the PET from virgin materials.[16] Again, recent increases in energy prices make this advantage even more significant.

In the near future, efforts to reduce emissions of greenhouse gases may become an important driver for use of plastics in general and for biobased and recycled plastics in particular. For example, a recent study by the Center for Packaging Technology (Cetea) in Spain reported that PET recycling reduces carbon dioxide emissions by 25 percent and methane emissions by 18 percent.[17] In the farther-term future, when oil supplies diminish significantly, production of plastics from renewable feedstocks will likely be critical.

8.2 RECYCLING COLLECTION

For plastics recycling to occur (or for recycling of other materials), three basic elements must be in place. First, there must be a system to collect the targeted materials, to gather them together. Second, there must be a facility capable of processing the materials into a form that permits them to be used to make a new product. Third, new products made in whole or part from the recycled materials must be manufactured and sold. A breakdown in any part of this system eventually stops the whole process. Because of this, efforts to increase recycling rates must pay attention to markets for the recycled materials as well as to the infrastructure to allow collection and processing of the materials.

Collection of plastics for recycling often occurs as part of a system designed to collect a variety of materials, not just plastics. Similarly, initial processing, in which collected materials are separated by generic type, often occurs in a multimaterial recycling facility.

8.2.1 Collection of Materials

For postconsumer materials, including plastics, the most difficult part of the recycling process may be getting the material collected in the first place. Industrial scrap is "owned" by the industrial entity that produced it. If the owners cannot get the scrap recycled, they will either have to dispose of it or pay some other business to do so. For much consumer scrap, there is little or no monetary incentive for its owner, the individual consumer, to direct it into a recycling system. Furthermore, industrial scrap tends to be concentrated, with substantial amounts of material in relatively few locations, making it relatively easy to collect. Postconsumer materials are typically very diffuse, so a more elaborate collection infrastructure is needed to get this material gathered together in quantities that make its processing economically viable.

There are three main approaches to collection: (1) go out and get the material, (2) create conditions such that the material will be brought to you, or (3) use a combined approach. There is a trade-off between motivation and convenience in getting people to participate in recycling by appropriately diverting the targeted recyclables from the waste stream into the recycle stream. Highly motivated individuals will participate in recycling even if they have to go to considerable effort to do so. If systems are set up to be very convenient, less motivation will be required to get people to participate. Therefore, increasing participation in recycling can be increased by providing greater motivation, by providing greater convenience, or by a combination of the two. Usually (although not always), systems that go out and get the materials provide greater convenience than those that require individuals to deliver the materials to a collection point.

When evaluating the success of recycling collection programs, authorities may report either participation rates or diversion rates. Participation rates reflect the proportion of people (often calculated by household rather than by individual) who actively participate in recycling. For curbside programs, a household is usually counted as participating if they put out (or deliver) any recyclables for collection any time in a one-month period. Diversion rates instead calculate the proportion of the targeted material(s) that is directed into the recycling stream rather than the waste stream. In principle, diversion rates are easier to calculate and more informative than participation rates, as they more directly get at the issue of how well a program is doing in acquiring materials for recycling. In practice, the only way to absolutely measure diversion is to do waste sorts to see what recyclables are left in the garbage stream. Due to the complexity and expense, not to mention the mess, of doing so, diversion rates are usually calculated rather than measured, based on amounts of targeted materials that are expected to be in the waste stream and measurements of the amounts that reach the recycling stream. This is the method used to obtain the recovery and recycling values in the series of EPA solid waste reports, for example.

8.2.1.1 Beverage Bottle Deposit Systems.   Recycling of postconsumer plastics in the United States got its start with the recycling of PET beverage bottles in states with bottle deposit legislation. The 5 or 10 cents per container deposit proved to be a sufficient incentive to get consumers to bring in 90 percent or more of the covered containers to centralized collection points (usually retail stores). This, in turn, spurred the development of effective reprocessing systems for these bottles and end markets for the recovered resin. In recent years, redemption rates (and therefore recycling rates) for containers covered by deposits have fallen. One reason may be that 5 cents is not as strong a motivation now as it was decades ago, when most of these laws were passed. Inflation has greatly decreased the real value of the deposit. A recent study reported that a 1981 nickel was worth only 2.5 cents in 2001.[18] One support for this view is that redemption rates in Michigan, the only state with a 10-cent deposit, have not fallen as much as those in most other states—remaining well above 90 percent.

Beverage bottle deposit programs are still relatively rare in the United States. Only 11 states have passed this type of deposit legislation (Table 8.3).[19] Initially, these laws were passed as litter-reduction measures. Therefore, they targeted beer and soft drink containers, since these represented a large and highly visible portion of litter. Later, several states recognized the value of these laws in achieving recycling, and both Maine and California amended their deposit laws to cover a wider variety of beverage containers in explicit efforts to increase beverage container recycling. The most recent deposit state, Hawaii, had recycling as an explicit goal when the law was initially passed.

The consumer pays a deposit, usually 5 cents, when buying the container and then receives a refund of that fee when the bottle is returned to a designated collection point. In most cases, any retailer that sells beverages of that type is obligated to accept the returns and refund the deposit. The majority of states provide for a handling fee for the retailer to at least partially offset the costs of managing the system.

Several years ago, Maine extended its early deposit law in an explicit attempt to increase recycling. The state now has deposits in place on most beverages, with the exception of milk.

TABLE 8.3 Bottle Deposit Legislation in the United States[19]

State	Containers covered	Characteristics
California	Beverages except milk, infant formula, distilled spirits, wine, medical foods, 100% fruit juice in 46 oz or larger containers, vegetable juice in containers larger than 16 oz, large refillable beverage containers; some other specified containers such as multimaterial pouches exempted	Refund value: 4 cents if capacity less than 24 oz, 8 cents if capacity 24 oz or greater; industry may be required to pay processing fee to cover part of recycling cost
Connecticut	Beer, malt beverages, carbonated soft drinks, soda water, mineral water	5¢ deposit
Delaware	Nonalcoholic carbonated beverages, beer, and other malt beverages	5¢ deposit, aluminum cans exempt
Hawaii	All nonalcoholic drinks except milk and dairy; beer, malt beverages, mixed spirits, mixed wine; aluminum, glass, PET, and HDPE containers only	5¢ deposit plus 1¢ nonrefundable container fee; containers over 64 oz exempt
Iowa	Beer, soda, wine, liquor	5¢ deposit
Maine	All beverages except dairy products and unprocessed cider	5¢ deposit, 15¢ on wine and liquor
Massachusetts	Carbonated soft drinks, mineral water, beer, and other malt beverages	5¢ deposit; containers 2 gal or larger exempt
Michigan	Beer, soda, canned cocktails, carbonated water, mineral water, wine coolers	10¢ deposit
New York	Beer, malt beverages, soda, wine coolers, carbonated mineral water	5¢ deposit
Oregon	Beer, malt beverages, soft drinks, carbonated and mineral water	5¢ deposit
Vermont	Beer and soft drinks, malt beverages, mixed wine drinks, liquor	5¢ deposit, 15¢ on liquor bottles, all glass bottles must be refillable

California's system is actually a refund value system rather than a true deposit. Until the law was changed effective January 2000, this cost was buried in the product price rather than charged as a separate item. Containers can be returned only to designated redemption centers unless there is not one within a specified distance, so return of containers is often significantly less convenient than it is in most deposit states, where containers can be returned to any retailer selling the covered beverage. It should also be noted that, in California, manufacturers must pay a processing fee to the state, in addition to the refund value, to cover the costs of recycling beverage containers covered by the refund value system.

California extended its refund value system, effective January 2000, to a wide variety of beverages, again in an explicit attempt to increase recycling of plastic bottles. Water and fruit juice containers are included, along with several other beverages (see Table 8.3). Initially, this produced a reduction in redemption rates. For the first half of 2000, the redemption rate fell to 70 percent from 80 percent in the first half of 1999. The decline was especially steep for PET, which dropped from 83 percent to 40 percent. The overall California redemption rate in 1999 for containers covered by the refund value system was 76 percent. The rate in the second half of the year was lower than in the first half, as has been the case for the last several years. The refund value started out at 1 cent. At the time of the expansion to cover additional containers, it was 2.5 cents per container for sizes less than 24 oz and 5 cents per container 24 oz or larger. Effective January, 2004, the refund value increased to 4 cents per container under 24 oz and 8 cents per container 24 oz or larger.[20] The result was an increase in recycling amounts and rates. The number of containers recycled reached 12 billion, up from 10.5 billion in 2003 and 10.6 billion in 2002. The beverage container recycling rate rose to 59 percent, compared to 55 percent in 2003, for the first increase in recycling rate since 1995.[21] The increase was attributed to efforts to increase public awareness, better customer service at recycling centers, a greater number of such recycling opportunities, and increased recycling at private businesses, in addition to the increased redemption value.[22] Recycling rates by material type continued to differ sharply, as shown in Fig. 8.14. It is important to note that beverage manufacturers must pay a processing fee to the Division of Recycling to cover a portion of the costs of processing the returned containers. For plastic containers, processing fees are lowest for PET (currently 12 percent of processing payments), somewhat higher for HDPE (20 percent), and significantly higher for other plastics (65 percent of processing payments).[23]

Hawaii's system is the newest, going into effect Jan. 1, 2005. It covers nonalcoholic drinks, except for milk and dairy products, and certain alcoholic drinks (beer, malt beverages, mixed spirits, and mixed wine) that are sold in aluminum, PET, or HDPE containers of 64 oz capacity or less. In addition to the 5-cent deposit, consumers pay a non-refundable 1-cent container fee. As in California, containers must be returned to redemption centers.[24]

Canada has deposit systems for most beverage containers except milk in 6 of its 10 provinces. Eight of the 10 provinces have deposits on soft drink and beer containers. Refillable (glass) beer bottles, which account for 75 percent of all beer containers sold in Canada, are subject to a mandatory or voluntary 10-cent deposit in all provinces. New Brunswick, Nova Scotia, and Newfoundland have half-back deposit systems for nonrefillable containers, and Prince Edward Island for alcohol containers, in which consumers get back only half of their original deposit when they return the empty container. This is intended to influence consumers to purchase refillable rather than nonrefillable bottles. Table 8.4 shows the deposit systems currently in place in Canada.[25] PET bottle recovery steadily increased through 2003, reaching a total of 110.0 million pounds, for a recovery rate of about 60 percent.[26] Recovery rates in western Canada, where deposits are high and the programs have a longer history, were 80 to 90 percent for large bottles and about

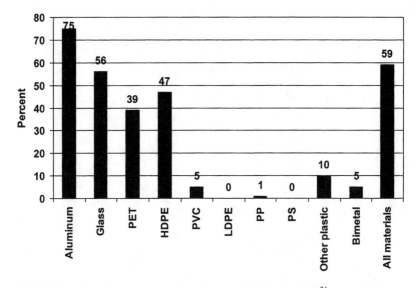

FIGURE 8.14   Beverage container redemption rates in California, 2004.[21]

70 percent for single-serving sizes. In Ontario, which is one of the two provinces without a deposit, about 45 percent of PET bottles were recovered, mostly through curbside collection. This is particularly of concern, because about a third of Canada's population lives in Ontario.[27] A report commissioned by the Environment and Plastics Industry Council (EPIC) determined that the overall recovery rate for plastic bottles was 36 percent in Canada in 2002, the average for beverage bottles in deposit programs was 75 percent, and the average in nondeposit programs was only 33 percent.[28]

A number of other countries also require deposits on beverage containers. For example, Austria requires a 40-cent deposit on refillable PET bottles. Belgium and Finland impose a tax on beer and soft drink containers that do not carry a deposit. Denmark has a mandatory deposit on imported glass and plastic soft drink and beer containers. The Netherlands requires deposits on PET and glass containers for soft drinks and waters. Sweden has voluntary deposits on PET beverage bottles. Norway has a deposit on beverage containers except those for milk, milk products, vegetable juices and water. Switzerland has a voluntary deposit on beverage containers.[19] Germany instituted a much-criticized deposit law that ran afoul of EU regulations, as it was found to interfere with trade. In January 2005, the German cabinet amended the rules in an effort to meet some of the objections. Nonexempt beverage containers will now carry a 25 cent deposit.[29] DPG GmbH has been founded to establish and administer the German deposit system.[30] South Australia has a deposit system that is about 30 years old and was originally modeled on that of Oregon.[31] Israel established a deposit, effective in 2001, on beverage containers under 1.5 l in size. The recovery rate in 2003 for covered containers was reported to be 60 percent. The Israel Union for Environmental Defense is trying to get 1.5-l bottles added to the deposit.[32] Korea, Japan, Taiwan, and India, among others, also have deposit programs for some beverage containers.[33]

From a recycling perspective, the most significant aspect of bottle deposit legislation is that, in most cases, the financial incentive provided does an excellent job of getting people to return their empty plastic bottles to appropriate places. Recycling rates for covered con-

TABLE 8.4 Beverage Container Deposit Programs in Canada[25]

Province	Deposit amount
Alberta	10¢ beer and wine 1 liter or less 5¢ other containers 1 liter or less 20¢ containers greater than 1 liter
British Columbia	10¢ beer and wine 1 liter or less 5¢ other containers 1 liter or less 20¢ containers greater than 1 liter
Manitoba	10¢ beer 1 liter or less 20¢ beer greater than 1 liter
New Brunswick	10¢ nonalcohol containers 10¢ wine and spirits 500 ml or less 10¢ beer 1 liter or less 20¢ wine and spirits greater than 500 ml 20¢ beer greater than 1 liter
Newfoundland and Labrador	8¢ nonalcohol containers 10¢ beer 1 liter or less 20¢ wine and spirits 20¢ beer greater than 1 liter
Nova Scotia	10¢ nonalcohol containers 10¢ wine and spirits 500 ml or less 10¢ beer 1 liter or less 20¢ wine and spirits greater than 500 ml 20¢ beer greater than 1 liter
Ontario	10¢ beer 1 liter or less 20¢ beer greater than 1 liter
Prince Edward Island	10¢ refillable soft drink and beer under 500 ml 10¢ wine and spirits 500 ml or less 10¢ beer 1 liter or less 20¢ wine and spirits greater than 500 ml 20¢ beer greater than 1 liter
Quebec	5¢ containers 450 ml or less 20¢ containers greater 450 ml
Saskatchewan	5¢ juice box and gable top 10¢ aluminum, steel, or bimetal cans and plastic bottles 1 liter or less 10¢ glass bottles 300 ml or less 20¢ cans, plastic bottles greater than 1 liter, glass bottles 301–999 ml 40¢ glass bottles greater than 1 liter, beer bottles greater than 1 liter

tainers are generally two to three times as high as those in nondeposit states. The redemption rate in Michigan with its 10-cent deposit is about 95 percent, compared to about 70 percent in all deposit states. In Sweden, where the value of a deposit is also about 10 cents, the aluminum can recycling rate is 86 percent. The Container Recycling Institute estimates that a national deposit system with a 10-cent deposit would achieve a recycling rate for covered containers of at least 80 percent.[34]

A negative aspect of deposit systems is that the per-container cost of managing these systems, as they are currently designed, is higher than the cost of alternative collection systems.[34] A national system rather than the current multiplicity of state systems could reduce some of this cost differential, however. There are also very real sanitary concerns, especially when deposits are expanded to noncarbonated beverages.

The labor requirements, and hence cost, for bottle deposit systems can be reduced using automated redemption systems. Reverse vending machines are often used to accept containers, usually designed to print a receipt for the consumer that can be redeemed at the store or other facility.

Businesses & Environmentalists Allied for Recycling (BEAR), a project of Global Green USA, in 2001 started the Multi-Stakeholder Recovery Project to examine beverage container recovery and recycling programs in the United States. The study was released in January, 2002, and was quickly attacked by industry groups. It concluded that container deposits result in 78 percent recovery, compared to 28 percent in nondeposit states. Drop-off programs have the lowest recovery—about 4.5 percent in nondeposit states. Curbside collection programs suffer a 13 percent yield loss due to handling.[35] The group originally wanted to produce a consensus-based approach for moving toward its goal of 80 percent recovery of beverage containers but was unable to progress and eventually disbanded.

More recently, the Association of Postconsumer Plastic Recyclers (APR) backed away from pushing for passage of bottle bills, in the face of opposition from the American Plastics Council and soft drink companies.[36]

A recent analysis of the ongoing controversy about bottle bills reported that the overall recycling rate for aluminum, glass, and PET beverage containers has fallen from 54 percent in 1992 to 37 percent in 2002, despite the tripling of the number of curbside collection programs during this period. The decline is attributed to increased beverage consumption away from home, stagnant scrap prices, a decreased emphasis on recycling by government and the media, and increasing public apathy about recycling. Beverage container recycling in deposit states remains much higher than in nondeposit states and, as mentioned earlier, is especially high in Michigan, the only state with a 10-cent deposit. The 10 deposit states had an aggregate beverage container recycling rate in 1999 of 71.6 percent, while the rate in the 40 nondeposit states (Hawaii did not have a deposit at that time) was only 27.9 percent. Using the costs from the BEAR report, it can be estimated that the average additional cost of $0.0168 per six-pack of containers in deposit states translates into a more than 1.5-fold increase in recovery rate.[37]

8.2.1.2 Other Deposit Systems. The idea behind beverage bottle deposits has also been applied to other products. Automobile batteries are subject to deposits in many states. While the primary motivation is to avoid the introduction of lead into landfills and incinerators, these systems have been very successful at facilitating the recycling of the polypropylene (PP) battery cases. Deposits are also common on tires.

8.2.1.3 Drop-Off Systems. Deposit systems are one variety of drop-off systems—systems in which consumers deliver the recyclables to the collection point. They achieve high levels of participation because of the strong motivation they provide: a monetary reward.

Drop-off systems encompass a wide range of designs, including barrels in supermarkets for people to place their plastic grocery sacks, roving multimaterial collection centers coming to a location once a month, permanent multimaterial centers in a centralized location in a community (or in an out-of-the-way location), collection bins in apartment building laundry rooms, and even sophisticated garbage and recyclables chutes in high-rise apartment buildings.

Many years ago, the aluminum industry began building a network of collection points for aluminum cans. These buyback centers also provided a monetary incentive for recycling, but at an average 1 cent per container, the incentive was much lower than with deposits. Convenience was also less. In most deposit states, containers can be taken for redemption to any retailer selling that product; in contrast, there was usually only one buyback center for a large geographical area, and it generally required a special trip rather than being in a place that consumers would routinely visit. Buyback centers never achieved the same level of participation or level of redemption as did deposit systems. There was a time period where buyback centers were seen as a viable alternative to deposit systems. The beverage industry, through the beverage industry recycling program (BIRP), supported efforts to build centers that accepted beverage containers, sometimes along with newspapers, paying consumers by weight of material they brought to the centers. While these centers were arguably effective at helping stave off deposit legislation in a number of states, they never achieved high recycling rates. With the growth of curbside recycling, most such multimaterial buyback facilities have closed. Even some of the aluminum-only centers have gone out of business.

Drop-off centers that do not buy back materials, and that often provide greater convenience (although less motivation) than the buyback centers, are more prevalent. These centers, often open 24 hours per day, 7 days per week, provide containers where sorted recyclables can be placed. While participation rates tend to be low, they can provide an important adjunct to curbside collection programs, discussed next. Such centers can relatively easily and inexpensively add new materials to collection programs just by adding another container. Collection costs are usually lower than for other types of recycling collection, since transportation of the materials is required only from the collection point to the processing location, and only limited labor is involved. On the other hand, such programs may have significant problems with contamination of the collected recyclables with undesirable materials.

In 2001, it was estimated that about 60 percent of the population of deposit states and 65 percent of the population of nondeposit states had access to residential drop-off programs, and they recovered 5 to 10 percent of the targeted beverage containers.[38]

8.2.1.4 Curbside Collection Systems.
In the United States, the most prevalent means of collecting recyclables is through curbside collection programs. These target primarily single family residences and involve picking up recyclables in much the same way as garbage is collected. Consumers place the recyclables at the curb, and they are picked up by the agency running the collection program.

Collection systems differ in design but fall into three general categories. Collection of commingled recyclables, often called *single-stream recycling,* refers to systems where the participant places all the recyclables together, usually in a container provided by the operator of the system. Other systems require consumers to separate the recyclables by type and thus use multiple containers, usually provided by the individual consumer, for set-out of the materials. Many systems are hybrids, with most materials collected in a commingled form and others collected separately. For example, a common design is a bin for commingled bottles and cans, with newspapers bundled separately. Virtually all collection systems accept multiple materials.

Systems in which consumers set out commingled recyclables at the curb can be further divided into three categories, depending on how the materials are handled in the collection vehicle.

In a few communities, recycled materials are placed into bags (usually blue in color) and collected in the same vehicles, standard compactor trucks, and at the same time as the garbage. When the load is dumped, the blue bags (and sometimes other readily identifiable recyclable materials) are sorted out. While some of these systems seem to work reasonably well, others have experienced significant contamination problems. Even without losses due to contamination, the yield of recyclables in general is lower than in systems that provide separate collection, simply because not all the bags are recovered intact. One frequently encountered problem is contamination of newspapers with broken glass. Some systems, therefore, require that newspaper be bagged separately from the other recyclables. The largest city to try this type of collection system is Chicago, where operation began in December 1995. While the city has repeatedly reported that the system was working well, there has been ongoing controversy. In late 2004, the city announced that it would launch a pilot program to study use of bins for recyclables.[39] In 2005, it was reported that the city's reported 25 percent diversion rate (required of the waste hauler by contract) was calculated by omitting more than 30 percent of the city's waste. The city has been allowing Allied Waste to bypass the sorting and recycling recovery facility, taking about 30 percent of the waste directly to landfill. If this added 325,000 tons were included in the recycling rate calculation, the 25 percent rate would fall to 17 percent.[40]

The second category includes systems that use a separate truck, or at least a separate compartment, for commingled recyclables. The recyclables are then delivered to a sorting facility called a materials recovery facility (MRF, pronounced "merf"), where they are separated by material type (and for plastics, sometimes by resin type as well, although this may take place at a separate facility dedicated to plastics only). While the first-generation MRFs relied almost exclusively on hand sorting, modern MRFs are becoming increasingly mechanized. The major advantages of this system are efficiency in the time on route and in the filling of the vehicle. Disadvantages include the need for a separate vehicle and crew, a dedicated sorting facility, and sometimes high residual levels of unwanted materials. A variation of this system uses ordinary garbage trucks for collection of recyclables rather than a specialized recycling vehicle but, unlike in the first category, garbage and recyclables are not collected together. Most of these systems use bins for set-out of the containers, but some use blue bags, and others use larger containers designed for automated dumping into the truck.

The third category includes systems in which the commingled recyclables are sorted at truck side into several categories. The separated streams may or may not require additional processing at a MRF before sale, depending on the materials included. The major advantage of this system is the quality control that can be practiced by the driver, coupled with ongoing education of consumers. If householders put unacceptable items into the bin, they will find the materials left there, ideally with an explanatory flyer, so they can learn from their errors. Another advantage is that less sorting is required after collection, and a dedicated processing facility may not be required. The major disadvantages are increased time per stop and the potential for the truck filling one compartment and therefore having to leave the route and off-load, even though other compartments are not full. The general recommendation is that truck-side sorting works well for moderate to small-sized communities, and commingled collection and a MRF work best for large communities.

General rules of thumb for effective design of curbside recycling systems are as follows:

- Collect the recyclables in at least partially commingled form and require little if any preparation beyond cleaning.
- Provide a readily identifiable container for use by the householder in putting out the recyclables.
- Collect recyclables weekly on the same day as garbage collection.
- Put considerable effort into ongoing education and publicity efforts.

Providing a container is particularly important. Often, this container is a blue, high-density polyethylene (HDPE) bin, frequently made from recycled HDPE. The container serves several functions. First, when a program is initiated, the delivery of a bin to all eligible households helps ensure that everyone knows about the existence of the recycling program. Thereafter, its presence in the home serves as a reminder. It increases convenience by providing a place for recyclables to be put until time for collection. It even brings peer pressure to bear, as its presence or absence at the curb on collection day lets the whole neighborhood know who is—and who is not—recycling. According to some reports, participation rates average 70 to 80 percent for curbside programs that provide containers and only 30 to 40 percent for programs that do not.[41]

Participation and diversion rates increase when collection is every week, on the same day as garbage collection. Allowing most recyclables to be mixed together in a single container (commingling) is preferred to requiring individuals to sort all materials by category. Minimizing the amount of preparation of materials that is required also increases convenience and therefore participation. Of course, some degree of preparation is essential to provide basic cleanliness and therefore ensure that odor and other problems do not render the collected material unusable.

In 2001, about 76 percent of the population of deposit states and 61 percent in nondeposit states had access to curbside recycling collection. About 62 percent of the targeted containers are unavailable for collection because they have been redeemed. Overall recovery through curbside programs averages about 9.5 percent in deposit states and 18.5 percent in nondeposit states.[38]

A study by WRAP in the UK found similar factors to be important. Curbside recycling that includes plastic bottles is less common there than in the United States, reaching 34 percent of households. Northern Ireland has the largest availability, with 48 percent of households covered, while Scotland has the lowest, with 23 percent of households. Collection rates are low, 7.9 percent of plastic bottles in 2004, for a total of 36,350 tonnes. Curbside collection was found to be about four times as effective as a drop-off system. Collection amounts are greater in systems that provide weekly (rather than every other week) collection, and systems with monthly collection fall still lower.[42]

8.2.1.5 Combination Systems.

Often, a combination of drop-off and curbside programs is very effective. Curbside programs work well for single-family residences but are not suited for high-rise apartment buildings, for example. Systems of distributed drop-off centers are especially suited for high-density housing. Locating recycling bins adjacent to garbage dumpsters or in laundry rooms or other community facilities can provide a significant level of convenience for residents.

In much of Europe, such distributed drop-off systems are a prevalent way of collecting recyclables. Many containers are collected through "bottle banks" placed on street corners, for example.

Even for communities with predominantly single-family housing, drop-off facilities can provide a useful supplement to curbside collection. They provide the ability to collect additional materials that are not economical to include in curbside programs, and they also provide an option for residents who for some reason do not want to wait for collection day to get rid of their accumulated recyclables, and for people who work and generate recyclables in the community but perhaps do not have recycling opportunities where they live.

8.2.1.6 Voluntary Versus Mandatory Systems.

Another difference between recycling programs is whether they are voluntary or mandatory. The majority of curbside programs are voluntary, but several states and a large number of municipalities have instituted mandatory programs. There seems to be general agreement that mandatory programs increase participation if enforcement efforts are included. If no enforcement takes place, results are not as clear. Typically, enforcement activities involve a series of warnings, ending in refusal to pick up the garbage for a set period of time. While fines and even jail terms may be permitted by the mandating ordinances, they are seldom employed.

Many programs find they can increase participation, with or without mandatory recycling, by charging a volume-based fee (often a per-bag charge) for garbage pickup but collecting recyclables for free.

8.2.1.7 Collection of Plastics.   Most of the approximately 8875[3] curbside collection programs in the United States include plastic bottles among the materials collected. A significant problem in including plastics in curbside programs is the space the containers take up in the truck, relative to their value. Educational programs urging consumers to compact the plastic bottles by stepping on them before placing them in the recycling bin can help. The use of on-truck compacting equipment is more effective in reducing volume. Disadvantages of on-truck compacting include the space consumed on the truck by the compactor itself as well as issues associated with more difficult sorting of compacted containers. On-truck compacting of commingled recyclables must also contend with problems caused by broken glass. In this regard, the presence of plastic bottles is an asset, since they reduce glass breakage. Shredding or chipping the plastic on the truck is generally not seen as viable, in large part because of the lack of reliable and efficient methods for separating chipped plastics by resin type.

Many collection systems, including curbside collection, accept only PET and HDPE bottles, which together represent about 96 percent of all plastic bottles (54 percent PET, 42 percent HDPE).[43] However, evidence that collection of these desired containers can be increased substantially by collecting all plastic bottles, and the support of the plastics industry for such programs, have led to an increasing number of communities accepting all types of plastic bottles.

A study carried out by the American Plastics Council found that programs targeting only PET and HDPE bottles received the same number of bottles of other resins and three times as many nonbottle containers as did programs targeting all plastic bottles. On average, PET and HDPE bottles made up 93 percent of the all-bottle program plastic stream but only 89 percent of the PET and HDPE-only program plastic stream. The difference is attributed to the higher incidence of nonbottle rigid containers such as trays, tubs, and cups in the PET/HDPE systems. When the city of Mesa, AZ, switched from HDPE and PET only to all plastic bottle collection in 1999, recovery of PET and HDPE bottles increased by 12.1 percent, and recovery rates for pigmented HDPE and custom PET bottles grew even more—36 percent and 18 percent, respectively. The percentage of non-PET and HDPE bottles collected actually declined. Windham County, VT, had a similar experience, with recovery of PET and HDPE bottles increasing from 61 to 63 percent, no increase in bottles other than HDPE and PET, and contamination from nonbottle containers cut in half—from 4.2 percent to 2.0 percent of the collected plastic in the curbside program. Drop-off sites in the county, serving the rural population, had a 24-percent increase in PET and HDPE bottles, and a 72-percent decrease in nonbottle plastic containers. The net result was that the PET recovery rate increased from 40 to 65 percent, and HDPE recovery increased from 59 to 64 percent. Results in a variety of other communities, in both bottle-bill and nonbottle bill states, showed substantial increases in plastic bottle collection on switching from HDPE and PET only to all plastic bottles. However, most communities were unable to find markets for the non-PET and HDPE bottles, disposing of them as residue.[44] Overall, APC estimates that all-bottle collection generates an average of 12 percent more PET and HDPE than collection targeting only these resins.[45]

By 2003, more than 1600 U.S. communities had adopted "all plastic bottle" collection rather than specifying no. 1 PET and no. 2 HDPE bottles. A survey of communities determined that 35 percent relied solely on curbside collection, 53 percent provided both curbside and drop-off, and 12 percent relied on drop-off recycling alone. Increasingly, these programs are expanding beyond just bottles. The survey found that 36 percent of the programs included other rigid containers such as trays and tubs, 4 percent accepted film, and 12 percent accepted "other" plastics such as 5-gallon buckets. As was the case in the earlier study, many programs discard the non-PET/HDPE plastics. Residents are not always aware that this is the practice. Only 35 percent of the communities surveyed reported that residents were aware of this fact, while 38 percent reported residents were not aware, and 7 percent were unsure. The lack of recycling of some bottles has not been an issue in most communities (55 percent) but has been an issue in 32 percent. Nearly half (49 percent) of the communities surveyed said that processing mixed bottles did not increase costs, and 18 percent said costs had actually decreased. Contaminant levels were not portrayed quite as favorably in this survey as in the first APC survey. Nearly half (45 percent) of the communities said contamination levels were unchanged, 23 percent said contamination levels declined, and 32 percent said levels rose. Recycling volumes increased in nearly all cases, with 91 percent of programs reporting increased PET recovery, 82 percent increased HDPE recovery, and 91 percent an increase in mixed bottle collection.[46]

Costs of curbside collection programs are generally intermediate between bottle deposit programs, which are the most costly, and drop-off programs, which cost the least.

8.2.1.8 Mixed Waste Processing.   Another approach to recycling plastics and other materials is not to ask consumers to do any special sorting or preparation but, instead, to recover recyclables from the garbage stream. The advantage of these systems is that, since they do not require any particular cooperation by consumers, they have the potential to recover the largest amount of recyclables.

The major disadvantages are the high cost of such systems and the low quality of the collected materials. The U.S. Bureau of Mines began experimenting with mixed-waste sorting facilities in the 1970s. Techniques employed were drawn from the mineral processing industries and included size reduction and various types of size- and density-based sortation methods. For plastics, the result of such processing is a mixed stream of plastics, with the resultant problems. Residual contamination is also a major concern.

Nevertheless, a number of mixed waste processing facilities (MWPFs) were built during the 1990s, with the total number in operation in the United States reaching 63 in 1997. By 2000, many of these had been closed, so the number was down to 52, with more closures forecast. Capital costs were often higher than anticipated, recovery rates were lower, and there were difficulties in operation. Some MWPFs changed to MRFs because of these difficulties. More than half of all U.S. MWPFs in 2000 were located in the western states, mostly in California. These facilities generally handle yard waste and construction and demolition debris in addition to municipal waste.[47]

While research on this type of recovery continues, the vast majority of plastics is recovered through source-separation-based programs, where the "free" labor of the individuals who keep the recyclable plastics separate from the garbage is crucial, both in terms of overall economics and in quality of the recovered materials.

8.2.1.9 Single-Stream Recycling.   In the United States, in recent years, there has been increased interest in collecting all recyclables commingled in a single stream. A number of communities have changed from collecting some materials separately to mixing them all together and performing all separation at a MRF. However, there is considerable disagreement about the desirability and effectiveness of this way of handling collection and sorting.

Proponents of single-stream recycling claim an increase in recycling rates and a decrease in costs compared to dual-stream (or more) recycling systems. A study by Skumatz Economic Research Associates, Inc. (SERA) reported that moving to single-stream recycling is the second or third most effective method of increasing collected recycling tonnage, behind pay as you throw (variable rate fees for garbage collection). Furthermore, it is the second most effective method of decreasing collection costs, second only to moving from every-week to every-other-week collection. The increase in collection amount is attributed to increased convenience provided by use of a single container and single collection day, a larger container that holds more material, and consumer education. It was acknowledged that residuals from processing increased, but there was still a net increase in recycling tonnage. Processing costs increased, but the savings from collection generally more than compensated. An increase in contamination of output streams was expected, but the study found considerable variability, with both good and bad single- and dual-stream systems. Glass contamination of paper was identified as a particular problem. A disconnect between material quality and material price can exacerbate quality problems, as recycling facilities have no incentive to invest in producing higher-quality materials if the marketplace does not value this effort.[48]

A 2004 study for the American Forest & Paper Association reported that single-stream collection increased collected tonnages by an average of 20 percent and decreased collection costs by 5 to 25 percent, for a total of $10 to $20 per ton, depending on the previous system characteristics. On the other hand, processing costs increased by $5 to $25 per ton. Because of increased contamination levels and more materials being removed during processing, the net increase in amount of recycled materials was only 1 to 3 percent. Average contamination levels were estimated to be 14.4 percent in single-stream MRF systems, compared to 6.8 percent in dual-stream MRFs.[49] Other studies have reported similar values.[50]

A related issue is decreased quality of some streams of recycled materials. Manufacturers of recycled paper seem to have the most problems, reporting that the additional contamination causes economic harm to their operations. Not only must the receiving facilities pay for unusable material, they also have to do more processing and use more chemicals to remove the contaminants.[51] It has been estimated that mill operation costs increase by about $8 per ton when fiber from single-stream rather than dual-stream MRFs is used.[49]

Some also charge that waste companies that favor the single-stream approach and dispute the charge that large amounts of residuals are produced focus on reporting "recovery" (input) rather than recycling (output) rates. In some cases, these residuals are being used as daily cover in landfills, for example.[51]

Much of the problem of increased level of residuals and contamination of paper streams in single-stream (and also two-stream) MRFs is directly attributable to broken glass. This fact, coupled with the relatively low value of collected glass containers, especially if mixed color, has led many programs to abandon collection of glass altogether.[52] In such cases, the cost comparisons for single-stream processing would differ. Glass contamination is also less in deposit states, since more than half of all glass containers are beer bottles, which are covered under deposit programs and hence recycled outside of the curbside system.[50]

Contamination of paper bales with plastic bottles does not produce the same level of concern from the paper industry as does glass contamination, but the plastics industry is concerned about the loss of millions of pounds of valuable plastic bottles from the recycling system due to this contamination. The American Plastics Council carried out a study of single-stream processing facilities to evaluate their ability to recover plastic bottles. They found that single-stream processing facilities typically have a higher percentage of plastic bottles and other containers in paper bales than do dual-stream facilities, but that this does not have to be the case. When markets offer better prices for cleaner materials, MRF operators are motivated to produce higher-quality materials. Loss of plastic materials to paper bales ranges from 10 to 200 tons per MRF per year, averaging about 70 tons per year. Another analysis found plastic bottles comprising 0.09 percent of bales of old corrugated containers and up to 0.96 percent of old newspapers, for a total of about 300 tons per year of lost containers for a 200 ton-per-day MRF operating 5 days per week. A series of recommendations for MRF operation were presented in the report. However, discounting of the value of paper bales that contain greater amounts of contamination was seen as key to improving performance.[53]

Even the claim that single-stream recycling leads to increased collection has been challenged. Usually, a switch to single-stream from dual-stream recycling also involves a change from the use of bins to the use of carts. A pilot study in Maryland found that changing from a bin to a cart for mixed paper, while keeping the bin for commingled containers, resulted in a 50 to 100 percent increase in collection without a change to single-stream collection.[50] Another factor that may account for some of the increases measured is that any program change can result in increased education of citizens about the program and, consequently, increased participation.[54]

At the same time as the United States is moving increasingly toward decreased sorting by consumers, in Japan, consumers are increasingly being asked to sort garbage into more and more categories so as to recover more materials. In Yokohama, ten categories of garbage are now collected, rather than the previous five. To support this extensive trash sorting, Yokohama has created a 27-page instruction manual for trash sorting. Proper sorting of the trash is seen as proof that a person is a responsible adult citizen. Some smaller communities collect more than 40 different categories of trash. In Kamikatsu, the garbage station has 44 bins for specific categories of trash.[55]

8.3 RECYCLING PROCESSES

Recycling processes for plastics can be classified in a variety of ways. One categorization differentiates between primary, secondary, tertiary, and sometimes quaternary recycling.

Primary recycling originally was defined as applications producing the same or similar products, whereas secondary recycling produces products with less demanding specifications. EPA's current definition considers use of in-plant scrap as primary recycling and use of postconsumer material as secondary recycling, regardless of the end products. Tertiary recycling uses the recycled plastic as a chemical raw material. Quaternary recycling uses the plastic as a source of energy. This last category is often not considered to be true recycling.

An alternative categorization that is gaining in popularity is mechanical and feedstock recycling. Mechanical recycling, as the name indicates, uses mechanical processes to convert the plastic to a usable form, thus encompassing the primary and secondary processes outlined above. Feedstock recycling is essentially equivalent to tertiary recycling, using the recycled plastic as a chemical raw material, generally (but not always) for the production of new plastics. The term *recovery* is often used to encompass mechanical and feedstock recycling plus incineration with energy recovery. This categorization is used in Europe in particular.

Plastic resins differ in terms of which recycling technologies are appropriate. Thermoplastics are more amenable to mechanical recycling than thermosets, which cannot be melted and reshaped. Typically, condensation polymers such as PET, nylon, and polyurethane are more amenable to feedstock recycling than addition polymers such as polyolefins, polystyrene, and PVC. Most addition polymers produce a complex product mixture that is difficult to use economically as a chemical feedstock, while condensation polymers usually produce relatively pure one- or two-component streams.

Germany does a substantial amount of feedstock recycling of plastics packaging, with a total of 330 thousand tonnes in 2002.[10] In the United States, feedstock recycling is mostly limited to nylon 6, polyurethane, and PET.

8.3.1 Separation and Contamination Issues

When plastics are collected for recycling, they are not pure. They contain product residues, dirt, labels, and other materials and often contain more than one type of plastic resin, resins with different colors, additive packages, and so on. This contamination is one of the major stumbling blocks to increasing the recycling of plastic materials. Usefulness of the recovered plastic is greatly enhanced if it can be cleaned and purified. Therefore, technologies for cleaning and separating the materials are an important part of most plastics recycling systems.

8.3.2 Separation of Nonplastic Contaminants

Since most household plastics are collected for recycling mixed with other materials, the first step in processing is usually to separate the plastics from these other materials. At drop-off facilities, this is usually accomplished when individuals place materials into the proper bins. For curbside collection, as discussed, the initial sorting is sometimes done at truck side by the driver of the recycling vehicle. A more common alternative is for the recycling truck to collect all or most materials in a commingled form and deliver them to a materials recovery facility (MRF) for initial sorting. Whether sorting is done at truck side or in a MRF, initial separation may simply separate plastics from nonplastics, with later separation by resin type, and sometimes by color as well, taking place in a dedicated plastics separation facility that receives mixed plastics from a number of MRFs (and likely from industrial facilities as well).

Waste plastics from commercial or industrial facilities are likely to be separated from nonplastic materials at the point of generation. For example, a facility that receives products in corrugated boxes, loaded on pallets and stabilized with stretch wrap, generates both corrugated and stretch wrap that can be recycled—and pallets as well unless they are reused. In such facilities, the waste streams should be managed in a way that provides for recyclables not only to be diverted from trash destined for disposal but to be kept separate from each other to minimize downstream processing requirements. The generating facility is likely to be able to reduce its disposal costs by doing so. Depending on the material and the circumstances, the company may simply be able to pay a reduced fee for disposal of its recyclables (compared to that for garbage) or may even be able to sell the recyclable materials to a user or processor.

Inclusion of nonplastic contaminants in recycled resins can affect both processing of the material and performance of the products manufactured from these materials. Presence of remnants of paper labels, for example, can result in black specks in plastic bottles, detracting from their appearance and rendering them unsuitable for some applications. These paper fragments can also build up in the screens in the extruder during processing, resulting in greater operating pressures (and energy use) and requiring more frequent screen changes. The presence of solid inclusions in the polymer can adversely affect the physical performance of the molded parts, resulting in premature failure. Mechanical properties can be decreased to the extent that thicker sections are required to obtain the desired performance.

Separation of plastics from nonplastic contaminants typically relies on a variety of fairly conventional processing techniques. Typically, the plastic is granulated, sent through an air classifier to remove light fractions such as labels, washed with hot water and detergent to remove product residues and dirt and to remove or soften adhesives, and screened to remove small heavy contaminants such as dirt and metal. Magnetic separation is used to remove ferrous metals, and techniques such as eddy current separators or electrostatic separators are also often used to remove metals. Many of these techniques were originally developed for mineral processing and have been adapted to plastics recycling.

8.3.3 Separation by Resin Type

Contamination of one resin with another can also result in diminished performance. One of the most fundamental problems is that most polymers are mutually insoluble. Thus, a blend of resins is likely to consist, on a microscopic scale, of domains of one resin embedded in a matrix of the other resin. While this sometimes results in desirable properties, more often it does not. To further complicate matters, the actual morphology, and thus the performance, will be strongly dependent not only on the composition of the material but also on the processing conditions. Therefore, for most high-value applications, it is essential to separate plastics by resin type.

Another problem arises from differences in melting temperatures. When PET is contaminated with polyvinyl chloride (PVC), for example, the PVC decomposes at the PET melt temperatures, resulting in black flecks in the clear PET. A very small amount of PVC contamination can render useless a large quantity of recovered PET. On the other hand, at PVC processing temperatures, PET flakes fail to melt, resulting in solid inclusions in the PVC articles that can cause them to fail. Again, a small amount of PET contamination can render recovered PVC unusable.

More subtle problems can arise, even from very similar resins. When injection-molded HDPE base cups from soft drink bottles were contaminated by newly developed blow-molded HDPE base cups, the recovered HDPE consisted of a blend of a high-melt-flow resin with a low-melt-flow resin that neither blow molders nor injection molders found usable. This caused serious difficulty for some recyclers. While this problem disappeared with the discontinuation of base cups, the difficulty in separating injection-molded HDPE bottles from blow-molded HDPE bottles is an ongoing concern.

Mixing resins of different colors can also be a problem. Laundry detergent bottle producers were able to fairly easily incorporate unpigmented milk bottle HDPE in a buried inner layer in detergent bottles, but they found it much more difficult to use recycled laundry detergent bottles. The color tended to show through the thin pigmented layer, especially in lighter-colored bottles. Motor oil bottlers who used black bottles had no such problem. As a general rule, the lighter the color of the plastic article, the more difficult it is to incorporate recycled content and, conversely, the lighter the color of the plastic article, the easier it is to find a use for it when it is recycled (and hence the higher its value).

8.3.4 Categories of Sorting Systems

Initially, many systems for separating plastics by resin type relied on hand sorting. Sorting by consumers delivering the appropriate types of plastics to the correct bins is nominally free of cost. However, costs do arise due to errors, and material is lost when consumers choose not to participate. Hand sorting at a MRF has also been common, but can be very expensive. Some systems have relied on less costly labor through training programs for the disabled, or sometimes by using prison labor. In these systems, also, human error is an issue.

Automated sorting systems use various technological devices to identify plastics in materials rather than relying on embedded codes. Capital investment is generally higher than for hand sorting, but operating costs are typically lower. Accuracy may be as good or better.

Sorting systems can be divided into macrosorting, designed to operate on whole or nearly whole plastic articles; microsorting, designed to operate on chipped plastics; and molecular sorting, designed to act on dissolved plastics. Hand sorting is always of the macrosorting variety. In principle, automated sorting can be of any type but is most often macrosorting or microsorting.

Various systems are now available to quickly and automatically identify plastics by resin type—or at least to distinguish between desired and undesired materials. Many of these rely on differences between resins in absorption or transmission of various wavelengths of electromagnetic radiation. Many can also separate plastics by color or color family in addition to sorting by resin type. Most of these systems work well with whole containers but are not effective in separating chipped plastics or multilayer materials.

8.3.5   Resin Identification Codes

The first control point for separation of plastics by resin can be at the time of collection. To facilitate identification of plastics packaging by resin type, and to satisfy pressure by states for such a system, the Society of the Plastics Industry (SPI) developed a coding system, comprised of a triangle formed by three chasing arrows, with a number inside and a letter code below (Fig. 8.15).[56] This system allows communities to tell consumers in a relatively simple way what materials are desired in the recycling system and what materials should not be included. In many cases, recycling programs collect only high-density polyethylene (HDPE) and polyethylene terephthalate (PET) bottles. They can communicate this to consumers by asking for no. 2 and no. 1 plastic bottles, respectively.

In most states where it has been adopted, the SPI code is required to be present on all bottles of 16 oz up to 5 gal capacity and on other containers of 8 oz up to 5 gal. In some states, there is no top limitation, so the code is required even on very large containers, including 55-gal drums. No states require its use on objects other than containers. Usually, the code is molded into the bottom of the container. When it is used on flexibles, it is most often printed.

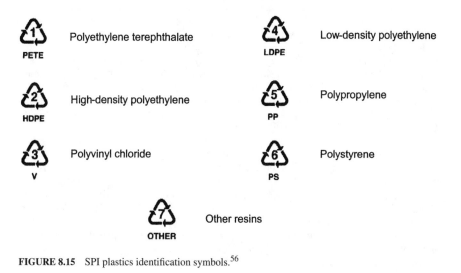

PETE	Polyethylene terephthalate	LDPE	Low-density polyethylene
HDPE	High-density polyethylene	PP	Polypropylene
V	Polyvinyl chloride	PS	Polystyrene
OTHER		Other resins	

FIGURE 8.15   SPI plastics identification symbols.[56]

This system has been widely adopted by U.S. states; as shown in Table 8.5, at least 39 states mandate its use.[57] However, it is also far from perfect. From the beginning, the system was challenged by groups that felt that the "recycling symbol" was misleading to consumers, as it was placed on plastic items that were not recyclable in most places, given that there were no systems in place to accept, process, and use the materials. This was aggravated by manufacturers use of the SPI coding system on a variety of products, including caps, trays, lids, bags, and others. For several years, there was discussion about modifying the system to eliminate the chasing arrows, but these efforts have now been abandoned.

TABLE 8.5 States that Require the SPI Coding System on Plastic Containers[57]

Alaska	Kansas	North Carolina
Arizona	Kentucky	North Dakota
Arkansas	Louisiana	Ohio
California	Maine	Oklahoma
Colorado	Maryland	Oregon
Connecticut	Massachusetts	Rhode Island
Delaware	Michigan	South Carolina
Florida	Minnesota	South Dakota
Georgia	Mississippi	Tennessee
Hawaii	Missouri	Texas
Illinois	Nebraska	Virginia
Indiana	Nevada	Washington
Iowa	New Jersey	Wisconsin

Another problem with the system is its lack of specificity. For example, it does not distinguish between high-melt-flow and low-melt-flow grades of resins. For example, most users of high-density polyethylene want extrusion grades rather than injection grades. While most HDPE bottles are extrusion blow molded, and hence have the desired low melt flow rates, some are produced by injection blow molding. Some recycling programs try to eliminate these containers for collection by specifying that they want only no. 2 bottles with a "seam" on the bottom—not an easy distinction for the average consumer, who probably won't be able to tell the difference between the parison cutoff line (which is what is really meant by the term "seam") and the mold parting line. Also, the coding system specifically identifies only six plastics, the most common packaging plastics, with everything else lumped in the no. 7 "other" category. As efforts to increase plastics recycling move beyond packaging, distinctions between other plastics become more crucial.

The SPI system has been adopted in many countries outside the United States as well. For example, Australia's Plastics and Chemicals Industries Association (PACIA) encourages manufacturers to voluntarily place the SPI code on plastic bottles and other containers, except those for poisonous or other dangerous substances. They also encourage use of the resin codes on flexible packaging, either by printing on the package or on an attached label.[58] Canada also uses the codes and encourages their use on other plastic articles, where it serves the same purpose as on containers, to encourage and facilitate the development of plastics recycling.[59] In the EU, the same basic system is used, with some modification of the letter symbols permitted. PET can be used instead of PETE, PE-HD instead of HDPE, PVC instead of V, PE-LD instead of LDPE, and O in place of Other.[60]

ISO 11469:2000, "Plastics—Generic identification and marking of plastics products," provides more specific identification of plastics in a variety of products, to facilitate their identification and decision making about their handling, waste recovery, and disposal. The standard provides for identification symbols to be placed between two angle brackets. For example, polyethylene would be labeled as >PE<, and nylon would be labeled >PA< (for polyamide). ISO 1043 provides symbols for plastics and additives. ISO 11469 provides for additional marking if desired, so that fillers, flame retardants, and so forth can also be identified.[61] Plastics components in vehicles that weigh more than 100 g are now required by EU regulation to be marked in accordance with this standard. The automotive standard allows brackets to be used in place of the > and < symbols.[62] A number of manufacturers are using the ISO codes on a variety of other types of products as well. For example, Hewlett Packard requires use of the ISO codes on all parts weighing 25 g or more if adequate space is available and the functionality of the part is not impaired. Inclusion of the codes on smaller parts is strongly encouraged. The preferred method is to mold the marking into the part, on an interior surface.[63]

ASTM International also has a standard for plastics identification, D1972, "Standard Practice for Generic Marking of Plastic Products." It provides for a triangle with the resin identification below. For the most part, standard resin abbreviations are used; these are tabulated in the standard. Copolymers and blends are identified with the two base polymers separated by a slash.[64]

8.3.6 Automated Macrosorting Systems

Much macrosorting, such as separating PET bottles from HDPE bottles, nylon carpet from polyester carpet, and so on, is still done by hand, often by workers picking materials off conveyor belts and placing them in the appropriate receptacle. However, mechanized means of sorting to make the process more economical and reliable continue to become more prevalent. The various devices commercially available to separate plastics by resin type typically rely on differences in the absorption or transmission of certain wavelengths of electromagnetic radiation, on machine vision systems that recognize materials by shape or color, or on some combination thereof. Many of these systems can separate plastics by color as well as by resin type.

These sorting systems can be divided into singulated-feed and mass-feed systems. In a singulated-feed system, objects are fed individually to the sensor, which then identifies each object (such as a plastic bottle) by type and directs it to the appropriate stream. In a mass-feed system, the recycling stream is spread out in a single layer across a wide belt. Singulated-feed systems can operate with a single sensor, although multiple lines (and therefore multiple sensors) are often required to increase capacity. The sort purity rate is typically 98 to 99 percent. In a mass-feed system, a separate sensor is required for each type of plastic to be sorted. A second sensor in a series can be used to increase the sorting purity. If a second sensor is not used, manual quality control is required as the final step. Such systems typically generate a purity of 90 to 95 percent, somewhat lower than singulated-feed systems.[65]

The Environment and Plastics Industry Council (EPIC) of Mississauga, ON, and Corporations Supporting Recycling commissioned a study of plastic sorting technologies in 2002. An update to the original study was released in 2005. It reported that the main manufacturers of mass-feed plastic bottle sorting equipment for the North American market were Magnetic Separation Systems (MSS), National Recovery Technologies Inc. (NRT), Pellenc Selective Technologies Inc., S&S Separation and Sorting Technology GmbH, and TiTech.[66]

Magnetic Separation Systems (MSS) of Nashville, TN, developed a system that sorts two to three plastic bottles per second, separating by resin type and color, using four sensors and seven computers. X-ray transmission is used to detect PVC, an infrared light high-density array separates clear from translucent or opaque plastics, a machine vision color sensor identifies bottle color (even ignoring the label), and a near-infrared spectrum detector identifies resin type.[67] A later MSS high-capacity plastic bottle separator uses a single sensor for both color and resin identification.[68] According to the EPIC report, the MSS Sapphire model can sort specific plastic resins, aseptic cartons, paper, and mixed plastics from a commingled stream of mixed plastics and paperboard, with the material positioned in a single layer on a conveyor. The system uses a combination of near-infrared detectors, high-speed microprocessors, and air jets. It cannot distinguish between colors or shapes. At each stage, the incoming stream is split into two output streams. It can handle 1500 to 3000 kg/hr. Reportedly, 18 such systems were installed in Germany, but none in North America. Typical product purity is more than 90 percent. The Aladdin system, designed for high-capacity MRFs, has all the capability of the Sapphire plus an integrated color sensor to allow separation between natural and colored containers. It can divide an input stream into three output streams and can count bottles separated by type and size. Seven such systems, two in Switzerland and five in North America, were reported to be in operation at four different facilities. Capacity is reported to be up to 4000 kg/hr.[66]

National Recovery Technologies, Inc. sells the MultiSort IR system, which uses near-infrared (NIR, also called shortwave infrared, SWIR) sensors to sort a designated polymer type from a commingled stream of mixed plastics and paperboard, with a throughput rate up to 4545 kg/hr. Like the MSS Sapphire system, it performs only one sort at a time and cannot sort by color. The company's MultiSort ES system can sort colors into groups at rates up to 3630 kg/hr. It also performs only one sort at a time. NRT's VinylCycle system was one of the original systems for separating PVC from PET, dating back to 1991. It is available in six bottle per second or ten bottle per second capacity and is widely used.[65,66] More recently, the NRT system was shown to be capable of sorting polylactide (PLA) bottles from PET bottles.[69]

Pellenc Selective Technologies (PST), based in France, has 100 machines installed around the world for various types of waste separation. It has two plastics sorting systems. The Mistral uses NIR to identify all materials in a single pass. The Sirocco uses a vision system to identify objects by location, shape, transparency, and color. Purity levels of 90 to 98 percent can be obtained; output varies from 2 to 10 tonnes/hr. The systems can handle plastic, paper, metal, and multilayer aseptic packaging. Ottawa recently installed a Mistral system in their MRF.[66]

Separation and Sorting Technology GmbH, based in Schönberg, Germany, has formed a joint venture with Tectron Engineering of Laguna Hills, CA. It has 500 machines around the world for sorting of plastics, glass, and metals. The Varisort system is designed for whole bottle sorting and accommodates a variety of sensors such as near-infrared, color-analysis, metal-detection, and X-ray. Different sensors can be placed on a single separation unit. Purity achieved is 99.5 percent. The company also sells SPEKTRUM color sorters.[66]

TiTech manufactures the TiTech Autosort beverage-carton sorter, originally developed to separate plastic-laminated paperboard. It uses near-infrared spectroscopy, particle detection, and selective impulses of compressed air. The position, size, and shape of the object are determined, in addition to determining the resin type. As with the systems described above, at each stage, only one sort is performed. Throughputs can be as much as 6000 kg/hr. Reportedly, two North American facilities are now using these systems.[65,66]

The only singulated sorting system identified in the EPIC report was the RapidSort system from Rofin Australia Pty. Ltd. It is reported to be capable of sorting both commingled and contaminated single polymer streams, achieving contamination levels of less than 50 ppm for PVC in PET, using high-resolution near-infrared and visible spectroscopy in a one-stage sensor. Each bottle is scanned several times to detect dirt, tops, labels, and other contaminants as well as to identify the bottle resin. Objects that are identified as nonplastics or that are not identified are separated from the streams of positively identified containers. The capacity of the system is 5 bottles per second, for a total of 800 to 900 kg/hr. Multiple lines can be used to increase overall throughput. Two facilities are using this system in Australia.[66]

A variety of other companies have developed similar equipment, some targeted at bottles and some at other plastics sorting tasks such as for carpet, automotive plastics, electronics, and so on.

Peter Walker Systems markets the Polyana and Tribopen systems in continental Europe. The Polyana system uses FT-IR spectroscopy to identify plastics. A special optical cell is used that identifies the plastic in a period of 4 to 6 s. It will identify plastics of all colors and can also identify blends by comparing the spectroscopic fingerprint with the spectra in its integrated database. The database can be modified, with compounds added or removed to suit the demands of a particular application. Two models are available, the portable 420 and the stationary 460 system, which provides more rapid identification. The Tribopen system, patented by Ford, identifies plastics by measuring the triboelectric (static) charge generated when its head is rubbed against the plastic to be identified. A single pen can be used only to separate two plastic groups from each other, but using two or three different pens can allow more complex separations to be made. Identification is negative only—that a tested polymer is not a specified plastic.[70]

Hamos GmbH of Penzberg, Germany, also has an IR-based plastics identification system for recycling applications. The company also manufactures equipment for color sorting that can be used in recycling processes.[71]

Sorting of black plastics has posed a difficult problem, as standard spectrometers cannot be used. Their high carbon content causes black plastics to absorb light to such a degree that, when intense light sources such as lasers are used to analyze them, they heat up and can emit light or even ignite. While Raman spectroscopy can be used to identify dark and intensely pigmented plastics, it requires low laser power, resulting in long measurement times on the order of 10 ss. In early 2001, SpectraCode, Inc., a manufacturer of spectrographic plastic identification devices, announced it had developed a new device that provides for instantaneous identification of postconsumer black plastics. It contains a modified probe that uses a sampling technique to test black samples at full laser power with no burning, allowing identification in half a second or less.[72] SpectraCode has targeted automotive and electronics plastics recycling for its technology.[73]

One way to simplify the identification task is to put taggants into the plastic, making it easy for automated systems to recognize them. Microtrace LLC, of Minneapolis, MN, manufactures Microtaggant identification particles. Currently, they are targeted mostly at preventing counterfeiting but also may be of use in plastics recycling.[74]

8.3.7 Microsorting Systems

Microsorting of plastics is commonly practiced for separation of lighter-than-water from heavier-than-water plastics. While such separation can be done in a simple float-sink tank, hydrocyclones are often used because of their advantages in size and throughput. Application of density-based separation for mixtures of plastics that are all heavier or all lighter than water is relatively rare.

Hamos GmbH of Penzberg, Germany, has developed triboelectric separation systems for plastic/plastic separations, and electrostatic separators for plastic/metal separation.[71]

Plas-Sep Ltd., of Canada, also sells electrostatic separators for mixed plastics. In 2003, it formed a joint venture with Sani Eco Inc., a recycling firm in Granby, Quebec, and now has four units in operation there, producing LDPE, HDPE, and PP. The system is based on putting a static charge on chopped pieces of plastic, 1 to 10 mm in size, by feeding them into a slightly tilted, slowly rotating drum. The particles are then dropped through an electric field in the separation tower, where separation occurs as pieces are attracted toward the electrode of the opposite polarity. If more than two plastics are included, multiple passes through the process are required. A second pass may also be used to increase purity. A laser-based monitoring system reads where the material is falling, allowing control over the material and correct placement of internal dividers to optimize separation. Energy costs for running the system are reported to be only about 8 cents per hour. The company reports that, from an initial 50/50 mix of plastics, purities of above 99.9 percent with yields of 80 percent can be achieved in a single pass. Materials to be separated must be clean and dry. Suggested applications include recycling of PET bottles, automotive parts recycling, recycling of post consumer plastics, and sorting of plastics from wire chopping operations.[75]

Argonne National Laboratory has developed a plastics separation system based on froth flotation. A series of six tanks is used, each with a specific function, depending on the plastic being recovered. The chemical solutions in each tank are chosen for the particular application. It has been used for recovering selected plastics from automobile shredder residue, disassembled car parts, industrial scrap plastics, and consumer electronics. Argonne claims it is the only technology that can successfully recycle ABS with a purity greater than 99 percent.[76]

Recovery Plastics is also reported to have a froth flotation process for plastics separation, targeting automotive plastics. Chemicals used for separation include surfactants, sodium hydroxide, and plasticizers.[77]

MBA Polymers of Richmond, CA, has developed plastics separation systems designed for auto shredder residue and electronics wastes. A series of proprietary processes are used to separate mixed resin flakes into pure streams as well as to separate plastics from other components in the waste stream.[77–79]

WIPAG Polymertechnik, of Germany, has developed proprietary separation processes for plastics, some of which are based on physical hardness. The technology is being used for recovering plastic from automobile bumpers and for sorting automobile instrument panels into hard flake, soft flake, and foam fractions. The company is involved in joint ventures with American Commodities in Flint, MI, and with PPR in Kent, England.[77,80]

Satake has developed machine vision systems to sort out colored from uncolored plastics flakes.[81]

The Salyp N.V. company, in Eiper, Belgium, developed an infrared sorting technology for separating various types of thermoplastics recovered from a system for recycling automotive shredder residue. The system uses infrared energy to heat and dry cleaned shredded plastics and to soften, but not melt, the targeted plastic. The mixed plastic stream is then fed through a set of rollers, and the softened plastic sticks to the roller and is removed. The remaining material is again heated, softening the next desired plastic, and the process is repeated until all the plastics have been separated. Since infrared radiators emit at different wavelengths, IR energy can provide selective heating of different thermoplastics by choosing appropriate emission wavelengths. Therefore, the system does not rely simply on differences in melting point for separating the plastics.[82] In addition to developing its own technology, Salyp has combined systems from other companies to produce mechanical sorting lines adopted for specific recycling requirements. Currently, the company offers two systems. The Salyp System A is designed for automotive shredder residue, and the WEEE System A is designed for mixed electronic scrap. In the electronic scrap system, separation of printed circuit boards, copper-rich, and pure plastics streams is the focus, with sorting of plastics by family as an option.[83]

8.3.8 Molecular Sorting

For multilayer materials such as coextrusions, plastic chips may themselves contain multiple materials. In such cases, a more complex approach may be needed if the materials are to be separated. In a few cases, chipping and grinding can be used to separate the materials. These fall into the microsorting category and have already been discussed. One example is the PET/EVOH ketchup bottles, which were designed without an adhesive layer between the PET and EVOH. When the bottle is chipped and the chips are washed, the PET and EVOH separate, and most of the EVOH is removed in the rinse step.

Another option is dissolving the plastics and later reprecipitating them. Either selective dissolution or selective reprecipitation in an organic solvent or a combination of solvents can be the basis of the separation. However, such systems are complex and seldom economical. Solvent retention is often a problem as well. While these systems have been the subject of research, no commercial systems currently use this approach.

A more promising option is to direct these materials to markets that are able to use them in commingled form. Sometimes, design changes can facilitate such use by selecting materials that are more compatible (or materials that will separate during processing). Compatibilizers can also be added to improve properties of such blends.

8.3.9 Safety Concerns

Even when plastics are sorted by type, the performance of recycled plastics may differ from virgin plastic because of the effects of the use cycle. These changes may be due to chemical changes within the polymer, sorption of materials into the polymer, or other factors. If materials are sorbed, there is potential for later release of these substances. For some critical applications, such possible or actual changes in behavior of recycled plastics pose unacceptable risks. For example, it is probably safe to conclude that recycled plastics will not be used for implantable medical devices. It is highly unlikely that recycled plastics will be used for the packaging of sensitive drugs. Other examples, of course, could also be cited where the small but real risk of unacceptable performance, or of release of some damaging substance coupled with the critical nature of the application, is likely to rule out the use of recycled plastics.

For less critical applications, such as the use of recycled packaging for food products, the conventional wisdom used to be that recycled plastics should not be used. This changed dramatically in the 1990s. In the United States, one of the earliest applications of recycled plastic for packaging of food products was recycled PET in egg cartons. The physical barrier of the egg shell provided a degree of added protection to the food that the FDA agreed was sufficient to allow ordinary recycled PET to be used. Next came use of recycled plastic in buried inner layers of packaging, such as a recycled PS clamshell used for hamburgers, in which the contact between the food and the recycled plastic was mediated by a layer of virgin plastic that acted as at least a partial barrier. Still later, repolymerized PET was used in direct contact with food (blended with virgin material). The repolymerization process, with its crystallization steps, provided assurance that any impurities present would be removed. Finally came FDA approval of specific systems for intensive cleaning of physically reprocessed PET, coupled with the limitation of incoming material to relatively pure streams of soft drink bottles returned for deposits. Then, production of 100 percent recycled content PET bottles using a process for physically reprocessing bottles collected from curbside was approved. Systems for processing recycled HDPE have been approved for limited direct food contact applications as well.

The concern over use of recycled plastics in food contact falls in two general areas. First is concern about biological contaminants. In most cases, the processing steps for production of plastic packaging materials provide a sufficient heat history to destroy disease-producing organisms. Therefore, this is not a major concern.

A more significant concern is the possible presence of hazardous substances in the recycled feedstock. FDA regulations require food packagers to ensure that the materials they use are safe for food contact—that they do not contain substances that might migrate into the food and cause deleterious effects on human health. Recycled resins, by their very nature, often have a somewhat unknown history. What if, for example, someone put some insecticide, some gasoline, weed killer, or any of a myriad of toxic substances into a soft drink bottle and later turned that bottle in for recycling? How can we prevent that container from contaminating new plastic packages? What we have seen in this, as in other areas, is movement at the FDA away from absolute prohibitions and toward a more reasonable evaluation of risk. In particular, the FDA has laid out guidelines for challenging recycling processes with known model contaminants and evaluating the ability of the process to remove those contaminants, thus providing some assurance that unacceptable levels of migration will not occur.

In October 2003, the FDA's Office of Food Additive Safety released guidance on *Recycled Plastics in Food Packaging*,[84] a supplement to its document *Points to Consider for the Use of Recycled Plastics in Food Packaging: Chemistry Considerations*.[85] Manufacturers wishing the FDA to consider the use of recycled plastic for a food-contact application were instructed to submit a complete description of the recycling process, including the source of the plastic and any source controls or other steps taken to ensure that the plastic is not contaminated; results of any tests showing the recycling process removes possible contaminants; and a description of proposed conditions of use for the plastic, including temperature, type of food, duration of contact, and whether it is single-use or repeated. PET and PEN recycled by a tertiary (feedstock) recycling process were determined to produce plastic of suitable purity for food-contact use, so no such submission was necessary for them. The FDA publishes a list of processes that, as of August, 2003, have received letters of nonobjection.[84]

8.3.10 Quality Issues

As discussed, the use history of a recycled plastic can affect its properties and performance. It is well known that plastics undergo chemical changes during processing and use that ultimately lead to deterioration in properties. In fact, much of the history of plastics is related to the development of appropriate stabilizing agents to prevent this degradation. We routinely stabilize plastics against thermo-oxidative degradation that would otherwise occur during processing. We know that some resins are much more sensitive than others. Depending on the amount of stabilizer initially present, the history of the resin, and the type of resin, a recycled plastic resin may or may not require additional stabilizer to be successfully utilized.[86]

Similarly, plastics that are designed to be used outdoors generally must be stabilized against photodegradation. Recycled materials are likely to need additional stabilizer to retain adequate performance.

When regrind began to be a common ingredient in plastics processing in the late 1970s, much effort was devoted to studying the effects of multiple processing cycles on polymer performance. For many polymers, three major types of chemical reaction occur. First is oxidation. Reaction of the polymer structure with oxygen results in the incorporation of oxygen-containing groups in the polymer, with concomitant changes in properties and increased potential for further reactions. Either with or without oxidation, chain cleavage can also occur. This results in a decrease in molecular weight, with a consequent decrease in many performance properties. Chain cleavage can be followed by cross-linking, the forming of new molecular bonds that increases molecular weight and also changes properties. In some polymers, one or the other of these reactions predominates. In others, such as polyethylene, the effects of one tend to be balanced by the effects of the other.

Some molecular structures are much more reactive than others. Polypropylene, for example, is significantly more susceptible to photo-oxidation than is polyethylene. Furthermore, for some materials, it is feasible to upgrade the material during reprocessing (such as in solid-stating of recycled PET), while for others it is not.

In summary, the general rule is that recycled polymers will have somewhat different properties than virgin polymers. These changes are usually detrimental and range in nature from virtually unnoticeable to major. Just as not all polymers are equally sensitive, not all properties are equally sensitive. It is not unusual, for example, for a recycled HDPE to have virtually the same tensile strength as virgin HDPE but at the same time have significantly decreased Izod impact strength.

8.4 POLYETHYLENE TEREPHTHALATE (PET) RECYCLING

The largest source of PET in the MSW stream in the United States is packaging, as shown in Fig. 8.16. For many years, PET was the most recycled plastic in terms of both total amount and recycling rate, largely due to recycling of soft-drink bottles. However, as the use of PET in other types of bottles grew, the recycling rate fell. Also, recycling rates for HDPE increased significantly with the addition of HDPE bottles to many curbside and drop-off recycling programs. Consequently, HDPE recycling began to exceed PET recycling. For 2003, American Plastics Council statistics show the total amount of PET bottles recycled again exceeding PET bottles.[43] EPA values, which report on containers rather

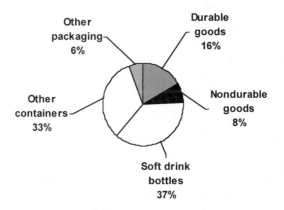

FIGURE 8.16   PET in U.S. municipal solid waste, 2003.[1]

than bottles only, still show HDPE with a larger amount of recycling in 2003.[1] In the United States, in 2003, 14 companies produced 517 million pounds of recycled PET flake from postconsumer bottles. The four largest companies accounted for nearly three quarters of the recycled PET. Capacity utilization was low, approximately 59 percent, illustrating the supply difficulties that continue to plague this industry. Nearly 40 percent of the total amount of bottles recovered for recycling, 321 million pounds of the 837.9 million pounds total, were exported, mostly to Asia.[43] NAPCOR, the National Association for PET Container Resources, reports very similar values: a total of 841 million pounds of postconsumer bottles recycled, 321 million of this sold to exporters, and 62 million pounds of postconsumer bottles imported. Of the exports, NAPCOR reports that 22.5 million pounds went to Canada and the remainder to China.[87]

Soft drink bottles remain the largest single use of PET in packaging, but nonbeverage bottle use continues to grow at a faster rate, so the aggregate use of PET in "custom" (non-soft-drink) bottles exceeds use in soft drink bottles, accounting for 57 percent of available PET bottles.[43] Use of small size soft drink bottles has grown, while use of liter and larger size bottles has declined. Fruit juice and water represent a sizeable segment of custom bottle use. Use of PET in beer bottles has grown less than had been anticipated. NAPCOR estimated that a total of 4.3 billion pounds of PET bottles and jars were available for recycling in the United States in 2003.[87] APC reports similar numbers, with a total of 4.0 million pounds of PET bottles.[43] The EPA reports somewhat different numbers for the soft drink/custom bottle split, estimating that 53 percent of the 4.0 billion pounds of PET bottles and containers in the MSW stream in 2003 were soft drink bottles.[1] The reason for this discrepancy in reported bottle types is not clear.

EPA reported an overall recycling rate for PET containers and packaging of 18.8 percent in 2003, a total of 820 million pounds; a PET container recycling rate of 18.4 percent, for 740 million pounds; and a PET soft drink bottle recycling rate of 25.2 percent, for 540 million pounds.[1] APC reports a PET soft drink bottle recycling rate of 30.2 percent in 2003, 531.8 million pounds, and a custom bottle recycling rate of 12.1 percent, 306.1 million pounds, for a total PET bottle recycling rate of 19.5 percent, 837.9 million pounds.[43] NAPCOR reports a PET bottle recycling rate of 19.6 percent in 2003, for a total of 841 million pounds.[87] Recycling rates for PET have been falling in the United States since 1994 (Fig. 8.17).

EPA reports a recycling rate for plastics in durables of 3.9 percent in 2003, a total of 330 thousand tons, but has not broken this figure down by resin type since 2000. That year, a PET recycling rate of 7.7 percent was reported for durable goods.[2] No significant recycling of plastics in nondurables is reported. In the noncontainer segment, EPA reported a recycling rate for PET packaging of 23.5 percent in 2003. The overall PET packaging recycling rate was 18.8 percent, and the overall recycling rate for PET in municipal solid waste was 14.3 percent.[1]

In contrast to the declining rates for PET bottle recycling in the United States, PET recycling continues to increase in much of the world. For example, Ontario reported a PET bottle recycling rate of 50 percent in 2003.[88] The Japanese Council for PET Bottle Recycling reports that the aggregate recycling rate for designated PET bottles (soft drinks, soy sauce, and liquors) reached 61.0 percent in 2003 after starting at only 0.4 percent in 1993. (Fig. 8.18). Designated bottles accounted for more than 93 percent of all PET bottles produced in Japan in 2003. The recycling rate calculated on the basis of all PET bottles, rather than designated bottles, was 56.7 percent in 2003. PET bottles are collected separately from other wastes in 2891 communities, 91.6 percent of all municipalities in the country.[89,90]

In Europe, Petcore reports that 665,000 tonnes of postconsumer PET were recycled in 2004, up 8.5 percent from 2003 and continuing a well established pattern of growth (Fig. 8.19). The recycling rate was reportedly 30.0 percent. Germany, France, and Italy ac-

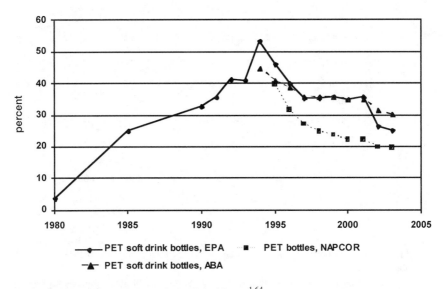

FIGURE 8.17 PET recycling rates in the United States.[1,64]

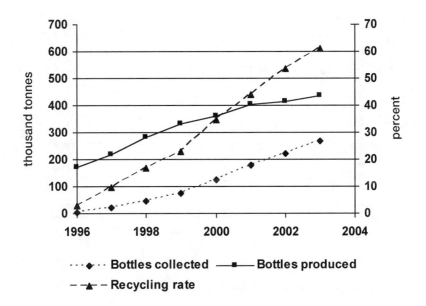

FIGURE 8.18 PET bottle recycling in Japan.[89]

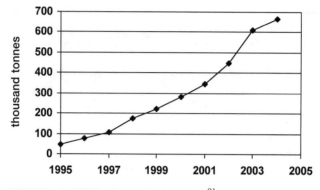

FIGURE 8.19   PET bottle recycling in Europe.[91]

counted for 60 percent of the total collected, but Ireland, Poland, and Spain had significant growth. By 2010, Petcore expects more than 1 million tons of European PET to be collected and recycled.[91]

Recycling rates vary significantly between EU countries. In 2005, the UK reported a recycling rate of only 7.5 percent for PET bottles.[92] Switzerland for many years claimed the highest recovery rate for PET beverage bottles in the world. From 1999 through 2001, the recovery rate was 82 percent (Fig. 8.20). Bottles are collected through a network of about 40,000 collection bins supplemented with return through the original distribution channels. About 90 percent of the collection occurs through the bins. The system is operated by the Swiss PET recycling organization, PET-Recycling Schweiz (PRS), which levies a prepaid recycling charge on participating members (about 85 percent of the total

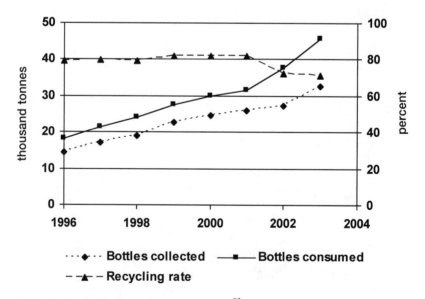

FIGURE 8.20   PET bottle recycling in Switzerland.[93]

retailers). However, the reported recycling rate for 2002 was only 72 percent and for 2003 was 71 percent. This falls short of the mandated 75 percent specified by Swiss law and allows the government to institute a deposit system. However, a decision was been made to give the industry some time to improve collection before imposing a mandatory deposit.[93]

In Brazil, the tonnage of PET recycled has continued to grow, reaching 120,000 tonnes and a recycling rate of 40 percent in 2003 (Fig. 8.21). PET packaging is collected using a combination of street collectors, factories, and municipal collection of separated recyclables.[94]

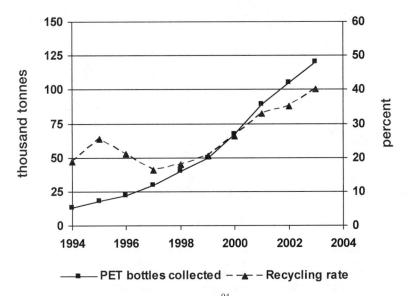

FIGURE 8.21   PET packaging recycling in Brazil.[94]

The overall PET recycling rate in Australia was reported to be 31.5 percent in 2003, up from 30.7 percent in 2002.[11]

8.4.1 Soft Drink Bottle Recycling

As mentioned above, soft drink bottle recycling got its start with the introduction of deposit legislation, which resulted in collection of significant volumes of material and recognition of the economic value embedded in them. In the United States, recycling rates for PET bottles grew until 1994 and then began to decline, as did recycling rates for PET soft drink bottles (Fig. 8.17). The decline in recycling rate for PET soft drink bottles is attributed in large part to a substantial increase in the tonnage of PET used in single-serving bottles, which are more likely to be consumed away from home and thus less likely to reach curbside recycling collection. Overall, between 1994 and 2003, the total tonnage and number of soft drink bottles recycled increased, even though the recycling rate fell. However, the tonnage of PET bottles (all types) collected for recycling declined from the previous year in 1996, 1997, 2000, and 2002, and the number of PET soft drink bottles recycled decreased in both 2002 and 2003 (Fig. 8.22). (The number can increase even if the

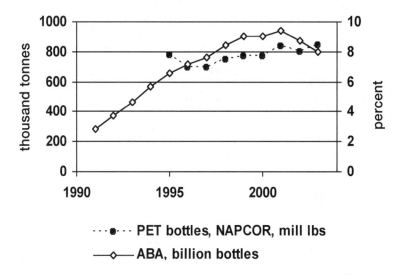

FIGURE 8.22   PET soft drink bottle recycling amounts in the United States.[95,96]

tonnage goes down due to the lighter average weight of the bottles collected.) The overall recycling rate for PET soft drink bottles in 2003 was 30.2 percent, for a total of 8 billion bottles, according to NDA, and 25.2 percent, for a total of 270 thousand tons, according to EPA.[1,95,96]

The Container Recycling Institute reports that the recycling rate for beer and soft drink containers (all types) in deposit states (including California) averages over 70 percent, 2 to 4 times that in nondeposit states.[97] Michigan, where the deposit is 10 cents rather than the 5 cents charged in most states, reports a 97 percent average redemption rate for covered containers for 1990 through 2003, and a 97 percent rate in 2003. This rate is slightly overestimated because of fraudulent returns; however, using the best estimate for this correction for 2003, the most recent year reported, decreases the redemption rate only to 95 percent.[98]

Maine estimates a 90 to 95 percent recycling rate for its expanded bottle bill, which covers all beverage containers except dairy products and cider. The state does not have specific figures by container type.[99] California reported a 39 percent recycling rate for PET bottles in 2004 under its expanded bottle bill, compared to 75 percent for aluminum cans.[100]

The decline in recycling rates for soft drink containers resulted in pressure on major soft drink manufacturers, first Coca-Cola and then PepsiCo, to increase PET soft drink bottle recycling and, in particular, to use recycled content in bottle manufacture. The GrassRoots Recycling Network (GRRN) and other organizations even took out full-page ads in the *New York Times* and *Wall Street Journal* blasting Coca-Cola for abandoning promises it made in 1990 to use recycled content in soft drink bottles. At least partially in response to consumer pressure, Coca-Cola began using recycled content in a portion of its U.S. soft drink bottles in 1998, although the company did not make any public announcement that it was doing so until 2000.[101] The company had been using recycled content in Australia and a few other countries for a number of years. Coca Cola, and later Pepsi as well, committed to work toward incorporation of 10 percent recycled content in their soft drink bottles by 2005. Coca Cola also participated in Businesses and Environmentalists

Allied for Recycling (BEAR) in its efforts to increase PET bottle recycling to 80 percent. However, the company later broke its ties to BEAR, in evident disagreement about BEAR's push for deposit systems as the most viable way to increase recycling. In 2002, Coke was reported to have achieved the 10 percent recycled content in 80 percent of its PET bottles, amounting to about 80 million pounds of recycled resin. Pepsi was far behind, using only about 1 million pounds of recycled PET that year. Dr. Pepper/Seven Up Inc. reported that it was not using recycled content because of high cost and shortage of material.[102]

The increased use of plastic bottles (mostly PET) at public events, coupled with the declining recycling rates for PET, have resulted in greater efforts to collect these bottles in such locations. For example, several major league baseball teams are making significant efforts to increase recycling. For PET bottles, such efforts usually involve placing recycling bins at various strategic locations around the ballpark and parking lots. Success has been mixed. The Seattle Mariners have 105 recycling bins for plastic bottles that are clearly marked and strategically placed, but they are reported to usually be contaminated with other kinds of trash, so the collected bottles are not usable. The Kansas City Royals and Colorado Rockies have had better success with bins around the concourse. The Rockies received special recognition from the Colorado state recycling alliance, Colorado Recycles, for using several reverse vending machines to collect plastic bottles, providing coupons redeemable for food and Rockies merchandise in exchange. The most successful recycling efforts, however, involve the cleanup crews. Evidently, so many people leave their trash in the seats that sorting at this time can result in high recovery rates.[103]

8.4.2 Recycling of Custom PET Bottles

Recycling rates for custom PET bottles are considerably lower than rates for soft drink bottles. Some custom bottles for beverages such as fruit juice and drinking water are covered by the expanded deposit legislation in Maine and in California. PET beer bottles are covered by deposits in all deposit states. However, nonbeverage containers such as peanut butter jars and shampoo bottles are not covered by deposits anywhere in the United States. The very rapid growth of bottled water, 23 percent in 2003, is pointed to as one reason for falling PET recycling rates.[6]

The introduction of PET bottles in colors other than green and clear has introduced new complications into PET recycling. For example, the blue color of some bottled water has sparked criticism. It appears that as long as the blue color remains a small fraction of collected materials, it can be incorporated into the green PET stream without problem. However, larger amounts would create problems. Many of the other colors of PET are handled by diverting the containers into disposal rather than recycling.

The introduction of PET bottles for beer sparked a new round of concern. The amber color of some of these containers means they must be separated from the clear and green PET. Furthermore, these bottles contain additional materials to provide the required level of protection against oxygen permeation. When ethylene vinyl alcohol (EVOH) is used as the barrier, the resulting structure is very similar to bottles for ketchup and some other foods. When the PET/EVOH ketchup bottle structure was introduced, it was shown that the bottle, which is designed to delaminate, is compatible with normal PET recycling systems, at least at current use and recovery levels. Most of the EVOH is removed during the washing and rinsing stages, and the small fraction that remains does not cause performance problems. For PET beer bottles containing a nylon barrier layer, estimates are that 30 to 40 percent of the barrier layer will be removed during processing, leaving the recycled PET containing about 3 percent nylon by weight. Manufacturers of some other bottle variants, such as those using an activated carbon coating, also claim either that the barrier material will be removed during the recycling operation, or that the tiny amount of material used in a bottle will not be enough to interfere with end uses of the recovered material.[104] In 2005, Petcore gave a limited two-year endorsement of the recyclability of ActiTUF resins, which incorporate an active oxygen scavenger, a gas barrier, or both, in colored PET bottles. Plasmax, a silicon oxide barrier coating from SIG Corpoplast, was endorsed for all applications.[105] However, concerns about PET recycling were stated as the reason for Asahi's decision in mid-2004 to drop plans to launch Japan's first PET beer container.[106]

8.4.3 Mechanical Recycling of PET Bottles

Most PET recycling processes use mechanical processing to convert the collected bottles into a usable form. While precise designs are proprietary, most operate similarly to the pilot recycling facility developed at Rutgers University under the sponsorship of the Plastics Recycling Foundation. This process began with color separation, followed by shredding the whole bottles, usually in two steps with initial shredding followed by a finer granulation step. Next, the shredded material was sent through an air classifier to blow off the light particles, which consist primarily of fines and label fragments. The material was then washed in hot detergent to remove product residues and soften and remove adhesive. Washing was followed by screening and rinsing. Next, a density-based separation using hydrocyclones separated the heavier-than-water PET from the lighter-than-water polyolefins, which consist predominantly of polypropylene from caps and, to a lesser extent, from labels. The PET was dried and then sent through a metal removal process, often using an electrostatic separator.

The Rutgers process was originally developed to handle beverage bottles at a time when nearly all bottles had paper labels, HDPE base cups, and aluminum closures. The disappearance of the base cups and change from aluminum to PP closures has greatly facilitated the recycling process. Aluminum, in particular, was difficult to remove and caused serious performance problems in the recycled material. All or nearly all PET containers now use plastic caps. A remaining source of aluminum in recycled PET is fragments of inner seal materials containing a foil layer that are sealed to the container during the packaging process and may not be removed completely when the consumer opens the container. Voluntary design guidelines discourage the use of aluminum inner seals that are not readily removed when the container is opened, but not all manufacturers adhere to these guidelines.[107] Changing the labels from paper to plastic has also facilitated recycling. In current processes, the PP label fragments that are not removed during previous process steps will be removed with the PP caps in the hydrocyclone.

Contamination of recycled PET with aluminum flakes remains an issue, with commonly used recycling processes producing average aluminum contamination of about 2000 parts per million (ppm). Electrostatic separation can reduce this level to 50 to 100 ppm, with PET recovery levels of 92 to 94 percent. However, electrostatic separation (ES) is vulnerable to fluctuations in atmospheric humidity and in temperature. Furthermore, even at these levels, aluminum tends to clog the screens of extruders, increasing maintenance requirements. Magnetic separation is effective at removing ferrous metals but not against aluminum. However, rare earth (RE) roll magnets, which generate a magnetic force many times that of conventional magnets, have been successful in producing less aluminum contamination and greater recovery of PET than electrostatic separation in some cases. While their advantages were not sufficient to replace ES, combining the two techniques, first proposed in 1990, was found to permit an increase in recovery to 96 to 98 percent, along with nearly doubling feed rates. Another alternative technology tried for aluminum separation was eddy current separation (ECS). Earlier versions of this technology did not provide acceptable performance, but now ECS systems based on rare earth magnetic rotors, resulting in rotating magnetic fields, have become available. In these systems, PET flakes are fed onto a conveyor belt, which moves them across a magnetic rotor. When a piece of aluminum or other nonferrous metal moves across the rotating separator, eddy currents are created in the metal, generating a magnetic field around the particle and causing it to be repelled from the magnet. PET particles are not affected and continue on their path. All ferrous metals must be removed before the separation, as they will generate very strong magnetic forces. High-grade PET can be produced at very high recovery rates using rare earth ECS systems. Even better results can be achieved by combining RE roll systems with ECS. A commercial operation using one RE roll followed by two ECS systems achieved a purity of less than 5.9 ppm aluminum and PET losses of only 0.5 to 0.9 percent.[108]

8.4.4 Recycling of Nonbottle PET

While most PET recycling processes in the United States handle only bottles, facilities in Germany handle mixed PET packaging, including bottles, tubs, dishes, and film, from yellow bag or bin collection of plastics through the Duales System Deutschland (DSD). Some PET recycling in Germany is through feedstock recycling but, increasingly, higher-value collected plastics such as PET bottles are diverted for mechanical recycling.

PET x-ray film represents another source of recycled material. Since these materials generally are coated with silver, there has long been a potent economic incentive for their recovery, and silver from x-ray film has been recovered since the early 1900s.[109] In such processes, recovered PET can be obtained as a by-product of silver recovery. Its recycling is complicated by the fact that it is generally coated with PVDC. Gemark is reported to have a proprietary process to remove the PVDC.[110,111] United Resource Recovery Corp. (URRC) of Spartanburg, SC, is another U.S. recycler of x-ray and other silver-coated PET film.[112]

DuPont operated a feedstock recycling facility, using its "Petretec" process, to recover PET materials such as x-ray film from 1995 to 1998 but discontinued the operation due to poor market conditions.[113]

8.4.5 Feedstock Recycling of PET

Recovered PET can be chemically broken down into small molecular species, purified, and then repolymerized. The two major processes for tertiary recycling of PET are glycolysis and methanolysis. Both result in PET that is essentially chemically identical to virgin resin and has been approved by the U.S. Food and Drug Administration, and similar regulatory organizations in some other countries, for food contact applications. However, PET produced by these processes is more costly than virgin resin, which significantly limits its use.

In 1991, Goodyear obtained a letter of nonobjection from the U.S. FDA for the use of its "Repete" tertiary recycled PET in food contact applications. The process, later sold to Shell Chemical Co., used glycolysis to partially break down PET, followed by purification and repolymerization. In tests using model contaminants, the contaminants were removed down to a 50 to 100 ppm level. That same year, both Eastman Chemical Co. and Hoechst-Celanese Corp. received letters of nonobjection from FDA for their methanolysis-based PET depolymerization processes.[114] Eastman's "Superclean" process reportedly can handle PET bottles of any color, including multilayer bottles such as those containing oxygen barriers, producing new PET equivalent to virgin.[115] Methanolysis processes provide full depolymerization and can remove colorants and certain impurities that cannot be removed by glycolysis.

Petrecycle Pty. Ltd., of Melbourne, Australia, announced in February 2001 that it would install a glycolysis system capable of processing more than 22 million pounds of postconsumer PET a year in the M&G Finanziaria Industriale SpA virgin PET production facility in Point Pleasant, WV. Petrecycle's "Renew" technology was designed to enable M&G to produce a blend of virgin and recycled PET, reportedly for lower costs than those associated with other recycling technologies.[116] However, this evidently never occurred. In 2004, Petrecycle opened a pilot scale facility in Painesville, OH, with a capacity of 34 kg per batch. In mid 2005, the company announced it would move the pilot facility back to Melbourne, Australia.[117]

Other tertiary recycling processes that have been developed include a Freeman Chemical Corp. process to convert PET bottles and film to aromatic polyols used for manufacture of urethane and isocyanurates.[114] Glycolized PET, preferably from film, since it is often lower in cost than bottles, can be reacted with unsaturated dibasic acids or anhydrides to form unsaturated polyesters. These can then be used in applications such as glass-fiber-reinforced bath tubs, shower stalls, and boat hulls. United States companies that have been involved include Ashland Chemical, Alpha Corp., Ruco Polymer Corp., and Plexmar.[110] Unsaturated polyesters have also been used in polymer concrete, where the very fast cure times facilitate repair of concrete structures. Basing polymer concrete materials, for repair or precast applications, on recycled PET reportedly leads to 5 to 10 percent cost savings and comparable properties to polymer concrete based on virgin materials. However, they are still approximately 10 times the cost of portland cement concrete.[118] There appears to be little commercial application of these processes at present.

8.4.6 Food Grade Mechanically Recycled PET

A variety of processes have received official nonobjection from the U.S. FDA for use of mechanically recycled PET in food packaging. The earliest processes relied on insensitive uses or on imposition of a physical barrier between the food and the recycled plastic. For example, the first approval was in 1989 for use of recycled PET in egg cartons.[110] Continental PET Technologies received approval in 1993 for a coinjected multilayer PET bottle with a one mil (0.001 in) layer of virgin PET between the core layer of recycled PET and the container contents. The approach was used initially for soft drink bottles in Australia, New Zealand, and Switzerland.[119–121] In 1994 and 1995, Wellman, Inc., obtained approval for use of mechanically recycled PET in multilayer packaging for a variety of food products.[122] Plastipak Packaging's Clean Tech affiliate also produces food-grade recycled PET.[123]

The first U.S. approval for use of mechanically recycled PET in direct contact with food came in 1994, for Johnson Controls' (later Amcor's) SuperCycle recycled PET. Amcor currently operates a PET recycling facility in Beaune, France, with a capacity of 15,000 tons per year. About 75 percent of production is food-grade SuperCycle resin, and the rest is nonfood-grade NuCycle.[124] Other companies subsequently also obtained FDA approval. For example, in 1996, Wellman received an FDA letter of nonobjection for their EcoClear resin made from 10 percent recycled PET from postconsumer bottles.[125]

In 2003, 6 of the 14 operating PET recycling plants used technologies that have received letters of nonobjection from FDA for use of the recycled material (RPET) in direct contact with foods and beverages. These processes rely on intensive cleaning, often in combination with control over the source material, and have been validated by challenge with known amounts of model contaminants. Use of RPET in food and beverage bottles was 106 thousand pounds in 2003, an increase of 23 percent over 2002. This is attributed directly to increased use by Coke and Pepsi.[87] In 2004, Plastipak Packaging announced that it would build a $13 million PET recycling plant in eastern Slovakia, in part to serve the demand for Coca-Cola and Pepsi-Cola for use of 10 percent recycled content in their bottles beginning in 2005.[126]

In 1999, Phoenix Technologies LP, a subsidiary of Plastic Technologies Inc., became the first company to receive FDA approval for use of 100 percent curbside recycled PET in food containers. In 2001, the company gained approval for use of this material in hot-filled bottles. Commercial production of food-grade curbside recycled PET began in December 2000. The process was already being used in Australia to make 25 percent recycled content bottles for Coca-Cola.[127] The company's facility in Ohio manufactures food-grade and nonfood-grade resins from 100 percent recycled content.[128] In 2004, the company received approval in Belgium for use of its recycled PET in food and beverage packaging, adding to its previous approvals in Australia, Canada, and New Zealand, in addition to the United States.[129]

United Resource Recovery Corp. (URRC) of Spartanburg, SC, became the second U.S. company to get approval for food-grade recycled PET from curbside collection in 2000, for its "Hybrid-UnPET" technology. The process is reported to use mechanical recycling without hot water, followed by a thermal treatment using sodium hydroxide, and a final stage for removal of residual contaminants.[130] The treatment with caustic soda results in a solid-phase reaction in which the outer surface of the PET chips is stripped off, and the resulting ethylene glycol and terephthalic acid are recovered as by-products. Any contamination adhering to the outer surface is removed during this stage. Residual contaminants are removed using a combination of air blowing and controlled temperature. The resultant mixture of salt and clean PET granules is separated by mechanical filtration followed by washing, and then removal of any small metal particles by a metal separator.[131] This process, therefore, could be regarded as a mix of mechanical and feedstock recycling, and URRC refers to it as *partial depolymerization*. The company reports that 90 to 95 percent of the polymer is preserved. This technology allows the company to use 100 percent curbside PET rather than the more expensive PET collected through deposit programs. Two large plants in Europe, one in Switzerland and one in Germany, are using this technology, along with the pilot-scale facility in Spartanburg.[112] In July 2005, the first bottle-to-bottle PET recycling facility in Latin America began operation in Toluca, Mexico. It has a capacity of 25,000 tons per year, matching the total amount of PET currently being recycled in Mexico. The plan is a joint effort of Coca-Cola de Mexico, Coca-Cola FEMSA and AL-PLA, a major PET bottle supplier.[132]

Another process producing food-grade recycled PET is the "Stehning BtoB Process," developed by OHL Apparatebau & Verfahrenstechnik GmbH, of Limburg, Germany, which received U.S. FDA approval in 1999. The first production unit began operation in October 1999 at PET Kunststoffrecycling GmbH (PKR) in Beselich, Germany. In this process, the clean PET bottle flakes, without preliminary drying, are fed into a modified twin screw extruder, where the PET is dried and degassed, and then melt-filtered and pelletized. The amorphous PET chips are fed into a discontinuous solid-stating process for crystallization and condensation/decontamination under vacuum. Reportedly, the sensory characteristics of the recycled material are superior to those of virgin PET. In particular, acetaldehyde and ethylene glycol levels are lower. The German facility has a production capacity of 7500 tonnes per year.[133] In 2002, Alba AG acquired a 50 percent share of PKR and took over its management. Alba and PKR together were reported to have a capacity of 7000 tonnes of PET per year.[134]

Erema North America received approval in 2002 for mechanical processing of PET from curbside collection systems into food-grade resin using the Vacurema technology that Erema, its parent company, first launched in 1997. In early 2003, it obtained additional approval for use of the resin in containers for hot-fill applications.[135] In 2005, the Czech company Silon was using the Vacurema system to produce fiber, and in a joint project, Kruschitz and Ovotherm were using a system to produce transparent egg packaging in Austria. The Swiss company ITW-Poly-Recycling uses a Vacurema system for bottle-to-bottle recycling. Snellcore, in the Netherlands, and the German company Texplast each have two systems.[136]

8.4.7 Properties of Recycled PET

Mechanically recycled PET usually retains very favorable properties. Some reduction in intrinsic viscosity is common, but it can be reversed by solid-stating. Residual adhesives from attachment of labels are a contaminant concern. Some of the adhesive residue can become trapped in the PET granules and is not removed by washing. Since these adhesives often contain rosin acids and ethylene vinyl acetate, the rosin acids plus acetic acid from ethylene vinyl acetate hydrolysis can catalyze hydrolysis of the PET during processing. A similar problem can be caused by residues of caustic soda or alkaline detergents from the wash step. Considerable loss of molecular weight can result, and darkening of the adhesive residues can cause discoloration.

PET is very susceptible to damage from PVC contamination, and vice versa. Contamination in the range of 4 to 10 ppm can cause serious adverse effects.[137] Because the densities of PET and PVC overlap, density-based separation methods are ineffective. Technologies have been developed for very effective sorting of whole-bottle PVC and PET. PVC contamination from materials such as coatings, closure liners, labels, and so on is more difficult to handle. Appropriate package design to avoid use of PVC or PVDC with PET containers is the most effective strategy. The Association of Postconsumer Plastic Recyclers issued a report detailing the effect of PVC contamination of PET, in which they estimated that the cost to the domestic PET recycling industry of addressing PVC contamination in 1998 totaled $6.5 million. Sorting accounted for 37 percent of the cost, with depreciated equipment, laboratory labor, and maintenance also representing major costs. The average cost was 1.67 cents per pound of PET produced. Because costs were lower in larger reclaimers, as well as in those specializing in deposit containers, the weighted average was 0.86 cents per pound.[138]

Repolymerized PET is essentially identical in performance to virgin PET.

8.4.8 Markets for Recycled PET

One of the earliest large-volume uses for recycled PET was as polyester fiberfill for applications such as ski jackets and sleeping bags. The range of applications has grown enormously and now includes items as diverse as carpet, automobile distributor caps, produce trays, and soft drink bottles. Fiber applications remain the largest market, although they have declined significantly in the last few years (Fig. 8.23).[87]

The first 100 percent recycled PET container in the United States was introduced in 1988 by Proctor & Gamble for household cleanser. Bottles, including those for food and beverages, are now a significant market for recycled PET, and pressure for bottle-to-bottle recycling continues. However, use of recycled PET in nonfood bottles has declined by nearly two-thirds since 1996 although, during this same period, use in food bottles has increased by more than 300 percent.[87]

In 2005, Marks & Spencer's became the first retailer in the UK to use recycled PET food packaging on a large scale. The company is encouraging closed-loop recycling, so it is marking the packaging with recycled content and notation that it is recyclable. Collection bins are being provided at the front of the stores. Items included are salad bowls, beverage bottles, recipe pots, and trays, with 30 to 50 percent recycled content.[139]

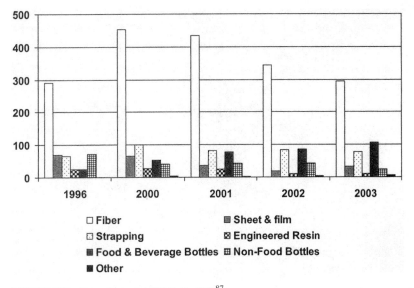

FIGURE 8.23 Uses of recycled PET in the U.S.[87]

Other markets for recycled PET include strapping, sheet, and film. A number of high-performance engineering alloys and compounds utilizing recycled PET have been developed, although this use has also decreased sharply since 1996.[87] Recently, a research project by Pera Innovation and Newcastle University showed that recycled PET could be treated and reused to manufacture high-value reinforced thermoplastic pipes for high-pressure engineering applications.[140] Another innovative use is uncleaned recycled PET being used as an adhesive to support steel rods holding up mine shafts.[141]

As is generally the case for recycled materials, the market situation is strongly affected by the supply and demand situation for virgin resin. The export market, as already discussed, is also a major factor. When virgin PET supply is low and prices are up, demand for recycled resin is strong. During the last half of the 1990s, there was a significant downturn in recycled PET demand caused by a large increase in production capacity for virgin resin that drove down prices. The situation was exacerbated by a temporarily plentiful supply of off-spec virgin resin from new facilities entering production. Some PET recyclers did not survive these lean years. Prices in 1995 for baled PET bottles were 27 to 35 cents per pound,[142] while prices in 2000 ranged from 7 to 20 cents per pound.[143] In the first half of 2005, prices were 20 to 26 cents per pound, largely due to strong demand from China and Vietnam.[144]

8.5 HIGH-DENSITY POLYETHYLENE (HDPE) RECYCLING

The sources of HDPE in the U.S. MSW stream are shown in Fig. 8.24. For many years, HDPE was the second most recycled plastic, but it overtook PET several years ago. In the United States, the EPA reports that the recycling rate for HDPE milk and water bottles was 31.9 percent in 2003, a total of 460 million pounds. Other HDPE containers were recycled

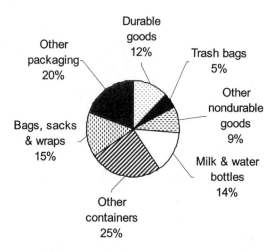

FIGURE 8.24   HDPE in the U.S. MSW stream, 2003.[1]

at a rate of 14.8 percent, for an overall bottle and container recycling rate of 21.0 percent. The recycling rate for HDPE packaging as a whole was 12.4 percent. The recycling rate for HDPE in durable and nondurable goods was negligible, for an overall HDPE recycling rate of 9.1 percent.[1]

The American Plastics Council (APC) reported a 2003 recycling rate for HDPE bottles of 24.8 percent, for a total of 823 million pounds—higher than the 19.5 percent reported for PET bottles.[72] APC reported rates of 24.2 percent in 2002 and 23.2 percent in 2001 for HDPE bottles.[43,145] As can be seen in Fig. 8.25, using values reported by the APC (using values for collected bottles), the recycling rate for HDPE bottles has been fairly steady since 1996, while the rate for PET bottles, and in particular for PET soft drink bottles, has been falling. The recycling rate for natural (unpigmented) HDPE bottles has fallen somewhat since 1998, from 31 percent that year to 27.3 percent in 2003, but the rate for pigmented bottles has increased from 19 percent in 2000 to 22.6 percent in 2003. The overall effect was a 2003 recycling rate for HDPE bottles of 24.8 percent.[43] As would be expected, this rate is higher than the EPA-reported rate for HDPE bottles and containers of 21.0 percent.[1] Most programs collect only bottles, not other containers such as margarine tubs.

APC reports that 30 companies engaged in HDPE bottle recycling in 2003, 1 more company than in 2002. Over 80 percent of the total recycled HDPE was handled by the eight largest companies. Capacity utilization was healthier for HDPE than for PET: 68 percent compared to 59 percent. In part, this was due to a lower percentage of exports for HDPE—12.1 percent of the total bottles collected in 2003, compared to 38.3 percent for PET.[43]

Ontario reported a recycling rate of 50 percent for HDPE bottles in 2003.[88] In Australia, the overall HDPE recycling rate was reported to be 23.1 percent in 2003, compared to 17.9 percent in 2002 and 19.0 percent in 2001.[11]

8.5.1   Recycling of Unpigmented HDPE Bottles

Unpigmented HDPE milk and water bottles are the most valuable type of HDPE for recycling. They are made from a high-quality fractional melt index homopolymer HDPE, usu-

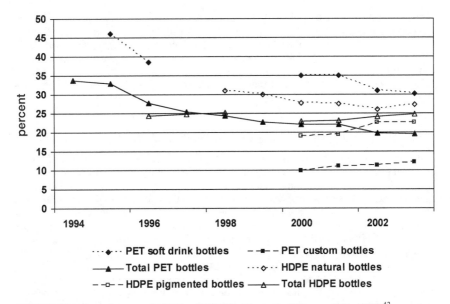

FIGURE 8.25 Recycling rates for PET and HDPE bottles in the U.S. reported by APC.[43]

ally unpigmented, that is suitable, when recycled, for a wide variety of uses. Considering that these bottles are generally not covered by deposits, their 27.3 percent recycling rate in the United States in 2003 is impressively close to the 30.2 percent rate for PET soft drink bottles.[43] Extrusion blow-molded HDPE bottles, both pigmented and unpigmented, are accepted for recycling in most community recycling programs. Injection blow-molded bottles, which are made from a high-melt-flow HDPE, are generally considered to be undesirable contaminants. Other containers that are unacceptable in many programs include motor oil bottles and those that contained caustic cleansers, insecticides, or other product residues that could pose a risk.

Ontario, Canada, has had a deposit system for HDPE milk bottles for a number of years, charging 25 cents per bottle, which is one of the reasons why milk sold in flexible pouches has been popular. The deposit system is mandated by Regulation 344, Disposable Containers for Milk, of the Environmental Protection Act.[146] When the government considered discontinuing the system in 1999, Ontario milk producers, retailers, and packaging suppliers objected.[147]

The Alberta (Canada) Dairy Council launched a voluntary recycling program in 1999 in an effort to increase the 35 percent recycling rate for HDPE milk containers to 70 percent. Municipalities and recycling programs receive a payment for collected and densified HDPE, subsidized by the Dairy Council, in the form of a guaranteed price of $400 (Canadian) per tonne. The three Alberta milk processors voluntarily pay two cents for each 4-l milk bottle and one cent for each 2-l bottle, into a Container Recovery Fund to support the program. In its first year of operation, July 1, 1999 to June 30, 2000, the program collected 1197 tonnes of material, a 32 percent increase over the previous year, bringing the province-wide recovery rate to 40 percent. Sixteen communities and recycling authorities achieved recovery rates of 70 percent or more.[148] By the 2002 to 2003 year, the most recent information available, the program had achieved a recovery rate of 48.3 percent for milk bottles. The recovery rate for polyethylene coated milk cartons, which were added to collection in January 2002, was 10.5 percent; the fee for these containers is also 1 cent per container. By 2003, 91 percent of Alberta residents had access to milk bottle recycling, and 76 percent to carton recycling, in their local communities. In 2002 and 2003, 1211 tonnes of HDPE milk bottles were collected, 2 percent higher than the previous year, along with 166 tonnes of milk cartons. Nova Scotia and Saskatchewan now operate similar collection programs, and British Columbia and New Brunswick are implementing programs.[149]

8.5.2 Recycling of HDPE Motor Oil and Pesticide Bottles

Motor oil bottle recycling is a significant issue, not only because of the volume of bottles involved but because of the potential adverse environmental consequences and the value of the oil remaining in the bottles. Honeywell Federal Manufacturing & Technologies of Kansas City, MO, developed a system for recycling motor oil bottles and recovering both plastic and oil. The system is licensed to ITec International Technologies, Inc., a subsidiary of Beechport Capital Inc., which is marketing the systems worldwide. ITec estimates that about 2 billion plastic motor oil containers are discarded each year in the United States, each containing, on average, an ounce of oil, for a total of 250 million pounds of plastic and more than 15 million gallons of motor oil. In 2005, the company announced that it intends to build two facilities in California to produce clean PET and HDPE flake using its ECO_2 system. Plastics, either oil-contaminated or others, are granulated, washed with an environmentally friendly solvent (which is reused), and rinsed to remove labels, glue, and most contaminants, including oil. Oil is separated and sent to a recycler. The flakes are then cleaned using pressurized carbon dioxide, which removes additional contaminants. The CO_2 is then distilled to remove contaminants and is reused. Residual paper is next removed from the cleaned plastic flakes using a mechanical vacuum system. In addition to oil containers, the technology has been tested successfully in ITec's Oakdale, CA, pilot plant for HDPE pesticide containers, mixed postconsumer HDPE containers, soft drink containers, and 5-gal HDPE paint buckets.[150]

In Canada, a National Used Oil Material Advisory Council was launched in 1997. This industry-led product stewardship program now includes the provinces of Alberta, British Columbia, Manitoba, Quebec, and Saskatchewan. The Alberta Used Oil Management Association (AUOMA) originated the program for used oil recycling, which also includes recycling of oil filters and motor oil containers under 30 l in size. Participating companies pay an Environmental Handling Charge (EHC) of 5 cents (Canadian) per liter of container size, at the wholesale level. The collectors of the used materials receive payment through the program for the returned materials. British Columbia reported 42.6 percent recovery of oil containers in 2004. Manitoba reported a 22 percent recovery plus 20 percent reuse rate, for a total of 42 percent. Alberta reported that 49.8 percent of used plastic oil containers were recovered. Saskatchewan's recovery rate was 25 percent recycled plus 24 percent reused, for a total rate of 49 percent. Quebec's program is new and did not report a recovery rate.[151] Some other used oil recycling programs around the world also recycle the oil bottles, although many do not do so.

Pesticide and similar containers, like oil bottles, are not accepted in most curbside and drop-off recycling programs because of the problems posed by contamination with these hazardous chemicals. Household pesticide containers usually must be disposed of in the regular trash or in special collection of hazardous wastes. However, there are opportunities for recycling such containers when they are generated in the agricultural sector, where the much larger volumes make collection more viable.

The Ag Container Recycling Council (ACRC) reports that 7.9 million pounds of plastic crop protection product containers were collected and recycled in 2004. Applications for these containers that are seen as acceptable include new pesticide containers, pallets, construction site mats, commercial truck or manure spreader decker boards, field drain tiles, speed bumps, parking stops, hazardous waste drums, scaffold nailing strips, and commercial truck subfloor support members. Some plastic lumber applications are also found acceptable—those that result in low human exposure, such as fence posts, marine pilings, railroad ties, landscape timbers, and sound barriers.[152]

Canada claims leadership in recycling of empty pesticide containers used for agriculture. The program began in 1989 and has collected about 55 million empty containers as of 2005. There are now about 1200 collection sites across Canada. Current recovery rates are about 70 percent, the highest return rate in the world. The recovered plastic is recycled into fence posts for use on farms and for highway guardrail posts. Some is used for energy recovery.[153]

8.5.3 Recycling of Nonbottle HDPE

While most recycled HDPE comes from bottles, limited recycling of other HDPE materials also occurs. Some recovery of HDPE film occurs, along with LDPE, when retail bags are collected for recycling. This is discussed in Sec. 8.6. This collection dropped precipitously between 1996, when it reached a high of 50 thousand tons, and 1998, when it fell to 10 thousand tons. EPA reports about 10 thousand tons of bags and sacks recycled in each year since 1998.[4]

DuPont operates a recycling program for its Tyvek envelopes. For small quantities, less than 25 envelopes per month, customers are instructed to turn any Tyvek envelope inside out, stuff it with other envelopes, and mail it to DuPont's Tyvek® recycling specialist in Virginia. Medium-quantity generators are eligible for a pouch program, with each pouch holding approximately 200 envelopes. Filled pouches are sent to a regional recycler. For generators of more than 500 envelopes per month, DuPont will set up a custom recycling program.[154]

In Canada, Haycore recycles tubs, lids, injection-molded plastics, and other non-HDPE and PET bottle materials from curbside collection program. The company produces high-quality clean flake.[155] Markets for the material include pallets made by Granville Composite Products, protective plastic spacers for shipping made by PBI Industries, and recycling containers made by Buckhorn Canada.[156]

HDPE drums are recycled by some companies, usually on a local or regional basis. Conigliaro Industries recycled drums of all types, including plastics.

8.5.4 Recycling Processes for HDPE

Recycling processes for HDPE bottles are similar to those for PET. Typically, the collected HDPE is first sorted to separate the higher-value unpigmented containers. In some cases, the pigmented HDPE is further separated into color families. Sorting of the unpigmented bottles is often done prior to initial baling but may be done at a later stage.

At the plastics processor, the baled HDPE containers are usually shredded, washed, and sent through either a float-sink tank or a hydrocyclone to separate heavy contaminants. Air classification may be done prior to washing. The clean materials are dried and then usually pelletized in an extruder equipped with a melt filter to remove residual nonplastic contaminants. If mixed colors are processed, the result from typical curbside or drop-off programs is a grayish-green color, which is most often combined with a black color concentrate to produce black products.

Several types of contamination are of concern in HDPE recycling. The first is contaminants that add undesired color to natural HDPE. A prime culprit is caps on bottles. While consumers are generally told to remove caps before turning the bottles in for recycling, a significant number arrive with the caps still in place, and the caps are generally brightly colored. Most of these are polypropylene, with the next largest fraction polyethylene. Neither of these materials is separable in the usual recycling systems, so they usually remain with the HDPE, where they result in discoloration of the resin. Typically, the amounts are low enough that mechanical properties are not affected, but they do impart a grayish color to the material. The introduction of pigmented HDPE milk bottles, which seem to periodically pop up in various places, is a concern to recyclers because, if they were widely adopted, they could significantly cut into the use of the more valuable natural bottles. Pigmented HDPE recycled resin typically sells for only 60 percent of the price of natural HDPE.[157]

A second type of contamination is mixing of high-melt-flow injection-molding grades of HDPE with low-melt-flow blow-molding grades. The result can be a resin with intermediate flow properties that are not desired by either injection or blow molders. Since the coding system for plastic bottles (see Sec. 8.3.5) does not differentiate between the two, it is difficult to convey to consumers in any simple fashion which bottles are desired in the recycling system and which are not. The recycling process does not separate the two grades, since their densities and most other physical properties are equivalent. Some programs simply accept the resulting contamination, while others try to get the message to consumers, sometimes by specifying bottles "with a seam." Fortunately, the vast majority of HDPE bottles, especially in larger sizes, are extrusion blow molded.

Mixing of polypropylene into the HDPE stream is also a concern. As discussed above, much of this arises from bottle caps left in place. Some also arises from fitments on detergent bottles and from inclusion of PP bottles in the recycling stream. Since both PP and HDPE are lighter than water, the density-based separation systems commonly employed will not separate the two resins. Fortunately, in most applications, a certain level of PP contamination can be tolerated. However, particularly in the pigmented HDPE stream, levels of PP contamination are often high enough to limit the amount of the recycled material that can be used, forcing manufacturers to blend the postconsumer materials with other scrap that is free of PP or with virgin (often off-grade) HDPE. While triboelectric systems for separating chipped PP from HDPE have shown some promise, they are not generally available or used.

Finally, contamination of the HDPE with chemical substances that may later migrate from a container with recycled content to the product can present problems. This is a more serious issue with HDPE than with PET for two primary reasons. First, the solubility of foreign substances of many types is greater in HDPE than in PET. Thus, the level of contamination that may be present is higher. Second, the diffusion of most substances is faster in HDPE than in PET. Combined, these factors create a significantly greater potential for migration of possibly hazardous contaminants out of recycled HDPE into container contents. Nevertheless, the use of some types of recycled HDPE for limited food contact applications has been approved by the U.S. FDA.

The strategies for dealing with potential migration of hazardous substances from recycled HDPE are essentially the same as with PET. First, a combination of the selection of starting materials and processing steps can be used to minimize the contamination levels that are present. The first company to obtain a letter of nonobjection from FDA for recycled HDPE in direct food contact was Union Carbide, which later sold the technology to Ecoplast, which also received a letter of nonobjection.[158]

Recycled HDPE can also be used in a multilayer structure that provides a layer of virgin polymer as the product contact phase. This is the standard approach for laundry products, where FDA approval is not an issue, but where consumer acceptability issues associated with objectionable odors in the product surfaced early in the development stage. In this case, the multilayer bottles used have a layer of virgin polymer on the outside as well as on the inside. This not only solved appearance problems that were also associated with the use of recycled plastic, it permitted a significant savings in the amount of fairly expensive colorants that are required. The middle layer in such structures is composed of a blend of recycled HDPE with process regrind. An additional benefit from the inner layer of virgin polymer is the better environmental stress crack resistance of the co-polymer virgin HDPE compared to the regrind/recycle mix.

8.5.5 Markets for Recycled HDPE

A major early market for recycled HDPE was agricultural drainage pipe. Pipe continues to be a significant market, but a number of additional markets have developed as well. In particular, coextruded bottles containing an inner layer of recycled HDPE have developed into a major market. Nearly all laundry products sold in plastic bottles in the United States use this structure, typically incorporating about 25 percent recycled content. Motor oil is often sold in single-layer bottles made from a blend of virgin and recycled HDPE. Figure 8.26 shows the proportion of recycled HDPE going into various market categories in the United States in 2003.[43]

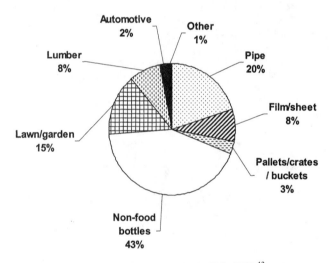

FIGURE 8.26   Uses of recycled HDPE in the U.S., 2003.[43]

As can be seen, while pipe remains a significant use of recycled HDPE, by far the largest use is in bottles (nonfood). United States companies that are major users of recycled HDPE include Procter & Gamble, which uses 25 to 100 percent recycled HDPE in most of its household products, and Clorox.

The use in plastic lumber and the related lawn and garden sector is increasing, as its benefits of long life compared to treated wood, freedom from the hazardous chemicals often used in outdoor grades of lumber, and maintenance of color without painting are recognized. Plastic lumber does carry a higher initial purchase price than wood equivalents, but life-cycle costing generally shows substantial benefits for plastic. Composites of wood fiber and plastic, either HDPE or LDPE, are also growing in use. Plastic lumber is discussed in more detail in Sec. 8.14.

Recycled HDPE is also used in manufacture of film, especially for merchandise bags. Often, the material used is recycled milk bottles. Other uses of recycled HDPE are manufacture of milk crates, curbside recycling bins, and other products. TransPac uses recycled HDPE in pallets, slip sheets, and pallet separator boards.[159] DuPont uses recycled HDPE in its Tyvek envelopes.[160]

8.6 RECYCLING OF LOW-DENSITY POLYETHYLENE (LDPE) AND LINEAR LOW-DENSITY POLYETHYLENE (LLDPE)

Because of the similarity in properties and uses of low-density polyethylene and linear low-density polyethylene, and because they are often blended in a variety of applications, use and recycling of LDPE and LLDPE are often reported and carried out together. Therefore, we will use the term LDPE to refer to both LDPE and LLDPE.

Nearly two-thirds of the LDPE found in municipal solid waste originates in packaging, as shown in Fig. 8.27.[1] Another sizable fraction comes from nondurable goods, especially trash bags. The two main sources of recycled LDPE are both in the bags, sacks, and wraps category: stretch wrap and merchandise bags. The U.S. EPA calculated that 150 thousand tons of LDPE/LLDPE bags, sacks, and wraps were recovered in 2003, for a recycling rate of 5.7 percent. The overall recycling rate for LDPE in MSW was 2.4 percent.[1] The overall recycling rate for LDPE and LLDPE in Australia was reported to be 12.2 percent in 2003—above the 2002 rate of 11.2 percent but lower than the 2001 rate of 13.4 percent. [11]

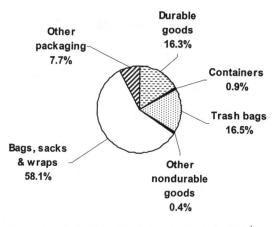

FIGURE 8.27   LDPE/LLDPE in MSW in the U.S., 2003.[1]

Recycling of grocery and other merchandise sacks is generally carried out at drop-off locations. At one time, there was a wide network of such sites across the United States, but many merchants discontinued the program due to contamination and other concerns. Bag recycling through schools has remained available in a significant number of locations.

In contrast to PET and HDPE, curbside recycling is not yet a significant factor in recycling of LDPE in the United States, although such recycling is increasing. San Juan Capistrano, CA, began collecting plastic bags at curbside in 2005, and other California cities are considering doing so. It reportedly is the first city in the country to collect bags curbside for use in new plastic bags. Residents put bags into a specially made plastic sack that goes in their curbside recycling bin. Increasing prices for oil have made plastic bags a more desirable feedstock.[161] The proposed bag taxes in California may also be a motivating factor. The American Plastics Council has set up a special web site for plastic bag and film recovery in California, which provides the general public, recycling coordinators, and businesses with information on recycling. Among other features, people can search for a drop off location in their area.[162]

Hilex Poly Co. announced in November 2004 that it was building a $13 million recycling plan in North Vernon, IN, in response to environmental challenges facing the plastic bag sector. The company is one of the largest plastic bag manufacturers in the United States, using primarily HDPE rather than LDPE. Its aim is to use the recycled material in the manufacture of new grocery bags, in a closed loop system it calls "Bag 2 Bag." The company also is working with its customers on reducing use of plastic bags through proper packing, and so on.[163]

Several bag-related companies have joined the NextLife Recycling Alliance to support film and bag recycling in the United States. It operates two recycling plants in Kentucky and Michigan and accepts truckloads of mixed material including plastic film and bags. Materials are sold to a variety of markets.[164]

Curbside collection of plastic film is much more common in Canada. The Plastic Film Manufacturers Association of Canada and EPIC have sponsored curbside recycling programs for plastic film of all types. By 1996, the program had grown to 146 communities in Ontario and 19 in the Montreal, QC, area. In 1998, EPIC published the *Best Practices Guide for the Collection and Handling of Polyethylene Plastic Bags and Film in Municipal Curbside Recycling Programs.* Source separation is recommended, but best practices for commingled collection that includes bags are also covered.[165] By 2005, 43 percent of the Canadian population had access to plastic bag recycling through either municipal curbside programs, drop-off sites, or in-store recycling programs.[166] Nonetheless, Ontario reported a recycling rate of only 5.6 percent for plastic film in 2003.[88]

Stretch wrap is collected primarily from establishments such as warehouses and retailers, where large quantities of goods arrive in pallet loads unitized with the wrap. Sending such material to be recycled, rather than paying to dispose of it, often makes economic sense for the companies involved. This is particularly true under present economic conditions, when companies may be able to sell their stretch wrap.[167]

Processing of film plastics is more difficult than processing of containers. The lower bulk density of the film leads to difficulty in handling the material, and it is more difficult to remove contaminants. Historically, a large fraction of merchandise bags collected in the United States were shipped to the Far East, where low labor costs permitted hand sorting. Stretch wrap is less contaminated, especially if paper labels are not used, and is often handled domestically.

EPIC has devoted considerable effort to identifying processing systems capable of handling postconsumer polyethylene film, as well as to collection of the material, as mentioned above. They chose to focus on dry processing as key to keeping the cost of pelletizing the materials down. EPIC found that pneumatic separation of polyethylene film can remove 98 percent of contaminants from film which has first been chopped, shredded, or granulated. For agglomerating the shredded material, continuous-feed agglomeration was found to be generally superior to pellet mill processing and batch-style agglomeration. While the equipment was more costly than batch agglomeration, labor costs were lower, and product quality was higher. EPIC has also produced a stretch wrap recycling guide.[168]

A major market for recycled plastic film and bags is manufacture of trash bags, typically in a blend with virgin resin. Recycled plastic has also been used in manufacture of new merchandise bags, bubble wrap, housewares, and other applications. In 2004, Dixons Group, a retailer, and Nelson Packaging, the UK's largest producer of plastic carrier bags, were recognized for the 2003 introduction of the first carrier bag with 100 percent recycled content to be adopted by a major UK retailer.[169] A rapidly growing application is the use of recycled LDPE in plastic lumber as a composite with wood fibers. Plastic lumber is discussed further in Sec. 8.14.

Agricultural film poses a special problem, since it may be heavily contaminated. On the other hand, its disposal can be costly, so users have motivation to find a less-expensive solution. A number of pilot projects have looked at recycling this film. Markets are often local, since it is expensive to collect from far away.

In 1997, New Jersey began a pilot recycling program targeted at nursery and greenhouse film. In the three-month project period, it collected nearly 450,000 pounds of film, about 45 percent of the total used by growers in the state. The biggest problem encountered was the dirt in the film. The program was successful enough to expand in 1998.[170] There are now two permanent collection sites in operation year round. One accepts film only from New Jersey, at a charge of $20 per ton. The other accepts film from out-of-state customers as well, at a charge of $25 per ton.[171]

EPIC has published a best-practices guide for agricultural films. It identifies plastic stretch wrap, silage bags, cover sheets for bunker silos, and hay bale wrap as viable sources of agricultural film for recycling. Film should be clean and dry, having less than 5 percent contamination. Air drying for one to three days is recommended. Shaking can be used to remove a lot of the dirt and other foreign objects. The film should be stored away from sunlight to prevent further degradation. Products from recycled film identified in the guide are plastic lumber, plastic plywood (puckboard or baleboard), horse fencing, and farm pens for dairy cattle, hogs, and poultry.[172]

Another specialty film is marine shrink film, used to wrap boats to protect them from winter weather. WasteCap, of Massachusetts, with funding from the American Plastics Council and the Massachusetts Marine Trades Association, started a recycling program for this film in 2003. Collection points are Massachusetts marinas, where significant amounts of the shrink film can be collected. During 2003 and 2004, approximately 35 tons of the white wrap was collected each year, out of about 130 tons that were sold in the area.[173,174]

Delta Plastics of the South, in Stuttgart, AR, recycles polyethylene tubing used for crop irrigation. Farmers often discard the tubing at the end of the year, because it becomes contaminated with soil and vegetation. Delta accepts used tubing from farmers who buy new tubing from the company. They operate over 100 collection sites in a 4-state area.[175] The collected material is cut, washed, and palletized. End uses include trash bags, construction film, and agricultural film. Between 1998 and 2005, Delta recycled over 69 million pounds of tubing.[176]

8.7 RECYCLING OF POLYPROPYLENE (PP)

Sources of PP in MSW in the United States are shown in Fig. 8.28. Packaging represents 38 percent of the total, and durable goods 37 percent. There is little recycling of PP from packaging reported by EPA, which noted a rate of only 0.7 percent in 2003, all from the "other plastics packaging" category.[1] It is likely that most recycled PP comes from durable goods, where the recycling rate was 12.4 percent in 1998.[2] As mentioned, EPA no longer divides plastic from durable goods by resin type, simply lumping it all in "other plastics"

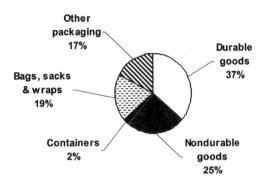

FIGURE 8.28   PP in MSW in the U.S., 2003.[1]

in the final compilation. A significant amount of recycled PP is known to come from automotive and other lead-acid battery cases. The overall recycling rate for PP in U.S. MSW in 2003 was reported as 0.3 percent, but this is likely an underestimate due to omission of battery cases.[1]

The overall recycling rate for PP in Australia was reported to be 9.9 percent in 2003, a significant increase from 7.0 percent in 2002 and 8.0 percent in 2001.[11]

Lead-acid batteries are prohibited from MSW disposal facilities in 37 U.S. states, primarily due to concern about the effects of lead. Several states impose deposits on batteries. Effective recycling programs for these batteries have existed for a number of years. The Battery Council International reports a 2003 recycling rate of more than 97 percent for lead-acid batteries. Polypropylene makes up about 7 percent of the battery, by weight, and is recovered along with the lead. The primary market for the recovered PP is new battery cases. A typical battery contains 60 to 80 percent recycled PP. The recycling rate for the cases can be assumed similar to that for batteries.[177]

Some PP is recovered from recycling of electronics and automobiles; see Secs. 8.15 and 8.16. Polypropylene hangers from department stores are also sometimes recycled.

Some PP bottles are collected in "all-bottle" collection programs; these may be recycled or may simply be discarded. Collection of other PP packaging is rare in the United States but more common in Europe and in some other parts of the world. Unfortunately, little information is available about recycling rates, products manufactured, and so forth. The American Plastics Council reported a 3.4 percent recycling rate for PP bottles in 2003, down from 3.9 percent in 2002.[43] The 2001 rate was 3.8 percent.[145]

A few innovative programs for recycling PP packaging do exist. Stonyfield Farm, a yogurt manufacturer, has entered into a partnership with Recycline to recycle the company's yogurt cups. All the handles of Recycline's products, such as toothbrushes and razors, are made from 100 percent recycled plastic, and at least 65 percent of that is recycled Stonyfield Farm yogurt cups. Furthermore, Recycline has a postage-paid recycling system so the end-of-life products can be sent back and again recycled, this time into plastic lumber.[178]

8.8 RECYCLING OF POLYSTYRENE (PS)

Nondurable goods represent by far the largest category of PS in U.S. MSW, 56 percent, with plastic plates and cups alone representing 31 percent (Fig. 8.29). Durable goods account for about 32 percent, with packaging amounting to about 10 percent.[1] A substantial

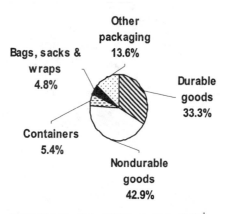

FIGURE 8.29   PS in MSW in the U.S., 2003.[1]

amount of PS is used in the building and construction industry, mostly for insulation materials, but these wastes are not considered part of the U.S. MSW stream. EPA reports no significant recycling of PS in 2003. The American Plastics Council reported a 1.1 percent recycling rate for PS bottles in 2001 and did not collect data on PS in 2003 or 2004.[43,145]

There is very little recycling of food-service PS in the United States. In 1989, the National Polystyrene Recycling Company (NPRC) was formed by several PS producers, with a goal of achieving a 25 percent recycling rate for food-service PS by 1995. It focused on institutional generators of such wastes, primarily schools and other cafeterias. However, the operation was plagued by high levels of contamination with food wastes and was unable to operate profitably. Recycling rates and amounts for PS food-service items, including packaging, declined after an initial period of growth. In 1999, NPRC and its remaining two recycling facilities were sold to Elm Packaging Co. and its name changed to Polystyrene Recycling Co. of America.[179] By late 2000, recycling of food-service PS had all but stopped.[180] However, Evergreen Partnering Group, Inc., in March 2005, announced plans to expand its small two-year-old pilot recycling program for polystyrene food-service items. The company recycles used PS trays, cups, and bowls in the Boston and Providence, RI, public schools and said it hoped to expand to Atlanta, Chicago, and Los Angeles.[181]

Nearly all recycling of PS from packaging in the United States now comes from recycling of foam cushioning materials. Such recycling has been much more successful than recycling of food-service PS. According to the Alliance of Foam Packaging Recyclers, recycling rates for EPS foam have ranged between 9 and 13 percent since 1992 (Fig. 8.30). In 2004, 25.0 million pounds of postconsumer EPS were recycled, for a rate of 12.0 percent, down from 26.2 million pounds and a 13.0 percent rate in 2003.[182] (It should be noted that these materials should be considered recycling by EPA, since all packaging is considered part of MSW; nevertheless, EPA does not report these amounts.) In addition to the postconsumer material, 32.9 million pounds of postindustrial EPS was recycled, mostly scrap from manufacturing facilities.[182] The Alliance provides a toll-free number for consumers to find whether EPS recycling is available in their area. It also sponsors a take-back program for small quantities of EPS, where consumers can mail the materials via the postal service or UPS to its national headquarters in Crofton, MD.[183]

In addition to recycling, there is considerable reuse of expanded polystyrene loose fill. The Plastic Loose Fill Council operates a toll-free "Peanut Hotline" to provide information to consumers about where to take EPS loosefill for reuse. This information is also

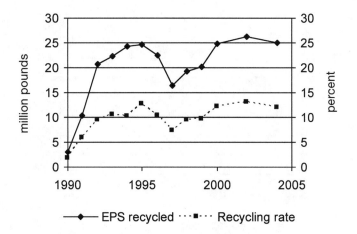

FIGURE 8.30 EPS recycling in the U.S.[182]

available via the organization's web page. PLFC reports that over 30 percent of all EPS loose fill is reused, and it averages 25 percent recycled content.[184]

One of the leaders in EPS recycling in the United States is FR International, which began PS recycling in 1990. The company recycled 10.9 million pounds of expanded polystyrene in 2000,[185] but only 8.5 million pounds in 2002, the latest figures available. The company has five manufacturing operations in the United States, and also has several facilities in Europe. It also manufactures loosefill EPS cushioning from recycled EPS.[186]

One of the problems in recycling EPS is the very low bulk density of the material, which makes shipping it over long distances uneconomical. International Foam Solutions, Inc. (IFS), of Boca Raton, FL, developed a process that dissolved EPS in a citrus-based solvent, producing a gel and eliminating 90 percent of the volume. The "Polygel" was stored in drums and shipped to IFS for processing. IFS further diluted the gel, filtered out contaminants, and produced new PS products. Contaminant levels were reportedly reduced to less than 1 ppm. The intent was to use the process for food-contaminated EPS as well as other EPS materials.[187] However, no current information about the company was found, so it may no longer be in operation.

Kodak operates a recycling program for PS in disposable camera bodies and also recycles film containers. Recovered camera bodies are ground, mixed with virgin resin, and used in the production of new disposable cameras. The PS internal frame and chassis of the cameras may be recovered intact and reused in new cameras. The cameras, from any manufacturer, are collected from photofinishers, who may be reimbursed for the cameras they return. The company's program is active in over 20 countries.[188]

In 1992, Japan, the United States, Germany, and Austria signed an international agreement to promote EPS recycling. By 2002, a total of 31 countries had signed. EPS Recycling International reported a global EPS packaging recycling rate of 45 percent in 2001 (no updates are currently available). The organization also provides information about member organizations in various countries.[189]

In Japan, one of the goals of the Japan Expanded Polystyrene Recycling Association (JEPSRA) is to increase recycling of expanded PS. Volume of collected material is reduced by pulverization, melting, or the use of solvents. With a network of more than 1000 recycling sites, Japan achieved a 39 percent recycling rate for EPS in 2002 (Fig. 8.31). JEPSRA's goal is to achieve a 40 percent recycling rate by 2005.[190]

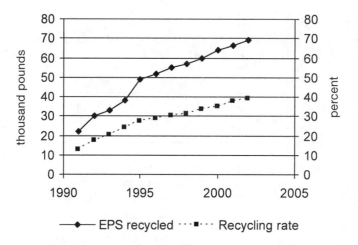

FIGURE 8.31   EPS recycling in Japan.[190]

Korea reports a recycling rate of 48.8 percent for EPS in 1999, excluding building insulation and food containers, for a total of 24,371 tonnes. The growth in EPS recycling is shown in Fig. 8.32.[191]

The Canadian Polystyrene Recycling Association (CPRA), with a plant in Mississauga, ON, accepts polystyrene from all over Canada. Over 95 percent of its revenue comes from the sale of recycled resins, mostly black high-impact polystyrene (HIPS). The remainder comes from member companies, representing manufacturers, distributors, and end users of polystyrene products. The plant, which has a capacity of 5000 tonnes of PS per year, accepts both food-service PS and cushioning materials. While it uses mostly postconsumer materials as feedstock, it also accepts obsolete PS materials from manufacturers, such as scrap from sign manufacturing.[192]

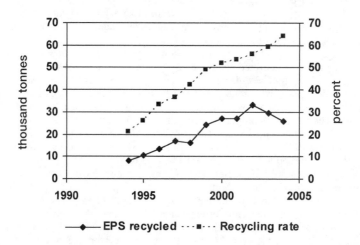

FIGURE 8.32   EPS recycling in Korea.[191]

In Australia, the overall recycling rate for PS (not including EPS) was reported to be 5.1 percent and for EPS was 7.0 percent in 2003. Recycling rates in 2002 and 2001 were 6.7 percent and 7.7 percent for PS, and 2.6 percent and 3.6 percent for EPS, respectively. As can be seen, the PS recycling rate continued to decline while the EPS rate rose significantly.[11]

Products manufactured from recycled polystyrene include cushioning materials, horticultural trays, video and audio cassette housings, rulers and other desktop accessories, clothes hangers, and other items.

8.9 RECYCLING OF POLYVINYL CHLORIDE (PVC)

Polyvinyl chloride in U.S. MSW comes mostly from nondurable goods, followed closely by packaging and durable goods (Fig. 8.33).[1] In western Europe, about 26 percent of PVC is used in pipes and fittings, 17 percent in film and sheet, 11 percent in window profiles, 9 percent in cables and wires, 5 percent in bottles, and 18 percent in floorings and coatings.[193]

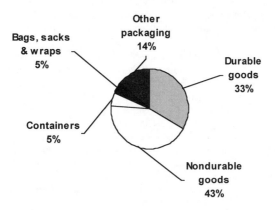

FIGURE 8.33   PVC in MSW in the U.S., 2003.[1]

The U.S. EPA reports no significant recycling of PVC from municipal solid waste in the United States.[1] The American Plastics Council reported a PVC bottle recycling rate in 2003 of only 0.2 percent, down from 0.4 percent in 2002.[43] There is limited recycling of PVC from other waste streams, especially construction and demolition material, which is not classified as MSW. Since non-MSW waste streams account for considerably more PVC and tend to be more uniform than MSW PVC waste streams, most efforts have focused on these materials.

The overall recycling rate for PVC was reported to be 4.0 percent in Australia in 2003, unchanged from 2002. The rate in 2001 was 4.9 percent.[11] One PVC recycler in Australia is Armstrong Australia, which uses both scrap from installation of flooring and recycled PVC bottles in manufacture of its commercial vinyl flooring. Amounts are relatively small—about 80 tonnes of PVC bottles (and 10 tonnes of HDPE shopping bags) in 2004.[194]

In the United States, the Vinyl Institute publishes a directory of North American companies involved in vinyl recycling, which is available in searchable on-line form. It currently lists 94 companies that accept postconsumer materials. The Institute also offers a database of U.S. and Canadian companies manufacturing products from recycled vinyl that currently lists 85 companies. A study commissioned by the Institute reported that about 18 million pounds of postconsumer PVC were recycled in 1997 from sources such as carpet backing, medical products, windows and siding, and packaging. Recycling of postindustrial materials was a much larger amount, representing about 78 percent of all recycled vinyl. Since nearly 70 percent of PVC production goes into products expected to last 10 years or more, recycling of postconsumer materials is expected to be more important in the future. The report found that demand for rigid postconsumer vinyl was smaller than the supply, despite the low recovery rates. The situation was somewhat better for flexible materials, but these were valued mostly for their plasticizer content, not the PVC itself. No updated information is currently available from the institute.[195,196]

PVC has been criticized by Greenpeace, and also by the Association of Postconsumer Plastic Recyclers, which in 1998 labeled PVC a recycling contaminant, pointing to the failure of a year-long effort to find markets for recycled PVC and the fact that many of their members had to landfill recovered PVC bottles because they could find no markets for them.[197] After that report, the Vinyl Institute funded a pilot project aimed to jump-start the PVC recycling market,[198] but, while recycling of PVC bottles increased slightly in 2002 from 2001, it decreased again in 2003. Overall use of PVC bottles continues to decline as well. In 2003, 208 million pounds of PVC were used in bottles, and only 0.2 million pounds were recycled.[43,145]

A number of projects have focused on recycling of vinyl siding, usually on scrap originating during building construction or remodeling, and often with financial support from the Vinyl Institute. As a result of one such pilot project in Grand Rapids, MI, recycling of vinyl siding waste is included in the EPA-funded publication, *Residential Construction Waste Management: A Builders' Field Guide.*[199] Among the most high-profile pilot projects are those involving Habitat for Humanity, which builds housing for low-income families. The Vinyl Institute and other PVC-related industry organizations have donated money and materials to some of these projects, in addition to supporting recycling efforts for the PVC scrap generated during construction. In 1997, Polymer Reclaim and Exchange, in Burlington, NC, was recycling about 300,000 pounds per month of vinyl siding scrap, through drop-off sites located at landfills and near manufacturers of mobile and manufactured homes.[200] However, the company soon thereafter went out of business. With the aid of a grant from the Vinyl Institute, it was purchased by a former employee and reopened as Reily Recovery Systems (RRS) in Sanford, NC. Within two years, it was recycling more than 2 million pounds per year, used to produce PVC pipe, mobile home skirting, and other products.[201]

In 2005, EPIC released a "best practices" guide for vinyl siding recovery from residential construction and demolition.[202]

Other recycling efforts that targeted building-related PVC wastes have focused on window profiles, carpet backing, and pipe. Often, these, along with vinyl siding scrap, are preconsumer rather than postconsumer wastes. There is also considerable interest in recycling of both preconsumer and postconsumer automotive scrap, especially in Europe, with legislative requirements for automobile recycling.

In fact, PVC recycling is much healthier in Europe than in North America. In Europe, Vinyl 2020 is a ten-year plan to put into action the Voluntary Commitment of the European PVC industry to achieve sustainability throughout the life cycle of PVC. Its two main foci are reduction in use of lead stabilizers and increase in recycling. The organization's 2005 progress report summarizes accomplishments during 2004. The European PVC Window

Profile and Related Building Products Association (EPPA) recycled 2865 tonnes of post-consumer waste, with preconsumer waste bringing the total to 5429 tonnes. The European Plastic Pipes and Fittings Association (TEPPFA) recycled 5640 tonnes of pipes and fittings. The European Single Play Waterproofing Association (ESWA) recycled 568 tonnes of PVC roofing membranes, more than 25 percent of the total available. The European PVC Floor Manufacturers (EPFLOOR) recycled 972 tonnes of waste. The EuPC PVC Coated Fabrics Sector Group (EPCOAT) started a trial collection project for coated fabrics. RGS of Denmark neared completion of a full-scale industrial demonstration project for recycling PVC waste into oil, salt, and minerals. It will have a capacity of 50,000 tonnes per year and is scheduled for completion in summer 2005.[203]

A leading PVC recycler is Solvay, which has a full-scale Vinyloop demonstration plant in Ferrara, Italy, based on solvent technology. The plant recycles vinyl cable waste and has also tested blisters, automotive wastes, roofing, and flooring, among other products. A modification in 2005 improved productivity and reduced steam consumption. In January 2005, a licensing agreement was signed with Kobelco Eco Solutions to build a Vinyloop facility near Tokyo to produce 18,000 tonnes per year of regenerated PVC from 20,000 tonnes of waste cable, greenhouse sheets, and wallpaper. Vinyloop Ferrara now is producing a recycled PVC resin, Vinyloop FC001, which contains PVC resin, plasticizer, stabilizer and filler. The company also has a subsidiary, Texyloop, which recycles old tarpaulins made of a composite of PVC and polyester.[204]

Hydro Polymers is installing a system to produce the UK's first recycled PVC resin, known as EcoVin. It will be made from postindustrial scrap and is intended for use in cable conduits, fencing, and window profiles, where it will have an outside layer of virgin PVC.[193]

In Japan, Sumitomo Corporation and Refineverse are building a PVC recycling company that will recycle a range of PVC materials, from tile office carpet, flooring, and wallpaper to construction and automotive materials.[205]

8.10 RECYCLING OF NYLON AND CARPET

Most nylon recycling projects target carpet as the largest volume of nylon in MSW. Occasional programs are directed toward other waste streams. The Monofilament Recycling Project, initiated in Florida in 2001, collected monofilament nylon fishing line in an effort to prevent some of the wildlife harm associated with discarded fishing line.[206]

About 4.7 billion pounds of carpet were discarded in the United States in 2002, about 1 percent of all MSW by weight and about 2 percent by volume, according to the EPA. Only 3.8 percent of it was recycled. As a consequence of pressure to do something about this large volume of waste, after two years of negotiation, carpet and fiber manufacturers, the Carpet and Rug Institute, state governments, nongovernmental organizations, and the EPA developed the National Carpet Recycling Agreement, signed in January 2002. This agreement established a ten-year plan to increase the amount of recycling and reuse of postconsumer carpet and decrease the amount going to landfill. The goals for 2012 are 3 to 5 percent reuse or at least 200 to 340 million pounds; 20 to 25 percent recycling or at least 1.4 to 1.7 billion pounds; 3 percent or 200 million pounds use in cement kilns as an alternative fuel source and an additive in cement production; and 1 percent or 67 million pounds energy recovery per year. The Carpet America Recovery Effort (CARE) was created as a third-party organization to achieve these goals. Total landfill diversion is targeted at 40 percent by 2012. Negotiations for goals for the next ten-year period are to start in 2010, with the goal of eventually eliminating land disposal and incineration of postconsumer carpet.[207,208]

CARE's 2004 Annual Report shows that progress is behind schedule, but continued increases in oil prices are expected to improve carpet recycling economics. The total amount of carpet recycled in 2004 was 98.4 million pounds, a 13.6 percent increase over 2003. This occurred despite the closure of the Polyamid 200 recycling plant in Germany in mid 2003. The recycling rates reported are lower than the EPA estimate given above: 0.99 percent for 2002, 1.79 percent for 2003, and 2.17 percent for 2004. CARE reports that recycling may be underestimated by as much as 60 percent, as response to their survey was only 7 percent. Despite a slow start, the organization still feels it has a very good chance of meeting or exceeding its 2012 goal.[209]

Market development is seen as a continuing challenge. CARE is currently sponsoring research on development of large-volume products capable of consuming significant quantities of recycled carpet. One effort is directed at railroad ties; tests are currently going on both in laboratories and in the field. Conigliaro Industries, of Massachusetts, is developing PlasCrete Wall Blocks, each block weighing 1850 lb, which can contain large quantities of mixed postconsumer carpet. Vortex Composites, of Chicago, has developed a retainer wall system designed to look like brick, but it is 40 percent lighter than concrete. Several efforts directed at recovery of nylon in a form usable for carpet face fiber are underway. Auburn University has developed a patented process that involves selective dissolution of nylon in formic acid, followed by precipitation in supercritical carbon dioxide. Both the formic acid and CO_2 are recovered and reused.[209]

About 25 percent of carpet is nylon, and many carpet recycling efforts have focused on it as the most recoverable and highest-value carpet material. BASF Corp. claims to have initiated the first comprehensive recycling program for used commercial carpet in the United States and Canada. Its 6ix Again program, which began in 1994, provides recycling of BASF nylon 6 carpet. At first, it provided recycling only for carpet manufactured after Feb. 1, 1994, and containing 100 percent Zeftron nylon 6 yarn. Carpets are sent to collection centers located throughout the United States and Canada. BASF recycles the carpet using chemical depolymerization into caprolactam and repolymerization, and it uses the material in new carpet, resulting in closed-loop recycling. In 1997, BASF expanded carpet recycling to customers replacing ineligible carpets with qualifying carpet, accepting all carpet yarn types and backing systems, and recycling them if a viable method is available—or incinerating the old carpet if recycling is not available. The program was later expanded again to include products containing Ultramid nylon 6 plastic with recycled content, and, in 2002, upholstery fabric made with Zeftron nylon upholstery yarn was added.[210,211]

DSM and Honeywell (formerly AlliedSignal) formed a joint venture, Evergreen Nylon Recycling (ENR), to recycle nylon 6 carpet. Like in the BASF process, the nylon 6 was depolymerized to caprolactam. The $85 million ENR plant in Augusta, GA, began operation in November 1999, with a capacity of more than 200 million pounds of carpet waste per year, producing over 100 million pounds of caprolactam. About 140 million pounds of the 200 million pounds processed each year was postconsumer carpet. The recovered caprolactam was used partly for production of new nylon 6 for carpet and other products and partly as a feedstock for production of other engineering plastics. Honeywell calculated that the process saved about 700,000 barrels of oil each year. The program accepted nylon 6 carpet made by any manufacturer.[212,213] In 2000, DSM and Honeywell announced plans to set up carpet recycling ventures in Europe and Asia but said that, first, an economically feasible way to collect the old carpet needed to be established.[214] The Evergreen process involved feeding entire carpets into the reactor. Polypropylene and latex backing and calcium carbonate filler from the carpets exited as a sticky brown substance that was sent to a cement kiln for incineration. The calcium carbonate in the mix became part of the cement.[210] However, in 2001, Honeywell and Evergreen suspended production of their Infinity recycled nylon resin due to higher than expected production and development costs, along with business and economic conditions. As of this writing, the company still maintains a web site, but there have evidently been no updates since 2001.[215]

C&A developed a closed-loop process for recycling postconsumer carpet and manufacturing waste into recycled content backing for new carpet. Recycled content ranges from 31 percent to 52 percent, and the material is recyclable back into carpet at the end of its life.[209]

Invista, based in Calhoun, GA, has been recycling carpet since 1991. It will collect and reclaim any used carpet, as well as installation scraps, from any manufacturer, and guarantees that no material will be sent to landfill. Recycled materials are used in manufacture of new carpet, carpet cushions, filtration devices, automotive parts, packaging materials, and furniture. Users must pay for the service.[209,216]

Wellman, Inc. of Johnsonville, SC, recycles nylon 6 and nylon 6,6 carpet, using a proprietary process, and formulates the product into moldable engineered nylon resins, with both filled and unfilled grades. Wellman received the CARE Recycler of the Year Award in 2005 for its developments in recycling nylon into automotive parts. Their EcoLon resin contains 25 percent postconsumer carpet.[68,217]

Shaw Industries has an environmental guarantee program, promising to collect, transport, and recycle any carpet tile made with their EcoWorx backing, at no cost to the customer. The company will recycle the material into new carpet. The company was awarded the Presidential Green Chemistry Challenge Award for development of that carpet tile as well as the Most Innovative Product Award at NeoCon. Shaw also manufactures Eco Solution Q nylon face fiber for carpets, which contains 25 percent recycled content.[209,218]

A number of other companies either are or have been engaged in nylon (or other) carpet recycling programs.

A limited amount of nylon recycling from appliances and automotive parts is taking place as well. For example, DuPont is working on recycling radiator tanks and air intake manifolds into nylon materials for new tanks and manifolds.[219,220]

Australia reported a 3.0 percent recycling rate for nylon of all types in 2003.[11]

8.11 RECYCLING OF POLYURETHANE

Recycling of polyurethane differs from recycling of thermoplastics, since the thermoset polyurethane cannot be melted and reformed. Most scrap from production and fabrication of soft polyurethane foam is shredded into small pieces and then rebounded into molded large blocks that are next laminated with film and used as carpet padding. Some is also used directly as filler for pillows or furniture. Used carpet padding can also be recycled in this way, making new carpet padding. Some mechanical recycling of polyurethane, especially harder foams, is being done using ground-up material as a filler in new materials. Other efforts have targeted chemical treatment to break down the material into polyol, which can then be used in production of new plastics.[221]

Primary potential sources of polyurethane foam for recycling include automobiles, upholstered furniture, mattresses, and refrigerators and freezers. In 2007, refrigerator disposal alone is expected to yield 126 million pounds of polyurethane foam.[222]

In the United States, polyurethane recycling, other than for production of carpet backing, has been quite limited. However, in Europe, the directive on waste from electric and electronic equipment (WEEE), by requiring the removal of ozone-depleting blowing agents that were once used in polyurethane foam production, has spurred development of treatment facilities for the material and thus is producing an increasing volume of rigid PU foam that is available for recycling.[223]

BASF teamed with Philip Services to open the first facility in North America for recycling rigid and semirigid polyurethane in 1997. This plant used a BASF-patented glycolysis process and was designed first for automotive waste, with the intention to expand to other waste streams.[224] However, the facility appears to be no longer in operation.

Mobius Technologies developed polyurethane foam recycling technology for PU scrap from slab and molded foam. The process pulverizes the scrap foam into powder of less than 50 μm and suspends it in polyol for use in production of new foam, replacing up to 10 to 12 percent of the new polyol and thus lowering costs. The scrap can be from either flexible or molded sources and can be production scrap or scrap from postconsumer applications such as old automobile seats. Mobius partnered with Dow Chemical to install a recycling system at Dow's Technical Development Center in Meyrin, Switzerland, in 2002.[225] Mobius also now has a strategic relationship with Energy and Environmental Ventures in China.[226]

ISOLA NV claims to be the largest rigid polyurethane recycling company in Europe, recycling 5000 tonnes of rigid polyurethane waste each year, including production waste from polyurethane panel and block industries and polyurethane recovered from end-of-life refrigerators in Germany, the Netherlands, Switzerland, Austria, and Scandinavia.[227] In 2004, it announced an intent to produce building blocks with up to 95 percent reclaimed polyurethane.[223]

The St. Vincent de Paul Society of Lane County, in Eugene, OR, in collaboration with the International Sleep Products Association and with additional funding from the city of San Francisco, opened a mattress components recycling facility in Alameda County, CA, designed to recycle about 500,000 pounds of polyurethane foam a year. The facility can handle 250 to 500 mattresses a day. It is designed to recover steel, wood, and cotton from the mattresses as well as polyurethane. The system first slices the incoming mattresses so that layers of polyurethane foam and cotton fiber can be removed from the steel framework. In the future, whole mattresses will be shredded.[228,229]

Conigliaro Industries, of Framingham, MA, in August 2001, received a grant from the Massachusetts Department of Environmental Protection for development and start-up of a mattress shredding and recycling facility. The company claims its plant, which opened in December 2002, was the first commercial plant for shredding and recycling mattresses. In addition to polyurethane foam, wood, cotton, and steel are recycled. The plant has a capacity of 1 mattress per minute, or 140,000 per year. Low-quality mattresses such as those from hospitals and schools are simply put through a shredder, and about 60 percent of the components are recycled. For brand-name mattresses, the mattress is first slit and then shredded, resulting in recovery of about 90 percent of the components.[230]

Troy Polymers is investigating glycolysis of mixed polyurethanes, reacting the polymer with diols at elevated temperature to yield materials that can be used to form new polyurethane or polyisocyanurate foams.[231]

RAMPF Ecosystems, in cooperation with FH Aalen and support from the EU LIFE-Programme, has developed three types of chemical recycling processes for PU foams: partial glycolysis, polyolysis, and acidolysis. The process targets both pre- and postconsumer polyurethane. The material is first sliced into pieces about 5 cm in size and then introduced to a depressurized reactor for chemical depolymerization. The result is a mixture of polyols and low-molecular-weight urethane. Filtration removes incidental foreign matter contaminants. The resultant polyol, Recypol, can be used alone or mixed with new polyol for production of foams. Yields are expected to be 97 to 99 percent when the process is fully developed. It is suitable for nearly all forms of polyurethane. Problems encountered are related mostly to the low quality of postconsumer residual materials.[232]

One focus of PU foam recycling efforts is motor vehicles. Such efforts are generally more prevalent in Europe, due to producer responsibility and recycling mandates.

In 1999, NV Salyp of Ieper, Belgium, began building the Salyp ELV Center in Belgium, which uses technology licensed from the Argonne National Laboratory, of Illinois, to recover polyurethane foam and other plastics from auto shredder residue (ASR). It also uses technology licensed from a German firm, KUTEC, for separating different types of thermoplastics from the Argonne technology reject stream. About 5 percent of automotive shredder residue, the material remaining after metals are recovered, is polyurethane. The Argonne technology separates the fluff into three streams: fines, foam, and a thermoplastic-rich stream. The foam stream is cleaned and sold for markets such as rebond foam in carpet underlay and for padding in automobiles. The thermoplastic stream is sorted further for recovery of a variety of pure resins. Even the fines stream may be recovered for applications such as cement, substituting for iron or mill scales that are now used to provide the needed iron. Salyp reports that the process can also recover PUR foam from shredded mattresses.[77,82,83]

In the United States, the Vehicle Recycling Partnership is supporting research efforts on ASR recycling. For example, Changing World Technologies has a pilot plant for recycling auto shredder residue, including polyurethane. Thermal depolymerization is used to convert hydrocarbons and other organics into fuels and specialty chemicals.[233]

The Alliance for the Polyurethanes Industry operates a recycled polyurethane markets database that allows searches for either buyers or sellers of polyurethane.[234]

Australia reported a relatively high 16.2 percent rate for polyurethane recycling in 2003.[11] In 2004, Coim Brasil, a subsidiary of the Italian company Chimica Organica Industrialle Milanese, began operation of the first polyurethane recycling plant in Brazil. It targets shoe soles and has the capacity to recover 200 tonnes per month.[235]

8.12 RECYCLING OF POLYCARBONATE (PC)

Some recycling of polycarbonate from products such as automobile bumpers, compact discs, computer housings, and telephones has been carried out.

Bayer built Europe's first polycarbonate CD recycling facility in Leverkusen, Germany, in 1995. The PC was separated from aluminum coatings, protective layers, and imprinting. The product was blended with virgin PC for use in a variety of products.[236] In Europe, Bayer recycles returnable polycarbonate bottles and jars used by dairies as well as compact discs.[237,238]

As the number of compact discs, and now also DVDs, has grown, so has interest in recycling this high-value engineering polymer. For example, Taiyo Yuden, in Japan, is engaging in a major effort to recycle used, defective, and obsolete optical discs.[239] In the UK, Polymer Reprocessors recycles CDs, producing injection-grade polycarbonate granulate.[240]

8.13 RECYCLING OF ACRYLONITRILE/ BUTADIENE/STYRENE COPOLYMERS (ABS) AND OTHER PLASTICS

Most ABS recycling is focused on appliances. Refrigerators alone are expected to yield 23 million pounds of ABS in 2007.[222] Telephones, especially cellular telephones, are also a significant source of ABS in the waste stream.

In Australia, the overall recycling rate for ABS/SAN was reported to be 8.5 percent in 2003, higher than the 7.2 percent rate in 2002 but lower than the rate of 9.8 percent in 2001.[11]

A few other types of plastics are recycled in small amounts. For example, the Thai Poly Acrylic Company Ltd. has depolymerization facilities for waste acrylic sheet, with the product used in production of new sheet.[241]

Many of these plastics are found in appliances, electronics, and automobiles. More information on electronics recycling can be found in Sec. 8.15, and automotive recycling in Sec. 8.16.

One of the most unusual plastics-related recycling programs is the KnoWaste LLC program for recycling disposable diapers and adult incontinence products. In addition to paper, the superabsorbent polymers used in the products are recovered, as are film materials. KnoWaste began with a pilot-scale facility in Mississauga, ON. It operates a facility in Arnhem, Netherlands, with a capacity of 70,000 tons per year. A successful nine-month demonstration diaper recycling program was conducted in Santa Clarita, CA, in 2002 and 2003. Most recently, it announced plans to provide diaper recycling in the Toronto, ON, metropolitan area.[242]

8.14 COMMINGLED PLASTICS AND PLASTIC LUMBER

When mixed plastics streams are collected, separation by resin type may not be feasible or economical. Similarly, heavily contaminated plastics streams may not be suitable for high-value end markets. Use of plastics in a commingled form, typically in applications where plastics replace wood or concrete, offers relatively low-value end uses that are much more tolerant of contamination than traditional plastic resin markets. In addition, the plastic substitutes for lumber and concrete tend to be considerably more durable than the materials they replace. This is especially true for plastic lumber, where the plastics also have the benefit of not employing some of the toxic chemical systems used to protect wood from degradation in moist outdoor environments. Many companies have been into and out of the commingled plastics business. The products typically sell for more than the wood or concrete items they replace, and convincing people to buy the products, even when they will last considerably longer than the cheaper alternatives, has been difficult.

A relatively new company in the business of recycling commingled plastics is ReSyk, of Utah. The company's patented process is claimed to be suitable for all kinds of incompatible plastics as well as contaminants such as motor oil, labels, aluminum, and copper. Products are compression molded. The company claims that, rather than just melting one of the plastics and having the others act as fillers, as was the case for many of the earlier commingled plastics processes, its process actually bonds the different types of plastics together. The company was formed in 1999 as a spin-off from Rotational Molding, of Utah, which had acquired the technology in the mid 1990s. At first, ReSyk made products including trash-can wheels, water-meter covers, and bumpers for loading docks.[243] Now, it evidently licenses its patented technology, manufactures prototypes, and serves as a consultant rather than being involved in product manufacture.[244]

As mentioned, one application for recycled commingled plastics is in plastic lumber. The plastic lumber business has grown rapidly in the last five years. Not all plastic lumber is made using recycled plastics, but much of it is. Some companies using recycled plastics for plastic lumber use mixed plastics, and others use relatively pure streams of material. Some use primarily postconsumer materials, while others use mostly industrial scrap.

Many manufacturers make composite plastic lumber out of wood fiber, with HDPE, LDPE, or PVC serving as the plastic component.

From 1992 to 1997, the U.S. Army Construction Engineering Research Laboratory (CERL) and Rutgers University, along with several recycled plastic lumber (RPL) manufacturers, carried out a project to optimize the production process for RPL and develop demonstration projects. In 1998 to 2000, the Plastic Lumber Trade Association and Battelle, with four RPL manufacturers, several government agencies, and others, completed a three-year cooperative effort to develop the use of RPL in structural applications in decking, marine waterfront, and material handling.[245] As a result of these and other projects, the usefulness of plastic lumber in several types of applications is now accepted.

In 1999, the Plastic Lumber Trade Association reported that RPL was growing at a 30 percent annual rate.[245] Growth of plastic lumber was been fueled by growing knowledge about performance properties of the material and how they relate to composition, as well as by the development of performance standards such as the standard test methods developed by ASTM.[246] By 2005, nine ASTM test methods for RPL had been established, along with two standards for plastic/fiber composites.[247] However, growth of plastic lumber has slowed, and the Association has not placed an annual report on their web site since 2002.

Residential decking is a large component of plastic lumber sales. Major manufacturers using recycled plastics include Trex and Advanced Environmental Recycling Technologies (AERT) with plastic/wood fiber composites, and U.S. Plastic Lumber Corp. with both structurally foamed HDPE and plastic/wood composites.[248]

Recycled plastic railroad ties have very large-scale potential. A number of test projects with a few ties each were carried out between 1996 and 1997. In 1998, the Chicago Transit Authority became the first commercial purchaser, buying 250 ties for a test on its elevated train line.[250] U.S. Plastic Lumber has been one of the major producers of these materials and claims they have twice the life span of wood ties. Polywood produces RPL for railroad ties as well as for bridges and boardwalks from a blend of polyethylene and polystyrene.[251] TieTek, Inc., a subsidiary of North American Technologies Group, Inc., produces railroad ties from a mixture of recycled plastics and recycled rubber, with a total recycled content of 75 percent. The company's plant facility in Houston, TX, began production in July 2000. In 2001, TieTek announced plans to build a production facility in New Zealand. In 2005, TieTek reported that more than 140,000 of its crossties have been installed in 18 U.S. states and 8 foreign countries.[252]

Plastic lumber has been used in constructing bridges, walkways, and piers. One of the first major uses was a bridge at Fort Leonard Wood in St. Robert, MO, that used 13,000 pounds of commingled recycled plastics from a number of different suppliers. The bridge is primarily for pedestrian use but can support light trucks. It is expected to last 50 years without maintenance—significantly longer than the 15-year life span of a treated wood equivalent.[253] Manufacturers include Polywood as well as U.S. Plastic Lumber, which manufactures fiberglass-reinforced RPL as well as foamed HDPE lumber.[254]

As mentioned, several companies have combined recycled plastic with wood fiber, sawdust, or recovered paper fiber. For example, Advanced Environmental Recycling Technologies Inc. (AERT) of Springdale, AR, has been producing window frames from recycled LDPE film (mostly pallet wrap) and wood fiber for a number of years.[255] Trex Co., of Winchester, VA, which also has production facilities in Nevada, manufactures recycled plastic/wood composites, primarily for decking, from LDPE and HDPE originating in pallet wrap and grocery bags, and waste wood fiber. Trex was selected by *Industry Week* as one of the 25 most successful small manufacturers in the United States.[68,256] A number of other companies manufacture plastic lumber as well.

In 2003, EPIC released a report on the potential for plastic lumber in marine applications in Canada. The higher cost of these materials compared to conventional ones is a major drawback at present. However, the report concludes that use will likely increase significantly in the future as price goes down.[257] A New Brunswick boat builder has made a barge that uses a considerable portion of plastic lumber, made from recycled bottles, shrink-wrap, and agricultural film. Recycled plastic 2 × 4s were used for the frame and 2 × 6-3/4-in tongue-and-groove planks cover wooden bulkheads. The hull is filled with foam blocks, making it unsinkable. In this application, using recycled plastic saved money; the barge cost about 20 percent less than a fiberglass-wood version.[258]

Another EPIC report looked at the potential for use of plastics in railway ties. Plastic railroad ties were reported to be superior in weather and abrasion resistance.[259]

8.15 RECYCLING PLASTICS FROM ELECTRONICS

Consumer electronics are getting increasing attention as a waste problem, as the boom in personal computers and their short lives before obsolescence result in a burgeoning number entering the waste stream. The hazardous heavy metals, such as lead and cadmium, present in some computer components make their disposal problematic, so there is a growing movement toward instituting recycling programs for these materials in particular.

In Europe, manufacturers will be held responsible for the ultimate recycling of consumer electronics, so they have an added financial incentive to improve the products' recyclability and to develop recycling systems. In the United States, several computer manufacturers have instituted voluntary take-back programs for computers. However, most charge customers for taking back the items, which significantly limits recovery. For example, in the United States, buyers of a new IBM PC can purchase the recycle of any manufacturer's computer system for a moderate fee. Customers send the systems via UPS to Envirocycle, a designated recycling center, where components will be either refurbished to benefit charities or recycled. A similar service is available in Canada. IBM also operates product take-back operations in Europe, Japan, and Taiwan.[260] For larger business customers, IBM operates an Asset Recovery Solutions program throughout the world that will pay participants for their outdated computer hardware or sell it for them.[261]

In 2001, a group of manufacturers, recyclers, and government representatives in the United States formed the National Electronics Product Stewardship Initiative (NEPSI), with the major goal of creating a framework for product stewardship for electronics for the whole United States. At this time, product stewardship initiatives have already been initiated in the northeast and the western regions of the country.[262] Efforts to reach agreement on the national initiative continued until 2005, when NEPSI was finally abandoned.[263]

Product take-back requirements for electronics continue to spring up around the world. In Ontario, the government added electronics and electrical equipment of all kinds to the list of items that are to be diverted from disposal. The plan is to require a company called Waste Diversion Ontario to develop a program to handle the diversion of waste electronics from the municipal waste stream.[264]

Because computers and electronics contain 15 to 20 different types of plastics, which in turn containing a variety of plasticizers, colorants, flame retardants, and fillers, recovery of plastics from these materials is complex.[265] Much of the research has been directed toward development of some of the microsorting techniques discussed in Sec. 8.3.7. Other efforts have been devoted toward design of computers and electronics to facilitate recycling, incorporating guidelines such as minimizing the use of different types of plastic resins, providing for identification of the types of resins used in computer parts, and designing items to be easily disassembled. Electronics manufacturers have also increased the use of recycled materials in their products. The Electronic Industries Alliance operates a Consumer Education Initiative, with a web site that directs consumers to electronics recycling and reuse opportunities.[266]

In 1992, IBM became the first major computer maker to code plastic parts for ease in identification for recycling. In 1999, IBM reported that 675,000 pounds of PVC, ABS, and PC/ABS resins recovered from old computers were reused in its products in 1998.[267] Nearly all the internal plastic parts in IBM PCs were reported to have recycled content at about this time.[68] The company's 2004 server brochure says that IBM uses recycled content resin into systems "where technically possible" and that several internal resin parts such as stiffeners, clips, and fillers use resin with pre- or postconsumer recycled content.[268]

Sony Electronics Inc., in cooperation with the State of Minnesota and Waste Management, Inc., initiated a five-year program in 2000 to take back all Sony electronic products in Minnesota at no cost to consumers.[269] It operates through drop-off sites, which also accept other brands of electronics for a small fee. Over 8000 pounds of electronics were collected through this program in 2002.[270]

Butler-MacDonald Inc., of Indianapolis, has recovered plastics from computer electronics for many years, initially concentrating on ABS and other plastics from telephones. The company has relied primarily on density-based separation systems but is adding new separation technologies.[271]

MBA Polymers, of Richmond, CA, with funding from the American Plastics Council, developed automated sortation techniques for plastics from computers and electronics. The mixture of resins is sorted by family, with the major products polypropylene, high-impact polystyrene, acrylonitrile/butadiene/styrene (ABS), polycarbonate (PC), and ABS/PC blends.[265,272]

In Japan, Kobe Steel Ltd. developed a sandwich technique to apply virgin resin to a core of recovered plastic, permitting increased use of recycled plastics in computers and other products.[273]

3M developed recycling-compatible label materials to enhance the recyclability of durable goods, including computers and electronics, by permitting the label to remain in place when the equipment housing is recycled, without detracting from resin performance.[274]

Mixed recycled plastics from computers and other electronics equipment are being used by Conigliaro Industries, of New England, in an asphalt road-paving mix.[275] The company also developed a system for recycling the plastic equipment housings from computer and electronic products and a unique cold patch asphalt/recycled plastic mix for filling potholes.[276]

In 1997, Dell announced that the computers in its OptiPlex PC line, sold to business, government, and education customers, would be designed to be recyclable by using an ABS chassis without fillers or coatings, and would also be designed to make the computer easier to maintain and upgrade.[277]

Several computer companies have minimized the number of types of plastics used in computers. Panasonic once used 20 different polystyrene resin grades but has cut that to 4. Hewlett-Packard (HP) tries to use a single polymer in products, if possible, preferably HIPS or ABS. HP also codes any piece of plastic greater than 25 g with a resin identification number.[278] HP also has an active return program for printer cartridges, through their Planet Partners program, incorporating prepaid mailing labels in new cartridge boxes so that the old ones can be returned at no cost to the user. The program is now in operation in 36 countries and recovered nearly 55 thousand tonnes of used products in 2004.[279]

Another design change is to facilitate disassembly by using standard screws and quick-release connections, and minimizing the number of fasteners. Molded-in ornamental components or information have replaced some painted decorations and adhesive labels. Many manufacturers are identifying the hazardous materials used in products and working toward reducing or eliminating them. On the other hand, some of the new, vibrant colors being used in electronics will make recycling more difficult. Also, smaller and less-expensive products, while achieving source reduction, can make repair, reuse, and upgrading more difficult.[278]

In 2000, the American Plastics Council published a report on recycling of plastics from residential electronics equipment. Plastics accounted for about one-third of materials collected in two pilot residential electronics recycling programs, which operated during 1997 and 1998 in Somerville, MA, and Binghamton, NY. Only about 25 percent of the plastics fraction was "clean plastics," homogeneous and free of contaminants. The report identifies 16 different generic plastic resins sold into the electrical and electronics sector (excluding wire and cable) in significant amounts in 1995. The six most common were polystyrene, acrylonitrile/butadiene/styrene (ABS), PP, PU, PC, and phenol formaldehyde (PF). Of plastic materials collected during a two-week period in Hennepin County, MN, through its curbside and drop-off collection programs for electronics, only 35 percent were found acceptable for further processing for recycling. Of these materials, high-impact polystyrene (HIPS) was the most common material, followed by ABS and polyphenylene oxide (PPO). There was considerable variation between categories of consumer electronics. Televisions had 75 percent HIPS, while computers contained mostly ABS and PPO. A similar project in San Francisco showed that recovered fans contained mostly polypropylene. APC concluded that, while televisions accounted for nearly half the electronics recovered, they were not a rich source of high-value engineering plastics. Computers were, however, and hence had a better chance of being effectively recycled.[280]

A report by researchers from Stockholm University drew attention to the problem of release of potentially toxic flame retardants during dismantling and recycling of consumer electronics. Airborne levels of flame-retardant additives were found to be two to three orders of magnitude higher in such facilities than in other work environments, and flame retardants were found in the blood of plant workers.[281] Subsequently, two of the three types (penta and octa) of polybrominated diphenyl ethers (PBDEs) that were widely used in fire retardants have been, or are about to be, banned in much of the world, as they were shown to bioaccumulate, and there is evidence that they are harmful. The PBDE most used in electronics (deca), which represents more than 95 percent of PBDEs used each year, is not affected by the bans. California became the first U.S. state to ban PBDEs, passing a law in August 2003 that takes effect in 2008. As originally written, the bill would have banned all three types, effective in 2006, but was changed after strenuous opposition from the electronics industry and others. The two compounds that are banned are used mostly in furniture.[282] The EPA has reached a voluntary agreement with manufacturers to remove penta and octa PBDEs from the U.S. market. The European Union is also banning PBDEs, and other countries either have done so or are considering similar action. The ban in Europe includes deca PBDEs, effective in 2006.[283]

Matsushita Electric Industrial Co., Ltd., of Osaka, Japan, in 2002 announced the development of what was claimed to be the first plastic recycling system capable of separating flame retardants from used plastic while maintaining its original physical properties. The plan was to release a commercial version of the system by March 2004, but no record of such a release was found.[284]

E-Scrap News conducted a survey of U.S. electronics scrap processors in 2004. More than one-third of respondents were in their first 5 years of operation, while 45 percent had been in operation for 11 years or longer. The majority were small organizations, with one or two facilities. Nearly half had fewer than 25 employees, and almost a third had $1 million or less in sales per year. When asked about the challenges they face, 62 percent saw legislation and regulation as positive for the overall industry, while only 22 percent felt they hurt the industry. Export markets were seen as hurting the industry by 64 percent of respondents and as helping by 31 percent. An overwhelming majority (92 percent) felt that additional research and development was either somewhat or very important. There was similar support for an environmental certification program to create system management standards, with 93 percent agreeing that there is a need for such a program. Cost of logistics, obtaining sufficient processing volumes, the cost of processing, and markets for plastics were rated as the biggest challenges facing the industry.[285]

For electronics, collection programs that charge participants have become the norm in the United States, with 53 percent of programs responding to the survey reporting that residents are charged for dropping off electronic waste, compared to 49 percent in the preceding year's survey. Fees are relatively stable, however, with only 25 percent of respondents reporting fee increases, and 7 percent reporting decreased fees. At the other end, fees charged by reclaimers are also relatively stable, with 50 percent of respondents reporting that these remained constant and 27 percent reporting decreases. More than 80 percent of programs accept CPUs, CRTs, computer peripherals, printers and scanners, TVs, and fax machines. About a quarter of all programs responding accept any item with a power cord. Periodic collection events are run by 45 percent of respondents, 31 percent have permanent drop-off collection facilities, and 22 percent have a combination of services. A number of programs plan to change from periodic collection to permanent facilities. Most programs report that participation in collection programs is increasing (79 percent) and that volume of collected materials is increasing (81 percent).[286]

A number of recent efforts are directed toward the rapidly increasing stream of discarded cellular phones. Unlike other electronics recycling programs, these usually do not charge a fee. Most of these operate through voluntary drop-off at participating retailers.

In Australia, the Australian Mobile Telecommunication Association (AMTA) has a mobile phone recycling program that is a voluntary, industry-led program, funded by a fee on sale of new mobile phones from participating manufacturers and network carriers. The phones, along with batteries and accessories, are collected from more than 1600 participating stores throughout Australia. At present, plastics from handset casings and housings are being stored until a processing facility in Melbourne is completed.[287]

ReCellular Inc., in 2004, signed a five-year exclusive agreement to provide cell-phone recycling and refurbishing in South America, Central American, and the Caribbean.[288]

The International Association of Electronics Recyclers has developed a searchable database that allows individuals to search for organizations that are involved in electronics recycling. The database includes organizations and companies that provide recycling services.[289]

8.16 RECYCLING AUTOMOTIVE PLASTICS

As is the case for electronics, regulations have spurred interest in recycling plastics from automobiles as well as in using recycled plastics and designing to facilitate recycling. European regulations are a major factor, as manufacturers will be responsible for achieving 95 percent recovery and 85 percent recycling of end-of-life vehicles by 2015.[77] Japan has a similar requirement. All major automobile manufacturers are working on design changes to improve recyclability and on incorporating more recycled materials in automobiles. For example, in late 1998, DaimlerChrysler set standards for recycled content that include asking suppliers of plastics parts to provide at least 20 percent recycled content in 2000, increasing to 30 percent beginning in 2002.[290] In 1999, it demonstrated two

Dodge Stratus sedans developed with 26 supplier companies to show the potential for use of recycled materials. More than 500 parts were modified to increase recycled content, and up to 40 percent of the plastic materials used were from recycled materials. Daimler-Chrysler is working toward making its vehicles 95 percent recyclable within the next few years.[291]

One of the problems with recycling automotive plastics is paint removal. This applies not only to end-of-life vehicles but also to thermoplastic olefin (TPO) painted parts, which have a high scrap rate. Polymer Sciences Inc. announced plans in late 1999 to build recycling facilities in Duncan, SC, and somewhere in Europe, using the company's mechanical process for stripping paint from TPO parts before they are ground and pelletized. However, the plant was not built, and the company now lists paint removal as a technology that it has available for sale or license.[292,293]

A mechanical paint-removal process is also used by American Commodities Inc., based in Flint, MI, which is reported to be the largest recoverer of postconsumer automotive plastics in North America. The company has capacity to recycle 40.5 million pounds of TPO a year. One user of the American Commodities process is Visteon, which recycles 1 million pounds of painted TPO parts each year, using about 15 percent recycled old TPO bumpers in manufacture of new bumpers. Even higher amounts of scrap can be used when it is available, as the company is approved to use 100 percent scrap. Visteon's Milan plant increased scrap use to 60,000 pounds per week of recycled TPO in 2000.[292,294,295] ACI specializes in bumper recycling, through its bumper buy-back program, but now recycles other materials as well. The company sells postconsumer Enviraloy plastics with a variety of base resins: ABS, nylon, polycarbonate, PC/acrylic alloys, PC/PBT alloys, and PPO-nylon alloys, in addition to TPO. The same group of resins are available in the Impact family of resins made from postindustrial materials.[296]

In early 2000, American Commodities, in a joint venture with Wipag Polymer Technique of Neuberg, Germany, began recycling scrap instrument panels, which contain styrene maleic anhydride (SMA), polyurethane foam, and PVC. The process separates the three materials so that each can be recovered. It can also be employed on door panels, where it also separates the textile inserts. Wipag has been using the process on instrument panels in Europe, where it has facilities in Neuberg, Germany, and Kent, England. Materials are returned to 99.98 percent purity at a cost savings of about 30 percent compared to virgin material. Recovered plastic is used in the production of new automotive parts.[295–297]

While many automotive recycling processes target disassembled components, including scrap parts produced during the manufacturing process, others are directed toward the more complex problem of recovering usable plastics from the residue from automobile shredder operations designed primarily for metals recovery. Auto shredder residue typically contains 15 percent to 27 percent plastics. The plastics fraction contains thermosets and thermoplastics. The thermosets, in addition to rubber, are mostly rigid polyurethane and fiberglass/polyester composites. The thermoplastics fraction includes nearly 50 different resins, but about 80 percent is PP and ABS.[77] One of the systems that has been developed for recovery of plastics from ASR was discussed in Sec. 8.11, since its major recovery target is polyurethane foam.

Another company that is working on recovery of plastics from ASR is Galloo Plastics, in France. According to a report in 2000, Galloo crushes the ASR into 1-in granules at the shredder and then shreds then to less than 1/2-in size at the recycling facility. Ferrous and nonferrous metals and small glass particles are removed, and then the ASR is aspirated to remove PUR foam. A drum tumbler removes wire and metal waste, leaving a plastics/wood mix. A proprietary float/sink process recovers filled and unfilled PP. The PUR foam fluff has been tested for sound-deadening applications but was not found to be economically viable. The PP is sold, and the company plans to add PE and ABS as products.[298]

Galloo is also working on recycling of electronic products and currently is expanding its facility.[299]

Argonne National Laboratory has a pilot facility for investigation of recycling from automotive shredder residue. It includes a mechanical separation facility and a wet-density/froth-flotation separation facility.[300]

8.17 DESIGN ISSUES

As mentioned, the design of items made of plastics can affect their recyclability. Obviously, design can also affect other environmental impacts associated with the item.

8.17.1 Design for Recycling

Obviously, product design can greatly affect the ease of the recycling the product at the end of its life. The original PET soft drink bottle had a PET body, HDPE base cup, aluminum cap, and paper label. All of these materials had to be separated to permit PET recycling to occur (and, of course, residual product and other contaminants also had to be removed). The current bottle design has a PET body, no base cup, and PP cap and label. Consequently, recovery of a pure PET stream for recycling is much easier.

Various organizations have produced guidelines for the design of products and/or packages, with the explicit goal of facilitating recycling. The general approach is to simplify identification of plastics by resin type and then to make it easier to separate the various plastic streams from each other and from nonplastic components, or to make the various plastics compatible with each other so that they do not need to be separated.

One of the early efforts to produce guidelines for "design for recycling" was the City/Industry Plastic Bottle Redesign Project, established in early 1994 to reach a consensus on design changes for plastic bottles so as to improve the economics of recycling. The "city" representatives included Dallas, Jacksonville, Milwaukee, New York, San Diego, and Seattle. "Industry" participants included Avery Dennison, Enviro-Plastic, Johnson Controls, Owens Illinois, Procter & Gamble, SC Johnson Wax, and St. Jude Polymer. The study received funding from the U.S. Environmental Protection Agency and the states of Wisconsin and New York. The focus was to assist plastic recyclers as well as to maximize the return to cities collecting postconsumer resin. Therefore, balancing costs of making package design changes against the recycling benefits was part of the effort. Recommendations for plastic bottle design included

- Making caps, closures, and spouts on HDPE bottles compatible with the bottles, ensuring that any aluminum seals used on plastic bottles pull off completely when the bottle is opened by the consumer
- Using unpigmented caps on natural HDPE bottles
- Phasing out the use of aluminum caps on plastic bottles and HDPE base cups on PET bottles
- Using water-dispersible adhesives for labels
- Not using metallized labels on plastic bottles with a specific gravity greater than 1.0
- Not printing directly on unpigmented containers
- Using PVC and PVDC labels only on PVC containers

- Making all layers in multilayer plastic bottles sufficiently compatible for use of the recyclate in high value end markets
- Avoiding the use of PVC bottles for products that are also packaged in other resins that look like PVC

The industry participants abstained from this final recommendation.[301] Some of these recommendations, such as phasing out use of aluminum caps on plastic bottles and HDPE base cups on PET bottles, have been nearly completely accomplished. Other recommendations, such as not using pigmented caps on unpigmented HDPE containers, have been almost totally ignored.

Other organizations have also issued plastic bottle design guides. The Association of Postconsumer Plastic Recyclers published a set of specific guidelines for PET, HDPE, PP, and PVC bottles.[107] It also issued a special advisory against the use of opaque white PET bottles, which were found to pose a special problem for recyclers.[302]

The Council for PET Bottle Recycling, in Japan, has published design guidelines for PET bottles.[303] The Association of Plastics Manufacturers in Europe also has guidelines for design of rigid plastic containers.[304] Petcore has published design guidelines for PET bottles that focus on the acceptability of additives and barrier materials.[305]

The automotive industry has also directed efforts toward improving the recyclability of automotive plastics by change in automobile design. Efforts include designing components for ease of disassembly, as well as efforts to minimize the use of different resins and to ensure that all resins in a component are compatible with one another. One of the reasons for these changes, in addition to voluntary actions by manufacturers, is requirements in Europe that mandate that the auto industry provide a system for collection and recycling of vehicles at the end of their life. Currently, 74 to 80 percent of the weight of a typical vehicle is reused or recycled; the End of Life Vehicle Directive requires that this amount be increased to 80 percent by 2006 and 85 percent by 2015. Furthermore, vehicles must increase their use of recycled materials.[306]

The computer and electronics industry is also being forced to increase recycling of materials, including plastics. Japan mandates certain percentages of recycled material in electronics.[307] Europe is instituting producer responsibility requirements for electronics that require specified recovery and recycling amounts be achieved beginning in 2006. (Precise requirements differ by category.)[308]

The American Plastics Council, with the Society of the Plastics Industry, published a set of design guidelines for information and technology equipment§[309] Fuji Xerox, in Japan, has a detailed set of design guidelines in support of its goal of zero waste.[310] IBM became the first computer manufacturer to code plastic components to promote their recycling, in 1992.[310]

Another of the early actions by computer manufacturers was by Dell Computer Corp., which announced in 1997 that it would make its computers marketed to business and government more recyclable by using plastic materials that do not contain fillers and coatings that can inhibit the recyclability of the materials. Dell also changed the chassis design for their computers, making metal and plastic parts easily separable. Plastic components are marked with international ISO standard codes. Dell also instituted a take-back program for its large corporate customers, accepting computers of any brand and giving discounts for the purchase of new Dell computers. Much of the equipment, instead of being recycled, was upgraded and resold in other countries.[311]

Apple Computer has banned the use of various chemicals in its products, including PBDEs. It prohibits any PVC parts or packaging of more than 25 g, except for closures for cables and wires. Several of its computers use recyclable polycarbonate for the enclosure. Large mechanical plastic parts are made of a single material or of compatible materials and marked with resin identification codes to facilitate recycling.[312]

8.17.2 Environmental Certification Programs

Another influence on design is the existence of programs to certify, in some fashion, the "environmental goodness" of a product or packaging. While recycling is not the only criterion, inclusion of recycled content is sometimes the basis for such certification.

Environmental certification programs include Green Cross and Green Seal in the United States, the Blue Angel program in Germany, the EcoMark in Japan, and the Environmental Choice label in Canada.

In all these programs, manufacturer participation is voluntary, with a fee generally charged to cover the cost of the program. The impetus for manufacturer participation is the marketing benefit derived from the eco-label.

8.17.3 Green Marketing

Significant marketing advantages can be derived from presenting products as having some positive environmentally related attributes, including recyclability or use of recycled content. When this phenomenon started, there was a great deal of misleading advertising. Such claims are subject to requirements under general legislation related to deceptive advertising, but, in a number of places, specific laws have been enacted to regulate environmental marketing claims. Companies considering making environmental claims about products need to be aware of the legal requirements. In the United States, laws differ somewhat from state to state. General guidance has been provided by the Federal Trade Commission, with its document "Complying with the Environmental Marketing Guides."[313] One of the requirements is that a claim of recyclability cannot be made about a product or package unless a substantial majority of consumers in the locations where the claim is made have access to recycling programs that accept the item for recycling. The U.S. Federal Trade Commission held that conspicuous display of the plastic resin codes constitutes a claim of recyclability, while inconspicuous use does not.[314]

8.17.4 Other Environmental Issues

Another important environmental issue facing plastics (of some types) at present is concern about components used in plastics, often as plasticizers, that are suspected of having hormone-mimicking or antagonizing effects. Initially, these concerns were directly predominantly to chlorine-containing chemicals, but a variety of nonchlorinated substances, many in the phthalate or related families, are now suspect as well. For example, one recent report linked prenatal exposure to phthalates to changes in the reproductive organs of male offspring.[315] In June 2005, the EU Commission voted to prohibit the use of three phthalate plasticizers in children's toys and to restrict the use of three others.[316]

Polyvinyl chloride (PVC) remains under attack, although the level of attention seems be less than it was in the 1990s. Recently, a life-cycle assessment of PVC and competing materials concluded that, in many applications, environmental impacts of PVC are comparable to those of alternatives. Furthermore, it found that the market share of PVC bottles even in Europe, where they once were very strong, is now minor.[317] This likely reflects the years of environmentally related pressure against PVC use as well as the positive attributes of PET, its main competitor.

Use of heavy metals in plastics as coloring agents, stabilizers, and so forth has decreased greatly in the last two decades. In the United States, the "Model Toxics Law," written by the Coalition of Northeastern Governors (CONEG) Plastics Task Force, has been adopted by at least 19 states (Table 8.6), resulting in nearly total elimination of use of lead, mercury, cadmium, and hexavalent chromium in packaging (including plastics packaging) in the United States. A number of other countries have now adopted similar poli-cies.[318] Currently, efforts are underway to mandate elimination of these heavy metals in other applications, including in plastics used in electronics and automobiles.

TABLE 8.6 States that Have Adopted the CONEG Model Toxics Law to Prohibit the Deliberate Introduction of Lead, Mercury, Cadmium, and Hexavalent Chromium in Packaging[57]

California	Iowa	New Hampshire	Vermont
Connecticut	Maine	New Jersey	Virginia
Florida	Maryland	New York	Washington
Georgia	Minnesota	Pennsylvania	Wisconsin
Illinois	Missouri	Rhode Island	

8.17.5 Life-Cycle Assessment and Sustainability

During the last two decades, there has been increasing interest in use of life-cycle assess-ment techniques to evaluate the environmental trade-offs associated with manufacturing and purchasing decisions. The philosophy behind life-cycle assessment is that the entire life cycle of a process or product, from acquisition of raw materials to eventual waste dis-posal, must be considered in evaluating the effects of that process or product on the envi-ronment. If only a portion of the life cycle is considered, then decisions about which of two alternatives has lesser adverse environmental impacts may be flawed, as looking at only a portion of the life cycle may result in ignoring serious impacts and lead to compari-sons that are not accurate.

A great deal of work has been done on trying to develop life-cycle assessment to the point that it is a useful tool. For example, the Industrial Designers Society of America, with funding from EPA, is currently developing a tool to produce a single score from a life-cycle assessment that can be used for comparison of alternative products. The Interna-tional Design Center for the Environment is also developing a system using streamlined LCA. The U.S. Green Building Council has developed the Leadership in Energy and Envi-ronmental Design (LEED) system for evaluating green buildings.[319]

Europe has generally been ahead of the United States in developing computerized models that can be used for life-cycle analysis as a way to cut down the complexity and expense associated with use of this tool. Some of these also produce single ratings; others produce a matrix.

Because of the complexity of the issues involved, life-cycle assessment has not yet reached the point at which it can be easily used to provide guidance in making complex production or purchasing decisions. Unfortunately, it still is too often true that "life-cycle analyses" issued by different groups about the same product choices come to diametrically opposing conclusions about which product is "best" for the environment. For example, in the 1980s, there were such analyses purporting to prove that disposable diapers were bet-ter than cloth diapers—and the reverse. In 2005, a newly issued life-cycle assessment, this time coming from the UK, may have signaled a reopening of the "nappy wars."[320,321] Life-cycle analysis may be a useful tool some day, but (certainly for now) both producers and consumers of goods should be very cautious in relying on life-cycle assessments, whether done by an organization with a particular point of view or even by supposedly unbiased commercial software programs, in determining whether product A or product B is environmentally "superior" to the other.

More recently, interest in life-cycle assessment is being superseded to some degree by interest in development of sustainable systems. Sustainability is perhaps a more useful concept than life-cycle assessment, since it avoids some of the complexities of determining, for example, whether x amount of air emission A is better or worse than y amount of water emission B. Rather, the focus is on whether goods are being produced, used, and disposed of in a manner that allows us to continue to produce and use them into the indefinite future. One intersection of sustainable development with plastics is increased interest in biobased plastics, since these likely come from renewable feedstocks, with potential for sustainable production, rather than from nonrenewable fossil fuels. Of course, recycling may also help make production systems more sustainable.

In 2005, K. Sonneveld et al. presented a discussion of how sustainable packaging can be defined and measured. The key factors are that packaging is effective, adding real value to society; efficient, using materials and energy as efficiently as possible throughout the life cycle; cyclic, with materials continuously cycled through the system, minimizing degradation or use of additives; and safe, not posing any risks to human health or ecosystems. The paper also reports on efforts of the Australian Sustainable Packaging Alliance to develop a Packaging Impact Quick Evaluation Tool (PIQET) to provide packaging technologists and managers with "hands-on input for defining company packaging strategies, selecting materials for packaging redesign or packaging innovation, and specifying packaging for procurement of incoming goods."[322]

Others have proposed the goal of zero waste, defined as waste reduction, clean production, and maximum recovery and use of materials. This idea has also begun to gain significant attention. In mid 2004, the Haut-Rhin Department, in the Alsace region of France, became the first locality in France to become a zero waste pilot program. The community had been considering an incinerator but opted for comprehensive recycling and composting instead.[323] As mentioned earlier, Fuji Xerox in Japan has adopted a zero waste goal.[310]

Another approach to the ideas of sustainability and zero waste is the "cradle-to-cradle" idea, where after use, the product or package becomes the same or another product, through reuse, technical nutrient flow (recycling), or biological nutrient flow (composting or resource growth) in a continuous cycle. This concept has led to some interesting ideas, such as using more packaging material, not less, and designing the best package possible rather than using the cheapest materials. Both of these approaches maximize the useful life of the package. The basic idea behind cradle-to-cradle design is to design waste out of the package life cycle. Another controversial idea is that "littering" can help the environment when the littered products are biodegradable.[324] A prime stimulus for this idea is the book by W. McDonough and M. Braungart, *Cradle-to-Cradle: Remaking the Way We Make Things*, published in 2002.[325]

8.18 LEGISLATION

One way to increase recycling of end-of-life plastics, and thus reduce their burden on disposal systems, is to institute legal requirements in the form of specific laws designed to aid recycling in some manner. Deposits on plastic bottles, discussed in Sec. 8.2.1.1, fall in this category. Several U.S. states and a large number of municipalities have instituted mandatory recycling for certain products, and plastic bottles are sometimes included. Many states have target levels for recycling or for reduction of waste going to disposal.

Another way to increase recycling is to require target levels of recycling and to mandate that the producers of the items take responsibility for ensuring that this occurs and that they bear the costs for doing so. This version of the "polluter pays" principle is known as *extended producer responsibility (EPR)*. A variant, also denoted EPR, is *extended product responsibility*. The difference between the two is that extended product responsibility spreads the burden past the producer to government entities and to users of the products involved, stating that they all must bear some of the responsibility. Most often, EPR is interpreted to mean extended producer responsibility in Europe, and extended product responsibility in the United States. A term that is being used increasingly in the United States and Canada is *product stewardship*, which is essentially equivalent to extended product responsibility and has the benefit of avoiding the confusion of the two different meanings that have been associated with EPR.

8.18.1 Recycling Rates and Recycled Content

Wisconsin, Oregon, and California have laws related to recyclability and recycled content of rigid plastic containers. In Wisconsin, plastic containers except those for food, beverages, drugs, and cosmetics, are required to contain 10 percent recycled content. The exemptions do not apply if the U.S. FDA has approved the use of recycled content. However, the law allows "remanufactured" material (regrind and so forth) to be counted as recycled, and it has had little effect.

Oregon requires that rigid plastic containers contain at least 25 percent recycled content, be recycled at an aggregate rate of at least 25 percent, be made of plastic that is recycled in Oregon at a rate of at least 25 percent, or be a reusable container that is made to be reused at least five times. Containers for medical foods, drugs, and devices, and containers of food other than beverages are exempt, as are packages that have been reduced in weight by at least 10 percent compared to packaging for the same product used five years earlier. Since the law went into effect, the recycling rate for covered containers has exceeded the 25 percent requirement, so no action by companies has been required. However, in determining the projected rate for 2000, the state warned that the recycling rate had been declining and might fall below the target 25 percent level as early as 2002. While the rate to date remains slightly above 25 percent (Fig. 8.34), the state has reiterated this warning each year.[326]

California had a law that was very similar to that in Oregon, except that the law set a higher recycling target for PET—55 percent. However, for most of the time since the law was passed, California's overall rigid plastic packaging container (RPPC) recycling rate was below the 25 percent required. Therefore, companies manufacturing, distributing, or importing nonexempt products in RPPCs had to comply with the law in some other way. Recycling rates in California are shown in Fig. 8.34. California recently revised its RPPC law, eliminating the state calculation of recycling rates and the option of manufacturers to comply via these rates. Manufacturers now have the following options for compliance:

- Containers made from at least 25 percent postconsumer resin
- Containers source-reduced by 10 percent
- Containers reused or refilled at least five times
- Containers having a 45 percent recycling rate (brand specific or for a particular type of RPPC) (product manufacturer must perform calculations to determine recycling rate)

Compliance may be achieved by averaging across an entire product line or sublines but can include only regulated RPPCs, not those that are nonregulated or exempt.[327] Industry

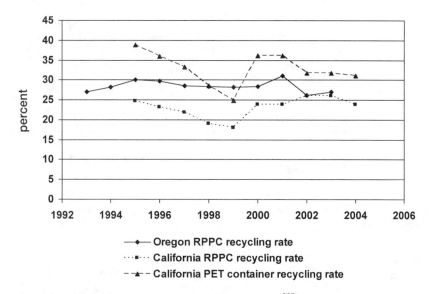

FIGURE 8.34   RPPC recycling rates in Oregon and California.[327]

generally opposed the changes in the law, feeling that the burden on them would increase. In 2004, California decided not to require that companies certify compliance for 2003, even though it determined that the RPPC recycling rate was only 23.9 percent for the state for that year.[328]

8.18.2   Plastic Bag Taxes and Bans

Over the last several years, plastic bags have become a particular focus of attention. For example, California for several years has had a requirement that plastic trash bags with a thickness of 0.7 mil (0.007 in) or greater contain at least 10 percent postconsumer recycled content, or that the manufacturer use at least 30 percent postconsumer recycled content averaged over all its plastic products. In 2002, the California Integrated Waste Management Board had recommended dropping this requirement except for bags purchased by state government, but this change was not made. Over the last few years, the amount of plastic used in garbage bags has dropped considerably, from about 8.4 million pounds in 1998 to about 2.6 million pounds in 2002. As a consequence, in 2003, the state banned two of the four largest trash bag manufacturers from getting state contracts because of their failure to comply. Bag makers attribute the decline to limited supplies because of competition from manufacturers of plastic lumber.[329]

Between 1989 and 1992, Italy had a tax on producers and importers of plastic bags equivalent to about 6 cents per bag, about five times the manufacturing cost. The tax greatly cut the use of plastic bags, as well as providing significant revenue for the government. However, it is no longer in effect.[330]

In 2002, Ireland implemented a tax on plastic bags equivalent to about 20 cents (U.S.) per bag. Proponents say the law has cut plastic bag use by 95 percent and dramatically reduced bag litter.[331] Opponents charge that plastic bags were never a large fraction of litter in the first place, cite the environmental benefits of lightweight, energy-efficient plastic bags that are frequently reused, and charge that the bag tax has increased theft.[332]

A number of other countries, including Australia, New Zealand, Hong Kong, Singapore, and India, have considered following Ireland's example in imposing plastic bag taxes. Taiwan implemented a plastic bag tax in 2002, and bag use fell by 80 percent.[333,334] Denmark has taxes on both paper and plastic bags. Hong Kong requires customers to be charged for bags over a minimum size.[335] In June 2005, plastic bag tax legislation was introduced in Scotland.[336] Kenya, in March 2005, proposed a ban on thin plastic bags (less than 30 μm) and a tax on heavier ones.[337]

Some locations have banned plastic bags altogether. In 2003, Coles Bay, Australia, a popular tourist spot, became the first community in Australia to ban plastic bags at all retail outlets.[338] Bangladesh banned them in 2002 because of flooding problems when the bags clogged drainage pipes.[334] South Africa banned thin plastic bags, requiring them to be thicker so they can be reused.[335]

8.18.3 Extended Producer/Product Responsibility and Product Stewardship

In many countries, concerns about the impact of plastics (as well as other materials) on the municipal waste stream are at a higher level than is currently the case is most of the United States. This may stem from a greater lack of disposal capacity, differences in the prevailing philosophical approach to waste management, more concern about resource use, and so on. Often, this means increased regulation of production, use, and/or disposal of products. The resulting legislative and societal pressure has led to increased growth in plastics recycling in many parts of the world during the period when plastics recycling (as well as that of some other materials) has declined in the United States.

Many countries have applied the concept of extended producer responsibility (EPR) to disposal of materials, including plastics. Under this philosophy, the manufacturer of a product (including packaging) is responsible for the ultimate recycling and/or disposal of that product. Government sets recycling requirements but leaves it to industry to formulate and manage the systems required to meet those requirements.

Extended producer responsibility got its start in Germany with take-back and recycling requirements for packaging. The government instituted requirements that the producers of products be responsible for recovery and recycling of product packages. The government set targets that must be met but left it up to industry to decide how to do so. What emerged, for consumer packages, was the Duales System Deutschland (DSD), or Green Dot system. It is an industry-funded system to collect and recycle packaging for consumer products. The current requirements call for 25 percent recycling and 50 percent recovery. By December 2008, the requirement will be recycling of 55 percent and recovery of 60 percent of waste packaging, with energy recovery through incineration counting towards the recovery target.[339] Recycling targets for individual materials are also being revised. For plastics, the requirement will be 22.5 percent recycling in December 2008, counting only materials that are recycled back into plastics. Current minimums are 15 percent for each packaging material. Furthermore, the targets are to be revised every five years. The 22.5 percent recycling target for plastics includes recycling of biodegradable plastics by composting.[340]

This system, while widely criticized at first, became policy for the whole European Union and is also in place in countries that are trying to become EU members. The EPR approach has also spread far beyond the borders of Europe. It has also expanded beyond packaging. For example, the EU has adopted EPR for automobiles and electronics.

The European Union's directive on Waste from Electrical and Electronic Equipment (WEEE) requires companies selling electronic products in Europe to set up end-of-life collection and recycling systems for these products by August 2005.[341]

An alternative to EPR are systems in which certain products seen as particularly problematic are subjected to fees or taxes, providing an economic incentive against their use. For example, as discussed above, plastic bag taxes have been springing up in a number of countries around the world. Another approach to minimizing the impact of plastics on systems for recycling or otherwise managing waste is laws that limit the types of products or packages that can legally be produced and sold and thereby limit what enters the waste stream. This may be directly in the form of bans on certain products or product components (such as bans on heavy metals in packaging), or it may be indirect in the form of requirements that certain types of products be manufactured from specific types of materials (such as biodegradability requirements).

Canada adopted a National Packaging Protocol in 1989 that required a 50 percent reduction in packaging waste going to disposal by 2000. The target was actually achieved four years ahead of schedule. However, most of the reduction was in transportation packaging rather than consumer packaging. Various provincial governments soon turned their attention to extended producer responsibility to achieve further reductions and to pass along part of the cost of managing packaging waste to the producing companies. All ten provinces in Canada now have EPR in place for one or more categories of products.[342]

A modification of extended producer responsibility, known as *extended product responsibility* (and also usually abbreviated EPR), asserts that the product manufacturer, the consumer, and government share responsibility for managing recycling and/or disposal of products and packaging. This is the basis for the fees levied in Ontario, Canada (and now being implemented in Quebec as well), on packaging and printed material, which are designed to require the producers of the materials to pay part of the cost of collecting and recycling them.

Environment Canada maintains an inventory of Extended Producer Responsibility and Product Stewardship programs in Canada on its web site. Users can search by region, program, sector, key word, and so on.[343]

EPR and product stewardship are being adopted in many other countries around the world. Examples include Japan, Korea, Brazil, Australia, and many more. In many cases, the systems are imposed by regulation. In others, governments and businesses are forging voluntary agreements to head off mandatory requirements. It seems certain that industry will be called on increasingly to assume some of the consequences of handling the wastes from the products they manufacture, and this means, of course, that industry will increasingly be paying attention to the disposal-related consequences of their product and package design decisions.

In Japan, the law provides for "shared responsibility" of consumers, local governments, and industry. Consumers must participate in the sorting and collection systems that are operated by local governments. The Japan Container and Packaging Recycling Association was established to carry out industry's obligation to recycled a percentage of the waste packaging that has been collected and sorted. Fees are charged to manufacturers on the basis of the quantity of packaging they generate; manufacturers, retailers, and wholesalers smaller than a designated limit are exempt.[344]

In the United States, extended producer responsibility has not yet been a major factor. Product stewardship, however, has been officially adopted as state policy in Minnesota, and other states are considering the approach. In Minnesota, product stewardship is currently being implemented for electronics, paint, and carpet. The state has set up voluntary programs to test collection strategies and recycling for household electronics waste and for paint. It has gone a step farther for carpet, negotiating to set national recovery and recycling goals and being instrumental in the carpet industry setting up a third party organization to achieve those goals. In all cases, cooperation and voluntary agreements with industry have been involved.[345] Minnesota's EPR program for carpet has been criticized for lacking consequences for not meeting recovery targets.[342]

In the United States, a major product stewardship effort was made for handling end-of-life computers and other electronics through the National Electronic Product Stewardship Initiative. However, after nearly four years of effort, it failed to reach its goal of "the development of a system, which includes a viable financing mechanism, to maximize the collection, reuse and recycling of used electronics, while considering appropriate incentives to design products that facilitate source reduction, reuse and recycling; reduce toxicity; and increase recycled content."[263,346] While participants were able to reach consensus on many program elements, such as the products to be included and the roles of various parties, they were unable to reach agreement on financing.[347] This has, at least for the time being, thrown the matter back to states, since many state governments have not been satisfied with voluntary actions by electronics manufacturers. States have taken two main approaches, advance recovery fees (ARF) and extended producer responsibility (EPR). Advance recovery fees are currently in place in two places in North America, the province of Alberta and the state of California. In both places, consumers pay a fee when they purchase certain electronic devices to finance the return and recycling of those products. In Alberta, the fees are equivalent to $7 to $37 (U.S. dollars) per item and apply to televisions, computers, and some computer accessories such as printers. In California, the fee is $6 to $10 and applies to televisions and cathode ray tube (CRT) display devices with display size greater than 4 in (diagonally). Both programs began collecting fees in 2005.[263] An industry-funded system will handle specific obsolete products, with fees to fund the system being charged to brand owners and importers. California also is implementing producer responsibility for cell phones under the Cell Phone Recycling Act of 2004, Section 42490 of the California Public Resources Code. Beginning in 2005, local governments in Maine will be required to collect computer monitors and televisions and deliver them to regional consolidations centers. Producers will be required to run the centers and fund the recycling of the products. Legislative study groups in Minnesota and Oregon have recommended establishment of advanced recycling fees in those states and legislation is being considered. Ontario is in the process of adding electronics and electrical equipment to its product stewardship program.[347] In 2005, Nova Scotia and Saskatchewan adopted requirements for province-wide electronics recycling programs to be financed by ARFs and managed by industry.[348]

While the product stewardship approach is still relatively uncommon, it seems likely that this approach to waste management and prevention will increase in the future.

8.19 BIODEGRADABLE PLASTICS

8.19.1 Overview of Plastics Degradation

Until the early 1970s, the focus was on preventing plastics degradation to avoid the loss in performance properties that resulted as plastics aged, were exposed to sunlight, and so on. The extent of degradation was generally measured by the percent loss of useful properties. One "rule of thumb" was that 90 percent loss in tensile strength was equivalent to total degradation, as this was sufficient to render the plastic object unusable.

In the mid 1980s, when concerns about solid waste disposal increased, interest in biodegradation intensified as some perceived it as a solution to the landfill "crisis." Plastics were attacked as major factors in this crisis, taking up landfill space for centuries or more, since nearly all synthetic plastics are not biodegradable. Biodegradation and other types of degradation were often not clearly distinguished from each other, and measures of degradation were equally unclear. This brought the initial generation of "biodegradable" plastics, which most often were mixtures of starch with low-density polyethylene.

As time went on, information increased about both the behavior and composition of solid waste in landfills. It became clear that the majority of material in landfills was, in fact, biodegradable, consisting of paper, food waste, and yard waste (see Fig. 8.4). Therefore, if biodegradation was the solution to landfill problems, there should be no problem! However, studies of landfill behavior showed clearly that, when landfills are kept dry, as is required by regulations designed to help prevent groundwater contamination, degradation is a very slow process, taking decades or more to be complete. Hence, simply making the plastics in a landfill biodegradable was not a real solution when conditions prevent rapid microbial destruction of landfilled materials. Furthermore, calling a plastic "biodegradable" on the basis of loss of tensile strength, or even complete loss of structural integrity, is questionable at best. The starch/polyethylene bags may disappear from view when the starch fraction is metabolized by microorganisms, but this is no guarantee that there has been any substantial change in the polyethylene fraction of the bag. The current view is generally that complete biodegradation means complete destruction of the molecular structure, a "return to nature" of the carbon content of the polymer rather than conversion of a plastic item to unidentifiable plastic powder. Implicit in labeling a polymer biodegradable is the assumption that this molecular destruction will take place in some "reasonable" time frame. Even the most recalcitrant plastic will probably biodegrade eventually—but eventually may mean centuries. Obviously, such slow biodegradation is of no practical value in the short term.

In recent years, there has been some tendency to stay away from terms like "biodegradable" that are fraught with uncertainty in meaning. Rather, plastics may be labeled "compostable" if they meet requirements set by national or international standards for this designation. Any material that is compostable is almost always biodegradable, but the reverse is not necessarily the case.

Another designation for plastics that is drawing increasing interest is sometimes related to, but by no means identical to, *biodegradable*. Biobased plastics are those that are formed from natural renewable feedstocks rather than fossil fuels. Biobased plastics may or may not be biodegradable (or compostable), and vice versa. In January 2005, the U.S. Department of Agriculture began setting guidelines for designating items made from biobased products that will be given preference in federal purchasing programs, much as products with recycled content are given preference.

Plastics may undergo a variety of other types of degradation. Of particular interest in the context of this discussion are photodegradation and hydrolytic degradation. Photodegradation is degradation that occurs as a result of exposure to light, generally ultraviolet radiation that is part of sunlight. During the 1980s, some plastics labeled "degradable" as a marketing tool were photodegradable rather than biodegradable. Photodegradation can be an important attribute for plastics that are frequently littered, but is of little or no value for plastics that are landfilled or composted. Hydrolytic degradation is degradation through the action of water, usually involving chemical reactions of the polymer with water, often in reactions that essentially reverse the polymerization process. A number of the commercially important biodegradable plastics degrade in part by hydrolysis. In some cases, the polymers are not biodegradable until they have first hydrolyzed enough to significantly reduce their molecular weight, making them susceptible to microbial attack.

As has been mentioned, degradation of plastics, as is the case for other materials, is affected by the conditions to which the plastic is exposed. Sunlight, mechanical stress, temperature, and humidity all affect the degradation rate.

The practical value of biodegradable plastics is a subject of some debate. As discussed, in a landfill, even foods degrade slowly, a fact amply illustrated by photographs that have been presented of grass, carrots, chicken, and other products still readily recognizable after ten years or more in a landfill.[349] Therefore, there is little value in using biodegradable plastics if a landfill will be their eventual destination. Nonetheless, some users perceive value in the use of biodegradable plastics as part of a "green" image for the company. This is especially true, for example, for products that are marketed as organic.

Composting, in contrast, is designed to accelerate biodegradation and serve as an alternative to landfilling. Use of biodegradable plastics permits disposal through composting and therefore can reduce the burden on landfill if systems to direct the product or package to composting are in place and utilized. In addition, for products that pose a litter problem, the use of biodegradable plastics can greatly reduce their prevalence and longevity in the environment. This can be of particular value for plastics that may reach water systems. Plastics in the marine environment are a significant problem. Even plastics improperly disposed of on land can eventually reach the ocean, where they pose significant problems to sea turtles and other marine life. Unfortunately, some compostable plastics do not biodegrade quickly in water, as they require the elevated temperature of a compost pile to cause sufficient hydrolysis to start the degradation process.

One way to structure composting programs that may accept biodegradable plastics is through curbside collection of a compostable fraction of waste, often termed *wet organics*. Such programs generally collect food wastes, yard wastes, and food-contaminated paper. Biodegradable plastics can also be accepted in such programs, at least in theory. (In practice, there may be concerns about the ability of individuals to discriminate properly between biodegradable and nondegradable plastics.) Biodegradable bags for collecting these organic wastes are already a substantial market in Europe.

In the United States, composting is limited almost entirely to yard waste. Food wastes are composted primarily in special programs targeted at institutions such as food-processing facilities, restaurants, and cafeterias. Recently, a few communities have started to institute residential composting programs. One of the first was in San Francisco, where what started as a pilot program in a few neighborhoods has now gone citywide and is expanding to neighboring communities. Freedonia forecasts that demand for biodegradable/compostable plastics will grow more than 16 percent per year between 2004 and 2008, reaching a total of over 290 million pounds in 2008.[350]

In Europe, composting has a much longer history and is more highly developed. The first composting plants for mixed municipal solid waste date to the 1970s. Collection of wet organics is commonplace in many countries. Expansion of composting is being driven by regulations requiring a reduction in landfilling of biodegradable municipal wastes. By 2006, the amount must be reduced to 75 percent of 1995 levels; additional targets culminate in reduction by 2016 to 35 percent of those levels.[351,352] With fees for nonbiodegradable plastics to ensure their collection and recycling, it is easier for compostable plastics, which generally cost more per pound, to compete economically. In part for this reason, interest in compostable plastics for European markets is generally greater than for U.S. markets. Canada also is actively increasing composting as a disposal alternative.

BASF, Cargill Dow, Novamont, and Rodenburg Biopolymers, in November 2004, signed a ten-year Environmental Agreement with the European Commission, committing themselves to using biodegradable and compostable polymers in the manufacture of packaging materials. The agreement includes a certification plan for quality control, and a labeling plan to facilitate waste handling. It will be managed by the International Biodegradable Polymers Association & Working Groups (IBAW). These four companies are estimated to represent more than 90 percent of the European market for biodegradable plastics and a similar share of the global market. As of mid 2005, supermarket carrier bags and bags for collection and composting of food waste were estimated to consume 38 percent of all biodegradable plastics.[353]

One concern that is frequently raised about degradable plastics is their effect on plastics recycling. As we have seen, in recycling, separation of plastics by resin type is key, at least for most high-value applications. If adequate separation is achieved, biodegradable plastics should not adversely impact recycling. Some biodegradable plastics can themselves be recycled if an infrastructure for doing so is developed. As a practical matter, it is unlikely that sufficient quantities will be available to make it worth setting up such systems, at least in the short term. Of course, some argue that composting itself is "nature's recycling." To the extent that biodegradable/compostable packages compete with alternatives that would be otherwise recycled, there is the additional issue of whether recycling or composting is more environmentally beneficial. A full answer to this question, which would require a thorough life-cycle assessment, is not available. Certainly it will be impacted by actual recycling rates and very possibly by local variables such as water availability, energy mix used for both corn (or other feedstock) cultivation and conversion and for plastic production, types and quantities of herbicides used, and a multitude of other variables.

Various organizations have issued standards for determining the biodegradability or compostability of plastics. For example, in Europe, EN 13432 describes methodology for evaluating the compostability of a polymer. In the United States, the Biodegradable Products Institute certifies materials that meet its requirements, and, in combination with the U.S. Composting Council, award their "Compostable Logo" to qualifying products. Japan's system for testing and certification of biodegradable plastics is GreenPla, managed by the Biodegradable Plastics Society. ASTM D6400, "Standard Specification for Compostable Plastics," is another commonly used evaluation procedure. There is ongoing international effort to standardize policies and procedures for certifying biodegradable products.[354]

Biodegradable plastics can be categorized in a number of ways. They can be divided into synthetic plastics and natural plastics, into biobased and nonbiobased plastics, or by polymer family. The reported chemical compositions of various biodegradable plastics, as compiled from various sources, are shown in Table 8.7.[355–358]

It should be noted that cellophane, the first transparent packaging material, is biodegradable, but it and other biodegradable cellulosic materials (such as paper) are not included in this discussion because they are not plastics. Biodegradable plastics designed primarily for medical applications, personal hygiene products, agricultural products, textiles, and so on also are not included.

8.19.2   Starch-Based Plastics and Other Polysaccharides

8.19.2.1   Starch-Based Plastics.   As can be seen in Table 8.7, a variety of starch-based plastics have been produced by several companies. Starch-based plastics are often water-soluble as well as biodegradable. Some contain almost entirely starch; others contain blends of starch with other biodegradable components.

An early producer of all-starch biodegradable plastics was Warner-Lambert. In 1990, it produced what was claimed to be the first biodegradable plastic from starch and sold it under the tradename Novon. The polymer contained about 70 percent branched starches and 30 percent linear starch, along with a glyceride as a plasticizer. In 1993, Warner-Lambert suspended operation of Novon after trying unsuccessfully to sell the business. In 1995, EcoStar International acquired the technology and formed Novon International, which soon thereafter was acquired by Churchill Technology. In 1996, Churchill Technology filed for bankruptcy, and production of Novon stopped.

StarchTech, Inc., of Golden Valley, MN, sells biodegradable starch-based resins for injection molding and other processes and licenses their technology to other producers. One

TABLE 8.7   Reported Composition of Selected Biodegradable Plastics[*355–358]

Polymer category	Trade name	Manufacturer
Starch	Mater-Bi	Novamont
Cellulose acetate	Acetate cellulose	Teijin
	CelGreen	Daicel Chemical Industry
	Eco-Excel	Ebara Jitsugyo Co. (EJ CO)
	Fasal	IFA
	Lunare	Nihon Shokubai
	Natureflex	Innovia Films
Cellulose acetate/polyethylene succinate	EnviroPlastic	Planet Polymer Technologies
Chitosan	Dolon	Aicello Chemical (Aicello Kagaku)
Polybutylene adipate copolymer	Bionolle	Showa Highpolymer
	Ecoflex	BASF
	EnPol	Ire Chemical
Polybutylene succinate	Bionolle	Showa Highpolymer
	SkyGreen BDP	SK Polymers
Polybutylene succinate copolymer	Biomax	DuPont
	Bionolle	Showa Highpolymer
	Eastar Bio	Novamont (was by Eastman Chemical)
	Ecoflex	BASF
	EnPol	Ire Chemical
	Iupec	Mitsubishi Gas Chemical
Polycaprolactone	CAPA	Solvay
	CelGreen	Daicel Chemical Industry
	Tone	Dow (was Union Carbide)
Polycaprolactone copolymer	Celgreen	Daicel Chemical Industries
Polyester, aliphatic-aromatic	Eastar Bio	Eastman Chemical Co. (Novamont)
	Ecoflex	BASF
Polyester carbonate	IUPEC	Mitsubishi Gas Chemical
Polyethylene sebacate	Eternacoll	Ube Industries
Polyethylene succinate	Lunare	Nippon Shokubai
Polyethylene succinate copolymer	Lunare	Nippon Shokubai
Polyethylene terephthalate copolymer	Biomax	DuPont
	Green Ecopet	Teijin
Polyhydroxyalkanoate	Nodax	Procter & Gamble
	Pullulan	Hayashibara Co
Polyhydroxybutyrate and copolymers	Biogreen	Mitsubishi Gas Chemical
	BioMer	Biomer
	Biopol	Metabolix
	PHBH, Nodax	Kaneka Corp.

TABLE 8.7 Reported Composition of Selected Biodegradable Plastics[*355–358] (Continued)

Polymer category	Trade name	Manufacturer
Polylactic acid	EcoPla	Cargill Dow Polymers
	Hycail	Hycail
	Lacea	Mitsui Chemicals
	Lactron	Kanebo Gohsen Ltd.
	Lacty (Lacti)	Shimadzu
	NatureWorks	Cargill
	Toyota Ecoplastics	Toyota Motor Corp.
	Vyloecol	Toyobo
PLA blend	Bio-Flex	Nature Compounds (FKuR Kunststoffe)
PLA copolymer	GS Pla	Mitsubishi Chemical
	Mazin	Gemplus, U. of Nebraska
	Plamate	Danippon Ink & Chemicals
	Vyloecol	Toyobo
Polysaccharide	Pullulan	Hayashibara Co.
Polytetramethylene adipate co terephthalate	Eastar Bio	Chemitech (Novamont)
Polyvinyl alcohol based	Aquarto	Planet Polymer Technologies
	Dolon	Aicello Chemical
	Ecomaty, Gosenol	Nihon Gohsei Kagaku Kogyo (Nippon Synthetic Chemical)
	Enpol	Polyval
	Elvanol	DuPont
	Erkol	Erkol
	Hydrolene	Idroplast
	J-Poval	Japan VAM & Poval
	Kuraray Poval, Kuraray Exeval, Kuralon	Kuraray
	MonoSol	MonoSol Div. of ChrisCraft
	Polinol	DC Chemical Co.
	Poval	ShinEtsu Chemicals, Kureha (Kuraray)
Starch-based	BIOPar	BIOP Biopolymer GmbH
	Bioska	Plastiroll Oy
	Clean Green	Clean Green Packaging, StarchTech
	Cohpol	VTT Chemical Technology
	Cornpol	Japan CornStarch
	Earthshell	EarthShell Corp.
	Eco-Foam	American Excelsior Co.
	Eco Ware	Nissei
	Envirofil	EnPac (DuPont & ConAgra)
	EverCorn	Evercorn (Japan Corn Starch & Grand River Technology)
	Flo-Pak Bio	Marfred Industries, Free-Flow Packaging Corp.
	Greenfill	Green Light Products Ltd., Heygates

TABLE 8.7   Reported Composition of Selected Biodegradable Plastics[*355–358] (Continued)

Polymer category	Trade name	Manufacturer
Starch-based (continued)	Greenpol	Greenpol Co
	Greensack	Convex Plastics
	Mater-Bi	Novamont
	Paragon	Avebe Bioplastics
	Placorn	Nihon Shokuhin-Kako
	Plantic	Plantic Technologies
	Renature	Marfred Industries, Storopack
	Solanyl	Rodenburg Biopolymers
	Star-Kore	Star-Kore Industries
	Supol	Supol GmbH
	SWIRL	Milleta
	Vegmat	Vegeplast

*Note: not all of these plastics are currently commercially available; reported compositions in some cases may be inaccurate.

of their products is Clean Green Packing.[359] American Excelsior, headquartered in Arlington, TX, manufactures several starch-based plastics, including Eco-Foam loose fill and sheet and laminated structures. These materials also are water soluble as well as biodegradable.[360]

Novamont makes Mater-Bi, another starch-based biodegradable polymer. This material is widely used for bags for collection of organic wastes for composting. Novamont claims programs serving over 15 million people use Mater-Bi bags and carriers for collection of organic wastes and grass clippings.[361]

Biotec is a German producer of starch-based polymers. In mid 2005, it announced plans to expand its production capacity from 2000 tonnes per year to 12,000 tonnes per year within the next 6 months, with the announcement of acquisition by Stanelco. Stanelco makes radio-frequency welding equipment and pioneered RF technology used to seal starch-based polymers without overheating or burning. Stanelco also recently acquired Adept Polymers, a manufacturer of water-soluble polymers.[362,363] Stanelco uses Biotec products to make food trays, air pillows, and edible packaging. The company hopes to bring the price for starch-based sheet for thermoforming down to less than 10 percent more than competitive PET sheet.[364]

EarthShell Corp. uses a combination of starch and limestone to make packaging products that are biodegradable and compostable. Products manufactured include cups, plates, bowls, sandwich wraps, and hinged-lid containers. The company's foam laminate is produced by mixing the starch and limestone with water and fiber and placing it in a heated mold. Vaporization of the water foams, forms, and sets the product. The material physically disintegrates in water when it is crushed or broken.[365]

Freedonia predicts demand for starch-based plastics will increase an average of 11.6 percent per year through 2008, reaching a total of 83 million pounds, compared to 48 million in 2003.[350]

8.19.2.2 Other Polysaccharides.

Some biodegradable or biobased plastics are made by starting with cellulose and modifying it to make it thermoplastic. For example, the Japanese company Ebara Jitsugyo, Ltd. (EJ CO), manufactures several products based on cellulose acetate that are claimed to be biodegradable.[366]

Some efforts are being made to develop biodegradable plastics based on chitosan, produced from shrimp and crab shells. One company involved in this effort was Aisero Chemical, in Japan. Other natural polysaccharides have also been investigated. Hayashibara produced Pullulan polysaccharide-based films, with properties similar to polystyrene. Pullulan is still used in cosmetics applications but does not appear to be used as film.

8.19.2.3 PHB, PHBV, and Other Bacterial Polyesters.
Currently, the leading company in the development of polyhydroxyalkanoates (PHAs), a type of polyester produced by biological fermentation, is Metabolix, based in Cambridge, MA. The company's PHAs are stable to water but biodegrade in fresh water, sea water, soil, and composting environments. They also degrade under anaerobic conditions such as in septic systems and municipal waste treatment plants. The polymers are also claimed to be recyclable. Properties of homopolymers and copolymers in this family vary from strong moldable thermoplastics to highly elastic to soft and sticky, depending on the chemical composition. Molecular weights range from about 1 thousand to 1 million. The polymers are produced in "biofactories" by accumulation inside microorganisms (produced by recombinant DNA) and later harvesting. The company has demonstrated its fermentation technology, based on renewable resources such as corn sugar and vegetable oil, on the tonnage scale and claims that commercial-scale trials indicate that production costs will be well under $1 per pound. This is much lower than was achieved by previous ventures into PHA production. Metabolix is also engaged in research to produce PHAs directly in nonfood crop plants. The U.S. Department of Energy has supported research on using native American prairie grass for this purpose, through genetic engineering. In 2005, Metabolix was awarded the 2005 Presidential Green Chemistry Award in the small business category for its progress in commercializing these materials.[367,368]

In 2004, Metabolix announced an alliance with Archer Daniels Midland to commercialize its fermentation technology. The two companies, in a 50/50 joint venture, will establish a state-of-the-art 50,000 ton production facility and will manufacture and market natural PHA polymers for a wide variety of applications, including coated paper, film, and molded products.[369] Also in 2004, Metabolix began a joint project with the U.S. Army Natick Soldier Center on development of PHA packaging film for the Navy. The project is being supported by the Navy's Waste Removal Afloat Protects the Sea (WRAPS) Program. It focuses on melt-processing PHA films used to enclose many of the fresh foods found in grocery stores and may also include exploration of nanocomposites and coextruded multilaminate systems as potential food packaging systems for both the Navy and the Army.[370] In 2005, Metabolix announced an alliance with British Petroleum to further develop direct production of its PHAs in switchgrass. The two-year agreement will research and develop grass crops containing high levels of naturally grown polymers that can be used to produce biodegradable plastics. A coproduct will be "advantaged biomass material which can be converted to energy."[371]

Recently, investigators at University College in Dublin, Ireland, found that a certain strain of bacteria can use waste styrene to make PHAs, opening the door for remediation of a toxic waste and its conversion into a useful plastic at the same time. The team is working on scale-up and increasing the efficiency of the bacterial action to make commercially useful amounts of PHA plastics.[372]

8.19.2.4 Polylactides.
Polymers based on lactic acid have a very long history. DuPont patented the ring-opening polymerization process for lactic acid polymers, following conversion of lactic acid to a cyclic dimer in the first reaction stage, back in 1954.[373] Applications were primarily in the medical area, including items such as absorbable sutures. High cost was a major deterrent to more widespread use. One of the key developments permitting more economical production of polylactides, and opening the door to larger-volume uses such as packaging, was development of the ability to control the ratio and distribution of the d- and l- forms of lactide in the polymer backbone. This is essential to

controlling crystallization and producing plastics with the desired combination of physical properties.

The development of polylactides for packaging and other nonmedical applications was spurred by a partnership between Dow Chemical Company and Cargill. Cargill had begun working on polylactides before 1987. They began production of pilot plant quantities in 1992, under the EcoPLA name, and built a 4000-tonne-per-year facility near Minneapolis. The 1997 joint venture with Dow, under the name Cargill Dow Polymers, enabled further commercialization of these materials.[374,375] In 2005, Cargill bought Dow out of the company, which is now operated as a wholly owned subsidiary of Cargill under the name NatureWorks LLC. In addition to NatureWorks PLA plastics, the new company sells PLA fiber under the Ingeo brand name. Cargill reports that PLA prices are now competitive with PET.[376]

The current production method for polylactide plastics begins with bacterial fermentation of carbohydrates, generally cornstarch that has first been fermented to sugar using a type of lactobacillus. In 1992, Cargill patented a polymerization process for polylactide production using prepolymerization to low-molecular-weight polylactic acid, catalytic conversion to lactide, and finally ring-opening polymerization to produce PLA. The properties of the resulting polymer are determined by its molecular weight and the proportion of d-, l-, and meso-lactide in the polymer and by processing conditions.[377] Cargill Dow built a large commercial facility for PLA production near Blair, NE, which opened in 2002.

PLA bottles are currently being used for Biota water bottles. Since the molding temperature for these bottles is lower than for PET, energy savings are claimed. The bottles are claimed to be the first in the world to be approved by the Biodegradable Products Institute. The bottles are substantially heavier than the PET alternative, but reportedly this was done by choice so that the bottle would have a premium feel. The bottles will disintegrate in 75 to 80 days in a composting environment that provides a temperature of 120 to 140°F, microorganisms, and moisture. They will not, however, degrade quickly in a backyard compost operation, as these do not reach as high a temperature.[378]

Alcas, of Italy, is using NatureWorks PLA for its ice cream cups and tubs as well as for drinking glasses, straws, and spoons in its "02 line." The name stands for "zero consumption × zero waste."[379] Sony reports it is using PLA with other materials, including an inorganic flame retardant, in casings for some of its electronics products.[380] In 2005, Fujitsu claimed to be the first company to use PLA in a large laptop computer; the material used is a blend of about 50 percent PLA with an amorphous plastic developed by Toray.[381] In 2004, the Belgian brewery Alken Maes became the first Belgian brewer to use PLA cups. Previously, the brewery used polycarbonate cups at public festivals but saw the PLA cups, made by Huhtamaki from Cargill Dow's NatureWorks PLA, as a more environmentally friendly option.[382]

In 2005, Sharp Corporation announced that it has developed technology to blend PLA with polypropylene recovered from electronics recycling, using a newly developed compatibilizer that greatly improves the properties of the blended materials. Its intent is to use the blend in new consumer electronics, thus significantly reducing their environmental impact as compared to using petroleum-based feedstocks.[383] NAT-UR, in 2005, became the first cutlery product to be granted the Biodegradable Products Institute compostable designation. The utensils are manufactured with NatureWorks PLA in combination with starch. They are reported to degrade in 30 to 60 days in a compost environment.[69]

Several companies have determined that IR sorting equipment will successfully separate PLA from a PET recycling stream.[384]

Freedonia predicts demand for PLA will grow at an annual rate of 24.6 percent through 2008, reaching a total of 135 million pounds that year, compared to 45 million in 2003.[350]

8.19.3 Other Biodegradable Polyesters

8.19.3.1 Polycaprolactone. Polycaprolactone has been used in relatively small quantities for a long time. The most well known supplier was Union Carbide, which manufactured Tone brand polymers, which are compostable. Dow Chemical acquired this technology when it acquired Union Carbide and continues to manufacture Tone polymers.[385]

8.19.3.2 Ecoflex. BASF manufactures Ecoflex, a synthetic aliphatic-aromatic copolyester. Ecoflex resins have been certified as compostable by the Biodegradable Products Institute, DIN CERTCO in Europe, and the Biodegradable Plastics Society in Japan (which describes the polymer as polybutylene adipate/terephthalate). When used for bags for collecting and composting food scraps and yard trimmings, disposable packaging or agricultural sheeting, it decomposes in compost within a few weeks without leaving any residues. Combinations of thermoplastic starch and Ecoflex are used for films and coatings for food packaging. BASF's plant in Ludwigshafen, Germany, has a production capacity of 8000 tonnes per year.[386] In 2005, the company announced that it will start up a new 6000-tonne-per-year plant in early 2006, at Schwarzheide in Germany, due to the expanding world market for biodegradable plastics. In particular, the company mentioned the amendment to the German packaging ordinance that exempts packaging certified as biodegradable from fees under the German DSD packaging ordinance until 2012. Ecoflex F is designed for flexible film applications, whereas Ecoflex S is designed for blends. Most current use of Ecoflex is in blends with other renewable materials, including starch, cellulose, and polylactic acid.[387,388]

8.19.3.3 Eastar Bio. Eastar Bio is a family of copolyesters developed by Eastman Chemical Company in 1998. Properties are similar to low-density polyethylene. The material meets requirements for food contact in a variety of applications and is compostable. In September, 2004, Eastman sold the business and technology to Novamont.[389]

8.19.3.4 Polybutylene Succinate (PBS). Manufacturers of polybutylene succinate, produced from polymerization of succinic acid and 1,4-butanediol, include Showa Highpolymer, which produces Bionolle polymers; SK Polymers, which makes SkyGreen BDP; and Mitsubishi Chemical. Normally, the source of both monomers is maleic anhydride. However, Mitsubishi is working with Ajinomoto to produce succinic acid by fermentation of sugar and starch, providing a biodegradable polymer that is partly biobased.[390]

8.19.3.5 Biomax. Biomax is a family of aliphatic/aromatic polyesters based on polyethylene terephthalate and manufactured by DuPont. A combination of hydrolysis and microbial action breaks down the polymer, and some grades have been certified as compostable. Reportedly, as many as three different proprietary aliphatic monomers may be incorporated into the polymer.[391]

8.19.3.6 Polyvinyl Alcohol and Other Water-Soluble Plastics. Several water-soluble polymers have a long history of use in niche applications. Many of these polymers are biologically stable when they are in the solid state but will biodegrade readily once they are dissolved. These include polyvinyl alcohol, cellulose esters and ethers, acrylic acid polymers, polyacrylamides, and polyethylene glycol, among others. Polyethylene oxide is biodegradable at low molecular weights.

8.19.3.7 Polyvinyl Alcohol. Polyvinyl alcohol is formed by hydrolysis of polyvinyl acetate. By controlling the degree of hydrolysis, the solubility can be modified, resulting in grades that will dissolve only in hot water or grades that dissolve in cold water as well. Some grades can be extruded, but others must be cast from solution.

One important application is in laundry bags and hamper liners for use in health care facilities. The filled bag is sealed shut with an attached adhesive strip. When placed in the washer, the adhesive and bag break down completely during the hot washing and disinfection. Any remaining polymer will biodegrade during the wastewater treatment process. The bags are impermeable to bacteria and viruses during normal use, as well as resistant to gases, solvents, and cool liquids, so they cut the risk of contamination, protecting hospital staff.[354]

PVOH films are also used to encapsulate agricultural chemicals to avoid human exposure when the chemicals are measured into water for application to crops. It is widely used as a binder and has other niche applications as well.

One of the first manufacturers of polyvinyl alcohol (PVOH) was Air Products & Chemicals, which manufactured it under the Airvol trade name. It appears that the product may no longer be available, however. Another early manufacturer was ChrisCraft Industrial products, Inc., which still makes PVOH through its Monosol division. Another major supplier is Kuraray, which makes a variety of water soluble plastics, including PVOH under the name Poval. DuPont sells PVOH under the Elvanol tradename and Polyval under the name Enpol. There are other suppliers as well.

8.19.4 Other Biodegradable Plastics

A number of investigators are working on the development of protein-based plastics. In some cases, these are being targeted as edible films. Of course, if a film is edible, it is generally biodegradable. Starting materials include zein, a corn protein; soy protein; and other materials. None of these materials has yet reached any large-scale commercial application.

Researchers at Cornell have created a biodegradable polymer called polylimonene carbonate, from limonene obtained from orange peel and converted to limonene oxide, plus carbon dioxide. The polymer has properties similar to those of polystyrene and has potential to use carbon dioxide that would otherwise be emitted into the atmosphere, adding to the greenhouse effect.[392]

8.19.5 Nonbiodegradable Biobased Polymers

As mentioned, there has also been interest in biobased polymers because of their basis on renewable feedstocks, regardless of whether they are biodegradable. One of these materials is Sorona™, manufactured by DuPont. In 2004, DuPont and Tate & Lyle PLC, a London-based renewable ingredients company, formed a joint venture, DuPont Tate & Lyle BioProducts LLC. A plant in Louden, TN, will produce the key polymer building block, 1,3-propanediol (PDO), by fermentation from corn sugar in place of the petroleum-based process currently used. The polymer will then be 37 percent based on corn. DuPont has a goal of deriving 25 percent of its revenue from nondepletable resources by 2010, and this will be a step toward that goal. In 2002, the company derived 14 percent of its revenue from such resources. Sorona is intended for a variety of applications, including textile apparel, interiors, engineering resins, and packaging. Sorona is not biodegradable, although DuPont says it has the technology to make biodegradable Sorona resins if the market grows for such materials, and mentions packaging as a potential application.[393,394]

Another market for biopolymers is production of polyurethanes. Soybean oil can be combined with an isocyanate to create soy polyol, which can then be used in all types of rigid and flexible polyurethane foam applications.[395]

8.19.6   Additives Claiming to Instill Biodegradability

As mentioned earlier, in the late 1980s, when the "solid waste crisis" hit the public consciousness, a number of manufacturers took advantage of consumers' desire to do something positive for the environment by marketing "biodegradable" merchandise sacks and garbage bags that contained about 6 percent starch and 94 percent polyethylene. The claims of the manufacturers that the polyethylene fraction of the bags would biodegrade once the starch portion was consumed was eventually disproved, and these products disappeared from the marketplace. The next generation of starch/polyethylene blends claimed to be biodegradable because the blend included pro-oxidants. The argument was that the polyethylene would be oxidized, causing fragmentation of the molecular structure, and the small polymer chains would then be biodegradable. In this case also, there was no evidence that the process actually worked as claimed. In particular, when these bags were sent to landfill, the anaerobic environment that soon develops would prevent any substantial oxidation, even if the claim that the plastic was biodegradable after oxidation were true. These products also disappeared from the marketplace, and that was where the matter of additives stood until recently.

Currently, Symphony Plastic Technologies Plc is selling additives and polyolefin resins that incorporate the additives, claiming that the materials produced with this additive are biodegradable. The claim is that incorporation of usually less than 4 percent of the additive (a metal salt) in any polyethylene or polypropylene resin will render it "totally degradable." The action of the additive is initiated by any combination of heat, light, and stress, and the material can be engineered to start to degrade in a time period of between 60 days and 5 to 6 years, depending on the formulation, amount of additive used, heat, light, stress, and so forth. The compostable version is claimed to "completely degrade between 60 to 90 days in a commercial compost environment." If the resins are later added to a recycling stream, the extrusion process reportedly deactivates the additive, so it will not harm the recycled resin (yet the company says it can be added to resins being recycled to make them degradable and also states that the action of the prodegradant additive is "triggered by the extrusion process").[396]

The "totally degradable" $d_2w^\circledR$ Symphony resins are even claimed to be superior to biodegradable plastics, because they do not need a biologically active environment to start degrading. A close look at the claims shows that the resins reportedly undergo molecular fragmentation due to the effect of the additive. Biodegradation is claimed because, as a result of the "oxidative action" of the additive, "the molecular 'backbone' collapses," and the "molecular chains become shorter and water 'wettable,' permitting the formation of a bio-film on the surface of the plastics, which allows microbial deterioration to take over."[396]

In the section on "Credentials" on the company web site, there is one report of testing by an independent laboratory—but this testing ("Migration Testing and Ageing Tests on Symphony EMc Film") provides the information that accelerated weathering reduces tensile strength and elongation more than in a control film, and that oxidation at 175°C is much more rapid than in the control. Biodegradation was not evaluated.[396]

EPI Environmental Plastics, Inc., markets a similar product. Their additive can reportedly make LDPE, LLDPE, HDPE, PP, and PS "that will be returned to the natural biocycle." The company says the products will "degrade in landfills, photodegrade (through exposure to sunlight) in city streets and in the countryside, and begin to biodegrade at compost sites and in soil." The polymers with the additive are said to retain the same properties as without the additive "for a controlled period of time (service life) until degradation is triggered by one or all triggers of heat, sunlight, or enhanced mechanical stress." Again, the mechanism appears to be oxidation, as the company says it is "Pioneering OXO-Biodegradable Technology,"[397]

In June 2005, the International Biodegradable Polymers Association & Working Groups (IBAW) issued a formal "Position on 'Degradable' PE Shopping Bags," talking about the appearance in the marketplace of plastic bags and other products made with polyethylene that claim to be "degradable" or "bio-, UV- or oxo-degradable," or even "compostable" and where the underlying technology is "based on special additives (master batch) which, if incorporated into standard PE resins, are purported to accelerate the degradation of the film products." IBAW concludes that this technology and the products are not new but appeared in the 1980s when "many doubts [were] expressed as to whether these products provide what they promise. Such doubts are still valid in the current context."[398]

8.20 REFERENCES

1. U.S. Environmental Protection Agency, *Municipal Solid Waste Generation, Recycling, and Disposal in The United States: Facts and Figures for 2003*, May 2005, http://www.epa.gov/msw/pubs/msw05rpt.pdf.
2. U.S. Environmental Protection Agency, *Municipal Solid Waste in the United States*, available at (links to current and previous reports).
3. Kaufman, S., N. Goldstein, K. Millrath, and N. Themelis, The State of Garbage in America, *BioCycle*, Jan. 2004, pp. 31–41.
4. U.S. Environmental Protection Agency, *Characterization of Municipal Solid Waste in The United States: 1998 Update*, EPA530-R-99-021, Washington, D.C., Sept. 1999.
5. 2004 Survey of N. American Plastics Recyclers & Brokers, *Plastics News*, May 24, 2004, p. 9.
6. Toloken, S., PET Bottle Use Up, Recycling Still Slipping, *Plastics News*, Nov. 8, 2004, p. 4.
7. Powell, J., The Squeeze is on in Plastics Recycling, *Resource Recycling*, Nov. 2003, pp. 19–22.
8. China Spells Crisis for Postconsumer Plastics, *C&EN*, May 17, 2004, p. 11.
9. Toloken, S., NAPCOR Laying Off President, Two Execs, *Plastics News*, May 24, 2004, p. 1, 27.
10. Association of Plastics Manufacturers in Europe, *Plastics in Europe; An Analysis of Plastics Consumption and Recovery in Europe, 2002 & 2003*, APME, Brussels, Belgium, 2004.
11. Nolan-Itu Pty. Ltd., *PACIA National Plastics Recycling Survey 2004*, Plastics and Chemicals Industries Association, Melbourne, Australia, 2004.
12. Toloken, S., Contaminants Seep into Rates for Recycling, *Plastics News*, Aug. 31, 1998, p. 1.
13. Denison, R.A., *Something to Hide: The Sorry State of Plastics Recycling*, Environmental Defense Fund, Washington, D.C., 1997.
14. Container Recycling Institute, Aluminum Beverage Can Waste Passes the One Trillion Mark, press release, May 24, 2004.
15. Fenton, R., Reuse Versus Recycling: A Look at Grocery Bags, *Resource Recycling*, vol. 11, no. 3, March 1992, pp. 105–110.
16. Miller, C., DOE Report: Recycle Plastic and Metal, Refill Glass, Burn Paper, *Waste Age's Recycling Times*, March 7, 1995, p. 10.
17. Recycling Helps to Control Planet's Temperature, *CEMPRE News*, May, 2005, p. 1.
18. Felder, M. and C. Morawski, *Evaluating the Relationship Between Refund Values and Beverage Container Recovery*, CM Consulting, April, 2003, http://www.bottlebill.org/assets/pdfs/geography/deposit%20levels.pdf.
19. Container Recycling Institute, *Bottle Bill Resource Guide*, http://www.bottlebill.org/.
20. California Dept. of Conservation, www.consrv.ca.gov.
21. California Department of Conservation, Calendar Year 2004 & Biannual Report of Beverage Container Sales, Returns, Redemption & Recycling Rates, May 10, 2005.
22. California Department of Conservation, Californians Set State's All-Time Record for Bottle and Can Recycling in 2004, press release, May 19, 2005.
23. California Department of Conservation, AB 28 Frequently Asked Questions, http://www.consrv.ca.gov/DOR/AB28FAQs.htm.
24. Hawaii State Department of Health, Hawaii Beverage Container Deposit Program, http://www.hawaii.gov/health/environmental/waste/sw/waste/sw/hi5/index.html.

25. CM Consulting, *Who Pays What—An Analysis of Beverage Container Recovery and Costs in Canada 2001–2002*, June, 2003, http://www.bottlebill.org/assets/pdfs/geography/WPW_FINAL_REPORT.pdf.
26. McKenney, M., PET Recovery Blossoms North of the Border, *Resource Recycling*, March 2004, pp. 18–22.
27. Morawski, C., Recovering Canada's Containers, *Resource Recycling*, Jan. 2004, pp. 9–16.
28. CM Consulting, *An Overview of Plastic Bottle Recovery in Canada*, Environment and Plastics Industry Council, Aug. 2004.
29. Verpackungsrundschau, German Cabinet Confirms Amended Deposit Rules, Jan. 12, 2005, http://www.verpackungsrundschau.de/archiv/news/2005/01/lang_e/12010501.html.
30. Jobwerx News, Beverage Container Security Solution in Germany, July 8, 2005, http://www.jobwerx.com/news/Archives/tomra_biz-id=947351_740.html.
31. Environment Protection Agency, Government of South Australia, http://www.parliament.sa.gov.au/Catalog/legislation/Regulations/e/1995.41.htm.
32. Israel Union for Environmental Defense, The Deposit Law in Jeopardy? IUED Goes to Court, March 2005, http://www.iued.org.il/text_item.aspx?tid=320&menu=19.
33. White, L., *Extended Producer Responsibility: Container Deposit Legislation Report*, Zero Waste New Zealand Trust, Auckland, New Zealand, Sept. 2002, http://www.zerowaste.co.nz/assets/Reports/Beveragecontainers.pdf.
34. Gitlitz, J. and P. Franklin, *The 10¢ Incentive to Recycle*, 3rd ed., Container Recycling Institute, Arlington, VA, Feb. 2004.
35. B.E.A.R., *Multi-Stakeholder Recovery Project, Stage One*, 2002, http://www.globalgreen.org/BEAR/Projects/FinalReport.pdf.
36. Toloken, S., APR Retreats from Lobbying for Controversial Bottle Bills, *Plastics News*, Oct. 18, 2004, p. 4.
37. Gitlitz, J., Are Bottle Bills Still Relevant? The Two Faces of Beverage Container Recycling in America, *Resource Recycling*, Sept. 2004, pp. 10–15.
38. Boisson, E., C. McLendon, B. Franklin, J. Stutz and J. Morris, Understanding Beverage Container Recycling, *Resource Recycling*, Feb. 2002, pp. 25–28.
39. Chicago to Try Bins, *Resource Recycling* electronic newsletter, Dec. 10, 2004.
40. More Bad News for Blue Bags, *Resource Recycling* electronic newsletter, March 29, 2005.
41. Moore, W., Collection and Separation, in R. Ehrig, Ed., *Plastics Recycling: Products and Processes*, Hanser Pub., Munich, 1992, pp. 45–72.
42. The Waste and Resources Action Programme, *UK Plastic Bottle Recycling Survey 2005*, Recoup, March 2005.
43. American Plastics Council, *2003 National Post-Consumer Plastics Recycling Report*, 2004, http://www.plasticsresource.com/s_plasticsresource/docs/1700/1646.pdf.
44. Perkins, R. and B. Halpin, Breaking Bottlenecks in Plastic Bottle Recovery, *Resource Recycling*, June 2000, pp. 28–33.
45. Toloken, S., Chicago Getting All-Bottles Recycling Site, *Plastics News*, June 2, 2003, p. 1.
46. Dunbar, J., The status of all-plastic-bottle recycling collection, *Resource Recycling*, June 2003, pp. 20–24.
47. Birenyi, E., State of MRFs: 2001, *Resource Recycling*, Jan. 2001, pp. 16–21.
48. Skumatz, L. and C. Bicknell, Single-Stream Recycling: Assessing the Tradeoffs, *Resource Recycling*, Aug. 2004, pp. 22–29.
49. One Size Won't Fit All, *Resource Recycling*, July 2004, pp. 47–50.
50. Stein, S., Single-stream: A Recycling Method that Cuts Both Ways, *Resource Recycling*, Oct. 2004, pp. 22–27.
51. Emerson, D., Single Stream vs. Source Separated Recycling, *BioCycle*, March 2004, pp. 22–25.
52. Pytlar, T., Trends in Materials Recovery Facility Modernization, *Resource Recycling*, Oct. 2004, pp. 15–21.
53. Perkins, R. and J. Dunbar, Optimizing Plastics Recovery from Single-Stream MRFs, *Resource Recycling*, Feb. 2005, pp. 12–18.
54. Hubbard, S., Talking Points: Single Stream; Analyzing Collection and Processing Costs, *Resource Recycling*, Oct. 2004, pp. 28–31.
55. Japan: Forget 3-Stream Sorts … How About 10 or More?, *BioCycle*, June 2005, p. 63.

56. Society of the Plastics Industry, SPI Resin Identification Code: Guide to Correct Use, http://www.plasticsindustry.org/outreach/recycling/2124.htm.
57. *Environmental Packaging: U.S. Guide to Green Labeling, Packaging and Recycling*, Thompson Publishing Group, Tampa, FL, 2000.
58. PACIA, Plastics Coding System, Plastics and Chemicals Industries Association, May 2001, http://www.pacia.org.au/index.cfm?menuaction=gen&page=2.plasticscodingsys.html.
59. Canadian Plastics Industry, Resin Coding, http://www.plastics.ca/newsroom/default.php?ID=344.
60. Envocare, Plastics: Recycling and Marking Codes, http://www.envocare.co.uk/plastics.htm.
61. International Standards Organization, ISO 11469:2000: Plastics—Generic Identification and Marking of Plastics Products, May 2000.
62. Commission Decision of 27 Feb. 2003 Establishing Component and Material Coding Standards for Vehicles Pursuant to Directive 2000/53/EC of the European Parliament and of the Council on End-of-Life Vehicles, *Official Journal of the European Union*, L 53/58, Feb. 28, 2003, http://europa.eu.int/eur-lex/pri/en/oj/dat/2003/l_053/l_05320030228en00580059.pdf.
63. Hewlett Packard, Plastic Part Marking Algorithm, March 29, 2005, http://www.hp.com/hpinfo/globalcitizenship/environment/pdf/plasticmarkspec.pdf.
64. ASTM International, Standard Practice for Generic Marking of Plastic Products, D1972–91(2001), West Conshohocken, Pennsylvania, 2001.
65. Cirko, C. and B. Graham, Plastic Sorting Technologies Under the Microscope, *Resource Recycling*, June 2003, pp. 16–19.
66. Graham, B., *A Review of Automated Technology to Sort Plastic & Other Containers*, Revised, Environment and Plastics Industry Council, Mississauga, Ontario, May 2005.
67. Powell, J., The PRFect Solution to Plastic Bottle Recycling, *Resource Recycling*, Feb. 1995, pp. 25–27.
68. Powell, J. (Plas'tik): in a Flexible or Changing State, *Resource Recycling*, Jan. 2001, pp. 32–37.
69. NatureWorks LLC, Leading Recycler Sorting System Separates NatureWorks PLA from PET, press release, Jan. 24, 2005.
70. Peter Walker Systems, http://www.walkersystems.de/.
71. Hamos GmbH, http://hamos.com/en/index.htm.
72. SpectraCode Breaks Black Plastics Recycling Barrier, *Purdue News*, Feb. 2001,
73. SpectraCode, http://www.spectracode.com/practical5.html.
74. Encoded particles identify plastic parts, *Plastics News*, Oct. 4, 2004, p. 12.
75. Plas-Sep Limited, London, Ontario, http://www3.sympatico.ca/jamesd.brown/new-site/mainenglish.html.
76. Award-Winning Researchers Introduce World's First Continuous, Multistage Plastics Separation Plant, TransForum, Vol. 4, No. 3, 2004, http://www.transportation.anl.gov/publications/transforum/v4n3/plastics_separation_plant.html.
77. Schut, J., Commingled Plastic Waste: New Gold Mine for Automotive Processors, *Plastics Technology*, May 2001, http://www.plasticstechnology.com/articles/200105fa1.html.
78. Henricks, M., Electronics Recycling Separation Systems, *American Recycler Newspaper*, March 2005, http://www.americanrecycler.com/0305spotlight.shtml.
79. MBA Polymers, What We Do, http://www.mbapolymers.com/whatwedo.htm.
80. WIPAG Polymertechnik, http://www.wipag.com/main-e.htm.
81. Satake USA, http://www.satake-usa.com/pages/products/vision/plastic_products.htm#.
82. DeGaspari, J., Infrared Recycler, *Mechanical Engineering*, 2000, www.memagazine.org/contents/current/feature/infrared/infrared.html.
83. Salyp, http://www.salyp.com.
84. U.S. Food and Drug Administration, Center for Food Safety and Applied Nutrition, Office of Food Additive Safety, *Recycled Plastics in Food Packaging*, Oct. 2003, http://www.cfsan.fda.gov/~dms/opa-recy.html.
85. U.S. Food & Drug Administration, Center for Food Safety & Applied Nutrition, Office of Premarket Approval, *Points to Consider for the Use of Recycled Plastics in Food Packaging: Chemistry Considerations*, Dec. 1992, http://www.cfsan.fda.gov/~dms/opa-cg3b.html.
86. Sitek, F., Restabilization Upgrades Post-consumer Recyclate, *Modern Plastics*, Oct. 1993, pp. 64–68.

87. NAPCOR, *2003 Report on Post Consumer PET Container Recycling Activity Final Report*, National Association for PET Container Resources, Charlotte, North Carolina, http://www.napcor.com/2003_Report.pdf.

88. Toloken, S., Canadian Packagers Paying Recycling Costs, *Plastics News*, Jan. 31, 2005, p. 5.

89. Council for PET Bottle Recycling, www.petbottle-rec.gr.jp/english/.

90. Japan for Sustainability, "Japan Achieves High PET Bottle Collection Rate at 61%, http://www.japanfs.org/db/database.cgi?cmd=dp&num=861&dp=data_e.html.

91. Petcore, http://www.petcore.org/news_press_01.html.

92. New Research into Food Grade Plastic Bottle Recycling, *Food Packaging Bulletin*, 13(10), 2005, p. 5–6.

93. SAEFL—Swiss Agency for the Environment, Forests and Landscape, A Guide to Waste: PET Beverage Containers, http://www.umwelt-schweiz.ch/imperia/md/content/abfall/separatsammlung_e.pdf.

94. CEMPRE (Compromisso Empresarial para Reciclagem), http://www.cempre.org.br/english/.

95. National Association for PET Container Resources (NAPCOR), 2003 Report on Post Consumer PET Container Recycling Activity, Final Report, Sept. 2004, http://www.napcor.com/2003_Report.pdf.

96. American Beverage Association, 1994–2003 Carbonated Soft Drink Container Recycling Rates, http://www.ameribev.org/environment/rateschart.asp.

97. Container Recycling Institute, Report Shows Plastic Bottle Waste Tripled Since 1995; Group Calls on Coke and Pepsi to Stop Attaching Bottle Bills, news release, Sept. 15, 2003.

98. Michigan Dept. of Environmental Quality, *Green Glass Task Force Final Report to the Legislature*, February 222, 2005, http://www.deq.state.mi.us/documents/deq-whm-stsw-greenglasstaskforcereport2-22-05.pdf.

99. State of Maine, *Final Report of the Joint Study Commission to Study Reimbursement Rates for Maine's Bottle Redemption Businesses and Other Issues Related to the Handling and Collection of Returnable Containers*, Dec. 2001, http://www.maine.gov/legis/opla/bottlerpt.PDF.

100. California Department of Conservation, Biannual Report of Beverage Container Sales, Returns, Redemption, and Recycling Rates, May 10, 2005.

101. Coca-Cola, Consumer Advisory: Coca-Cola System Using Recycled Plastic, press release, March 1, 2000.

102. Toloken, S., PTI Banks on Growing Recycled PET Market, *Plastics News*, May 26, 2003, p. 9, 21.

103. Santosusso, A., Stadiums Hit a Homerun with Recycling Programs, *Resource Recycling*, Nov. 2004, pp. 27–29.

104. The Plastic Redesign Project, The Potential Impacts of Plastic Beer Bottles on Plastics Recycling, working paper, Sept. 1999, http://www.plasticredesign.org/files/plasticbeerbottles.html.

105. Petcore, Petcore Announces New Test Results for Recyclable Barrier Technologies, news release, Jan. 25, 2005, http://www.petcore.org/news_press_01.html.

106. Asahi Drops PET Bottle Plans, *Food Production Daily.com*, July 10, 2004, http://www.foodproductiondaily.com/news/news-ng.asp?id=55214-asahi-drops-pet.

107. Association of Postconsumer Plastic Recyclers, http://www.plasticsrecycling.org/.

108. Heubel, D., The Magnetic Duo, *Resource Recycling*, May 2004, pp. 24–28.

109. DeWinter, W., Poly(ethylene terephthalate) Film Recycling, in F. LaMantia, Ed. *Recycling of Plastic Materials*, ChemTec Pub., Ontario, 1993, pp. 1–15.

110. Milgrom, J., Polyethylene Terephthalate (PET), in R. Ehrig, Ed. *Plastics Recycling: Products and Processes*, Hanser Pub., Munich, 1992, pp. 17–44.

111. Gemark Corp., http://www.gemark.com/gemark_corporation.htm.

112. United Resource Recovery Corp. (URRC), http://www.urrc.net/.

113. DuPont Ends Recycling Experiment, *Plastics News*, Jan. 25, 1999.

114. Bisio, A. and M. Xanthos (eds.) *How to Manage Plastics Waste: Technology and Market Opportunities*, Hanser Pub., Munich, 1994.

115. PET Container Recycling Europe, Eastman Announces Breakthrough in PET Packaging Recycling, *PETCORE*, 1(2), Dec. 1999.

116. Tilley, K., PET Resin Giant Investing in U.S. Recycling, *Plastics News*, Feb. 12, 2001, p. 4.

117. Petrecycle, http://www.petrecycle.com/.

118. Rebeiz, K., D. Fowler, and D. Paul, Recycling Plastics in Construction Applications, *J. of Resource Management and Tech.*, vol. 21, no. 2, 1993, pp. 76–81.

119. Bakker, M., Using Recycled Plastics in Food Bottles: The Technical Barriers, *Resource Recycling*, May 1994, pp. 59–64.
120. Myers, J., Coca-Cola Closes the Loop with Multi-layer PET Bottle, *Modern Plastics*, April 1994, pp. 28–30.
121. Rabasca, L., Coca-Cola Introduces Recycled PET Bottle in Switzerland, *Waste Age's Recycling Times*, Feb. 21, 1995, p. 1.
122. Ford, T., FDA Clears Wellman's Recycled Sheet, *Plastics News*, Aug. 28, 1995, p. 6.
123. Busard, T., Use of "PCR" PET & HDPE Materials in Bottle Manufacturing, Presented at California RMDZ Administrators Workshop, April 7, 2005, http://www.longbeach-recycles.org/zoneworks_presentations/4_05_busard.pdf.
124. Amcor, http://www.amcor.com.
125. Wellman, Inc., www.wellmaninc.com/PETResins/.
126. Pryweller, J., Plastipak Adds Plants in Europe, *Plastics News*, March 8, 2004.
127. Toloken, S., FDA Approves Phoenix Process, *Plastics News*, April 30, 2001, p. 3.
128. Phoenix Technologies International LLC, http://www.phoenixtechnologies.net/.
129. Phoenix Wins Belgian OK, *Plastics News*, May 24, 2004, p. 19.
130. PET on Everybody's Lips, *PETplanet Insider*, April 2000.
131. Swiss Recycling Plant Inaugurated, *PETplanet Insider*, Aug. 2000.
132. Coca Cola, Opening of the First Nutrition-Grade Bottle-to-Bottle PET Recycling Plant in Latin America, press release, July 13, 2005, http://www2.coca-cola.com/presscenter/nr_20050713_americas_pet_recycling.html.
133. PET Bottle-to-Bottle with the Stehning Process, *PETplanet Insider*, Aug. 2000.
134. Committed to Recycling, *PETplanet Insider*, Sept., 2002.
135. Toloken, S., Erema Unveils Recycling System, *Plastics News*, July 7, 2003, p. 19.
136. Vacurema Makes its Mark in the Field of Post-consumer PET, *PETplanet Insider*, April, 2005.
137. Babinchak, S., Current Problems in PET Recycling Need to be Resolved, *Resource Recycling*, Oct. 1997, pp. 29–31.
138. PVC Costs in PET Bales, *Resource Recycling*, April 2000, p. 56.
139. Marks & Spencer Project Closes the Plastics Packaging Recycling Loop, *WRAP*, June 6, 2005, http://www.wrap.org.uk/waste_minimisation/retailer_initiative_innovation_fund/news_events/news/marks_spencer.html.
140. Project Delivers High Value Engineering Use for Recycled Plastics, *Omnexus*, Nov. 10, 2004.
141. Toloken, S., Uncleaned PET Bottles Put to Use in Coal Mines, *Plastics News*, Aug. 2, 2004, p. 10.
142. Apotheker, S., The Bottle Is the Bottleneck, *Resource Recycling*, Sept. 1994, pp. 27–42.
143. NAPCOR, 2000 Report on Post Consumer PET Container Recycling Activity, http://www.napcor.com.
144. *Plastics Recycling Update*, Jan.–June, 2005 issues.
145. American Plastics Council, 2002 National Post-Consumer Plastics Recycling Report, 2003.
146. Ontario, Environmental Protection Act, R.R.). 1990, Regulation 344, Disposable Containers for Milk.
147. Milk Run, *Solid Waste & Recycling*, Feb./March 1999.
148. Alberta Dairy Council, *Plastic Milk Jug Recycling Program, Annual Report 1999–2000*, http://www.milkcontainerrecycling.com.
149. Alberta Dairy Council, Alberta Dairy Council Milk Container Recycling Program, Annual Report, 2002–2003, http://www.milkcontainerrecycling.com/AB/documents/annualreport2003_lowres.pdf.
150. Itec Environmental Systems, http://www.iteceg.com/index.htm.
151. Used Oil Recycling.com, http://www.usedoilrecycling.com/index.cfm.
152. Ag Container Recycling Council, http://www.acrecycle.org/.
153. Croplife Canada, Crop Protection Stewardship, http://www.cropro.org/english/aboutcpi/stewardshipfirst.cfm.
154. DuPont, Tyvek Envelopes, Recyclability, http://envelopes.tyvek.com/en/aboutUs/faqs.shtml#1.
155. Haycore Canada, http://www.haycore.ca/indexen.html.
156. Feature Story: Recycling Tubs and Lids, *EPIC News & Views*, June 2004.
157. Toloken, S., Recyclers Worried over Opaque Milk Bottles, *Plastics News*, Oct. 27, 1997, p. 7.
158. Ecoplast Gets FDA Nod, *Plastics News*, April 6, 1998.
159. TransPac, http://www.gotranspac.com/.

160. Tyvek, Frequently Asked Questions, http://envelopes.tyvek.com/en/aboutUs/faqs.shtml#1.
161. Toloken, S., Film Recycling is Gaining Momentum, *Plastics News*, May 16, 2005.
162. PlasticBagRecycling.org—The online resource for film recovery in California, http://ww.plasticbagrecycling.org/.
163. Hilex Poly Co., Hilex Poly Announces "Bag 2 Bag" Recycling Facility and Program, news release, Nov. 5, 2004, http://www.hilexpoly.com/news112904.htm.
164. Nextlife Recycling, http://www.nextlife-recycle.com/.
165. Environment and Plastics Industry Council, *Best Practices Guide for the Collection and Handling of Polyethylene Plastic Bags and Film in Municipal Curbside Recycling Programs*, http://www.cpia.ca/files/files/files_BestPracticesGuide.pdf.
166. EPIC, The Truth About Plastic Bags, July 7, 2005, http://www.cpia.ca/epic/media/news.php?ID=997.
167. EPIC, Plastic Stretch-wrap in Big Demand, http://www.cpia.ca/epic/media/news.php?ID=995.
168. EPIC, Stretch Wrap Recycling, a How To Guide, http://www.plastics.ca/epic/.
169. Dixons Group, Dixons Group and Nelson Packaging Bag Environmental Award, press release, May 17, 2004.
170. Statewide Program to Recycle Nursery Greenhouse Film Expands, *BioCycle*, Sept. 1998, p. 19.
171. New Jersey Department of Agriculture, New Jersey Nursery and Greenhouse Film Collection Sites (Year-Round), http://www.state.nj.us/agriculture/filmrecycling.htm.
172. Environment and Plastics Industry Council (EPIC), *Best Practices Guide for Agricultural Plastic Film*, http://www.cpia.ca/files/files/files_best_practices2.pdf.
173. Geiselman, B., Mass. Floats Plastic Recycling Program, *Plastics News*, May 26, 2003, p. 22.
174. WasteCap of Massachusetts, 2005 Marine Shrink Wrap Recycling Program, http://www.wastecap.org/wastecap/Programs/shrinkwrap/shrinkwrap.htm.
175. Johnson, J., Delta Recycles Irrigation Pipe in 4 States, *Plastics News*, Oct. 13, 2003.
176. Delta Plastics, Recycling, http://www.deltapl.com/recycling.htm.
177. Battery Council International, Battery Recycling, http://www.batterycouncil.org/recycling.html.
178. Recycline, Inc., Recycling Partnership with Recycline Gives New Life to More Than 1 Million Stonyfield Farm Yogurt Cups, press release, Oct. 29, 2004.
179. Toloken, S., Thermoformer Elm Packaging buys NPRC, *Plastics News*, July 5, 1999, p. 1.
180. Ehrlich, R.J., The Economic Realities of Recycling, Polystyrene Packaging Council, http://www.polystyrene.org/environment/econ.html.
181. Toloken, S., Evergreen Expanding Program to Recycle PS Cafeteria Trays, *Plastics News*, March 7, 2005, p. 7.
182. Polystyrene Packaging Council, 2004 EPS Recycling Rate Report, May 2005, http://www.epspackaging.org/pdf/2004RecyclingRateRptWeb.pdf.
183. Alliance of Foam Packaging Recyclers, EPS Recycling, http://www.epspackaging.org/info.html.
184. Plastic Loosefill Council, http://www.loosefillpackaging.com/default.asp.
185. FP International, FP International Announces 23% Increase in Annual Polystyrene Recycling Totals, press release, Jan. 31, 2001.
186. FP International, http://www.fpintl.com/.
187. Alliance of Foam Packaging Recyclers, EPS Recycling—What's Next? *Molding the Future* 5(2), Oct. 1998, pp. 1,4.
188. Kodak, http://www.kodak.com/eknec/PageQuerier.jhtml?pq-path=2879/4191&pq-locale=en_US.
189. EPS Recycling International, http://www.epsrecycling.org/.
190. Japan Expanded Polystyrene Recycling Association, http://www.jepsra.gr.jp/en/.
191. Korea Foam-Styrene Recycling Association, http://www.eps.or.kr/english/status.html.
192. Canadian Polystyrene Recycling Association, http://www.cpra-canada.com/.
193. Hydro Polymers: Searching for a More Sustainable PVC, Environmental Expert.com, http://www.environmental-center.com/articles/article1358/article1358.htm.
194. Infolink Australia, Why Armstrong's Floors are Always Green, http://www.infolink.com.au/articles/e5/0c032ae5.asp.
195. Principia Partners, *Post-industrial and Post-consumer Vinyl Reclaim: Material Flow and Uses in North America*, July 1999.
196. The Vinyl Institute, Recycling Vinyl, http://www.vinylinfo.org./recycling/index.html.
197. Toloken, S., Recyclers Tag PVC as Contaminant, *Plastics News*, April 20, 1998.

198. Toloken, S., Recycling Program in the Works for PVC, *Plastics News*, Aug. 31, 1998, p. 5.
199. Wisner, D., Recycling Post-consumer Durable Vinyl Products, presented at The World Vinyl Forum, Sept. 7–9, 1997, Akron, Ohio.
200. Alaisa, C., Giving a Second Life to Plastic Scrap, *Resource Recycling*, Feb. 1997, pp. 29–31.
201. American Recycler, Virginia Business Finds Niche in Recycling Vinyl Siding, http://www.americanrecycler.com/1virginia.html.
202. EPIC, *Best Practices Guide for the Recovery of Vinyl Siding from Residential New Construction and Demolition Projects*, Aug. 2005, http://www.cpia.ca/files/files/files_best_practives2.pdf.
203. Vinyl 2010, Progress Report 2005, http://www.vinyl2010.org.
204. Vinyloop, news, http://www.vinyloop.com/services/news/0,,2044–2-0,00.htm.
205. Sumitomo Opens PVC Recycling Operation, *Recycling Today*, Dec. 10, 2004, http://www.recyclingtoday.com/news/news.asp?ID=6852.
206. Environmental News Service (ENS), Anglers Recycling Fishing Line Before it Strangles Wildlife, Nov. 29, 2001.
207. U.S. EPA, Product Stewardship: Carpet, http://www.epa.gov/epr/products/carpet.htm.
208. Carpet America Recovery Effort, Memorandum of Understanding for Carpet Stewardship (MOU), http://www.carpetrecovery.org/about/mou.asp.
209. Carpet America Recovery Effort, Annual Report 2004, http://www.carpetrecovery.org/annual_report/04_CARE-annual-rpt.pdf.
210. Tullo, A., DuPont, Evergreen to Recycle Carpet Forever, *Chemical & Engineering News*, Jan. 24, 2000, pp. 23–24.
211. Zeftron Nylon, 6ix Again, http://www.zeftronnylon.com/main/enviro/6ixagain.cfm.
212. Evergreen Nylon Recycling, Inc., press release, Nov. 15, 1999.
213. Honeywell Nylon Products, press release, March 5, 2001.
214. Steady Progress Seen for Nylon Chemical Recycling, *Modern Plastics*, June 2000, p. 17.
215. Honeywell, Infinity, http://www.infinitynylon.com/usa/main/newsstory/news.asp.
216. Antron, http://antron.invista.com/content/sustainability/ant08_04.shtml.
217. Carpet America Recovery Effort, *CARE Quarterly Newsletter*, June 2005.
218. Shaw Contract Group, http://www.shawcontractgroup.com/html/html/capabilities/cap_sustain1.shtml.
219. DuPont, DuPont Composite Recycle Technology Makes Like-New Nylon from Scrapped Car parts, Helps Meet ELV Regulations, news release, June 24, 2003.
220. DuPont, Toyota, DuPont Engineering Polymers Show Value of Composite Recycle Technology in Automotive Air Intake Manifolds, news release, March 8, 2004.
221. Stone, H., Basics of Polyurethane Foam and the Use of Combustion Modifying Additives, Polyurethane Foam Association, presentation at San Francisco meeting, April 29, 2003, http://www.pfa.org//EFC9_Handout.html.
222. American Plastics Council, *Composition, Properties and Economic Study of Recycled Refrigerators*, http://www.plasticsresource.com/s_plasticsresource/doc.asp?TRACKID=&CID=174&DID=381.
223. Polyisocyanurate Insulation Manufacturers Association, Recycling of Rigid Polyurethane Foam Increases for European Firm With New Procedure, *PIMA Newsletter*, Aug. 2004, www.pima.org/newsletters/newsletter_080004.html.
224. BASF Corp., Philip Services Corp., press release, Sept. 16, 1997.
225. Dow Chemical Company, Dow and Mobius Technologies Open Demonstration Recycling System, press release, May 15, 2002.
226. Mobius Technologies, http://www/mobiustechnologies.com/.
227. Isola NV, http://www.isola.be/eng/index2.htm.
228. Alliance for the Polyurethanes Industry, press release, Oct. 18, 2000.
229. St. Vincent de Paul Society of Lane County, DR3 Mattress Recycling, http://www.svdp.us/mainsite/recycling/dr3.html.
230. McNulty, M., Conigliaro Expands Its Mattress Recycling Program, *BEDtimes*, April 2003.
231. Alliance for the Polyurethane Industry, *2004 API Year in Review*, http://www.apcnewsmedia.com/docs/2200/2124.pdf.
232. RAMPF Ecosystems, Layman's report LIFE ENV/D/000398, Large Scale Polyurethane Recycling, http://www.rampf-ecosystems.de/download/layman_report_engl.pdf.

233. Winslow, G. and T. Adams, Recycling Automotive Shredder Residue and Plastics Using Thermal Depolymerization Process, *GPEC 2004*, SPE, 2004, U.S. DOE, Energy Efficiency and Renewable Energy, http://www.eere.energy.gov/vehiclesandfuels/pdfs/alm_04/6c_daniels.pdf.
234. Alliance for the Polyurethane Industry, Recycled Polyurethane Markets Database, http://www.polyurethane.org/recycling/markets/markets_search.asp.
235. Wagner, T., Subsidiary of Italian Company Inaugurates the First Polyurethane Recycling Plant in Brazil, *Strategis*, http://strategis.ic.gc.ca/epic/internet/inimr-ri.nsf/en/gr125173e.html.
236. CD Recycling Plant is Europe's First, *Modern Plastics*, Sept. 1995, p. 13.
237. Bayer Plastics, Recycling—Household, http://plastics.bayer.com/plastics/emea/en/literature/3955/article.jsp?docId=3938.
238. Bayer Plastics, Recycling—Information and Communication Technology, http://plastics.bayer.com/plastics/emea/en/literature/3955/article.jsp?docId=3948.
239. Taiyo Yuden, Taiyo Yuden: Recycling Business of Polycarbonate Substrates for Optical Disc Launched, press release, Oct. 27, 2004.
240. Polymer Reprocessors, http://www.polymer-reprocessors.co.uk/.
241. Thai Poly Acrylic Company Limited, SHE Policy and Product Stewardship, http://www.thaipolyacrylic.com/english/coporateImage/coporateImage.htm.
242. Knowaste, http://knowaste.com.
243. Toloken, S., Recycling Technology Handles Contaminants, *Plastics News*, Dec. 2, 2002, p. 10.
244. Resyk, http://www.resyk.net.
245. Robbins, A., 2000 State of the Recycled Plastic Lumber Industry, Plastic Lumber Trade Association, http://www.plasticlumber.org.
246. Standards Boost an Industry, *ASTM Standardization News*, July 1999, pp. 22–26.
247. Robbins, A., 2001–2002 State of the Recycled Plastic Lumber Industry, March 11, 2002, presented at the Annual Meeting of the Plastic Lumber Trade Association, http://www.plasticlumber.org/srplpdfs/srpl01.pdf.
248. Sparks, K. Stiff Competition: Plastic Lumber Makers Shape the Market, *Resource Recycling*, May 1999, pp. 33–38.
249. Powell, J., Plastics Recycling: Changing Markets are the News, *Resource Recycling*, Oct. 2000, pp. 19–23.
250. Bregar, B., Plastic Rail Ties Gaining Favor, *Plastics News*, May 11, 1998.
251. Black, M., Crossties: New Technology and Old Standbys That Still Get the Job Done, *Railway Track & Structures*, Sept. 1999, pp. 17–23.
252. TieTek, http://www.tietek.com/html/installations.html.
253. Urey, C., Uncle Sam Recruits Recycled Plastic Lumber, *Plastics News*, July 13, 1998.
254. U.S. Plastic Lumber, www.usplasticlumber.com/.
255. Bregar, B., AERT Wins Major Contract for its Recycled Material, *Plastics News*, May 20, 1991, p. 1.
256. Trex, http://www.trex.com/.
257. CIC Innovation Consultants Inc., *A Custom Market Research Study Reviewing the Potential for: Plastic Lumber in Canadian Marine Applications*, Environment and Plastics Industry Council (EPIC), 2003.
258. New Brunswick Boat Builder Shows Off New Recycled Plastic Barge, *EPIC News & Views*, Dec. 2003, p. 2.
259. CIC Innovation Consultants Inc., *A Custom Market Research Study Reviewing the Potential for: Plastic Railroad Ties in Canada*, Environment and Plastics Industry Council (EPIC), 2003.
260. IBM, http://www.ibm.com.
261. CNET News, IBM's Recycling Program Goes Global, Nov. 9, 2004, http://news.com.com/IBMs+recycling+program+goes+global/2100-7341_3-5445207.html.
262. Duff, S., Group Addresses Electronics Recycling, *Plastics News*, May 7, 2001, p. 16.
263. Holmes, K., Filling the Policy Gap: Trends in E-scrap Recycling, *Resource Recycling*, April, 2005, pp. 32–35.
264. Canada Embraces E-Waste Recycling, *Environment News Service*, Oct. 27, 2004, http://www.ens-newswire.com/ens/oct2004/2004-10-27-01.asp.
265. J. Porter, Computers & Electronics Recycling: Challenges and Opportunities, *Resource Recycling*, April 1998, pp. 19–22.
266. Electronic Industries Alliance Consumer Education Initiative, http://www.eiae.org/

267. Durables Recycling Is Surprisingly Healthy, *Modern Plastics*, March 1999, p. 14.
268. IBM, Designing with a Green Pen: IBM E.server Environmentally Conscious Products Program, http://www.ibm.com/ibm/environment/products/greenpensummary2004.pdf.
269. Hansen, B., Minnesota Offers Unique Electronics Recycling Plan, *Environment News Service*, Oct. 18, 2000.
270. Plug in to Recycling, http://www.moea.state.mn.us/plugin/sonyevents.cfm.
271. Toloken, S. and J. Doba, Industry Wrestles with Electronics Recycling, *Plastics News*, Nov. 27, 2000.
272. MBA Polymers, http://mbapolymers.com/recyclingtoday.htm.
273. Japan Firms Reusing Plastics in Electronics, *Plastics News*, Oct. 5, 1998.
274. 3M, press release, Nov. 1998.
275. Old Computers Turn up in Road-paving Product, *Modern Plastics*, June 1999, p. 12.
276. Dunbar, J. and G. Conigliaro, Plastics from Recovered Electronics Pave the New Information Highway, *Resource Recycling*, May 1999, p. 43.
277. Dell Computer Corp.,
278. T. Krause, Design for Environment: A Last Will and Testament for Scrap Electronics, *Resource Recycling*, March 2001, pp. 14–23.
279. Hewlett Packard, Recycling Program Overview, http://www.hp.com/hpinfo/globalcitizenship/environment/recycle/index.html.
280. American Plastics Council, Plastics from Residential Electronics Recycling, Report 2000, Arlington, VA, 2000.
281. K. Betts, Could Flame Retardants Deter Electronics Recycling? *Environmental Science & Technology*, 35(3), pp. 58A–59A, 2001.
282. Herrick, T., As Flame Retardant Builds Up in Humans, a Ban is Debated, *Wall Street Journal*, Oct. 8, 2003.
283. Toxic-Free Legacy Coalition, Toxic Flame Retardants (PBDEs), A Priority for a Healthy Washington, http://www.environmentalpriorities.org/toxics/PBDE_fact_sheet-NG_112004.pdf.
284. Matsushita Electric Industrial Co, Ltd., Matsushita Electric (Panasonic) Develops Advanced Plastic Recycling System, press release, Nov. 5, 2002.
285. Holmes, K. Emerging Trends in E-scrap Processing, *Resource Recycling*, Jan. 2005, pp. 14–19.
286. Holmes, K., Collecting Scrap Electronics: End-of-life or Growth of an Industry? *Resource Recycling*, Dec. 2004, pp. 18–22.
287. Australian Mobile Telecommunications Association, The Mobile Phone Industry's Recycling Program, http://www.amta.org.au/default.asp?id=117.
288. Cellular Online, ReCellular to Provide Cell-phone Recycling in Latin America, http://www.cellular.co.za/news_2004/july/072204-recellular_to_provide_cell.htm.
289. International Association of Electronics Recyclers, IAER Directory of the Electronics Recycling Industry, http://www.iaer.org/search/
290. Pryweller, J., DaimlerChrysler Sets Recycling Standards, *Plastics News*, Jan. 4, 19999, p. 1.
291. DaimlerChrysler Corporation Working with Recyclers to Increase Recovery, Reuse of Plastics from Automobiles, *Environment News Service*, Sept. 19, 2000.
292. J. Pryweller, "New Plants Will Recycle Painted TPO Parts," *Plastics News*, Nov. 15, 1999, p. 12.
293. Polymer Sciences, Inc., http://www.polymersciences.com/.
294. Milan Injection Moulding Plant—USA, *Plastics Technology*,
295. Brooke, L., Plastics Recycling Goes Global, *Automotive Industries*, Feb., 2000.
296. ACI Plastics, http://www.aciplastics.com/.
297. Miel, R., ACI 'Liberates' Instrument-panel Plastics, *Plastics News*, April 24, 2000, p. 20.
298. DaimlerChrysler Tests Recycling System, *American Metal Market*, Sept. 25, 2000.
299. Galloo Plastics Recycling Plastics, *Actuscience*, March 2004, http://ambafrance-ca.org/hyperlab/actualite/archive-US/US-galloo.htm.
300. Daniels, E., Postshred Materials Recovery Technology Development and Demonstration, FY 2004 Progress Report, http://www.petcore.org/news_press_01.html.
301. Anderson, P., S. Kelly, and T. Rattray, Redesigning for Recycling, *BioCycle*, July 1995, pp. 64–65
302. Toloken, S., APR Urges Against the Use of Opaque White PET, *Plastics News*, April 29, 2002, p. 13.

303. The Council for PET Bottle Recycling, *Design Guideline for Recycling Designated EPT Bottles*, April 1, 2001, http://www.petbottle-rec.gr.jp/english/en_design.html.
304. Association of Plastics Manufacturers in Europe (APME), *Design for Recycling of Rigid Plastics Containers*, Brussels, June, 1996.
305. Petcore, *Guidelines on Acceptability of Additives and Barrier Materials in the PET Waste Stream for an Effective Recycling of PET*, http://www.petcore.org/chargement/publications/Guidelines.pdf.
306. Europa, Management of End of Life Vehicles, http://europa.eu.int/scadplus/leg/en/lvb/l21225.htm.
307. SpecialChem, Environmental and Recycling Issues for Plastics and Additives, Sept. 16, 2003, http://www.specialchem4polymers.com/resources/articles/article.aspx?id=1327.
308. Europa, Waste Electrical and Electronic Equipment, http://europa.eu.int/scadplus/leg/en/lvb/l21210.htm.
309. American Plastics Council, *A Design Guide for Information and Technology Equipment*, http://www.plasticsresource.com/s_plasticsresource/docs/800/745.pdf.
310. Fuji Xerox Co., Recycling Systems, http://www.fujixerox.co.jp/eng/ecology/report2000/pdf/14-20.pdf.
311. Goldsberry, C., OEMs Undertake Quest for Computer Afterlife, *Plastics News*, Jan. 13, 1997, p. 9.
312. The MPW 50, *Modern Plastics*, July 1, 2005.
313. U.S. Federal Trade Commission, Complying with the Environmental Marketing Guides, http://www.ftc.gov/bcp/conline/pubs/buspubs/greenguides.htm.
314. Toloken, S., "FTC Cracks down on Resin Code Placement," *Plastics News*, May 4, 1998, pp. 5, 24.
315. Phthalates and Male Babies, *C&EN*, June 6, 2005, p. 8.
316. Hileman, B., EU Bans Three Phthalates From Toys, Restricts Three More, *C&EN*, July 11, 2005, p. 11.
317. Baitz, M. et al, *Life Cycle Assessment of PVC and of Principal Competing Materials*, European Commission, July 2004, http://europa.eu.int/comm/enterprise/chemicals/sustdev/pvc-final_report_lca.pdf.
318. Toxics in Packaging Clearinghouse, Fact Sheet, Jan., 2005, http://www.toxicsinpackaging.org/adobe/TPCH-fact-sheet.PDF.
319. Toloken, S., Groups Strive to Simplify Life-cycle Rating, *Plastics News*, Oct. 25, 2004, p. 12.
320. Nappy Clash Leaves Consumers Confused, *Edie News Network*, May 20, 2005.
321. Environment Agency, *Life Cycle Assessment of Disposable and Reusable Nappies in the UK*, Bristol, UK, May 2005.
322. Sonneveld, K. K. James and H. Lewis, Sustainable Packaging: How Do We Define and Measure it? presented at 22nd Iapri Symposium, May 22–24, 2005.
323. Seldman, N., Creating a Zero Waste Future in Europe, *BioCycle*, Aug. 2004, p. 66–67.
324. Newcorn, D., Cradle-to-cradle: the next Packaging Paradigm? *Packaging World*, May 2003, pp. 62–65.
325. Braungart, M. and W. McDonough, *Cradle to Cradle: Remaking the Way We Make Things*, North Point Press, New York, 2002.
326. Oregon Department of Environmental Quality, Minimum Content Requirements, http://www.deq.state.or.us/wmc/solwaste/mincont.html.
327. California Integrated Waste Management Board, http://www.ciwmb.ca.gov.
328. Toloken, S., Calif. Law Could Tighten Requirements, *Plastics News*, Sept. 27, 2004, p. 39.
329. Toloken, S., California Targets Trash Bags, *Plastics News*, Nov. 24, 2003, p. 3.
330. International Institute for Sustainable Development, Instruments for Change: Compendium of Instruments, http://www.iisd.org/susprod/displaydetails.asp?id=148.
331. Friends of the Irish Environment, http://www.friendsoftheirishenvironment.net/.
332. British Plastics Federation, The Full BPF Position on the Issue of a Plastic Bag Tax, http://www.bpf.co.uk/bpfissues/plastic_bag_tax_bpf_position.cfm.
333. Hui, S., Greens Step Up Fight for Plastic Bag Tax, *The Standard*, Nov. 27, 2004, http://www.the-standard.com.hk/stdn/std/others.
334. Zero Waste, Plastic Shopping Bag Report, July, 2002, http://www.zerowaste.co.nz/assets/Reports/PlasticShoppingBagsandbiodegradablepackaging.pdf.

335. Smith, S., Plastic Bags, Briefing Paper No. 5/2004, Parliament of New South Wales, May 3, 2004, http://www.parliament.nsw.gov.au/prod/parlment/publications.nsf/0/33469EB37225F1F8 CA256ECF00077479.

336. Scotland Could Bag New Eco-tax, *Edie News Network*, June 24, 2005, http://www.edie.net/.

337. Kenya Proposes Plastic Bag Ban in New Waste Strategy, *Edie News Network*, March 4, 2005, http://www.edie.net/news/news_story.asp?id=9618&channel=5.

338. Tilley, K., Australian Town Bans PE Bags from Outlets, *Plastics News*, Sept. 22, 2003, p. 13.

339. Burning Argument over Packaging Waste Bill, *Edie Weekly Summaries*, Dec. 12, 2003.

340. European Commission, Packaging and Packaging Waste, http://europa.eu.int/comm/environment/waste/packaging_index.htm.

341. U.S. Electronics Businesses Face New European Union Compliance Directives, *American Recycler*, July 2005, http://www.americanrecycler.com/0705us.shtml.

342. Sheehan, B. and H. Spiegelman, EPR in the U.S. and Canada, *Resource Recycling*, March 2005, pp. 18–21.

343. Environment Canada, Extended Producer Responsibility & Stewardship, http://www.ec.gc.ca/epr/inventory/en/index.cfm.

344. Japan Expanded Polystyrene Recycling Association (JEPSRA), Law for Promotion of Sorted Collection and Recycling of Containers and Packaging, http://www.jepsra.gr.jp/en/j/j03.html.

345. Minnesota Office of Environmental Assistance,

346. NEPSI, http://eerc.ra.utk.edu/clean/nepsi/.

347. Powell, J., Stewardship of Electronic Products: Successes and Failures, *Resource Recycling*, Jan. 2005, pp. 24–27.

348. Saskatchewan Adopts E-waste Program, *Resource Recycling* electronic newsletter, March 29, 2005.

349. *Degradable Plastics and the Environment*, Mobil Chemical Company, 1988.

350. Freedonia, US Degradable Plastics Demand to Reach 370 Million Pounds in 2008, news release, Nov. 2004.

351. Barth, J. and B. Kroeger, Marketing Compost in Europe, *BioCycle*, Oct. 1998, pp. 77–78.

352. De Bertoldi, M., Composting in the European Union, *BioCycle*, June 1998, pp. 74–75.

353. Four Plastics Companies Commit to Biodegradable Plastics, *Environmental News Service*, Feb. 16, 2005.

354. EPIC, Degradable Plastics Gaining Favour in Niche Applications, Special Report *EPIC News & Views*, Environment and Plastics Industry Council, March 2005.

355. Biodegradable Plastics Society, http://www.bpsweb.net/02_english/.

356. List of Companies Involved in Producing Biodegradable Packaging material, Friendly Packaging, http://www.friendlypackaging.org.uk/materialslist.htm.

357. PRA Inc., Bioplastics, Agriculture and Agri-Food Canada, Aug. 2003, http://www.agr.gc.ca/misb/spec/bio/pdf/plast2_e.pdf.

358. Types, International Biodegradable Polymers Association & Working Groups, http://www.ibaw.org/eng/seiten/basics_types.html.

359. StarchTech, Inc., http://www.starchtech.com/index.html.

360. American Excelsior Company, http://www.amerexcel.com/.

361. Mater-Bi (Novamont), http://www.materbi.com/.

362. Cundy, C., Biotec to Expand Starch-based Polymer Production, *Plastics & Rubber Weekly*, July 26, 2005.

363. Cundy, C. Stanelco Acquires Starch Polymer Maker Biotec, *Plastics & Rubber Weekly*, June 6, 2005.

364. Packager Moves: Stanelco Buys Biotec; Berry Buys Kerr, *Food Production Daily.com*, June 7, 2005.

365. EarthShell, http://www.earthshell.com.

366. Eco-Series Products, EJ CO, http://www.ej-p.co.jp/english/index.html.

367. Metabolix, *Where Nature Performs*, undated brochure, Metabolix, http://www.metabolix.com/resources/brochure.pdf.

368. Metabolix, Metabolix's Natural Plastics Win Presidential Green Chemistry Challenge Award, press release, June 20, 2005.

369. Metabolix, Metabolix, Inc. and Archer Daniels Midland Company Enter Strategic Alliance to Commercialize PHA Natural Polymers, press release, Nov. 4, 2004.

370. Miller, M., Metabolix Wins Grant to Explore PHA Bioplastics for Packaging Film, Metabolix press release, March 2, 2004.
371. Barber, J. and O. Peoples, BP and Metabolix Agree to a Joint Development Program for Renewable Plastics, Metabolix press release, March 22, 2005.
372. Biodegradable Plastics are Made from Toxic Waste, *Food, Cosmetics and Drug Packaging*, Feb. 2005, pp. 28–29.
373. Kharas, G., F. Sanchez-Rivera, and D. Severson, Polymers of Lactic Acids, in D. Mobley, Ed., *Plastics from Microbes: Microbial Synthesis of Polymers and Polymer Precursors*, Hanser Pub., Munich, 1994, pp. 93–137.
374. Cargill, Cargill Developing Degradable Polymers Made from Corn, press release, Minneapolis, Minn., Oct. 15, 1991.
375. Thayer, A., Polylactic Acid is Basis of Dow, Cargill Venture, *C&EN*, Dec. 8, 19997, pp. 14–16.
376. NatureWorks LLC is New Name for Cargill's Corn-based Plastic Business, *Omnexus*, Feb. 4, 2005.
377. Auras, R., B. Harte and S. Selke, An Overview of Polylactides as Packaging Materials, *Macromol. Biosci.*, 2004, pp. 835–864.
378. Lingle, R., BIOTA's High-water Mark in Sustainable Packaging, *Packaging World*, Jan. 2005, pp. 62–64.
379. Ice Cream Cups Use Corn-based Plastic, *Food, Cosmetics and Drug Packaging*, Feb. 2005, p. 23.
380. Sony Corp., Sony Develops Flame-retardant Vegetable-based Plastic for the Casing of DVD Player, press release, February 12, 2004, http://www.sony.net/SonyInfo/Environment/news/2004/01.html.
381. Vink, D., Fujitsu Launches Large Notebook with Biodegradable Housing, *Plastics & Rubber Weekly*, Jan. 14, 2005, http://www.prw.com/main/newsdetails.asp?id=3601.
382. Vink, D., Summer Beer Drinkers Get Biodegradable Cups, *Plastics & Rubber Weekly*, Aug. 5, 2004, http://www.prw.com/main/newsdetails.asp?id=3059.
383. Sharp Develops New Technology to Blend Plant-Based Plastic with Waste Plastic, *Omnexus*, July 12, 2005.
384. xx NatureWorks LLC, Leading Recycler Sorting System Separates NatureWorks PLA from PET, press release, Jan. 24, 2005.
385. Tone, Dow Chemical, http://www.dow.com/tone/.
386. BASF's Ecoflex, Resins Earn the Biodegradable Products Institute's Emblem, press release, *E-wire*, June 23 2003.
387. BASF, BASF Increases Capacity for Biodegradable, news release, April 26, 2005, http://www2.basf.de/basf2/html/plastics/englisch/pages/presse/05_225.htm.
388. BASF, *Ecoflex Biodegradable Plastic*, undated brochure, http://www2.basf.de/basf2/html/plastics/englisch/pages/presse/05_225.htm.
389. Novamont, http://www.novamont.com/.
390. Spotlight on Specialty Chemical—Polybutylene Succinate, *Nadini Chemical Journal*, June 2003, http://www.nandinichemical.com/online_journal/may03.htm.
391. DuPont Biomax Resins, http://www.dupont.com/packaging/products/biomax.html.
392. Sweet and Environmentally Beneficial Discovery: Plastics Made from Orange Peel and a Greenhouse Gas, *Omnexus*, Jan. 18, 2005.
393. DuPont, DuPont and Tate & Lyle Form Bio-Products Joint Venture, press release, May 26, 2004, http://www.dupont.com/sorona/news/052604.html.
394. DuPont Sorona, Frequently Asked Questions, DuPont, http://www.dupont.com/sorona/faqs.html.
395. United Soybean Board, New Life-Cycle Data Quantifies Soy Advantages, *Biobased Solutions*, Nov. 2003.
396. Symphony Plastics, http://www.degradable.net/.
397. EPI Environmental Technologies, Inc., http://www.epi-global.com/en/Index-e.htm.
398. Position on "Degradable" PE Shopping Bags, International Biodegradable Polymers Association & Working Groups, http://www.ibaw.org/eng/downloads/050606_Position_Degradable_PE.pdf.

CHAPTER 9
PLASTICS AND ELASTOMERS: AUTOMOTIVE APPLICATIONS

K. Sehanobish and Tom Traugott
The Dow Chemical Company
Midland, Michigan

9.1 ELASTOMERIC MATERIALS

9.1.1 Introduction

This section discusses elastomeric materials such as thermoplastic elastomers (TPEs), TPVs, and other rubber systems such as thermoset elastomers/rubbers (TSRs) invoked in automotive applications apart from their use as impact modifiers in polymer blends. If one starts from under the hood, elastomers are used primarily in belts and hoses, bellows, and gaskets. At the separation between engine compartment and the interior, elastomers are used for sound management. Inside the car, they are used in floors, instrument panel skins, instrument panels for soft touch, gaskets for side mirrors, and so on. Outside the car, they can be found in tires (base tire, treads, side walls) and, finally, they are used in wire and cables and coatings in almost all parts of the car as needed.

Worldwide consumption of TPE was estimated to be about 3.0 billion pounds for automotive applications (including use as impact modifiers) in 2005. Some of the major producers of elastomeric materials are Dupont, Advanced Elastomers Systems, Bayer, BASF, and Dow. The automotive market represents 65 percent of the total TPE demand in North America and is equivalent in the rest of the world. Two major performance segments that drive TPE volume growth in automotive are (1) energy absorption and (2) acoustics. In the energy absorption segments, TPE is a dominant player in the form of neat materials, blends, and foams. Thermosetting elastomers are predominant in applications where acoustics is the driver. Polyurethane (PU) foams dominate the market in all sorts of acoustic-related applications, including flooring and carpets. Polyolefin (PO) foams will continue to challenge this segment in future.

9.1.2 Thermoplastic Elastomers

Let us first review the various thermoplastic elastomers used in automotive applications. These are styrenic block copolymers (SBCs), thermoplastic olefins (TPOs) (cross-linked and uncross-linked), thermoplastic vulcanizates (TPVs), thermoplastic polyurethane (TPU), copolyester, polyamide, polyisoprene, polybutadiene, and last but not the least, natural rubber.[1] Various types of SBC include styrene-butadiene rubber (SBR), SB(butadiene)S, SI(isoprene)S, SE(ethylene)BS, SEP(propylene), and SEB. Thermoplastic olefins are ethylene-α-olefin copolymers (α-olefins range from C3 to C8), EPD(iene)M(monomer), and their blends with polymers such as polypropylene (PP). TPVs are mostly made from EPDM and PP through a dynamic vulcanization process with some type of crosslinker. One can replace EPDM with nitrile or butyl rubber as well TPV. Thermoplastic polyurethanes are usually made using aliphatic or aromatic isocyanates and in large part are used as coatings, adhesives, and dispersions. Most of the TPO elastomers are used as impact modifiers in polypropylene and will also be covered separately in the section on engineering applications of TPO. About 80 percent of TPO is consumed by the automotive industry.

9.1.2.1 Styrene Butadiene Copolymer (SBC).   In the automotive industry, SBCs are often referred to as f-TPVs (fully cross-linked TPVs) as well. Most of them are SEBS blocks, and total global demand is approximately 30 million pounds. Global usage of SEBS is roughly expressed in Fig. 9.1. Most of these resins end up as air bag covers. Due to the possible litigation resins associated with airbags, SEBS compounds are designed to have superior elongation, tensile, and tear strength over other thermoplastic elastomers. Figure 9.2 displays a relative comparison of SEBS properties with a fully cross-linked TPV.

FIGURE 9.1   Global SEBS demand in terms of the major auto producers.

SEBS Compound	f-TPV Compound
• Tensile strength to 5000 psi	• Tensile strength to 1200 psi
• Elongation to 750%	• Elongation to 375%
• Shore A 70 ± 5	• Shore A 45 to 50 D
• Limited resistance to hydrocarbons	• "Some" resistance to hydrocarbons
• Flexible at ≤70°C	• Flexible at ≤ –40°C

FIGURE 9.2   Comparison of mechanical properties of SEBS compound vs. f-TPV.

Some of the SEBS grades in the market are Kraton G, Europrene, Bergaflex, and Multiflex G. The safety cushion provided by its properties makes it difficult for other elastomers to penetrate in this market, and SEBS compounds have a high price, in the range of $2.00/lb. Other application of SBCs are in window seals, gasketing and other noise, vibration, and harshness (NVH) mitigation. Its interesting that, although polystyrene is not the best in terms of squeak and rattle performance, the rubbery block structure of the copolymers makes it suitable for dealing with the dissipation characteristics needed in NVH.

SBCs are characterized by their molecular architecture, which has a "hard" thermoplastic segment and a "soft" elastomeric segment that alternate in many different ways. Figure 9.3 shows a typical molecular architecture of SBS rubber. In a very crude sense, SBCs have strength properties equivalent to vulcanized elastomer systems without vulcanization.

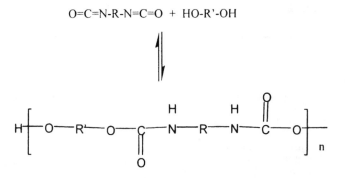

FIGURE 9.3   SBS molecule showing alternate blocks of sizes a and b.

9.1.2.2 Thermoplastic Olefin (TPO) Elastomers.   Elastomeric TPOs (soft TPOs) are increasingly used in special automotive applications because of their lower specific gravity, injection moldability, economics, recyclability, and noise performance. Essentially, elastomeric TPO is some type of blend of PP, filler, additives (slip agent, antioxidant, and so on) and a thermoplastic elastomer. Some variations of these essential ingredients are either compounded in extruders (other mixers) or prepared in some combination of reactors (often referred to as *reactor TPO*) that can make both PP and the rubber. TPOs are offered in Japan and sold under the name Toyota Super Olefin (TSOP). It is differentiated in its flow characteristics, modulus, and balance of low temperature properties due to some unusual co-continuous morphology achieved through some unique fabrication route of the PP and elastomer blend.[2] The elastomer-to-PP ratio can be adjusted to control the modulus and elastic recovery. These elastomers are not designed for high elastic recovery or compression set performance. They can be formulated at low cost and can be injection molded for applications that do not require high recovery. They are often used in such applications as battery covers or dash mats, as they provide adequate energy absorption characteristics at low cost. Certainly, the higher melting temperature of the PP component over typical elastomers is a positive attribute for some applications. Consumption of TPO in these applications is much less than TPO as a thermoplastic in interior and exterior applications such instrument panels, fascia, and so on.

9.1.2.3 Thermoplastic Vulcanizate (TPV).   TPVs are generally classified as f-TPV and p-TPV, wherein the prefixes indicate fully and partially vulcanized, respectively. Compounds of f-TPVs are typically made with EPDM and PP. In few special TPV compounds, EPDM is replaced by nitrile rubber, and natural rubber as well. The industry benchmark is TPV compounds, and they are sold under the trade name of Santoprene™ and marketed by Advanced Elastomer Systems (AES) Ltd. Mitsui is also dominant in the TPV based interior skins market in Europe. Seventy-five to 90 percent of TPV compounds are fully crosslinked with EPDM and are priced at approximately $1.80/lb. Two to 10 percent are partially cross-linked TPV compounds that sell for approximately $1.47/lb. Another 8 to 15 percent of the TPV compounds are fully cross-linked with nitrile rubber and cost more than $2.00/lb. Fully cross-linked TPVs are positioned for complex injection molded components, as they have good oil and fluid resistance and compression set. Typical automotive requirements that a fully cured TPV can meet are shown in Table 9.1.

TABLE 9.1 Typical Automotive Requirements Met by Fully Cross-Linked TPV

Fully cured EPDM in a polyolefin base			
	Grade B	Grade C	
Original durometer	67 ± 3	73 ± 3	Shore A
Original elongation	300	340	min. %
Compression set			
22 hr @ 70°C	35	35	max. %
70 hr @ 125°C	50	55	max. %
Allowable changes to original values			
After 168 hr @ 150°C			
Durometer	8	10	pt
Elongation	−35	−35	@
After 1008 hr @ 135°C			
Durometer	10	10	pt
Elongation	−30	−40	@
After 70 hr in oil at 125°C			
Durometer	−40	−40	pt
Elongation	−70	−70	@
Volume change	120	95	max. %

A. Y. Coran and a few other researchers discovered a new route to thermoplastic elastomers based on dynamic vulcanization—the process of vulcanizing an elastomer during its melt-mixing with molten plastic. The resulting compositions comprise completely vulcanized micron-size particles of rubber dispersed in a thermoplastic matrix. The elastomer phase can be quite voluminous; thus, very rubbery products can be produced that can be processed as thermoplastic materials. Commercial elastomers that were the result of the work are Santoprene® and Geolast® thermoplastic elastomers. Such products are called *thermoplastic vulcanizates,* a term coined by Coran. Thermoplastic elastomers are processed into finished parts at far less expense than are conventional TSRs. They are not vulcanized in the mold. (Conventional "in-the-mold" vulcanization requires long periods of time for the vulcanization process, or cross-linking, to take place. During this time, expensive molding equipment is engaged.). The Santoprene and Geolast thermoplastic elastomers can be rapidly and directly fabricated into finished parts using techniques (injection molding, calendaring, extrusion, and so forth) that are generally used with thermoplastics [e.g., polyethylene, polypropylene, poly(vinyl chloride), polystyrene, and others].[3]

TPV growth in automotive applications will be stimulated by its penetration into body/ glazing seals, interior skins, and belting. TPVs ability to be blow molded is a major reason why it is replacing TSRs, mainly neoprene and copolyesters, in automotive uses. It can easily be made into boots and bellows to act as fluid seals, acoustic seals, lubricant containers, and dust protectors.

9.1.2.4 Thermoplastic Urethane (TPU). Automotive TPU applications are rather limited. Mostly, it appears as aqueous dispersions for the coating industry. There is a very small market for molded goods, but the higher price makes TPU uninteresting even with some differentiated performance over other TPEs. TPU elastomer is renowned for its high tensile strength; abrasion resistance; elongation; tear strength; resilience; resistance to fuels, oils, and oxygen; good hydrolytic stability; excellent impact resistance; good vibration damping; and excellent low-temperature properties (usually the caprolactone-based TPUs).

TPUs are generally high-molecular-weight linear polymers that exhibit room-temperature elastomeric properties and are thermoplastic in nature. The aqueous dispersions are created by incorporating carboxyl group in the chain to provide water affinity. A typical polycaprolactone or polyether polyols with pure methylene-diphenylene isocyanate, and the procedure of polymerization includes reacting long chain diols such as liquid polyester, then the NCO terminated prepolymer, and adding a chain extender such as 1,4 butanediol.

9.1.3   Thermoset Elastomers

A substantial amount of these elastomers, in cross-linked form, also end up in the tire industry, in belts and hoses, and in gaskets. They are formulated and vulcanized in almost all cases. Most treads are made of isoprene rubber, polybutadiene, or SBR. Base tires and the side walls are generally made from natural rubber and polybutadiene. While tires are the biggest segment of TSRs, the nontire industry represents about 25 percent of the total value of the rubber consumed in North America and Europe. EPDM is the primary ingredient for the nontire TSRs. Others are natural rubber, neoprene, nitrile rubber, and others. The belt industry is another area in which TSR dominates. Belt manufacturing is a multistep process that may face some pressure from TPV in future. The major belt manufacturers (Gates, Dayco, and Goodyear) control 75 percent of the market. Nitrile rubber, neoprene, and EPDM are the usual ingredients for belt compounds. The automotive body mount segment is primarily controlled by butyl rubber. Globally, roughly 350 million pounds of EPDM ends up in body seals and glazing seals applications. TPVs are attempting to penetrate this segment, as they offer some performance advantages (compression set) and the opportunity to utilize thermoplastic processing tools. The compression set advantage over EPDM occurs only after long-term exposure at 70°C. However, the price of TPV still remains a challenge to its rapid penetration in the TSR market.

9.1.4   Conclusions

Elastomers serve a large potion of the overall need for plastics in the automotive industry. Their prime applications are in energy absorption, followed by noise, vibration, and harshness. However, the industry is crowded with many equivalent choices, and competition is growing fiercely. Sometimes, as a result of regulatory changes in areas such as safety and fuels, the need arises for either more elastomers or novel ones with differentiated properties.

9.2 OTHER THERMOPLASTICS IN AUTOMOTIVE APPLICATIONS

9.2.1 Introduction

This section discusses the more rigid non-elastomeric thermoplastics used in automotive applications like body panels, fenders, under-the-hood connectors, instrument panels, knee bolsters, radiator end cap etc. Generally, moduli of these resins are higher than elastomeric materials. Thermoset resins that are often used as fiber reinforced composites for high stiffness applications through reaction injection molding, transfer molding, liquid molding etc. will not be covered here.

9.2.2 Thermoplastics

Let us first review various thermoplastics used in automotive applications. These include nylon 6,6-based blends (e.g., nylon 6,6-PPO), glass-filled nylon 6,6 with without impact modifiers, homo- and copolymers of PP, polybutylene terephthalate (PBT), polyethylene (PE), bis-phenol A polycarbonate (PC), acrylonitrile-butadiene-styrene (ABS), PC-ABS blends, glass-filled PP, and ABS.

9.2.2.1 Nylon 6,6. Nylon 6,6 appears in automotive applications mostly in impact-modified, blended, and filled (mostly glass reinforced, with or with out impact modifiers) form. Impact modification of nylon 6,6 is of significant commercial importance, especially for automotive uses. Typical applications include fasteners for interior and exterior components, a host of connectors, holding fixtures, and radiator end caps, and a fairly common application is for air intake manifolds (Fig. 9.4). These applications rely on the ductility and toughness of impact-modified nylon 6,6 as a mechanism to connect assemblies or join components.

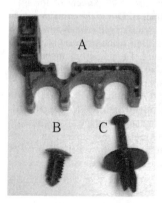

FIGURE 9.4 Photo of typical injection-molded impact-modified nylon 6/6 components including (A) fuel line connector, (B) christmas tree fastener, and (C) tufflock fastener. These applications require high toughness, so modifier selection and attention to detail during production are critical.

Fuel vapor canisters represent a relatively new automotive use of molded impact-modified nylon 6,6, and it is experiencing rapid growth (Fig. 9.5). It's an excellent choice for the application, given its inherent toughness and resistance to the combined effects of temperature and gasoline vapor. The Clean Air Act of 1990 mandated that all cars, beginning with the 1998 model year, must make provisions to trap hydrocarbon fumes emitted from the fuel tank and related to the refueling process.

FIGURE 9.5 Photo of two different fuel vapor canisters molded from impact-modified nylon 6,6. These applications generally require minimal toughness, so lower-cost impact modifiers and compounding technology are employed to produce the resultant products.

DuPont leads the market in impact-modified nylon 6,6, with products that exhibit a range of impact and mechanical performance. The market is segmented into three performance categories: super tough, moderately tough, and low toughness. The biggest breakthrough in relatively brittle nylon 6 and nylon 6,6 systems occurred in 1979 with the issuing of a key U.S. patent to Epstein.[4] The patent claimed that maleic anhydride or fumeric acid grafted EPDM rubber, among a variety of rubbers, after being blended with nylon, can provide a high degree of toughness. Epstein believed that the soft phase of the polymer only has to adhere with the polyamide matrix at the interface and that adhesion may be achieved by hydrogen or covalent bonding (Fig. 9.6). There were also claims with regard to the modulus of the rubber phase and rubber size. Mechanical performances of DuPont's major products that serve these markets are summarized in Table 9.2.

Historically, a number of different impact-modification technologies have been used. These include various maleated olefinic rubber such as EB, EP, and so on; SBS; brominated isobutylene-para-methyl styrene elastomers produced by Exxon; and many others. Dow's metallocene-based ethylene-α-olefin elastomers were found to be very effective as well. The rheology of toughened nylon 6,6 is usually directly related to the maleic anhydride graft level of the impact modifier. Rubber particle size averages of greater than 0.25 μm and less than 0.5 μm are required to achieve the required balance of mechanical performance. Optimum particle size varies with the percentage of rubber.

Nylon 6,6 is often blended with PPO (Noryl GTX) to make extremely rigid polymers for exterior body panels. Most of its applications are in Europe, in automobile fenders. This product is offered by General Electric Plastics, which currently is developing a commercial conductive polymer solution to facilitate production-line painting. Absorption of moisture by nylon remains a continuing critical issue to be dealt with to meet the performance requirements.

9.2.2.2 Homo, Co-, and Impact Copolymers of Propylene (H-PP, Co-PP, ICP).   PP can be divided into two types based on stereo regularity: isotactic and syndiotactic. The designation depends on the placement of CH_3 and H units along the backbone. Absence of regularity and random placement results in atactic PP. Both iso- and syndio-polymers are

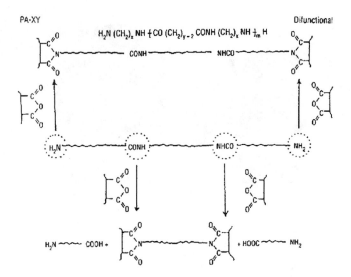

FIGURE 9.6 Maleic anhydride grafted rubber and its possible chemical interaction with PA6 or PA 6,6.

TABLE 9.2 Mechanical Performance of Dupont's Two Major Zytel Product Lines

Performance test	Units	Super tough Dupont Zytel ST801	Moderate tough Dupont Zytel 408
DTUL @ 66 psi	°F	420	420
Izod @ 72°F	ft-lb/in	17	5
Izod @ 0°F	ft-lb/in	6	2
Izod @ –40°F	ft-lb/in	2	1.2
Flex strength	psi	9800	–
Flex modulus	ksi	245	160
Tensile strength	psi	7500	7500
Tensile elongation	%	60	270

crystalline, whereas atactic is amorphous. Sundiotactic PP has not made a measurable entry in automotive applications and will not be discussed. Crystallinity in polypropylene can also be disrupted by incorporating α-olefin comonomers ranging anywhere from ethylene to octene, resulting in copolymers (co-PPs). Commercially, ethylene to butene comonomers are available. The absence of comonomers usually results in homopolymer polypropylene (H-PP) with predominantly isotactic conformation. Stereo defects in a predominantly isotactic structure result in homopolymers with varying crystallinity and crystal morphology. On top of stereo (tactic) defects, one can also introduce regio defects, depending on the catalyst used. Metallocene catalyst tends to introduce regio defects.

Impact copolymers (ICPs) can be made in one or more reactors by incorporating ethylene into the reaction process along with polymerization of homo- and copolymers of PP. Most commercial products end up incorporating approximately 50:50 weight ratio of ethylene and propylene. Thus, most ICP structure essentially contains a h-PP or co-PP with a certain weight percent of EP rubbers. Each producer has variations in the molecular weight and molecular weight distribution of these components, resulting in performance differentiation. Some uncontrolled levels of propylene-ethlyene (PE) and atactic PP are often formed in the reactors, which also results in variation in ICPs among various producers. One can also consider other ethylene-α-olefin rubbers to differentiate their ICP structures.

Due to lower specific gravity, injection moldability, economics, recyclability, and gloss, TPOs are continuously replacing ABS and PC/ABS polymers. As the price gap closes, disadvantages like weatherability, higher gloss, and noise characteristics still plague ABS-type polymers. Most automotive applications are for interior components such as instrument panels, interior trims, airbag doors, some interior skins, and so on— with the exception of bumper fascia, which is the most common exterior application. All TPOs used in automotive applications also require the balance of low-temperature toughness, density, a certain elastic modulus, coefficient of linear thermal expansion (CLTE), and shrinkage to fit into a specific part and tool design. Scratch and mar resistance of the final molded part can become very important, depending on the application (e.g., instrument panels). Crystallinity of the PP matrix plays a dominant role in determining the overall modulus. Thus, the high-stiffness TPOs always start from a very high-crystallinity, high-isotacticity homo- or copolymer or sometimes a reactor TPO that has a very high-modulus PP component. On the other hand, the high-toughness TPOs tend to contain a higher proportion of rubber. Fillers are mainly introduced for reasons of cost, CLTE, and (to a lesser extent) for modulus. The final performance of a TPO part is strongly dependent on mixing as well as the subsequent molding conditions. Dow Chemical Co. has the largest participation in the elastomer component of the compounded PP, while Exxon, Basell, and others are the biggest suppliers of the PP component of the TPO. While producers try to differentiate in the ingredients, tiers and compounders try to create differentiation through their processing and molding technology.

Although H-PP is often used as a blend for making TPO, or filled with short or long fibers, and so forth, there is very little direct H-PP application in the automotive industry. These applications take advantage of the lower cost and scratch resistance of H-PP. PP may also go into some of the fibers used in automotive components. In contrast, ICPs, with their excellent balance of toughness, rigidity, and processability, have made enormous inroads into automotive applications over the last two decades, becoming a dominant interior thermoplastic. (Clearly, the economics of hydrocarbon-based resins plays an important role.) Table 9.3 shows some characteristic properties of homo, copolymer, and impact copolymer PPs.

Automotive applications for PP, especially ICPs, are numerous and varied. Functional requirements that call for the use of ICPs include moderate impact, easy molding, moderate modulus, noise absorption, UV resistance, and a low-gloss surface. Most applications in which ICPs can be found are in the interior—trim, consoles, door panels, and instrument panel trim. Exterior parts are dominated by highly modified PPs, but ICPs can be found in less challenging applications such as wheel well covers.

In general, they can be classified into two broad categories: high stiffness and high toughness. TPO compound specifications are proliferated by the compounders, tiers, and OEMs into more than five types to bring differentiation to the consumer, but there is a

TABLE 9.3 Typical Properties of Homo- and Copolymer Polypropylenes

Property (ASTM)	Homo-PP	Co-PP	ICP
MFR	20	35	35
Flexural modulus (kpsi)	250	200	170
Izod impact strength (ft-lb)	0.8	0.7	2.9
Tensile yield strength (kpsi)	5.2	5.1	3.2
Percent elongation at yield	9	–	4
DTUL (264 psi, °F)	244	216	216

strong underlying current from the OEMs to use fewer specifications to meet all needs. Some typical TPO properties in the high-stiffness and high-toughness applications are listed in Table 9.4.

9.2.2.3 Polybutylene Terephthalate (PBT).   PBT is another polymer used mostly under the hood or close to the engine electrical and electronic connectors, for smart network in-terface devices, power plugs and electrical components, switches and controls, circuit breaker enclosures, and various housings. GE, Dupont, and Ticona are the biggest manu-facturers of PBT, which is sold under the brand names Valox® (GE), Crastin® (DuPont), and Celanex® and Vandar® (Ticona). Usually, these products range from 100 percent un-modified PBT resins to combinations of glass-fiber reinforced, mineral-filled, mineral/glass-reinforced, and flame-resistant grades. In automotive use, there is no known applica-tion of the unmodified PBT. The base PBT resin is made by reacting butane diol with dim-ethyl terephthalate or terephthalic acid, and this material is the foundation for grades that are reinforced with fiberglass, minerals such as mica or wollastonite, stainless steel fibers, or carbon fibers, or that are filled with glass beads or other essentially nonorienting fillers (Fig. 9.7).

PBT and modified resins offer chemical resistance, outstanding dielectric strength, out-standing electrical properties, low-temperature performance down to –40°F (–40°C), strength and modulus at elevated temperatures, very good processability (long flow in thin sections), and last but not the least, flame resistance.

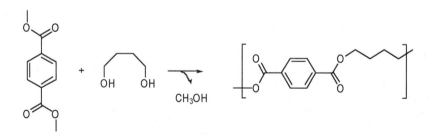

FIGURE 9.7   Linear PBT molecule polymerized by reacting butanediol with terephthalic acid. Conventional commercial production of polybutylene terephthalate (PBT) involves a condensa-tion reaction between dimethyl terephthalate and 1,4-butane diol.

TABLE 9.4 Typical Properties of High-Stiffness and High-Toughness TPOs

In or at reactor TPO	MFR	Flex mod., MPa	D/B transition instrumented falling dart, 3 mm	Application and global volume, million lb	CCR	Typical commercial products
High-toughness	>15	800–1000	Min. –40°C	Fascia and airbag, critical interior, >500	Paint adhesion (CLTE 80–120)	Stamylan P108 M97 Basell Hifax E308G9 Exxon 8224/8114 Dow and others
High-stiffness	>15	1200–1500	Min. –20°C	Instrument panels (IPs) and trim, >400	Scratch res. (CLTE 80–120) Gloss 60=1.5	Basell PPU9057HSHGL23 Exxon Exxtral BNU011 Dow and others

9.2.2.4 Polyethylene (PE). Polyethylene covers a huge segment of the polymer industry and can be classified into mostly linear high-density polyethylene (HDPE), substantially linear ethylene-α-olefin copolymers [also known commercially as linear low-density polyethylene (LLDPE)], and mostly long-chain branched low-density polyethylene (LDPE). Another subsegment of the ethylene-α-olefin copolymers was discussed earlier in terms of their elastomeric performance. HDPE is used predominantly as fuel tanks and is supplied by companies such as BASF, Solvay, and Phillips. Since these products are blow molded, they require a unique balance of high melt strength and toughness. LLDPE and LDPE are used in many packaging applications in shops, essentially for shipping and transportation of parts. They do not play any primary role in automotive applications.

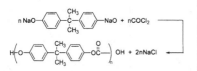

FIGURE 9.8 Polycarbonate polymerization via interfacial process.

9.2.2.5 Polycarbonate (PC). Although the first reported synthesis of aromatic polycarbonates from bisphenol-A can be traced back to Einhorn[5] in 1898, there were no additional investigations into this polymer for the next 50 years. Despite numerous investigations into polycarbonates derived from other aromatic diols, the foundation of the polycarbonate industry lies on bisphenol-A. Hereafter, the term "PC" will refer only to bisphenol-A polycarbonate. Figure 9.8 shows the structures of the PC polymer and its monomer precursors.

As in many such technical breakthroughs, the development of high-quality PC resins was aided by related activities, and acceptance of PC relied on advancements in high-purity bisphenol-A processes.[6] Early in its development, polymer chemists realized the importance of monomer purity to reaching ultimate color and mechanical properties in PC. In this connection, Kissinger and Wynn[7] have been credited with process innovations that eliminated the need for water, organic solvents, distillation, or extraction and put the commercial manufacture of this monomer in an accessible position.

Today, it is estimated that the global PC manufacturing capacity is about 2.6 million Mton. The major producers are General Electric Plastics and Bayer, with a combined 60 percent of global capacity. Global capacities can be seen in Table 9.5. Capacities are relatively balanced between western Europe and the United States. Japan's share of the global capacity is about half that of the western regions. Between 2000 and 2010, global capacity growth is projected to be >50 percent; however, the regional shares should be the same except for some relatively increased capacity in western Europe.

Over the last 35 years, the process for manufacturing PC has undergone significant modernization and evolution. Early process attempts at melt transesterification and solution polymerization with pyridine were both deemphasized based on equipment limitations and economics, respectively. Both batch and continuous processes are practiced today. Solution polymerization in methylene chloride in contact with aqueous sodium hydroxide has become the preferred process by some major producers. Melt polymerization has also regained attention, facilitated by improvements in polymerization equipment. Melt polymerization also addresses recent concerns with potential release of volatile organic compounds into the environment.

The commercial polycarbonate polymerization processes can be categorized according to whether phosgene ($COCl_2$) or its derivative is utilized as a raw material. Thus, the processes are distinguished as employing either phosgenation or transesterification.

Phosgenation is most often implemented via interfacial polymerization. In this process, PC polymer is formed at the interface of an aqueous sodium alkoxide bisphenol-A salt solution and phosgene organic solvent mixture. The PC polymer becomes soluble in the organic solvent phase as it is formed. The phosgene remains partitioned between aqueous

TABLE 9.5 PC Capacity and Demand by Region

	1999	2000	2001	2002	2003	2004	2005	2010
Capacity								
U.S.	666	666	666	815	815	865	915	1029
W. Europe	575	575	705	845	845	885	885	955
Japan	347	347	347	395	395	395	395	445
Total	1588	1588	1718	2055	2055	2145	2195	2429
Demand								
U.S.	409	428	424	441	472	509	550	736
W. Europe	400	440	435	453	480	516	555	760
Japan	201	226	182	191	199	208	217	271
Total	1010	1094	1041	1085	1151	1233	1322	1767

10^3 Mton

and organic layers. The appeal of this process is the low operating temperature, the fact that no drying of monomers is required, and its relative insensitivity to impurities. Polymers of high molecular weights are readily achieved. The disadvantages of this approach are the rather involved isolation and purification steps and the health and safety aspects of the large amounts of phosgene and organic solvent (typically methylene chloride) required.

Nonphosgene routes to PC have been extensively developed and are in commercial use. Companies operating such processes are GE Plastics, Bayer, and Asahi/Chi Mei. The literature indicates that other producers are investigating this area. The approach being used by these companies relies on the transesterification of diphenyl carbonate with bisphenol-A. This process takes place in two major steps. In the initial step, bisphenol-A and diphenyl carbonate are reacted, and phenol is liberated. The phenol is removed to produce a prepolymer. This prepolymer can then be polymerized via an ester disproportionation reaction whereby diphenylcarbonate is formed and volatilized. A challenge of this process is reported to be the low pressures and high temperatures required (<1 mm Hg and 300°C, respectively) to promote the reaction. An additional complexity of the nonphosgene route is the preparation of the diphenyl carbonate. This can be generated from phosgene, which doesn't address the hazards previously discussed. A popular route to diphenyl carbonate is via the reaction of dimethylcarbonate with phenol. Although many factors go into the selection of the PC process utilized by resin manufacturers, some analyses report that there is a capital cost advantage in the nonphosgene approach; however, it involves a higher cost for raw materials.

Polycarbonate was the first amorphous engineering polymer commercialized, and it possesses a very enviable combination of performance attributes. Figure 9.9 illustrates how PC compares to other amorphous commercial resins with respect to heat and impact resistance. Clearly, PC has the most advanced combination of these two properties among these significant polymers. Useful molecular weights are in the range 20,000 to 35,000 amu. Resins above 35,000 amu are difficult to process, whereas molecular weights below 20,000 amu lose toughness. Molecular weight distributions are typically narrow.

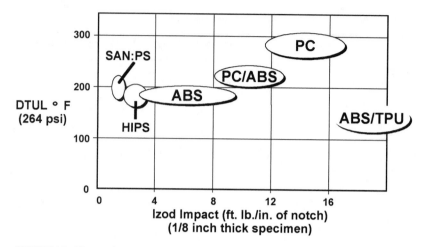

FIGURE 9.9 Heat vs. impact resistance for amorphous thermoplastic resins.

Some of PC's strengths and weaknesses are listed below:

Strengths

- Impact resistance, ductility
- Clarity
- Dimensional stability
- Inherent ignition resistance
- High temperature resistance
- Rigidity
- Creep resistance

Weaknesses[*]

- Solvent resistance (organics)
- Abrasion resistance
- UV resistance (limited)
- Notch sensitivity
- Hydrolytic stability (limited)

PC has good resistance to UV light but displays yellowing. UV stability can be extended with absorbers, coatings, and cap layers. Typical automotive exterior lighting applications rely on UV protection via proprietary coatings. Solvent resistance is an issue for some automotive applications that otherwise would use PC. For example, the use of PC and its blends in parts that can come in contact with gasoline is discouraged because of potential stress cracking after exposure. Table 9.6 shows some typical properties of PC resins.

---

[*]Relative to a single property of an alternative material. The overall balance of properties is excellent for many applications.

TABLE 9.6 Properties of Three PC Resin Types

	General-purpose	Ignition resistant	10% glass-filled
Melt flow rate @ 300°C/1.2 kg (g/10 min)	10	10	—
Light transmittance (%)	87–91	—	—
DTUL AT 264 psi (°F)	263	266	280
Tensile strength (psi)	9000	8700	8500–9500
Flexural modulus (× 100,000 psi)	3.5	3.6	4.6–5.5
Notched Izod impact (ft-lb/in)	17	12	2–4
UL-94 rating @ 1/16 (0.0625) in	HB	V-0	V-2
@ 1/8 (0.125) in	HB	V-0/5V	V-0

Although the PC displays ductile behavior in many applications, there are cases in which ductility needs to be extended to very low temperatures. Many parts that must conform to government energy management standards, such as instrument panels or airbag covers, must fail in a ductile manner with no cracking or shards at or below −20°C. PC's low-temperature ductility can be improved markedly by incorporating low levels of elastomeric impact modifiers. The principles of impact modification can be applied to PC with a number of rubbery dispersed phases, depending on the overall performance and economic balance sought. Methacrylate-butadiene-styrene copolymer can be incorporated into a PC matrix to give resins that meet many low-temperature impact requirements. Table 9.7 compares the properties of two typical impact-modified PCs. Note that some degree of impact is sacrificed in the easier-flow grade.

TABLE 9.7 Properties of Two Impact-Modified PC Resins

Grade	MFR, g/10 min	Izod, ft-lb/in	Temperature, °C
Impact-modified	11	10	−30
Impact-modified, high-flow	18	8	−30

9.2.2.6 ABS. ABS polymers are prepared by the polymerization of acrylonitrile, butadiene, and styrene (Fig. 9.10). This polymer has a very useful combination toughness, stiff-

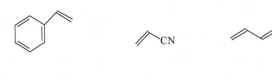

Styrene   Acrylonitrile   Butadiene

FIGURE 9.10 ABS monomers.

ness, solvent resistance, and processability. The amounts of these three monomers play an important role in the ultimate properties of the copolymer. Resin suppliers can tailor resins for particular applications and markets by varying the ratio of these three monomers and the morphology of the dispersed phase.

In 2005, the worldwide capacity of ABS has been estimated to be 8.4 million Mton/yr (18.5 billion lb/yr). There has been significant capacity expansion, from about 6.2 million Mton in 2000. This amounts to an increase of 35 percent in 5 years (Fig. 9.11). By region, one can see that significant growth in supply of this material has occurred between the years 2000 and 2005. A clear trend is a large increase in capacity in the Pacific, followed by North America and Europe. No doubt this growth has been driven by the strong, emerging economies of China and Korea. Incremental expansion has been the trend in North America and Europe led by BASF, Bayer, and Dow Chemical.

Any process to manufacture ABS resins must fulfill these functions economically:

1. A two-phase structure must result, consisting of styrene-acrylonitrile high polymer (SAN) with a dispersed phase of polybutadiene (PBD).
2. The PBD must be well grafted by SAN and cross-linked.
3. The graft-PBD must be dispersed in the SAN matrix at micron scale.

Two processes have emerged that fulfill these qualifications and have distinct advantages in terms of process economies and product characteristics: *emulsion* and *mass* (also referred to as *solution*) ABS.

The emulsion ABS process was the initial industrial process and is still used today because of its flexibility. The process borrows from rubber latex manufacture by forming and cross-linking the PBD phase in a water/emulsifier system. Independent control of particle size and cross-link level is achieved by initiator and temperature settings. After the PBD latex is formed, the next step of the process involves polymerizing styrene and acrylonitrile monomers in the presence of the latex to form the grafted rubber concentrate. At this point, the final ABS composition is formed by either mixing concentrate with additional SAN latex and coagulated or by first isolating the PBD concentrate and subsequently compounding it with SAN. Coagulation of the lattices is very important to the final stability of the resin, since salts, soaps, and other residues can subsequently lead to poor thermal stability. Separate control of each step of the manufacture allows for numerous opportunities to tailor the structure of the final product for diverse applications.

In contrast to the emulsion process, the mass process uses only monomers and low diluent levels to effect the conversion of monomer to polymer. It is generally accepted that the mass ABS process has less complexity than emulsion but is arguably less versatile. An additional point of distinction is that preformed, uncross-linked PBD rubber is dissolved in the monomers and solvents. The solution is then continuously pumped to a series of stirred reactors where the polymerization and grafting is advanced. Phase inversion from PBD to SAN continuous occurs when the phase volume of SAN exceeds that of the rubber. Peroxide initiators and chain transfer agents are introduced at strategic stages of the process for rate and molecular weight control. Agitation rate, especially in the first half of the reaction, is critical to the proper formation of the rubber phase. The last major unit operation is devolatilization, which entails flash removal of unconverted monomers and diluent. It is well understood that the majority of the PBD grafting occurs in the initial part of the process, whereas rubber cross-linking is achieved at the final stages of polymerization and upon devolatilization. The majority of ABS made in this process is ready for sale and requires no additional modification via compounding.

The structure of ABS is more complex than the designation of copolymer indicates. A morphology is developed in the polymerization processes whereby two phases are pro-

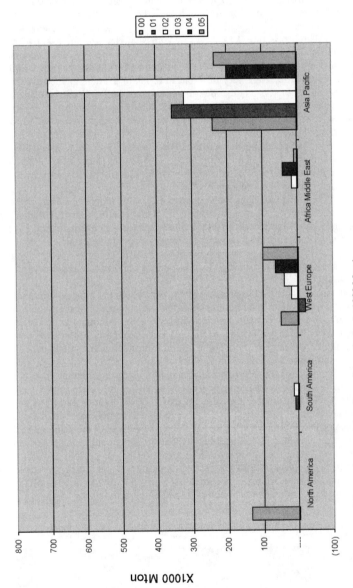

FIGURE 9.11 ABS capacity expansions by region and year (× 1000 Mton).

9.17

duced: a styrene-acrylonitrile (SAN) continuous phase and a discrete grafted polybutadiene one. The rubber particles are finely dispersed in the rigid phase. Thus, a situation is created in which copolymerizing acrylonitrile and styrene in the presence of PBD results in an amorphous molding polymer that is much better suited to automotive applications than the polystyrene homopolymer. Although the addition of the more polar monomer brings about improvements in modulus, equally important is a step improvement in resistance to impact damage. Figure 9.12 shows transmission electron microscopy images of two distinct morphologies present in commercial ABS resins. The darker domains are droplets of grafted polybutadiene that have been stained by osmium tetraoxide. One can see that the rubber particles are very different in terms of average diameter and appearance. Particles with diameters on the order of the wavelength of light can be made with the emulsion process, whereas larger particles with occluded matrix resin is very typical of mass. In mass ABS, control of the particle size and particle density allows for a broad range of gloss.

The key function of the dispersed, grafted PBD phase is to dissipate energy in the case of an impact event. There are two types of energy absorption found in ABS: crazing and shear banding. In the case of crazing, rubber particles can dissipate energy by initiating and terminating this type of microcracking. The initial step in the dissipation process is the deformation of the rubber particles to the point of void formation. This void formation, in turn, initiates additional crazes that are terminated at neighboring rubber particles. Crazes characteristically have high levels of surface area in the form of fibrils spanning the craze direction. This is the dominant mechanism for lower AN containing ABS resins. In higher AN containing ABS copolymers, the dominant mechanism becomes shear, yielding evidenced by the appearance of banding. Submicron-size particles are reported to facilitate this energy dissipation mode.

The composition of the ABS has a very large bearing on the final performance of the resin. Figure 9.13 illustrates the relationship between composition and structure in ABS. Polymer scientists carefully balance the proportions of these monomers when designing ABS resins for particular applications. Table 9.8 shows properties typical of some commercially available classes of ABS. These material properties are suitable for injection molding automotive parts.

ABS copolymers can be used in a wide variety of applications ranging from plated exterior grills to molded-in-color glove box doors. However, advances in vehicle aerodynamics have resulted in much more sunlight exposure for interior parts and an overall increase in maximum temperatures. As a rubber modified amorphous resin, the glass transition temperature (T_g) of the rigid phase and the phase volume of the dispersed elastomer have a very large influence on its modulus response to temperature. There are numerous approaches to extending the upper service temperatures of ABS. These include alloying with higher-heat polymers, reinforcing with fibrous fillers, and terpolymerization. This group of chemically modified resins is commonly referred to as high-heat ABS (HHABS).

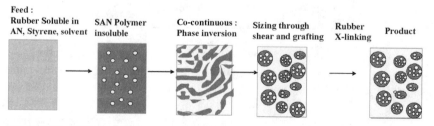

FIGURE 9.12 ABS phase development schematic.

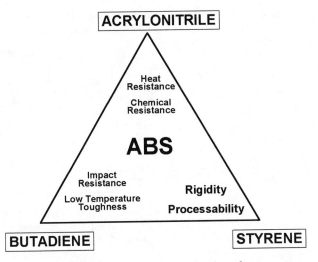

FIGURE 9.13 Monomer contribution to ABS resin performance.

TABLE 9.8 Typical Properties of ABS Resins for Automotive Applications

Property	Test method (ASTM)	High-flow mass	General-purpose mass	High-impact mass	General-purpose emulsion
Specific gravity	D-792	1.05	1.04	1.03	1.04
Coefficient of linear thermal expansion (cm/cm, °C)	D-696	7.6E-05	9.3E-05	9.3E-05	8.8E-05
Notched Izod impact strength (23°C, J/m)	D-256	160	310	553	203
Flexural modulus (MPa)	D-790	2170	2070	1980	2620
Tensile yield strength (MPa)	D-638	39	42.2	37	43
Ultimate elongation (%)	D-638	60	25	30	40
Melt flow rate (230°C, 3.8 kg)	D-1238	6.5	2.5	0.9	8
Heat deflection temperature (unannealed, °C)	D-648	80	83	82	79

Terpolymerization is a popular method of raising the heat resistance of ABS, and these resins account for a large portion of the volume used in automotive applications. That is, an additional monomer is added to the polymerization. These monomers can act in two ways when incorporated in the rigid phase: chain stiffening and/or modification of the cohesive energy density. Table 9.9 lists the T_g of some common styrenic polymers. You can see that the influence of comonomer AN is a modest 0.3 to 0.4°C/percent. The three most commonly used monomers that increase the T_g of ABS are substituted styrenes, maleimides, and maleic anhydrides (MAs). The majority of HHABS uses α-methylstyrene (αMS) or N-phenylmaleimide (NPMI) for enhancing the T_g. The efficiency with which these monomers raise the T_g depend on the chemistry employed. The most common monomer added to HHABS is αMS. Reaction kinetics change significantly with the introduction of

TABLE 9.9 Properties of Some HHABS Resins, Medium and High Heat Range

Heat range	Grade	Supplier	Technology	MFR (g/10 min, 220°C, 10 kg)	Charpy impact (kJ/m^2)	Vicat (°C)
Medium heat	MAGNUM™ 3325MT	Dow Chemical Co.	Mass, low residuals	10	18	101
	BDT 5510	GE Plastics	αMS, mass	15	13	100
High heat	MAGNUM™ 3416SC	Dow Chemical Co.	NPMI, mass	6.5	18	108
	Ronfolin HX-10	BASF Corp.	αMS, emulsion	3.5	12	110

this substituted styrene, requiring that the emulsion process be used for higher levels of monomer incorporation. Maleic anhydride (MAH) is an excellent monomer for modifying styrenics for heat resistance; however, it can't be used in the polymerization of ABS because of its tendency to induce cross-linking of the rigid phase. This can be avoided by blending a styrene-MAH copolymer with ABS in a compounding step. These two polymers are highly compatible. Finally, NPMI has recently found its way into HHABS in both the emulsion and mass process. This monomer is especially reactive with styrene and compatible with multiple processes.

9.2.2.7 PC/ABS Blends. Polymer blends have an important role in bringing property options for automotive applications that can't be reached with a single material. Unfortunately, due to unfavorable thermodynamics, only a few examples of miscible blends have found their way into commercial use in this industry. In contrast, many compatible (or partially miscible) blends have reached commercial importance and can be found in key applications. Numerous examples are impact modified resins: ABS, TPO, PVC, IM-PC, and so forth. However, as a binary mixture of resins approaches equal volume fraction in the blend, the requirements for compatibility at the phase interfaces increases dramatically and excludes the majority of combinations. One such blend having properties that make it unique and useful in automotive applications is PC/ABS. The concept of blending PC and ABS dates back to 1964 with U.S. patent no. 3130177, Borg Warner's grandfather patent in the area. Additional inventions have ensued: a patent was granted to Teijin in 1974 that covers impact strength and rubber location in PC/ABS/MBS blends (U.S. patent no. 3582394). U.S. patent no. 3880783 was issued in 1975 to Bayer, which demonstrates that polycarbonate and ABS (emulsion polymer) blends give high gloss and good impact. U.S. patent no. 4098734 was issued in 1978 to Monsanto, showing that improved properties are obtained for PC/ABS alloys when bimodal rubber particles are employed. Based on its versatility, this blend is considered a mainstay engineering material for applications such as automotive instrument panels, body panels, and wheel covers.

Although ABS and PC are extremely useful amorphous resins in their own right (see the ABS and PC sections), they have some limitations that can be solved by their blends. Although PC has exceptional clarity, toughness, and heat resistance, it is notch sensitive and more difficult to process than ABS resins. Similarly, ABS is a tough material, very processable, and adheres to paint and foams well. Due to exceptional compatibility between the phases, alloys of these two resins result in a resin that has a unique combination of their properties. PC/ABS blends are noted for having high heat properties, stiffness, and toughness with less notch sensitivity, improved processability, and versatile surface characteristics.

Since there are at least two major, immiscible phases in the PC/ABS blends, it is not surprising that careful attention has to be paid to developing an optimum morphology during processing. Figure 9.14 shows the morphology developed in a commercial grade of PC/ABS. Note that the darkened phase is ABS surrounded by PC. Thus, scales as diverse as molecules through to phase morphology, and part scale millimeters to meters, contribute to useful parts. Numerous studies have been done to understand the role that composition of the neat polymers, phase volumes, compatibilization, and processing play in the ultimate performance of this blend.

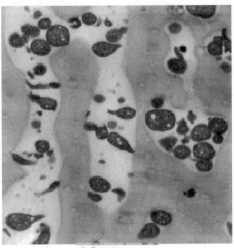

1.0 μm ——

FIGURE 9.14 Transmission electron micrograph of a PC/ABS blend (phases light to dark: SAN, PC, polybutadiene).

The thermodynamics of polymer blends, although semiquantitative, play a key role in understanding phase morphology. For the ABS composition, Callaghan et al.[8] determined that an optimum AN level exists at about 25 percent. This gives the highest level of phase adhesion and compatibility. Morphology development at the micron scale must also be optimized so as to get the most desirable toughness. Figure 9.15 shows the response of impact resistance to the weight percent of PC in the blend. Clearly, toughening is optimum near the composition at which the ABS phase first becomes co-continuous (or 65/35 percent by weight PC/ABS).

Stability of both the morphology and the components must be considered in this system. Although it can undergo oxidative chain scission under very extreme temperatures, mass ABS is known to be robust during molding processes. Polycarbonate, with its active carbonyls, is known to undergo hydrolysis and can be attacked by basic impurities resulting in molecular weight losses. Figure 9.16 shows the effect of time and temperature on the molecular weight of the PC portion of the blend. Note that, if PC molecular weight falls below 20,000 amu, then embrittlement of that phase occurs. The rate of degradation is a function of some the common impurities and even moisture. Depending on the level of these agents, degradation can be changed dramatically. Plastics producers understand the

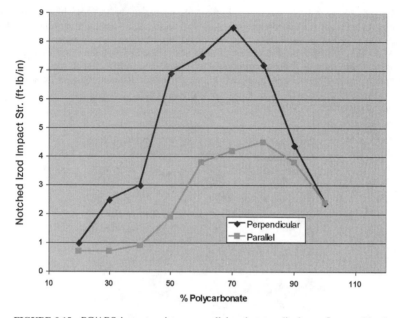

FIGURE 9.15 PC/ABS impact resistance parallel and perpendicular to flow vs. blend composition.

need to start with PC and ABS with a minimum of impurities to have molecular weight control through compounding and processing into parts. Impurities in ABS are known to be strongly dictated by the polymerization process employed. The process to manufacture emulsion ABS involves substances that are detrimental at trace levels (surfactants, flow aids, heat stabilizers, and so on) compared to mass ABS. In addition, the mass process exposes the polymer to less heat history during its manufacture than emulsion. Likewise, impurities in polycarbonate or PC blends, particularly those capable of an alkaline reaction, can reduce its resistance to moisture. Stability of the melt is especially significant as resins suppliers push to minimize viscosities during molding via lowering the feedstock molecular weights. High-flow PC/ABS resins must start with high-purity feedstocks to assure that molecular weight attrition doesn't lead to brittle parts.

Table 9.10 shows the properties typical of commercial PC/ABS resins. Four performance types are listed: general-purpose, high-flow, blow molding and low-gloss grades. Some important characteristics of PC/ABS can be learned from this table. For example, these PC/ABS resins are all engineered with low-temperature ductility and a robust modulus. The rheology is modified to meet the forming applications.

9.2.2.8 LGF PP and ABS.   Estimates are that about 30 percent of the 2 million Mton of e-glass fiber consumed globally for polymer reinforcement is used in thermoplastic composites. Glass-filled thermoplastic composites have been growing at a very healthy pace of 15 to 20 percent per year, largely fueled by automotive applications. Two key reasons that glass-reinforced thermoplastics are becoming so important are their recyclability and compatibility with the injection-molding process. One of the most revolutionary technologies to come to the forefront is the long glass fiber-reinforced (LGF) resins. Specifically, the use of long glass fiber reinforcement in polypropylenes has allowed the use of a lower-cost

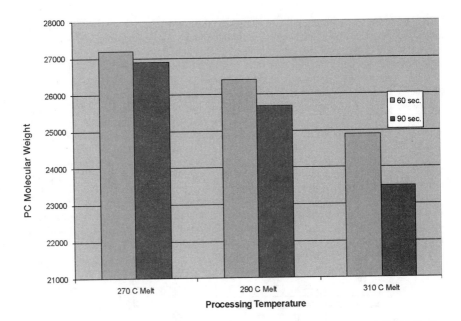

FIGURE 9.16   Effect of molding time and temperature on PC molecular weights in PC/ABS blends.

TABLE 9.10   Properties of Some Typical Automotive Grades of PC/ABS

Properties	ISO test method	Unit	General-purpose	Easy-processing	Blow molding	Low-gloss
Specific gravity	1183 B; 1987	g/ml	1.13	1.14	1.13	1.13
Tensile modulus	527-2; 1993	MPa	2500	2390	2350	2500
Tensile yield strength	527-2; 1993	MPa	53	55	54	51
Elongation at break	527-2; 1993	%	125	110	115	120
Flexural modulus	178; 1993	MPa	2375	2375	2400	2300
Notched Izod impact strength (23°) C (−30°) C (−40°) C	180-4A; 1993 180-4A; 1993 180-4A; 1993 180-4A; 1993	kJ/m^2 kJ/m^2 kJ/m^2 kJ/m^2	— 50 36 24	— 46 34 25	— 53 25 20	— 52 32 24
DTUL (0.45 MPa), unannealed	75A; 1987	° C	131	130	132	128
DTUL (1.8 MPa), unannealed	75A; 1987	° C	109	110	109	106
Melt flow rate (MFR), 260°C 3.8 kg	1133; 1991	g/10 min	7	18	4	4.5

polymer to be used in structural, engineered applications. LGF-PP is especially interesting, since the glass fiber's enhanced reinforcing abilities and the cost/performance of PP make it a very economical engineering material. This material is quickly gaining in importance, as evidenced by annual growth rates of >35 percent over the last four years.

Long glass fiber PP derives its unique properties by artfully combining a low-density, semicrystalline PP resin with compatibilized e-glass fibers in such a way as to preserve the filler aspect ratio. All references to fiber length will be for the molded part, not the starting fiber length. It has long been understood that this is a key to maximizing the potential of glass fiber reinforcement of PP; however, the combination of material and transformation science has just now gelled to make this a popular commercial option. Composites theory can be used to describe the effect of fillers on modulus as a function of aspect ratio.[*] (Note that this predicts modulus in the direction of fiber orientation.) These relationships can be seen in Fig. 9.17. Another important property improvement seen in LGF-PP is impact resistance. Figure 9.18 compares the falling dart impact (FDI) strength and tensile properties of long and short glass-reinforced PP. Clearly, an impressive level of toughening is achieved with long glass fiber reinforcement. Material scientists attribute this drastic improvement[†] in impact energy management to a mechanism whereby energy is dissipated due to slippage at the long glass/PP interface.

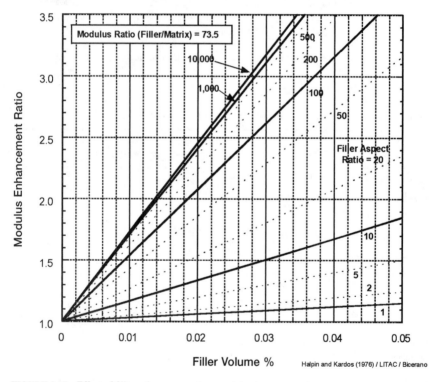

FIGURE 9.17   Effect of filler volume percent on modulus improvement.

---

[*]Fiber length/diameter ratio.
[†]Relative to short glass-reinforced PP.

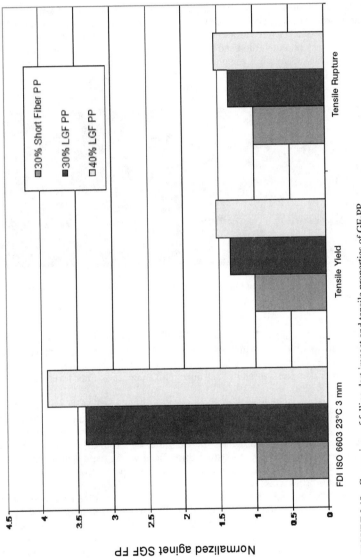

FIGURE 9.18 Comparison of falling dart impact and tensile properties of GF-PP.

Another important element to attaining optimum properties via LGF-PP is the addition of a compatibilizing, grafted additive. Because of the chemical mismatch at the interface of PP and e-glass fibers, without additional treatments, there is the potential for very poor energy transfer between fiber and matrix. Numerous microscopic studies have verified poor adhesion of glass fibers to PP and motivated the development of grafted polypropylenes to improve the adhesion. The addition of maleic anhydride grafted PP (MA-g-PP) has been shown to improve the adhesion between the two major phases. Figure 9.19 compares the relative adhesion of a fracture surface with and without modification by MA-g-PP. Smooth fibers indicate very poor adhesion and stress transfer, whereas a roughened glass surface is indicative of cohesive failure at the glass-PP interface. Better interfacial bonding has been shown to be a strong contributor to modulus in these types of composites.

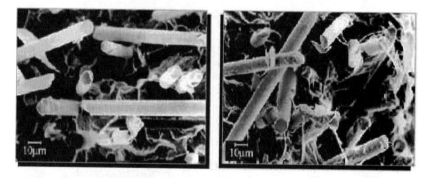

FIGURE 9.19 Scanning electron micrograph of PP-glass fracture surfaces. The sample on the left had no compatibilizer, sample on right had MA-g-PP added.

LGF-PP properties are very much dependent on the conversion process used to incorporate the fibers and form the parts. Thus, representative properties should be reported for each of the major processes and glass levels. Table 9.11 gives a summary of properties from both direct compression and pellet injection. Three glass fiber levels were chosen: 20, 30, and 40 percent.

PP can be modified with long glass fibers and formed into articles in a number of ways. However, two very distinct methods are used to incorporate PP and long glass fibers: direct and pellet processes. In the direct process, glass roving is fed to a portion of the forming process, and the fiber filled melt is transferred to either an injection or compression molding process. Alternatively, if existing injection molding capital must be used, then molders have the option of purchasing a material from the pellet process.

In the direct process, glass roving is added to a stage of the process where shear can be carefully controlled and fiber lengths are maximized. Two major direct processes are commercially practiced: injection and compression. The direct injection process uses a coupled process whereby a compounding extruder is coupled to an injection molding press. Typically, the extruder is operated continuously and fills an accumulating tank that, in turn, feeds the molding machine. In this case, the extruder can clearly be seen mounted on the top of the press. Direct compression is also linked to the forming process, and glass modified, molten "buns" are transferred to a compression press.

Pellet processes are practiced by suppliers who may also do more conventional molding. The original process, begun by Fiberfill Co. in the 1950s, consisted of simply pulling glass rovings through wire coating dies. This incomplete wetting of the glass fibrils re-

TABLE 9.11 Example Properties of LGF-PP at Three Levels of Glass Reinforcement

Process	Pellet		Direct	
	Injection	Injection	Compression	Compression
Long glass fiber, %	20	30	40	30
Specific gravity	1.03	1.12	1.22	1.12
Tensile strength (MPa)	96	120	78.6	54.5
Tensile elongation, 5 mm/min, break	2.7	2.6	2	1.5
Flexural modulus (GPa)	5.0	6.7	5.7	5.0
Flexural strength (MPa, 2 mm/min)	140	170	146.9	108.3
Izod impact @ 23°C (kJ/m^2)			15.3	15.1

sulted in poor dispersion and suboptimum part properties. Although this process was ultimately abandoned, development began again in the 1980s, fueled by the rapid growth of advanced composites. As a result of their intense efforts, pultrusion processes to produce LGF-PPs with wetted fibrils and good dispersion were developed. Factors that influence the quality of the product and output of the process include impregnation chamber design, die geometry, line speed, number of strands, and extruder output. Line speed must be balance against the desired degree of fiber wet-out with minimal free fibers.

Long glass fiber PP offers a unique engineering option of automotive engineers in place of metal or high-performance engineering thermoplastics (ETPs). Compared to alternatives, LGF-PP has the following very compelling advantages:

Metal

- Cost
- Corrosion resistance
- Weight reduction
- Intricate design capability
- Thermal and sound insulation

ETPs

- Cost
- Density (vs. filled ETPs)
- Processability
- Sound insulation
- Moisture absorption

Numerous applications for LGP-PP have been developed. The most popular applications in automotive are: instrument panels, underbody shields and front end carriers. Figure 9.20 shows the relative share of each application.

As LGF-PPs find their way in to new applications, it is only natural that additional performance requirements are requested by automotive molders. One such development is for exterior, unpainted panels and steps. This requires materials for these applications not only to have sufficient modulus and processability but also the ability to withstand impact (e.g., stones and grocery carts) and maintain acceptable appearance for the life of the vehicle. Although various methods for impact modifying LGF-PPs are possible, a novel approach using masterbatches was recently reported by Richardson et al. (WO 2004/035295 A1). She

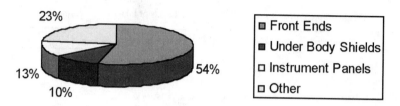

FIGURE 9.20   Automotive applications for LGF-PP.

demonstrates that, by adding a specific levels of elastomer impact modifier (i.e., ethylene/octene copolymer) directly to the process, impact strengths can be greatly increased. This masterbatch approach could also bring other functionality to the part, such as weather resistance, thermal stability, and colorant. Table 9.12 shows the performance one can achieve with LGF masterbatches that are designed to bolster impact performance of the part.

TABLE 9.12   Properties of Direct Compression LGF-PP with Impact Modifier Added On-Line (WO 2004/035295)

	1	2	3	4
Composition				
PP	67	65	63	61
POE	0	2	4	6
Glass	30	30	30	30
Color	2	2	2	2
SA	1	1	1	1
Properties				
F_M, MPa	4760	5330	5000	4820
F_S, MPa	97	113	115	108
T_S, MPa	52	55	53	58
T_E, %	1.4	1.6	1.4	1.5
IDI @ 3 mm, J	15	14	14	17
IDI @ 4 mm, J	18	20	22	24
IDI @ 5 mm, J	29	33	33	34

IDI = instrumented dart impact; POE = polyolefin elastomer; SA == glass/PP-compatibilizer.

9.3 REFERENCES

1. Market Analysis of Thermoplastics Elastomers, Robert Eller Associates, Inc., September 2000.
2. N. Kawamura., et al., Super Olefin Polymer for Material Consolidation of Automotive Interior Plastic Parts, SAE Technical Paper 960296, 1996.
3. A. Y. Coran and R. Patel, Rubber-Thermoplastic Compositions. Part I. EPDM-Polypropylene Thermoplastic Vulcanizates, *Rubber Chem. Technol.*, 5, 141, 1980.
4. B. N. Epstein, U.S. patent no. 4,174,358, 1979.
5. A. Einhorn, Annahlin, 300,135,1898.
6. J. E. Jansen, U.S. patent no. 2,468,982, 1949.
7. G. M. Kissinger and Nicholas P. Wynn, U.S. patent no. 5,362,400, 1994.
8. T. A. Callaghan, K. Takakuwa, D.R. Paul and A.R. Padwa, *Polymer* 34, 3796, 1993.

INDEX

ABOUT THE EDITOR

Charles Harper, president of Technology Seminars, is a leading authority on plastics, plastics fabrication, and plastics applications. He has written numerous other references for the field, including *Modern Plastics Handbook; Handbook of Plastics, Elastomers, and Composites;* and *Handbook of Materials in Product Design* (all published by McGraw-Hill).